Consumers' surplus

$$\text{CS} = \int_0^{x*} (D(x) - p_*)dx$$

Producers' surplus

$$\text{PS} = \int_0^{x*} (p_* - S(x))\, dx$$

PROBABILITY THEORY

Sample mean $\bar{x} = \frac{1}{n}\sum_{i=1}^{n} x_i$

Sample variance $s^2 = \frac{1}{n-1}\sum_{i=1}^{n} (x_i - \bar{x})^2$

Discrete random variables

$$\mu = E(X) = \sum_{i=1}^{n} x_i p_i$$

$$\sigma^2 = \text{Var}(X) = \sum_{i=1}^{n} (x_i - \mu)^2 p_i$$

Continuous random variables

$$\mu = E(X) = \int_A^B xp(x)\, dx$$

$$\sigma^2 = \text{Var}(X) = \int_A^B (x - \mu)^2 p(x)\, dx$$

$$= \int_A^B x^2 p(x)\, dx - \mu^2$$

MATHEMATICAL FORMULAS

Average value $\bar{f} = \frac{1}{b-a}\int_a^b f(x)\, dx$

Area between curves $A = \int_a^b [f(x) - g(x)]\, dx$

Area of a circle $A = \pi r^2$

Circumference of a circle $C = 2\pi r$

Volume of revolution $V = \pi \int_a^b f(x)^2\, dx$

Volume of a cylinder $V = \pi r^2 h$

Volume of a cone $V = \frac{1}{3}\pi r^2 h$

Volume of a sphere $V = \frac{4}{3}\pi r^3$

Surface area of a sphere $A = 4\pi r^2$

Approximation formula $f(a + h) \approx f(a) + f'(a)h$

Newton's method $x_n = x_{n-1} - \frac{f(x_{n-1})}{f'(x_{n-1})}$

Trapezoidal rule

$$\int_a^b f(x)\, dx \approx \Delta x \left[\frac{y_0}{2} + y_1 + y_2 + y_3 + \cdots + y_{n-1} + \frac{y_n}{2}\right]$$

Simpson's rule

$$\int_a^b f(x)\, dx \approx \frac{\Delta x}{3} [y_0 + 4y_1 + 2y_2 + 4y_3 + 2y_4 + \cdots + 2y_{n-2} + 4y_{n-1} + y_n]$$

Method of least squares

$$m = \frac{n\Sigma x_i y_i - (\Sigma x_i)(\Sigma y_i)}{n\Sigma x_i^2 - (\Sigma x_i)^2}$$

$$b = \frac{1}{n}(\Sigma y_i - m\Sigma x_i)$$

APPLIED CALCULUS

APPLIED CALCULUS

an intuitive approach for management, life, and social sciences

RICHARD L. FABER
Boston College

MARVIN I. FREEDMAN
Boston University

JAMES L. KAPLAN
Boston University

WEST PUBLISHING COMPANY
ST. PAUL • NEW YORK • LOS ANGELES • SAN FRANCISCO

COPYEDITOR: Constance Day
DESIGNER: Deborah Schneider
ART STUDIO: ANCO/Boston
COMPOSITOR: Polyglot Compositors
PRODUCTION COORDINATOR: Schneider & Co.
COVER PHOTOGRAPH: J. Lurie/FPG

Library of Congress Cataloging-in-Publication Data

Faber, Richard L.
Applied calculus.

Includes index.
1. Calculus. I. Freedman, Marvin I. II. Kaplan, James L. III. Title.
QA303.F22 1986 515 85-20317
ISBN 0-314-85235-2

Contents

* A computer symbol accompanying a section title indicates that optional exercises (entitled "If You Work with a Computer") requiring the use of a computer occur at the end of that section's exercise set.

Preface

Does the world need another calculus book? The three authors of *Applied Calculus: An Intuitive Approach for Management, Life, and Social Sciences* answer with a definite yes!

Now that you have our text in your hands, let us tell you what we have tried to achieve in writing this book.

The typical student taking a course based on this text is likely to be preparing for a professional career in business, economics, or in the life or social sciences. This student is studying calculus for important practical reasons, so the goal for any text meant for this audience must be to emphasize, in its examples and exercises, applications meaningful to these disciplines. We have met this goal by carefully selecting applications that are of intrinsic interest in themselves and, in addition, will prove to have the greatest relevance to the student's educational objectives. Thus, the key concepts of calculus—the derivative and the integral—are introduced informally and intuitively by relating them to ideas with which the student is already familiar.

We found in our own teaching experience that texts written for an "applied calculus audience" often appear lifeless and convey little of the sense of discovery that should accompany the learning of any new subject. Indeed the discovery of the calculus by Sir Isaac Newton and Baron Gottfried Wilhelm von Leibniz stands as one of the great triumphs of human thought. Although mathematical rigor is inappropriate for the applied calculus audience, there is no reason that the "spirit" of calculus cannot be adequately conveyed.

We have therefore written a spirited text—one that both the student and the instructor will enjoy working with in the coming months. We have emphasized readability and "teachability" and have provided an informal, intuitive, and lively introduction to the calculus.

FEATURES

Spiral Approach Many topics are first introduced at an elementary or intuitive level early in the text and then reappear later at a more advanced level, as the reader gains more mathematical sophistication. For example, differential equations, a key concept of applied calculus, are introduced in Chapter 4 as a tool used to solve problems involving exponential growth and decay. In Chapter 7 more general separable differential equations are studied and applied to problems in physics, chemistry, sociology, and economics. Still later, differential equations appear a final time in Chapter 11, where they are solved with the aid of power series.

Linear regression, in the special case of a simple proportionality relationship (that is, a line through the origin), is introduced in Chapter 3 (single variable optimization) through examples taken from ecology and biology. Then, in Chapter 8, the general problem of linear regression (the method of least squares) is treated as an illustration of multivariable optimization.

Examples and Exercises We have included a large number of worked examples, varying from the purely mechanical, or computational, to specialized applications of interest to specific disciplines. In many cases, worked examples are followed immediately by a practice exercise that reinforces the skills introduced. The end-of-section exercises are graded in level of difficulty and offer a wide variety of applications.

Chapter Tests Each chapter is followed by two examinations. The first, which we call the "Warm-Up Test," is accompanied by fully worked solutions at the end of the text. The second, the "Final Exam," is similar to the first, but the solutions are to be found in the *Instructor's Guide*. The Final Exam may be used, at the instructor's option, as either a class test or a set of chapter review exercises.

Computer Programs (Optional) The pervasiveness of the use of the computer in our society needs no elaboration. To demonstrate the interplay between calculus and the use of the computer, we have included, at the end of several exercise sets, machine-independent computer programs in both BASIC and Pascal, together with exercises on various numerical approximation and iterative techniques. However, these programs and exercises are entirely optional and may be omitted without loss of continuity. They are included for the benefit of those with access to a computer and some knowledge of BASIC or Pascal. A software disk is available free to adopters for either the IBM PC™ (and compatibles) or the Apple Macintosh.™

Algebra Review In our experience, most students have little difficulty understanding the basic concepts of calculus. Many, however, struggle with the mechanical aspects of the subject because of dimly remembered prerequisites from high-school algebra. Accordingly, we have included, in Appendix A, a review of algebraic techniques that should serve as a convenient reference. It includes examples and exercises that can be used for student drill or self study.

Trigonometry We have chosen to collect all the information about the trigonometric functions into a single chapter, where we provide a self-contained review of the basic definitions, functions, and identities before taking up the calculus of these functions.

Flexible Curriculum Development Finally, we have tried to arrange the subject material to allow for flexible curriculum development. The first flow chart shows the logical interdependence among the chapters. The second indicates a possible syllabus for a one-semester course.

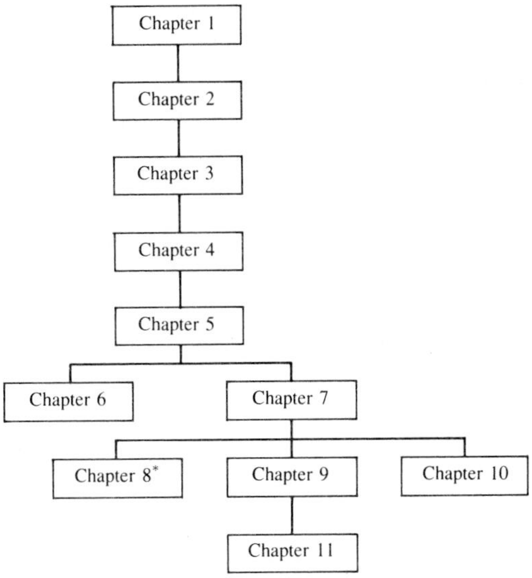

Interdependence of chapters

* Only Section 8.6 (Double Integration and Its Applications) requires Chapter 7.

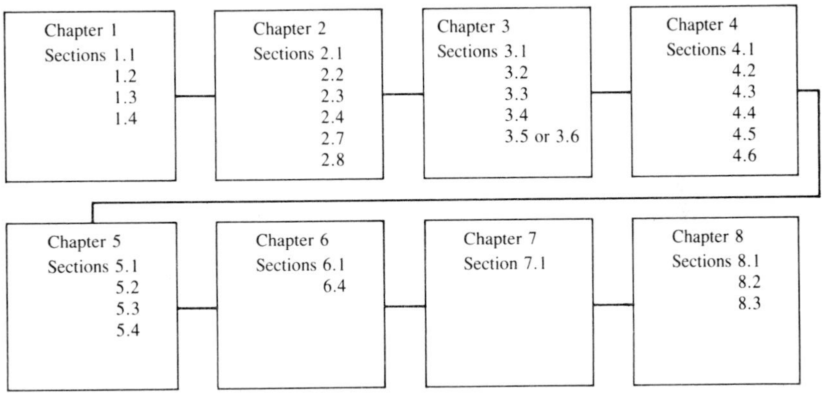

A possible one-semester course

ACKNOWLEDGMENTS

The writing of this book required the assistance and support of numerous people. This brief expression of thanks hardly does justice to our deep appreciation for their contributions to this project.

First and foremost, we would like to thank our wives, Susan Faber, Corey Freedman, and Ellen Kaplan, who provided continuous encouragement, despite sacrificing a seemingly endless number of evenings and weekends.

We would like to thank the following reviewers, who thoughtfully provided both praise and criticism:

John Bosovicki
Indiana University of Pennsylvania

James Burgmeier
University of Vermont

Sam Councilman
California State University, Long Beach

Arthur Crummer
University of Florida

Richard Davitt
University of Louisville

Gary Etgen
University of Houston

Murray Gerstenhaber
The University of Pennsylvania

Shirley Goldman
University of California, Davis

John Hardy
University of Houston

Kevin Hastings
University of Delaware

Charles Himmelburg
University of Kansas

Joyce Longman
Villanova University

Stanley Lukawecki
Clemson University

Richard Marshall
Eastern Michigan University

Lyle Mauland
University of Indiana

Carl Minda
University of Cincinnati

Robert Moreland
Texas Technical University

David Morency
University of Vermont

David Pentico
Virginia Commonwealth University

Anthony Peressini
University of Illinois

Manley Perkel
Wright State University

Dix Petty
University of Missouri at Columbia

Richard Porter
Northeastern University

Stephen Rodi
Austin Community College

Linda Sherrell
Louisiana State University

Clifford Sloyer
University of Delaware

Dale Thoe
Purdue University

Thomas Woods
Central Connecticut State University

We are especially grateful to our editor, Patrick J. Fitzgerald, whose constant encouragement and strong vision of the shape of the final product resulted in a far better text than we would ever have written without him.

We would like to thank Debbie Schneider of Schneider & Co., our production coordinator, whose creative flair and coordinating talents transformed a visually bland typed manuscript into a highly attractive volume. We are also grateful to the members of the West Publishing Co. staff for their efforts: Nancy Hill-Whilton, our developmental editor; Sherry H. James, our production editor; and Kristi Shuey, our promotion manager. We also want to thank Constance Day, our copyeditor; Phyllis Coyne, our proofreader; and Susan Vayo and Gloria Langer for accuracy checking.

Our sincere thanks to Tom Orowan and Ellen Rigoli for their ability and patience in typing from handwritten pages only slightly more legible than the Dead Sea Scrolls, and to John Aversa for assisting with the art work.

Finally, we appreciate the assistance of the many students who checked the answers to every exercise in the text: Mako Haruta, Dan Leonard, Susan Sadofsky, Peter Senak, Patricia Sniffen, Beverly Stratton, and Amy Waldhauer.

We would also like to thank Edwin Weiss of Boston University for his valuable contribution during the early stages of this endeavor.

R.L.F.
M.I.F.
J.L.K.

APPLIED CALCULUS

ONE Preliminaries

Outline

In Context

The purpose of this chapter is twofold: first, to review some essential facts from analytic geometry concerning points, lines, distance, and circles; and second, to introduce the concept of a *function*, which is central to all of mathematics.

SECTION 1.1

Analytic Geometry

This section is designed to provide a quick review of some elementary facts from analytic geometry that play an important role in developing and motivating the calculus.

We assume you are familiar with the introduction of coordinates in the plane. We start with two perpendicular straight lines; the horizontal one is called the **x-axis**, and the vertical one is called the **y-axis**; their intersection is called the **origin**. We set up a coordinate scale on each. With every point P in the plane, we then associate an ordered pair of numbers (a, b) in the usual way. These numbers are called the **coordinates** of P; the first number is known as the **x-coordinate** and the second number is known as the **y-coordinate**. In Figure 1.1 we plot a few points in the plane to illustrate how this goes. The coordinates just introduced are known as **Cartesian coordinates**, after the French philosopher and mathematician René Descartes (1596–1650). In general, we shall denote a point P by $P = (x, y)$ or simply by $P(x, y)$. The essential feature here is that we have associated a geometric object in the plane, namely a point, with an ordered pair of numbers. Another important geometric object in the plane is the straight line. How can we describe a straight line in terms of numbers (that is, algebraically)? Given a straight line l, choose two points on the line and denote them by $P_1 = (x_1, y_1)$ and $P_2 = (x_2, y_2)$. If $x_1 \neq x_2$, let us divide the difference in the y's by the difference in the x's (see Figure 1.2). The quotient,

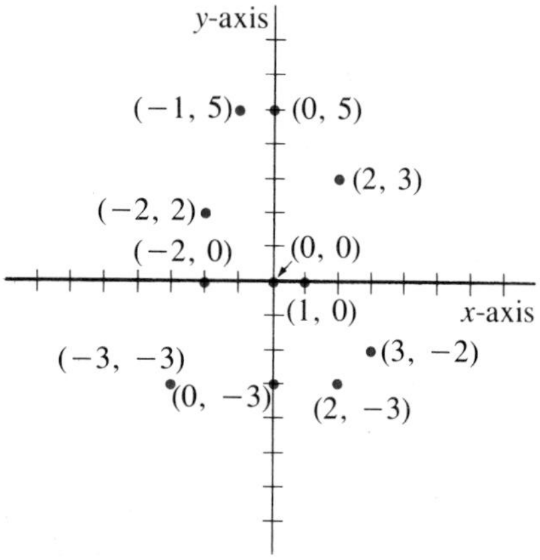

FIGURE 1.1

(1.1)
$$m = \frac{y_2 - y_1}{x_2 - x_1}$$

is called the **slope** of the line l.

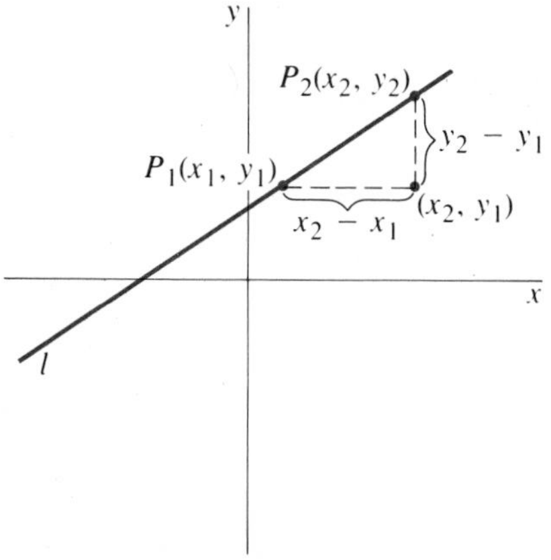

FIGURE 1.2

By using similar triangles, it is easy to demonstrate that, for any choice of the points P_1, P_2 on l, we always get the same value for m. (In mathematical terminology, the slope m is independent of the choice of the points P_1, P_2 on the line.) Note also that

$$\frac{y_2 - y_1}{x_2 - x_1} = \frac{y_1 - y_2}{x_1 - x_2}$$

so it makes no difference how the points are labeled (that is, which one is P_1 and which is P_2).

The slope is a measure of the "tilt" of a line. In Figure 1.3 we illustrate this by sketching a number of lines with their corresponding slopes. Observe that a horizontal line has slope $m = 0$, whereas a vertical line has **undefined** slope (for a vertical line we must have $x_1 = x_2$, so computation of m would involve division by 0, which is meaningless—that is, undefined). It is also worth noting, as illustrated in Figure 1.4, that a line with positive slope tilts upwards (is rising) as we go from left to right, whereas a line with negative slope tilts downwards (is falling) as we go from left to right.

Let us return to the nonvertical line l with slope m. Take a point on the line; denote it by $P = (x_1, y_1)$. Consider now any point $Q = (x, y)$ in the plane. When is point Q on line l? As Figure 1.5 indicates, Q is on l if and only if the slope of the line QP is the same as the slope of line l. In other words,

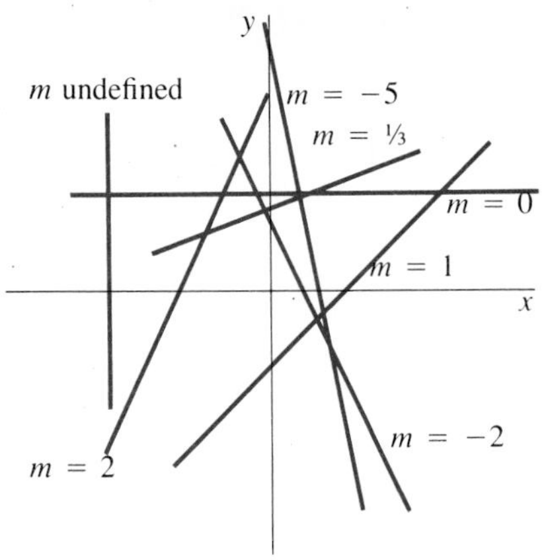

FIGURE 1.3

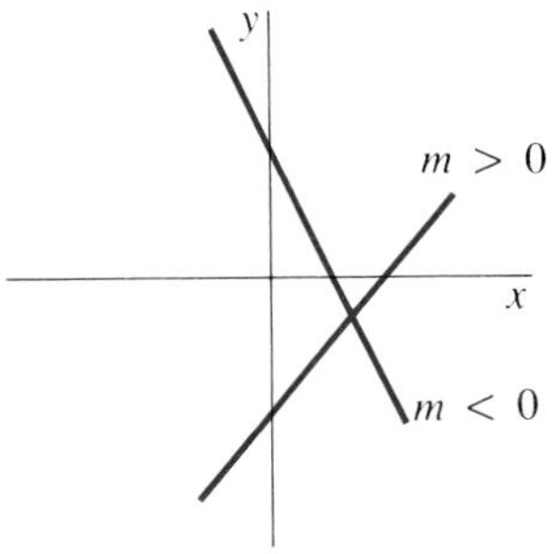

FIGURE 1.4

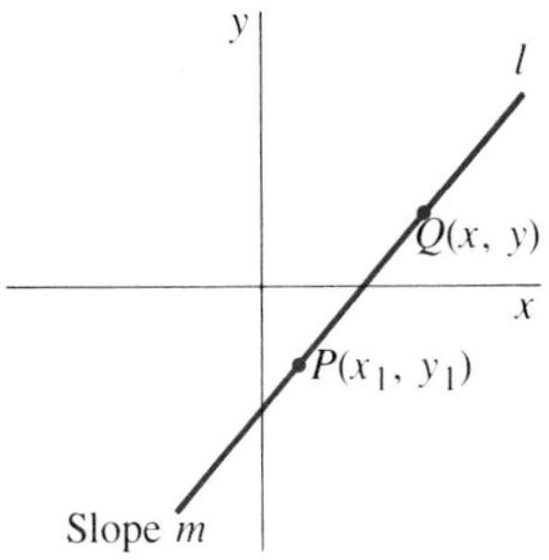

FIGURE 1.5

$$\frac{y - y_1}{x - x_1} = m$$

or

$$y - y_1 = m(x - x_1) \tag{1.2}$$

This is known as the **point–slope form** of the equation of a straight line. Specifically, it is the equation of the line of slope m that passes through the point (x_1, y_1). An arbitrary point in the plane is on this line if and only if its coordinates satisfy Equation (1.2).

EXAMPLE 1 Find an equation of the line of slope 3 that passes through the point $(-1, -5)$.

SOLUTION If we apply the point–slope form, Equation (1.2), with $m = 3$, $x_1 = -1$, and $y_1 = -5$, we have

$$y - (-5) = 3[x - (-1)]$$

or, equivalently,

$$y + 5 = 3x + 3$$

which, in turn, is the same as

$$y = 3x - 2$$

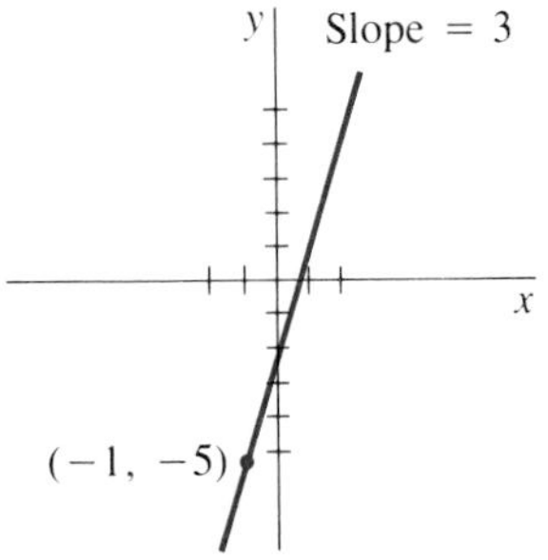

FIGURE 1.6
Line with slope 3 through the point $(-1, -5)$

The point (x_1, y_1) in the point–slope form, Equation (1.2), could be *any* point on the line. One point that is frequently used is the point where the line crosses the y-axis. The x-coordinate of this point is 0; the y-coordinate, called the y-intercept, is generally denoted by b. If we use the point–slope form for the equation of the line through the point $(0, b)$ with slope m, the result is

$$y - b = m(x - 0)$$

or

$$y = mx + b \tag{1.3}$$

This familiar form of the equation of a straight line is called the **slope–intercept form**.

In applications, letters other than x and y are frequently used to name the variables of interest. In the next example we consider the changing value of a building over time. Here V replaces y and t replaces x as the names of the axes.

EXAMPLE 2 A building was purchased new for $149,400, and it depreciates at a constant rate of $8300 per year until its value is reduced to zero. What is the building's value 5 years after purchase?

SOLUTION Let V denote the building's value t years after purchase. Because V decreases at $8300 per year, the plot of V versus t is a straight line with slope -8300, passing through (0, 149,400). By the slope–intercept formula,

$$V = -8300t + 149{,}400$$

After $t = 5$ years, $V = -8300(5) + 149{,}400 = \$107{,}900$.

EXAMPLE 3 A manufacturing company's costs consist of *fixed costs* plus *variable costs.* Fixed costs (such as rent, heat, and license fees) are independent of how many units are produced and are the same even if no units are produced.

Suppose the Acme Company's fixed costs are $1500 per day and that, when 300 units are produced per day, the total daily cost C is $8700. Assuming that C and the number x of units produced per day are linearly related, find an equation for C in terms of x.

SOLUTION The desired equation has the form $C = mx + b$. The total cost when $x = 0$ consists only of fixed costs, so $b = 1500$:

$$C = mx + 1500$$

Moreover, the point (300,8700) must satisfy this equation, so

$$8700 = m(300) + 1500$$

Solving for m, we obtain $m = 24$. Consequently,

$$C = 24x + 1500$$

EXAMPLE 4 A business's energy expenditures were $83,000 in 1980 and $115,000 in 1985. Assuming a linear relationship between energy expenditures and time, estimate the company's energy costs in 1990.

SOLUTION The arithmetic will be a little easier if we choose the year 1980 as the origin of our t-axis, as shown in Table 1.1.

TABLE 1.1

t	*Year*	E
0	1980	83,000
5	1985	115,000
10	1990	?

In other words, let t be the number of years since 1980. Then

$$m = \frac{115000 - 83000}{5 - 0} = 6400$$

and the point–slope formula gives

$$E - 83000 = 6400(t - 0)$$

or

$$E = 6400t + 83000$$

(We could have used the slope–intercept formula instead.)

For 1990 $t = 10$, so $E = \$147{,}000$ is our estimate of energy costs in 1990.

EXAMPLE 5 Find the equation of the line passing through the points $(3, -1)$ and $(-2, -1)$.

SOLUTION The situation is depicted in Figure 1.7.

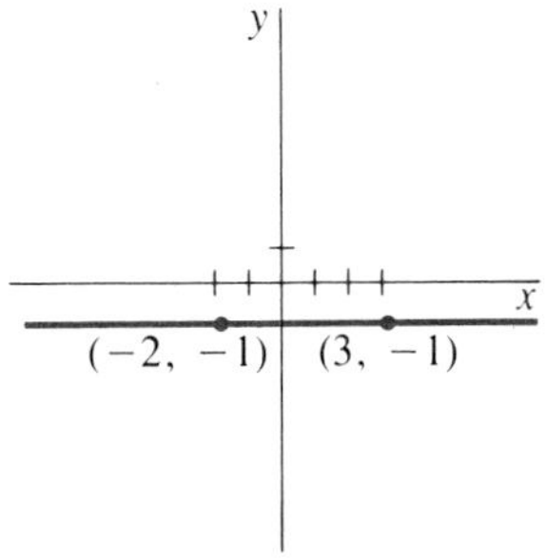

FIGURE 1.7
Line through $(-2, 1)$ and $(3, -1)$

The line has slope

$$m = \frac{(-1) - (-1)}{3 - (-2)} = 0.$$

Because the line is horizontal (parallel to the x-axis), all its points have the y-coordinate -1, including the intersection of the line with the y-axis. Hence $b = -1$, and the point–slope formula, $y = mx + b$, becomes $y = -1$. In general every horizontal line has an equation of the form $y = b$.

EXAMPLE 6 Find the equation of the line determined by the points $(3, -1)$ and $(3, 4)$. See Figure 1.8.

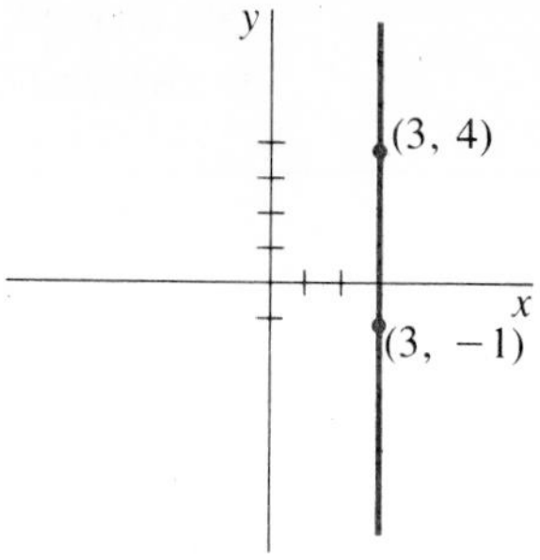

FIGURE 1.8
Line through (3, 4) and (3, −1)

SOLUTION The two points have the same x-coordinate, so the line is vertical and the slope is undefined (Figure 1.8). In this case the line has the equation

$$x = 3$$

Note that this equation is not of the standard form $y = mx + b$. As a matter of fact, one purpose of this example is to emphasize that equations of the form $y = mx + b$ represent only nonvertical straight lines. Vertical lines have equations of the form $x =$ constant.

In general, consider an equation of the form

(1.4) $$ax + by = c$$

where a, b, and c are constants.

1 If $b \neq 0$, we may divide through by b and solve for y. This gives

$$y = \left(\frac{-a}{b}\right)x + \left(\frac{c}{b}\right)$$

which tells us that, in this case, Equation (1.4) represents the nonvertical line with slope $-a/b$ and the y-intercept c/b.

2 If $b = 0$, Equation (1.4) becomes $ax = c$. So, if $a \neq 0$, Equation (1.4) represents the vertical line

$$x = \frac{c}{a}$$

whereas, if $a = 0$, there is nothing to talk about because we have $0 = c$.

We conclude that the equations of form $ax + by = c$ represent *all* straight lines, vertical and nonvertical. A word of caution: Two different equations of this form may represent the same straight line. For example, $2x - 3y = 5$ and $-4x + 6y = -10$ both represent the same line $y = \frac{2}{3}x - \frac{5}{3}$.

EXAMPLE 7 Find the point of intersection of the lines $2x - 3y = 5$ and $3x + 4y = 9$ (Figure 1.9).

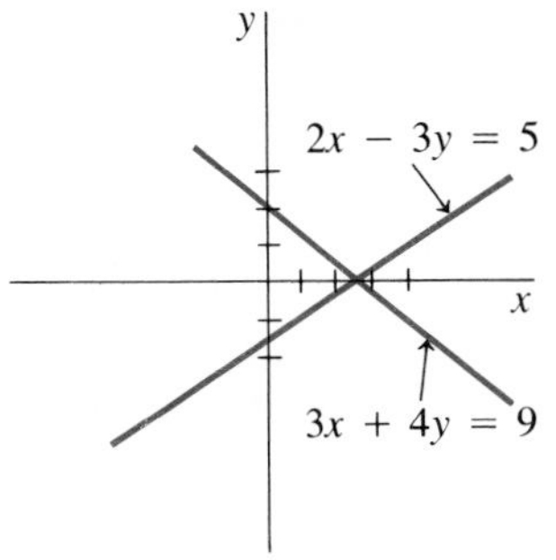

FIGURE 1.9

SOLUTION The point of intersection is on both lines, so it satisfies both equations. Thus the way to find it is to solve the two equations simultaneously. One way to do this is to multiply the first equation by 4 and the second equation by 3, which gives

$$8x - 12y = 20$$
$$9x + 12y = 27$$

Adding these equations eliminates y and gives $17x = 47$, so $x = \frac{47}{17}$. Substituting this value of x in either of the original equations, we get $y = \frac{3}{17}$, so the point of intersection is $(\frac{47}{17}, \frac{3}{17})$. This problem is just like those considered in Section 3 of Appendix A.

Now suppose we are given straight lines l_1 and l_2 with slopes m_1 and m_2, respectively. It is clear that l_1 and l_2 are parallel if and only if their slopes are equal. Thus

$$l_1 \parallel l_2 \leftrightarrow m_1 = m_2$$

It is also known that nonvertical lines l_1 and l_2 are perpendicular if and only if their slopes are negative reciprocals of each other (which is equivalent to saying that the product of their slopes is -1). Thus

(1.5) $$l_1 \perp l_2 \leftrightarrow m_2 = -\frac{1}{m_1} \leftrightarrow m_1 m_2 = -1$$

The case where a vertical line appears can be handled directly.

EXAMPLE 8 Given the line $3x - 5y = 12$ (call it l) and the point $P = (2, 1)$ (see Figure 1.10), find the equation of the line that passes through P and is:

(a) parallel to l. (b) perpendicular to l.

SOLUTION If we solve the equation of l for y, we obtain $y = \frac{3}{5}x - \frac{12}{5}$. Thus l has slope $\frac{3}{5}$.

(a) We want the line with slope $\frac{3}{5}$ that passes through (2, 1). Using the point–slope formula and simplifying, we get $3x - 5y = 1$.

(b) We want the line with slope $-\frac{5}{3}$ that passes through (2, 1). By using the same procedure, we get $5x + 3y = 13$.

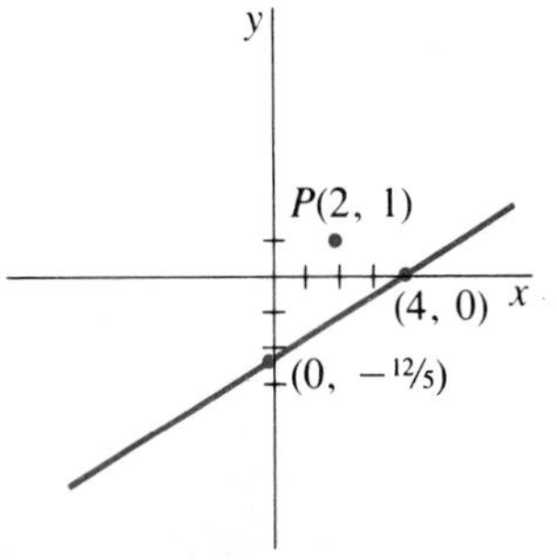

FIGURE 1.10

Let us consider some further applications of these ideas.

EXAMPLE 9 Suppose that a manufacturer of a certain product can sell 80 units if she prices them at \$10 each. Experience tells us that an increase in the price usually results in a reduced demand for the product, whereas a price cut usually results in increased sales. An equation that relates the demand for a product, q_d (quantity demanded), to the price p at which it can be sold is called a **demand equation**. For instance, imagine that our manufacturer finds that the demand for her product declines to only 60 units at a price of \$20. If the demand equation is assumed to be linear, what is the demand equation?

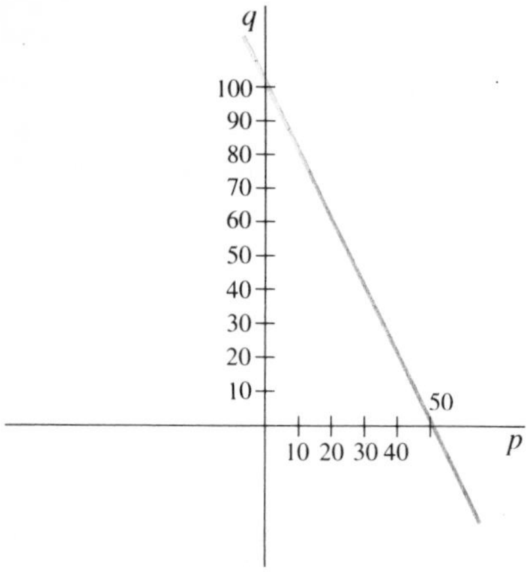

FIGURE 1.11

SOLUTION We set up a pair of axes with the price, p, on the horizontal axis, and the demand, $q = q_d$, on the vertical axis (Figure 1.11). The description of our problem tells us that the points $(p_1, q_1) = (10, 80)$ and $(p_2, q_2) = (20, 60)$ must lie on the line. The slope of the line (with p playing the role of x and q playing the role of y) is given by

$$m = \frac{q_2 - q_1}{p_2 - p_1} = \frac{60 - 80}{20 - 10} = -2$$

The point–slope equation of the demand equation is then

$$q_d - 80 = -2(p - 10)$$

or

$$q_d = 100 - 2p$$

EXAMPLE 10 In the preceding example, we introduced the concept of a demand equation, which relates the price of an item to the number of that item consumers will buy. But the supply of a product also depends on its price. Manufacturers are eager to produce additional units as the price at which they can sell them increases, but they cut back as the selling price drops. An equation relating the supply of a product, q_s (quantity supplied), to its price p is called a **supply equation**.

Suppose, for example, that the supply equation for the product discussed in Example 9 is

$$q_s = 3p - 50$$

This means that, if the product sells for \$20, manufacturers will produce only $q_s = 3(20) - 50 = 10$ units, whereas, if the price increases to \$100, the supply will be $q_s = 3(100) - 50 = 250$.

In a free market there is a continual interplay between supply and demand. If the price is very low, demand is high and supply is low. This excessive demand tends to increase the price, stimulating manufacturers to produce additional supplies. Conversely, if the price is very high, manufacturers produce more but demand is small. This excess supply tends to depress prices, and soon demand increases. In a purely competitive economy, **equilibrium** is achieved at a price where supply and demand are in balance. Mathematically, the equilibrium price is found as the simultaneous solution of the supply and demand equations.

Let us find the equilibrium price for the equations

$$q_d = 100 - 2p \quad \text{and} \quad q_s = 3p - 50$$

SOLUTION We set the equations equal to one another.

$$100 - 2p = 3p - 50$$
$$150 = 5p$$
$$30 = p$$
$$q = 100 - 2(30) = 40$$

Thus both supply and demand are 40 units at a price of $30 (Figure 1.12).

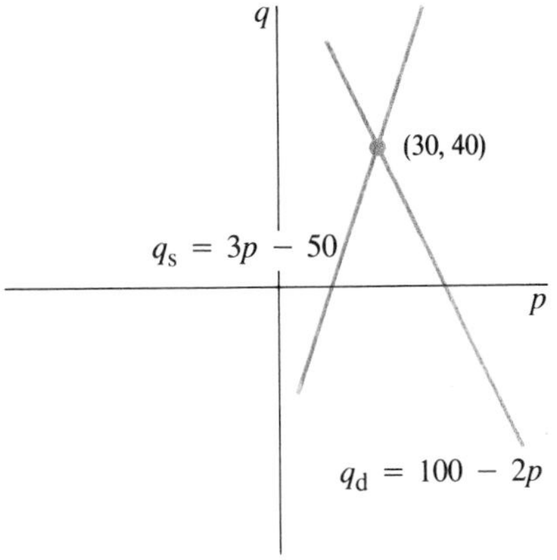

FIGURE 1.12

EXAMPLE 11 Suppose the government decides to raise revenue by imposing a tax of $5 per unit on the sellers in the previous example. What will be the effect of this policy on the equilibrium price?

SOLUTION The demand equation remains unchanged:

$$q_d = 100 - 2p$$

The supply equation must be modified to take the tax into account. The sellers will receive $5 less on each sale, so that they will produce goods according to the formula

$$q_s = 3(p - 5) - 50$$

(Do you see why?) Thus we must find the point of intersection of the lines

$$q = 100 - 2p$$

and

$$q = 3p - 15 - 50 = 3p - 65$$

Setting these equations equal results in

$$100 - 2p = 3p - 65$$
$$165 = 5p$$
$$p = 33$$

The new equilibrium price is $33. Of this, the seller must give the government $5.

It is interesting to note that the seller will keep \$28 (compared to \$30 without any tax), while the consumer will pay \$33 (compared to \$30). Thus, despite the government's attempt to tax sellers, the greater part of the tax burden is passed along to the consumer in a free market.

PRACTICE EXERCISE Suppose that, in Examples 10 and 11, the government decides to impose a 5% sales tax rather than a flat \$5 per item. Find the new equilibrium price. How much of this does the seller keep? *Note*: The before-tax price must be multiplied by 1.05 to get the price p including tax, so p must be *divided* by 1.05 to get what the seller receives.

[Your answer should be \$30.88, of which the seller keeps \$29.41 and the government receives \$1.47. Thus the seller winds up receiving 59¢ less while the buyer pays 88¢ more.]

Thus far we have been concerned with points and lines. We conclude this section with a few comments about circles. Here the key role is played by the notion of distance.

EXAMPLE 12 Find the distance between the points $P_1 = (1, -3)$ and $P_2 = (4, 2)$.

SOLUTION We wish to find the length of the line segment $\overline{P_1P_2}$; call it d. As Figure 1.13 illustrates, we draw lines parallel to the x- and y-axes through P_1 and P_2, respectively. Their intersection point Q has coordinates $(4, -3)$. Then $\overline{P_1Q}$ has length 3 and $\overline{QP_2}$ has length 5; so, according to the Pythagorean theorem, $d^2 = 3^2 + 5^2$ and thus $d = \sqrt{34}$.

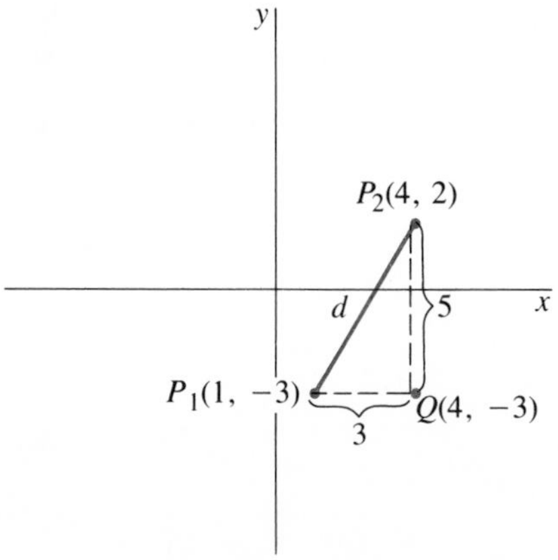

FIGURE 1.13

In the same way, if we are given two arbitrary points in the plane $P_1 = (x_1, y_1)$ and $P_2 = (x_2, y_2)$, by using the Pythagorean theorem (as Figure 1.14 illustrates), we find that the distance d between them is

$$d = \sqrt{(x_2 - x_1)^2 + (y_2 - y_1)^2} \tag{1.6}$$

Note that in the *distance formula*, Equation (1.6), it does not matter which

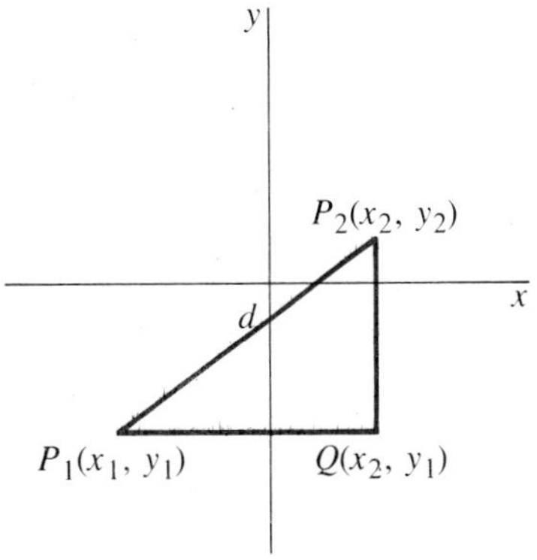

FIGURE 1.14

point is called P_1 because $(x_2 - x_1)^2 = (x_1 - x_2)^2$ and $(y_2 - y_1)^2 = (y_1 - y_2)^2$.

EXAMPLE 13 Describe (that is, characterize) all points on the circle of radius 5 whose center is at (2, 3).

SOLUTION In Figure 1.15, let $P = (x, y)$ be an arbitrary point in the plane. When is P on the circle in question? Clearly, if and only if the distance from P to the center (2, 3) is 5. Therefore the condition is

$$\sqrt{(x - 2)^2 + (y - 3)^2} = 5$$

which, when we square both sides, becomes

$$(x - 2)^2 + (y - 3)^2 = 5^2$$

We refer to this as the equation of the given circle; a point is on the circle if and only if its coordinates satisfy this equation.

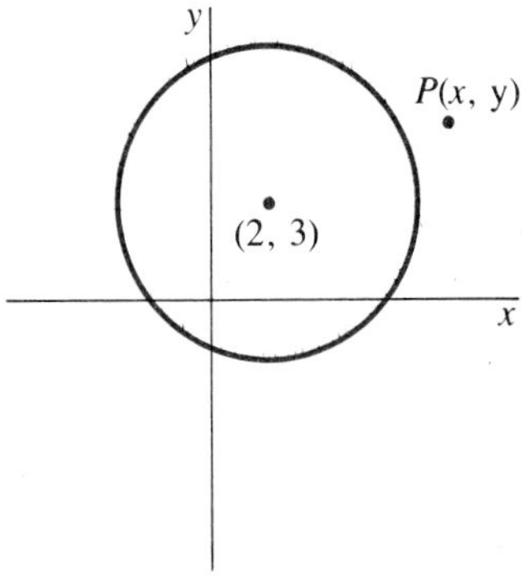

FIGURE 1.15
Circle of radius 5 centered at (2, 3)

In exactly the same way, the circle with center at (a, b) and radius r (Figure 1.16) has the equation

(1.7)
$$(x - a)^2 + (y - b)^2 = r^2$$

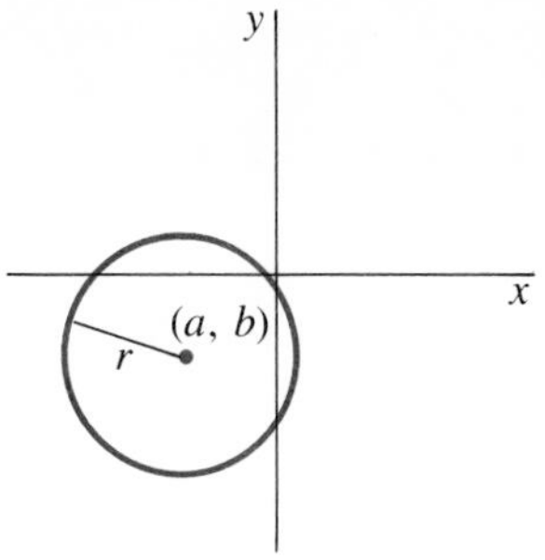

FIGURE 1.16

EXAMPLE 14 What can you say about all points that satisfy the following equation?

$$x^2 + y^2 - 2x + 6y - 20 = 0$$

SOLUTION We rewrite this equation by grouping the terms in x and the terms in y separately.

$$x^2 - 2x \qquad + y^2 + 6y = 20$$

Then we take $x^2 - 2x$ and "complete the square," which means turn it into a perfect square. To do this, we must add 1.

$$x^2 - 2x + 1 = (x - 1)^2$$

Note that, starting from $x^2 - 2x$, we added the square of half the coefficient of the x-term. In the same way, adding half the coefficient of the y-term, we arrive at

$$y^2 + 6y + 9 = (y + 3)^2$$

Thus the original equation may be written as

$$(x - 1)^2 - 1 + (y + 3)^2 - 9 = 20$$

Because we added 1 and 9 in completing squares, we must subtract 1 and 9 in order to obtain an expression that is equivalent to the left side of the original equation. We now have

$$(x - 1)^2 + (y + 3)^2 = 30$$

Consequently we are dealing with the circle of radius $\sqrt{30}$ and center $(1, -3)$!

Section 1.1 EXERCISES

1. Plot each of the following points in the plane. Use the same set of axes.
   (a) $(4, 3)$ (b) $(-1, 2)$ (c) $(3, -3)$
   (d) $(-2, -1)$ (e) $(5, 0)$ (f) $(0, -1)$

2. Give the form of any point:
   (a) on the x-axis (b) on the y-axis

3. In each case, find the slope of the straight line passing through the two given points.
   (a) $(1, 2)$ and $(3, 7)$ (b) $(-3, 7)$ and $(1, -1)$
   (c) $(5, 7)$ and $(-2, 7)$ (d) $(-3, 4)$ and $(-3, -7)$

4. Find the slope and y-intercept (if they exist) of each of the following straight lines.
   (a) $y = 3x + 4$ (b) $y = 3 - 4x$
   (c) $2x - 3y = 5$ (d) $x = 3y + 2$
   (e) $x = 2$ (f) $y = -3$
   (g) $y = \dfrac{2 - 3x}{5}$ (h) $y = -7x$

In Exercises 5–16 find an equation of the given line.

5. Slope 3 and passing through $(-1, 4)$
6. Slope -2 and passing through $(3, -2)$
7. Slope -1 and y-intercept $(0, 5)$
8. Slope 0 and y-intercept $(0, -3)$
9. Passing through $(1, 3)$ and $(2, 7)$
10. Passing through $(3, 1)$ and $(7, 2)$
11. Passing through $(3, 6)$ and $(3, -2)$
12. Passing through $(6, 3)$ and $(-2, 3)$
13. Passing through $(-1, -2)$ and parallel to $y = 2x - 3$
14. Passing through $(2, 3)$ and parallel to $x = 5$
15. Passing through $(-4, 3)$ and parallel to $2x - 5y = 4$
16. Passing through $(-4, 3)$ and parallel to $y = 3$

In Exercises 17–22 determine where each of the pairs of straight lines meet.

17. $y = 2x + 5$ and $y = -3x + 7$
18. $y = 2x + 5$ and $y = 3x + 5$
19. $2x + 3y = 5$ and $3x + 4y = 8$
20. $x + 3y = 5$ and $3x - 2y = 2$
21. $3x - 2y = 2$ and $9x - 6y = 6$
22. $2x + 3y = 5$ and $6x + 9y = 10$

In Exercises 23–28 find the equation of the straight line that passes through the given point and is perpendicular to the given line.

23. Point: $(-3, 4)$ line: $2x + 3y = 5$
24. Point: $(0, 0)$ line: $x - y = 2$
25. Point: $(1, 1)$ line: $y = -3x + 4$
26. Point: $(2, -3)$ line: $3x = 4$
27. Point: $(3, 0)$ line: $\sqrt{2}y + 3 = 0$
28. Point: $(-1, 2)$ line through $(3, -3)$ and $(0, 4)$

29. Find the distance between each of the following pairs of points.
   (a) $(3, 0)$ and $(7, 0)$ (b) $(0, -2)$ and $(0, 5)$
   (c) $(3, 4)$ and $(4, 3)$ (d) $(0, 0)$ and $(7, -24)$
   (e) $(3, 1)$ and $(-7, 1)$ (f) $(-2, 6)$ and $(4, 7)$
   (g) $(\sqrt{2}, -\sqrt{2})$ and $(3, 4)$ (h) (a, b) and (x, y)

In Exercises 30–34 find the equation of each of the indicated circles.

30. Center at $(2, 2)$; radius 3
31. Center at $(-2, -5)$; radius $\sqrt{2}$
32. Center at $(-3, 4)$; passing through $(-5, 7)$
33. Center at $(-5, 2)$; tangent to the y-axis
34. Center at $(-3, 4)$; tangent to the x-axis

Each of the equations in Exercises 35–40 is the equation of a circle. Find the center and the radius.

35. $(x - 2)^2 + (y - 3)^2 = 16$
36. $(x + 3)^2 + (y - 4)^2 = 17$
37. $x^2 + y^2 - 4x - 6y = 3$
38. $x^2 + 4x + y^2 - 10y + 12 = 0$
39. $2x^2 + 2y^2 + 8x + 4y = 10$
40. $2x^2 + 2y^2 + 8x + 4y + 10 = 0$
41. Consider the circle with center at $(3, 4)$ and passing through the point $(5, 7)$. Does it pass through $(1, 1)$? Through $(6, 6)$?
42. Which of the following points lie on the unit circle (the circle of radius 1 centered at the origin)? If a point does not lie on the unit circle, determine whether it is inside the unit circle or outside.
   (a) $(1, 0)$ (b) $(-\frac{3}{5}, \frac{4}{5})$ (c) $(-1, 1)$
   (d) $(\frac{1}{2}, \frac{1}{2})$ (e) $\left(\frac{\sqrt{2}}{2}, \frac{-\sqrt{2}}{2}\right)$ (f) $\left(\frac{\sqrt{2}}{2}, 1\right)$
   (g) $(0, -1)$ (h) $(\frac{8}{17}, \frac{15}{17})$ (i) $\left(\frac{-\sqrt{3}}{2}, \frac{1}{2}\right)$
   (j) $\left(\frac{1}{2}, \frac{\sqrt{2}}{2}\right)$

In Exercises 43–46 decide whether the three given points are *collinear*. (That is, do they lie on the same straight line?)

43. $(5, 3), (2, 2), (8, 4)$
44. $(-5, 3), (7, -4), (-4, 4)$
45. $(1, 4), (2, 6), (-1, -2)$

46. $(0, 3), (2, 1), (-3, -9)$

47. Is the triangle with vertices at $(2, 5), (-4, -1)$, and $(6, -5)$ isosceles?

48. Is the triangle with vertices at $(3, 4)$, $(7, 7)$, and $(6, 8)$ isosceles?

49. Is the triangle with vertices at $(2, 3), (-3, 0)$, and $(-1, 8)$ a right triangle?

50. Find algebraically all points in the plane that are equidistant from the points $(2, 5)$ and $(4, 8)$. (*Hint*: This is the perpendicular bisector of the segment joining the two points.)

51. Are the points $(-2, -3)$, $(6, 5)$, $(3, -4)$, and $(1, 6)$ the vertices of a parallelogram?

52. A parallelogram has vertices at $(0, 0), (8, 2)$, and $(3, 6)$. Find the fourth vertex. How many answers are there?

53. Show that the straight line with x-intercept $a \neq 0$ and y-intercept $b \neq 0$ has the equation

$$\frac{x}{a} + \frac{y}{b} = 1$$

This is known as the **two-intercept form** of the equation of a line.

54. Find algebraically all points in the plane whose distance from $(-2, 3)$ is twice their distance from $(4, -5)$.

55. For which values of m will the two straight lines $4x - 3y + 11 = 0$ and $mx - y + 5 = 0$ be:
(a) parallel? (b) perpendicular?

56. Find all values of m for which the straight lines $2x - my + 3 = 0$ and $x + my + 5 = 0$ are:
(a) parallel (b) perpendicular

57. Find the equation of the line tangent to the unit circle at the point $(\frac{4}{5}, \frac{3}{5})$.

58. Consider the circle $(x - 2)^2 + (y + 3)^2 = 73$. Find the equation of the tangent to the circle at the point $(-1, 5)$.

59. A book publisher believes that he can sell 25,000 copies of a forthcoming novel if the price is \$12 and 31,000 copies if the price is \$9. Find the demand equation (assuming that it is linear).

60. A supermarket manager can sell 1040 qt of milk per day if the price is 57¢/qt and 1280 qt if the price is 51¢. Assuming a linear demand equation, how many can she sell at 55¢/qt? What is the demand equation?

61. Suppose milk producers are willing to provide the supermarket described in Exercise 60 with 1100 qt of milk per day when the retail price is 57¢/qt and with 1400 qt when the retail price is 60¢/qt. Assuming that the supply equation is linear, find it.

62. In connection with Exercises 60 and 61, find the equilibrium price. (Your answer may not be an integer.)

63. What happens to the equilibrium price in Exercise 62 if the government imposes:
(a) a tax of 1¢ per quart on the producer?
(b) a 4% sales tax (on the consumer)?
(See Example 11 and following Practice Exercise.)

64. A bus tour company finds that it can sell 120 tickets at \$5 each and that, for each 25¢ reduction in the price, an additional 10 tickets can be sold. Find the demand equation.

65. A woman sets up a word processing business in her home. She figures that her fixed costs (for equipment and furniture amortization, utilities, insurance, supplies, and so on) amount to \$240 per week. Her operating costs are \$5/hr, and she bills her clients at the rate of \$20/hr. Find her cost and gross income from working x hr per week. How many hours a week must she work to break even?

SECTION 1.2

The Concept of a Function

Researchers in almost every field are constantly striving to establish quantitative relationships. Demographers want to know how reproductive rates depend on population density. Biologists want to know how many mutations are associated with a particular radiation dosage, or how cell metabolism depends on cell size. Economists ask how demand for a product depends on its price or how changes in interest rates affect the availability of mortgage money. Psychologists study the relationship between learning time and the size of the task being learned. Sociologists examine the dependence of suicide rates on such factors as age and economic status. The number of such relationships is literally infinite, and you can no doubt think of others from your own experience.

As an example of such a quantitative relationship, consider the well-known formula

$$F = \tfrac{9}{5}C + 32$$

which relates a temperature C, measured in degrees Celsius, to the equivalent temperature F, measured in degrees Fahrenheit. We formalize this concept of relationship into the following definition.

DEFINITION 1.1 A **function** is a relationship between two variables such that a unique value of the second variable is associated with each value of the first variable. The set of values that can be taken on by the first variable is called the **domain** of the function; the set of values of the second variable that are associated with these by the function is called the **range** of the function. □

It is customary to call the first variable (C in our example) the **independent variable** and to call the second variable (F in the example) the **dependent variable**. Of course, other letters could have been used for the variables, so

$$y = \tfrac{9}{5}x + 32$$

describes the same function.

Often we give a name to a function (such as f, g, or h). Then, if x is a number in the domain of a function f, the value associated with or "assigned" to x is denoted $f(x)$, which is read "f of x." Thus our temperature example involves the function f defined by

$$f(x) = \tfrac{9}{5}x + 32$$

Although x is most commonly used to name the independent variable any other letter could be used; thus $f(t) = \tfrac{9}{5}t + 32$ is the same function. The key idea is the rule by which values are assigned, not the labels given the variables.

EXAMPLE 15 Consider the square root function,

$$f(x) = \sqrt{x}$$

Whenever we write $\sqrt{\ }$, we mean the nonnegative square root. Thus we have, for example, $f(4) = \sqrt{4} = 2$, $f(7) = \sqrt{7}$, $f(0) = 0$, $f(t) = \sqrt{t}$, and $f(x - 1) = \sqrt{x - 1}$.

The domain of a function is often stated explicitly, as in

$$g(t) = -16t^2 + 64t + 192 \qquad 0 \leq t \leq 6$$

which signifies that the domain is the closed interval [0, 6]. In other cases, the domain is implied by the formula, as in the case of the square root function. Negative numbers do not have real square roots, so the domain of the square root function consists of all numbers greater than or equal to zero. The range is the same set of numbers, because every nonnegative number is the square root of some nonnegative number.

EXAMPLE 16 As another example, consider the function

$$h(u) = (4 - u^2)^{1/2}$$

Because of the $\frac{1}{2}$ power (square root), we must have $4 - u^2 \geq 0$, or $-2 \leq u \leq 2$. Thus the domain of h is $[-2, 2]$, and a moment's thought will reveal that the range is the interval [0, 2].

In general, a function is usually defined by simply giving the rule (in most cases, a formula) with no explicit mention of the domain. In such a case, as in the preceding example, the domain of the function is the set of all numbers for which it "makes sense" to apply the rule or formula.

Often the domain is determined by some limitation of the real-world situation being modeled. For example, quantities such as length and area must be positive, hourly wages must not be below the legal minimum wage, and one cannot consume more resources than one has. In our initial temperature example, physicists tell us that C cannot be less than "absolute zero," or approximately $-273.16°$. Thus the domain is $[-273.16, \infty)$ and the range is $[-459.69, \infty)$.

EXAMPLE 17 The Music Tent Summer Theater offers a tour operator the following deal. A minimum of 80 tickets must be purchased, for which the price will be \$15 per ticket. However, for each additional ticket purchased above 80, the price for *all* tickets purchased will be reduced by 10¢. Find the price $p = p(x)$ (that is, the *demand function*) and the revenue $R = R(x)$ when x tickets are purchased (assume $x \geq 80$).

SOLUTION The excess over 80 is $x - 80$. Because the price of every ticket is reduced \$0.10 for each of these $x - 80$ tickets, the price is

$$p = 15 - 0.1(x - 80) = 23 - 0.1x$$

The revenue is the product of the price by the number of tickets sold, or

$$R = px = 23x - 0.1x^2$$

EXAMPLE 18 Let $f(x) = x^2 - 3x$. Evaluate: (a) $f(-x)$; (b) $f(1 - t)$; (c) $f(x + h)$.

SOLUTION We can write the function as $f(\ ) = (\ )^2 - 3(\ )$, meaning that whatever is placed within the parentheses on the left must be placed within each pair of parentheses on the right. Thus:

(a) $f(-x) = (-x)^2 - 3(-x) = x^2 + 3x$

(b) $f(1 - t) = (1 - t)^2 - 3(1 - t) = 1 - 2t + t^2 - 3 + 3t = t^2 + t - 2$

(c) $f(x + h) = (x + h)^2 - 3(x + h) = x^2 + 2xh + h^2 - 3x - 3h$

An expression involving functions that we shall explore in great detail in Chapter 2 is

(1.8) $$\frac{f(x+h)-f(x)}{h}$$

This expression lies at the heart of the calculus (as a matter of fact, one of the main purposes of our detailed discussion of functions is to enable us to understand and handle this expression). It is called the **difference quotient**, or the **Newton quotient**, of f. The difference quotient is relatively easy to compute if the function f is not too complicated. Let us do this for a few simple functions.

EXAMPLE 19 (a) The function f is given by

$$f(x) = 2x + 3$$

which may write as

$$f(\quad) = 2(\quad) + 3$$

Thus $f(x+h) = 2(x+h) + 3 = 2x + 2h + 3$, and the difference quotient becomes

$$\begin{aligned}\frac{f(x+h)-f(x)}{h} &= \frac{(2x+2h+3)-(2x+3)}{h}\\ &= \frac{2h}{h}\\ &= 2\end{aligned}$$

(b) The function f is given by

$$f(x) = x^2 + 1$$

Alternatively, we may write this as

$$f(\quad) = (\quad)^2 + 1$$

Then

$$f(x+h) = (x+h)^2 + 1 = x^2 + 2xh + h^2 + 1$$

and the difference quotient becomes

$$\begin{aligned}\frac{f(x+h)-f(x)}{h} &= \frac{(x^2+2xh+h^2+1)-(x^2+1)}{h}\\ &= \frac{2xh+h^2}{h}\\ &= 2x + h\end{aligned}$$

(c) The function f is given by

$$f(x) = x^3 - x^2 + 5x - 2$$

Then

$$\begin{aligned} f(x+h) &= (x+h)^3 - (x+h)^2 + 5(x+h) - 2 \\ &= (x^3 + 3x^2h + 3xh^2 + h^3) - (x^2 + 2xh + h^2) \\ &\quad + (5x + 5h) - 2 \\ &= x^3 + 3x^2h + 3xh^2 + h^3 - x^2 - 2xh - h^2 \\ &\quad + 5x + 5h - 2 \end{aligned}$$

and the difference quotient becomes

$$\begin{aligned} &\frac{f(x+h) - f(x)}{h} \\ &\quad = \frac{(x^3 + 3x^2h + 3xh^2 + h^3 - x^2 - 2xh - h^2 + 5x + 5h - 2)}{h} \\ &\qquad - \frac{(x^3 - x^2 + 5x - 2)}{h} \\ &\quad = \frac{3x^2h + 3xh^2 + h^3 - 2xh - h^2 + 5h}{h} \\ &\quad = 3x^2 + 3xh + h^2 - 2x - h + 5 \end{aligned}$$

(d) The function f is given by

$$f(x) = \frac{1}{x}$$

which may be written as

$$f(\quad) = \frac{1}{(\quad)}$$

Observe that $f(0)$ is undefined, so the domain of f consists of all real numbers except 0. Because

$$f(x+h) = \frac{1}{x+h}$$

the difference quotient becomes

$$\begin{aligned} \frac{f(x+h) - f(x)}{h} &= \frac{\dfrac{1}{x+h} - \dfrac{1}{x}}{h} \\ &= \frac{\dfrac{x}{x(x+h)} - \dfrac{x+h}{x(x+h)}}{h} \\ &= \frac{\dfrac{-h}{x(x+h)}}{h} \\ &= \left(\frac{-h}{x(x+h)}\right)\left(\frac{1}{h}\right) \\ &= \frac{-1}{x(x+h)} \end{aligned}$$

Section 1.2 EXERCISES

In each of Exercises 1–6 a function* f is given. Compute (provided the computation is permissible):

(a) $f(1)$ (b) $f(-3)$ (c) $f(0)$ (d) $f(2.5)$ (e) $f(\pi)$ (f) $f(-\sqrt{2})$ (g) $f(t)$

1. $f(x) = 7x$

2. $f(x) = 5x - 2$

3. $f(x) = |1 + x|$

4. $f(x) = 2x - 3$

5. $f(x) = \sqrt{2 - x}$

6. $f(x) = \dfrac{|x|}{x}$

In each of Exercises 7–12 find the value of the given function f for the following values of the independent variable:

(a) $-x$ (b) $-2 + h$ (c) $h^2 - 1$ (d) $x + h$ (e) $\dfrac{1}{x - 1}$

7. $f(x) = 4 - 3x$

8. $f(x) = (x - 1)(2x + 3)$

9. $f(x) = x^2 + x + 1$

10. $f(x) = 1 + x^2$

11. $f(x) = x^3 - x + 1$

12. $f(x) = \dfrac{x - 1}{x^2 + 1}$

In each of Exercises 13–18, find the domain of the given function f.

13. $f(x) = x - 3$

14. $f(x) = 5 - 2x$

15. $f(x) = \dfrac{1}{x - 3}$

16. $f(x) = \sqrt{x - 3}$

17. $f(x) = \sqrt{3 - x}$

18. $f(x) = 4 + 2x - x^2$

In each of Exercises 19–24, find the range of the given function f.

19. $f(x) = x + 1$

20. $f(x) = 5 - 2x$

21. $f(x) = |1 - x|$

22. $f(x) = \sqrt{1 + x}$

23. $f(x) = \dfrac{1}{x - 2}$

24. $f(x) = \sqrt{x - 2}$

In each of Exercises 25–33, describe the domain and range of the given function f.

25. $f(x) = x|x|$

26. $f(x) = \dfrac{|x|}{x}$

27. $f(x) = x^3$

28. $f(x) = -5x^2$

* The absolute value function, $|x|$, is reviewed in Appendix A.

29. $f(x) = x^3 + 2$

30. $f(x) = \sqrt{x^2 + 9}$

31. $f(x) = \sqrt{x^2 - 4}$

32. $f(x) = \sqrt{4 - x^2}$

33. $f(x) = mx + b$ (m, b constant)

34. The function f assigns to a circle of radius r the area of the circle. Find a formula for $f(r)$. What is the domain of f? What is the range of f?

35. The function g assigns to a circle of radius r the circumference of the circle. Find a formula for $g(r)$. What is the domain of g? What is the range of g?

36. You are told to pick any number and then perform the following operations in sequence: (i) add 1, (ii) square what you have, (iii) subtract four times the number you started with, (iv) take the square root, (v) add 1. This process determines a function; find a formula for this function.

37. The current cost of mailing a first-class letter is 22¢ for the first ounce and 17¢ for each additional ounce or fraction thereof. Let f represent the function that assigns to each letter of weight x (in ounces) the postage required to mail it. (a) Find $f(1)$, $f(2.3)$, and $f(\pi)$; (b) What are the domain and the range of f?

38. Consider the function $f(x) = x^2 - 2x - 2$. What is the domain of f? Does 4 belong to the range of f? Does -4 belong to the range of f? What is the range of f?

39. The U-Drive-It car rental agency charges \$15 per day plus 10¢/mi for a subcompact car. Write an expression for the cost of renting a subcompact for three days and driving a total of x mi.

40. The price p of each theater ticket for a group sale is related to the number x of people in the group by the demand equation

$$p = 30 - \frac{x}{20} \qquad 50 \leq x \leq 500$$

The theater holds only 500 people.

(a) What is the revenue R as a function of x?

(b) The theater owner's fixed costs amount to \$350 per show, and the variable cost is \$8 per ticket. Find the total cost C and profit P ($= R - C$) for a performance given for a group of x people, who have booked the theater exclusively.

41. An open box is to be formed from a 12″ × 12″ square of cardboard by cutting $x'' \times x''$ squares from the four corners and then folding up the resulting flaps (see the accompanying figure). Find the volume of the box as a function of x. What is the domain of this function?

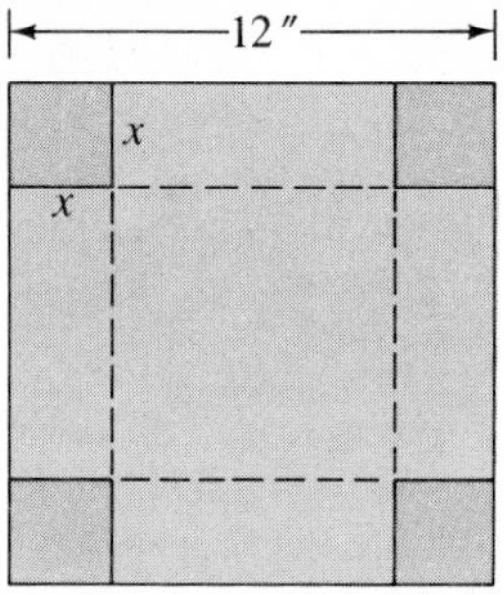

FIGURE FOR EXERCISE 41

42. A power station is on one side of a river that is 2 mi wide. Five miles upstream on the opposite shore is a lumber mill. A power line is laid underwater from the power station to a point on the opposite shore x mi downstream from the mill, and from there overground to the mill (see the accompanying figure).

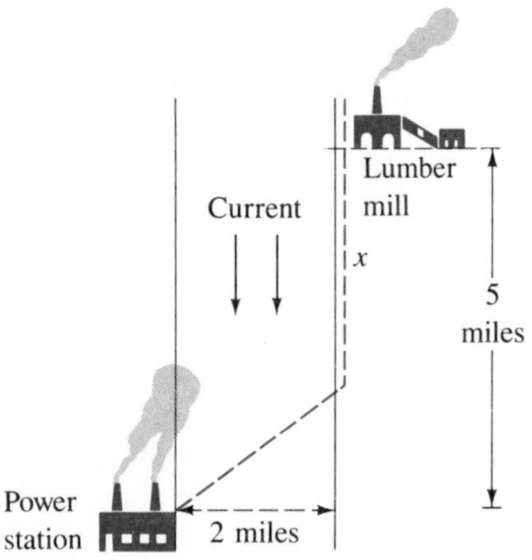

FIGURE FOR EXERCISE 42

(a) Find the total length of the line as a function of x.

(b) It costs \$4200/mi to lay a power line under water and only \$2600/mi to lay it overland. Find the total cost of the power line as a function of x.

43. Consider the function f defined by $f(x) = 2x + 1$. Are there any real numbers a for which $f(a^2) = [f(a)]^2$?

In each of Exercises 44–49 compute:

(a) $\dfrac{f(1 + h) - f(1)}{h}$

(b) $\dfrac{f(-3 + h) - f(-3)}{h}$

for the given function f.

44. $f(x) = 3x - 2$

45. $f(x) = x^2$

46. $f(x) = 1/x$

47. $f(x) = \dfrac{x}{x + 1}$

48. $f(x) = x^2 - \dfrac{1}{x}$

49. $f(x) = (2x + 1)^3$

In each of Exercises 50–57 a function f is given.

(a) Write the expression for $\dfrac{f(x + h) - f(x)}{h}$.

(b) Simplify and evaluate this expression (as far as possible).

50. $f(x) = 2 - 3x$

51. $f(x) = mx + b$ (m, b constant)

52. $f(x) = x^2$

53. $f(x) = 3x^2 - x + 1$

54. $f(x) = 1/x$

55. $f(x) = x^3$

56. $f(x) = |x|$

57. $f(x) = \sqrt{x}$

SECTION 1.3

The Graph of a Function

For each number x in the domain of a given function f, we can plot a point (x, y) in the coordinate plane by taking $y = f(x)$. The collection of all such points is defined to be the *graph of f*.

The graph of a function provides a picture, or geometric representation, of the function. In this section we take the first step in the study of graphs by seeing what the graph looks like for a few simple functions.

How can we draw the graph of a function? Typically, a function f provides us with an infinite number of pairs $(x, f(x))$. It is, however, humanly impossible to plot all of these points in the plane. Instead, the standard procedure is to construct a representative table of points $(x, f(x))$ associated with the function and then sketch a curve that accommodates the points of the table in a reasonable manner. How many pairs $(x, f(x))$ should we tabulate and plot? This is a matter of judgment and experience. Obviously enough points should be plotted so that the graph, with all its significant features, may be sketched with confidence.

EXAMPLE 20 Sketch the graph of the function

$$f(x) = x^2$$

SOLUTION We make a table that lists a few points $(x, f(x))$ on the graph (see Table 1.2).

TABLE 1.2

x	0	1	−1	2	−2	3	−3	4	−4
$f(x) = x^2$	0	1	1	4	4	9	9	16	16

Then a sketch of the graph looks like Figure 1.17. You should check that this sketch is correct. You are probably aware that this curve is known as a **parabola**.

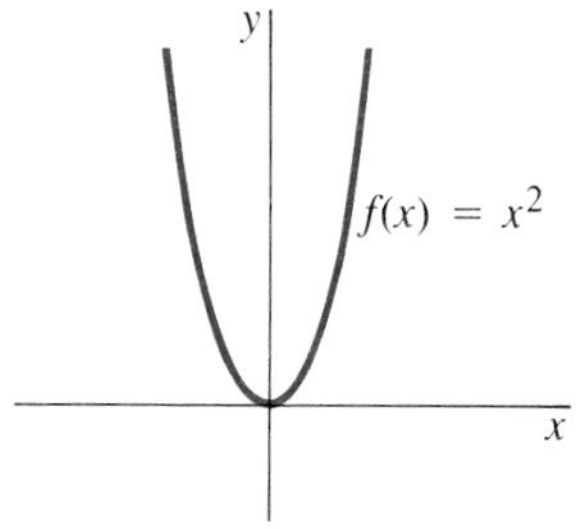

FIGURE 1.17

When we plot points on the graph, $f(x)$ plays the role of the y-coordinate. For this reason, instead of referring to the "graph of the function f," we often speak of the "graph of $y = f(x)$." When we write $y = f(x)$ for any given function f, the value of y depends on the choice of x, and a value of x determines the corresponding y-value unambiguously. This is why x is often called the **independent variable** and y the **dependent variable**.

EXAMPLE 21 Sketch the graph of the function

$$y = f(x) = \tfrac{9}{5}x + 32$$

SOLUTION Recall, from Section 1.2, that f is the rule for converting degrees Celsius to degrees Fahrenheit. In Table 1.3, we have computed several points, and the resulting graph is shown in Figure 1.18.

TABLE 1.3

x	0	5	10	20	30	37	40	100	-5
$y = \frac{9}{5}x + 32$	32	41	50	68	86	98.6	104	212	23

Of course, from our work on lines in Section 1.1, we know that $y = \frac{9}{5}x + 32$ is the equation of a straight line with slope $\frac{9}{5}$ and y-intercept 32. It was therefore not necessary to plot all of these points (any two would have sufficed).

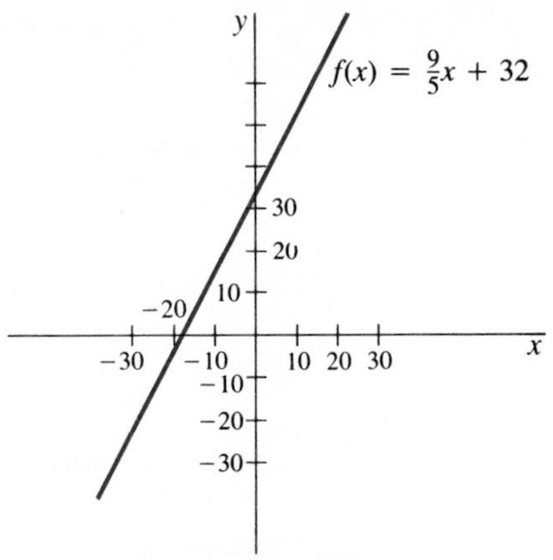

FIGURE 1.18

EXAMPLE 22 Sketch the graph of the square root function

$$y = f(x) = \sqrt{x}$$

SOLUTION Table 1.4 lists a few values. (Note that we may use only x's that are ≥ 0.)

TABLE 1.4

x	0	1	2	3	4	8	9
$y = \sqrt{x}$	0	1	$\sqrt{2}$	$\sqrt{3}$	2	$2\sqrt{2}$	3

The graph takes the form shown in Figure 1.19.

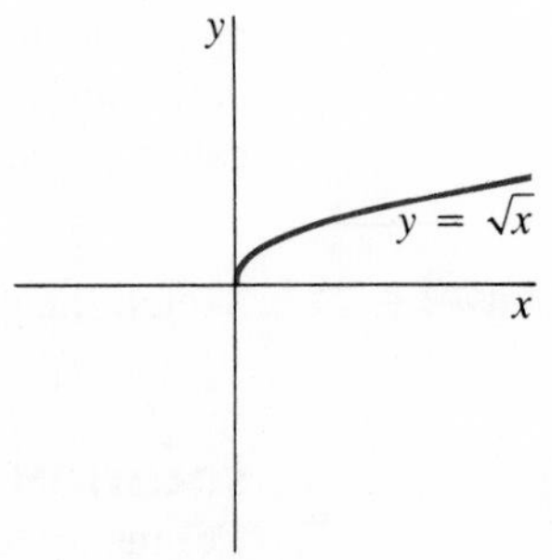

FIGURE 1.19

EXAMPLE 23 Sketch the graph of $y = f(x) = 1/x$.

SOLUTION We make a table such as Table 1.5 (the domain contains all numbers but zero).

TABLE 1.5

x	1	-1	2	-2	3	-3	4	-4	$\frac{1}{2}$	$-\frac{1}{2}$	$\frac{1}{3}$	$-\frac{1}{3}$	$\frac{1}{4}$	$-\frac{1}{4}$
$y = 1/x$	1	-1	$\frac{1}{2}$	$-\frac{1}{2}$	$\frac{1}{3}$	$-\frac{1}{3}$	$\frac{1}{4}$	$-\frac{1}{4}$	2	-2	3	-3	4	-4

and arrive at the graph shown in Figure 1.20. This curve is called a **hyperbola**; it has two halves or "branches."

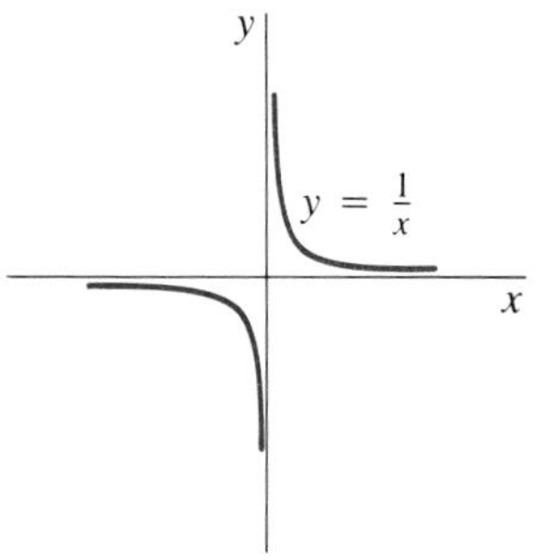

FIGURE 1.20

EXAMPLE 24 Older readers may remember when the cost of a local telephone call was 10¢ for the first 5 minutes and 5¢ for each additional 3 minutes or fraction thereof. Let x denote the duration of a call, and let f be the function that computes the cost of that call. Some representative values of this function are given in Table 1.6.

TABLE 1.6

x	1	2	4	5	6	7.4	8.1	12
$f(x)$	10	10	10	10	15	15	20	25

We can write this function in the form

(1.9)
$$f(x) = \begin{cases} 10 & \text{if} \quad 0 < x \le 5 \\ 15 & \text{if} \quad 5 < x \le 8 \\ 20 & \text{if} \quad 8 < x \le 11 \\ 25 & \text{if} \quad 11 < x \le 14 \\ 30 & \text{if} \quad 14 < x \le 17 \\ 35 & \text{if} \quad 17 < x \le 20 \\ \vdots & \qquad \vdots \end{cases}$$

The graph of this function is shown in Figure 1.21. For obvious reasons, f is said to be a **step function**. Observe that this graph has several breaks (jumps, gaps, holes, or any word you choose to convey the idea). We say that there is a **discontinuity** at each point where the function jumps. More generally, we have the following rough definition.

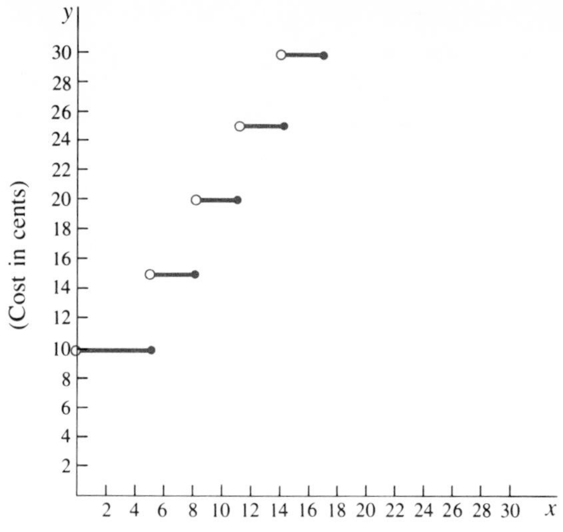

FIGURE 1.21

DEFINITION 1.2 If the graph of the function f can be drawn in one piece (that is, without taking the pencil point off the paper), we say that the function f is **continuous**. If the graph of the function f cannot be drawn in one piece, the function f is said to be **discontinuous**.

Thus the discontinuities of f in Equation (1.9) occur at $x = 5, 8, 11, 14, \ldots$. ⊟

EXAMPLE 25 In Figure 1.22, we have reproduced the Tax Rate Schedule for individual taxpayers for 1984. Let us try to reformulate this information using function notation.

Let x denote an individual's taxable income for 1984 and let $y = f(x)$ represent the corresponding federal income tax liability on that income. Then, as is easily checked, $f(x)$ is given by Table 1.6. For example, if an individual has a taxable income of \$20,000, his tax liability is

$$\begin{aligned} f(20{,}000) &= \$2737 + 0.26(20{,}000 - 18{,}200) \\ &= 2737 + 468 = \$3205 \end{aligned}$$

A portion of the graph at this function appears in Figure 1.23.

Given a function f, we have learned to associate with it a set of points in the plane; this set is called the graph of the function. A set of points in the plane may also arise in other ways. In such a situation, one may ask: Is this set of points the graph of some function?

Schedule X
Single Taxpayers

Use this Schedule if you checked **Filing Status Box 1** on Form 1040—

If the amount on Form 1040, line 37 is: Over—	But not over—	Enter on Form 1040, line 38	of the amount over—
$0	$2,300	**—0—**	
2,300	3,400	**.......11%**	$2,300
3,400	4,400	**$121 + 12%**	3,400
4,400	6,500	**241 + 14%**	4,400
6,500	8,500	**535 + 15%**	6,500
8,500	10,800	**835 + 16%**	8,500
10,800	12,900	**1,203 + 18%**	10,800
12,900	15,000	**1,581 + 20%**	12,900
15,000	18,200	**2,001 + 23%**	15,000
18,200	23,500	**2,737 + 26%**	18,200
23,500	28,800	**4,115 + 30%**	23,500
28,800	34,100	**5,705 + 34%**	28,800
34,100	41,500	**7,507 + 38%**	34,100
41,500	55,300	**10,319 + 42%**	41,500
55,300	81,800	**16,115 + 48%**	55,300
81,800	....	**28,835 + 50%**	81,800

FIGURE 1.22
1984 Tax rate schedules

TABLE 1.6

$$f(x) = \begin{cases} 0 & \text{if} \quad 0 < x \le 2{,}300 \\ 0.11(x - 2{,}300) & \text{if} \quad 2300 < x \le 3{,}400 \\ 121 + 0.12(x - 3{,}400) & \text{if} \quad 3400 < x \le 4{,}400 \\ 241 + 0.14(x - 4{,}400) & \text{if} \quad 4400 < x \le 6{,}500 \\ 535 + 0.15(x - 6{,}500) & \text{if} \quad 6500 < x \le 8{,}500 \\ 835 + 0.16(x - 8{,}500) & \text{if} \quad 8500 < x \le 10{,}800 \\ 1{,}203 + 0.18(x - 10{,}800) & \text{if} \quad 10{,}800 < x \le 12{,}900 \\ 1{,}581 + 0.20(x - 12{,}900) & \text{if} \quad 12{,}900 < x \le 15{,}000 \\ 2{,}001 + 0.23(x - 15{,}000) & \text{if} \quad 15{,}000 < x \le 18{,}200 \\ 2{,}737 + 0.26(x - 18{,}200) & \text{if} \quad 18{,}200 < x \le 23{,}500 \\ 4{,}115 + 0.30(x - 23{,}500) & \text{if} \quad 23{,}500 < x \le 28{,}800 \\ 5{,}705 + 0.34(x - 28{,}800) & \text{if} \quad 28{,}800 < x \le 34{,}100 \\ 7{,}507 + 0.38(x - 34{,}100) & \text{if} \quad 34{,}100 < x \le 41{,}500 \\ 10{,}319 + 0.42(x - 41{,}500) & \text{if} \quad 41{,}500 < x \le 55{,}300 \\ 16{,}115 + 0.48(x - 55{,}300) & \text{if} \quad 55{,}300 < x \le 81{,}800 \\ 28{,}835 + 0.50(x - 81{,}800) & \text{if} \quad 81{,}800 < x \end{cases}$$

The simple answer, once you think about it, is the following criterion:

(1.10) A set of points in the plane is the graph of a function if and only if no two points of the set have the same x-coordinate—or, to put it another way, if no two distinct points of the set lie on the same vertical line.

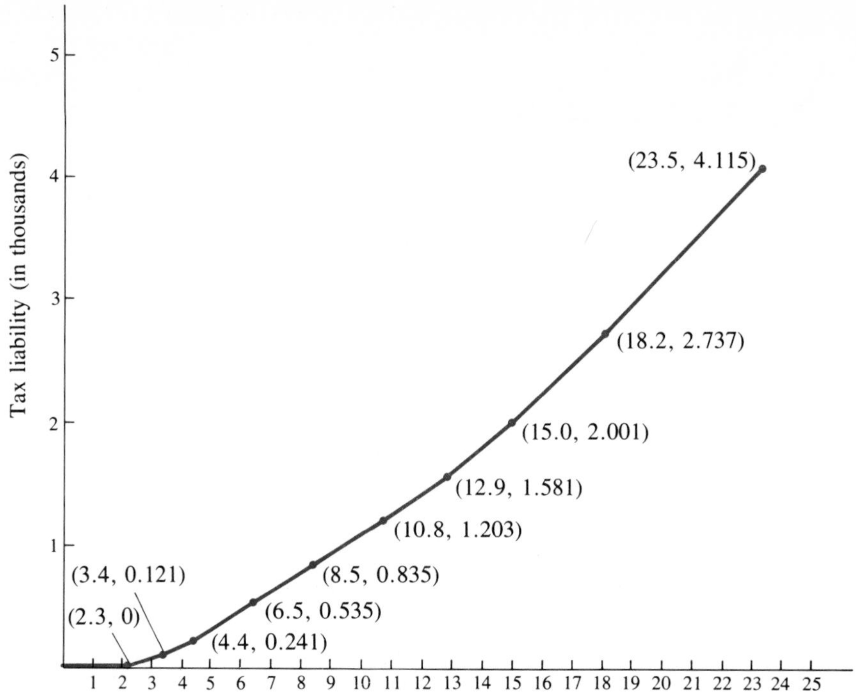

FIGURE 1.23

EXAMPLE 26 A common way in which a set of points in the plane may arise is from an equation. Consider, for example, the equation $x^2 + y^2 = 1$ and the set of all points in the plane whose coordinates satisfy this equation. This set consists of all points whose distance from the origin is 1. It is called the **unit circle** (Figure 1.24).

The unit circle is not the graph of any function because it does not satisfy Criterion (1.10). For example, the vertical line through $(\frac{1}{2}, 0)$ contains two

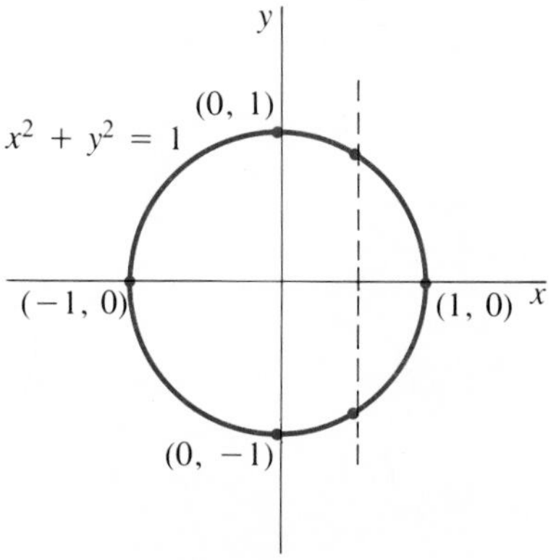

FIGURE 1.24

distinct points of the unit circle, namely

$$\left(\frac{1}{2}, \frac{\sqrt{3}}{2}\right) \quad \text{and} \quad \left(\frac{1}{2}, -\frac{\sqrt{3}}{2}\right)$$

In general, when we have an equation connecting x and y, the set of all points whose coordinates satisfy the equation is called the **graph of the equation.** The example of the unit circle shows that the graph of an equation need not be the graph of a function.

EXAMPLE 27 Sketch the graph of the equation

$$y = \sqrt{1 - x^2}$$

where, as usual, $\sqrt{\ }$ means the nonnegative square root.

SOLUTION Let us square both sides of the equation to obtain

$$y^2 = 1 - x^2$$

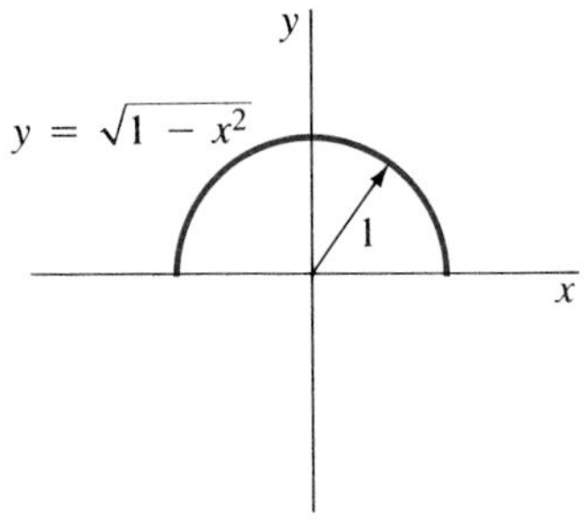

FIGURE 1.25

This is nothing but the equation $x^2 + y^2 = 1$ of the unit circle. However, not every point (x, y) on the unit circle satisfies $y = \sqrt{1 - x^2}$. In order to satisfy this equation, we are permitted (because $\sqrt{1 - x^2}$ is never negative) to consider only points for which y is nonnegative. Our equation, therefore, has as its graph the upper half of the unit circle. This graph is shown in Figure 1.25. Note also that the lower half of the unit circle is the graph of the function

$$g(x) = -\sqrt{1 - x^2}$$

The domain of g is $\{x: -1 \le x \le 1\}$ and the range of g is $\{y: -1 \le y \le 0\}$.

Section 1.3 EXERCISES

In each of Exercises 1–18 sketch the graph of the given function f.

1. $f(x) = 3x - 5$
2. $f(x) = 5 - 3x$
3. $f(x) = 3x^2$
4. $f(x) = x^4$
5. $f(x) = 3x^2 - 4$
6. $f(x) = -x^2 + 2$
7. $f(x) = 4 - 3x^2$
8. $f(x) = (x - 1)^2 = x^2 - 2x + 1$
9. $f(x) = (x + 1)^2 + 3 = x^2 + 2x + 4$
10. $f(x) = (x - 2)^2 + 1$
11. $f(x) = x^2 - 5x + 8$
12. $f(x) = 1/x^2$
13. $f(x) = 1/(x - 2)$
14. $f(x) = |x|$
15. $f(x) = |x + 1|$
16. $f(x) = x + |x|$
17. $f(x) = \sqrt{x - 1}$
18. $f(x) = \sqrt{9 - x^2}$
19. Find the linear function f whose graph passes through the points (1, 3) and (2, −5).
20. Find the linear function f whose graph passes through the points (1, 3) and (4, 3).
21. Find the quadratic function f ("quadratic" means that f is of form $f(x) = ax^2 + bx + c$) whose graph passes through the points (1, 4), (0, 1), and (3, 4).
22. Find the quadratic function f whose graph passes through the points (−1, −2), (0, −4), and (2, 4).
23. What happens in Exercise 19 if the points are (1, 3) and (1, −5)?
24. Sketch the graph of $f(x) = x - |x|$.

In each of Exercises 25–29 sketch the graphs of the given pair of equations on the same set of axes. Determine the points of intersection (if any) of the graphs.

25. $2x + y = -1, \quad x - 3y = 2$

26. $3x + 2y = 1, \quad 2x - 3y = 0$

27. $y = x^2, \quad y = 2x + 3$

28. $x = 1 - y^2, \quad y = 3x - 1$

29. $x^2 + y^2 = 25, \quad y = 7x + 25$

In each of Exercises 30–36 make a rough sketch of the graph of the given equation and describe the graph as best you can. Is it the graph of some function?

30. $x^2 - y = 0$

31. $x - y^2 = 0$

32. $xy = 1$

33. $x^2 - y^2 = 1$

34. $(x - 3)^2 + y^2 = 36$

35. $x^2 + (y - 3)^2 = 36$

36. $|y| = |x|$

37. Sketch the graph of the "postage function" given in Exercise 37 of Section 1.2.

38. A particular savings bank pays 6% interest per year. The interest is credited each December 31. Suppose that on January 1 a depositor opens an account with \$1000.00. Let f denote the function that determines the dollar value of the account at each future time t (measured in years). Sketch the graph of f.

39. Determine which of the following sets of points represent the graph of a function. Explain.

(a)

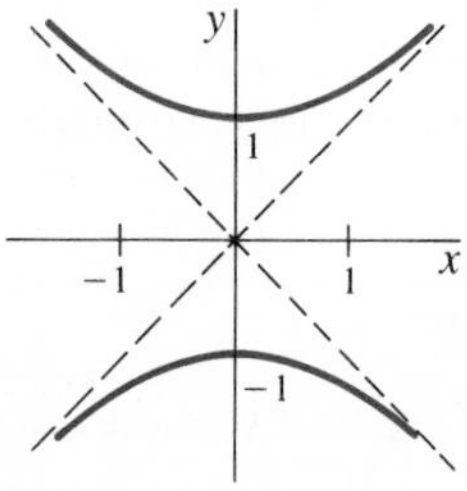

(b)

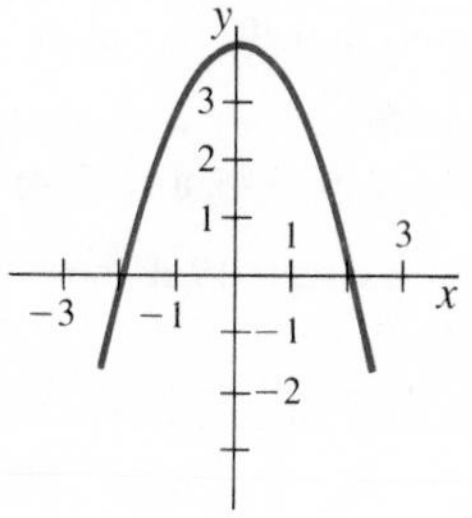

(c)

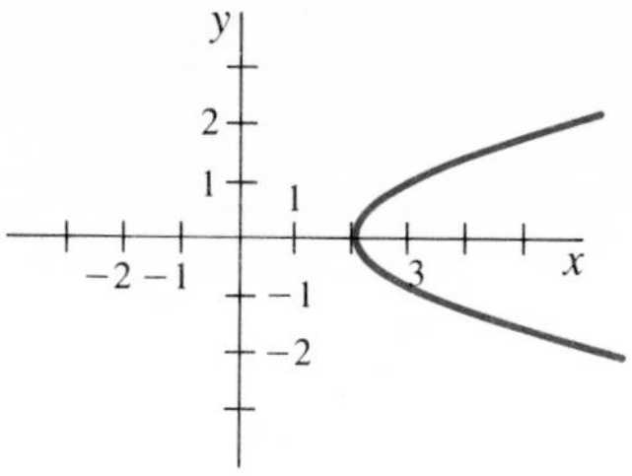

(d)

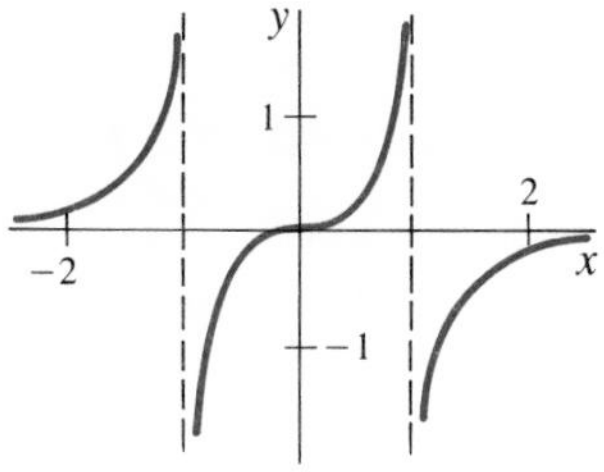

(e)

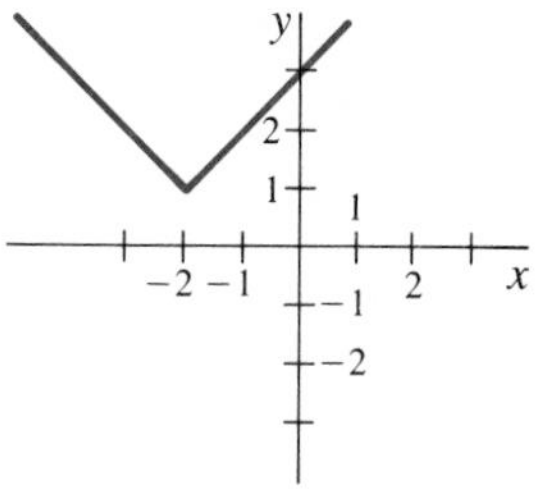

(f)

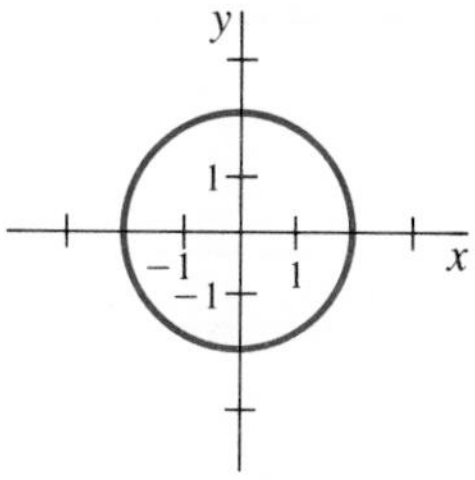

(g)

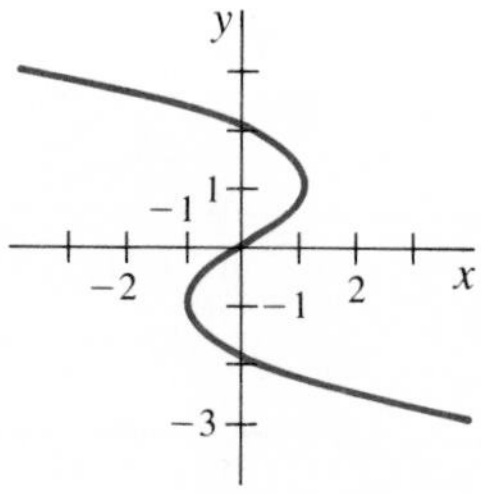

(h)

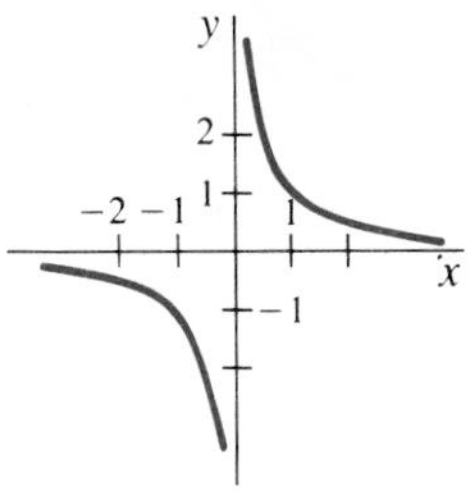

(i)

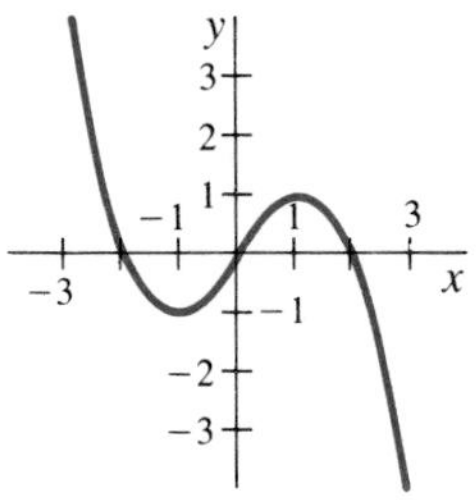

(j)

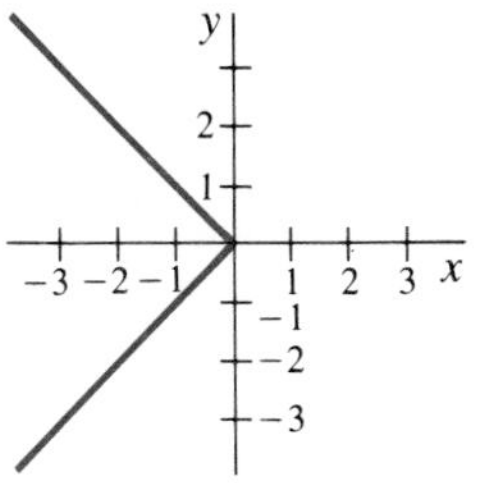

(k)

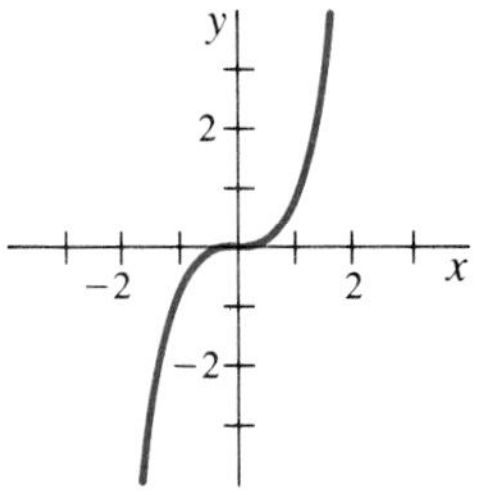

(l)

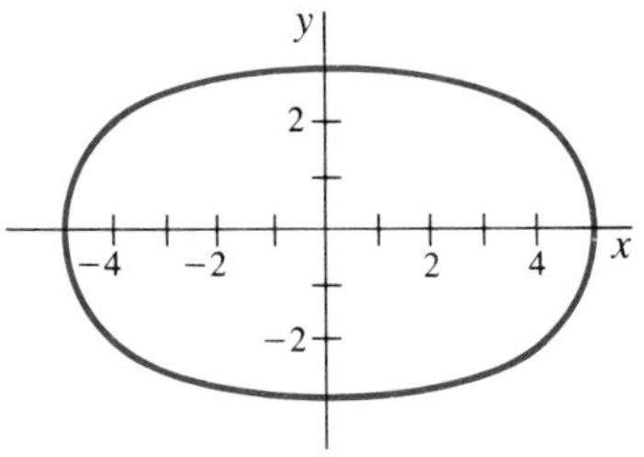

40. Which of the following points lie on the graph of the function $f(x) = \dfrac{x^2 + x + 1}{x^2 - 1}$?

(a) $(0, -1)$ (b) $(1, 0)$ (c) $(2, \frac{7}{3})$
(d) $(-1, 0)$ (e) $(-2, 1)$

41. Which of the following points lie on the graph of the function $f(x) = \sqrt{\dfrac{x}{x - 3}}$?

(a) $(0, 0)$ (b) $(1, \sqrt{1/2}$ (c) $(-1, -\frac{1}{2})$
(d) $(-2, \sqrt{2/5})$ (e) $(2, +\sqrt{2})$

42. Which of the following points are on the graph of the equation $x^2 + xy + y^2 = 3$?

(a) $(0, 1)$ (b) $(-\sqrt{3}, 0)$ (c) $(-1, 2)$
(d) $(2, -1)$ (e) $(-1, -1)$

43. Which of the following points are on the graph of the equation $x^2 + y^2 - 4x + 6y + 9 = 0$?

(a) $(2, 3)$ (b) $(3, 2)$ (c) $(4, 3)$
(d) $(0, 3)$ (e) $(2, 1)$

44. A sociologist has determined (see R. Cargo and J. N. Yanouzas, *Formal Organization: A Systems Approach*, Homewood, IL: Richard D. Irwin, 1967, p. 489) that, for a person whose predetermined goal level is $g \geq 0$ and whose perceived justice per unit of reward is $a > 0$, satisfaction S is related to total reward R by the formula

$$S(R) = \frac{aR}{g - R}$$

Sketch the graph of this function when $a = 2, g = 6$.

45. A library book sale starts at 10 A.M. At noon the salespeople drop the prices 25%. At 2 P.M. they drop the prices 50% of the current price, and at 4 P.M. they drop them an additional 50%. If books originally sell for \$4.00, sketch the function that describes their price throughout the day.

46. A property owner wants to fence off a rectangular piece of land bordered on one side by a freeway. If she has 240 ft of fencing for the remaining three sides, what dimensions will maximize the area of the rectangle? (*Hint:* You should find a quadratic equation for area whose vertex will give a maximum.)

47. What if the property owner in Exercise 46 decides to put a fence on all four sides? What dimensions will maximize the area of the rectangle?

48. Car rental costs \$25 per day or fraction thereof, and in addition there is a \$30 drop-off charge for not returning the car to the original location. Let $C(x)$ represent the cost of renting a car for x days. If you want to drive from Boston to San Francisco, it will take you from 4 to 8 days. Sketch the graph of $y = C(x)$.

49. The demand for toaster ovens is given by $p(x) = 40 - x$ where $p(x)$ is the price when x units are demanded.

Revenue, $R(x)$, is the product of the price and the number of units sold.

(a) Find the function $R(x)$.

(b) Graph $y = R(x)$.

(c) From the graph, determine the maximum revenue.

(d) What price corresponds to the maximum revenue?

50. Suppose that the unit price of a commodity when x units are sold is given by the demand equation $p = 200 - 10x$ and that the cost of producing x units is $C(x) = 480 + 40x$. Sketch the graph of the revenue function and the cost function using the same set of axes. Find the *break-even points*—that is, the points where $R(x) = C(x)$.

IF YOU WORK WITH A COMPUTER (OPTIONAL)

In each of Exercises 51–57, use either of the programs following Exercise 57 to sketch the graph of $f(x)$. Use the x and y intervals given.

51. $f(x) = x^3 - 3x^2 + 5;\ -2 \le x \le 3,\ -15 \le y \le 5$

52. $f(x) = \sqrt{25 - x^2};\ -5 \le x \le 5,\ 0 \le y \le 5$

53. $f(x) = \dfrac{16}{1 + x^2};\ -5 \le x \le 5,\ 0 \le y \le 18$

54. $f(x) = |x|;\ -1 \le x \le 1,\ 0 \le y \le 1$

55. $f(x) = 3x^4 - 12x^3 + 40x - 5;\ -2 \le x \le 4,\ -40 \le y \le 160$

56. $f(x) = \sqrt{4 - x};\ -6 \le x \le 4,\ 0 \le y \le 4$

57.* $f(x) = 2^x - x^2 - 1;\ -1 \le x \le 5,\ -2.5 \le y \le 6$

```
10    REM - Primitive graph plotter - BASIC version

20    INPUT "Domain end points a,b";a,b
30    INPUT "No. of steps on x-axis";M
40    INPUT "Min and Max y values: c,d";c,d
50    INPUT "Number of steps on y-axis";N

60    REM - define function to be graphed
70    DEF FNf(x)=x*x
80    REM - function to round x to nearest integer
90    DEF FNr(x)=INT(x+.5)

100   REM - h is x increment, k is y increment
110   h=(b-a)/M
120   k=(d-c)/N

130   PRINT
140   PRINT , "Y-AXIS FROM ";c;" TO ";d;
150   PRINT " IN STEPS OF ";k
160   REM - draw y-axis
170   PRINT ,
180   FOR i=0 TO N
190     PRINT "-";
200   NEXT i
210   PRINT
```

```
(* Primitive graph plotter - Pascal version *)
program Plot (input, output);

 var
  a, b, c, d, h, k, x, y : real;
  M, N, i, j, y1 : integer;

 function f (x : real) : real;
 begin
  f := 16 / (1 + sqr(x * x))
 end;

begin
 write('Domain left endpoint? ');
 readln(a);
 write('Domain right endpoint? ');
 readln(b);
 write('No. of steps on x-axis? ');
 readln(M);
 write('Minimum y value? ');
 readln(c);
 write('Maximum y value? ');
 readln(d);
 write('Number of steps on y-axis? ');
 readln(N);
```

(*Programs continue at top of next page*)

* In Pascal, 2^x is entered as exp(x*ln(2)). The functions exp and ln are discussed in Chapter 4.

```
220  x=a
230  FOR i=0 TO M
240    y=FNf(x)
250    PRINT USING "##.##";x;
260    PRINT ,
270    IF y<c OR y>d THEN PRINT : GOTO 330
280    REM - get y-c as a multiple of k
290    y1 = FNr( (y-c)/k )
300    FOR j=1 TO y1
310      PRINT " ";
320    NEXT j
330    PRINT "*"
340    x = x + h
350  NEXT i

360  PRINT
370  END
```

```
h := (b - a) / M;  (* compute x and y increments *)
k := (d - c) / N;
writeln;
write('Y-AXIS FROM ', c : 5 : 2, ' TO ', d : 5 : 2);
writeln(' IN STEPS OF ', k : 5 : 2);
writeln;
write('             ');    (* draw y-axis *)
for i := 0 to N do
  write('-');
writeln;

x := a;
for i := 0 to M do
  begin
    y := f(x);
    write(x : 5 : 2, '        ');
    if (y < c) or (y > d) then
      writeln
    else
      begin
        y1 := round((y - c) / k);
        for j := 1 to y1 do
          write(' ');
        writeln('*');    (* plot point *)
      end;
    x := x + h
  end;

end.
```

SECTION 1.4

Composite Functions

Sometimes it is useful to think of a function as a kind of "machine" that can accept any number x (from its domain) as input, process this number in some manner, and produce as output the value $f(x)$ (see Figure 1.26).

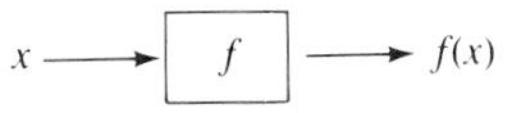

FIGURE 1.26

If we have two such "machines," f and g, we can feed the output from f into g, just as on an assembly line, where the result of one manufacturing process is the input to the next (see Figure 1.27). All that is required is that x belong to the domain of f and that $f(x)$ belong to the domain of g. The resulting function, which takes x into $g(f(x))$ is called the **composition** of g and f and is denoted $g \circ f$. Thus

$$(g \circ f)(x) = g(f(x))$$

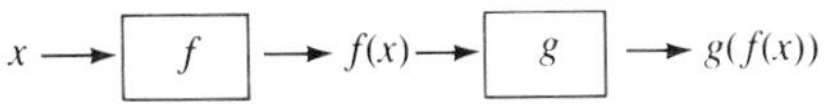

FIGURE 1.27

For example, suppose that

$$f(x) = 2x + 1 \qquad g(x) = x^2$$

Then

$$(g \circ f)(x) = g(f(x)) = g(2x + 1) = (2x + 1)^2$$

On the other hand, we may place the g machine ahead of the f machine in our assembly line (that is, compose f and g in the other order) to get the function $(f \circ g)(x) = f(g(x))$. For f and g as above,

$$(f \circ g)(x) = f(g(x)) = f(x^2) = 2x^2 + 1$$

Note that, in general, $f \circ g \neq g \circ f$.

EXAMPLE 28 Let $f(x) = \sqrt{x}$ and let $g(x) = x^3 + 1$. Find $f \circ g$ and $g \circ f$, as well as their domains and ranges.

SOLUTION

$$(f \circ g)(x) = f(g(x)) = f(x^3 + 1) = \sqrt{x^3 + 1}$$

Because of the square root, we must have $x^3 \geq -1$ or $x \geq -1$. Accordingly, $f \circ g$ has domain $[-1, \infty)$, and the range is easily seen to be $[0, \infty)$, the same as the range of f itself.

$$(g \circ f)(x) = g(f(x)) = g(\sqrt{x}) = x^{3/2} + 1$$

Because x cannot be negative, $g \circ f$ has domain $[0, \infty)$ and range $[1, \infty)$.

PRACTICE EXERCISE Let $f(x) = x^3 + 1$ and let $g(x) = \sqrt[3]{x - 1}$. Compute $f \circ g$ and $g \circ f$.

(Your answer should be $f(g(x)) = x$ and $g(f(x)) = x$.)

The property illustrated by the preceding exercise—namely, $f(g(x)) = x$ and $g(f(x)) = x$—will be investigated further in Section 4.3.

EXAMPLE 29 During the course of a day, the temperature in a certain city varies according to the formula

$$C(t) = 20 - (t - 12)^2/24 \qquad 0 \leq t \leq 24$$

where t is in hours past midnight and $C(t)$ is in degrees Celsius. Using the formula

$$F = \tfrac{9}{5}C + 32$$

express the temperature at time t in degrees Fahrenheit.

SOLUTION We substitute $C = C(t)$ into the function $F(C) = \frac{9}{5}C + 32$ and obtain the composition

$$\begin{aligned} F = F(C(t)) &= \frac{9}{5}\left[20 - \frac{(t - 12)^2}{24}\right] + 32 \\ &= 36 - \frac{9(t - 12)^2}{5 \cdot 24} + 32 \\ &= 68 - \frac{3(t - 12)^2}{40} \end{aligned}$$

EXAMPLE 30 Express the function

$$f(x) = \frac{3}{\sqrt{x^2 + 4}}$$

as a composition of two simple functions.

SOLUTION The function $f(x)$ has the form $\frac{3}{\sqrt{(\ )}}$, where $x^2 + 4$ is within the parentheses. Therefore, if we let $g(x) = 3/\sqrt{x}$ and let $h(x) = x^2 + 4$,

$$g(h(x)) = \frac{3}{\sqrt{h(x)}} = \frac{3}{\sqrt{x^2 + 4}} = f(x)$$

Section 1.4 EXERCISES

In Exercises 1–4 find the expressions for $(f \circ g)(x)$ and $(g \circ f)(x)$. Give the domains of all compositions.

1. $f(x) = 4x - 2, \quad g(x) = \dfrac{1}{x + 2}$

2. $f(x) = x^2 - 2, \quad g(x) = \dfrac{1}{x + 1}$

3. $f(x) = x^2 - x, \quad g(x) = \sqrt{x + 2}$

4. $f(x) = \dfrac{x - 1}{x + 1}, \quad g(x) = \dfrac{x + 1}{x - 1}$

In Exercises 5–8 express each of the given functions as a composition of two simple functions.

5. $f(x) = (x^2 - x - 3)^2$

6. $f(x) = \sqrt{x^2 + x + 1}$

7. $f(x) = \dfrac{1}{\sqrt{3x^2 + 4}}$

8. $f(x) = (3x^2 - 2)^{3/2}$

9. After t hr of operation a manufacturing plant has produced $N(t) = 60t - \frac{3}{2}t^2$ computer circuit boards, where $0 \le t \le 8$. The cost of producing x of these boards is $C(x) = 200 + 5x$.

(a) Express the cost as a function of t.

(b) What is the cost for an 8-hr day?

10. Suppose a corporate vice president's income $I(P)$ is 3% of the corporation's profit P. Suppose also that the profit $P(S)$ is given in terms of the corporation's sales S by the formula

$$P = 0.42S - 5000$$

(a) Express I as a function of P, and express I as a function of S.

(b) For each dollar increase in profit, how much does the vice president's income change?

(c) For each dollar increase in sales, how much does the corporation's profit increase?

(d) What is the effect of a dollar increase in sales on the vice president's income?

Key Mathematical Concepts and Tools

Slope	**Domain**
Function	**Range**
Graph of a function	**Composition of functions**

Applications Covered in This Chapter

Demand function
Supply function
Cost of a telephone call as a function of call duration
Income tax liability as a function of income
Cost function
Demand-supply equilibrium
Effect of tax changes on demand
Conversion of Celsius to Fahrenheit

ONE Warm-Up Test

1. In each case find the slope of the straight line passing through the two given points.
   (a) $(-1, 2)$ and $(3, 5)$ (b) $(1, 5)$ and $(2, 0)$
   (c) $(-3, 0)$ and $(5, 0)$ (d) $(1, 8)$ and $(4, 8)$

2. Find an equation for the straight line:
   (a) Passing through $(2, 5)$ and parallel to the line
   $$y = 4x + 5$$
   (b) Passing through $(1, 3)$ and $(-4, -22)$
   (c) Having slope $-\frac{1}{2}$ and y-intercept $(0, 6)$
   (d) Passing through $(2, 3)$ and perpendicular to the line
   $$y = -2x + 5$$

3. At a price of 86¢ per dozen, a supplier is willing to provide 1500 dozen eggs per day to a supermarket. At 70¢ per dozen, she will supply only 1200 dozen eggs. Assuming a linear relationship, find the supply equation.

4. In each case find the domain of the given function f.
   (a) $f(x) = \sqrt{x^2 - 9}$ (b) $f(x) = \dfrac{1}{x - 4}$
   (c) $f(x) = x^2$ (d) $f(x) = \dfrac{3x + 4}{5 - x^2}$

5. Describe in words the graph of each of the following equations.
   (a) $x^2 - 8x + y^2 + 10y = 128$
   (b) $x^2 - 8x + y^2 + 10y = -41$

6. Express each of the following functions as a composition of two given functions.
   (a) $f(x) = \sqrt{x^2 - 4x + 5}$
   (b) $f(x) = \dfrac{1}{(x^2 + 1)^3}$
   (c) $f(x) = (8x^2 + 5)^{1/4}$

7. (a) Work out the linear formula for converting temperature in degrees Fahrenheit F to temperature in degrees Celsius C.
   (b) Suppose Boston's temperature varies over a certain 24-hour period according to the formula
   $$F(t) = 70 + \frac{(t - 8)^2}{30} \qquad 0 \le t \le 24$$
   where $F(t)$ is the Fahrenheit temperature at time t. Express the formula for the temperature at time t in Celsius.

8. Let $f(x) = 2x - 3$ and let $g(x) = 3x + B$. What value can we assign to B so that we will have $f(g(x)) = g(f(x))$?

9. Find the perpendicular distance from the point $(-1, 4)$ to the straight line with the equation $3y - 2x = 6$.

10. A car rental agency at a local airport charges \$22 per day plus 14¢/mi for rental of a subcompact car. In addition, there is an \$18 drop-off charge if the car is not returned at the airport. A customer plans to use the car for 5 days and to return it at city center. Write an expression for that customer's cost for driving x mi.

1. Assume that $f(x) = x^2 + 1$ and that $g(x) = 2x$. Find each of the following functions.
   (a) $f(x)g(x)$ (b) $f(g(x))$ (c) $g(f(x))$
   (d) At what point(s) do the graphs of $f(x)$ and $g(x)$ intersect?

2. Find an equation in slope–intercept form for the straight line:
   (a) Passing through $(2, -2)$ and parallel to the line $y = -3x + 2$
   (b) Passing through $(8, 2)$ and $(-6, 4)$
   (c) Passing through the origin and perpendicular to the line $y = 3x - 4$

3. The two straight lines $y = 3x + 6$ and $y = 3x + 10$ are parallel. Find the perpendicular distance between them.

4. Sketch the graph of each of the following functions.
   (a) $f(x) = |x - 3|$
   (b) $f(x) = \sqrt{x - 1} \quad 1 \le x < \infty$
   (c) $f(x) = x^2 - 3x + 2$

5. A cafeteria finds that, if it prices its "special" at \$2.50, it will serve 180 meals and that, for each 15¢ reduction in the special price, it can sell 25 extra meals. (a) Find the demand equation. (b) Write down a formula for the total revenue produced if the price per meal is p.

6. The following equations represent circles. Find their centers and radii.
   (a) $2x^2 - 20x + 2y^2 + 8y = 14$
   (b) $100x^2 + 100x + 100y^2 + 40y = 371$

7. In each case, find the domain and range of the given function f.
   (a) $f(x) = \dfrac{3 - x}{x^2 + 9}$ (b) $f(x) = \dfrac{3 - x}{x^2 - 9}$
   (c) $f(x) = \sqrt{4 - x^2}$ (d) $f(x) = x^{1/3}$

8. A parcel of real estate in a given locale was purchased initially for \$342,000 in 1980. It has been found to appreciate in value in each succeeding year by \$40,000.
   (a) Find a function $V(t)$ that describes its value t years after 1980.
   (b) What will its value be in 1995?

9. Supply and demand functions are given below. Find the equilibrium value of price in each case.
   (a) $q_d = 80 - 3p \quad q_s = p + 20$
   (b) $q_d = 170 - 5p \quad q_s = 3p + 10$
   (c) $q_d = 65 - p \quad q_s = 3.5p + 20.9$

10. In each of the following situations, determine whether the three points lie on a straight line. If your answer is yes, give an equation for the line.
    (a) $(1, 7)$, $(2.5, 13.5)$, $(4, 22)$
    (b) $(2, 2)$, $(-3, -13)$, $(1.2, -0.4)$
    (c) $(4, 12.8)$, $(0, 0)$, $(\frac{5}{16}, 1)$

TWO Differentiation

Outline

In Context

The purpose of this chapter is to begin the study of one of the principal concepts of elementary calculus — namely, the *derivative.* Our approach is based on the fact that this notion arises quite naturally (as it did historically) from a careful consideration of certain simple physical problems.

In Sections 2.1 and 2.2 we will motivate the derivative by studying in detail the problems of the "broken speedometer" and the "slope of a graph at a point." We hope this discussion will convince you that the derivative is a useful and important concept that deserves further study.

Later in this chapter we will learn how to compute the derivative of many of the basic functions that arise in the application of mathematics to various disciplines. We will also develop certain rules for computing the derivatives of more complicated functions that are constructed out of these basic functions.

SECTION 2.1

The Case of the Broken Speedometer

Let us begin by examining a simple situation. A man decides to take a trip by car along a straight highway from, say, Boston to New York; the distance is 220 miles. He leaves at noon and expects to arrive at 4:00 P.M. The car is equipped with the usual gauges and dials on the dashboard. We focus on three of them:

1 A clock: It tells time!

2 An odometer: It measures the total distance, in miles, that the car has traveled from the moment it rolled off the assembly line. For example, the odometer reading shown in Figure 2.1 signifies that the car has traveled a total of 12,000.0 miles.

FIGURE 2.1

3 A speedometer: At each moment it measures and displays the speed of the car in miles per hour—as, for example, in Figure 2.2.

During the trip, the driver notices that the speed limit on a particular stretch of highway has been reduced to 45 miles per hour because of construction. He glances at the dashboard and finds that his speedometer is out of order. However, the other dials seem to be functioning correctly. The clock reads 3:00 P.M. and the odometer reading is 12,000.0 miles.

We wish to consider the following question: Using only information obtainable from the clock and the odometer, is it possible to decide whether the car is traveling within the speed limit? In more detail, can the driver determine what the speedometer would read (if it were working) at 3:00 P.M.?

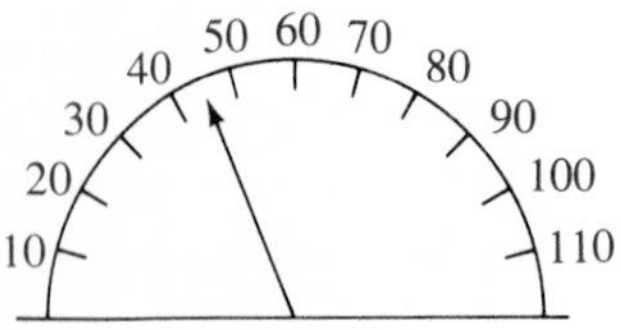

FIGURE 2.2

A very primitive answer to this problem is provided by the following reasoning. Because the trip is 220 miles long and it requires 4 hours, the speed of the car is "most likely" to be

$$\frac{\text{Distance traveled}}{\text{Time elapsed}} = \frac{220 \text{ miles}}{4 \text{ hours}} = 55 \text{ miles per hour}$$

Of course, this answer has almost no chance of being correct. It represents the *average speed* during the period from 12:00 noon to 4:00 P.M. (if the trip turns out as originally anticipated). The correct speedometer reading at 3:00 P.M. may well be much larger or much smaller than the value of 55 miles per hour that we just computed. Can we do better? Yes—but only if we have more information.

One way to proceed is as follows: Suppose that, starting at 3:00 P.M., the driver records the odometer reading every minute until 3:05 P.M. And suppose these mileage readings are as shown in Table 2.1.

TABLE 2.1

Time	*Odometer reading*
3:00 P.M.	12,000.0
3:01	12,000.7
3:02	12,001.5
3:03	12,002.3
3:04	12,003.2
3:05	12,004.2

Thus we may compute the average speed for the 5 minutes ($\frac{1}{12}$ hour) from 3:00 P.M. to 3:05 P.M. It is given by

$$\frac{12{,}004.2 - 12{,}000.0 \text{ miles}}{\frac{1}{12} \text{ hour}} = \frac{4.2 \text{ miles}}{\frac{1}{12} \text{ hour}} = 50.4 \text{ miles per hour}$$

This is a better estimate than the previous one of 55 miles per hour. However, it may still contain a large error because there could have been very wide fluctuations in the actual speed during this 5-minute period.

PRACTICE EXERCISE Use the data in Table 2.1 to find: (a) the average speed over the 3-minute period from 3:00 P.M. to 3:03 P.M.; (b) the average speed over the 2-minute period from 3:00 P.M. to 3:02 P.M.

[Your answers should be (a) 46 miles per hour and (b) 45 miles per hour.]

Another, more accurate, estimate of the correct speedometer reading at 3:00 P.M. is surely given by the *average speed* for the 1-minute ($\frac{1}{60}$-hour) period from 3:00 P.M. to 3:01 P.M.:

$$\frac{12{,}000.7 - 12{,}000.0 \text{ miles}}{\frac{1}{60} \text{ hour}} = \frac{0.7 \text{ miles}}{\frac{1}{60} \text{ hour}} = 42.0 \text{ miles per hour}$$

This answer is clearly the best estimate that can be made on the basis of the data given in Table 2.1. Thus it seems reasonable to say that the correct speedometer reading (the actual speed of the car) is safely below the posted speed limit. In order to compute a more accurate answer, it would be necessary to check the odometer reading at more frequent intervals—for example, at 10-second intervals instead of at 1-minute intervals.

A lesson we may draw from the preceding discussion is as follows:

(2.1) The speed of a moving car at a given instant of time can be estimated by computing the average speed of the car over shorter and shorter time intervals (where the time intervals contain the instant in question).

A question immediately arises: Does this method ever enable us to determine the *exact* speed of the car, or will our answer always be just an approximation? Surprisingly, perhaps, the answer is the former (provided we have enough information). Let us illustrate this contention by an example.

Suppose again that a car is being driven from one city to another. Denote the total distance the car has traveled by s (measured in miles) and the time elapsed by t (measured in hours), with both s and t measured (for convenience) from the moment when the trip begins. Let us assume for illustrative purposes that

$$s = 10t^2$$

Thus s is a function of time t that, at each time t, tells us the distance traveled, s, with complete accuracy. This particular formula was chosen for its simplicity; it is not intended to describe the motion of any real car. In particular, this equation says that, when $t = 0$ (which is the starting time of the trip), we have s (which is the odometer reading) equal to 0.

In this situation, let us consider a closely related question: Using only the information provided by the equation $s = 10t^2$, can the driver determine the speed of the car at $t = 3$? In other words, how fast is the car moving at the instant of time exactly 3 hours after the start of the trip?

Let us approach this problem by generalizing the method used earlier. We computed the average speed of the car over several specific time intervals, each interval starting at $t = 3$. Rather than restricting ourselves to a single specific time interval, let us in this situation compute the average speed of the car during the time interval of length h that starts at $t = 3$ and ends at $t = 3 + h$. (For example, if the value of h happens to be 0.5, we are computing the average speed during the time interval from $t = 3$ to $t = 3.5$.) In general, h can take on any value other than 0. Thus, when we work with the time interval from $t = 3$ to $t = 3 + h$, we are actually working with an arbitrary time interval that starts at $t = 3$; h is the length of this time interval.

To fix these ideas, let us examine Table 2.2, which displays the values of s at the times $t = 3$ and $t = 3 + h$.

TABLE 2.2

t	$s = 10t^2$
Time in hours	*Odometer reading*
3	$10(3)^2 = 90$
$3 + h$	$10(3 + h)^2 = 90 + 60h + 10h^2$

The average speed from $t = 3$ to $t = 3 + h$ is computed as follows:

(2.2)
$$\frac{(90 + 60h + 10h^2) - 90 \text{ miles}}{h \text{ hours}} = \frac{60 + 10h^2 \text{ miles}}{h \text{ hours}}$$
$$= 60 + 10h \text{ miles per hour}$$

This expression gives us, in one fell swoop, the average speed over any time interval that starts at $t = 3$.

Now comes the crucial step, the step that enables us to arrive at an exact answer for the speed at $t = 3$. We allow the length of the time interval (from $t = 3$ to $t = 3 + h$) over which we compute the average speed to get shorter and shorter. To put it another way, we imagine that the number h is getting closer and closer to 0. As h becomes very small (very near 0 but never equal to 0), what happens to the expression for the average speed, Equation (2.2)? Clearly, as h becomes very small, so does $10h$, and consequently $60 + 10h$ gets very close to 60. We say that 60 is the **limiting value** of the expression $60 + 10h$ as h approaches 0, and we write this as

$$\lim_{h \to 0} (60 + 10h) = 60$$

(The way to read the left side here is "the limit as h approaches 0 of the expression $60 + 10h$.") This tells us that the speed of the car at the moment when $t = 3$ is precisely 60 miles per hour. This is the number that a properly operating speedometer would show at $t = 3$; in fact, a speedometer is designed to determine and display this number.

The procedure just employed is an example of what is called a **limit**

process. The use of limit processes is the special property that distinguishes calculus from the branches of mathematics that you have studied earlier.

Because the number 60 represents the precise speed of the car at the instant of time $t = 3$, it is generally referred to as the **instantaneous speed** or **instantaneous velocity** at $t = 3$. If we denote the instantaneous velocity at time t by $v(t)$, our previous result is $v(3) = 60$.

PRACTICE EXERCISE Consider the car whose motion is given by the equation $s = 10t^2$.

(a) Find the average speed over the interval from $t = 2$ to $t = 2 + h$ and the instantaneous speed at $t = 2$ [that is, $v(2)$].

(b) Find the average speed over the interval from $t = 4$ to $t = 4 + h$ and the instantaneous velocity $v(4)$ at $t = 4$.

[Your answers should be (a) $40 + 10h$ miles per hour and 40 miles per hour and (b) $80 + 10h$ miles per hour and 80 miles per hour.]

Now let us consider this very general question: Using the information provided by the function $s = 10t^2$, can we determine the instantaneous velocity $v(t)$ at any arbitrary time t? In other words, can we determine what the correct speedometer reading should be at each instant of time t?

As before, our approach to this problem is to compute the average speed of the vehicle over the time interval from t to $t + h$ and then to let the length of the time interval (which is h) get shorter and shorter. Let us construct Table 2.3.

TABLE 2.3

t	$s = 10t^2$
Time in hours	*Odometer reading*
t	$10t^2$
$t + h$	$10(t + h)^2 = 10t^2 + 20th + 10h^2$

We have used the formula $s = 10t^2$ to compute the distance traveled by the car up to time t (which is $10t^2$) and the distance traveled by the car up to time $t + h$ (which is $10(t + h)^2$). Now the **average speed** from time t to time $t + h$ is given by

$$\frac{(10t^2 + 20th + 10h^2) - 10t^2 \text{ miles}}{h \text{ hours}} = \frac{20th + 10h^2 \text{ miles}}{h \text{ hours}} \tag{2.3}$$

$$= 20t + 10h \text{ miles per hour}$$

Next we examine this expression and seek its limit (if one exists) as h gets closer and closer to 0. As h becomes very small, so does $10h$, and consequently $20t + 10h$ becomes very close to $20t$. We write this as

$$\lim_{h \to 0} (20t + 10h) = 20t$$

and conclude that the instantaneous velocity at any time t is given by

$$v(t) = 20t$$

Note the great generality of what we have done. With a single computation we have found the velocity $v(t)$ at every time t. Moreover, we can find the velocity at any specific time t by substituting the value of t in the formula $v(t) = 20t$. In particular, if $t = 3$, this formula yields $v(3) = 60$, which agrees with our previous result. We also have $v(2) = 40$ and $v(4) = 80$. These should be the results you have already found in the last practice exercise.

We close this section by restating Equation (2.3), using the function notation of Chapter 1. We may reformulate Table 2.3 as Table 2.4.

TABLE 2.4

t	$s(t) = 10t^2$
t	$s(t) = 10t^2$
$t + h$	$s(t + h) = 10(t + h)^2$

Then the left side of Equation (2.3) becomes

$$\frac{s(t + h) - s(t)}{h}$$

which we recognize as the difference quotient for the function $s(t) = 10t^2$. The instantaneous velocity was obtained by computing the limit of this expression as $h \to 0$. Thus

(2.4)
$$v(t) = \lim_{h \to 0} \frac{s(t + h) - s(t)}{h}$$

We will consider this expression further in Section 2.3.

PRACTICE EXERCISE Replace the distance function $s = t^2$ by the new function $s(t) = t^3$. For this equation of motion, find the instantaneous velocity $v(t)$ at an arbitrary time t.

[Your answer should be $v(t) = 3t^2$.]

Section 2.1 EXERCISES

1. Is $s = t^3$ a reasonable equation for the motion of a car? Explain.
2. Mention some properties that a function f should have in order for $s = f(t)$ to give a reasonable description of the motion of a car.
3. Find the instantaneous speed at time $t = 2$ of a car whose motion is given by $s = 3t$.
4. Find the instantaneous speed at time $t = 2$ of a car whose motion is given by $s = 3t + 5$.
5. Find the instantaneous speed at $t = 3$ of a car whose motion is given by $s = 3t^2$.
6. Find the instantaneous speed at $t = 3$ of a car whose motion is given by $s = 3t^2 + t + 5$.

In each of Exercises 7–13 find the instantaneous velocity $v(t)$ at any time t of a car whose motion is according to the given rule.

7. $s = 3t + 5$
8. $s = at + b$ $\quad$ (a, b constant)
9. $s = 3t^2 + t + 5$
10. $s = at^2 + bt + c$ $\quad$ (a, b, c constant)
11. $s = \dfrac{1}{t}$
12. $s = \dfrac{1}{t + 1}$
13. $s = \dfrac{1}{t^2}$

14. We have found the instantaneous velocity of a car at time t [that is, $v(t)$] by computing the average velocity over the time interval from t to $t + h$ (with $h > 0$) and then taking the limit of the average velocity as h approaches 0. Investigate what happens if we work with time intervals ending at t rather than starting at t. In more detail, what happens if we first compute the average velocity from $t - h$ to t (with $h > 0$) and then take the limit as h approaches 0? Examine a few specific examples, such as those in the text or in the foregoing problems.

15. In our discussion in the text, we defined the velocity $v(t)$ of a car at time t as the limit of the rate of change of distance $s = s(t)$ with respect to time (that is, as the limit of the average speed). The notation for this is

$$\lim \frac{\text{Change of distance}}{\text{Change of time}} = \lim_{h \to 0} \frac{s(t + h) - s(t)}{h}$$

In similar fashion, the *acceleration* $a(t)$ of a car at time t is defined as the limit of the rate of change of velocity $v = v(t)$ with respect to time. The notation for $a(t)$ is

$$\lim \frac{\text{Change of speed}}{\text{Change of time}} = \lim_{h \to 0} \frac{v(t + h) - v(t)}{h}$$

Thus acceleration "$a = a(t)$" stands in the same relation to velocity "$v = v(t)$" as velocity "$v = v(t)$" stands in relation to distance "$s = s(t)$". Compute the acceleration, at any time t, for (a) $s = t^3$ and for (b) $s = 3t^2$.

In each of Exercises 16–20 find the velocity $v(t)$ and the acceleration $a(t)$ at any time t of a car whose motion is according to the given rule.

16. $s = t^4$ **17.** $s = 5t^3$ **18.** $s = 4t^2$

19. $s = 5t$ **20.** $s = \dfrac{1}{t + 3}$

21. The accompanying table shows the unit price of a certain commodity over a 10-year period. What is the average rate of change (in dollars per year) in the commodity's price over this period? Over the first 5 years?

22. At noon a 250-gal water tank springs a leak. The water volume V in the tank is observed at 1-min intervals, as indicated in the accompanying table. What is the average rate of change of volume during the 5-min period from noon to 12:05 P.M.? During the first minute? During the interval from 12:04 P.M. to 12:05 P.M.? (Your answers will be negative because the volume is decreasing.)

TABLE FOR EXERCISE 21

Year	*Unit*
1975	\$ 5.65
1976	5.89
1977	6.92
1978	8.50
1979	10.42
1980	11.83
1981	14.72
1982	16.04
1983	16.95
1984	17.46
1985	18.02

TABLE FOR EXERCISE 22

Time	*Volume*
12:00	250.0
12:01	244.8
12:02	240.2
12:03	236.2
12:04	232.8
12:05	230.0

23. Refer to Exercise 22. If we let t be minutes past noon, the function

$$V = 250 - 5.5t + 0.3t^2$$

fits the data given in the table. Assuming that this function actually does describe the changing volume:

(a) Find the average rate of water flow from the tank during the interval from $t = 3$ to $t = 3 + h$.

(b) Find the instantaneous rate of flow at 12:03 P.M. ($t = 3$).

SECTION 2.2

The Tangent Line

In Chapter 1 we saw that the slope of a straight line is a measure of the line's steepness or, equivalently, of how fast the y-coordinate changes compared to the x-coordinate. As the previous section suggested, we are interested in rates of change also in the more general situation where y is an arbitrary (nonlinear) function of x. Here the steepness of a function's graph at a particular point is measured by the slope of what is called the tangent line to the graph at that point.

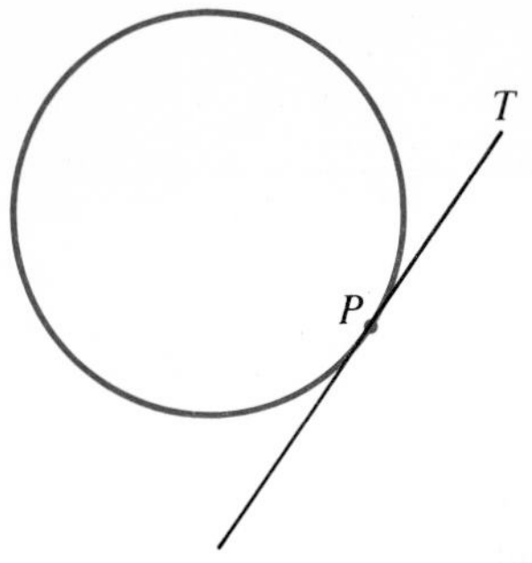

FIGURE 2.3

In plane geometry we encountered the notion of the tangent line T to a circle at a point P (see Figure 2.3). There are several ways in which this **tangent line** can be defined. For example, it can be viewed as the line through P that is perpendicular to the radius of the circle that ends at P. Alternatively, it can be viewed as the straight line through P that intersects (or meets) the circle in exactly one point. Intuitively, of course, the tangent line is a straight line that just "grazes" the circle at P.

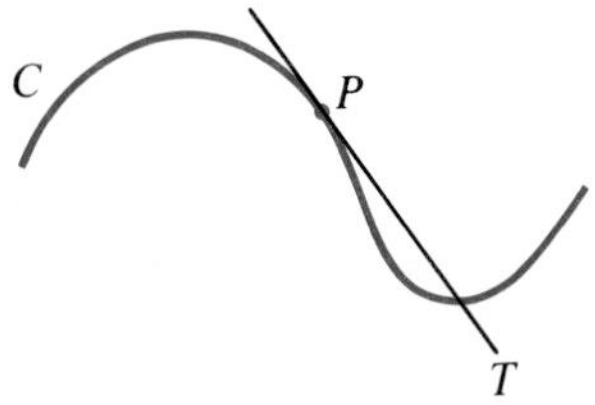

FIGURE 2.4

The question naturally arises: What do we mean by a tangent to an arbitrary curve in the plane? Clearly, for curves other than circles, it is meaningless to use any definition of the term *tangent* that mentions a "radius." Also, the requirement that a tangent line meet the curve at exactly one point is not quite what we want. For a curve such as the one shown in Figure 2.4, we would like to call T the tangent line to the curve C at the point P, in spite of the fact that line T intersects the curve at two points. It turns out that the intuitive notion of tangent line is the most useful because, as we are about to show, it can be extended to general curves.

Given an arbitrary curve C and a point P on the curve, we would like to determine the tangent line to the curve at the given point (assuming that there is such a thing). Of course, we need to clarify what is meant by "determining" the tangent line. It is reasonable to say that the tangent line is determined as soon as we have enough information to write down its equation. Now, one form of the equation of a straight line is the so-called *point–slope* form:

(2.5) $$y - y_1 = m(x - x_1)$$

This is an equation for the straight line of slope m that passes through the point (x_1, y_1). In our situation, the point P is given [which means that its coordinates are known; we denote them by (x_1, y_1)], and the tangent line must pass through it. It follows that the tangent line is determined as soon as we know its slope.

Let us pursue these ideas in a concrete situation. We wish to consider the following question: Can we find the slope of the tangent line to the curve $y = 10x^2$ at the point (3, 90) on the curve?

Recall first of all that, if any two points P and Q in the plane are given and we write their coordinates as (x_1, y_1) and (x_2, y_2), respectively, the slope of the straight line joining P and Q is given by

(2.6) $$\frac{y_2 - y_1}{x_2 - x_1} = \frac{\text{difference in the } y\text{'s}}{\text{difference in the } x\text{'s}}$$

Let us fix P as the point on the curve $y = 10x^2$ whose coordinates are $(x_1, y_1) = (3, 90)$. If we choose Q to be another point on our curve, then for different choices of Q we can compute, from Equation (2.6), the slopes of various lines through P. (Any such line joining two points P and Q on the same curve is often called a **secant line**.) To be specific, suppose Q is the point with coordinates $(x_2, y_2) = (4, 160)$. Then Equation (2.6) shows that the slope of the secant line PQ is

$$\frac{160 - 90}{4 - 3} = 70$$

Unfortunately, there is no reason to expect that the slope of the secant line PQ has any bearing on the solution of our question, which requires that we find the slope of the tangent line to the curve at the point $P = (x_1, y_1) = (3, 90)$. In fact, Figure 2.5 shows that the slope of secant line PQ and the slope of the tangent line T at P are clearly different.

Is there anything we can do to correct or improve this situation? One possibility is that, instead of using the point $Q = (4, 160)$, we choose another point Q' on the curve $y = 10x^2$ that is closer to P than the point Q. For example, suppose we take $Q' = (3.5, 122.5)$. A look at Figure 2.6 makes it apparent that secant line PQ' is closer than secant line PQ to the tangent line T. The slope of PQ', which is given by

$$\frac{122.5 - 90}{3.5 - 3} = 65$$

is therefore closer to the slope of the tangent line T than the previously computed value of 70. It seems plausible that, the closer we choose Q to the fixed point P, the closer the slope of secant line PQ will be to the slope of the tangent line.

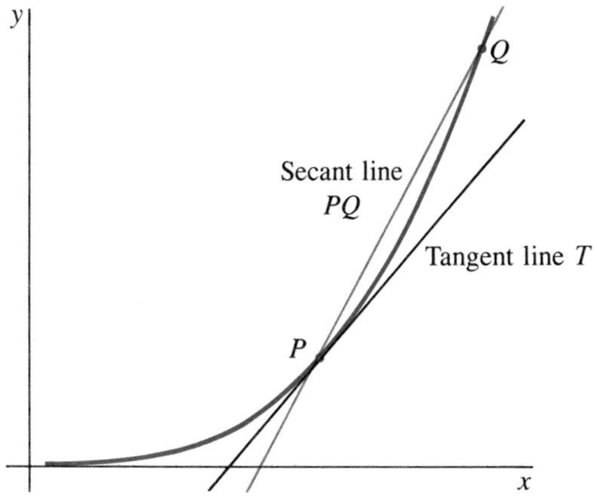

FIGURE 2.5

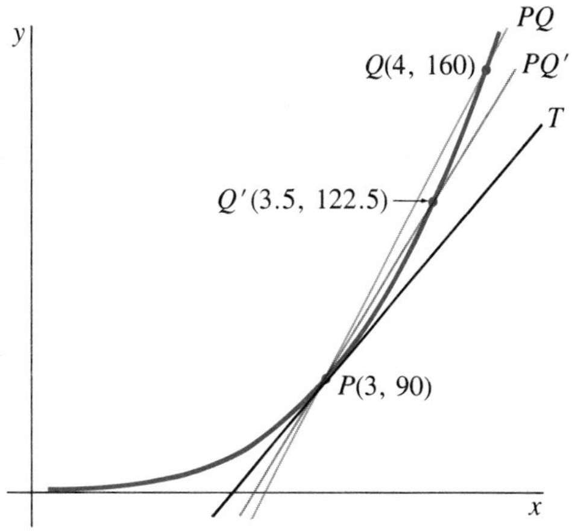

FIGURE 2.6

PRACTICE EXERCISE For the points $Q_1 = (3.1, 96.1)$, $Q_2 = (3.01, 90.601)$, and $Q_3 = (3.001, 90.06001)$, find the slopes of the secant lines PQ_1, PQ_2, and PQ_3, respectively.

[Your answers should be 61, 60.1, and 60.01.]

From the preceding, we see that as the point Q moves toward P along the curve $y = 10x^2$, we generate a set of numbers representing the slopes of the various secant lines PQ. In our specific case, these slopes are

$$70, \quad 65, \quad 61, \quad 60.1, \quad 60.01$$

Because these numbers appear to be tending toward 60, and because we feel that these numbers provide better and better approximations to the actual slope of the tangent, we *guess* that the slope of the tangent line at the point (3, 90) is 60.

Is there any way to confirm our guess? Let us reformulate our procedure in more general terms. As before, we fix $P = (3, 90)$. But, rather than select specific coordinates for the point Q, let us simply require that Q be a point on the curve $y = 10x^2$ whose x-coordinate is h units to the right of P. For example, if we choose $h = 1$ then $Q = (4, 160)$, whereas if $h = 0.5$ then $Q = (3.5, 122.5)$. In particular, because we are selecting different values for h, the corresponding points Q will be distinct points on $y = 10x^2$. In general, if the x-coordinate of Q is $3 + h$, its y-coordinate is given by

$$10(3 + h)^2 = 90 + 60h + 10h^2$$

so
$$Q = (3 + h, 90 + 60h + 10h^2)$$

Applying Equation (2.6), we find that the slope of the secant line PQ (see Figure 2.7) is given by

(2.7)
$$\frac{(90 + 60h + 10h^2) - 90}{(3 + h) - 3} = \frac{60h - 10h^2}{h} = 60 + 10h$$

Now let us "picture" the point Q moving along the curve $y = 10x^2$ toward the point P. This means that the coordinates of Q are getting closer and closer to those of P. It is equivalent to letting the number h get closer and closer to 0. We expect that, as this occurs, the slope of the secant line PQ will approach (get closer and closer to) the slope of the tangent line at P. As h becomes very small, so does $10h$, and therefore $60 + 10h$ gets very close to 60.

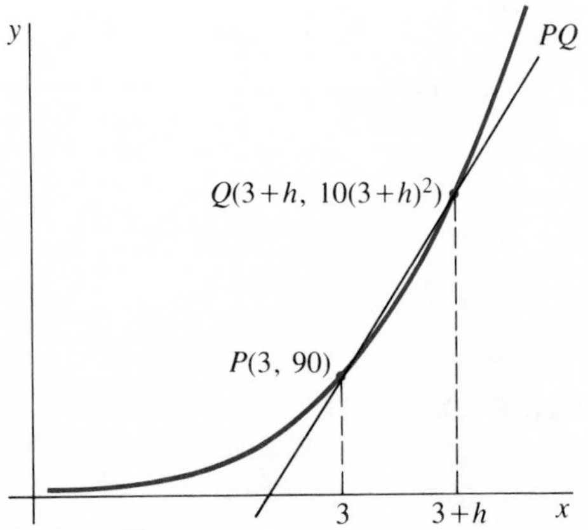

FIGURE 2.7

We say that 60 is the limiting value of the left side of Equation (2.7) as h approaches 0, and we write

$$\lim_{h \to 0} (60 + 10h) = 60$$

This confirms our earlier guess.

We can generalize the foregoing procedure in order to answer the following more general question: Can we find the slope of the tangent line to the curve $y = 10x^2$ at an arbitrary point P on the curve?

Let $P = (x, 10x^2)$ be any point on our curve. As before, let Q be any other point on the curve, with Q not far from P. Its x-coordinate may be written as $x + h$. Thus

$$\begin{aligned} Q &= (x + h, 10(x + h)^2) \\ &= (x + h, 10x^2 + 20xh + 10x^2) \end{aligned}$$

Then, according to Equation (2.6), the slope of secant line PQ (see Figure 2.8) is given by

$$\frac{(10x^2 + 20xh + 10h^2) - 10x^2}{(x + h) - x} = \frac{20xh + 10h^2}{h} = 20x + 10h$$

We then take the limit of this expression as h approaches 0. Now, as h becomes very small, $10h$ becomes small, and $20x + 10h$ gets very close (as close as we want) to $20x$. We say that the slope of the tangent line to the curve $y = 10x^2$ at the point $P = (x, 10x^2)$ is given by $20x$. (Observe that, if we choose $x = 3$, the slope is $20x = 20(3) = 60$, in accordance with our previous calculation.) Thus, by a straightforward procedure, based on the idea that the slope of the tangent line is the limit of the slopes of the secant lines, we have computed the slope of the tangent at *any* point on the curve. Thus we can readily find the equation of the tangent to the curve at any given point.

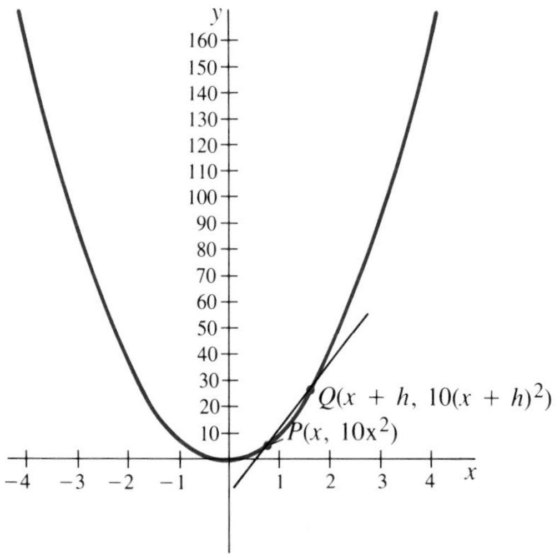

FIGURE 2.8

PRACTICE EXERCISE Give an equation for the tangent line to the graph of $y = 10x^2$ at (3, 90).

[Your answer should be $y = 60x - 90$ (or the equivalent)].

As we did at the end of Section 2.1, let us conclude by reformulating our computations using function notation. Let f represent the function $f(x) = 10x^2$. Then $f(x + h) = 10(x + h)^2 = 10x^2 + 20xh + 10h^2$. It follows that the point P on the curve in Figure 2.8 has coordinates $(x, f(x))$, whereas the coordinates of Q are given by $(x + h, f(x + h))$. Then the slope of the secant line joining P and Q (see Figure 2.9) is given by

$$\frac{f(x + h) - f(x)}{h} \tag{2.8}$$

which we recognize as the difference quotient for the function f. In order to obtain the slope of the tangent line to the curve at P, we let $h \to 0$, which is equivalent to sliding Q along the curve toward P. Thus the slope of the tangent line at an arbitrary point P is

$$\lim_{h \to 0} \frac{f(x + h) - f(x)}{h} \tag{2.9}$$

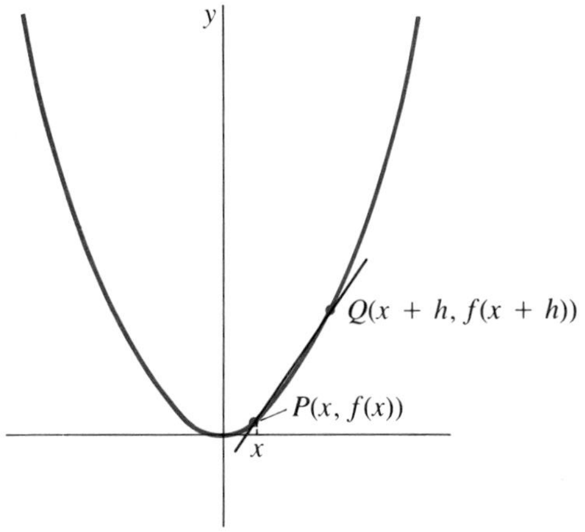

FIGURE 2.9

Section 2.2 EXERCISES

1. On the curve $y = f(x) = 3x$, consider the point whose x-coordinate is 2. Find the slope of the tangent to the curve at this point. Interpret your result.
2. Do Exercise 1 for the curve $y = f(x) = 3x + 5$.
3. On the curve $y = f(x) = x^3$, consider the point whose x-coordinate is 3. Find the slope of the tangent line to the curve at this point.
4. Do Exercise 3 for the curve $y = f(x) = 2x^3$.

5. Consider the point $(-1, 7)$ on the curve $y = f(x) = 3x^2 + x + 5$. Find the slope of the tangent to the curve at this point.

6. Find the slope of the tangent to the curve $y = f(x) = 3x^2 - x + 5$ at the point on the curve whose x-coordinate is -1.

7. Do Exercise 6 for the curve $y = f(x) = \frac{3}{x} + 2x + 5$.

In each of Exercises 8–20 a curve $y = f(x)$ is given. Find the slope of the curve (that is, the slope of the tangent to the curve) at any point on the curve.

8. $y = f(x) = 3x + 5$

9. $y = f(x) = ax + b$ (a, b constant)

10. $y = f(x) = 3x^2 + x + 5$

11. $y = f(x) = ax^2 + bx + c$ (a, b, c constant)

12. $y = f(x) = \frac{1}{x}$

13. $y = f(x) = x^2 + 2x$

14. $y = f(x) = 3x^2 + 5$

15. $y = f(x) = \frac{2}{x}$

16. $y = f(x) = \frac{1}{2x}$

17. $y = f(x) = x^3 - 1$

18. $y = f(x) = x^4$

19. $y = f(x) = \frac{1}{x - 3}$

20. $y = f(x) = x - x^2$

21. Suppose the cost of producing x commercial air conditioning systems is $C = 2x^2 + 60x + 1200$.
   (a) What is the cost to produce 5 air conditioning systems?
   (b) What is the average cost per system when 5 systems are produced?
   (c) Find the rate of change of the cost with respect to x when $x = 5$. (This is the slope of the tangent line.)

SECTION 2.3

Derivatives and the Marginal Concept

In Section 2.1 we examined in detail the motion of a car along a straight road. We used the notation $s(t)$ to denote the distance of the car at time t from a fixed reference point. Our analysis of this "broken speedometer" problem led us to consider the difference quotient

$$\frac{s(t + h) - s(t)}{h}$$

[See Equation (2.4).] This represented the car's average speed from time t to time $t + h$. We were forced to compute the limit of this expression as h approached 0. The limit represented the car's speedometer reading, or instantaneous velocity, at time t.

In Section 2.2 we considered a different problem: how to determine the slope of a curve at some point $(x, f(x))$ on it. Again we were led to examine the difference quotient

$$\frac{f(x + h) - f(x)}{h}$$

[See Equation (2.8).] This represented the slope of the secant line joining the points $P(x, f(x))$ and $Q(x + h, f(x + h))$. We found the desired slope of the curve (or, equivalently, the slope of the tangent line to the curve) by allowing the point Q to move along the curve toward the point P. This is the same as evaluating the limit of the expression for the slope of the secant line as h approaches 0.

If you were reading the previous two sections attentively, you probably noticed the similarities between the calculations done to solve each problem. In fact, the broken speedometer problem and the slope problem are the same problem under different disguises. Let us now formalize what we have done in the preceding two sections.

DEFINITION 2.1 (The Derivative) Given a function $y = f(x)$ and a number x_0 in the domain of f, we say that $y = f(x)$ is **differentiable** at x_0 if the limit as h tends toward 0 of the difference quotient

$$\frac{f(x_0 + h) - f(x_0)}{h}$$

exists*. The value of this limit

$$\lim_{h \to 0} \frac{f(x_0 + h) - f(x_0)}{h}$$

is called the **derivative of the function $f(x)$ at x_0** and is denoted by

$$f'(x_0), \qquad \left.\frac{dy}{dx}\right|_{x_0}, \qquad y'(x_0), \quad \text{or} \quad \left.\frac{dy}{dx}\right|_{x=x_0}$$

Thus

$$f'(x_0) = \lim_{h \to 0} \frac{f(x_0 + h) - f(x_0)}{h}$$ □

The process of computing a derivative is called **differentiation**. If a function is differentiable at x_0 for every choice of x_0 in its domain, we simply say that $y = f(x)$ is differentiable, and we denote its derivative by

$$f'(x), \quad \frac{dy}{dx}, \quad \text{or simply} \quad y'$$

You may be puzzled about the reason for the alternative notations y' and dy/dx for the derivative. The first notation (involving the apostrophe or prime symbol) is called **Newton notation**,† whereas the latter is referred to as **Leibniz notation**. Sir Isaac Newton (1642–1727) and Baron Gottfried Wilhelm Leibniz (1646–1716) independently developed the theory of calculus and, for historical reasons, both notations are still in use.

In our new terminology, we may summarize the results of Sections 2.1 and 2.2 by saying that, if $y = f(x) = 10x^2$, then

$$f'(x) = y' = \frac{dy}{dx} = 20x$$

---

* We adopt the standard convention that, if an expression tends toward $+\infty$ or $-\infty$, the limit does *not* exist.

† Newton actually used a dot rather than an apostrophe.

and, in particular,

$$f'(3) = y'(3) = \left.\frac{dy}{dx}\right|_{x=3} = 60$$

EXAMPLE 1 Let $f(x) = 1/x$. Compute $f'(x)$.

SOLUTION We observe first that $x = 0$ does not belong to the domain of f because division by 0 is meaningless. Therefore we must require that $x \neq 0$. The Newton quotient of f is

$$\frac{f(x+h) - f(x)}{h} = \frac{\dfrac{1}{x+h} - \dfrac{1}{x}}{h} = \frac{\dfrac{x - (x+h)}{(x+h)(x)}}{h} = -\frac{1}{x(x+h)}$$

Now, as h tends toward 0, $x + h$ tends toward x. Therefore the limit of the Newton quotient is

$$f'(x) = \lim_{h\to 0} -\frac{1}{x(x+h)} = -\frac{1}{x(x)} = -\frac{1}{x^2}$$

EXAMPLE 2 If $y = 3x - 1$, compute dy/dx.

SOLUTION We write $y = f(x) = 3x - 1$. Then

$$\begin{aligned}\frac{f(x+h) - f(x)}{h} &= \frac{[3(x+h) - 1] - (3x - 1)}{h} \\ &= \frac{3x + 3h - 1 - (3x - 1)}{h} \\ &= \frac{3x + 3h - 1 - 3x + 1}{h} \\ &= \frac{3h}{h} = 3\end{aligned}$$

According to the definition of the derivative (Definition 2.1),

$$\frac{dy}{dx} = \lim_{h\to 0} \frac{f(x+h) - f(x)}{h} = \lim_{h\to 0} 3$$

But, as the value of the number h changes, the number 3 remains constant, so $\lim_{h\to 0} 3 = 3$ and

$$\frac{dy}{dx} = 3$$

For instance, this means that

$$\left.\frac{dy}{dx}\right|_{x=1} = 3, \quad \left.\frac{dy}{dx}\right|_{x=\sqrt{7}} = 3, \quad \text{and} \quad \left.\frac{dy}{dx}\right|_{x=-\pi} = 3$$

In fact,

$$\left.\frac{dy}{dx}\right|_{x=x_0} = 3$$

regardless of what value of x_0 we choose. This is to be expected, of course, because the graph of $y = 3x - 1$ is a straight line with slope 3. At *any* point on the line, the tangent line is the given line itself.

There is an aspect of our computation of derivatives from "first principles"—that is, via the definition of the derivative, that often causes some confusion and is perhaps worth explaining. In order to compute the derivative, we must evaluate the limit of the difference quotient at x_0:

$$\lim_{h \to 0} \frac{f(x_0 + h) - f(x_0)}{h}$$

But if we allow h to tend toward 0, it looks as though we will be dividing by a number that is very close to 0, and we know that division by 0 is not permitted.

This difficulty is illusory, however. As h tends toward 0, $x_0 + h$ tends toward x_0, which in turn implies (for most functions we shall encounter) that $f(x_0 + h)$ tends toward $f(x_0)$. Thus $f(x_0 + h) - f(x_0)$ tends toward 0 as h tends toward 0. Both the numerator and the denominator of the difference quotient will therefore be small, and it is their *ratio* (more precisely, the limit of their ratio) that we are trying to determine.

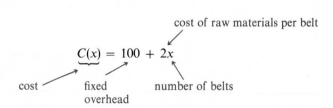

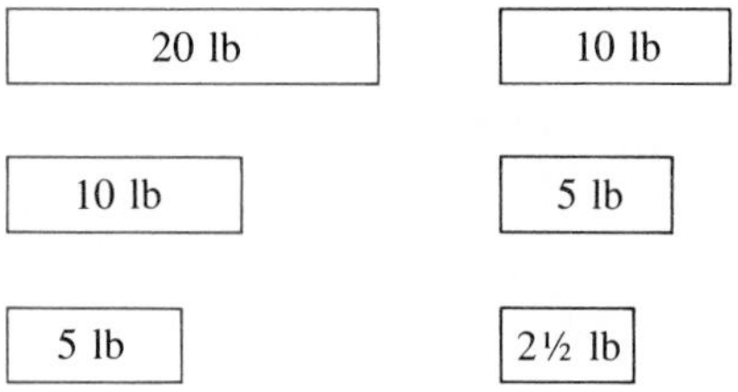

FIGURE 2.10

Let us try to explain this idea by analogy with another situation. Imagine that we have two steel bars, one of which weighs 20 pounds and the other 10 pounds. The ratio of their weights is 20/10, or 2. If we cut each bar in half, the first bar will weigh 10 pounds, and the second will weigh only 5 pounds. Each bar is smaller, but the ratio is 10/5, which is still 2. No matter how much we cut the first bar, as long as we cut the second bar proportionally, the ratio of the weights remains fixed (Figure 2.10). We encounter the same idea in the evaluation of the derivative. Both the numerator and the denominator of the difference quotient are tending toward 0, but it is their ratio that we want. Of course, in general, the ratio (the difference quotient) does not remain constant as the numerator and denominator tend toward 0.

Let us consider some practical applications of the concept of the derivative.

An enterprising student is helping to finance the cost of her college education by making and selling hand-tooled leather belts in a small leather shop. Her fixed overhead—which includes such items as rent and electricity—is \$100 per month. In addition, it costs her \$2 for the raw materials to produce each belt. If we let x denote the number of belts she produces in a month, then the total cost of producing those x belts must be given by

cost of raw materials per belt

$$C(x) = 100 + 2x$$

cost — fixed overhead — number of belts

This is an example of a **cost function**. For example, it tells us that the cost of producing 40 belts in one month is

$$C(40) = 100 + 2(40) = 180$$

If, in the next month, the student produces 50 belts (cost \$200), the *change in cost* to produce 10 extra belts will be

$$C(50) - C(40) = 200 - 180 = 20$$

Therefore, the *average rate of change* as x goes from 40 to 50 is

$$\frac{C(50) - C(40)}{10} = \frac{20}{10} = 2$$

dollars per belt. In general, the average rate of change from x to $x + h$ is given by

$$\frac{C(x + h) - C(x)}{h} = \frac{[100 + 2(x + h)] - [100 + 2x]}{h} = \frac{2h}{h} = 2$$

We recognize this as simply the difference quotient for the cost function C. Of course, for $C(x) = 100 + 2x$, the instantaneous rate of change (or derivative) is given by

$$C'(x) = 2$$

This is hardly surprising. For any linear function, the average change (which represents the slope of a secant line) agrees with the instantaneous rate of change (which is the slope of the tangent line).

Now let's modify our problem somewhat. Suppose that the leather wholesaler agrees to offer a discount depending on how much leather is ordered. To be specific, let's suppose that the cost of raw materials per belt is now

$$2 - 0.001x$$

rather than a flat \$2. Thus, the more belts produced, the cheaper the cost of leather per belt. Then the total cost for producing x belts is

$$C(x) = 100 + (2 - 0.001x)x = 100 + 2x - 0.001x^2$$

Now, for example, the average increase in cost per belt for producing 50 belts as opposed to 40 is

$$\frac{C(50) - C(40)}{10} = \frac{197.50 - 178.40}{10} = \frac{19.10}{10} = \$1.91$$

That is, when 50 belts are made in a month, as opposed to 40, raw material for each additional belt is \$1.91 (on the average), as opposed to \$2.00 without a discount.

We could similarly ask for the increase in cost that would be incurred for producing 41 belts instead of 40 belts. In this case

$$\frac{C(41) - C(40)}{1} = \$1.919$$

Thus it costs almost \$1.92 to increase production from 40 to 41 belts. Note that unlike the linear cost function we considered previously, the average rate of change of cost is no longer a constant. In the linear case, the average rate of change of cost was \$2 per additional belt, and that value agreed with the derivative. What does the first derivative tell us when the cost function is nonlinear? As you can easily verify,

$$C'(x) = 2 - 0.002x$$

and $$C'(40) = 2 - (0.002)(40) = \$1.92$$

Thus the derivative at $x = 40$ is approximately the cost of producing one additional belt. Economists call it the marginal cost at $x = 40$.

In general, in theoretical work, economists define **marginal cost** (at any x) as the derivative of the cost function:

$$\text{Marginal cost at } x = C'(x) = \lim_{h \to 0} \frac{C(x + h) - C(x)}{h}$$

It is the (instantaneous) rate of change of cost with respect to production (at the production level x), and it approximates the additional cost of increasing production from x to $x + 1$:

$$C'(x) \approx C(x + 1) - C(x)$$

Geometrically, the marginal cost is the slope of the tangent line to the graph $y = C(x)$ at the point $(x, C(x))$.

The marginal concept is a very important one in economics, and it can be applied to many other functions. Another illustration is provided by examining Example 25 of Chapter 1 once more. We reproduced the tax-rate schedule for individual taxpayers for 1984 in Figure 1.22. This schedule was formulated as a function in Table 1.6, and its graph appears in Figure 1.23.

Consider a taxpayer earning \$16,000 per year. According to the schedule, his federal income tax for the year was

$$f(16{,}000) = 2001 + 0.23(16{,}000 - 15{,}000) = 2231$$

Suppose our taxpayer gets a raise of \$1. How much does his tax bill increase? Again, from Table 1.6 we compute the tax liability on an income of \$16,001 as

$$f(16{,}001) = 2001 + 0.23(16{,}001 - 15{,}000) = 2231.23$$

Thus the additional tax per unit of additional income was

$$\frac{f(16{,}001) - f(16{,}000)}{1} = 0.23$$

We recognize 0.23 as simply the slope of the graph of the tax function at $x = 16{,}000$. If we repeat this analysis for a raise of \$$h$, where h is presumed to be very small, the additional tax per additional unit of income will be

$$\lim_{h \to 0} \frac{f(16{,}000 + h) - f(16{,}000)}{h} = 0.23$$

This is called the **marginal tax rate** at an income level of \$16,000.

Note that, if we have a second taxpayer with a higher income of \$25,000, her tax bill is

$$f(25{,}000) = 4115 + 0.30(25{,}000 - 23{,}500) = 4565$$

Her marginal tax on an additional \$1 of income is

$$f'(25{,}000) = 0.30$$

(the derivative, or slope, at $x = 25{,}000$). Thus \$1 of additional income increases her tax bill by a greater amount than it does for someone with a smaller income.

In the problems at the end of this section, we explore the marginal concept further.

Section 2.3 EXERCISES

In each of Exercises 1–5 $s = s(t)$ and $y = y(x)$ are given. Compare the work involved in computing the instantaneous speed $v(t) = s'(t)$ of a car and the slope $y'(x)$ of a curve. In each exercise find the derivative of $s(t)$ or, equivalently, of $y(x)$.

1. $s = s(t) = t^3$ and $y = y(x) = x^3$

2. $s = s(t) = t^4$ and $y = y(x) = x^4$

3. $s = s(t) = 3t^2 - t + 5$ and $y = y(x) = 3x^2 - x + 5$

4. $s = s(t) = at + b$ and $y = y(x) = ax + b$ (a, b constant)

5. $s = s(t) = at^2 + bt + c$ and $y = y(x) = ax^2 + bx + c$ (a, b, c constant)

6. Can you give any examples of situations in which the notion of a derivative arises?

In Exercises 7–10 apply "first principles" (that is, compute the limit of the appropriate difference quotient) to find:

7. $f'(7)$ when $f(x) = x^2$

8. $f'(-3)$ when $f(x) = 3x^2 - 5x + 1$

9. $f'(-2)$ when $f(x) = x^3 - 5$

10. $f'(-1)$ when $f(x) = \dfrac{1}{x}$

In Exercises 11–12 apply "first principles" (that is, compute the limit of the appropriate Newton quotient) to find:

11. $f'(x)$ when $f(x) = 3x^2 - 5x + 1$

12. $f'(x)$ when $f(x) = \dfrac{x^2 - 1}{x}$

In each of Exercises 13–20 evaluate the given limit and interpret what you have accomplished by this computation (that is, what derivative does the limit represent?).

13. $\displaystyle\lim_{h \to 0} \frac{(2 + h)^2 - 2^2}{h}$

14. $\displaystyle\lim_{h \to 0} \frac{(-1 + h)^3 - (-1)^3}{h}$

15. $\displaystyle\lim_{h \to 0} \frac{3(2 + h) - 3(2)}{h}$

16. $\displaystyle\lim_{h \to 0} \frac{\dfrac{1}{3 + h} - \dfrac{1}{3}}{h}$

17. $\displaystyle\lim_{h \to 0} \frac{\dfrac{1}{(2 + h)^2} - \dfrac{1}{2^2}}{h}$

18. $\displaystyle\lim_{h \to 0} \frac{(x + h)^2 - x^2}{h}$

19. $\displaystyle\lim_{h \to 0} \frac{(x + h)^3 - x^3}{h}$

20. $\displaystyle\lim_{h \to 0} \frac{\dfrac{1}{(x + h)^2} - \dfrac{1}{x^2}}{h}$

21. A steel manufacturer believes that his cost C of producing steel is given by the formula

$$C(x) = x^2 - x + 10$$

where x is measured in tons and C is measured in dollars. (Why the manufacturer uses this formula is a question that does not concern us.)

(a) The cost of producing 2 tons of steel is $C(2) = 2^2 - 2 + 10 = 12$ dollars. Similarly, $C(5) = 30$. Find $C(3)$, the cost of producing 3 tons; and $C(2 + h)$, the cost of producing $2 + h$ tons.

(b) The additional cost incurred by the manufacturer in producing 5 tons of steel instead of 2 tons is

$$C(5) - C(2) = 30 - 12 = 18 \text{ dollars}$$

(In general, by "additional cost incurred" we mean the difference between the costs.) The additional cost of producing 4 tons instead of 2 tons is then 10 dollars. Find the additional cost of producing 3 tons instead of 2 tons and the additional cost of producing $2 + h$ tons instead of 2 tons.

(c) The additional cost per ton of producing 5 tons instead of 2 tons is

$$\frac{C(5) - C(2)}{5 - 2} = 6 \text{ dollars per ton}$$

(In general, by "additional cost per ton" we mean the difference between the costs divided by the difference in tonnage.) The additional cost per ton of producing 4 tons instead of 2 tons is 5 dollars. Find the additional cost per ton of producing 3 tons instead of 2 tons. Find the additional cost per ton of producing $2 + h$ tons instead of 2 tons.

(d) Compute the limit as $h \to 0$ of the expression you just obtained. (The answer is 2.) As we have seen, economists call this the marginal cost at $x = 2$; it measures the rate of change of cost with respect to a "small" increase in production when $x = 2$. Find the marginal cost at any x for the cost function of this problem.

22. The steel manufacturer of Exercise 21 believes that the revenue R that he receives from the sale of his steel is given by the formula

$$R(x) = 100x - x^3$$

where x is measured in tons and R is measured in dollars.

(a) The revenue from selling 2 tons of steel is $R(2) = (100)(2) - 2^3 = 192$ dollars. Similarly, $R(5) = 375$. Find the revenue from the sale of 3 tons and of $2 + h$ tons, respectively.

(b) The "additional revenue" received from selling 5 tons instead of 2 tons is

$$R(5) - R(2) = 375 - 192 = 183 \text{ dollars}$$

The additional revenue from the sale of 4 tons instead of 2 tons is then 144 dollars. Find the additional revenue from the sale of 3 tons instead of 2 tons. Find the additional revenue from the sale of $2 + h$ tons instead of 2 tons.

(c) The "additional revenue per ton" from the sale of 5 tons instead of 2 tons is

$$\frac{R(5) - R(2)}{5 - 2} = 61 \text{ dollars}$$

The additional revenue per ton from the sale of 4 tons instead of 2 tons is 72 dollars. Find the additional revenue per ton from the sale of 3 tons instead of 2 tons. Find the additional revenue per ton from the sale of $2 + h$ tons instead of 2 tons.

(d) Compute the limit as $h \to 0$ of the expression you just obtained.

(e) In theoretical work, economists define **marginal revenue** (at any x) as the derivative of the revenue function. Marginal revenue at x:

$$R'(x) = \lim_{h \to 0} \frac{R(x + h) - R(x)}{h}$$

It measures the "instantaneous" rate of change of revenue with respect to sales (at the sales level of x tons). Find the marginal revenue at any x for the revenue function of this problem.

23. During the course of a 2-hour storm, it is observed that the barometric pressure varies according to the formula

$$P(t) = 3t^2 - 6t + 32$$

where P represents pressure and t represents time measured in hours (with the storm beginning at $t = 0$ and ending at $t = 2$).

(a) Compute the pressure at $t = \frac{1}{2}$, at $t = 1$, and at $t = \frac{1}{2} + h$.

(b) Compute the change in pressure from $t = \frac{1}{2}$ to $t = 1$ and from $t = \frac{1}{2}$ to $t = \frac{1}{2} + h$.

(c) Compute the change in pressure per unit of time from $t = \frac{1}{2}$ to $t = 1$ and from $t = \frac{1}{2}$ to $t = \frac{1}{2} + h$.

(d) Compute the limit as $h \to 0$ of the expression you just obtained. This is called the *pressure gradient* at $t = \frac{1}{2}$.

(e) In similar fashion, compute the pressure gradient at $t = \frac{1}{4}$, at $t = 1$, and at $t = \frac{3}{2}$.

(f) Compute the pressure gradient at any time t_0 (where $0 < t_0 < 2$). In other words, compute

$$\lim_{h \to 0} \frac{P(t_0 + h) - P(t_0)}{h}$$

Thus the pressure gradient is simply the derivative (that is, the instantaneous rate of change) of the pressure.

(g) Sketch a graph of P against t (that is, plot t on the horizontal axis and P on the vertical axis). Can you explain why the pressure gradient is 0 at the precise instant when the pressure is lowest?

24. Is it possible that a function does not have a derivative? In the case of a moving car, what would this indicate? In the case of a curve in the plane, what would this indicate?

25. A magazine publisher finds that the demand q_d (number of subscribers) for his magazine is given by the formula

$$q_d = 25{,}000 - 0.1p^2 \qquad 100 \le p \le 500$$

where p is the magazine's price in cents.

(a) Compute the magazine's circulation at prices of $p = 200$ and $p = 250$. How many subscribers are lost due to a price increase from \$2.00 to \$2.50?

(b) Compute the average number of subscribers lost for each cent of price increase in going from $p = 200$ to $p = 250$.

(c) Compute the average number of subscribers lost per cent of price increase in going from $p = 200$ to $p = 200 + h$.

(d) Compute the limit of the latter expression as $h \to 0$. This is the instantaneous rate of change in demand at $p = 200$. How would you interpret it in colloquial English?

IF YOU WORK WITH A COMPUTER (OPTIONAL)

(Exercises 26–30 involve the use of a computer for approximating derivatives.)

Difficulties may arise in computing the difference quotient $[f(x + h) - f(x)]/h$ when h is small. First of all, because the numerator involves the difference of two quantities that are very close to one another, the subtraction may result in the loss of several significant digits. Subsequent division by the small number h may magnify this loss of significance, with the result that the difference quotient is not a very good estimate of $f'(x)$, even though h may be quite small (perhaps on the order of 10^{-7}).

An expression that often gives more accuracy without requiring h to be quite so small is the so-called symmetric difference quotient,

$$\frac{f(x + h) - f(x - h)}{2h}$$

It can be shown that this expression is the derivative at x of a parabola fitted to the three points $(x - h, f(x - h))$, $(x, f(x))$, and $(x + h, f(x + h))$. For most functions we shall encounter, $h = 10^{-4}$ is usually sufficient to give good accuracy without the loss-of-significance problems described above. For more information, see David A. Smith, *Interface: Calculus and the Computer*, New York: Saunders College Publishing, 1984.

For Exercises 26–30, use either of the following computer programs to estimate f' at several points of the given function on the given interval, using n subintervals.

26. $f(x) = x^4 + 3x^3 - 4x^2 + 2$ on $[-1, 3]$ $\quad n = 8$

27. $f(x) = (x^2 - 1)/(x^2 + 1)$ on $[-1, 1]$ $\quad n = 5$

28. $f(x) = \sqrt{\dfrac{x + 2}{x^2 + 1}}$ on $[0, 2]$ $\quad n = 4$

29. $f(x) = \dfrac{1}{x^2}$ on $[\frac{1}{2}, 3]$ $\quad n = 5$

30. $f(x) = \dfrac{2}{\sqrt{x}}$ on $[1, 9]$ $\quad n = 16$

```
10  REM - Derivatives - BASIC version

20  REM - estimates f' at several pts. in an interval

30  REM - Define f(x)
40  DEF FNf(x) = x^4

50  INPUT "End points of interval";a,b
60  INPUT "How many subintervals";n
70  PRINT
```

```
(* Derivatives - Pascal version *)
program Derivatives (input, output);

 var
  a, b, x, h, increment : real;
  i, n : integer;

(* Define function *)
 function f (x : real) : real;
 begin
  f := 3 * x * x - 5 * x + 2
 end;
```

(*Programs continue at top of next page*)

```
80  PRINT " x", "f(x)", "f'(x) est."

90  h=.0001
100 increment=(b-a)/n
110 x=a

120 FOR i = 0 TO n
130    REM - compute "symmetric diff. quotient"
140    PRINT x, FNf(x), (FNf(x+h)-FNf(x-h))/(2*h)
150    x = x + increment
160 NEXT i

170 END
```

```
begin
  write('Interval left endpoint a? ');
  readln(a);
  write('Interval right endpoint b? ');
  readln(b);
  write('How many subintervals? ');
  readln(n);

  writeln;
  writeln('  x', '            ', 'y', '           ', 'dy/dx');
  h := 0.0001;
  increment := (b - a) / n;

  for i := 0 to n do
   begin
     write(x : 8 : 5, '    ', f(x) : 8 : 5, '    ');
(* compute symmetric difference quotient *)
     writeln((f(x + h) - f(x - h)) / (2 * h) : 8 : 5);
     x := x + increment
   end

end.
```

SECTION 2.4

Some Simple Derivatives

We have computed the derivative of a few simple functions in the text, and we hope you have computed others in doing the problems. It should be apparent that the computation of derivatives can be a very tedious procedure, even for relatively simple functions. Imagine how messy the algebra would be if we were to attempt to differentiate a function like

$$y = \sqrt{\frac{x}{1 + x^2}}$$

directly from the definition of derivative. Nevertheless, it is possible (and, in fact, not even terribly difficult) to differentiate this function. To do so, it is necessary to have at our disposal certain rules and formulas for the differentiation of those simpler functions out of which the given one is built (such as $\sqrt{x}$ and $1 + x^2$), as well as a working knowledge of how to differentiate combinations of such basic functions (such as products, quotients, and compositions).

To begin with, let us consider the simplest class of functions that we know, functions whose graphs are straight lines: $f(x) = mx + b$.

The difference quotient is

$$\frac{f(x + h) - f(x)}{h} = \frac{m(x + h) + b - mx - b}{h} = \frac{mh}{h} = m$$

which is a constant. Because the difference quotient is the same for all $h \neq 0$, its limit as $h \to 0$ is m. Hence, for $f(x) = mx + b$, $f'(x) = m$.

Of course, this result could have been anticipated because a line is its own tangent line at each of its points.

As a special case ($m = 0$), we note that the derivative of a constant function, $f(x) = b$, is 0 for every x. Let us now pass to quadratic functions.

EXAMPLE 3 Let us find $f'(x)$ when $f'(x) = x^2$.

SOLUTION Again we must evaluate the limit of the difference quotient as h tends toward 0. This time $f(x) = x^2$, so

$$\frac{f(x+h) - f(x)}{h} = \frac{(x^2 + 2xh + h^2) - x^2}{h} = \frac{2xh + h^2}{h}$$

Because h is not 0, we may carry out the indicated division to obtain

$$\frac{2xh + h^2}{h} = 2x + h$$

All that is left for us to do is to examine this expression as h tends toward 0.

As h takes on smaller and smaller values tending toward 0, the expression $2x + h$ tends toward $2x$. Thus we conclude that, for $f(x) = x^2$, the derivative is $f'(x) = 2x$.

This tells us, for instance, that on the parabola $y = x^2$, the slope of the tangent line at the point (3, 9) on the curve is given by $f'(3) = 2(3) = 6$. Similarly, at $(-1, 1)$ the slope is $f'(-1) = -2$, and in general, the slope at any point of the curve is twice the x-coordinate (Figure 2.11).

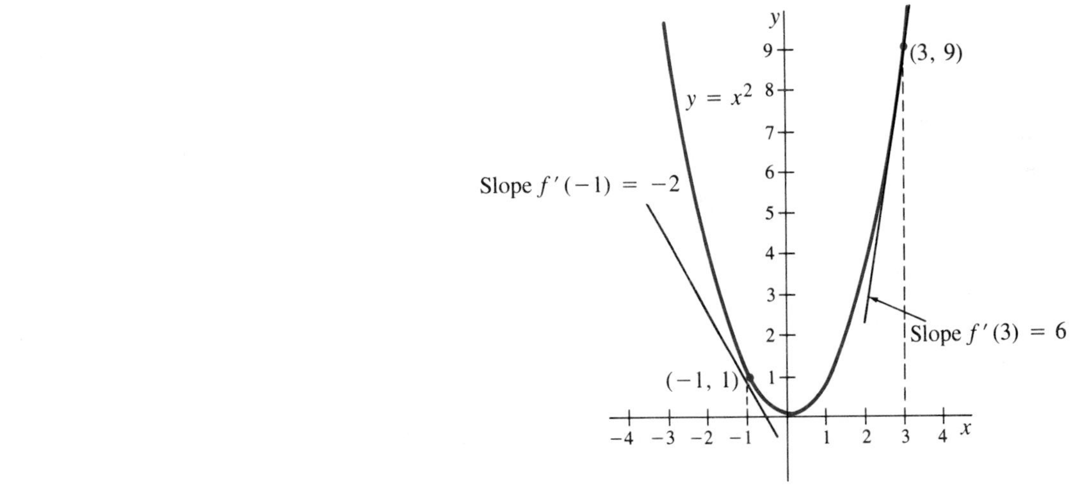

FIGURE 2.11

PRACTICE EXERCISE Compute $f'(x)$ for (a) $f(x) = 5x^2$; (b) $f(x) = -9x^2$; (c) $f(x) = cx^2$ for any constant c.

[Your answers should be (a) $f'(x) = 10x$, (b) $f'(x) = -18x$, and (c) $f'(x) = 2cx$.]

PRACTICE EXERCISE Guess (on the basis of our previous computations) the derivative of $f(x) = x^3 - 9x^2 + 8x + 2$?

[Your answer should be $f'(x) = 3x^2 - 18x + 8$.]

In an earlier practice exercise we found that $f(x) = x^3$ has derivative $f'(x) = 3x^2$. Consider now Table 2.5, which summarizes some of our results.

TABLE 2.5

$f(x)$	$f'(x)$
$x^0 = 1$	0
$x^1 = x$	1
x^2	$2x$
x^3	$3x^2$

What would you guess the derivative of $f(x) = x^4$ to be? A reasonable guess would be $f'(x) = 4x^3$ if the pattern of Table 2.5 is to be maintained. For $f(x) = x^5$, $f(x) = x^8$, and $f(x) = x^{100}$, we would probably guess $f'(x) = 5x^4$, $f'(x) = 8x^7$, and $f'(x) = 100x^{99}$, respectively.

This suggests that we make the following very reasonable conjecture.

THEOREM 2.2 (The Power Rule) If n is any nonnegative integer, the polynomial $f(x) = x^n$ has derivative

$$f'(x) = nx^{n-1}$$

That is, if we wish to differentiate the nth power of x, we simply reduce the exponent by 1 and then multiply the result by n. Of course, we must realize that all of our examples merely suggest that this rule is a correct one; a proof is still necessary.

PROOF We must evaluate

$$\lim_{h \to 0} \frac{f(x+h) - f(x)}{h}$$

where $f(x) = x^n$, n a nonnegative integer. In this case, the difference quotient becomes

(2.10)
$$\frac{(x+h)^n - x^n}{h}$$

But now we encounter a difficulty. In order to simplify this expression, we must expand $(x+h)^n$. This poses no problem when $n = 2$, in which case

$$(x+h)^2 = x^2 + 2xh + h^2$$

or when $n = 3$, in which case

$$(x+h)^3 = x^3 + 3x^2h + 3xh^2 + h^3$$

But what if $n = 100$? Certainly it would spoil the day to have to multiply $(x + h)$ by itself 99 times!

To resolve this difficulty, let us write down some of the powers of $(x + h)$ to see whether we can perceive some pattern.

$$(x + h)^1 = x + h$$

$$(x + h)^2 = x^2 + 2xh + h^2$$

$$\begin{aligned}(x + h)^3 &= x^3 + 3x^2h + 3xh^2 + h^3 \\ &= x^3 + 3x^2h + (3x + h) \cdot h^2\end{aligned}$$

$$\begin{aligned}(x + h)^4 &= x^4 + 4x^3h + 6x^2h^2 + 4xh^3 + h^4 \\ &= x^4 + 4x^3h + (6x^2 + 4xh + h^2) \cdot h^2\end{aligned}$$

It should now be clear that the first term in the expansion of $(x + h)^n$ is x^n and that the second is $nx^{n-1}h$. All succeeding terms have either h^2 or higher powers of h in them. Thus we can factor h^2 out of each of these terms, lump all that "stuff" together, and write

$$(x + h)^n = x^n + nx^{n-1}h + (\text{stuff}) \cdot h^2$$

Now Expression 2.10 becomes

$$\frac{x^n + nx^{n-1}h + (\text{stuff}) \cdot h^2 - x^n}{h} = \frac{(nx^{n-1} \cdot h) + [(\text{stuff}) \cdot h^2]}{h}$$

Dividing by h (which is permissible because $h \neq 0$), yields

$$nx^{n-1} + (\text{stuff}) \cdot h$$

Now, as h tends toward 0, all the "stuff" in the parentheses is multiplied by the small number h, so that term goes to 0. Thus

$$f'(x) = \lim_{h \to 0} \frac{f(x + h) - f(x)}{h} = nx^{n-1}$$

This completes the proof. ⊟

In our earlier work, we found that the derivative of $f(x) = 10x^2$ is $20x$, which is just 10 times the derivative of x^2. This was generalized in part (c) of the practice exercise following Example 3. These relationships are examples of the following theorem.

THEOREM 2.3 If $f(x) = cx^n$, where c is any constant and n is a nonnegative integer, then

$$f'(x) = ncx^{n-1}$$

The proof is left for you to complete. It is straightforward, amounting to reproducing the proof of Theorem 2.2 with the additional factor of c in every term. ⊟

Although we have proved the power rule (Theorem 2.2) and Theorem 2.3 only for the case where n is a positive integer, they in fact hold when n is *any* real number. We shall take this on faith for now and prove these theorems for rational (fractional) values of n later.

Section 2.4 EXERCISES

In each of Exercises 1–11 compute the indicated derivative of the given function from first principles (that is, by taking the limit of the difference quotient).

1. $f'(x)$ when $f(x) = 3x^4 + 5$

2. $\left.\dfrac{dy}{dx}\right|_{x=2}$ when $y = f(x) = 3x^2 - 8x^3$

3. $f'(-3)$ when $f(x) = 3x^2 + \dfrac{1}{x}$

4. $\left.\dfrac{dy}{dx}\right|_{x=-1}$ when $y = f(x) = x^2 - \dfrac{1}{x}$

5. $f'(2)$ when $f(x) = (2x + 3)^2$

6. $\left.\dfrac{dy}{dx}\right|_{x=-2}$ when $y = f(x) = (2x + 3)^2$

7. $f'(0)$ when $f(x) = 5x^6 + 8x^4$

8. $\left.\dfrac{dy}{dx}\right|_{x=1}$ when $y = f(x) = (x + 1)^3$

9. $f'(x)$ when $f(x) = 2x^3 - x^2 + 5x + 1$

10. $f'(x)$ when $f(x) = ax^2 + bx + c$ (a, b, c *constant*)

11. $f'(x)$ when $f(x) = \dfrac{1}{x} - \dfrac{1}{x^2}$

In each of Exercises 12–15 find the equation of the tangent line to the given curve at the given point.

12. $y = f(x) = x^2 + 3x + 5$ at the point $(1, 9)$

13. $y = f(x) = x^3 + x + 4$ at the point $(-2, -6)$

14. $y = f(x) = \dfrac{1}{x}$ at the point where $x = -3$

15. $y = f(x) = x^{10}$ at the point $(-1, 1)$

In each of Exercises 16–20 evaluate the given limit and explain what it represents.

16. $\lim\limits_{h\to 0} \dfrac{(2 + h)^5 - 2^5}{h}$

17. $\lim\limits_{h\to 0} \dfrac{(1 + h)^n - 1^n}{h}$

18. $\lim\limits_{h\to 0} \dfrac{\dfrac{1}{(x + h)^3} - \dfrac{1}{x^3}}{h}$

19. $\lim\limits_{h\to 0} \dfrac{[(x + h)^2 + 5(x + h)] - [x^2 + 5x]}{h}$

20. $\lim\limits_{h\to 0} \dfrac{\sqrt{x + h} - \sqrt{x}}{h}$

21. Given a function $y = f(x)$, we can find $f'(x_0)$ by computing
$$\lim_{h\to 0} \frac{f(x_0 + h) - f(x_0)}{h}$$
Do we get the same result by first finding the derivative $f'(x)$ and then substituting x_0 in it? Why?

In each of Exercises 22–30 use the power rule (Theorem 2.2) and Theorem 2.3 to find the derivative of the given function.

22. $f(x) = x^{365}$

23. $f(x) = x^{256}$

24. $f(x) = 3x^{10}$

25. $f(x) = \dfrac{5}{x^2}$

26. $f(x) = 4\sqrt{x}$

27. $f(x) = 13x^{13}$

28. $f(x) = -5x^4$

29. $f(x) = \dfrac{2}{x}$

30. $f(x) = 9\sqrt[3]{x}$

31. A bowling ball manufacturer estimates that the profit from selling x hundred bowling balls is $800x - 2x^3 - 800$. Find the marginal profit:
(a) When $x = 10$; (b) When $x = 12$.

32. Suppose the revenue from producing x units of a certain product is $0.02x^2 - 4x$. What is the marginal revenue at a production level of 500?

33. If the demand function is $p = 960 - 3x$, what is the marginal revenue when $x = 120$?

SECTION 2.5

What Can Go Wrong (Optional)

You may have noticed that our definition of the derivative, Definition 2.1, included the qualifying statement

> if the limit as h tends toward 0 of the difference quotient $\dfrac{f(x_0 + h) - f(x_0)}{h}$ exists

Why did we include such a qualification? After all, in every situation considered, the limit did exist. We found its value by simply observing the behavior of the difference quotient as h tended toward 0.

Unfortunately, there are examples of functions for which the limit of the difference quotient does not exist. Intuitively, because the derivative of a function $y = f(x)$ at a point x_0 represents the slope of the tangent line at $(x_0, f(x_0))$, the failure of the limit to exist amounts to the fact that the curve $y = f(x)$ may fail to have a clearly defined tangent line at the point $(x_0, f(x_0))$.

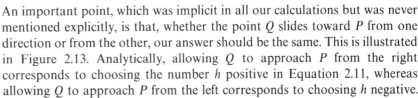

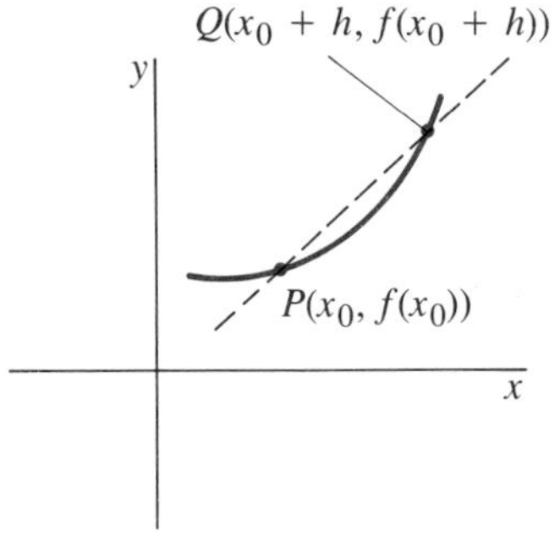

FIGURE 2.12
The slope of PQ is $\dfrac{f(x_0 + h) - f(x_0)}{h}$

Before looking at some specific instances in which this occurs, let us digress momentarily to recall some of our earlier work. We considered the problem of determining the slope of the tangent line to a curve $y = f(x)$ at some point $P(x_0, f(x_0))$ on it. To do so, we chose another point on the curve near P, with coordinates $Q(x_0 + h, f(x_0 + h))$, and then we found the slope of the secant line PQ (see Figure 2.12). This slope turned out to be

$$\frac{f(x_0 + h) - f(x_0)}{h}$$

Next we reasoned that, if we allow the point Q to slide toward P along the curve, then the slope of the secant line PQ should approach that of the tangent line. That is,

$$\underbrace{f'(x_0)}_{\substack{\text{slope of tangent}\\ \text{line at } (x_0,\, f(x_0))}} = \underbrace{\lim_{h \to 0}}_{\substack{\text{limit as we let}\\ Q \text{ approach } P}} \underbrace{\frac{f(x_0 + h) - f(x_0)}{h}}_{\substack{\text{slope of secant line}\\ \text{joining } P \text{ and } Q}} \tag{2.11}$$

An important point, which was implicit in all our calculations but was never mentioned explicitly, is that, whether the point Q slides toward P from one direction or from the other, our answer should be the same. This is illustrated in Figure 2.13. Analytically, allowing Q to approach P from the right corresponds to choosing the number h positive in Equation 2.11, whereas allowing Q to approach P from the left corresponds to choosing h negative.

Frequently, students ask why we don't write

$$f'(x_0) = \lim_{-h \to 0} \frac{f(x_0 - h) - f(x_0)}{-h}$$

if we want to allow h to be negative in Equation (2.11). The answer is that we could, but it is easier to simply let the symbol h represent both positive and negative numbers. You have done this before—for example, in elementary algebra. If we write $2x = -6$, the symbol x represents the negative number -3. The symbol x can represent both positive and negative numbers.

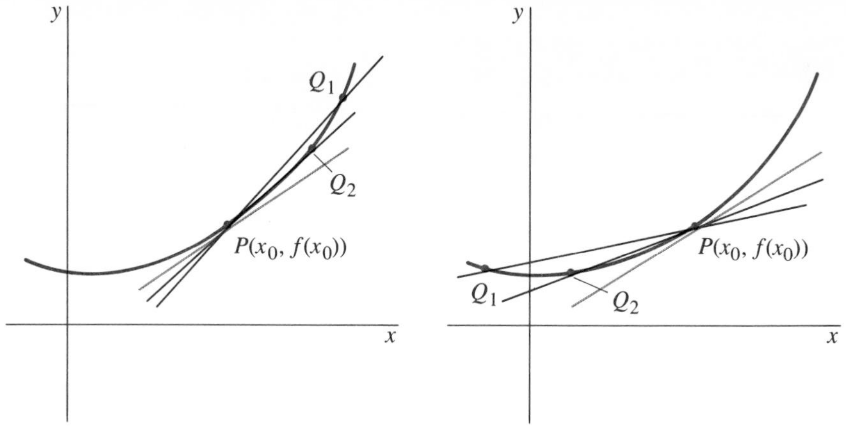

(a) Q approaches P from the right (b) Q approaches P from the left

FIGURE 2.13

EXAMPLE 4 Consider the function $f(x) = |x|$. Let us see whether this function is differentiable at a point x_0 by evaluating the limit of the difference quotient. As is clear from the graph of the absolute value function (see Figure 2.14), there are three cases to consider: (a) x_0 is positive; (b) x_0 is negative; (c) $x_0 = 0$.

SOLUTION

(a) Suppose x_0 is any positive number. Then, because

$$f(x) = |x|$$

and the absolute value of a positive number is itself, we have

$$f(x_0) = |x_0| = x_0$$

Next, if we choose h to be a number close enough to 0, either positive or negative, $x_0 + h$ is close to x_0 and hence positive (see Figure 2.14). Thus

$$f(x_0 + h) = |x_0 + h| = x_0 + h$$

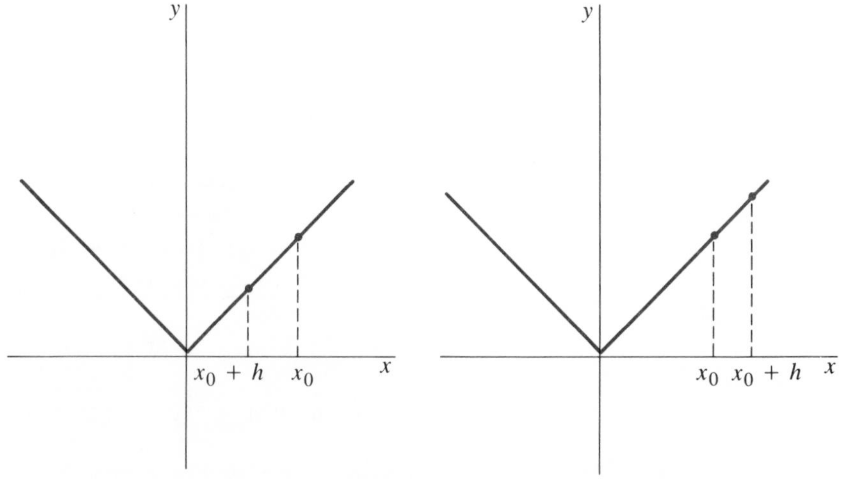

(a) $y = |x|, x_0 > 0, h < 0$ (b) $y = |x|, x_0 > 0, h > 0$

FIGURE 2.14

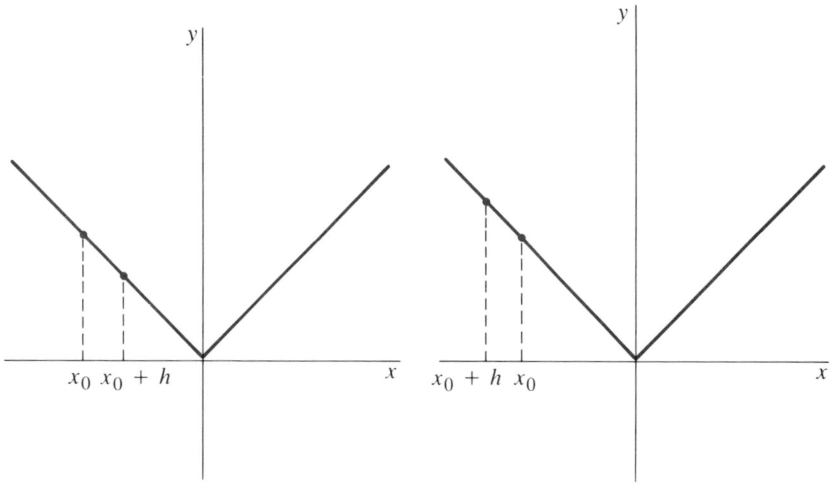

(a) $y = |x|, x_0 < 0, h > 0$ (b) $y = |x|, x_0 < 0, h < 0$

FIGURE 2.15

It follows that

$$f'(x_0) = \lim_{h \to 0} \frac{f(x_0 + h) - f(x_0)}{h} = \lim_{h \to 0} \frac{|x_0 + h| - |x_0|}{h}$$

$$= \lim_{h \to 0} \frac{x_0 + h - x_0}{h} = \lim_{h \to 0} \frac{h}{h} = 1$$

This is what we expect because, for x positive, $y = |x|$ is the same as $y = x$, which is a straight line having slope 1.

(b) Now suppose x_0 is negative. Then

$$f(x_0) = |x_0| = -x_0$$

If h is very small (again, it can be either positive or negative), $x_0 + h$ is negative (see Figure 2.15). Thus

$$f(x_0 + h) = |x_0 + h| = -(x_0 + h)$$

We have then

$$f'(x_0) = \lim_{h \to 0} \frac{f(x_0 + h) - f(x_0)}{h} = \lim_{h \to 0} \frac{-(x_0 + h) - (-x_0)}{h}$$

$$= \lim_{h \to 0} \frac{-h}{h} = -1$$

This too seems reasonable. For x negative, $y = |x|$ is the same as $y = -x$, which is a straight line having slope -1.

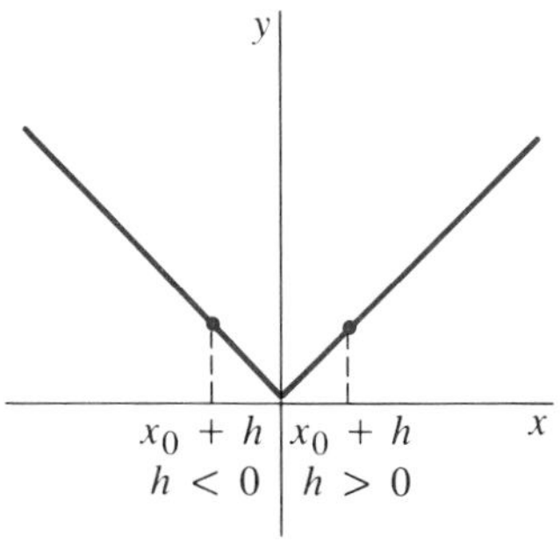

FIGURE 2.16
$y = |x|, x_0 = 0$

(c) Now let $x_0 = 0$ (see Figure 2.16). Then

$$f(x_0) = |x_0| = |0| = 0$$

If h is very small and positive, $x_0 + h = 0 + h = h$ is also positive, so

$$f(x_0 + h) = |h| = h$$

Therefore

$$\frac{f(x_0 + h) - f(x_0)}{h} = \frac{h - 0}{h} = 1$$

On the other hand, if h is small and negative, $x_0 + h$ is negative and

$$f(x_0 + h) = |h| = -h$$

Then
$$\frac{f(x_0 + h) - f(x_0)}{h} = \frac{-h - 0}{h} = -1$$

We have obtained different values depending on whether we chose h positive or negative. If h tends toward 0 through positive values of h, the difference quotient approaches 1; if h tends toward 0 through negative values of h, the difference quotient approaches -1. So the limit of this difference quotient as $h \to 0$ does not exist. In other words, the function $y = f(x) = |x|$ is *not* differentiable at $x_0 = 0$; $f'(0)$ does not exist. In such a situation, as is clear from the Figure 2.16, the curve has no clearly defined tangent line at the point in question.

EXAMPLE 5 Suppose $f(x) = \sqrt{x}$, $x \geq 0$. Let us try to compute $f'(x_0)$.

Assume for the moment that $x_0 > 0$. Then

$$\frac{f(x_0 + h) - f(x_0)}{h} = \frac{\sqrt{x_0 + h} - \sqrt{x_0}}{h}$$

Multiplying top and bottom by $\sqrt{x_0 + h} + \sqrt{x_0}$, we obtain

$$\begin{aligned}\frac{f(x_0 + h) - f(x_0)}{h} &= \frac{\sqrt{x_0 + h} - \sqrt{x_0}}{h} \\ &= \frac{\sqrt{x_0 + h} - \sqrt{x_0}}{h} \cdot \frac{\sqrt{x_0 + h} + \sqrt{x_0}}{\sqrt{x_0 + h} + \sqrt{x_0}} \\ &= \frac{(x_0 + h) - x_0}{h(\sqrt{x_0 + h} + \sqrt{x_0})} = \frac{1}{\sqrt{x_0 + h} + \sqrt{x_0}}\end{aligned}$$

Now, if we let h tend toward 0, $\sqrt{x_0 + h}$ tends toward $\sqrt{x_0}$, so that

$$\begin{aligned}f'(x_0) &= \lim_{h \to 0} \frac{f(x_0 + h) - f(x_0)}{h} = \lim_{h \to 0} \frac{1}{\sqrt{x_0 + h} + \sqrt{x_0}} \\ &= \frac{1}{\sqrt{x_0} + \sqrt{x_0}} = \frac{1}{2\sqrt{x_0}}\end{aligned}$$

The foregoing result holds when $x_0 > 0$, but what happens when $x_0 = 0$? Unfortunately, our formula

$$f'(x_0) = \frac{1}{2\sqrt{x_0}}$$

makes no sense in this case, because division by 0 is impossible. How can we interpret this result? The graph of $y = \sqrt{x}$ is depicted in Figure 2.17. If

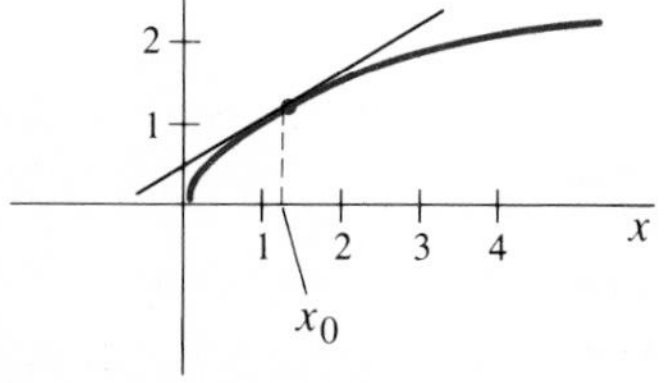

FIGURE 2.17
$y = \sqrt{x}$

$x_0 > 0$, the slope of the tangent line at the point $(x_0, \sqrt{x_0})$ is given by

$$\frac{1}{2\sqrt{x_0}}$$

For instance, at the point (1, 1) the slope is

$$f'(1) = \frac{1}{2\sqrt{1}} = \frac{1}{2}$$

At $x_0 = 0$, however, the graph of the curve possesses a vertical tangent line, and the slope of a vertical line is undefined. We conclude that $f(x) = \sqrt{x}$ is *not* differentiable at $x_0 = 0$.

In summary, we expect a function to be nondifferentiable at any point where its graph has a sharp corner or a vertical tangent.

In spite of the fact that we have spent a great deal of time discussing when differentiability of a function fails to hold, most of this calculus course and almost all applications you are likely to encounter involve functions that pose no problems. That is, the limit of the difference quotient will exist as $h \to 0$. In the next section, we shall begin the systematic computation of derivatives of some elementary functions.

Section 2.5 EXERCISES

In each of Exercises 1–8 show that the given function $y = f(x)$ is differentiable at every $x \neq 0$, and then find the derivative $f(x)$ at such an x. Decide whether each function is differentiable at $x = 0$; if it is, find its derivative at $x = 0$ [that is, $f'(0)$].

1. $y = f(x) = x + |x|$

2. $y = f(x) = |x| - x$

3. $y = f(x) = |x^2|$

4. $y = f(x) = (x)|x|$

5. $y = f(x) = 2x$

6. $y = f(x) = \dfrac{|x|}{x}$

7. $y = f(x) = x^2|x|$

8. $y = f(x) = x|x^2|$

In each of Exercises 9–18 sketch the graph of the given function. Then decide where the function is differentiable and where it is not.

9. $y = f(x) = |x - 4|$

10. $y = f(x) = |3 - x|$

11. $y = f(x) = |2x - 1|$

12. $y = f(x) = |3 - 2x|$

13. $y = f(x) = \sqrt{3 - x}$

14. $y = f(x) = \sqrt{x - 4}$

15. $y = f(x) = \sqrt{2x - 1}$

16. $y = f(x) = \sqrt{3 - 2x}$

17. $y = f(x) = |x| + |x - 2|$

18. $y = f(x) = |2x - 1| + |3x - 5|$

19. In Example 5 we computed the derivative of $f(x) = \sqrt{x}$ at any $x_0 > 0$ by evaluating the limit of the difference quotient as $h \to 0$. For the case where $x_0 = 0$, set up the difference quotient and examine its behavior as $h \to 0$. (Of course, h must be taken positive, because $f(x)$ is not defined for negative x.) Interpret what you find geometrically. What is happening to the curve and its tangent near $x_0 = 0$?

SECTION 2.6

A Discussion of Limits

In order to derive additional differentiation formulas, we will have to utilize some elementary properties of limits. In this section, therefore, we discuss the notation of a limit in a nonrigorous way and then summarize the properties that we shall need later.

DEFINITION 2.4 We say that the limit of the function $f(x)$, as x approaches a, is L, or

$$\lim_{x \to a} f(x) = L$$

if the values of $f(x)$ approach L as x approaches (but does not equal) a. (The function need not be defined at a.) ⊟

EXAMPLE 6 It is easy to see that:

(a) $\lim_{x \to 5} x^2 + 1 = 26$

(b) $\lim_{h \to 0} \dfrac{h^2 + 8}{h - 1} = -8$

(c) $\lim_{z \to 9} (z^3 + 3z - 1) = 755$

EXAMPLE 7 Let us try to evaluate $\lim_{h \to 1} \dfrac{h^2 - 1}{h - 1}$.

SOLUTION Observe first that, if $h = 1$, the denominator of

$$u(h) = \frac{h^2 - 1}{h - 1}$$

is 0, so the function $u(h)$ is undefined there. In particular, it makes no sense to evaluate this limit by substituting $h = 1$. Suppose, however, that h is near 1. For instance, choose $h = 0.9$. Then

$$u(0.9) = \frac{(0.9)^2 - 1}{(0.9) - 1} = \frac{-0.19}{-0.1} = 1.9$$

If $h = 0.99$, which is even closer to 1, then

$$u(0.99) = \frac{(0.99)^2 - 1}{(0.99) - 1} = \frac{-0.0199}{-0.01} = 1.99$$

Similarly, if h is larger than 1 but near 1, such as $h = 1.01$, then

$$u(1.01) = \frac{(1.01)^2 - 1}{(1.01) - 1} = \frac{0.0201}{0.01} = 2.01$$

The values we obtain in this way certainly appear to cluster around the number 2. In fact, the closer h is chosen to 1, the closer our answer appears to be to the number 2, so we say

$$\lim_{h \to 1} \frac{h^2 - 1}{h - 1} = 2$$

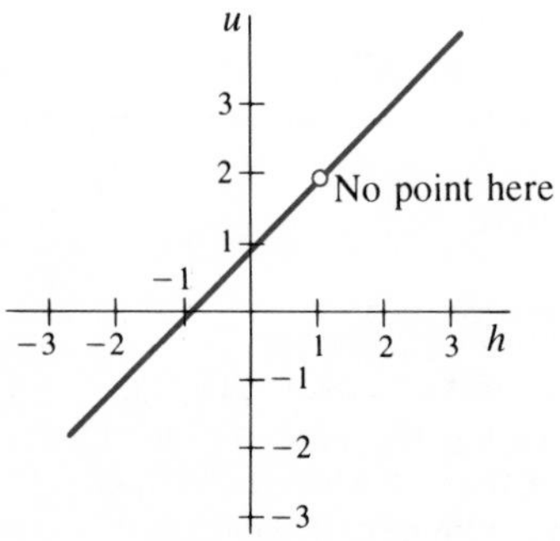

FIGURE 2.18 $u(h) = \dfrac{h^2 - 1}{h - 1}$

This answer becomes even more plausible when we observe that, for $h \neq 1$,

$$\frac{h^2 - 1}{h - 1} = \frac{(h - 1)(h + 1)}{h - 1} = h + 1$$

Thus, if h is near 1 but not equal to it, then $u(h) = \dfrac{h^2 - 1}{h - 1} = h + 1$, which is near 2. The graph of $u(h)$ (see Figure 2.18) is the same as that of the function $v(h) = h + 1$, except that $u(h)$ is not defined at $h = 1$ whereas $v(h)$ is.

EXAMPLE 8 In answering the question posed on page 41, we determined the instantaneous velocity of a car moving according to the rule $s = 10t^2$ at $t_0 = 3$. To do this, we considered (without writing it explicitly there) the difference quotient

$$\frac{s(t_0 + h) - s(t_0)}{h}$$

which represented the average velocity of the vehicle from time t_0 to time $t_0 + h$. The difference quotient was found to have the value

$$u(h) = \frac{60h + 10h^2}{h}$$

The instantaneous velocity, or speedometer reading, was then determined as

$$\lim_{h \to 0} \frac{60h + 10h^2}{h}$$

If we choose values for h "in the neighborhood of 0" and tending toward 0, such as $h = 0.1$, $h = 0.01$, or $h = 0.001$, we obtain $u(h) = 61$, $u(h) = 60.1$, and $u(h) = 60.01$, respectively. Because these numbers appear to be getting closer to 60, we say that

$$\lim_{h \to 0} \frac{60h + 10h^2}{h} = 60$$

Of course, if we note that

$$\frac{60h + 10h^2}{h} = 60 + 10h$$

for $h \neq 0$, it is clear that when h is "almost" 0 this value is "almost" 60.

We now state the properties of limits that we shall use in Section 2.7.

THEOREM 2.5 Let $u(h)$ and $v(h)$ be two functions of h. Suppose that $\lim_{h \to a} u(h)$ and $\lim_{h \to a} v(h)$ both exist. Then:

(i) $\lim_{h \to a} [(u(h) \pm v(h)]$ exists and

$$\lim_{h \to a} [u(h) \pm v(h)] = \lim_{h \to a} u(h) \pm \lim_{h \to a} v(h)$$

That is, the limit of a sum or difference equals the sum or difference of the limits.

(ii) $\lim_{h \to a} [u(h)v(h)]$ exists, and

$$\lim_{h \to a} [u(h) \cdot v(h)] = (\lim_{h \to a} u(h)) \cdot (\lim_{h \to a} v(h))$$

That is, the limit of a product equals the product of the limits.

(iii) $\lim_{h \to a} [u(h)/v(h)]$ exists and

$$\lim_{h \to a} [u(h)/v(h)] = \lim_{h \to a} u(h)/\lim_{h \to a} v(h) \qquad \text{provided that } \lim_{h \to a} v(h) \neq 0$$

That is, the limit of a quotient equals the quotient of the limits. ⊟

Section 2.6 EXERCISES

Evaluate the limits given in Exercises 1–44 if they exist (*a*, *b*, *c*, *d*, *m* constant).

1. $\lim_{h \to 0} (3h + 5)$

2. $\lim_{h \to 0} (-3h)$

3. $\lim_{h \to 0} (\sqrt{2h})$

4. $\lim_{h \to 0} (2)$

5. $\lim_{x \to 0} (-2)$

6. $\lim_{x \to 0} (0)$

7. $\lim_{x \to 2} (3x + 5)$

8. $\lim_{x \to 2} (3x - 6)$

9. $\lim_{h \to 0} (h^2 - 3h + 2)$

10. $\lim_{h \to 0} (7 - 5h - h^2)$

11. $\lim_{x \to 0} (x^3 - 2x^2 + 5x - 9)$

12. $\lim_{x \to -1} (x^3 - 2x^2 + 5x - 9)$

13. $\lim_{x \to 3} (mx + b)$

14. $\lim_{x \to -2} (ax^2 + bx + c)$

15. $\lim_{h \to 0} (h - 2)(3 - h)$

16. $\lim_{h \to 0} (2h + 3)(5h - 7)$

17. $\lim_{x \to 0} (ax + b)(cx + d)$

18. $\lim_{x \to 3} (ax + b)(cx + d)$

19. $\lim_{h \to 0} \sqrt{h^2 + 4}$

20. $\lim_{x \to -1} \sqrt{x^2 + 4}$

21. $\lim_{h \to 0} \sqrt{2h + 3}$

22. $\lim_{x \to 2} \sqrt{2x - 3}$

23. $\lim_{h \to 0} (1 + h)^{10}$

24. $\lim_{h \to 0} (1 - h)^{1982}$

25. $\lim_{h \to 0} (1 + 3h)^{1982}$

26. $\lim_{h \to 0} (h^3 - 2h^2 + 13h - 4)^7$

27. $\lim_{h \to 0} \dfrac{1}{h - 5}$

28. $\lim_{h \to 0} \dfrac{2h + 3}{5h + 7}$

29. $\lim_{h \to 0} \sqrt{\dfrac{2h + 3}{5h + 7}}$

30. $\lim_{x \to 0} \dfrac{ax + b}{cx + d}$

31. $\lim_{x \to 0} \sqrt{x^2 - 3x + 4}$

32. $\lim_{x \to 0} (\sqrt{5 + x} - \sqrt{5})$

33. $\lim_{h \to 0} (-1 + h)^n$

34. $\lim_{x \to 0} (2\sqrt{3x + 5} - 3\sqrt{5x + 7})$

35. $\lim_{x \to 0} \dfrac{x^2 - 9}{x + 3}$

36. $\lim_{x \to 3} \dfrac{x^2 - 9}{x + 3}$

37. $\lim_{x \to 0} \dfrac{x^2 - 5x + 6}{x - 2}$

38. $\lim_{x \to -3} \dfrac{x^2 - 9}{x + 3}$

39. $\lim_{x \to 2} \dfrac{x^2 - 5x + 6}{x - 2}$

40. $\lim_{x \to 2} \dfrac{2x - 3}{x^2 - 5x + 6}$

41. $\lim_{x \to 0} \dfrac{x + 2}{(x - 5)^2}$

42. $\lim_{x \to 5} \dfrac{x + 2}{(x - 5)^2}$

43. $\lim_{x \to 2} (3x^2 - 4x + 1)^{13}$

44. $\lim_{x \to 9} \dfrac{\sqrt{3x - 2} + x}{\sqrt{x + 7}}$

In Exercises 45–51 evaluate the given limit (if it exists); it is understood that *x* is constant.

45. $\lim_{h \to 0} (x^2 + 10xh +. h)$

46. $\lim_{h \to 0} (2xh^2 + 10h + 20)$

47. $\lim_{h\to 0} (x + h)(x^2 + h)$

48. $\lim_{h\to 0} (\sqrt{x + h} - \sqrt{x})$

49. $\lim_{h\to 0} (x^3 + 3x^2h + h - 5)$

50. $\lim_{h\to 0} \dfrac{1}{x^2 + 10xh + h^2}$

51. $\lim_{h\to 0} \dfrac{\sqrt{x^2h + x + h}}{\sqrt{h^2 + 4x + xh}}$

SECTION 2.7

Differentiating Sums, Products, Quotients, and Powers

If you compute the derivative of $f(x) = 5x^2 + 8x$, you can easily show, by taking the limit of the difference quotient, that its derivative is $f'(x) = 10x + 8$. Now $10x$ is the derivative of $5x^2$, while 8 is the derivative of $8x$. This suggests that the derivative of a polynomial is the sum of the derivatives of its separate monomial terms. Let us prove that differentiation does "work" in this manner.

THEOREM 2.6 If $f(x)$ and $g(x)$ are two differentiable functions of x, then $f(x) + g(x)$ is also differentiable and its derivative, which we denote by $[f(x) + g(x)]'$ is given by

$$[f(x) + g(x)]' = f'(x) + g'(x)$$

□

Before presenting the proof, let's make sure that we understand exactly what the theorem says.

EXAMPLE 9 Let us find $\phi'(x)$ if $\phi(x) = x^4 + 7x^2$. We may proceed as follows. Let $f(x) = x^4$, and let $g(x) = 7x^2$. Then

$$f'(x) = 4x^3$$

and

$$g'(x) = 14x$$

Now Theorem 2.6 tells us that

$$\phi'(x) = f'(x) + g'(x) = 4x^3 + 14x$$

Of course, it was not necessary for us to write down $f(x) = x^4$ and $g(x) = 7x^2$ explicitly. In practice, we simply differentiate term by term by inspection.

An important point worth mentioning about Theorem 2.6 is that it does not require $f(x)$ and $g(x)$ to be polynomials. This means that, once we complete the proof of this theorem, we shall be able to differentiate such expressions as

$$\sqrt{x} + \sqrt{x^2 + 1}$$

provided that we can individually differentiate the function $\sqrt{x}$ and the function $\sqrt{x^2 + 1}$.

Let us turn to the proof of our theorem.

PROOF OF THEOREM 2.6 Let $\phi(x) = f(x) + g(x)$. To find the derivative of $\phi(x)$, if it exists, we must evaluate the limit of the difference quotient for ϕ.

Thus
$$\phi'(x) = \lim_{h \to 0} \frac{\phi(x + h) - \phi(x)}{h}$$

Now
$$\phi(x + h) = f(x + h) + g(x + h)$$

The difference quotient may therefore be written as

$$\frac{\phi(x + h) - \phi(x)}{h} = \frac{[f(x + h) + g(x + h)] - [f(x) + g(x)]}{h}$$

We can regroup the terms to obtain

$$\frac{f(x + h) - f(x) + g(x + h) - g(x)}{h} = \frac{f(x + h) - f(x)}{h} + \frac{g(x + h) - g(x)}{h}$$

Thus

$$\phi'(x) = \lim_{h \to 0} \left[\frac{f(x + h) - f(x)}{h} + \frac{g(x + h) - g(x)}{h} \right]$$

This is the limit of a sum, and by part (i) of Theorem 2.5, it must equal the sum of the individual limits, provided that these individual limits exist. Therefore

$$\phi'(x) = \lim_{h \to 0} \left[\frac{f(x + h) - f(x)}{h} \right] + \lim_{h \to 0} \left[\frac{g(x + h) - g(x)}{h} \right]$$

But, by assumption, the functions f and g are both differentiable. This means precisely that these individual limits exist, and in fact

$$\lim_{h \to 0} \frac{f(x + h) - f(x)}{h} = f'(x)$$

$$\lim_{h \to 0} \frac{g(x + h) - g(x)}{h} = g'(x)$$

Substitution yields

$$[f(x) + g(x)]' = \phi'(x) = f'(x) + g'(x)$$ ⊟

PRACTICE EXERCISE Accept it as a fact (we will prove it later) that the function $f(x) = \sqrt{1 + x^2}$ is differentiable and that its derivative is $f'(x) = \dfrac{x}{\sqrt{1 + x^2}}$. Using this information, differentiate

$$g(x) = x^3 + \sqrt{1 + x^2}$$

$\left[\text{Your answer should be } g'(x) = 3x^2 + \dfrac{x}{\sqrt{1 + x^2}}\right]$.

Theorem 2.6 can readily be extended to a sum of several functions.

COROLLARY 2.7 If $f_1(x), f_2(x), \ldots, f_n(x)$ are n differentiable functions of x with derivatives $f'_1(x), f'_2(x), \ldots, f'_n(x)$, respectively, the sum $[f_1(x) + \cdots + f_n(x)]$ is differentiable and

$$[f_1(x) + f_2(x) + \cdots + f_n(x)]' = f'_1(x) + f'_2(x) + \cdots + f'_n(x)$$

In other words, the derivative of a sum is the sum of the separate derivatives of the terms, regardless of how many terms the sum contains. ⊟

EXAMPLE 10 If

$$f(x) = x^3 - 9x^2 + 8x + 2$$

then

$$f'(x) = 3x^2 - 18x + 8 + 0$$

Compare this answer with your guess in answer to the second practice exercise following Example 3.

Let's now see whether we can guess at a rule for the differentiation of a function that is the product of two simpler expressions. The most logical and obvious guess might be that, to differentiate the product of two functions, we differentiate each one separately and multiply the results together. (Thus we might guess that the derivative of a product is the product of the derivatives.) Before trying to prove this fact, let's test it on an example.

We select two functions—say, $f(x) = x^2$ and $g(x) = x^3$. Then

$$\phi(x) = f(x) \cdot g(x) = x^2 \cdot x^3 = x^5$$

and we know that $\phi'(x) = 5x^4$. We know also that $f'(x) = 2x$ and that $g'(x) = 3x^2$. Unfortunately, $f'(x)g'(x) = (2x)(3x^2) = 6x^3$, which is not equal to $5x^4$. Our rule doesn't seem to work in this case: $[f(x)g(x)]' \neq f'(x)g'(x)$.

We are forced to accept the conclusion that the derivative of a product is somewhat more complicated than that of a sum. Fortunately, a rule for the differentiation of a product does exist.

THEOREM 2.8 (The Product Rule) If $f(x)$ and $g(x)$ are two differentiable functions, the product $f(x)g(x)$ is differentiable and its derivative is given by

$$[f(x)g(x)]' = f(x)g'(x) + g(x)f'(x)$$

In other words, the derivative of a product of two functions equals the first function times the derivative of the second plus the second function times the derivative of the first. ⊟

Before plunging into a proof of this rule, it seems only fair to subject it to the same scrutiny that proved so disastrous to our previous attempt.

EXAMPLE 11 Let $f(x) = x^2$ and let $g(x) = x^3$, as before. Thus, if $\phi(x) = f(x) \cdot g(x) = x^5$, then $\phi'(x) = 5x^4$. We know that $f'(x) = 2x$ and that $g'(x) = 3x^2$. Substituting into the product-rule formula (Theorem 2.8) yields

$$5x^4 = (x^2 \cdot x^3)' = x^2(3x^2) + x^3(2x) = 3x^4 + 2x^4 = 5x^4$$

PRACTICE EXERCISE Suppose you are told that the function

$$f(x) = \frac{1}{x + 1}$$

is differentiable and that its derivative is

$$f'(x) = \frac{-1}{(x + 1)^2}$$

Use this fact to find the derivative of

$$g(x) = \frac{x^2}{x + 1}$$

$\left[\text{Your answer should be } g'(x) = x^2 \cdot \frac{-1}{(x + 1)^2} + 2x \cdot \frac{1}{x + 1} = \frac{2x + x^2}{(x + 1)^2}.\right]$

EXAMPLE 12 In Example 5 we found that, if $f(x) = \sqrt{x}$, then at any $x > 0$, $f'(x) = 1/(2\sqrt{x})$. Let us use this fact to derive a formula for the derivative of $g(x) = x^{3/2}$, $x > 0$.

We note first that $\sqrt{x}$ means the same thing as $x^{1/2}$. Thus we can write g as

$$g(x) = x^{3/2} = x \cdot x^{1/2}$$

which is the product of the functions x and $x^{1/2}$. The derivative of x is 1 and that of $x^{1/2}$ is $1/(2\sqrt{x})$. Making use of the product rule, we find that

$$g'(x) = (x^{3/2})' = x\left(\frac{1}{2\sqrt{x}}\right) + \sqrt{x}(1)$$

$$= x\left(\frac{1}{2x^{1/2}}\right) + x^{1/2}(1) = \frac{x^{1/2}}{2} + x^{1/2} = \frac{3}{2}x^{1/2}$$

Now that you know how to use the product rule, let's prove it.

PROOF OF THEOREM 2.8 Let $\phi(x) = f(x)g(x)$. Our goal is to show that $\phi(x)$ is differentiable and to determine its derivative. As always, we must consider the limit as h tends toward 0 of the difference quotient. In this case, because

$$\phi(x + h) = f(x + h)g(x + h)$$

we have

$$\frac{\phi(x+h)-\phi(x)}{h}=\frac{f(x+h)g(x+h)-f(x)g(x)}{h} \tag{2.12}$$

Now comes a trick from the mathematician's bag. It is a simple step, really, but it is probably not the kind of thing you would think of doing yourself. We add and subtract the expression $f(x+h)g(x)$ in the numerator of Equation (2.12)! This does not change the value of the difference quotient (because $+f(x+h)g(x)$ and $-f(x+h)g(x)$ add up to 0), although the result looks somewhat different. We obtain

$$\frac{f(x+h)g(x+h)-f(x+h)g(x)+f(x+h)g(x)-f(x)g(x)}{h}$$

We then factor and regroup terms to write

$$f(x+h)\left[\frac{g(x+h)-g(x)}{h}\right]+g(x)\left[\frac{f(x+h)-f(x)}{h}\right] \tag{2.13}$$

Note that the terms in the square brackets are precisely the difference quotients of f and g, respectively. Now we want to take the limit as h tends toward 0. According to Theorem 2.5, we may consider the limit of the terms individually.

By assumption, $f(x)$ and $g(x)$ are differentiable functions. Thus

$$f'(x)=\lim_{h\to 0}\frac{f(x+h)-f(x)}{h} \quad \text{and} \quad g'(x)=\lim_{h\to 0}\frac{g(x+h)-g(x)}{h}$$

Now as h tends toward 0, $f(x+h)$ must tend toward $f(x)$, whereas $g(x)$ remains unchanged because it does not depend on h. [The fact that $f(x+h)\to f(x)$ as $h\to 0$ is a consequence of the continuity of f at x. It is a fact that differentiability implies continuity.] Combining all this information, we see that, as h tends toward 0, the limit of Expression (2.13) exists and so

$$\underbrace{f(x+h)\left[\frac{g(x+h)-g(x)}{h}\right]}+\underbrace{g(x)\left[\frac{f(x+h)-f(x)}{h}\right]}$$

approaches $\quad f(x)\cdot g'(x) \quad + \quad g(x)\cdot f'(x)$

This completes our proof. ⊟

An easy consequence of the product rule is the following corollary:

COROLLARY 2.9 Let $g(x)$ be a differentiable function with derivative $g'(x)$, and let c denote any constant real number. Then the function $cg(x)$ is differentiable, and its derivative is $cg'(x)$. That is,

$$[cg(x)]'=cg'(x)$$

PROOF In the Theorem 2.8, put $f(x)=c$. Then $f'(x)=0$ (the derivative of any constant is 0), so

$$[cg(x)]' = cg'(x) + g(x)c' = cg'(x) + 0 = cg'(x)$$ ⊟

PRACTICE EXERCISE Suppose you know somehow or other that, if $f(x) = \sqrt{x+1}$, then

$$f'(x) = \frac{1}{2\sqrt{x+1}}$$

Find the derivative of $g(x)$ when:

(a) $g(x) = 5\sqrt{x+1}$ (b) $g(x) = x\sqrt{x+1}$ (c) $g(x) = 5x\sqrt{x+1}$

[Your answers should be: (a) $g(x) = \dfrac{5}{2\sqrt{x+1}}$; (b) $g(x) = \dfrac{x}{2\sqrt{x+1}} + \sqrt{x+1}$; (c) $g(x) = \dfrac{5x}{2\sqrt{x+1}} + 5\sqrt{x+1}$]

By this time we know how to differentiate the sum of two functions and the product of two functions. The natural next step for us to take is to develop a rule for differentiating a quotient $f(x)/g(x)$ of two differentiable functions.

We first look at the special case where the function $f(x)$ is constant and equal to 1. That is, let's first learn how to differentiate $1/g(x)$.

THEOREM 2.10 If $g(x)$ is a differentiable function with derivative $g'(x)$, then for any value of x for which $g(x) \neq 0$, the derivative of $1/g(x)$ exists and is given by

$$\left[\frac{1}{g(x)}\right]' = -\frac{g'(x)}{g(x)^2}$$ ⊟

Let's work a few examples before proving our theorem.

EXAMPLE 13 To compute the derivative of $1/(x^2 + 1)$, we get $g(x) = x^2 + 1$. Then $g'(x) = 2x$ and our theorem says that

$$\left(\frac{1}{x^2+1}\right)' = -\frac{2x}{(x^2+1)^2}$$

wherever $g(x) \neq 0$. But $x^2 + 1$ is always positive, so the derivative makes sense for any value of x.

EXAMPLE 14 Differentiate $\dfrac{1}{x-1}$.

SOLUTION Here $g(x) = x - 1$, which is 0 if $x = 1$. Also $g'(x) = 1$. Thus, by

Theorem 2.10, the derivative of $\dfrac{1}{x-1}$ is given by

$$\left(\frac{1}{x-1}\right)' = -\frac{1}{(x-1)^2}$$

which is valid whenever $x \neq 1$.

Let's prove our theorem now.

PROOF OF THEOREM 2.10 Let $\phi(x) = \dfrac{1}{g(x)}$. Then $\phi(x+h) = \dfrac{1}{g(x+h)}$, so the difference quotient takes the form

$$\frac{\phi(x+h) - \phi(x)}{h} = \frac{\dfrac{1}{g(x+h)} - \dfrac{1}{g(x)}}{h}$$

Combining fractions by using the least common denominator, $g(x+h)g(x)$, yields

$$\frac{\dfrac{g(x) - g(x+h)}{g(x+h)g(x)}}{h}$$

which, according to the rules for working with fractions (see Appendix A), becomes

$$\frac{g(x) - g(x+h)}{hg(x+h)g(x)}$$

We may write this as

$$-\frac{g(x+h) - g(x)}{hg(x+h)g(x)} = -\frac{1}{g(x+h)g(x)}\left[\frac{g(x+h) - g(x)}{h}\right]$$

Now we take the limit as h tends toward 0 in this expression. Because g is assumed differentiable, $\dfrac{g(x+h) - g(x)}{h}$ tends toward $g'(x)$. Also $g(x+h)$ tends toward $g(x)$. Thus

$$-\underbrace{\frac{1}{g(x+h)g(x)}}\underbrace{\left[\frac{g(x+h) - g(x)}{h}\right]}$$

$$-\frac{1}{g(x)g(x)} \cdot g'(x) = -\frac{g'(x)}{g(x)^2}$$

□

There is a very important consequence of Theorem 2.10. Let us set $g(x) = x^n$, where n is a positive integer. Then $g'(x) = nx^{n-1}$. Provided that x is not 0, Theorem 2.10 tells us that

$$\left(\frac{1}{x^n}\right)' = \frac{-nx^{n-1}}{(x^n)^2} = -\frac{nx^{n-1}}{x^{2n}} = -nx^{n-1-2n} = -nx^{-n-1}$$

Therefore

$$\left(\frac{1}{x^n}\right)' = -nx^{-n-1}$$

Now $1/x^n$ means exactly the same thing as x^{-n}. We have just discovered, therefore, that $(x^{-n})' = -nx^{-n-1}$.

If we let $m = -n$, then this becomes $\frac{d}{dx} x^m = mx^{m-1}$. This is the same rule that worked for positive integer exponents.

We can therefore summarize our discussion as follows:

COROLLARY 2.11 If m is any nonzero integer (positive or negative),

$$\frac{d}{dx}(x^m) = mx^{m-1}$$

(If m is negative, both x^m and its derivative are undefined at $x = 0$.) □

Later we shall prove that this power rule holds for fractional exponents also (assume it for now).

EXAMPLE 15 If $f(x) = x^{-3}$, then $f'(x) = -3x^{-4}$.

We are now in a position to formulate the general quotient rule.

THEOREM 2.12 (The Quotient Rule) If $f(x)$ and $g(x)$ are two differentiable functions, then the quotient $f(x)/g(x)$ is differentiable for each x for which $g(x) \neq 0$, and its derivative is given by

$$\left[\frac{f(x)}{g(x)}\right]' = \frac{g(x)f'(x) - f(x)g'(x)}{g(x)^2}$$

In other words, the derivative of a quotient equals

> the bottom times the derivative of the top, minus the top times the derivative of the bottom, all divided by the bottom squared.

Say this to yourself ten times!

PROOF The quotient $f(x)/g(x)$ is just another way of writing the product

$$f(x)\left[\frac{1}{g(x)}\right]$$

Now $f(x)$ has derivative $f'(x)$, whereas $\dfrac{1}{g(x)}$ has derivative $-\dfrac{g'(x)}{g(x)^2}$ (according to Theorem 2.10). Consequently,

$$
\begin{aligned}
\left[\frac{f(x)}{g(x)}\right]' = \left[f(x)\frac{1}{g(x)}\right]' &= f(x)\left[\frac{-g'(x)}{g(x)^2}\right] + \frac{1}{g(x)}\cdot f'(x) \\
&= \frac{-f(x)g'(x)}{g(x)^2} + \frac{f'(x)}{g(x)} \\
&= \frac{-f(x)g'(x)}{g(x)^2} + \frac{g(x)f'(x)}{g(x)^2} \\
&= \frac{g(x)f'(x) - f(x)g'(x)}{g(x)^2}
\end{aligned}
$$

which is valid provided that $g(x) \neq 0$. ⊟

EXAMPLE 16 To differentiate $x/(x^2 - 1)$, we set $f(x) = x$ and $g(x) = x^2 - 1$ in the quotient rule (Theorem 2.12). Then, keeping in mind that $f'(x) = 1$ and $g'(x) = 2x$, observe that the quotient rule gives

$$
\left(\frac{x}{x^2 - 1}\right)' = \frac{\overset{g(x)}{(x^2 - 1)}\overset{f'(x)}{(1)} - \overset{f(x)}{x}\overset{g'(x)}{(2x)}}{\underset{g(x)^2}{(x^2 - 1)^2}} = \frac{x^2 - 1 - 2x^2}{(x^2 - 1)^2} = \frac{-x^2 - 1}{(x^2 - 1)^2}
$$

which is valid whenever $x \neq 1$ and $x \neq -1$.

EXAMPLE 17 Find $f'(x)$ when $f(x) = \dfrac{x^3 + 3x + 1}{x - 1}$.

SOLUTION The derivative of the numerator $x^3 + 3x + 1$ is $3x^2 + 3$, while the derivative of the denominator $x - 1$ is 1. Substituting these into the quotient rule produces

$$
f'(x) = \frac{(x - 1)(3x^2 + 3) - (x^3 + 3x + 1)}{(x - 1)^2} = \frac{2x^3 - 3x^2 - 4}{(x - 1)^2}
$$

which is valid whenever $x \neq 1$.

EXAMPLE 18 Find the derivative of $\dfrac{(x^2 - 1)(2x + 3)}{x}$.

SOLUTION This example is slightly more involved. In order to compute the derivative of the numerator, we must make use of the product rule:

$$
((x^2 - 1)(2x + 3))' = (x^2 - 1)(2) + (2x + 3)(2x) = 6x^2 + 6x - 2
$$

The denominator, x, has derivative 1. Thus

$$\left[\frac{(x^2 - 1)(2x + 3)}{x}\right]' = \frac{x(6x^2 + 6x - 2) - (x^2 - 1)(2x + 3)(1)}{x^2}$$
$$= \frac{4x^3 + 3x^2 + 3}{x^2}$$

which is valid whenever $x \neq 0$.

Of course, the sum, product, and quotient rules can be restated in equivalent Leibniz notation. For u and v differentiable functions of x, we have

(2.14)

$$\frac{d}{dx}(u + v) = \frac{du}{dx} + \frac{dv}{dx}$$
$$\frac{d}{dx}(uv) = u\frac{dv}{dx} + v\frac{du}{dx}$$
$$\frac{d}{dx}\left(\frac{u}{v}\right) = \frac{v\dfrac{du}{dx} - u\dfrac{dv}{dx}}{v^2}$$

EXAMPLE 19 Suppose that a manufacturer has a cost function given by

$$C(x) = 1000 + 4x + \frac{625}{x + 1}$$

What is the marginal cost when $x = 50$? (Refer to Section 2.3 and its problems for a discussion of the marginal concept.)

SOLUTION Differentiating, we have

$$C'(x) = 4 - \frac{625}{(x + 1)^2}$$

Thus

$$C'(50) = 4 - \frac{625}{(51)^2} \approx 3.76$$

This means that, if our manufacturer increases his production from 50 to 51 units, his cost will increase by approximately \$3.76.

THE GENERALIZED POWER RULE Before continuing, there is one additional differentiation rule we shall need for some of the applications in the next chapter. This rule will be deduced from the product rule,

$$[f(x)g(x)]' = f(x)g'(x) + g(x)f'(x)$$

If f and g are the same function, and we denote it by $u(x)$, the product rule becomes

$$[u(x)u(x)]' = u(x)u'(x) + u(x)u'(x)$$

or

$$[u^2(x)]' = 2u(x)u'(x)$$

Having done this, if we now set $f(x) = u^2(x)$ and $g(x) = u(x)$ in the product rule, then

$$[u^2(x)u(x)]' = u^2(x)u'(x) + u(x)[2u(x)u'(x)]$$

or

$$[u^3(x)]' = 3u^2(x)u'(x)$$

In general, this procedure can be continued indefinitely to show that

$$[u^n(x)]' = n[u(x)]^{n-1}u'(x)$$

or, in Leibniz notation,

(2.15)
$$\frac{d}{dx}(u^n) = nu^{n-1}\frac{du}{dx}$$

for n a nonnegative integer. Later we shall prove this rule's validity for fractional n as well. In fact, it is valid for all real numbers n, and we shall henceforth assume this. In the special case $u = x$, $du/dx = 1$, Equation (2.15) becomes

$$\frac{d}{dx}x^n = nx^{n-1}$$

which we recognize as the power rule. For this reason, Equation 2.15 is often called the **generalized power rule**. A few examples should serve to illustrate its use.

EXAMPLE 20 Find $g'(x)$ when $g(x) = (x^2 - 1)^4$.

SOLUTION Let $u = x^2 - 1$, so that $g(x) = u^4$. Then

$$g'(x) = 4u^3\frac{du}{dx} = 4(x^2 - 1)^3 2x = 8x(x^2 - 1)^3$$

EXAMPLE 21 Find $g'(x)$ for $g(x) = \sqrt{x + x^3} = (x + x^3)^{1/2}$.

SOLUTION We set $u = x + x^3$, so that $g(x) = u^{1/2}$. Then $du/dx = 1 + 3x^2$ and

$$g'(x) = \frac{1}{2}u^{-1/2}\frac{du}{dx} = \frac{1}{2}(x + x^3)^{-1/2}(1 + 3x^2) = \frac{1 + 3x^2}{2\sqrt{x + x^3}}$$

EXAMPLE 22 Find $\dfrac{dy}{dx}$ when $y = \left(\dfrac{x^2 - 1}{x^2 + 1}\right)^3$.

SOLUTION Let $u = \dfrac{x^2 - 1}{x^2 + 1}$, so that $y = u^3$. Then, by the quotient rule,

$$\frac{du}{dx} = \frac{(x^2 + 1)2x - (x^2 - 1)2x}{(x^2 + 1)^2} = \frac{4x}{(x^2 + 1)^2}$$

Thus

$$\frac{dy}{dx} = 3u^2\frac{du}{dx} = 3\left(\frac{x^2 - 1}{x^2 + 1}\right)^2\frac{4x}{(x^2 + 1)^2} = \frac{12x(x^2 - 1)^2}{(x^2 + 1)^4}$$

Section 2.7 EXERCISES

In Exercises 1–10 find $f'(x)$ when the function $f(x)$ equals:

1. $5x^3 - 3x + 2$
2. $\sqrt{2}x^5 - \pi x^4 + x^2$
3. $(x^2 + x + 1)(x^3 - x + 1)$
4. $(3x^5 - 7x^3 + x)(3x^4 - x^2 - 1)$
5. $\dfrac{1}{x^2}$
6. $\dfrac{5}{x^3}$
7. $\dfrac{-3}{x + 3}$
8. $\dfrac{x}{x + 3}$
9. $\dfrac{x^2 + x + 1}{x^3 - x + 1}$
10. $\dfrac{3x^5 - 7x^3 + x}{3x^4 - x^2 - 1}$

In Exercises 11–18 find dy/dx when y equals:

11. $x^{82} - x^{81} + x^{80}$
12. $(\sqrt{x} + 1)(x^2 + 1)$
13. $x(x - 2) + (x - 1)(x + 3)$
14. $3x^7 + 8x^3$
15. $\left(\dfrac{x - 1}{x + 1}\right)^2$
16. $\dfrac{3x + 5}{\sqrt{x}}$
17. $\dfrac{5}{x^2 + x + 1}$
18. $\sqrt{\dfrac{3x - 2}{x}}$

In Exercises 19–30 differentiate as indicated.

19. $\dfrac{d}{dx}[(3x + 5)(5x + 7)]$
20. $\dfrac{d}{dx}\left(\dfrac{3x + 5}{5x + 7}\right)$
21. $\dfrac{d}{dx}(-5)$
22. $\dfrac{d}{dx}[(x^2 + x + 1)(x^3 - 1)(5x - 3)]$
23. $\dfrac{d}{dx}\left[\dfrac{(x^2 + x + 1)(5x - 3)}{x^3 - 1}\right]$
24. $\dfrac{d}{dx}[(3x^2 - 2x + 1)^2]$
25. $\dfrac{d}{dx}\left(\dfrac{2x - 3}{3x + 5} + \dfrac{x + 7}{5x - 1}\right)$
26. $\dfrac{d}{dx}\left[\dfrac{x^3 + 2x + 1}{(x - 3)(x + 2)}\right]$
27. $\dfrac{d}{dx}\left[(3x - 2)\dfrac{1}{x^3} - \dfrac{1}{x^2}\right]$
28. $\dfrac{d}{dx}[(\sqrt{x} + 2)^2]$
29. $\dfrac{d}{dx}[(x^2 + 1)^3]$
30. $\dfrac{d}{dx}[(x^2 + x + 1)^n]$
31. Find $\left.\dfrac{dy}{dx}\right|_{x=-2}$ when $y = \dfrac{1}{x^2 + 1}$.
32. Find $\left.\dfrac{dy}{dx}\right|_{x=-1}$ when $y = (x + 1)^7$.
33. Find $\left.\dfrac{dy}{dx}\right|_{x=2}$ when $y = \dfrac{\sqrt{x}}{x^2 + 1}$.
34. Find $\left.\dfrac{dy}{dx}\right|_{x=3}$ when $y = \dfrac{1}{(x^2 - 6)^5}$.
35. Compute $f(1)$ and $f'(1)$ when $f(x) = \dfrac{x}{3} - \dfrac{3}{x}$.
36. Compute $f(4)$ and $f'(4)$ when $f(x) = \dfrac{\sqrt{x}}{1 + \sqrt{x}}$.
37. Compute $f(1)$ and $f'(\frac{1}{2})$ when $f(x) = \dfrac{1}{x^2 + 1}$.
38. Compute $f(-1)$ and $f'\left(\dfrac{b}{a}\right)$ when $f(x) = ax^2 + bx + c$ (a, b, c constant, $a \neq 0$)

In Exercises 39–42 find the slope of the given curve (that is, the slope of the tangent to the curve) at the specified point.

39. $y = f(x) = (2x + 3)(3x - 5)$ at the point $(-1, -8)$
40. $y = f(x) = \dfrac{2x + 3}{3x - 5}$ at the point $(0, -\frac{3}{5})$
41. $y = f(x) = (x^2 - 2)^3$ at the point where $x = 2$
42. $y = f(x) = ax^2 + bx + c$ (a, b, c constant) at the point where $x = -3$

In Exercises 43–48 find the equation of the tangent line to the given curve at the indicated point.

43. Curve: $y = f(x) = \frac{1}{x}$ point where $x = 3$

44. Curve: $y = f(x) = 4$ point where $x = -1$

45. Curve: $y = f(x) = \sqrt{x}$ point $(4, 2)$

46. Curve: $y = f(x) = \frac{2x + 3}{3x + 5}$ point $(1, \frac{5}{8})$

47. Curve: $y = f(x) = x^2$ point where $x = x_0$

48. Curve: $y = f(x) = ax^2 + bx + c$ (a, b, c constant) point where $x = x_0$

49. Find all points where the curve $y = \frac{x^3}{3} + \frac{x^2}{2} - 2x + 1$ has slope 0.

50. Find all points where the slope of the curve $y = \frac{1}{x}$ is -2.

51. Find all points on the curve $y = x^2 + 1$ where the tangent is parallel to the line $y = 3x - 2$.

52. Find all points on the curve $y = \frac{x + 1}{x - 1}$ where the tangent line is: (a) parallel to the line $2x - y = 5$; (b) perpendicular to the line $x - 2y = 5$.

53. Consider the chord (line segment) connecting the points $(0, -6)$ and $(5, 9)$ on the curve $y = f(x) = x^2 - 2x - 6$. Find a point on the curve, with x-coordinate between 0 and 5, at which the tangent to the curve is parallel to the chord.

54. Consider the chord (line segment) connecting the points $(1, 0)$ and $(4, \frac{15}{17})$ on the curve

$$y = f(x) = \frac{x^2 - 1}{x^2 + 1}$$

Find an equation for a point on the curve at which the tangent to the curve is parallel to the chord.

55. A ball is thrown upward so that its height $s = s(t)$ above the ground, starting at time $t = 0$, is given by the formula

$$s(t) = 160 + 48t - 16t^2$$

(a) What is $s(0)$? Interpret your answer.

(b) Find $s(1)$ and $s(2)$. What conclusion can you draw from the fact that they are equal?

(c) What meaning do you attach to $s(-2)$?

(d) For which values of t does $s(t) = 0$?

(e) When does the ball hit the ground?

(f) For which values of t does the formula for $s(t)$ really give the height of the ball?

(g) Find the velocity $v(t) = s'(t)$.

(h) When is $v(t) = s'(t)$ equal to 0? Explain what is happening at this instant.

(i) Verify that $s'(2) = -s'(1)$. Interpret your answer.

In each of Exercises 56–63 compute the marginal cost for the given cost function $C(x)$.

56. $C(x) = \frac{x^3}{x^2 + 3x + 4}$

57. $C(x) = 350 + 5x$

58. $C(x) = 400 + 6x^2$

59. $C(x) = x(x + 5x^2)$

60. $C(x) = \frac{x^4 - 3x^2 + 2}{x}$

61. $C(x) = \frac{x^3 - 4x^2 + 3}{x^2}$

62. $C(x) = x^{3/2}$ (*Hint*: $x^{3/2} = x\sqrt{x}$, so you can use the product rule.)

63. $C(x) = 8x^{3/2}$

64. A population of 1000 bacteria is placed in a culture medium and grows such that the number of bacteria after t hr is given by

$$N(t) = 1000\left(1 + \frac{2t}{40 + t^2}\right)$$

Find the (instantaneous) rate at which the population is growing when $t = 2$ (in bacteria per hour).

65. Suppose the price p charged for a certain product, when x units are sold per month, is given by the demand function

$$p = 50(30 - \sqrt{x})$$

(a) Write an expression for the monthly revenue R as a function of x. (Revenue = price times number sold.)

(b) Find the price and marginal revenue when: (i) $x = 100$; (ii) $x = 400$.

66. A company's annual cost C for inventory storage and management is given by

$$C = \frac{240{,}000}{x} + 2.4x$$

where x is the order size when inventory must be replenished from the company's supplier.

(a) Find the exact change in C when order size is increased from $x = 150$ to $x = 151$.

(b) Compute the marginal cost when $x = 150$ and compare this with your answer to part (a).

67. Suppose it is known that a beginning typist can type

$$W(x) = \frac{60x^2}{x^2 + 15} + 25$$

words per minute after x weeks of training.

(a) How many words per minute can the typist type after 5 weeks of training?

(b) What is the typist's rate of improvement (in words per minute per week) after 5 weeks of training?

(c) Do parts (a) and (b) for 10 weeks.

68. The number x of quarts of oil a gas station sells in a month is related to the selling price p by the demand equation

$$p = \frac{1350 - x}{500}$$

(a) Find the revenue R from the sale of x quarts in a month.

Suppose the monthly cost of x quarts of oil is

$$C = 1.6x + 50$$

(b) Find the monthly profit P as a function of x.

(c) Find the marginal profit $P'(x)$ when $x = 50$.

(d) At what sales level is the marginal profit 0?

(Later we shall see that this value of x yields the largest possible profit.)

69. As we know, the difference quotient

$$\frac{f(x + h) - f(x)}{h}$$

gives the average rate of change of $f(x)$ as x goes from x to $x + h$. A related concept, which is often more revealing, is the **percentage average rate of change** given by

$$\frac{f(x + h) - f(x)}{h} \cdot \frac{100}{f(x)}$$

The limit of this quantity as $h \to 0$, namely

$$100 \frac{f'(x)}{f(x)}$$

is called the **percentage rate of change of $f(x)$** at x.

(a) The Consumer Price Index (CPI) is the present cost of a prescribed collection of goods and services that cost \$100 in 1967. If the CPI was 125.3 in 1972 and 181.5 in 1977, find the percentage average rate of change in the CPI (or the **average inflation rate**) over the 5-year periods 1967–1972 and 1972–1977. (By definition, the CPI in 1967 was 100.) *Note*: The percentage rate of change of the CPI is called the **inflation rate**.

(b) Consider the typist of Exercise 67. What is this typist's percentage rate of improvement after 5 weeks of training?

SECTION 2.8

Higher Derivatives

From our discussion of the "broken speedometer" problem we learned that, given a vehicle moving according to some distance function $s = s(t)$, its derivative

$$s'(t) = \lim_{h \to 0} \frac{s(t + h) - s(t)}{h}$$

represents the speedometer reading, or instantaneous velocity, at time t. Thus we view the velocity as the instantaneous rate of change of distance.

Although cars generally are not equipped with devices to measure the rate of change of velocity (that is, how fast the speedometer needle changes), airplanes are. Such devices are called accelerometers; they measure **acceleration**, which is defined as the instantaneous rate of change of velocity. In other words, the acceleration is simply the derivative of the velocity function.

This seems reasonable. After all, given a function $s(t)$, its derivative is another function that we denote by $s'(t)$. It is quite possible that this function has a derivative in its own right. This latter is called the second derivative of $s(t)$.

In general, we make the following definition:

DEFINITION 2.13 Given a differentiable function $y = f(x)$, we say that the function is twice differentiable if its derivative $f'(x)$ is differentiable. In this case we call the derivative of $f'(x)$ the **second derivative** of f and denote it by

$$f''(x) \quad \text{or} \quad \frac{d^2f}{dx^2} \quad \text{or} \quad \frac{d^2y}{dx^2} \quad \text{or} \quad y''$$ ⊟

EXAMPLE 23 Suppose a particle moves along a straight path according to the distance-versus-time formula $s(t) = 18t^3 + 5$. Let us find the acceleration of the particle at $t = 4$.

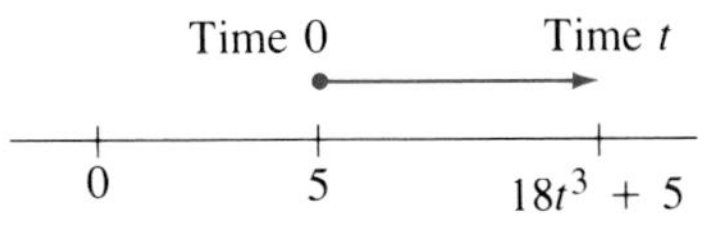

FIGURE 2.19

SOLUTION The formula $s(t) = 18t^3 + 5$ means that at time t our particle is $18t^3 + 5$ units from some fixed reference point (see Figure 2.19). Note that $s(0) = 5$ at time $t = 0$, so the particle begins its motion 5 units from our reference point. Then $s'(t) = 54t^2$ represents the velocity of this particle at time t.

The derivative of $s'(t) = 54t^2$ is

$$s''(t) = 108t$$

Thus $s''(4) = (108)(4) = 432$, and this is the acceleration at $t = 4$.

In summary, the procedure for obtaining the acceleration function from the distance function is simply to differentiate the distance function once to obtain the velocity function and then to differentiate the velocity function to obtain the acceleration.

EXAMPLE 24 Given $y = x^2 + x^{-1}$, find $\dfrac{d^2y}{dx^2}$.

SOLUTION We note first that $\dfrac{dy}{dx} = 2x - x^{-2}$ and that therefore

$$\frac{d^2y}{dx^2} = 2 + 2x^{-3} \qquad (\text{valid for } x \neq 0)$$

EXAMPLE 25 It can be shown (we will learn how to do so at a later stage) that, if a ball is dropped from a building 100 feet high, its distance from the ground t seconds later is given by

$$s(t) = -16t^2 + 100 \text{ feet}$$

Of course, this formula is valid only until the ball hits the ground.

(a) Find the velocity of the ball at $t = 2$.

(b) Find the acceleration of the ball at $t = 2$.

(c) At what time does the ball strike the ground?

SOLUTION

(a) If $s(t) = -16t^2 + 100$, then $s'(t) = -32t$. At $t = 2$,

$$s'(2) = -32(2) = -64 \text{ feet/second}$$

(b) Because $s'(t) = -32t$, $s''(t) = -32 \text{ feet/second}^2$. Observed that the acceleration is 32 feet/second^2 downward, regardless of the time. This constant acceleration is due to the pull of gravity on the ball. This fact is presented early in any physics course and was discovered by Isaac Newton.

(c) At the time that the ball strikes the ground, its distance from the ground (which is our reference point) is 0. Thus we must find the time t at which $s(t) = 0$. Setting

$$s(t) = -16t^2 + 100 = 0$$

we find that $16t^2 = 100$, or $t = \frac{5}{2}$.

There are situations other than physical-motion problems in which the second derivative plays an important role. The next example illustrates a subtle application of the second derivative that you have encountered many times in everyday life, probably without knowing it. Because this application is not obvious, it requires a bit of preparation.

EXAMPLE 26 Consider the function $f(x) = 10x - x^2$. Let us tabulate the value of this function and its derivative $f'(x) = 10 - 2x$ for $x = 0, 1, 2, 3, 4, 5$.

TABLE 2.6

x	$f(x) = 10x - x^2$	$f'(x) = 10 - 2x$
0	0	10
1	9	8
2	16	6
3	21	4
4	24	2
5	25	0

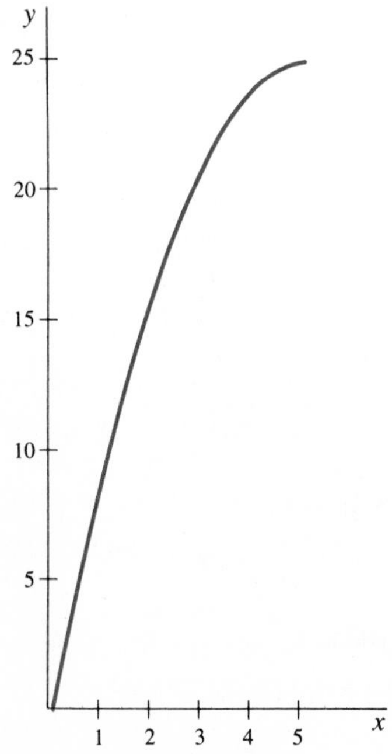

FIGURE 2.20
$f(x) = 10x - x^2$

What does Table 2.6 tell us? First we see that, as x increases, so does $f(x)$. However, the function appears to be increasing more and more slowly with each unit change in x. It changed from 0 to 9, from 9 to 16, from 16 to 21, from 21 to 24, and finally from 24 to only 25. The function is increasing, but its *rate* of increase is decreasing. This is emphasized by the values in the last column. The first derivative measures the change in the function. If the derivative is large, then the tangent line to the curve (and hence the curve itself) is rising steeply; if the derivative is small and positive, the curve is rising slowly; if the derivative is negative, the curve is falling. For the function $f(x) = 10x - x^2$ the derivative is getting smaller and smaller, so that, even though the curve is rising, it is rising more and more slowly. Its graph is shown in Figure 2.20.

This entire discussion can be summarized by one simple statement! That statement is

$$f''(x) = -2$$

After all, the second derivative measures the rate of change in the first derivative. If it is negative, the first derivative is getting smaller. This means that the function is increasing more and more slowly (or possibly decreasing more and more rapidly). Observe that, if the second derivative is negative, as it is in our example, the function can still be increasing (although more and more slowly).

What has this to do with real life? Government officials frequently employ a device known as the "decreasing rate of increase." This is a trick whereby a rising figure can be made to look like a decline. Take, for example, an imaginary catastrophe like a monthly rise of 10¢ in the price of a 5-pound bag of sugar. How can that be made to appear almost like good news, even to sugar lovers? Simple, as shown in Table 2.7.

TABLE 2.7

Date	*Increase*	*Price*	*Percent increase*
Jan.		\$1.00	
Feb.	10¢	1.10	10%
Mar.	10¢	1.20	9.1
Apr.	10¢	1.30	8.3
May	10¢	1.40	7.7
June	10¢	1.50	7.1
July	10¢	1.60	6.7
Aug.	10¢	1.70	6.3
Sept.	10¢	1.80	5.9
Oct.	10¢	1.90	5.6
Nov.	10¢	2.00	5.3
Dec.	10¢	2.10	5.0

As you can see, in a period of under a year, the price of sugar has more than doubled. But the *rate* of increase has steadily declined, so that in December it (the rate) is barely half what it was in February (from 10% down to 5%). Picture the headlines: "Sugar Inflation Seen Checked—Rate of Price Rise Cut in Half!"

By the time the price of sugar reaches \$5 for a 5-pound bag, the rate of increase will have dropped to only 2%. The trick is, of course, that the headline is really reporting the fact that the second derivative is negative (that is, the percentage increase is diminishing). The headline is not saying anything about the price.

Incidentally, if the rate of increase were to have remained at a steady 10% per month (as opposed to a rise of 10¢ per month), the price of sugar over the same period would have risen to \$2.85 instead of to \$2.10. And there still would have been no increase in the "rate of inflation" to report. (In this case, the second derivative would be 0.)

Obviously the same technique can be used in other settings. For example, the crime rate and the inflation rate are already first derivatives (the crime rate is the rate of change of crime, and the inflation rate is the rate of change of prices). You might read in the papers that "The rate of increase of serious crime is declining" or "The inflation rate dropped this month." Both sound like hopeful notes. In both cases, however, the headlines are actually reporting a negative second derivative. It is probable that both the number of crimes and the cost of living are continuing to increase, just like the price of sugar.

One final note: Just as our examples have exhibited situations wherein the second derivative naturally arises, you may encounter situations wherein higher derivatives are required. Let's formulate the appropriate notation.

Given a function $y = f(x)$ and a positive integer $n > 2$, we denote by

$$f^{(n)}(x) \quad \text{or} \quad \frac{d^n y}{dx^n}$$

the nth derivative of $y = f(x)$, if it exists. This nth derivative is defined as the derivative of the $(n - 1)$th derivative:

$$\frac{d^n y}{dx^n} = \frac{d}{dx}\left(\frac{d^{n-1}y}{dx^{n-1}}\right)$$

or

$$f^{(n)}(x) = [f^{(n-1)}(x)]'$$

EXAMPLE 27 To compute $f^{(4)}(x)$ if $f(x) = x^6 - 3x^2$, we observe that

$$f'(x) = 6x^5 - 6x$$
$$f''(x) = 30x^4 - 6$$
$$f^{(3)}(x) = 120x^3$$
$$f^{(4)}(x) = 360x^2$$

Section 2.8 EXERCISES

Find the first four derivatives of the functions given in Exercises 1–6.

1. $x^8 - 3x^6 + x^2 - 2$
2. $x^5 + x + 1$
3. $3x^2 - 4x + 5$
4. $x^4 - 3x^2 - 1$
5. x^{25}
6. $(x^2 - 1)^3$

Find the 4th derivative, $f^{(4)}(x)$, of the functions given in Exercises 7–14.

7. $f(x) = x^9$
8. $f(x) = x^{10}$
9. $f(x) = x^{11}$
10. $f(x) = x^{15}$
11. $f(x) = x^9 + x^7 - x^3 + 1$
12. $f(x) = \sqrt{x}$
13. $f(x) = -x^{11} + x^8 + x^4$
14. $f(x) = (x^3 - 1)^3$

Find the 99th derivative, $\dfrac{d^{99}y}{dx^{99}}$, of the functions given in Exercises 15–20.

15. $y = x^{98}$
16. $y = x^{99}$
17. $y = x^{100}$
18. $y = x^{200}$
19. $y = x^{75} + x^{50}$
20. $y = (x^3 - 1)^{30}$

Find the derivatives requested in Exercises 21–24.

21. $f''(-1)$ when $f(x) = \dfrac{x^3 + 1}{x}$
22. $\left.\dfrac{d^2}{dx^2}\left(\dfrac{1}{x - 3}\right)\right|_{x=5}$
23. $\left.\dfrac{d^2}{dx^2}(2x^4 - x^3 + 3x)\right|_{x=2}$

24. $f''(0)$ when $f(x) = (x^2 + 1)^3$

25. Consider the function $y = f(x) = \frac{1}{x}$. Find the first 5 derivatives. Can you find a formula for the nth derivative, $f^{(n)}(x)$, for any positive integer n?

26. Compute $f(0)$, $f'(0)$, and $f''(0)$ when $f(x) = ax^2 + bx + c$.

27. Compute $f(0)$, $f'(2)$, and $f''(3)$ when $f(x) = \frac{1}{1 - x^2}$.

28. Compute $f(-1)$, $f'(1)$, and $f''(0)$ when $f(x) = \frac{2x + 3}{3x + 5}$.

29. Compute $\left.\frac{dy}{dx}\right|_{x=4}$ and $\left.\frac{d^2y}{dx^2}\right|_{x=-1}$ when $y = \frac{x}{x^2 - 4}$.

30. The height of a ball t sec after it is dropped from the top of a 20-story office building is given (in feet) by

$$h(t) = 240 - 16t^2$$

(a) How high is the building?

(b) When does the ball strike the ground?

(c) At what speed does the ball strike the ground?

31. An object is thrown upward from the edge of the roof of a tall building. The object's height (in feet) t sec later is given by

$$h = -16t^2 + 64t + 192$$

(a) What is the object's initial velocity?

(b) What maximum height does the object reach? (*Hint*: What is the object's velocity when it reaches its maximum height?)

(c) With what speed does the object hit the ground?

Key Mathematical Concepts and Tools

Derivative
Limit
Power rule
Product rule
Quotient rule
Higher derivatives

Applications Covered in This Chapter

Broken speedometer and velocity
Tangent to a curve
Marginal cost of production
Marginal tax rate
Marginal revenue
Pressure gradient
Growth rate of bacteria
Acceleration
Inflation rate

TWO Warm-Up Test

1. Using the $\lim_{h \to 0}$ definition of the derivative, find the instantaneous velocity $v(t)$, at the given time, for an object whose motion is given by the rule:

(a) $s = 4t + 2, t = 3$ (b) $s = t^2 + t, t = 2$

(c) $s = \frac{3}{t + 1}, t = 1$

2. The accompanying table shows the unit price of a certain commodity over a 5-year period. What is the average rate of change (in dollars per year) in the commodity's price over this period? Over the first 3 years?

TABLE FOR EXERCISE 2

Year	1980	1981	1982	1983	1984	1985
Unit price	\$108.21	109.46	112.32	123.81	132.14	155.17

3. For the curve $y = f(x)$, find the slope of the tangent line at the point on the curve as given.

(a) $y = f(x) = 8x + 1$; point (2, 17)

(b) $y = f(x) = x^2 - 4x$; point $(-1, 3)$

(c) $y = f(x) = x^2 - 2x + 3$; point (0, 3)

4. For $y = f(x) = 3x^2 - 4x + 6$, compute:

(a) $f'(x)$ (b) $f'(3)$

(c) $f'(2 + z)$ (d) $\left.\dfrac{dy}{dx}\right|_{x=-3}$

5. A software manufacturer is able to realize a profit of $75x^3 - 84$ dollars if he sells x units. Find the marginal profit:

(a) when $x = 7$; (b) when $x = 11$.

6. Use the $\lim_{h\to 0}$ method to compute $f'(3)$ for:

(a) $f(x) = \sqrt{x + 1}$ (b) $f(x) = \dfrac{1}{x + 1}$

7. Evaluate each of the following limits.

(a) $\lim_{h\to 0} \sqrt{h^2 + 5}$ (b) $\lim_{x\to 4} \dfrac{x^2 - 16}{x - 4}$

(c) $\lim_{x\to 4} \dfrac{x + 7}{(2x - 1)^2}$

8. Consider two differentiable functions $u = f(x)$ and $v = g(x)$.

(a) State the product rule in both Newton and Leibniz notation.

(b) Exhibit the use of this rule to differentiate $y = (x^2 + x - 1)(x^3 + 4)$.

9. Find dy/dx when y equals:

(a) $x^{74} - 2x^{16} + 5$ (b) $(3x^4 - 2x^2 + 1)^2$

(c) $\dfrac{x^3 - x}{2x - 1}$

What rules have you used in each of parts (a) through (c)?

10. Write the equation of the tangent line to the curve $y = 1/(x - 3)$:

(a) At the point (4, 1) (b) At the point $(5, \frac{1}{2})$

TWO Final Exam

1. Find the equation of the tangent line to the given curve at the indicated point.

(a) $y = f(x) = (x^2 - 2)^3$ point where $x = 3$

(b) $y = f(x) = \sqrt{x}$ point (9, 3)

2. Compute the marginal cost for each of the following cost functions $C(x)$.

(a) $C(x) = \dfrac{x^2 + 1}{(x - 1)^2}$ (b) $C(x) = x(x^2 + 4)$

(c) $C(x) = \dfrac{2x^2 + 1}{x}$

3. Use the $\lim_{h\to 0}$ method to compute $f'(2)$ for:

(a) $f(x) = \sqrt{x^3 + 1}$ (b) $f(x) = 5$

4. A ball is thrown upward. We are told that its height $s = s(t)$ above the ground after t seconds is given by the formula

$$s(t) = 80t - 16t^2$$

(a) Find its instantaneous velocity at $t = 2$; at $t = 3$.

(b) At what time t is the ball's instantaneous velocity equal to 0? How high is the ball at this time? Can you interpret the significance of this height?

5. Water is flowing from a tank in such a way that, at the end of t min, the volume of water in the tank is $V(t) = (10 - t)^2$ gal, where $0 \le t \le 10$. Note that initially ($t = 0$) there are 100 gal of water in the tank and that, at the end of 10 min ($t = 10$), the tank is empty.

(a) How fast is the water flowing out at $t = 3$?

(b) How fast is the water flowing out at $t = 5$?

6. For the function $y = f(x) = \dfrac{x}{x^2 + 1}$, compute:

(a) $f'(x)$ (b) $f'(2)$

(c) $\left.\dfrac{dy}{dx}\right|_{x=-1}$ (d) $f'(2 + h)$

7. Consider two differentiable functions $u = f(x)$ and $v = g(x)$.

(a) State the quotient rule in both Newton and Leibniz notation.

(b) Exhibit the use of this rule to differentiate

$$y = \frac{x^2 + 4}{x^3 - 3x + 1}$$

8. For each of the following compute dy/dx, indicating, as you go along, which differentiation rules you are using.

(a) $y = \dfrac{(x^2 + 1)(x - 4)}{x}$ (b) $y = (3x^2 + 7)^{11}$

(c) $y = (x^2 - 8)^{2/3}$

9. Compute $f''(x)$ for:

(a) $f(x) = \dfrac{x}{x + 1}$ (b) $f(x) = (x^2 - 1)(3x + 2)$

(c) $f(x) = (x + 2)^{5/2}$

10. Evaluate each of the following limits.

(a) $\lim\limits_{h \to 0} \dfrac{h^2 - 25}{h - 5}$ (b) $\lim\limits_{x \to -3} \dfrac{x^2 - 9}{x + 3}$

(c) $\lim\limits_{x \to 5} \dfrac{x - 7}{(3x + 1)^2}$

THREE

Curve Sketching and Applied Optimization

Outline

In Context

In this chapter we will study two types of applications of differentiation.

First we develop techniques that enable us to sketch accurately the graph of a function. Our methods are based on information that we can derive from a careful analysis of the first and second derivatives of the given function, and they do not require extensive plotting of individual points.

We then turn our attention to the important practical problem of determining the maximum or minimum value of a particular quantity. Problems of this type are known collectively as *optimization problems.* Our solution of this class of problems draws on the techniques developed in the earlier sections on curve sketching.

SECTION 3.1

The "Ups and Downs" of a Graph

Suppose we are given the problem of drawing the graph of the function $y = x^2$. Of course we know in advance what the graph should look like; this is an equation that we have encountered many times. But let us draw the graph from scratch. The easiest way to proceed is probably to construct a table of points of the graph, plot these points on graph paper, and then connect them with a smooth curve. This procedure is illustrated in Figure 3.1.

x	$y = x^2$
−2	4
−1.5	2.25
−1	1
−0.5	0.25
0	0
0.5	0.25
1	1
1.5	2.25
2	4

(a) Construct a table of points on the graph of the function.

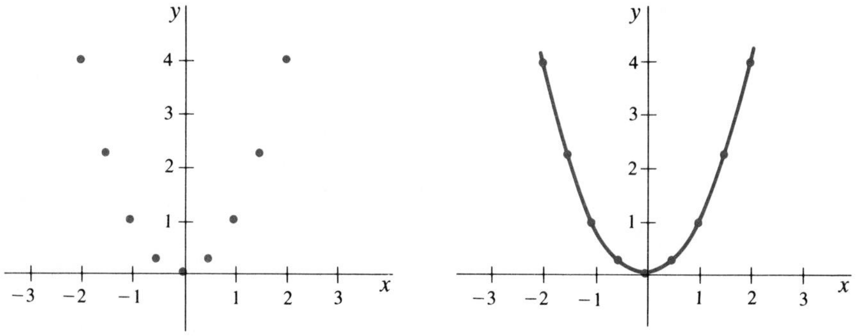

(b) Plot the points on the graph paper.

(c) Connect the points.

FIGURE 3.1
Steps in graphing the function $y = x^2$

Let us also draw the graph of the function $y = x^3 - 3x + 1$ in the same manner (see Figure 3.2).

x	$y = x^3 - 3x + 1$
−2	−1
−1.5	2.125
−1	3
−0.5	2.375
0	1
0.5	−0.375
1	−1
1.5	−0.125
2	3

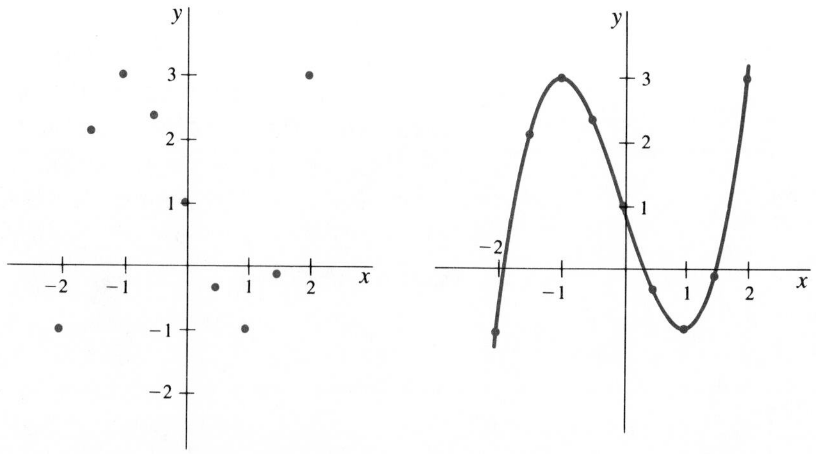

FIGURE 3.2
Steps in graphing the function $y = x^3 - 3x + 1$

PRACTICE EXERCISE Use the primitive procedure given above to sketch the graph of the function $y = 3x^4 - 4x^3$.

[Refer to Figure 3.37 on page 113.]

If you have conscientiously done this exercise, or if you have checked the computations of Figures 3.1 and 3.2, it will be apparent that the technique of drawing a graph by constructing a table of values is a very tedious procedure. In addition to the difficult computations we are likely to encounter, we can imagine serious theoretical difficulties that might arise when employing this technique. First, we can expect our graph to be reasonably accurate only for those values of x that are close to the ones in our table. For example, if the function whose graph we wish to draw has a graph that makes an unexpected turn near $x = 2000$, we will never find out about this simply by plotting points whose x coordinates lie between -2 and 2, as we did in Figures 3.1 and 3.2. Second, even after we have computed a table of values, such as the one shown in Figure 3.2, we are not really *sure* how these points should be connected. For example, in Figure 3.3 we have plotted the same points as in Figure 3.2 but then connected them in two different ways. How do we really know that the first picture is correct, and not the second?

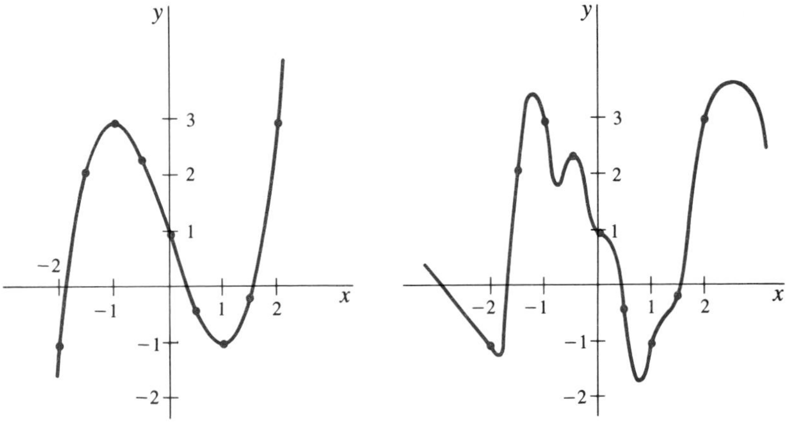

FIGURE 3.3

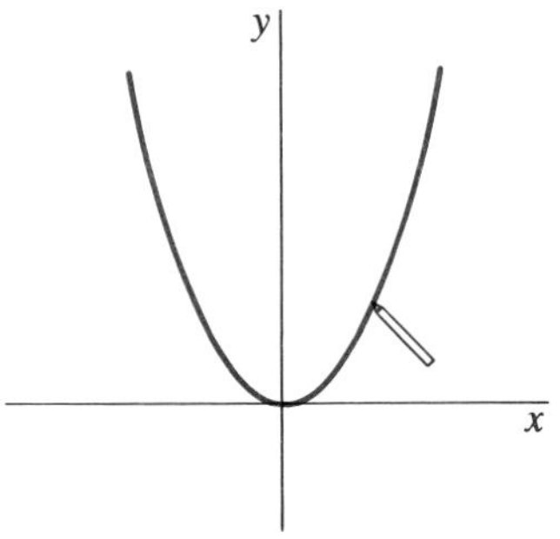

FIGURE 3.4

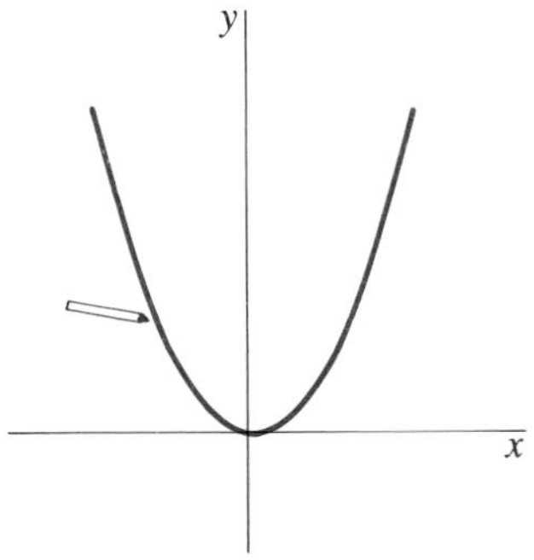

FIGURE 3.5

Our goal is to develop methods that will enable us to sketch the graph of a function in a way that incorporates all the significant features of the graph and at the same time minimizes computation.

Let us try to give a rough verbal description of the graph of the function $y = x^2$ (see Figure 3.4). If we place our pencil at a point on the graph where $x > 0$ and move it along the curve toward the right, we will touch points with increasing y-coordinates. On the other hand, if we place our pencil at a point on the graph where $x < 0$ and move it along the curve toward the right, we will touch points with decreasing y-coordinates (see Figure 3.5). The transition from points with decreasing y-coordinates to those with increasing y-coordinates takes place at the origin, where the graph has a "low point," or minimum. This discussion suggests the following definitions.

DEFINITION 3.1 The function f is *decreasing* on an interval if its graph is falling as we move along the graph from left to right. Similarly, the function f is *increasing* on an interval if its graph is rising as we move along the graph from left to right. ⊟

Figure 3.6 illustrates the graphs of functions that are decreasing and increasing, respectively, on the interval $[a, b]$.

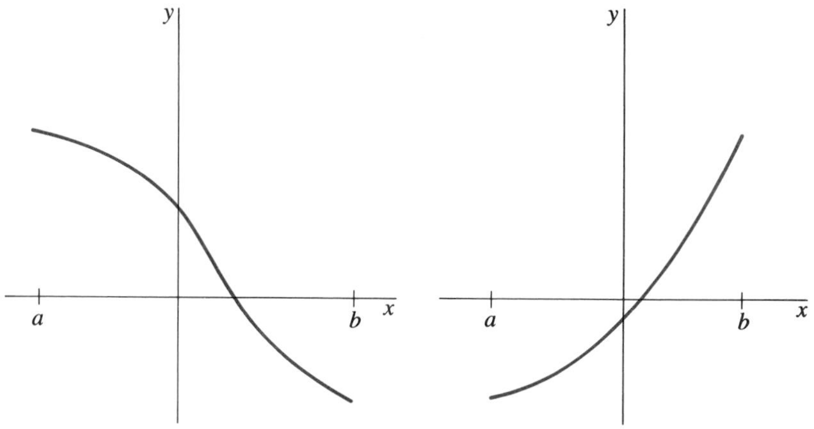

(a) f decreasing on [a, b] (b) f increasing on [a, b]

FIGURE 3.6

EXAMPLE 1 Consider the linear function

$$y = f(x) = mx + b$$

Its graph is, of course, a straight line (see Figure 3.7). It is easy to see from the graph that:

(i) If $m > 0$, the function is increasing.

(ii) If $m < 0$, the function is decreasing.

(iii) If $m = 0$, the line is a horizontal line, which is neither increasing nor decreasing.

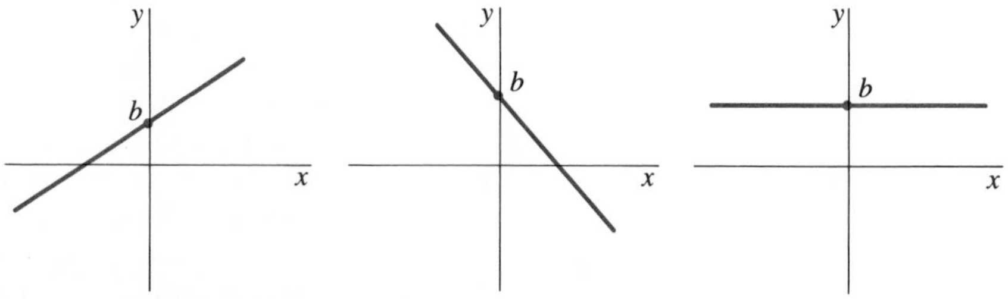

(a) $m > 0$ (b) $m < 0$ (c) $m = 0$

FIGURE 3.7

DEFINITION 3.2 The function f has a **minimum** at $x = c$ if $f(c) \leq f(x)$ for all x in the domain of f. Equivalently, there is no point on the graph that is lower than the point $(c, f(c))$. (Note that such a smallest value can occur at more than one point.)

The function has a **maximum** at $x = c$ if $f(c) \geq f(x)$ for all x in the domain of f. Thus f has a maximum at $x = c$ if $f(c)$ is the largest value the function may assume. (Note that such a largest value can occur at more than one point.)

A point at which either a minimum or a maximum occurs is called an **extreme point**. ⊟

EXAMPLE 2 We consider again the graph of the parabola $y = x^2$ plotted in part (c) of Figure 3.1. We may summarize our description of the graph by saying that it is decreasing on the interval $(-\infty, 0)$, is increasing on the interval $(0, \infty)$, and has a minimum at $x = 0$.

EXAMPLE 3 Consider again the function $y = x^3 - 3x + 1$, whose graph is shown in Figure 3.8. A brief examination indicates that the function is increasing on $(-\infty, -1)$, is decreasing on $(-1, 1)$, and is increasing on $(1, \infty)$. At first glance our intuition suggests that the point $(-1, 3)$ represents a maximum point [and that the point $(1, -1)$ represents a minimum]. This is an illusion, however. A more careful examination reveals that points sufficiently far to the right on the graph are higher than the point $(-1, 3)$. For example, the point $(5, 111)$ on the graph is higher than the point $(-1, 3)$. Thus we *cannot* call the point $(-1, 3)$ a maximum point of the function $y = x^3 - 3x + 1$. In fact, the graph does not possess a maximum (the values of y get arbitrarily large). Similarly the point $(1, -1)$ is not a minimum point. Nevertheless, these points are somehow distinguished. This discussion points out the need for some additional terminology.

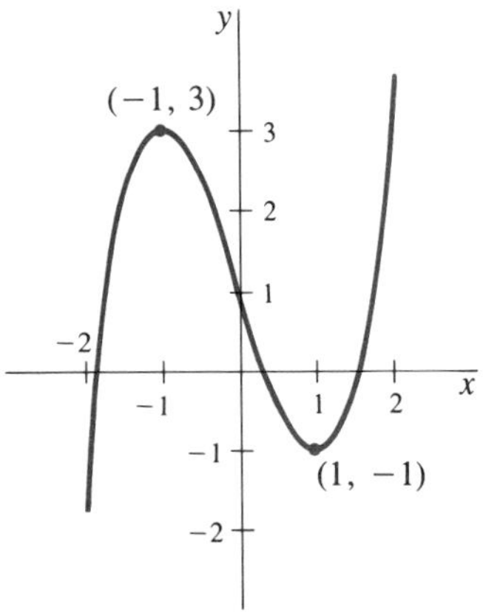

FIGURE 3.8
Graph of $y = x^3 - 3x + 1$

DEFINITION 3.3 The function f has a **local minimum** (also called a **relative minimum**) at $x = c$ if $f(c) \leq f(x)$ for all x near c. Alternatively, f has a **local minimum** at $x = c$ if the point $(c, f(c))$ is the lowest point on a small portion of the graph containing that point (not as an endpoint). ⊟

Operationally, if the graph is decreasing on some interval just to the left of $x = c$ and increasing on some interval just to the right of that point, then f has a local minimum at $x = c$. See Figure 3.9(a).

By analogy, f has a **local (relative) maximum** at $x = c$ if $f(c) \geq f(x)$ for all x near c. If the graph of f is increasing just to the left of $x = c$ and decreasing just to the right, f has a local maximum at $x = c$. See Figure 3.9(b). A point at which either a local maximum or a local minimum occurs is called a **local extremum** or a **local extreme point**.

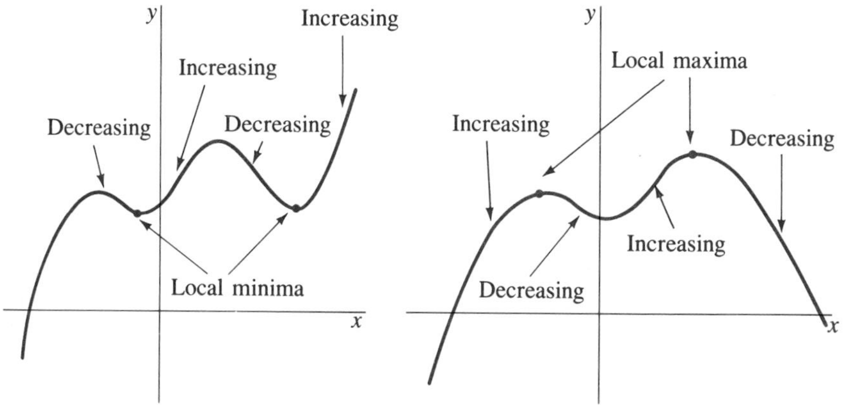

(a) Local minima (b) Local maxima

FIGURE 3.9

If you examine the sketch of the mountain range shown in Figure 3.10, you will see that each peak corresponds to a local maximum. The tallest mountain in the range is a maximum.

Applying our newly developed terminology to Example 3, we can conclude that $(-1, 3)$ is a local maximum and that $(1, -1)$ is a local minimum.

In the preceding examples we have obtained our graph by plotting a good many points and then drawing a curve connecting them. Then we looked at the resulting graph and labeled the regions where the curve was rising or falling and where the extrema and local extrema occurred.

The objective of the next two sections is to reverse the procedure we have just described. That is, we will learn (1) to determine at the outset (before we have the sketched graph) where a curve is increasing, where it is decreasing, and where its extrema and local extrema occur, and (2) to use this information as an aid in plotting the desired graph, showing all relevant features.

FIGURE 3.10

Section 3.1 EXERCISES

For each of the following functions: (a) make a rough sketch of the graph; (b) try to determine where the curve is rising and where it is falling; (c) try to locate all extrema and local extrema.

1. $y = (x - 2)(x - 3)$

2. $y = 3x^2 + x - 1$

3. $y = \dfrac{x^3}{3} - 2x^2 + 4x + 1$

4. $y = -x^3 + 6x^2 + 6$

5. $y = \dfrac{x}{x + 1}$

6. $y = \dfrac{x - 2}{x - 3}$

7. $y = |9 - x^2|$

8. $y = x^2 - 5x + 6$

9. $y = \dfrac{8}{4 - x^2}$

10. $y = \dfrac{x}{x^2 - 4}$

11. $y = \dfrac{x^2}{1 - x^2}$

12. $y = \dfrac{x^2 - 1}{x^2 + 1}$

13. $y = \sqrt{x}$

14. $y = x^{2/3}$

15.* $y = e^x$

16.* $y = xe^{-x}$

17.* $y = \ln x$

18.* $y = \dfrac{\ln x}{x}$

19. $y = \dfrac{x}{1 + |x|}$

20. $y = \dfrac{x^2 - 4}{x}$

SECTION 3.2

Use of the First Derivative in Curve Sketching

Given any function $y = f(x)$, we know that the derivative $f'(c)$ represents the slope of the tangent line to this curve at $x = c$. If this number, $f'(c)$, is positive, the tangent line is tilting upward near the point $(c, f(c))$. Our intuition tells us that the graph of f moves in the same direction as its tangent line, at least for a little while. This is indeed the case if the derivative f' is continuous at c: $y = f(x)$ increases on some interval containing $x = c$ whenever $f'(c) > 0$. See Figure 3.11(a). Similarly, Figure 3.11(b) reveals that f decreases near $x = c$ if $f'(c) < 0$ (again assuming that f' is continuous at c).

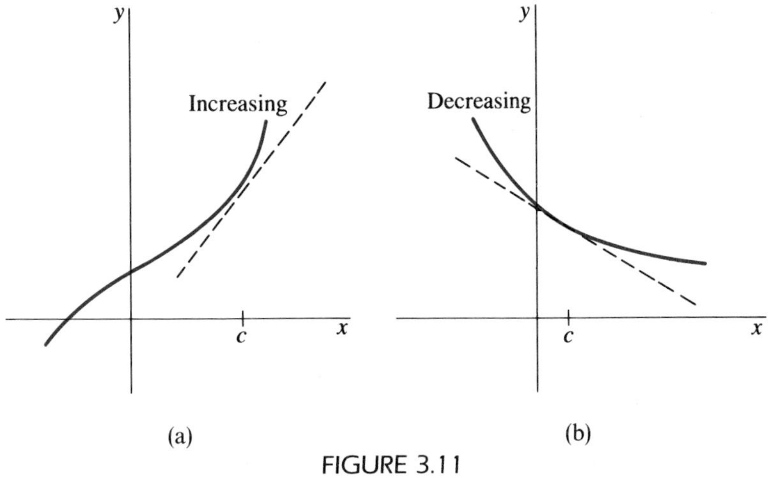

FIGURE 3.11

* Do only at this time if you have a calculator with exponential and logarithm buttons. Discussion of these functions is in the following chapter.

The values of x at which the derivative is zero are of crucial importance in curve sketching.

DEFINITION 3.4 A point $x = c$ for which $f'(c) = 0$ is called a **critical point** for f. ⊟

The behavior of a graph near a critical point $x = c$ can be determined by constructing a **sign pattern chart** for the derivative near $x = c$. Typical sign pattern charts are shown in Figures 3.12 and 3.13.

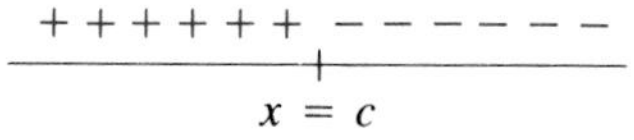

FIGURE 3.12
Sign of $f'(x)$ near local maximum $x = c$

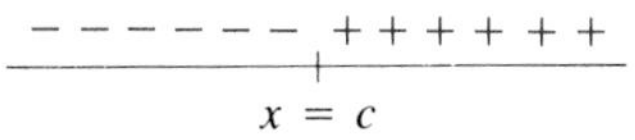

FIGURE 3.13
Sign of $f'(x)$ near local minimum $x = c$

FIRST DERIVATIVE TEST FOR CRITICAL POINTS

Let $f'(x)$ be defined in an open interval containing the critical point $x = c$. Then

1 If $f'(x)$ is positive to the left of $x = c$ and negative to the right of $x = c$, f has a local maximum at $x = c$ (see Figure 3.12).

2 If $f'(x)$ is negative to the left of $x = c$ and positive to the right of $x = c$, f has a local minimum at $x = c$ (see Figure 3.13).

Let us go back and apply these ideas to some of the functions we considered in Section 3.1.

EXAMPLE 4 Consider again the parabola $y = f(x) = x^2$. Its derivative is $f'(x) = 2x$. If we set this equal to 0, we find that $x = 0$ is the only critical point. Moreover, if $x < 0$, then $\frac{dy}{dx} = 2x < 0$; whereas if $x > 0$, then $\frac{dy}{dx} = 2x > 0$. We may summarize this by writing

$$f'(x) \text{ is } \begin{cases} <0 & \text{if} \quad x < 0 \\ =0 & \text{if} \quad x = 0 \\ >0 & \text{if} \quad x > 0 \end{cases}$$

which results in the sign pattern chart shown in Figure 3.14. This, of course, agrees with the sketch shown in Figure 3.1.

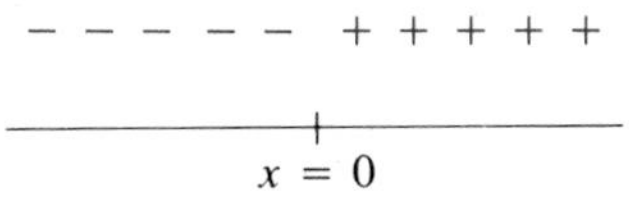

FIGURE 3.14
Sign pattern chart for $\frac{dy}{dx} = 2x$

EXAMPLE 5 Let us perform an analysis of the function $y = f(x) = x^3 - 3x + 1$ (see Figure 3.2). For this function

$$f'(x) = 3x^2 - 3 = 3(x^2 - 1) = 3(x + 1)(x - 1)$$

Setting $f'(x) = 3(x + 1)(x - 1) = 0$ reveals that the derivative has two roots, $x = -1$ and $x = 1$. These are the critical points of f. These roots divide the x-axis into three regions: $x < -1$, $-1 < x < 1$, and $x > 1$ (see Figure 3.15). Choose a value of x in each of the three regions and substitute it into f' to determine the sign of the derivative there. For instance, in the region $x < -1$

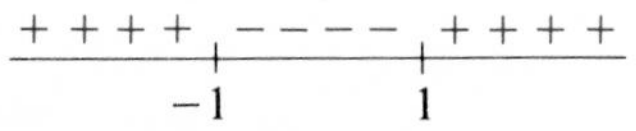

FIGURE 3.15
Sign pattern chart for $\frac{dy}{dx} = 3(x + 1)(x - 1)$.

you might select $x = -2$. Then $f'(-2) = 3(-2)^2 - 3 = 9 > 0$ and we conclude that $f' > 0$ in this region. It doesn't matter which number you choose. The value of f' will be different, but the sign depends only on the region.

To illustrate further, in the region $-1 < x < 1$ you might choose $x = 0$. Then

$$f'(0) = 3(0)^2 - 3 = -3 < 0$$

so $f' < 0$ in this region. You should convince yourself that $f' > 0$ in the region $x > 1$.

Thus we conclude that the graph of $y = x^3 - 3x + 1$

is increasing for $x < -1$

has a local maximum at the critical point $x = -1$

decreases for $-1 < x < 1$

has a local minimum at $x = 1$

increases for $x > 1$

Once again, this analysis corroborates the point-by-point plot shown in Figure 3.2.

EXAMPLE 6 Consider the cubic equation $y = f(x) = x^3$ whose graph is shown in Figure 3.16(a). Here $\dfrac{dy}{dx} = 3x^2$ and $3x^2 = 0$ only if $x = 0$. The presence of the x^2 term in the derivative indicates that the derivative will be positive whether x is positive or negative. The sign pattern chart for the derivative is shown in Figure 3.16(b). The critical point $x = 0$ is thus neither a local maximum nor a local minimum. The graph

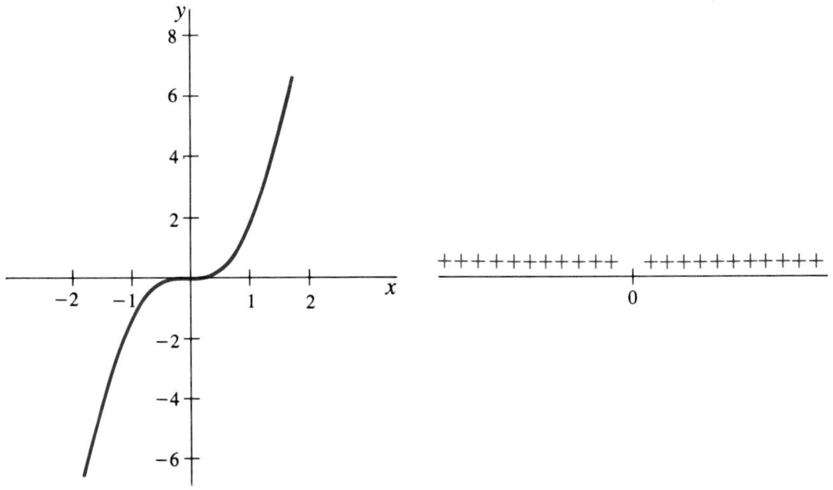

(a) $y = f(x) = x^3$ (b) Sign pattern chart for $dy/dx = 3x^2$

FIGURE 3.16

increases for $x < 0$

has a horizontal tangent line at $x = 0$

increases for $x > 0$

We will discuss what goes on at such a critical point in the next section.

EXAMPLE 7 Sketch the graph of $y = -x^3 + 6x^2 + 6$.

SOLUTION We first compute $dy/dx = y' = -3x^2 + 12x$. This factors into

$$y' = -3x(x - 4)$$

Setting $y' = 0$, we find that there are two critical points: $x = 0$ and $x = 4$. The value of y corresponding to $x = 0$ is $y = 6$, and the value of y corresponding to $x = 4$ is $y = -4^3 + 6 \cdot (4)^2 + 6 = 38$. Thus the points (0, 6) and (4, 38) must lie on our sketch (see Figure 3.17). Then we construct the sign pattern chart, which is shown in Figure 3.18. We determine that the derivative is negative to the left of $x = 0$ by trying, for example, the value $x = -1$ and computing $dy/dx = -3(-1)(-5) < 0$. Trying $x = 2$, which is between $x = 0$ and $x = 4$, gives $dy/dx = -3(2)(-2) > 0$. Thus we have the $++++$ pattern between $x = 0$ and $x = 4$. Similarly, for any $x > 4$ (try $x = 5$, for example), we have $dy/dx < 0$, as indicated by the $----$ pattern.

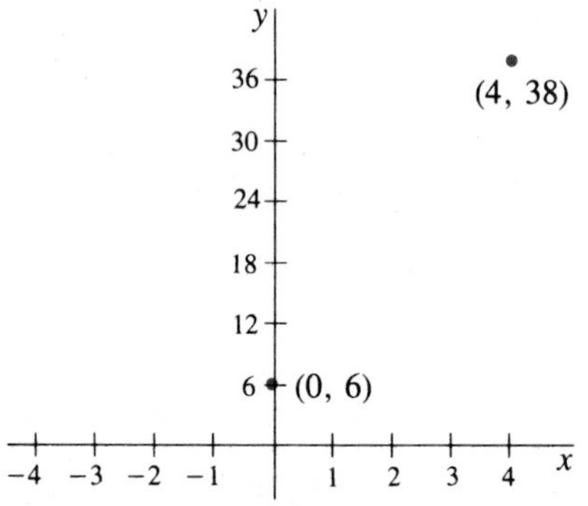

FIGURE 3.17
Critical points of $y = -x^3 + 6x^2 + 6$

Thus we know that the curve falls to the left of $x = 0$ as we move toward (0, 6), then rises between (0, 6) and (4, 38), and finally falls to the right of $x = 4$. Furthermore, the

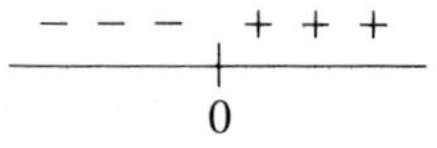

pattern indicates that $x = 0$ gives a local minimum, whereas the

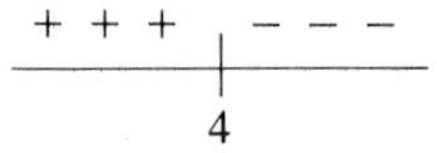

pattern indicates that $x = 4$ gives a local maximum. This information is incorporated in Figure 3.19.

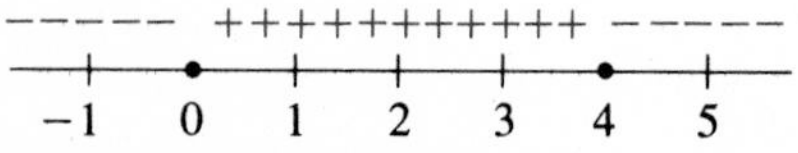

FIGURE 3.18
Sign pattern chart for $\frac{dy}{dx} = -3x(x - 4)$

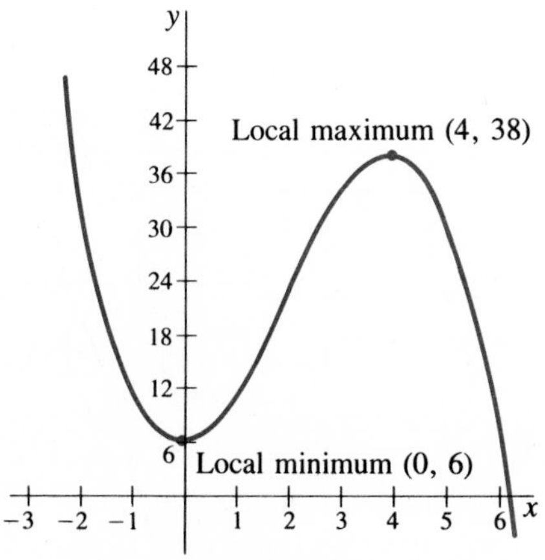

FIGURE 3.19
$y = -x^3 + 6x + 6$

Finally, we complete our sketch by seeing what happens to y as $x \to \pm\infty$. For x large and positive, say $x = 10^6$ (one million), $y = -10^{18} + 6 \cdot 10^{12} + 6$, which is very large in magnitude but negative. Thus we conclude that

$$\lim_{x \to \infty} (-x^3 + 6x^2 + 6) = -\infty$$

However, for x large and negative, say $x = -10^6$,

$$y = -(-10^6)^3 + 6(-10^6)^2 + 6 = 10^{18} + 6 \cdot 10^{12} + 6$$

which is large and positive. Thus we conclude that

$$\lim_{x \to -\infty} (-x^3 + 6x^2 + 6) = \infty$$

We summarize our methodology in the form of a guide. We shall modify it later and formulate a more complete approach.

PRELIMINARY GRAPHING GUIDE

Given the differentiable function $y = f(x)$:

1 Compute $y' = \dfrac{dy}{dx}$.

2 Factor y', if possible.

3 Set the factored $dy/dx = y'$ equal to 0 to find the critical points; these are the possibilities for local extrema. Compute the values of y corresponding to these x's and plot the points.

4 Construct a sign pattern chart for $\dfrac{dy}{dx}$.

5 Begin to plot the curve. Remember that, if the sign chart is $+ + + +$ in a certain region, the curve rises there, whereas a $- - - -$ pattern indicates a region where the curve is falling. Be sure that your curve goes through any points already plotted.

6 Label the critical points on your graph as local maxima or local minima if the sign pattern charts so indicate.

7 Examine the behavior of the curve $y = f(x)$ when x is large and positive and when x is large and negative. Indicate this information in the way you complete your sketch.

EXAMPLE 8 Sketch the graph of the function

$$y = \frac{x}{1 + x^2}$$

SOLUTION As in the previous example, we compute y' (this time making use of the quotient rule).

$$y' = \frac{(1 + x^2)(1) - x(2x)}{(1 + x^2)^2} = \frac{1 - x^2}{(1 + x^2)^2}$$

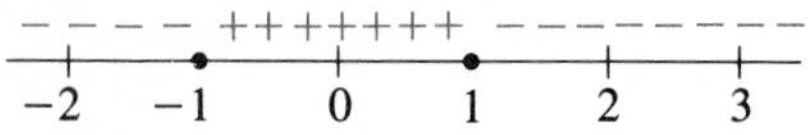

FIGURE 3.20

Sign pattern chart for $y' = \dfrac{1 - x^2}{(1 + x^2)^2}$

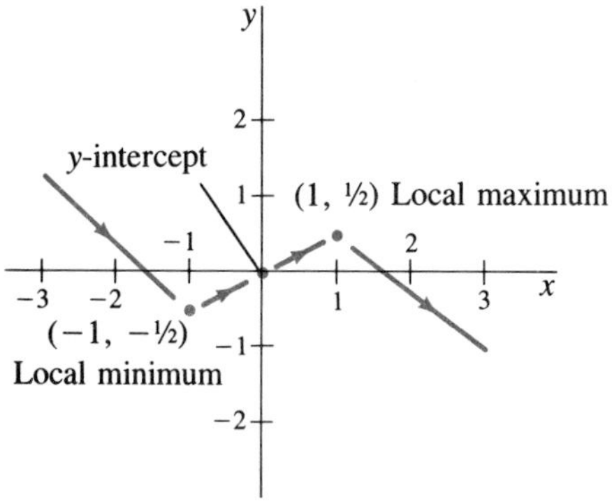

FIGURE 3.21

Initial sketch of $y = \dfrac{x}{1 + x^2}$

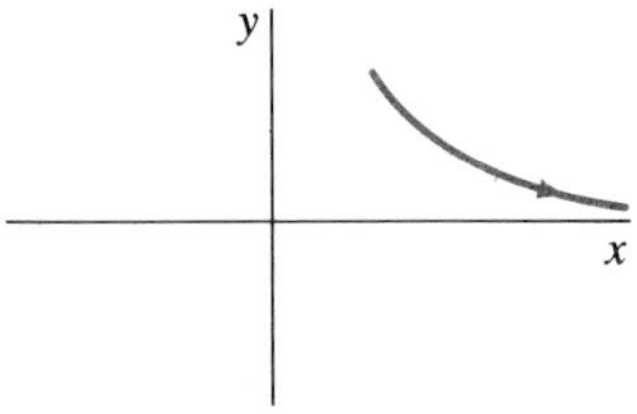

FIGURE 3.22

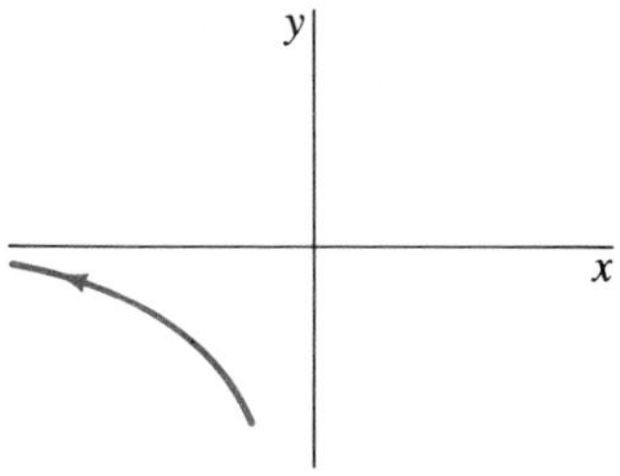

FIGURE 3.23

The denominator of this fraction is always positive, so the fraction will be positive, negative, or zero according to whether its numerator $(1 - x^2)$ is positive, negative, or zero, respectively. Clearly $1 - x^2 = (1 - x)(1 + x)$ equals 0 when $x = -1$ or $x = 1$. When $x = -1$, $y = -\frac{1}{2}$; and when $x = 1$, $y = \frac{1}{2}$.

Our next step is to construct a sign pattern chart for dy/dx. The sign of $(1 - x^2) = (1 - x)(1 + x)$ in the various regions determined by the critical points $x = -1$ and $x = 1$ gives the pattern shown in Figure 3.20.

We can now begin to plot the curve. Note that we have also labeled the point $(x, y) = (0, 0)$ where the graph crosses the y-axis (see Figure 3.21). This is called the ***y*-intercept** of the curve. It is usually helpful (and never harmful) to add this point to the sketch.

It remains for us to examine the behavior of $y = \dfrac{x}{1 + x^2}$ as $x \to +\infty$ or $x \to -\infty$. As indicated in item 7 of our preliminary graphing guide, we first try a large positive value of x (say $x = 10^6$) and compute y. We have

$$y = \frac{x}{1 + x^2} = \frac{10^6}{1 + 10^{12}}$$

which is approximately $\dfrac{1}{10^6}$. Thus we see that, for x large and positive, y is still positive but very small in magnitude. Thus we write

$$\lim_{x \to +\infty} \frac{x}{1 + x^2} = +0$$

where the $+0$ indicates that, as x tends toward ∞ through positive numbers, y tends toward 0 through positive values. This behavior is illustrated in Figure 3.22.

To see what happens to our curve as $x \to -\infty$, let us try a negative value for x that is very large in magnitude, say $x = -10^6$. Then

$$y = \frac{x}{1 + x^2} = \frac{-10^6}{1 + (-10^6)^2} = -\frac{10^6}{1 + 10^{12}}$$

which is approximately $-1/(10^6)$. Thus we see that, for x large but negative, y is negative but very close to 0. We write

$$\lim_{x \to -\infty} \frac{x}{1 + x^2} = -0$$

indicating that, as x tends toward $-\infty$ through negative numbers, y tends toward 0 through negative values. This behavior is illustrated in Figure 3.23.

A straight line such as $y = 0$ in the foregoing example, which a curve $y = f(x)$ approaches as $x \to +\infty$ or as $x \to -\infty$, is called a **horizontal asymptote**. Figure 3.24 incorporates our finding of a horizontal asymptote at $y = 0$ into the initial sketch of Figure 3.21. This example illustrates that item 7 of our preliminary graphing guide involves the search for and labeling of possible horizontal asymptotes.

There is also another type of asymptotic behavior that frequently occurs: vertical asymptotes.

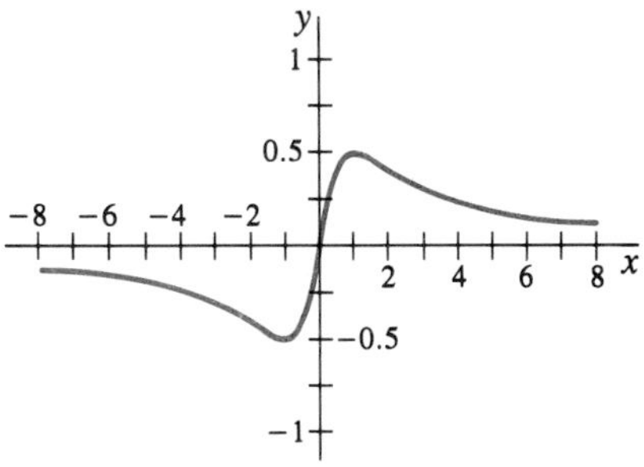

FIGURE 3.24

$$y = \frac{x}{1 + x^2}$$

EXAMPLE 9 Sketch the graph of

$$y = f(x) = \frac{1}{x - 4}$$

SOLUTION Before proceeding, we note that this function exhibits a property we have not encountered in the previous graphs we have drawn. It is not defined for all values of x. In particular, because the denominator of the fraction $1/(x - 4)$ is 0 for $x = 4$, we must exclude this number from the domain of f. Our analysis of the graph will have to incorporate this fact.

As a first step, let's find the critical points, if any. To do this, we compute

$$\frac{dy}{dx} = -\frac{1}{(x - 4)^2}$$

Now dy/dx can vanish (that is, it can equal 0) only when the numerator of $-1/[(x - 4)^2]$ is 0. But this cannot occur; hence there are no critical points.

Also, at every point of our function's domain the sign of dy/dx is negative. Thus the graph must always be falling to the left of $x = 4$ (as we move from left to right) and must be falling to the right of $x = 4$ (as we move from left to right) as well.

Next we investigate what happens to our curve as $x \to \pm\infty$. First, for x large and positive, let's try $x = 10^7$. Then $y = 1/(10^7 - 4)$, which is a very small positive number. Thus we write

$$\lim_{x \to +\infty} \left(\frac{1}{x - 4}\right) = +0$$

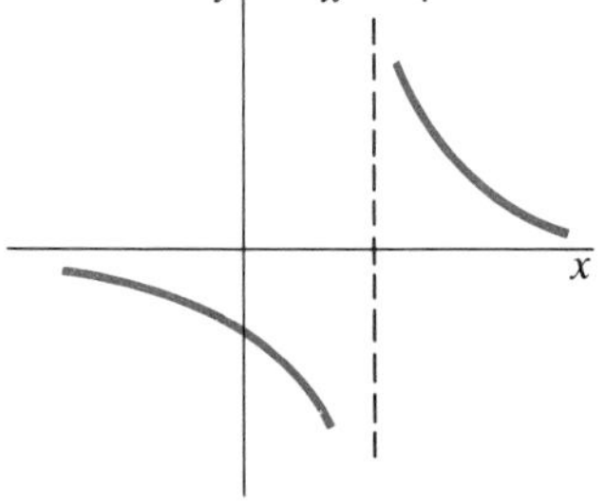

FIGURE 3.25

Therefore, as x tends toward $+\infty$, the graph approaches the horizontal asymptote $y = 0$ (the x-axis) from above. Similarly, we could demonstrate that

$$\lim_{x \to -\infty} \left(\frac{1}{x - 4}\right) = -0$$

Figure 3.25 illustrates the information we have gathered up to this point. It also includes a dotted vertical line at $x = 4$. This is to remind us that our

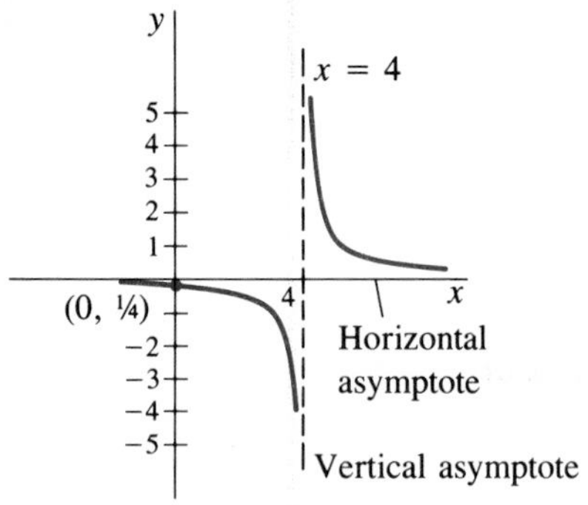

FIGURE 3.26

equation $y = 1/(x - 4)$ makes no sense for this value of x. However, we *can* ask how our curve behaves as x tends toward $x = 4$. We note first that, for x near 4 but very slightly to the left of it (say, $x = 3.99$), we have

$$y = \frac{1}{3.99 - 4} = \frac{1}{-.01} = -100$$

which is large and negative. Furthermore, as x gets closer to $x = 4$ from the left (as it reaches $x = 3.999$, for example), y remains negative but its magnitude is even larger.

Next we consider values of x near 4 but slightly to the right of it, say $x = 4.01$, yielding $y = \dfrac{1}{4.01 - 4} = 100$. We see that, as x tends toward $+4$ from the right, y is positive and is getting larger and larger.

Our final sketch, Figure 3.26, includes the y-intercept—that is, the point $x = 0, y = -\frac{1}{4}$. The vertical line $x = +4$ is called a **vertical asymptote** of the curve $y = \dfrac{1}{x - 4}$.

In sketching, it is generally good practice to label the vertical asymptotes, if any.

Section 3.2 EXERCISES

Find all critical points for each of the functions in Exercises 1–6.

1. $8 + 2x - x^2$

2. $(x - 2)(x + 3)$

3. $x^4 - x^2$

4. $(x + 1)(x - 2)^2$

5. $\dfrac{x + 1}{x}$

6. $\dfrac{x}{(x - 1)^2}$

For each of the functions $y = f(x)$ in Exercises 7–12, draw the sign pattern chart for $dy/dx = f'(x)$ and interpret what it says.

7. $x^3 - x^2$

8. $(x - 1)(x - 2)(x - 3)$

9. $\dfrac{1}{x^2 + 4}$

10. $(x - 1)^3 + 2$

11. $\dfrac{x^2}{4 - x^2}$

12. $x^2(x - 3)^2$

How does each of the functions in Exercises 13–18 behave as $x \to \infty$ and also as $x \to -\infty$? (*Hint*: In Exercises 16–18, divide the numerator and the denominator by the highest power of x present.)

13. $1 + 4x - 3x^2$

14. $3x^5 + 4x^4$

15. $5x^3 + x - 10^9$

16. $\dfrac{3x + 4}{4 - 7x}$

17. $\dfrac{3x^2 + x + 1}{5x^2 - x + 4}$

18. $\dfrac{1 + 4x - 3x^2}{x^3 + x + 1}$

As carefully as possible, sketch the graph of each of the functions $y = f(x)$ in Exercises 19–30. Your discussion should include the sign pattern chart for dy/dx, critical points, relative maxima and minima, rising and falling, behavior of the graph as $x \to \pm\infty$, asymptotes, and so on.

19. $2x^3 + 3x^2 - 12$

20. $(x - 3)(x^2 + 4)$

21. $x^4 - 4x + 8$

22. $(x + 2)^3(x - 3)$

23. $\sqrt{4 - x^2}, \quad -2 \le x \le 2$

24. $x^{2/3}$

25. $\dfrac{x + 1}{x - 1}$

26. $\dfrac{1}{x^2 + 4}$

27. $\dfrac{2}{x - 3}$

28. $\dfrac{1}{(x - 2)^2}$

29. $x\sqrt{x - 3}$

30. $\dfrac{x + 3}{x^2 + x - 6}$

SECTION 3.3

Use of the Second Derivative

Let us apply our graphing techniques to the function $y = 3x^4 - 4x^3$. In this case $y' = 12x^3 - 12x^2 = 12x^2(x - 1)$. Thus $y' = 0$ when $x = 0$ or $x = 1$. The corresponding function values are

$$y = 0 \quad \text{at} \quad x = 0$$

and

$$y = -1 \quad \text{at} \quad x = 0$$

An analysis of the sign of $y' = 12x^2(x - 1)$ yields the information shown in Figure 3.27. The y-intercept is the point (0, 0). We also have

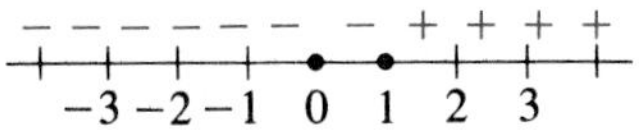

FIGURE 3.27
Sign of $y' = 12x^2(x - 1)$

$$\lim_{x \to -\infty} (3x^4 - 4x^3) = \lim_{x \to -\infty} (3x^4 - 4x^3) = \infty$$

This information yields Figure 3.28.

If you now go back to the introductory remarks to this chapter, you will note that the practice exercise on page 97 asked you to draw the graph of the same function, $y = 3x^4 - 4x^3$. At that time, however, the problem was to be done by constructing a table of values and then plotting them on graph paper. If you performed this exercise conscientiously, using x-values spaced $\frac{1}{2}$ unit apart, as was done in the text, you should have obtained a graph like the one shown in Figure 3.29. In Figure 3.30 these two graphs are shown on the same set of axes. The curve of Figure 3.28 is a solid line; that of Figure 3.29 is a dashed line. What went wrong? Why are these two curves different in that portion of Figure 3.30 which is boxed in? Roughly, we may say that the solid curve appears to be "cupped upward," whereas the dashed curve is "cupped downward."

It appears that the manner in which a graph is "cupped," which is called its **concavity**, is of considerable importance in determining certain aspects of its appearance. Our goal in this section is to show how examining the sign of the second derivative of $y = f(x)$, much as we examined the sign of the first derivative, enables us to understand this important idea.

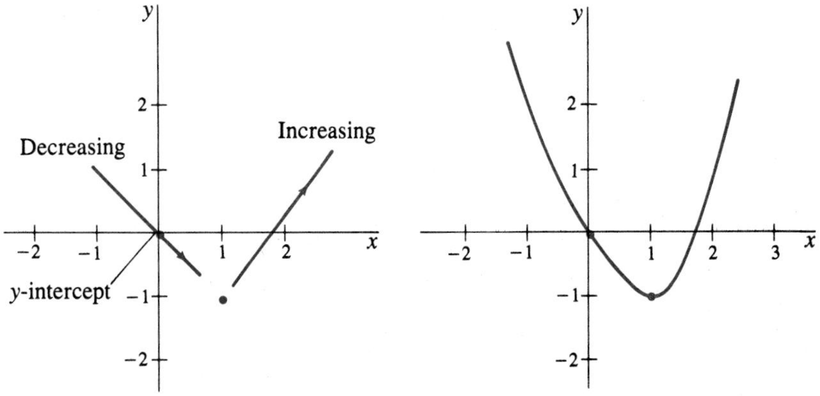

FIGURE 3.28
Initial sketch of $y = 3x^4 - 4x^3$

x	$y = 3x^4 - 4x^3$
-2	80
-1.5	$\frac{459}{16} \approx 29$
-1	7
-0.5	$\frac{11}{16} \approx 0.7$
0	0
0.5	$-\frac{5}{16} \approx -0.3$
1	-1
1.5	$\frac{27}{16} \approx 1.7$
2	16

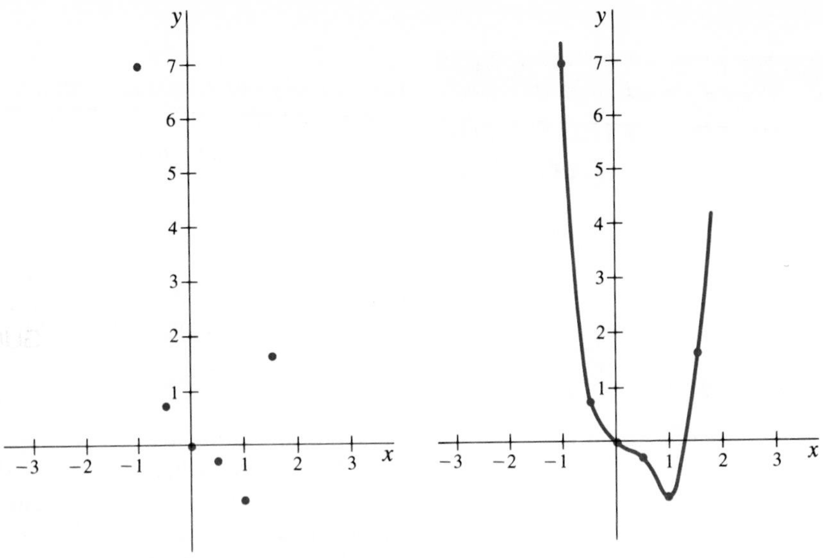

FIGURE 3.29

Plot of selected points of $y = 3x^4 - 4x^3$

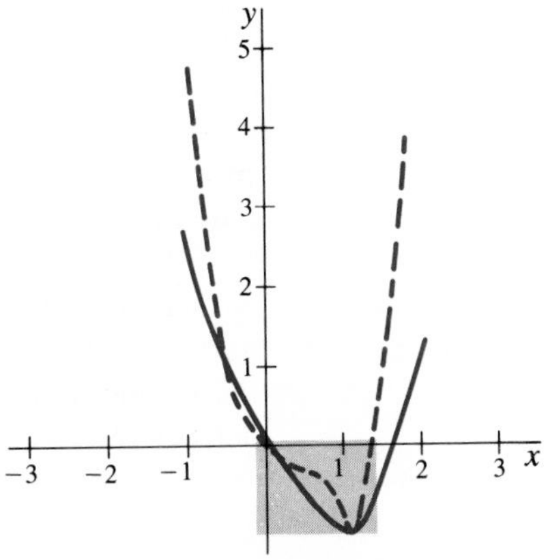

FIGURE 3.30

Preliminary graph

DEFINITION 3.5 Assume that the derivative of f exists at $x = a$.

(i) We will call the function f **concave up** at the point $x = a$ if, near the point $(a, f(a))$, the graph of $y = f(x)$ lies *above* its tangent line. See Figure 3.31(a).

(ii) We will call the function f **concave down** at the point $x = a$ if, near the point $(a, f(a))$, the graph of $y = f(x)$ lies *below* its tangent line. See Figure 3.31(b).

(iii) We will call the point $x = a$ an **inflection point** of the function f if the graph of $y = f(x)$ crosses its tangent line at $(a, f(a))$. See Figure 3.31(c). Thus f changes its concavity at a point of inflection. ⊟

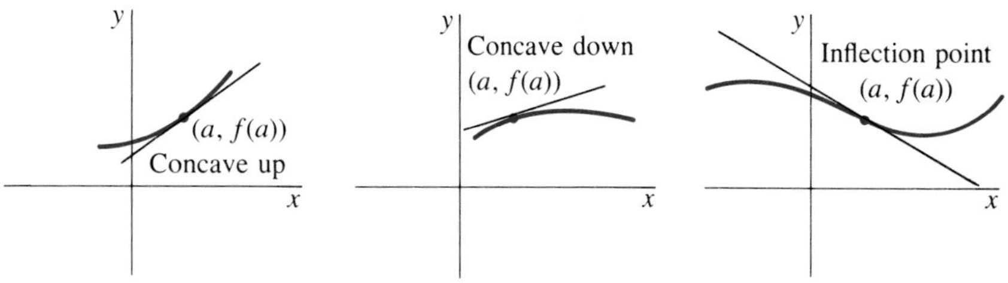

(a) The function f is concave up.

(b) The function f is concave down.

(c) An inflection point of the function f.

FIGURE 3.31

Consider now the case of a curve that is concave up at each point of an interval containing the point a, as shown in Figure 3.32. An examination of tangent lines drawn to the curve at several points would show that the slope of those lines [that is, $f'(x)$] is increasing as we pass from left to right. Note that it is not necessary for f to be increasing near $x = a$ [it is not doing so in Figure 3.32(a); it *is* in Figure 3.32(b)]. Rather, it is $f'(x)$ that is getting larger. From our study of derivatives we know that the second derivative of f,

$$f'' = \frac{d^2y}{dx^2}$$

measures the change in the first derivative of f (just as the first derivative measures the change in f). Thus we see that, if $f''(a) > 0$, then f is concave up at $x = a$.

A similar argument shows that, if $f''(a) < 0$, then f is concave down at $x = a$. We emphasize again that the curve itself may be either increasing, as shown in Figure 3.33(a), or decreasing, as shown in Figure 3.33(b).

In summary, we can characterize concavity as follows.

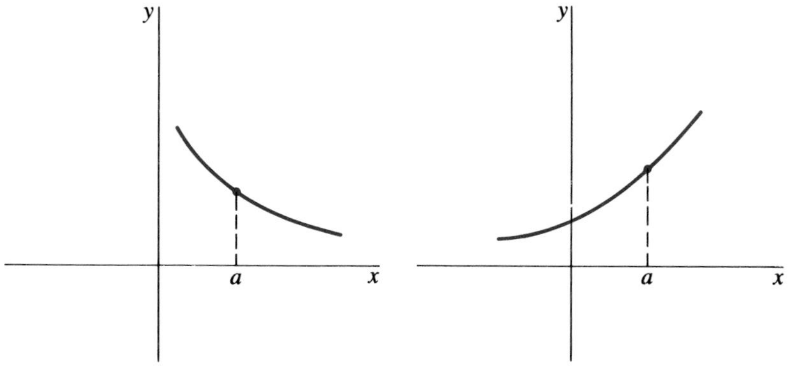

(a) f concave up at $x = a$, curve decreasing at a

(b) f concave up at $x = a$, curve increasing at a

FIGURE 3.32

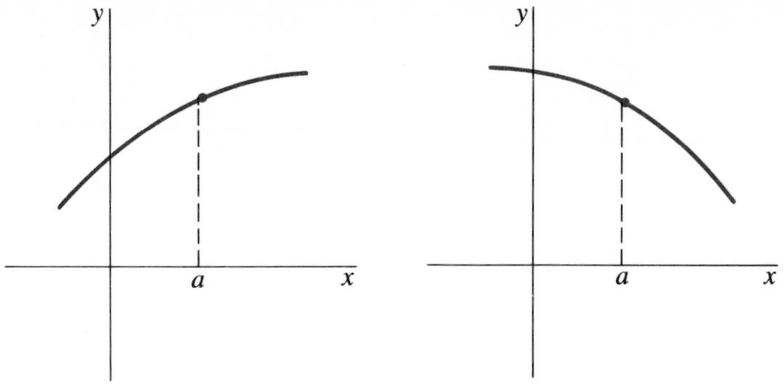

(a) f concave down at $x = a$, curve increasing at a

(b) f concave down at $x = a$, curve decreasing at a

FIGURE 3.33

CHARACTERIZATION OF CONCAVITY

1 If $f''(a) > 0$, then $y = f(x)$ is concave up at $x = a$.

2 If $f''(a) < 0$, then $y = f(x)$ is concave down at $x = a$.

3 If $f''(a) = 0$, no conclusion is possible. The graph may be either concave up or concave down at $x = a$, or it may have a point of inflection.

The second derivative provides an easy test to determine whether a critical point is a maximum or a minimum.

SECOND DERIVATIVE TEST FOR CRITICAL POINTS

Let $x = a$ be a critical point of $y = f(x)$. Thus $f'(a) = 0$. Then

1 If $f''(a) > 0$, $x = a$ is a local minimum.

2 If $f''(a) < 0$, $x = a$ is a local maximum.

3 If $f''(a) = 0$, the test fails. In this case you should rely on the first derivative sign pattern chart to determine the character of the critical point.

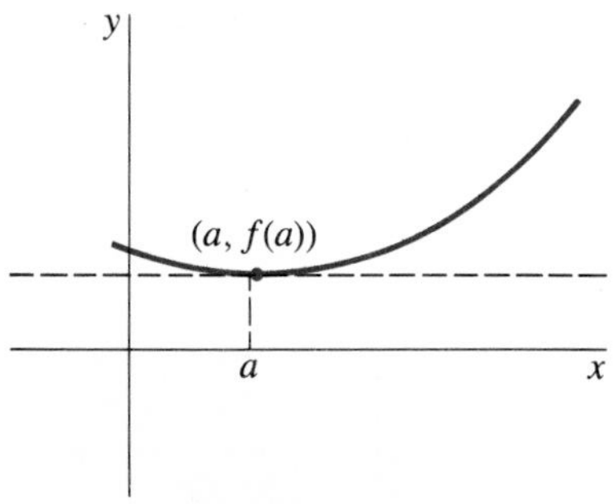

FIGURE 3.34

If $f'(a) = 0$ and $f''(a) > 0$, the curve has a horizontal tangent at $x = a$ and is concave up. Thus the picture must look somewhat like Figure 3.34. Clearly the function f has a local minimum at $x = a$. This proves item 1. Item 2 is proved in analogous fashion.

If you apply the second derivative test at a critical point $x = a$ and find that $f''(a) = 0$, the test is inconclusive. What you should do is go back to the first derivative test, which is always conclusive, and decide whether you have a local minimum $(----0++++)$, a local maximum $(++++0----)$. It is known that $(++++0++++)$ or $(----0----)$ indicates a point of inflection. (See Exercise 43.)

Because the first derivative test always gives an answer whereas the second derivative test is sometimes inconclusive, you may ask why one should ever use the second derivative test? The answer is expediency. It is often faster to compute $f''(x)$ and plug in a critical value a than to construct the sign pattern for $f'(x)$. So we often try the second derivative test first. If we get a conclusive answer, fine; if not, we must fall back on the sure-fire first derivative test.

Let us illustrate the use of the second derivative in curve sketching.

EXAMPLE 10 Sketch the graph of the equation

$$y = f(x) = 3x^4 - 4x^3$$

SOLUTION This is the same example we used to motivate the results of this section. We compute

$$f'(x) = 12x^3 - 12x^2 = 12x^2(x - 1)$$

$$f''(x) = 36x^2 - 24x = 12x(3x - 2)$$

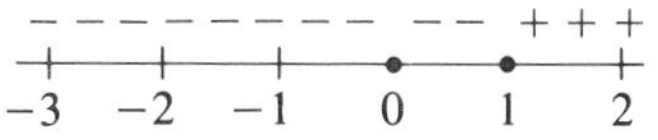

FIGURE 3.35
Sign pattern chart for f'

We see that

$$f' = 0 \quad \text{if } x = 0 \quad \text{or} \quad x = 1$$

These are the only critical points. Moreover

$$f''(0) = 0 \qquad f''(1) = 12 > 0$$

By the second derivative test, $x = 1$ must be a local minimum. The test yields no information about the point $x = 0$.

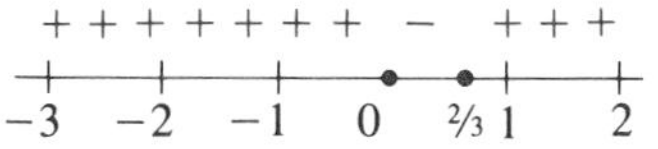

FIGURE 3.36
Sign pattern chart for f''

Our previous analysis determined the sign pattern chart of $f'(x) = 12x^2(x - 1)$ to be as shown in Figure 3.35. Because the curve is falling on both sides of $x = 0$, this point is neither a relative maximum nor a relative minimum. We also know that $f(0) = 0$, $f(1) = -1$, and the y-intercept is the point $(0, 0)$.

What additional information does $f''(x)$ give us? Let us construct a sign pattern chart for the second derivative in the various regions determined by its zeros in a manner similar to what we did for the first derivative. We find that

$$f'' = 0 \quad \text{if } x = 0 \quad \text{or} \quad x = \tfrac{2}{3}$$

We also find that f'' has the sign pattern shown in Figure 3.36. Thus f is

concave up if $x < 0$
concave down if $0 < x < \frac{2}{3}$
concave up if $x > \frac{2}{3}$

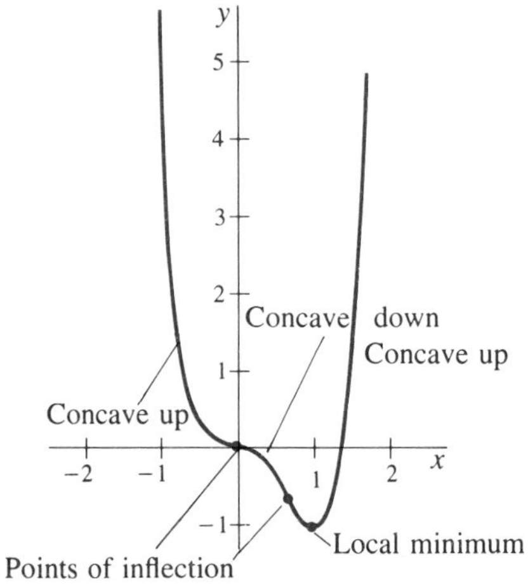

FIGURE 3.37

The points $x = 0$ and $x = \frac{2}{3}$ are therefore points of inflection. We find that $f(\frac{2}{3}) = 3(\frac{2}{3})^4 - 4(\frac{2}{3})^3 = -\frac{16}{27} \approx -0.6$. Combining all this into a single graph and noting that

$$\lim_{x \to \infty} (3x^4 - 4x^3) = \infty \quad \text{and} \quad \lim_{x \to -\infty} (3x^4 - 4x^3) = \infty$$

we have Figure 3.37. The dotted graph of Figure 3.30 is thus fairly accurate, except that we now know that $f' = 0$ when $x = 0$.

EXAMPLE 11 Sketch the graph of the function

$$f(x) = \frac{4}{4 + x^2}$$

SOLUTION

$$f'(x) = \frac{-4(2x)}{(4 + x^2)^2} = \frac{-8x}{(4 + x^2)^2}$$

$$f'' = \frac{(4 + x^2)^2(-8) - (-8x)2(4 + x^2)2x}{(4 + x^2)^4}$$

$$= \frac{24x^2 - 32}{(4 + x^2)^3}$$

Because $4 + x^2 > 0$ for all x, it follows that $f'(x) = 0$ only if $-8x = 0$; that is, if $x = 0$. Thus $x = 0$ is the only critical point.

We find

$$f(0) = \frac{4}{4 + 0^2} = 1$$

Moreover,

$$f''(0) = \frac{-32}{4^3} = -\frac{32}{64} = -\tfrac{1}{2} < 0$$

so that $x = 0$ is a local maximum according to the second derivative test.

The sign of

$$f'(x) = -\frac{8x}{(4 + x^2)^2}$$

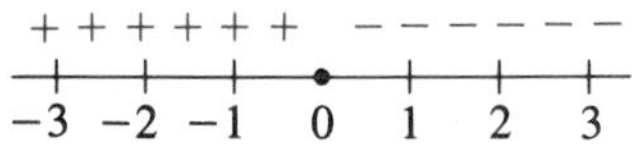

FIGURE 3.38
Sign pattern of f'

is determined by the sign of $-8x$ because $(4 + x^2)^2 > 0$. We therefore find the sign pattern shown in Figure 3.38 for f'. Thus f is increasing on $(-\infty, 0)$ and decreasing on $(0, \infty)$. Figure 3.39 indicates what we know about the graph up to this point.

The sign of

$$f''(x) = \frac{24x^2 - 32}{(4 + x^2)^3}$$

is determined by the sign of $24x^2 - 32 = 8(3x^2 - 4)$ because $(4 + x^2)^2 > 0$. So $f''(x) = 0$ only when $3x^2 - 4 = 0$; that is, when $x = \pm\sqrt{\frac{4}{3}} = \pm 2\sqrt{3}/3$. By computing values of $f''(x)$ for x's in the interval $(-\infty, -2\sqrt{3}/3)$, $(-2\sqrt{3}/3, 2\sqrt{3}/3)$, and $(2\sqrt{3}/3, +\infty)$, we can construct the sign pattern chart for $f''(x)$ shown in Figure 3.40.

Thus we see that $f'' > 0$ in the interval $(-\infty, -2\sqrt{3}/3)$ and in the interval $(2\sqrt{3}/3, +\infty)$, indicating that f is concave up on these intervals while $f'' < 0$ in the interval $(-2\sqrt{3}/3, +2\sqrt{3}/3)$, indicating that on this interval f must be concave down. The two points $x = \pm 2\sqrt{3}/3$ are points of inflection.

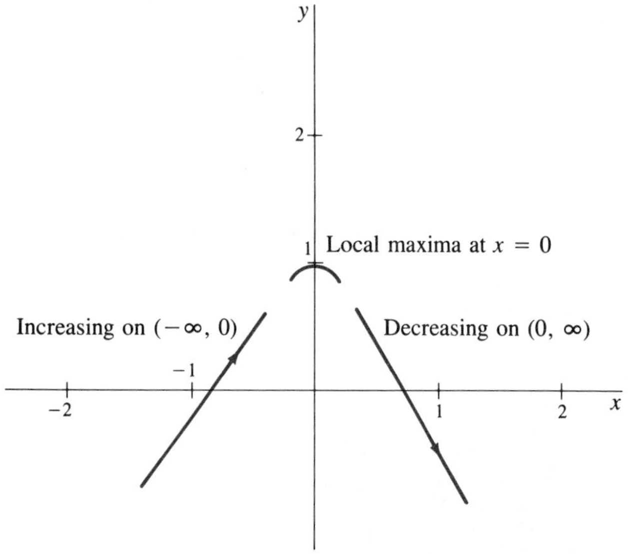

FIGURE 3.39

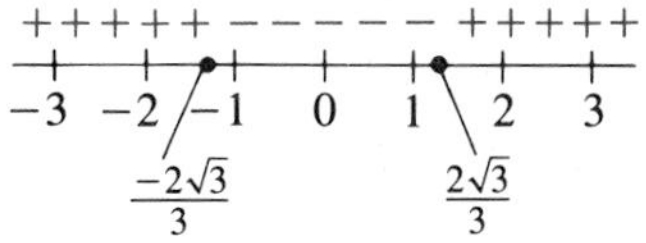

FIGURE 3.40
Sign pattern of f''

We easily compute

$$f\left(-\frac{2\sqrt{3}}{3}\right) = \frac{3}{4} \qquad f\left(\frac{2\sqrt{3}}{3}\right) = \frac{3}{4}$$

Before producing our final sketch, we investigate the asymptotic behavior of f. We have

$$\lim_{x \to +\infty} \frac{4}{4 + x^2} = +0$$

and

$$\lim_{x \to -\infty} \frac{4}{4 + x^2} = +0$$

Thus f has $y = 0$ as a horizontal asymptote on both the left and the right. Incorporating these features into our sketch yields Figure 3.41.

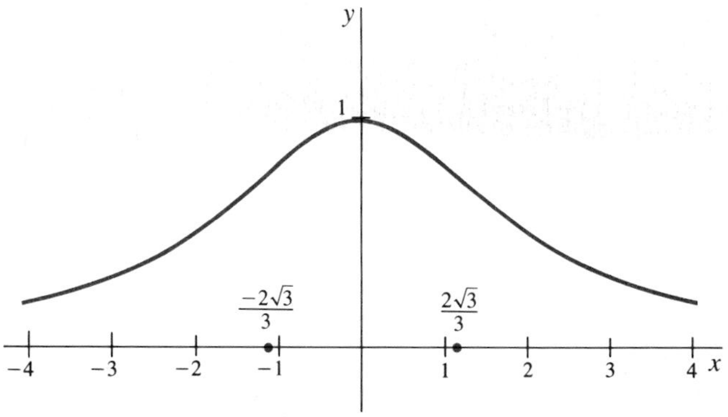

FIGURE 3.41

The next example introduces two features we have not encountered before.

EXAMPLE 12 Sketch the graph of the function

$$f(x) = x^{2/3}$$

SOLUTION We note first that

$$f(x) = x^{2/3} = (x^{1/3})^2 = [(-x^{1/3})]^2 = f(-x)$$

Thus f has the property that it produces the same value when evaluated at x and at $-x$. For example,

$$f(8) = 8^{2/3} = 4 = (-8)^{2/3} = f(-8)$$

From this we see that whatever the graph of f looks like to the right of the y-axis (that is, for $x \geq 0$) will be the mirror image of the graph to the left of the y-axis. The function f is said to be **symmetric** with respect to the y-axis. Thus it suffices to sketch the graph for $x \geq 0$ only. The rest of the graph can then be obtained by the use of symmetry.

Now

$$f'(x) = \frac{2}{3}x^{-1/3} = \frac{2}{3x^{1/3}}$$

$$f''(x) = -\frac{2}{9}x^{-4/3} = -\frac{2}{9x^{4/3}}$$

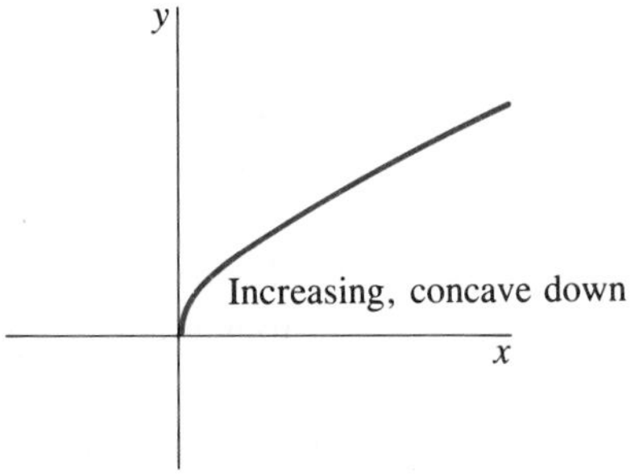

FIGURE 3.42
$y = x^{2/3} \quad x \geq 0$

Observe that, even though the function f is defined for all values of x, neither f' nor f'' is defined for $x = 0$. If $x > 0$,

$$f'(x) = \frac{2}{3x^{1/3}} > 0$$

so that f is increasing on $(0, \infty)$. As x approaches 0, $f'(x) \to \infty$. Also, if $x > 0$,

$$f''(x) = -\frac{2}{9x^{4/3}} < 0$$

so that f is concave down $(0, \infty)$. Furthermore,

$$\lim_{x \to \infty} x^{2/3} = \infty$$

so that f has no horizontal asymptotes. And, because f is defined for all x, there are no vertical asymptotes.

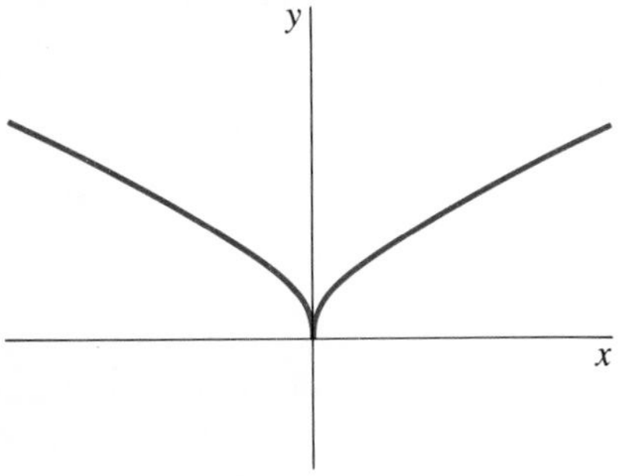

FIGURE 3.43
$y = x^{2/3}$

The y-intercept occurs at $x = 0$: $f(0) = 0$. Combining these into a single sketch, we obtain Figure 3.42. By symmetry, the complete graph is as shown in Figure 3.43. Observe that the graph has a minimum at $x = 0$. We were not able to determine this by using the second derivative test because $f'(0)$ and $f''(0)$ do not exist. In fact, the graph has a sharp spike at $x = 0$, where the tangent line is vertical. This is called a **cusp**.

In closing this section, let us assemble a complete graphing recipe that includes both our preliminary guide and the second derivative test.

COMPLETE GRAPHING GUIDE

To sketch the graph of $y = f(x)$

1. Compute $y' = f'$ and $y'' = f''$.
2. Factor y' if possible, and find the critical points—that is, those values of x for which $y' = 0$. Compute the value of y corresponding to each critical value $x = c$ and plot the points. Also plot points, if any, where f is defined but f' is not.
3. Examine each critical point $x = c$ to determine whether it gives a maximum or minimum (it may give neither). Do this either by the first derivative sign pattern or by the second derivative test. If you use the second derivative test, note that, if $f''(c) > 0$, then c is a local minimum and that, if $f''(c) < 0$, then c is a local maximum. But if $f''(c) = 0$, the test fails and you *must* go back to the first derivative test in order to classify $x = c$.
4. Find the sign of f' in each of the regions of the x-axis determined by the critical points, and use this information to indicate where $y = f(x)$ is rising and where it is falling.
5. Find the sign of f'' in each of the regions determined by its zeros. In any interval where f'' is >0, the curve is concave up; where f'' is <0, the curve is concave down. A point where the concavity changes is called an inflection point.
6. See what happens to the curve $y = f(x)$ as x tends toward $+\infty$ or x tends toward $-\infty$. The curve may tend toward $\pm\infty$ (as a polynomial does), or it may tend toward a horizontal asymptote.
7. Find the vertical asymptotes, if any, by looking for those values of x at which f is undefined. [This almost invariably occurs when the denominator of $f(x)$ vanishes.]
8. If necessary, compute some extra points on the graph.
9. Draw a smooth curve incorporating what you learned in steps 1 through 8.

One last example may help to synthesize these ideas.

EXAMPLE 13 Sketch the graph of a function $y = f(x)$ that satisfies all of the following conditions (a) through (f).

(a) $f(-3) = -2, f(-2) = 0, f(-1) = 1, f(0) = 0, f(1) = 2, f(2)$ does not exist, $f(4) = 0$

(b) $f'(-1) = 0$, $f'(0)$ does not exist, $f'(1) = 0$

(c) $f'(x) > 0$ if $x < -1$

$f'(x) < 0$ if $-1 < x < 0$

$f'(x) > 0$ if $0 < x < 1$

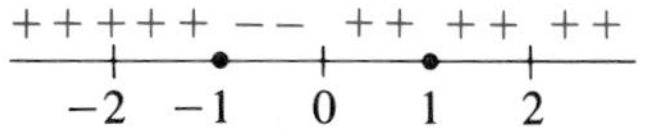

FIGURE FOR EXAMPLE 13, PART (c)
Sign pattern of f'

$f'(x) > 0$ if $1 < x < 2$

$f'(x) > 0$ if $2 < x$

(d) $f''(1) = 0$

$f''(x) < 0$ if $x < 0$

$f''(x) < 0$ if $0 < x < 1$

$f''(x) > 0$ if $1 < x < 2$

$f''(x) < 0$ if $2 < x$

(e) $\lim_{x \to \infty} f(x) = 1$, $\lim_{x \to -\infty} f(x) = -\infty$

(f) $\lim_{x \to 2^-} f(x) = \infty$, $\lim_{x \to 2^+} f(x) = -\infty$

The notation $\lim_{x \to c^-} f(x)$ means the limit as x approaches c from the left.

Similarly, $\lim_{x \to c^+} f(x)$ means the limit as x approaches c from the right.

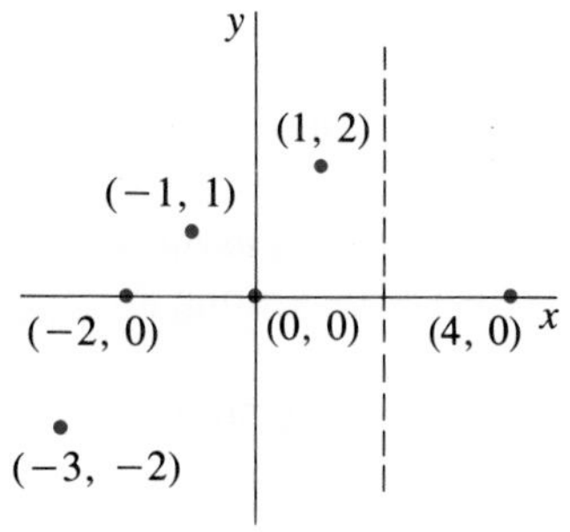

FIGURE 3.44

SOLUTION Condition (a) locates six points on the graph and indicates a probable vertical asymptote at $x = 2$ (see Figure 3.44).

Condition (b) tells us that $x = 1$ and $x = -1$ are critical points and that the graph probably has a sharp spike at $x = 0$. From condition (d) we know that $f''(x) < 0$ if $x < 0$. In particular, $f''(-1) < 0$. By the second derivative test, $x = -1$ is a local maximum. Also from condition (d), we know that $f''(1) = 0$, so that $x = 1$ is a possible point of inflection. This is confirmed by the fact that $f''(x) < 0$ if $0 < x < 1$ [f is concave down on $(0, 1)$] and that $f''(x) > 0$ if $1 < x < 2$ [f is concave up on $(1, 2)$]. See Figure 3.45.

Condition (c) reveals that f is increasing on $(-\infty, -1)$, $(0, 1)$, $(1, 2)$, and $(2, \infty)$ and that f is decreasing on $(-1, 0)$. See Figure 3.46.

We have already mentioned some of the information that we glean from condition (d). Additionally, it tells us that f is concave down for $x > 2$. See Figure 3.47.

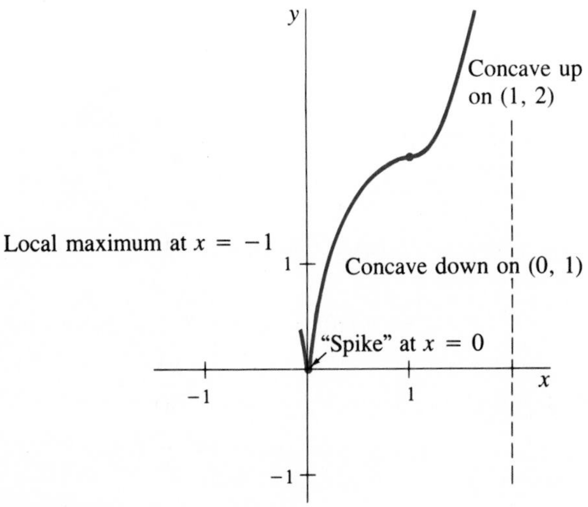

FIGURE 3.45

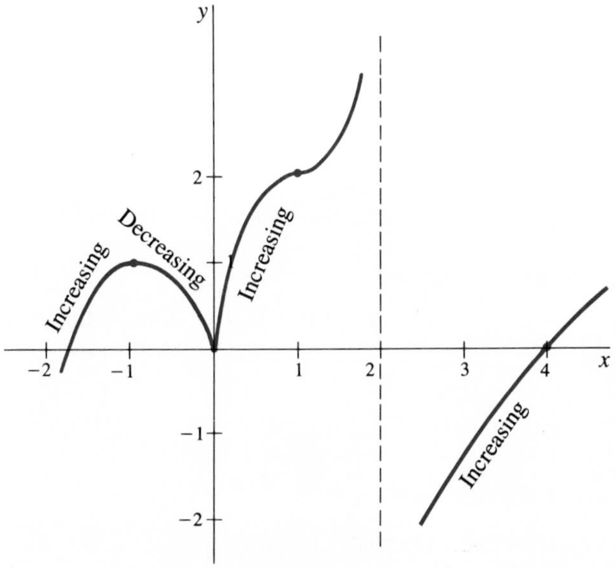

FIGURE 3.46

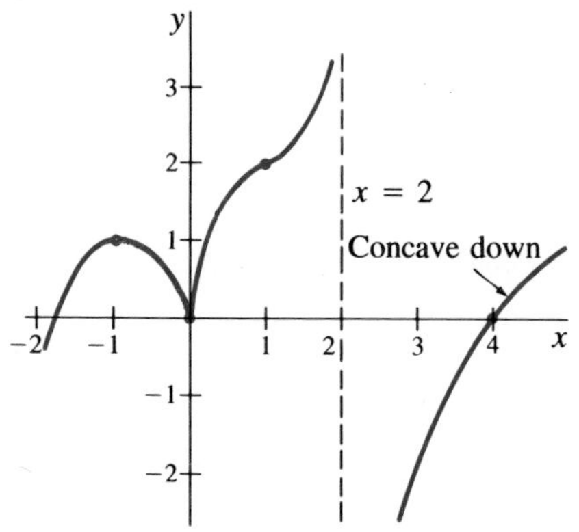

FIGURE 3.47

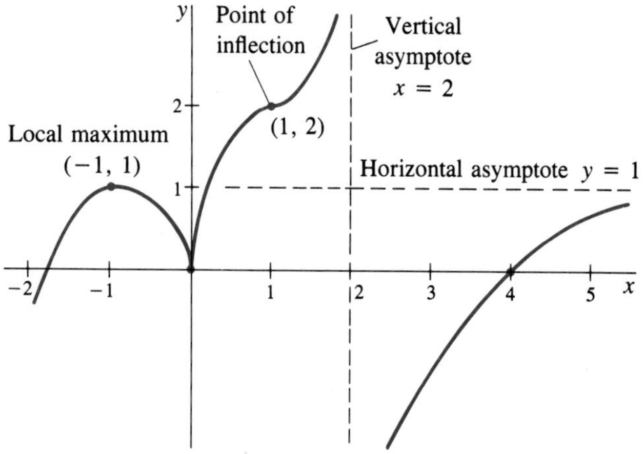

FIGURE 3.48

Condition (e) tells us that the graph has a horizontal asymptote to the right at $y = 1$ and that it moves downward as we move far along the graph to the left.

Condition (f) confirms the existence of a vertical asymptote at $x = 2$. Figure 3.48 shows the completed sketch.

Section 3.3 EXERCISES

For each of the functions in Exercises 1–29, sketch the curve carefully and discuss all its relevant features. You should deal with such items as rising and falling, maxima and minima, concavity up or down, inflection points, and asymptotes.

1. $2x^3 + 3x^2 - 12x - 10$

2. $\frac{x^3}{3} - \frac{x^2}{2} - 2x - 6$

3. $10 + x - 3x^3$

4. $x^2(4 - x)$

5. $(x + 3)(x^2 + 1)$

6. $(x - 1)(x - 2)(x - 3)$

7. $(x + 2)^3 - 1$

8. $x^4 + 2x^2 + 4$

9. $12 + 2x^2 - x^4$

10. $x^4 - 4x^3 - 2x^2 + 12x + 1$

11. $5x^3 - 3x^5$

12. $3x^8 - 4x^6$

13. $\frac{x - 2}{x + 1}$

14. $\frac{2x + 3}{5x + 7}$

15. $(x - 1)^2(x - 2)$

16. $\frac{(x - 1)}{(x - 2)^2}$

17. $\frac{6 + x - x^2}{(x - 2)^2}$

18. $\frac{x}{x^2 + 1}$

19. $\frac{x^2 + 1}{x^2 - 1}$

20. $\frac{x}{x^2 - 4}$

21. $\frac{(x - 2)^2}{(x - 3)^2}$

22. $\frac{x - 1}{(x - 2)(x - 3)}$

23. $\frac{8}{1 - x^2}$

24. $\frac{x^2}{1 - x^2}$

25. $x^2 + \frac{1}{x}$

26. $\frac{x^2}{(x - 2)(x + 3)}$

27. $x + \sqrt{x^2 - 1}$

28. $x\sqrt{x^2 - 4}$

29. $x^2 - \frac{4}{x}$

The curves in Exercises 30–34 approach slanted lines; that is, they have oblique asymptotes. To find this asymptote, use long division to rewrite the equation in the form

$$y = \frac{a}{x - c} + mx + b$$

Then, for large x, the line $y = mx + b$ is approached as an oblique asymptote. Find the oblique asymptote for each function.

30. $\dfrac{x^2 + x + 1}{x - 1}$ **31.** $\dfrac{(x - 1)^2}{x - 2}$ **32.** $\dfrac{x^2 - 4}{x}$

33. $\dfrac{x^2 + 1}{x}$ **34.** $\dfrac{x^2 + x - 5}{x + 1}$

In Exercises 35–42, sketch the graph of a function $y = f(x)$ that is consistent with the given data.

35. $f(3) = 2, f'(x) < 0$ for $x < 3, f'(x) > 0$ for $x > 3$

36. $f(-3) = 2, f'(x) > 0$ for $x < 3, f'(x) < 0$ for $x > 3$

37. $f(4) = 1, f'(x) > 0$ for all x

38. $f(0) = 1, f''(x) < 0$ for all x

39. $f(0) = 3, f(5) = -2, f''(x) < 0$ for $x < 3, f''(x) > 0$ for $x > 3$

40. $f(1) = 2$, $f(0)$ does not exist, $f(-x) = -f(x)$ for all x, $f'(1) = 0$, $f''(x) > 0$ for $x > 0$, $\lim_{x \to 0+} f(x) = \infty$, $\lim_{x \to \infty} f(x) = \infty$

41. $f(-3) = 5, f(-1) = 3, f(1) = 1, f'(-3) = 0, f'(1) = 0, f'(x) < 0$ for $-3 < x < 1, f'(x) > 0$ for $x < -3, f'(x) > 0$ for $x > 1, f''(x) < 0$ for $x < 0, f''(x) > 0$ for $x > 0$

42. $f'(x) = 2$ for $x > -1, f'(x) = 1$ for $x < -1, f(-1) = 0$

43. Consider the following functions:

(i) $f(x) = x^4$, (ii) $f(x) = -x^4$, (iii) $f(x) = x^3$

(a) Verify that in each case we have $f'(0) = 0$ and $f''(0) = 0$.

(b) Verify that: in case (i) there is a relative minimum at $x = 0$; in case (ii) there is a relative maximum at $x = 0$; in case (iii) there is neither a relative minimum nor a relative maximum at $x = 0$ (in fact, there is an inflection point at $x = 0$).

This illustrates why the second derivative test fails when $f''(a) = 0$ (see item 3 of the second derivative test for critical points) and also why $f''(a) = 0$ gives no information about concavity (see item 3 of the characterizations of concavity).

SECTION 3.4

Applied Optimization

Probably the most common objective of business management is to maximize profits and minimize losses. Problems in which we wish to maximize one quantity and/or minimize another are known as **optimization problems**. Other such problems spring to mind readily in economics: Executives strive to minimize taxes, maximize worker productivity, minimize transportation costs, and so on. But economics is by no means the only source of optimization problems. Architects wish to maximize the strength of structures; engineers try to minimize the failure rates of components in electrical devices; chemists attempt to maximize the rate of chemical reactions. Even nature pursues economy—the path of a ray of light through different media is such that its propagation time is minimized.

Thus determining maxima and minima can be an activity of great practical importance. When we are able to find a formula for the quantity to be optimized in terms of other variables, the techniques of calculus developed earlier in this chapter for curve sketching provide a method for locating extrema. Let us illustrate this contention by means of a few examples.

EXAMPLE 14 An open box is to be made from a square sheet of tin that is 12 inches on each side by cutting out a small square from each corner and bending up the sides. How large a square should be cut from each corner if the volume of the box is to be a maximum?

SOLUTION Let us denote by x the length of the sides of the small squares we are cutting out. Figure 3.49(a) shows the sheet of tin before it has been bent; Figure 3.49(b) shows the box formed by folding along the dotted lines.

From high school geometry we know that the formula for the volume of a rectangular box is

$$V = lwh$$

where l, w, and h represent the length, width, and height of the box, respectively. In our problem the length and width of the original flat sheet of tin are equal to 12 inches. After we fold up x inches from each side, the length and width of the box shown in Figure 3.49(b) are

$$l = w = 12 - 2x$$

The height of the box is simply the height of the portions we have folded up, so $h = x$. Substituting these values in the formula for V yields

$$V = (12 - 2x)(12 - 2x)(x) = 144x - 48x^2 + 4x^3$$

It is important to note that, even though this formula makes sense for any value of x, physically it makes sense only to consider

$$0 < x < 6$$

That is, it makes no sense to try to cut away a square whose sides have negative length, nor can we cut away more than we have to start with.

We can now rephrase our problem in the following way: Determine the value or values of x between 0 and 6 at which the function $V = 144x - 48x^2 + 4x^3$ assumes a maximum. When so formulated, this is a problem of the type we encountered when sketching curves. Taking the first derivative and setting it equal to 0 gives

$$V' = 144 - 96x + 12x^2 = 0$$

or

$$12 - 8x + x^2 = 0$$

This can be factored into

$$(x - 6)(x - 2) = 0$$

so that

$$x = 2 \quad \text{and} \quad x = 6$$

are critical points for V. We discard the value $x = 6$ as physically meaningless and retain $x = 2$ as the only critical point. Now,

$$V'' = -96 + 24x$$

so that

$$V''|_{x=2} = -96 + 24(2) = -48 < 0$$

Thus the graph of V as a function of x is concave downward at $x = 2$, so $x = 2$ is a relative maximum. It is also easy to convince yourself that $x = 2$ is actually an absolute maximum (see Figure 3.50). Our conclusion, then, is that we should cut a 2-inch square from each corner in order to achieve maximum volume.

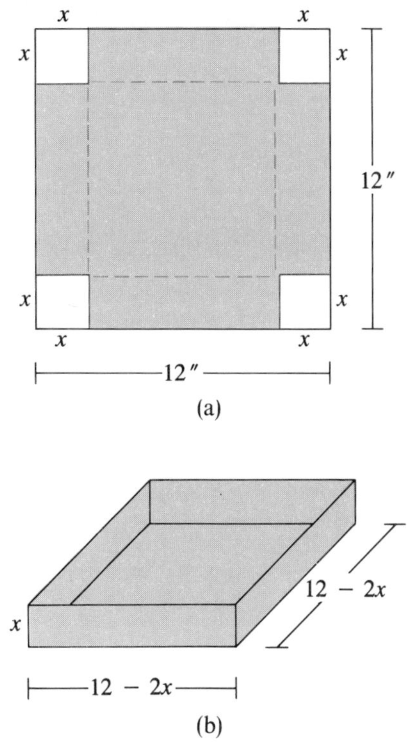

FIGURE 3.49

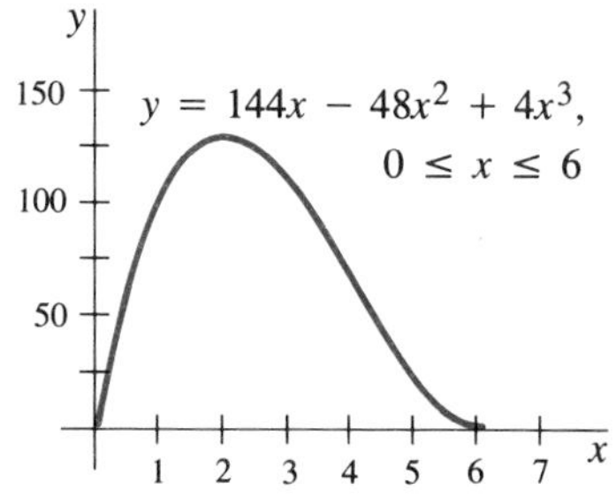

FIGURE 3.50

Let us abstract the steps in the solution of a maximum or minimum problem.

OPTIMIZATION GUIDE

1 If possible, draw a sketch illustrating the problem. Label the dimensions, assigning variables to those quantities that are free to vary.

2 Determine from the statement of the problem what quantity is to be maximized or minimized, and then write an equation for that quantity in terms of the other variables. (In Example 14 we were trying to maximize the volume V and wrote the equation $V = lwh$.)

3 Use any relationships you can find between the variables to write the quantity to be optimized in terms of a single variable. Determine any limitations on the size of that variable from the specific nature of the problem. (In our previous example, we expressed the length l, the width w, and the height h in terms of the single variable x:

$$l = 12 - 2x, \qquad w = 12 - 2x, \qquad h = x$$

Thus we wrote V as a function of x alone. Then we noted that $0 < x < 6$.)

4 Compute the first derivative of the resulting formula, and set it equal to 0 to find the critical points. Determine which of the points represent a relative maximum and which a relative minimum for the function. This may require using the first derivative test or the second derivative test, or it may be obvious from the concrete situation.

5 Keep in mind that the goal is to find an absolute extremum, not a relative extremum. It will sometimes be necessary to check the end points of the variable's range (which you determined in step 3) for end point extrema. Then you can determine an absolute maximum or minimum from among the relative and end point maxima and minima, respectively.

EXAMPLE 15 A homeowner has decided to curtail his spiraling food costs by planting a vegetable garden. He has chosen to locate his garden along the edge of a straight brook that runs through his property. He buys 50 feet of fencing to enclose his garden along the three sides that do not border the brook. What should be the dimensions of the garden if its area is to be as large as possible?

SOLUTION Denote by l and w the length and width, respectively, of the proposed vegetable garden. A sketch is shown in Figure 3.51. We wish to make the area A as large as possible. It is fairly simple to express the area in terms of the other variables:

$$A = lw$$

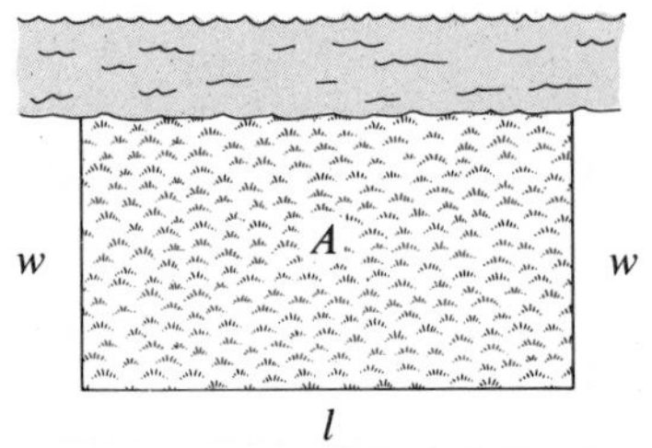

FIGURE 3.51

Our next step is to try to express the area as a function of one variable. This will require that we find an additional relationship between l and w. Now, because we know that our gardener has 50 feet of fencing, we must have

$$l + 2w = 50 \quad \text{or} \quad l = 50 - 2w$$

Substituting this into the formula for area yields

$$A = lw = (50 - 2w)w = 50w - 2w^2$$

Clearly the variable w is limited by the requirements that

$$0 < w < 25$$

Next we locate the critical points of A.

$$A' = 50 - 4w = 0$$

so that

$$w = \frac{50}{4} = 12.5$$

is the only critical point. From the relationship $l + 2w = 50$ we find that

$$l = 50 - 2(12.5) = 25$$

It remains only for us to determine whether the dimensions we have found furnish a maximum area. Because

$$A'' = -4$$

the graph of A as a function of w is concave downward. Thus $w = 12.5$ gives both a relative and an absolute maximum for A.

EXAMPLE 16 A lumber mill is being located at the east bank of a river that is 2 miles wide. A power station is located 5 miles downstream on the west bank. Power lines must be connected from the power station to the mill. The plan devised is to lay these lines underwater diagonally from the power station on the west bank to some point on the east bank and then overland northward to the mill. The cost of laying cable underwater is estimated at \$6000 per mile; the overland cost is \$3600 per mile. Locate the optimal point on the east bank for the cable to emerge from underwater so as to minimize the total cost.

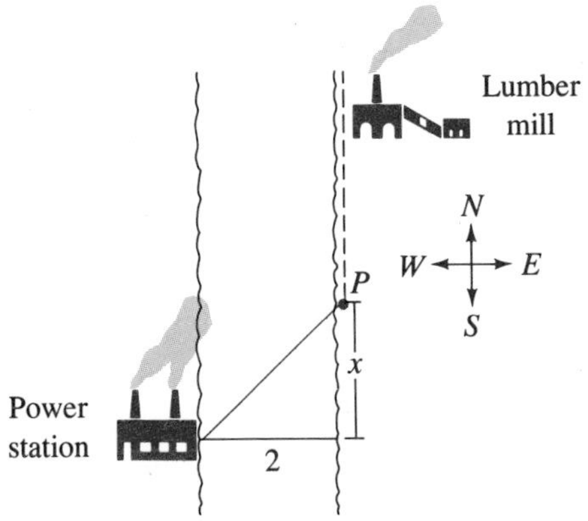

FIGURE 3.52

SOLUTION Let P denote the point on the east bank at which the cable emerges from the water. Let x denote the distance that point P is north of that point on the east bank directly opposite the power station. (see Figure 3.52). At \$3600 per mile the cost of laying the power line overland from P to the mill will then be \$3600 $(5 - x)$.

The underwater distance from the power station to the point P may be computed by the Pythagorean theorem as $\sqrt{2^2 + x^2} = \sqrt{x^2 + 4}$. At \$6000 per mile underwater, the cost of laying this portion of the power line is $\$6000\sqrt{4 + x^2}$.

Thus the total cost of laying the cable must be

$$C(x) = 3600(5 - x) + 6000\sqrt{4 + x^2}$$

In order to minimize this cost we first compute its derivative (with respect to x).

$$C'(x) = -3600 + \frac{1}{2}(6000)\frac{2x}{\sqrt{4 + x^2}}$$

$$= -3600 + \frac{6000x}{\sqrt{4 + x^2}}$$

Setting $C'(x)$ equal to 0, we obtain $3600 = \dfrac{6000x}{\sqrt{4 + x^2}}$ or, on simplification,

$$3\sqrt{4 + x^2} = 5x$$

Squaring both sides, we find that

$$9(4 + x^2) = 25x^2$$

so that $36 = 16x^2$. Hence

$$x^2 = \frac{36}{16}$$

or $x = \frac{6}{4} = \frac{3}{2} = 1.5$ miles. (The negative solution for x is discarded.)

We conclude that the optimal point for the cable to emerge from underwater is on the east bank $1\frac{1}{2}$ miles north of the point directly opposite the power station—that is, $3\frac{1}{2}$ miles south of the lumber mill.

One question still remains: How can we tell whether our answer represents a maximum cost or a minimum cost? Of course, we could resort to the second derivative test. In this problem, however, we can rely on our physical intuition to guide us. It is clear that our answer cannot represent a maximum because we can always incur *more* cost by locating point P far to the north or far to the south. Thus we are safe in saying that the *sole* critical point of C yields a minimum. (The first derivative test could also be used to corroborate our intuition.)

EXAMPLE 17 A tin can is to hold 12 cubic inches of a liquid. What dimensions for the can will require the smallest amount of tin? (Ignore the thickness of the can.)

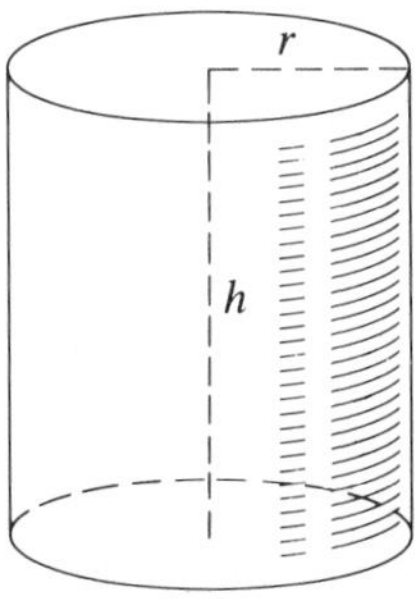

FIGURE 3.53

SOLUTION Let r = radius of the can, and let h = height of the can (see Figure 3.53). We want to minimize the amount of tin used to make the can. Because we are assuming that the can has negligible thickness, the amount of tin needed is proportional to the surface area. The total surface area of the can is the area of its top, bottom, and sides. The top and bottom are circles of radius r, so each has area

$$\pi r^2$$

In order to determine the surface area of the rest of the can, imagine that, after removing its top and bottom, we cut it vertically and then unroll it into a flat rectangular sheet (see Figure 3.54). The height of the resulting sheet is h, and its length is the circumference of the can, which is $2\pi r$. Thus the sheet has area

$$(2\pi r)h$$

Combining these various areas, we find that the surface of the can is

$$A = \underbrace{\pi r^2}_{\text{area of top}} + \underbrace{\pi r^2}_{\text{area of bottom}} + \underbrace{2\pi rh}_{\text{total area of side}} = 2\pi r^2 + 2\pi rh$$

In order to minimize A, we must write it as a function of a single variable rather than as a function of both r and h. This necessitates finding another

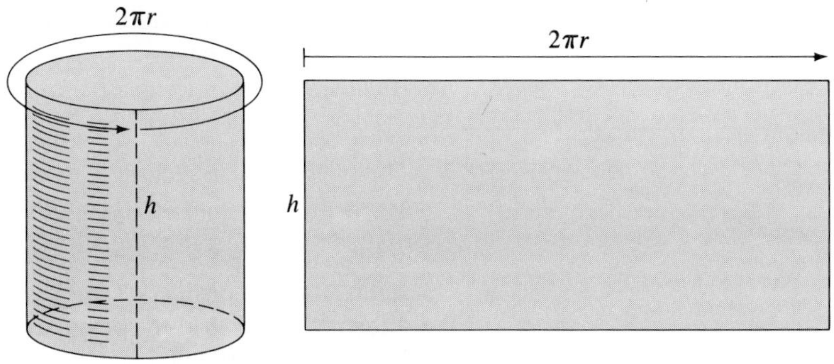

FIGURE 3.54

relationship between the variables. Well, we know that the can is to have a volume of 12 cubic inches. And because the volume of a right circular cylinder is given by $V = \pi r^2 h$, we have

$$\pi r^2 h = 12$$

or

$$h = \frac{12}{\pi r^2}$$

Substituting this into the formula for area yields

$$A = 2\pi r^2 + 2\pi r\left(\frac{12}{\pi r^2}\right) = 2\pi r^2 + \frac{24}{r}$$

Now, setting $A' = 4\pi r - \frac{24}{r^2}$ equal to 0 gives

$$r^3 = \frac{24}{4\pi}$$

and

$$r = \sqrt[3]{\frac{24}{4\pi}} \approx 1.24 \text{ inches}$$

We can check that this is a minimum by computing

$$A'' = 4\pi + \frac{48}{r^3}$$

which is > 0 for all $r > 0$. Finally,

$$h = \frac{12}{\pi r^2} \approx \frac{12}{\pi(1.24)^2} \approx 2.48 \text{ inches}$$

Thus a tin can of volume 12 cubic inches that will use the least amount of tin should have a radius $r \approx 1.24$ inches and height ≈ 2.48 inches.

You may be curious as to whether tin can manufacturers really design cans (which are actually made of steel, not tin) in this way. The answer, in general, is that they do not! That's because cans are usually packed in cardboard cartons that come in certain standard sizes. The cost to the manufacturer for the production of special-sized cartons would often more than offset the cost of additional raw material for a less than optimal can. Thus the dollars-and-cents reasoning of business frequently dictates practices that are not optimal in terms of the conservation of natural resources.

EXAMPLE 18 The strength of a rectangular beam of a given length is proportional to its width and the square of its depth. Find the dimensions of the strongest beam that can be cut from a log whose cross section is a circle 1 foot in diameter.

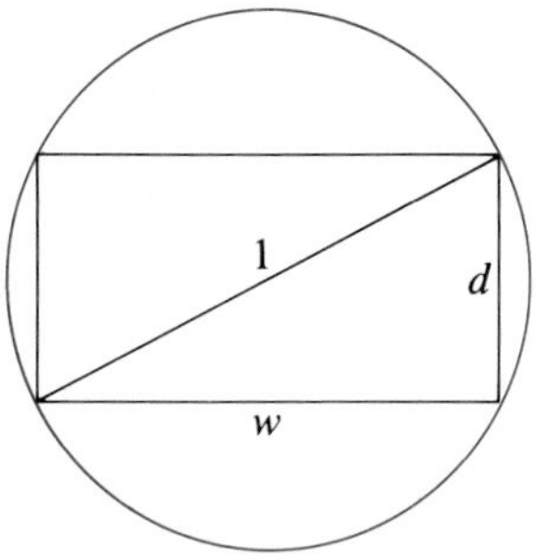

FIGURE 3.55

SOLUTION Consider a cross section perpendicular to the axis of the log. Let d denote the depth of the proposed beam, and let w denote its width (see Figure 3.55). We wish to maximize the strength of the beam S, which is given by

$$S = kwd^2$$

where k is the positive proportionality constant. We have written S as a function of the two variables w and d and would like to find another relationship between these variables. From the Pythagorean theorem,

$$d^2 + w^2 = 1$$

so that

$$d^2 = 1 - w^2$$

Substituting this into the formula for S yields

$$S = kw(1 - w^2) = kw - kw^3$$

where we must have $0 < w < 1$. Differentiating, we have

$$S' = k - 3kw^2 = 0$$

which implies that

$$w^2 = \frac{1}{3}$$

so

$$w = \sqrt{\tfrac{1}{3}} \approx 0.58 \text{ feet}$$

Because

$$S'' = -6kw$$

is negative for all w between 0 and 1, the graph of S as a function of w is concave downward on this interval, and our function has a maximum when $w = \sqrt{\frac{1}{3}}$. We may find the depth of the proposed beam by making use of the auxiliary equation

$$d^2 = 1 - w^2 = 1 - \frac{1}{3} = \frac{2}{3}$$

$$d = \sqrt{\frac{2}{3}} \approx 0.82 \text{ feet}$$

Note that the optimal dimensions in no way depend on the proportionality constant k.

Section 3.4 EXERCISES

1. The sum of two positive numbers is 24. What is the biggest value their product can take on? What are the two numbers in this situation?

2. Find the smallest value taken on by the sum of a positive number and its reciprocal.

3. Find the number that is greater than its square by the largest amount.

4. The sum of two positive numbers is to be 6. How should they be chosen in order to minimize the sum of the first and the square of the second?

5. The product of two positive numbers is a. What is the maximal value of their sum? What is the minimal value of their sum?

6. A rectangular plot for a small garden is to have a perimeter of 48 ft. What should its dimensions be in order to maximize the area?

7. A farmer has 500 yd of fencing material and wants to fence in a rectangular field, one of whose sides is to run along a straight stream (and hence will require no fencing). How big an area can he fence in?

8. Do the preceding problem when the given length of fencing is the constant c.

9. Prove that, of all rectangles with a given perimeter, the square has maximal area.

10. Prove that, of all rectangles with a given area, the square has minimal perimeter.

11. You have 150 in. of molding to go around the outer edge of a window whose shape is to be a rectangle topped by a semicircle. (Such a window is sometimes called a Norman window.) Show that, if you wish to maximize the area of the window, then r, the radius of the semicircle, should be $r = \dfrac{150}{4 + \pi}$. What is the height of the rectangle in this situation?

12. Show that, if the given length of molding in Exercise 11 is p inches, the maximal area occurs when $r = \dfrac{p}{4 + \pi}$.

13. What happens in Exercise 11 (and in Exercise 12) if you wish to minimize the area of the window?

14. Consider a semicircle of radius a, and inscribe a rectangle in it (with two vertices on the diameter and two vertices on the semicircle). What is the biggest area this rectangle can have?

15. Find the dimensions of the rectangle of maximal area that can be inscribed in the circle of radius a. What is its area?

16. Find the maximal area of a rectangle that has its base on the x-axis and the two upper vertices on the parabola $y = 16 - x^2$.

17. Do Example 17 under the assumption that the top and bottom of the can (each of them) are cut from a square piece of tin, and the waste material is counted as being used to make the can.

18. What dimensions of a tin can (in the usual shape of a right circular cylinder) minimize the amount of tin required to enclose a fixed volume V? (Show that $h = 2r$.)

19. A string of length l is cut in two. One piece is shaped into a square and the other into a circle. Where should the string be cut if the sum of the two areas is to be a maximum? A minimum?

20. Do Exercise 19 when the two pieces of string are shaped into a square and an equilateral triangle.

21. Do Exercise 19 when the two pieces of string are shaped into a circle and an equilateral triangle.

22. A straight line of negative slope is to be drawn through the point (2, 3) so that the triangle cut off in the first quadrant has minimal area. Find the straight line.

The distance from a given point to a given curve is defined as the shortest distance from the given point to an arbitrary point on the curve.

23. Using calculus, find the distance from the point $(-1, 2)$ to the straight line $y = 3x + 2$.

24. Using calculus, find the distance from the point $(-1, 2)$ to the straight line $2x + 3y = 12$.

25. What is the distance from each of the following points to the parabola $y = x^2$? In each case, find the nearest point on the parabola.
(a) $(0, -1)$ (b) $(0, 1)$ (c) $(0, \frac{1}{2})$ (d) $(0, \frac{1}{4})$
(e) (0, b) (You may find it necessary to distinguish the cases $b < \frac{1}{2}$ and $b \geq \frac{1}{2}$.)

26. Suppose the Post Office will accept only packages whose length plus girth is at most 72 in. How big a volume can you mail in a single rectangular carton? (Assume a square cross section. This exercise will reappear in Sections 8.3 and 8.5 *without* this assumption.)

27. Find the length of the shortest ladder that will reach from the ground to the side of a building if there is a wall 8 ft high parallel to the building and 5 ft from it.

28. You are given a fixed amount of material to use in constructing a sand box with a square base. What should the dimensions be in order to maximize the amount of sand it can hold?

29. A sheet of metal of area A is to be made (without waste) into a closed box whose base is square. What should the dimensions be if we want to maximize the volume?

30. You are given a square sheet of aluminum whose side is a and are asked to cut out equal-sized squares from each of the four corners. The sides are then to be bent up to form a box with no top. If the volume of this box is to be maximized, what size squares should you cut out?

31. We are given a piece of cardboard whose area is a square units, and we want to construct a carton. The base of the carton is required to be a rectangle whose length is 50% larger than its width. Find what the dimensions should be in order to maximize the volume if the carton: (a) has no top; (b) has a top.

32. A publisher decides that, for esthetic reasons, the pages of a book should have margins of $1\frac{1}{2}$ in. each on top and bottom and 1 in. on each side. A psychologist advises the publisher that the optimal amount of printed matter per page is 35 in.^{2}. Paper is rather expensive, so the publisher seeks to minimize the amount of paper used. What should the publisher choose as the dimensions of a page?

33. Let us find the distance from the point (x_0, y_0) to the line $y = mx + b$. To do this via calculus, we set up the distance s from (x_0, y_0) to an arbitrary point (x, y) on the line and then minimize s.
   (a) Show that s^2 is minimized (this is clearly equivalent to minimizing s) when the point (x, y) satisfies the condition $\dfrac{y - y_0}{x - x_0} = \dfrac{-1}{m}$—in other words, when (x, y) is also on the line through (x_0, y_0) with slope $\dfrac{-1}{m}$. (This line is, of course, the line through (x_0, y_0) that is perpendicular to the given line $y = mx + b$.)
   (b) Show that for this line we have
$$s^2 = (x - x_0)^2\left(1 + \frac{1}{m^2}\right)$$
and that, because this x equals $\dfrac{x_0 + my_0 - mb}{1 + m^2}$, it follows that the distance s from (x_0, y_0) to the line $y = mx + b$ is
$$s = \frac{|mx_0 - y_0 + b|}{\sqrt{1 + m^2}}$$

34. Prove, via calculus, that the distance from the point (x_0, y_0) to the line $Ax + By + C = 0$ is given by the formula
$$\frac{|Ax_0 + By_0 + C|}{\sqrt{A^2 + B^2}}$$

SECTION 3.5

Applications to Business and Economics

In Section 2.3 we introduced the "marginal" concept from economics. It may be helpful to reread that material now.

EXAMPLE 19 The **total revenue function** for a seller is the function that gives the total dollar receipts if a certain quantity x of the commodity is sold. Suppose the revenue function for a particular manufacturer is given by
$$R(x) = 500x + x^2 - 0.1x^3$$
Compute the marginal revenue when $x = 10$.

SOLUTION $R'(x) = 500 + 2x - 0.3x^2$ so $R'(10) = 490$

This means that the additional revenue to the manufacturer for one additional unit of production, provided that the manufacturer is currently producing 10 items, will be approximately \$490.

Marginal rates of change, by themselves, are of limited usefulness. Their importance lies in the fact that they enable the businessman to make management decisions in such a way as to maximize profits. In any company,

$$\text{Profit} = \text{revenue} - \text{cost}$$

or

$$P(x) = R(x) - C(x)$$

In order to maximize profit, we set its first derivative equal to 0. (Of course, we then check to see that the critical points we determine represent a maximum, not a minimum.) Thus

$$\frac{dP}{dx} = \frac{dR}{dx} - \frac{dC}{dx} = 0$$

or, equivalently,

$$\frac{dR}{dx} = \frac{dC}{dx}$$

This is a well-known theorem in economics: Profit is maximized when marginal revenue equals marginal cost.

EXAMPLE 20 Recall that an equation that relates the market price per unit, p, to the quantity demanded, x, is called a **demand function**. Typically, the greater the number of units produced, the cheaper they must be priced in order to sell. For example, a typical demand function might be

$$p(x) = 80 - 0.2x$$

Hence a manufacturer who wishes to produce and sell $x = 20$ units must price each of them at

$$p(20) = 80 - (0.2)(20) = 76$$

The revenue produced by the sale of x units priced at $p(x)$ dollars per unit is simply

$$R(x) = x \cdot p(x)$$

Suppose that the corresponding cost function is

$$C(x) = 200 + 20x$$

Find the production level that maximizes profits.

SOLUTION We know that

$$\underset{\substack{\uparrow \\ \text{profit}}}{P} = \overset{\substack{\text{revenue} \\ \downarrow}}{R} - \underset{\substack{\uparrow \\ \text{cost}}}{C}$$

and we are given the cost function $C(x) = 200 + 20x$. Our revenue function is

$$R(x) = x(80 - 0.2x) = 80x - 0.2x^2$$

Therefore, the profit function $P = R - C$ is

$$P = P(x) = (80x - 0.2x^2) - (200 + 20x)$$
$$= 60x - 0.2x^2 - 200$$

Setting the derivatives of P equal to 0, we have

$$P' = 60 - 0.4x = 0$$

and $$x = \frac{60}{0.4} = 150$$

If our manufacturer produces 150 units, the profit will be

$$P(150) = 60(150) - 0.2(150)^2 - 200 = \$4300$$

This must be a maximum, because

$$P''(x) = -0.4 < 0$$

which says that the graph of the profit equation is always concave downward.

EXAMPLE 21 The demand equation for a particular manufacturer is given by

$$p(x) = 500 + x - 0.1x^2$$

and the cost equation is determined to be

$$C(x) = 5000 + 200x + x^2$$

What output should our manufacturer produce and sell in order to maximize profits?

SOLUTION As in Example 20,

$$R(x) = (\text{number of units sold})(\text{price per unit})$$
$$= xp(x) = 500x + x^2 - 0.1x^3$$

The profit P is given by

$$P(x) = R(x) - C(x)$$
$$= (500x + x^2 - 0.1x^3) - (5000 + 200x + x^2)$$
$$= 300x - 0.1x^3 - 5000$$

Thus

$$P'(x) = 300 - 0.3x^2 = 0$$

gives

$$x^2 = 1000$$

and $$x = \sqrt{1000} \approx 31.6 \text{ units}$$

Of course, you should verify that this gives a maximum.

PRACTICE EXERCISE (a) In Example 20, if we set $x = 0$ in the cost equation, we find that

$$C(0) = 200$$

This represents the manufacturer's fixed overhead, which is a cost incurred whether or not the manufacturer makes anything. Suppose that the fixed overhead is \$500 instead of \$200. Find the production level that maximizes profit, and find the maximum profit.

(b) Repeat part (a) when the fixed overhead is \$1000. What can you infer from this about the danger of having large fixed operating expenses?

[Your answers should be: (a) $x = 150, P = 4000$; (b) $x = 150, P = 3500$.]

EXAMPLE 22 A theater owner charges \$4 for a ticket to see a movie. He averages 1000 customers per week. He has experimented with increasing his ticket prices and has found that each \$1 increase in the price results in a drop of 150 patrons. What ticket price will maximize revenue?

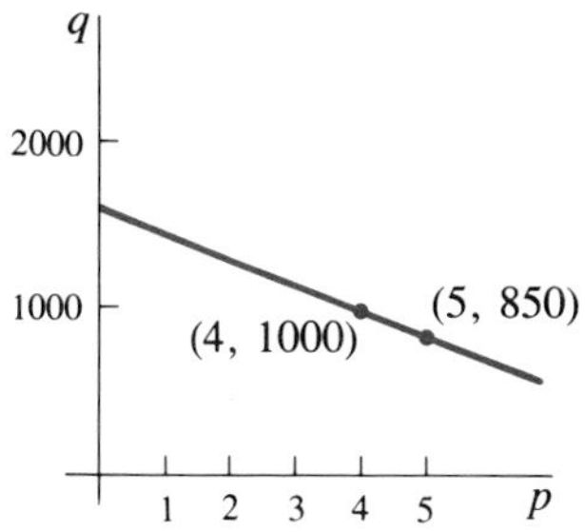

FIGURE 3.56

SOLUTION As a first step, let us determine the demand function for movie tickets. We know that 1000 customers attend the show weekly when the price is \$4 and that only 850 people attend if the price is increased to \$5. Utilizing the methods of Section 1.1, we can determine the equation of the straight line through the points (4,1000) and (5,850). See Figure 3.56. Its slope is

$$\frac{850 - 1000}{5 - 4} = -150$$

and its equation is given by

$$q - 1000 = -150(p - 4)$$

or

$$q = -150p + 1600$$

Here we denote the quantity of tickets demanded by q and consider p the independent variable. We know that revenue R is bound by multiplying the number of units sold by the price per unit, so

$$R = -150p^2 + 1600p$$

Thus (keep in mind that p is the variable),

$$R' = \frac{dR}{dp} = -300p + 1600 = 0$$

which gives $p = 5\frac{1}{3} \approx \$5.33$. This price maximizes revenue. The number of patrons at this price is

$$q(16/3) = -150\left(\frac{16}{3}\right) + 1600 = 800$$

ELASTICITY OF DEMAND The concept of **elasticity of demand**, which is related to the concept of marginal revenue, is a very useful one in economic analysis. It is basically a measure of the responsiveness of the quantity of goods sold to a change in the price of those goods.

Suppose we are given a demand curve for a particular good or service. At first glance it might seem that the slope of the demand curve describes the degree to which the level of demand responds to price changes. Further reflection, however, reveals that this is not the case. For example, suppose we are comparing the demand curve for automobiles with the demand curve for wheat, and we want to know which good is more responsive to a change in price. Comparing the slopes of the two demand curves tells us nothing. A $1 drop in the price of wheat may increase the quantity sold by 20 million bushels per month. A $1 decrease in the price of automobiles may increase sales by only one automobile per month. But this does not mean that wheat is more responsive to price changes than automobiles. After all, a $1 change in the price of wheat is a very large relative change, whereas a $1 change in the price of automobiles is of little consequence. Furthermore, a bushel of wheat and an automobile are very different entities, and there is little basis for comparing a unit of one with a unit of the other.

To overcome this difficulty, the economist Alfred Marshall defined **(price) elasticity of demand** as the *percentage* change in demand divided by the *percentage* change in price, when these changes are small. If Δq represents a small change in the demand q, and Δp a small change in the price p, then the elasticity of demand E is given by

$$E = \lim_{\Delta p \to 0} \frac{\dfrac{\Delta q}{q}}{\dfrac{\Delta p}{p}} = \lim_{\Delta p \to 0} \frac{\dfrac{p}{q}}{\dfrac{\Delta p}{\Delta q}} = \frac{\dfrac{p}{q}}{\dfrac{dp}{dq}}$$

EXAMPLE 23 The demand equation for sugar in the United States from 1915 to 1929 was estimated* to be given by

$$p = 14.7 - 0.1q$$

Find the elasticity of the demand for sugar when:

(a) $q = 80$; (b) $q = 100$; (c) $q = 50$

SOLUTION We readily compute $dp/dq = -0.1$. Thus the elasticity of demand, E, is given by

$$E = \frac{p/q}{-0.1} = \frac{14.7 - 0.1q}{-0.1q} = 1 - \frac{147}{q}$$

Thus:

(a) $E(80) = -0.8375$

(b) $E(100) = -0.47$

(c) $E(50) = -1.94$

Observe that the elasticity of demand in this example varies at different levels of demand. In general, if $|E| > 1$, the demand is said to be **elastic**; if $|E| < 1$ it is **inelastic**, and if $|E| = 1$ it is **unitary**.

* Henry Schultz, *Statistical Laws of Demand and Supply with Special Applications to Sugar* (Chicago: University of Chicago Press, 1928).

An example of an elastic good might be low-priced ball point pens. A given type of pen might sell quite well when priced at 49¢ each. But, because of competition from a large number of similar pens, it might sell quite poorly at 69¢ each. On the other hand, the demand for appendectomies is quite inelastic. The number of such operations does not depend strongly on the price that is charged to perform the surgery.

EXAMPLE 24 The demand equation for potatoes in the United States from 1915 to 1929 was estimated* to be given by

$$p = 2630/q^3$$

Find the elasticity of demand when $q = 3$.

SOLUTION We compute

$$\frac{dp}{dq} = \frac{-7890}{q^4} \quad \text{so that} \quad E = \frac{\dfrac{p}{q}}{\dfrac{dp}{dq}} = \frac{2360/q^4}{-7890/q^4} = -\frac{1}{3}$$

We close this section by showing the relationship between marginal revenue and elasticity of demand. Recall that the total revenue R is defined as

$$R = qp$$

Because p is a function of q, we can compute the marginal revenue by applying the product rule:

$$R' = \frac{dR}{dq} = q\frac{dp}{dq} + p$$

This expression may be rewritten in the form

$$R' = \left[\frac{q}{p}\frac{dp}{dq} + 1\right]p$$

But

$$\frac{q}{p}\frac{dp}{dq} = \frac{\dfrac{dp}{dq}}{\dfrac{p}{q}} = \frac{1}{E}$$

so that

(3.1) $$R' = p\left(1 + \frac{1}{E}\right)$$

* Henry Schultz, *The Theory and Measurement of Demand* (Chicago: University of Chicago Press, 1938).

In other words, the marginal revenue is the price times (1 plus the reciprocal of the elasticity of demand).

EXAMPLE 25 The demand function for cotton in the United States from 1915 to 1929 was estimated to be given by

$$p = 8 - q$$

Verify Equation (3.1) for this demand function.

SOLUTION The total revenue R is given by

$$R = qp = q(8 - q) = 8q - q^2$$

Thus the marginal revenue is

$$R' = 8 - 2q$$

On the other hand, $dp/dq = -1$, so the elasticity of demand is given by

$$E = \left(\frac{p}{q}\right)(-1) = -p/q$$

Equation (3.1) requires that

$$R' = p\left(1 + \frac{1}{E}\right)$$

Computing the right-hand side of this equation, we obtain

$$p\left(1 + \frac{1}{E}\right) = (8 - q)\left(1 - \frac{q}{p}\right) = (8 - q)\left(1 - \frac{q}{8 - q}\right)$$
$$= 8 - q - q = 8 - 2q = R'$$

as required.

Section 3.5 EXERCISES

1. A sand box, which is to contain 64 ft^3, is required to have a square base (and, of course, no top). What should its dimensions be if the cost of material used to build it (that is, for the base and sides) is to be a minimum?

2. Do the preceding problem if the material for the base costs $\$2/ft^2$ and the material for the sides costs $\$3/ft^2$.

3. A discount store owner can buy color televisions at \$200 each. He normally sells 20 per week at \$400 each. Furthermore, for each \$5 reduction in selling price, he can sell an additional one per week. What selling price will maximize his profit per week?

4. You need to have a wall of area 10,000 ft^2 painted. A painter can paint 50 ft^2/hr and he gets paid \$8/hr. It is also necessary to have a supervisor (who does no painting); he gets paid \$12/hr. And you have additional expenses of \$7 for each painter hired. How many painters should you hire in order to minimize your total cost?

5. A printing company has 10 presses each of which can print 10,000 copies/hr. It costs \$25 to set up a single press for a run and $50 + 5x$ dollars to run x presses for 1 hr. How many presses should be used to print 500,000 copies of a booklet most cheaply?

6. An archeological team has reached a layer of rock at a depth of 75 ft. From this point the team wants to dig a tunnel that will reach a point 100 ft down and 300 ft north of the given point. If the cost of digging a tunnel is \$120/ft

(of length) through ordinary earth and \$200/ft through rock, what is the minimal cost for digging the tunnel?

7. Suppose a company is strip-mining coal. Experience indicates that the cost is proportional to the depth at which the mining is done and that the value of the mined product is proportional to the square root of the depth. If the cost is \$2.40/yd^3 at a depth of 16 ft, and the value of the product is \$6.00/yd^3, at which depth should a profit-oriented company mine?

8. A business that produces hand-crafted items has a capacity of 250 per week. The owner estimates that he can sell x of them at a price of $1000 - 3x$ each and that the cost of producing these x items is $500 + 167x + 2x^2$. How many should he produce per week in order to maximize his profit? (The final answer must, of course, be an integer.)

9. A retail store manager can buy a certain gadget for 17¢ each. At what price, p¢, should she mark them if she estimates that the number she will sell (which depends on p) is

$$\frac{10{,}000}{p - 17} + 2(75 - p)$$

10. A pipe manufacturer can produce x first-quality pipes per week and y seconds per week, where x and y are related by the equation

$$y = 120 + \frac{40{,}000 - 80x}{600 - x} \qquad 0 \le x \le 560$$

If the profit to be made on a first-quality pipe is 3.2 times the profit to be made on a second, how many pipes of each kind should the manufacturer make?

11. A finance company borrows money from the public and can then lend everything it borrows. It cannot legally lend money at an interest rate of more than 20% per year, and the amount it can borrow from the public is proportional to the rate of return that it offers the investor. What interest rate (that is, rate of return) should it offer the investing public in order to maximize its profit?

12. A motel with 100 rooms can fill all of them if it charges \$120 per day. For every \$10 that the motel raises its price, 5 rooms become vacant. What is the optimal (for the owner) charge per day if the upkeep is \$30 per day for a vacant room and \$50 per day for an occupied room?

13. It costs a manufacturer $C(x)$ hundred dollars to produce x hundred items, and they can all be sold for a total revenue of $R(x)$ hundred dollars. Which choice of x maximizes profit when:

$$C(x) = x^3 - x^2 + 1 \qquad R(x) = 3 + 15x$$

A manufacturer estimates his cost $C(x)$ of producing x items per day (or week, or month, or any other appropriate unit of time) and the unit price $p(x)$ at which he can sell each of these items. Find the production level (the number of items produced per unit time) and the unit price (the price of a single item) at which he maximizes profit if $C(x)$ and $p(x)$ are as given in Exercises 14–16. Here (a, b, c, and d are positive constants.)

14. $C(x) = 100x + 10{,}000$ cents,
$p(x) = 750 - 0.05x$ cents

15. $C(x) = 400 + 10x + \dfrac{x^2}{4}$ dollars,
$p(x) = 150 - \dfrac{x}{2}$ dollars

16. $C(x) = a + bx$ monetary units,
$p(x) = c - dx$ monetary units

17. Suppose that the manufacturer in Exercise 14 has a new tax of 15¢ per item that he must pay. What is his optimal production level now?

18. You have advertised successfully for someone to drive your car from Boston to Washington, D.C., a distance of 450 mi. The following facts are known. To stay within the speed limits, the constant speed x (in miles per hour) at which the car is driven must satisfy $35 \le x \le 55$. Tolls average 1.5¢/mi. Gasoline costs \$1.20/gal. Mileage per gallon (which depends on speed) is given by a linear function (that is, one of form $y = ax + b$) that satisfies

x (*in miles per hour*)	40	50	60
y (*in miles per gallon*)	30	25	20

(*Note*: you should find that $a = -0.5$, $b = 50$.) What constant speed x should you instruct the driver to maintain, in order to minimize the total cost, if you pay the driver at the following rate in dollars per hour?

(a) 0 (b) \$3

19. You intend to drive from Boston to Miami and want to choose a constant speed of s mph in such a way as to minimize the total cost of the trip. What should be your choice of s if the cost of running your car per hour is as follows? (a, b, and c are constant.)

$$a + bs + cs^2$$

20. A real estate developer is preparing to put up an office building. It costs her \$1,000,000 to buy the land, hire the architect, and get the work started. She estimates that it will cost \$250,000 to construct the ground floor and that each additional floor will cost \$20,000 more than the preceding one. This implies that the cost to build x floors is $C = 10{,}000x^2 + 240{,}000x + 1{,}000{,}000$. If each floor is designed to produce an annual profit of \$30,000, how many floors should she build in order to maximize the **rate of return** on her investment (= annual profit/cost)?

21. A corn silo is to be in the shape of a hemisphere sitting on top of a right circular cylinder, and its capacity (volume) is fixed in advance. Find the relationship between the height of the cylinder and the radius of the hemisphere if the cost of construction per unit area for the hemisphere is 3 times that for the cylinder and we want to minimize the total cost of construction. (*Note*: If the cylinder has height y and the radius of its base is x, the surface area of the hemisphere is $2\pi x^2$ and the cylinder—which has no top—has surface area $\pi x^2 + 2\pi xy$.)

22. Using Equation (3.1), show that marginal revenue is positive if demand is elastic and that marginal revenue is negative if demand is inelastic. (*Note*: Because $dp/dq < 0$, $E < 0$.)

23. Find the price elasticity of demand for each of the following demand curves.

(a) $p = \dfrac{40}{\sqrt{q}}$

(b) $p = 300 - 0.1q$

(c) $p = 100 - 2q^{1/2}$

24. Consider the demand curve

$$p = 300 - 0.1q \quad 0 < q < 3000$$

Determine for what values of q demand is: (a) elastic; (b) unitary; (c) inelastic.

SECTION 3.6

Applications to Biology

In this section we will consider some maximum and minimum problems arising from biological applications. Some of these are problems of design; some involve the allocation of a limited resource. Others arise because nature itself seems to pursue economy. A hibernating animal, for example, should require minimal energy; the leaves of a plant should be arranged so as to receive maximal sunlight. Of course, we cannot conclude from this that nature does calculus! Rather, organisms tend to adapt in a way that is most likely to ensure their own survival; they evolve towards an optimal configuration. (This is simply a restatement of the principle of "survival of the fittest.") In such situations, calculus can enable us to understand why certain biological systems are the way they are rather than some other way.

Because of the complexity of many biological systems, some of our solutions will entail more involved analysis than those of the previous section. The basic principles, however, are the same.

EXAMPLE 26 Let us consider a problem of vascular branching. We will assume that an artery of radius 0.1 centimeter (cm) runs along a line from point B to point A, a distance of 6 centimeters. A smaller branch artery of radius 0.08 cm is to

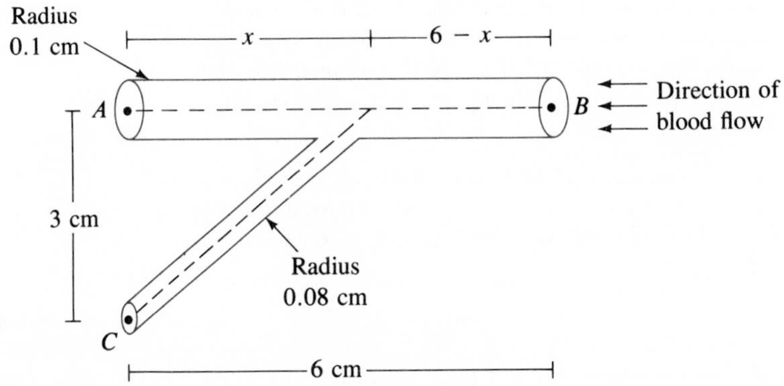

FIGURE 3.57

deliver blood to an organ located at a point C, which is 3 centimeters from point A. It is known empirically that the resistance R encountered by a fluid in a tube is proportional to the length l of the tube and inversely proportional to the fourth power of its radius r. This rule, known as Poiseuille's law, can be written as

(3.2)
$$R = k\frac{l}{r^4}$$

where k is a positive constant of proportionality. Making use of this fact, determine the point along the main artery from which the smaller artery should branch in order that the total resistance of the blood flowing from B to C should be minimized.

SOLUTION Let x denote the horizontal distance from point A to the axis of the branching blood vessel. We wish to minimize the total resistance encountered by the blood in flowing from B to C. This, in turn, consists of the resistance encountered by the blood in the main artery, R_1, plus the resistance encountered in the smaller artery, R_2. Now, by Poiseuille's law,

$$R_1 = k\frac{(6 - x)}{(0.1)^4}$$

In order to compute R_2 we must first find the length of the branching vessel, which we denote by l_2. From the Pythagorean theorem,

$$(3)^2 + x^2 = l_2^2$$

so that

$$l_2 = \sqrt{9 + x^2}$$

Thus

$$R_2 = k\frac{\sqrt{9 + x^2}}{(0.08)^4}$$

Combining these formulas yields

$$R = R_1 + R_2 = \frac{k(6 - x)}{(0.1)^4} + \frac{k(9 + x^2)^{1/2}}{(0.08)^4}$$

in which R is expressed as a function of the single variable x. We note that our description of the problem requires that $0 \le x \le 6$. Now

$$\begin{aligned} R' &= -\frac{k}{(0.1)^4} + \frac{k}{(0.08)^4}\left[\frac{1}{2}(9 + x^2)^{-1/2}(2x)\right] \\ &= -\frac{k}{(0.1)^4} + \frac{k}{(0.08)^4}\left(\frac{x}{(9 + x^2)^{1/2}}\right) \end{aligned}$$

Setting this equal to 0 yields

$$\frac{(0.1)^4}{(0.08)^4}x = (9 + x^2)^{1/2}$$

Using the approximation $(0.1)^4/(0.08)^4 \approx 2.44$, we find on squaring that

$$5.95x^2 = 9 + x^2 \quad \text{or} \quad 4.95x^2 = 9$$

Thus $$x = \sqrt{\frac{9}{4.95}} \approx 1.35 \text{ cm}$$

As in our previous work, you can show that $x = 1.35$ cm actually minimizes the total resistance. This is done by first using the second derivative test to demonstrate that $x = 1.35$ gives a local minimum and then examining the function R for possible end-point minima. The details are left to you.

EXAMPLE 27 Scientists in many countries are currently experimenting with fish farming. In these experiments fish are bred in man-made lakes and then harvested for food. A few countries (Israel, for example) already have commercial fish farms. It is desirable to harvest the fish at approximately the same rate as new fish are born, thus maintaining a roughly constant population. If too many are removed, the population of fish eventually dies out; if too few are taken, the lake will become overcrowded.

Suppose that a lake has sufficient natural resources to support 100,000 fish. (This is known as the **carrying capacity** of the environment.) Suppose further that the rate of growth R of the fish population is given by

$$R = aN(100{,}000 - N) \tag{3.3}$$

where N represents the number of fish in the pond and a is a positive constant representing the difference between the birth rate and the death rate of the fish, assuming unlimited resources. Find the size of the fish population that the "farmer" should try to maintain to produce maximum annual yield.

SOLUTION From a mathematical point of view, this problem is remarkably simple. We simply wish to determine the value of N that maximizes the rate of growth of the fish population. Thus, differentiating R with respect to N and setting the result equal to 0, we have

$$R' = 100{,}000a - 2aN = 0$$

and $$N = \frac{100{,}000}{2} = 50{,}000$$

Because $R'' = -2a < 0$ regardless of the value of N, the graph of R is always concave downward, and $N = 50{,}000$ yields an absolute maximum for R.

Incidentally, our assumption that the growth rate R is given by Equation (3.3) is not necessarily far-fetched. The form $aN(100{,}000 - N)$ is an example of **logistic growth**. There is a large body of experimental data for numerous organisms (especially unicellular ones) that apparently show that their populations grow logistically. We will examine this idea in more detail when we discuss mathematical modeling in Chapter 7.

EXAMPLE 28 In studies of nutrient recycling in forests, **litter** is defined to be plant debris that has fallen to the ground. Once on the ground, it is subject to attack by microbes, bacteria, yeasts, and fungi. The end result of this decomposition is a layer of humus on the forest floor. This layer remains relatively constant in depth from year to year.

It has been shown* that, in any given geographical region, the annual litter production (which we denote by y) is proportional to the accumulation of humus on the forest floor (which we denote by x). Thus

$$y = kx$$

for some constant of proportionality k. (The unit of measurement of both x and y is the *langley*, which is defined to be 1 gram calorie/meter^2; it is the unit of energy commonly used in ecology.) The constant k, known as the **rate parameter**, measures the effectiveness of microbial activity on the breakdown of litter. In a warm, moist, tropical rain forest, where rapid decomposition takes place, k might have a value near 4; in a pine forest, k might be as small as $\frac{1}{64}$.

An ecologist wishes to compute the rate parameter k for temperate forests in the eastern United States. In four separate such forests (all assumed to have the same rate parameter k), he collects the data given in Table 3.1. Find the best estimate for the value of k that is obtainable from the given data.

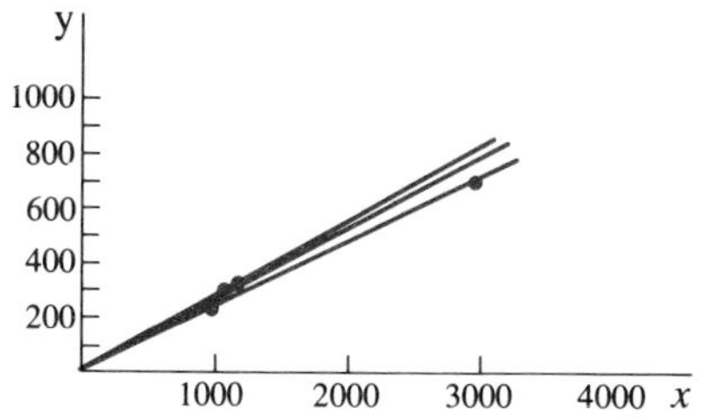

FIGURE 3.58

TABLE 3.1

	Humus accumulation x	*Annual litter production* y
Forest 1	1000	240
Forest 2	1200	320
Forest 3	1100	300
Forest 4	3000	700

SOLUTION Let's make sure we understand the problem at hand. In Figure 3.58 we have plotted the (x, y) pairs from Table 3.1. Now, we know that any equation of the form $y = kx$ has as its graph a straight line passing through the origin. Several such lines are also shown in Figure 3.58. Observe that no one line contains all four data points, because the four given points are not collinear. Our problem is to find that value of k for which the equation $y = kx$ best characterizes the Eastern timber forests under consideration.

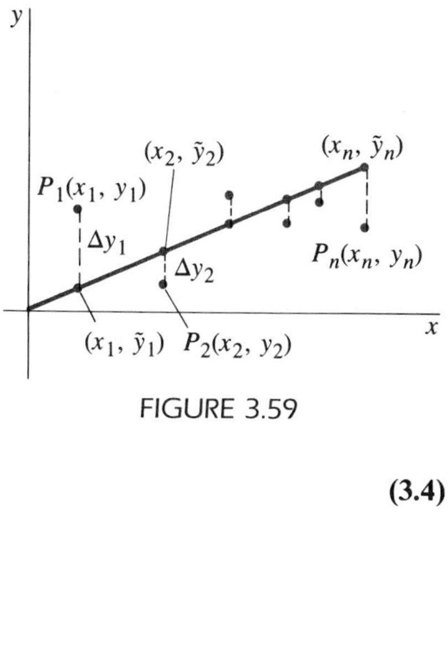

FIGURE 3.59

In order to solve this problem, we must select some criterion that will enable us to decide when one line fits the data better than another—and ultimately which line fits the data best of all. In general, suppose we have n data points, which we denote by $P_1(x_1, y_1), P_2(x_2, y_2), \ldots, P_n(x_n, y_n)$. Let some line $y = kx$ be given, and denote the points on the line with the same x-coordinates as the points $P_1, \ldots, P_n$ by $(x_1, \tilde{y}_1), (x_2, \tilde{y}_2), \ldots, (x_n, \tilde{y}_n)$. See Figure 3.59. The vertical deviations of the data points from the line are denoted by $\Delta y_1, \Delta y_2, \ldots, \Delta y_n$ and are given by

$$\Delta y_i = y_i - \tilde{y}_i \qquad i = 1, 2, \ldots, n \tag{3.4}$$

It is tempting to guess that, if the sum of the deviations is small, the given line fits the data well. This is not a very good criterion, however, for the fit of a line. The deviations given by Equation (3.4) can be either positive or negative. It is possible that some of the deviations may be large and positive while others are

* J. S. Olson, "Energy Storage and Balance of Decomposers in Ecological Systems," *Ecology*, 44(1963), pp. 322–332.

equally large and negative. Thus a line that is not near any of our data points might have the sum of its deviations 0 because of the cancellation of positive and negative terms. Instead, the criterion that is generally accepted is to define the line of best approximation to be the line that results in the smallest value for the sum of the squares of the deviations. That is, we want to minimize

(3.5) $$E = (\Delta y_1)^2 + (\Delta y_2)^2 + \cdots + (\Delta y_n)^2$$

Note that, if all our data points lie on the same line, then $\Delta y_i = 0$, $i = 1, 2, \ldots, n$, so $E = 0$ in this case. In general $E \geq 0$.

In our problem $n = 4$. Consider our first data point (1000, 240). Given any line $y = kx$, if we set $x = 1000$, then

$$\tilde{y}_1 = 1000k$$

Thus

$$\Delta y_1 = 240 - 1000k$$

Similarly,

$$\Delta y_2 = 320 - 1200k$$

$$\Delta y_3 = 300 - 1100k$$

$$\Delta y_4 = 700 - 3000k$$

Therefore

$$E = (240 - 1000k)^2 + (320 - 1200k)^2 + (300 - 1100k)^2 + (700 - 3000k)^2$$

and we want to select k so as to minimize E. Proceeding in our usual manner, we compute dE/dk and set it equal to 0.

$$\frac{dE}{dk} = -2(240 - 1000k)(1000) - 2(320 - 1200k)(1200)$$
$$- 2(300 - 1100k)(1100) - 2(700 - 3000k)(3000) = 0$$

which, upon simplification, yields

$$12{,}650{,}000k = 3{,}054{,}000$$

or

$$k = \frac{3{,}054{,}000}{12{,}650{,}000} \approx 0.241$$

It is easily verified that E has a minimum for this value of k.

This example is an illustration of the **method of least squares**. It is a special case of the more general problem of trying to find the equation of a straight line $y = mx + b$ that best fits a given set of data (in our case, we had $b = 0$). A procedure of this type is called a **simple linear regression**, and it is an important statistical procedure in every discipline in which we try to draw conclusions from data. We will consider the more general problem in Chapter 8, when we discuss functions of several variables.

Another example of the application of this idea is contained in the following:

EXAMPLE 29 An important biological constant for a given species is called the **respiratory quotient**. It is defined to be the ratio of the volume of carbon dioxide expelled during breathing to the volume of oxygen taken in.

$$\text{Respiratory quotient} = \frac{\text{volume of } CO_2 \text{ eliminated}}{\text{volume of } O_2 \text{ consumed}}$$

If we let k = respiratory quotient, y = volume of CO_2 eliminated, and x = volume of O_2 consumed, then

$$y = kx$$

as in Example 28. Suppose a biologist takes measurements of the values of x and y and gathers the following data:

x	y
7.0	6.2
8.4	7.6
10.1	9.0

where x and y are measured in liters per minute. Determine the respiratory quotient k for this organism, using the technique of Example 28. (*Note:* Your answer should be $k = 0.89$.)

The significance of the respiratory constant lies in what it tells us about the metabolic processes of the animal. A respiratory constant near 1 implies that the organism is metabolizing carbohydrate as its energy source. To see this, note that the chemical formula for a carbohydrate is

$$C_nH_{2n}O_n$$

For instance, if $n = 6$, then $C_6H_{12}O_6$ is glucose. The chemical reaction

$$C_nH_{2n}O_n + nO_2 = nCO_2 + nH_2O$$

indicates that the ratio of the number of molecules of carbon dioxide to the number of molecules of oxygen is 1. The respiratory constant is defined by gas volumes, not molecules, but Avogadro's law insures that the number of molecules in a given volume of gas is constant at standard temperature and pressure. Hence the respiratory constant is 1.

A similar analysis of the general formula for saturated fats, $(CH_2O)_3(CH_2)n(CO_2H)_3$, shows that the respiratory constant must be approximately 0.7 for an organism metabolizing mostly fats. The corresponding number in humans is 0.8, indicating a mixed source of energy. (It changes in humans only under conditions of starvation.) For mammals preparing for hibernation the respiratory constant can exceed 1; during hibernation it can be as low as 0.3.

Section 3.6 EXERCISES

1. Find a real number x that minimizes:

(a) $(x + 2)^2 + (x - 3)^2 + (x - 7)^2 + (x - 10)^2$

(b) $(x-1)^2+(x-2)^2+(x-3)^2+(x-4)^2+\cdots+(x - 10)^2$

2. Given n real numbers $a_1, a_2, a_3, \ldots, a_n$, find an x that minimizes

$$(x - a_1)^2 + (x - a_2)^2 + \cdots + (x - a_n)^2$$

3. Apply the method of least squares to find the straight line $y = mx$, through the origin, that most closely fits the points (1, 2), (3, 9), and (4, 10).

4. Do Exercise 3 for the points $(1, \frac{1}{2})$, (2, 1), (5, 2), and (11, 4).

5. Do Exercise 3 for the points (3, 1), (4, 2), (10, 6), (13, 7), and (19, 9).

6. Find the straight line of form $y = mx - 3$ that most closely fits the points $(-3, -7)$, $(0, -1)$, (2, 2), and (4, 4).

7. Can you find a straight line $y = 3x + b$, not necessarily passing through the origin, that most closely fits the points $(-2, -10)$, $(0, -4)$, (2, 2), and (5, 11)?

8. Do Example 26 in the following circumstances: The distance between points B and A is 7 centimeters, and the artery connecting them has radius 0.15 cm. The smaller branching artery has radius 0.06 centimeters and is to deliver blood to an organ at C, which is 4 centimeters from A.

9. Complete the details that were left to the reader at the end of the discussion of Example 26. That is, show that $x = 1.35$ cm. actually minimizes the total resistance.

10. Apple pie of standard thickness has 75 calories per square inch (of area of the top surface). What should be the radius of a circular apple pie if a piece of pie cut from it (that is, a wedge of the usual shape), of perimeter 16 inches, has a maximal number of calories?

11. A sculptor wishes to insert (inscribe) a wooden right circular cylinder inside a hollow, plastic right circular cone. Show that the volume of the largest such cylinder is $\frac{4}{9}$ the volume of the cone.

12. Find the volume of the biggest right circular cylinder that can be inscribed in a sphere of radius a. What are its dimensions?

13. A room has the following dimensions: length l, height h, and width w, with $h < w < l$. An ant starts on the floor at one of the corners and crawls on the walls until it reaches the diagonally opposite corner on the ceiling. What is the shortest path? Where does the ant cross from one wall to the other?

14. In Exercise 13, what happens if the ant is also permitted to crawl on the floor and the ceiling?

15. A group of archaeologists in the desert is 7 mi from a straight road (point P on the road is nearest to them) and needs to reach point A on the road as rapidly as possible. If the band can move 8 mph along the road and 2 mph in the desert, determine what point on the road it should head toward if the distance between P and A is:

(a) 1 mi; (b) 2 mi; (c) 3 mi; (d) $\frac{1}{2}$ mi; (e) 5 mi

Key Mathematical Concepts and Tools

Increasing and decreasing function
Critical points
Extrema (local and absolute maxima and minima)
Concave up, concave down
Inflection point
First derivative test
Second derivative test
Curve sketching
Optimization
Elasticity of demand

Applications Covered in This Chapter

Efficient design of a box
Maximal area garden
Least-cost cable configuration

Optimal design of a tin can of given volume

Maximal strength of a beam

Optimal location for arterial bypass

Maximal harvest rate

Simple linear regression for data concerning litter on forest floor

Method of least squares for data for respiratory capability

Optimal factory production level

Best price for a theater ticket

Elasticity of demand for potatoes

Elasticity of demand for cotton

THREE Warm-Up Test

1. Consider the function

$$y = f(x) = \frac{x - 1}{(x + 1)^2}$$

Its derivative is

$$f'(x) = \frac{3 - x}{(x + 1)^3}$$

and its second derivative is

$$f''(x) = \frac{2x - 10}{(x + 1)^4}$$

Sketch the graph showing where the curve is rising and where it is falling. Locate and classify all extrema and local extrema.

2. Sketch the graph of the function

$$y = -\frac{x^3}{8} + \frac{3}{2}x^2 + 6$$

Test for maxima and/or minima using the first derivative test. Where does the curve rise and where does the curve fall?

3. How does each of the following functions behave as $x \to -\infty$?

(a) $2 + 3x - x^3$ (b) $\dfrac{2x^3 + 4x + 5}{x^3 - 8x}$

(c) $\dfrac{3 + 2x}{x + 7}$ (d) $\dfrac{1 + x^3}{x^4 + x^2}$

4. A university is trying to determine what price to charge for football tickets. At a price of \$6 per ticket, attendance averages 20,000 fans per game. For every increase of \$1, attendance drops 10,000 people; each decrease of \$1 brings in 10,000 additional fans.

(a) Assuming that the demand function is linear, write a formula relating the price of a ticket p to the demand x.

(b) Fans at the game spend an average of \$1.00 on concessions. What price per ticket should be charged to maximize revenue? How many people will attend at that price?

5. A rectangular schoolyard is to be enclosed with a fence. One side of the schoolyard is against the school building, so no fence is needed on that side. If material for the fence costs \$2/ft for the two ends and \$5/ft for the side parallel to the school, find the dimensions of the field of largest area that can be enclosed for \$1000.

6. Use calculus techniques to sketch the graph of the function $y = \dfrac{x}{x + 2}$. Show all significant details.

7. Given the demand curve

$$p = 450 - 0.2q \qquad 0 < q < 2250$$

compute the elasticity of demand E for: (a) $q = 180$ (b) $q = 2000$ (c) an arbitrary q

8. Find the straight line $y = mx$, through the origin, that most closely fits the points (1, 3), (3, 11), and (5, 17).

9. Find a real number x that minimizes

$$(x + 5)^2 + 2(x - 3)^2 + 8(x - 4)^2$$

10. Sketch the curve $y = 2x^3 - 3x^2 - 12x + 12$, using calculus techniques. Be sure to show and label all extrema and inflection points. Discuss concavity via the second derivative test.

1. The Acme Television Rental Company can rent 120 color televisions to its customers if it prices them at $50 per month each. For each dollar that it increases the monthly rental price, it finds that it rents 2 fewer sets. That is, at $51 per month 118 sets will be rented, at $52 per month only 116 sets will be rented, and so on. What rental charge will produce the greatest gross income?

2. Let $f(x) = x^3 + \frac{3}{2}x^2 - 6x$.

 (a) On what interval(s) is f an increasing function? A decreasing function? What are the maximum and minimum points?

 (b) Where is the curve concave upward? Downward? Where is the point of inflection?

 (c) Draw a sketch of $f(x)$ between $x = -4$ and $x = +4$.

3. A rectangular corral is to contain 150 square meters. On three sides, it is to be built with wooden fencing that costs $10/m$. On the fourth side metal fencing, which costs $20/m$, will be used. What should the dimensions of the corral be to minimize the cost? Justify your answer.

4. Use the information given to sketch the curve that satisfies all of the following properties.

 (a) $f(-3) = 4$, $f(-2) = 7$, $f(-1)$ does not exist, $f(0) = -1$, $f(1) = 0$, $f(2)$ does not exist, $f(3) = 5$

 (b) $f'(0) = 0$, $f'(1) = 0$ and

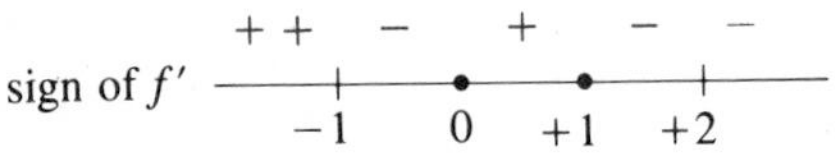

 sign of f' $\underset{\quad -1 \quad\quad 0 \quad\quad +1 \quad +2}{+\ +\quad -\quad\quad +\quad\quad -\quad\quad -}$

 (c) $f''(\tfrac{1}{2}) = 0$ and

 sign of f'' $\underset{\quad -1 \quad\quad 0 \ \ \tfrac{1}{2}\ \ 1 \quad\quad 2}{+\ +\quad +\ +\quad +\ -\quad -\ -\quad +\ +}$

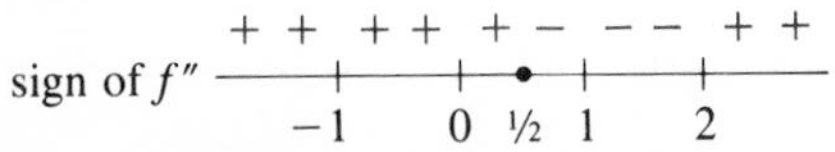

 (d) $\lim_{x \to \infty} f(x) = 0$ $\qquad \lim_{x \to -\infty} f(x) = 2$

 (e) $\lim_{x \to -1^-} f(x) = +\infty$ $\qquad \lim_{x \to 2^-} f(x) = -\infty$

 $\lim_{x \to -1^+} f(x) = +\infty$ $\qquad \lim_{x \to 2^+} f(x) = +\infty$

5. We want to build a wooden box with a square bottom. The box will have no top. If the volume of the box is to be 4 ft^3, find the dimensions (length, width, and height) of the box that will require the least amount of material. Justify your answer.

6. A company wishes to run a utility cable from point A on the shore to an installation at point B on an island. Point B is 6 mi from the nearest point C on the shore (see the Figure for Exercise 6). It costs $400/mi$ to run the cable over land and $500/mi$ to run it underwater. Assume that the cable starts at A and runs along the shoreline and then angles and runs underwater to the island. Find the point at which the line should begin to angle in order to yield the minimal total cost.

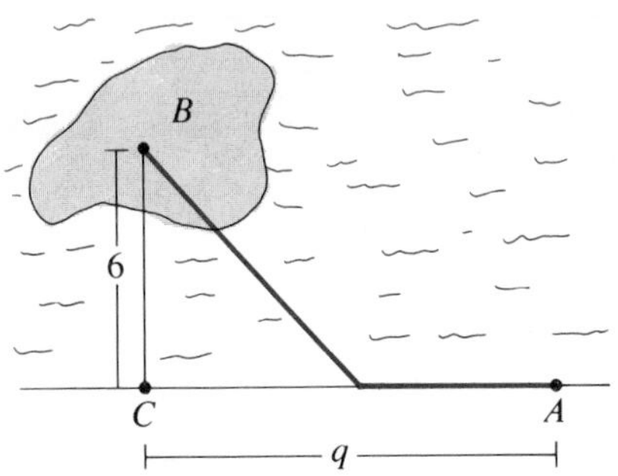

FIGURE FOR EXERCISE 6

7. A rectangular poster is to contain 50 in.2 of printed material and is to have a margin of 1 in. on each side and 2 in. on the top and bottom. Find the dimensions of the poster that will require the least amount of paper.

8. Find the straight line in the form $y = mx - 5$ that most closely fits the points $(-2, 6)$, $(0, 2)$, $(2, 5)$, and $(3, 8)$.

9. Find the real number x that minimizes

$$(2x - 3)^2 + (x - 8)^2 + 4(3 - x)^2 + x^2$$

10. Consider the demand curve

$$p = 900 - 0.3q \qquad 0 < q < 3000$$

Compute the elasticity of demand E for: (a) $q = 100$ (b) $q = 1500$ (c) an arbitrary q, $0 < q < 3000$

FOUR

Exponential and Logarithmic Functions

In Context

In this chapter we introduce two of the most important functions of the calculus: the *exponential* and *logarithmic functions*. We develop the algebraic properties of these functions and learn how to compute their derivatives. More significantly, we study the application of the exponential function to problems of growth and decay arising in several disciplines and introduce the important notion of a *differential equation*.

In order to accomplish these goals, we discuss the chain rule (Section 4.2) and the method of implicit differentiation (Section 4.4). In Chapter 2, we developed rules for differentiating combinations of functions formed via the arithmetic operations of addition, subtraction, multiplication, division, and raising to a power. The chain rule enables us to differentiate functions formed via composition of functions.

SECTION 4.1

The Exponential Function

We have studied the notion of differentiation and have developed rules for computing the derivative of certain simple functions. Let us digress for a while to consider a question of the type that mathematicians like to ask: Does there exist a function $E(x)$, not identically zero, that is equal to its own derivative? In symbols, we want to know whether there is a function $E(x)$ for which

$$E'(x) = E(x)$$

This is a natural question for a mathematician. After all, differentiation is a process that transforms one function into another; $f(x)$ goes in and $f'(x)$ comes out. If we visualize some sort of a "differentiation machine," it might look as represented in Figure 4.1. Believe it or not, electrical engineers actually have machines called analog computers that can do just this! Our question amounts to asking whether there exists a function that is unchanged by such a differentiation machine.

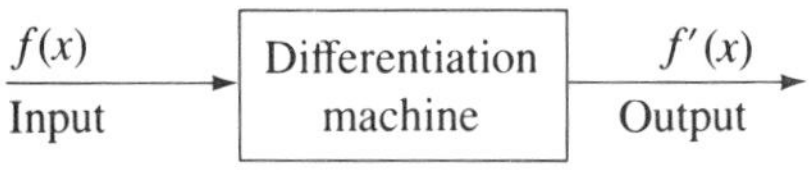

FIGURE 4.1

At this point you are probably saying to yourself, "Who cares about the answer to this question anyway? Of what use will the answer be to me?" Surprisingly enough, in spite of the fact that the problem we are considering is an abstract one, its answer serves as the key with which we can unlock numerous profound applications to biology, economics, sociology, physics, and chemistry. In fact, the function $E(x)$, which we shall soon construct, is probably the single most important function you will encounter in calculus—from the point of view of its significance in other disciplines. In subsequent sections we will pursue some of these applications.

First let us proceed in a nonrigorous way. Let us assume that there *does* exist a function $E(x)$ whose derivative is itself and try to find it. To start with, the simplest functions we know are probably the polynomials. Is it possible that there is a nonzero polynomial $E(x)$ with $E'(x) = E(x)$? Well, a typical polynomial is an expression of the form

$$p(x) = a_0 + a_1 x + a_2 x^2 + \cdots + a_n x^n$$

for some choice of coefficients $a_0, a_1, \ldots, a_n$. It is, of course, very easy to compute the derivative of $p(x)$. It is

$$p'(x) = a_1 + 2a_2 x + \cdots + na_n x^{n-1}$$

Now the derivative $p'(x)$ cannot possibly be equal to $p(x)$. Why? Because $p(x)$ is a polynomial of degree n, and differentiation reduces its degree by 1 so that $p'(x)$ is a polynomial of degree $n - 1$. The conclusion we draw is that any function $E(x)$ for which $E'(x) = E(x)$ cannot be a nonzero polynomial.

We could now try guessing at more complicated types of functions, such as trigonometric functions or rational functions, to see whether we could find one whose derivative is itself. Unfortunately, there is no clear-cut procedure for making these guesses. We might spend the rest of our lives guessing and never find one that works. We need a different approach.

Let us ask ourselves once again what went wrong with our attempt to find a polynomial equal to its own derivative. The problem was that differentiation of a polynomial of degree n produces a derivative of degree $n - 1$. But what if we take n to be infinite? That is, suppose we try a "polynomial of infinite degree." (Alternatively, think of a polynomial whose terms "go on forever.") Then reducing the degree by 1 by differentiating will still leave an infinite number of terms. Perhaps this might work. Accordingly, let us try

(4.1)
$$E(x) = a_0 + a_1x + a_2x^2 + a_3x^3 + a_4x^4 + \cdots$$

Such an expression is called an **infinite series**. When infinite series make sense and when they don't is an interesting and sometimes difficult mathematical question. At this stage, however, we will not worry about a rigorous justification of our ideas but will simply let our intuition guide us. Let us simply assume that Expression (4.1) for $E(x)$ is meaningful. What is its derivative? If we simply differentiate term by term, in the same way that we differentiate a polynomial, we get

(4.2)
$$E'(x) = a_1 + 2a_2x + 3a_3x^2 + 4a_4x^3 + \cdots$$

The expression for $E(x)$ and the expression for $E'(x)$ *both* go on indefinitely. Because we want to have $E'(x) = E(x)$, let us set these two expressions equal to one another. If they are equal, the constant term in one should equal the constant term in the other, and the coefficient of x in one should equal the coefficient of x in the other; in general, the coefficients of like powers of x in the expressions for $E(x)$ and $E'(x)$ should be identical. Writing Expressions (4.1) and (4.2) one beneath the other, we can see what must be true if corresponding coefficients are to be equal.

$$\begin{array}{rccccccc}
E(x) = & a_0 & + & a_1 \cdot x + & & a_2 \cdot x^2 + & & a_3 \cdot x^3 + \cdots \\
E'(x) = & \underbrace{a_1}_{\substack{\text{constant}\\ \text{terms}}} & + & \underbrace{2a_2}_{\substack{\text{coefficients}\\ \text{of } x}} \cdot x + & & \underbrace{3a_3}_{\substack{\text{coefficients}\\ \text{of } x^2}} \cdot x^2 + & & \underbrace{4a_4}_{\substack{\text{coefficients}\\ \text{of } x^3}} \cdot x^3 + \cdots
\end{array}$$

This yields

(4.3)
$$\begin{aligned}
a_1 &= a_0 \\
2a_2 &= a_1 \\
3a_3 &= a_2 \\
4a_4 &= a_3 \\
&\vdots
\end{aligned}$$

In order to determine the coefficients, let's start by choosing, arbitrarily,

$$a_0 = 1$$

Then, from the first Equation (4.3) we know that $a_1 = a_0$, so

$$a_1 = 1$$

The next Equation (4.3) tells us that $2a_2 = a_1 = 1$, so that

$$a_2 = \frac{1}{2} = \frac{1}{(2)(1)}$$

Similarly, $3a_3 = a_2 = \frac{1}{2}$, so

$$a_3 = \frac{1}{6} = \frac{1}{(3)(2)(1)}$$

Next, $4a_4 = a_3 = \frac{1}{6}$, so

$$a_4 = \frac{1}{24} = \frac{1}{(4)(3)(2)(1)}$$

You can probably guess that

$$a_5 = \frac{1}{(5)(4)(3)(2)(1)}$$

and that, in general, a_n must be given by

$$a_n = \frac{1}{(n)(n-1)\cdots(3)(2)(1)}$$

The expression $(n)(n-1)\cdots(3)(2)(1)$, which is the product of the first n positive integers, arises very frequently in mathematics. We use a shorthand notation to represent this number, namely $n!$. This is read as "n-factorial." Thus, for example,

$$4! = (4)(3)(2)(1) = 24$$

Using our notation, we have discovered that

$$a_n = \frac{1}{n!}$$

so that

(4.4)
$$E(x) = 1 + \frac{x}{1!} + \frac{x^2}{2!} + \frac{x^3}{3!} + \frac{x^4}{4!} + \cdots + \frac{x^n}{n!} + \cdots$$

is a function for which $E'(x) = E(x)$. Check it!

Let us try to analyze the function $E(x)$ in more detail, now that we have a formula for it. To start with, let us compute $E(0)$. From Equation (4.4) we know that

$$E(\quad) = 1 + \frac{(\quad)}{1!} + \frac{(\quad)^2}{2!} + \frac{(\quad)^3}{3!} + \frac{(\quad)^4}{4!} + \cdots$$

It follows that

$$\begin{aligned} E(0) &= 1 + \frac{0}{1!} + \frac{0}{2!} + \frac{0}{3!} + \frac{0}{4!} + \cdots \\ &= 1 + 0 + 0 + 0 + 0 + \cdots = 1 \end{aligned}$$

Let us compute $E(1)$ similarly.

$$\begin{aligned} E(1) &= 1 + \frac{1}{1!} + \frac{1}{2!} + \frac{1}{3!} + \frac{1}{4!} + \cdots \\ &= 1 + 1 + \frac{1}{2} + \frac{1}{6} + \frac{1}{24} + \cdots \end{aligned}$$

Using your hand calculator, start to add these terms together. An interesting thing happens. For example, the first six terms add up to

$$1 + 1 + \frac{1}{2} + \frac{1}{6} + \frac{1}{24} + \frac{1}{120} = \frac{326}{120} \approx 2.7167$$

The first seven terms add up to

$$1 + 1 + \frac{1}{2} + \frac{1}{6} + \frac{1}{24} + \frac{1}{120} + \frac{1}{720} = \frac{1957}{720} \approx 2.7181$$

As we continue, despite the fact that we are adding a large number of terms, the sum does not grow much larger. Rather, the more terms we add up, the closer our sum gets to a particular number, which is called e. It is one of the most important numbers in mathematics. To five decimal places its value is 2.71828 . . . . The more terms of $E(1)$ that we add up, the more decimal places we determine in the decimal expansion of e. In fact, with the help of computers, this sum has been accurately computed for 100,000 decimal places! The number e can never really be "known" with complete accuracy. By this we mean that e is an infinite decimal with no repeating pattern. It is an example of a **transcendental number**.

We have just said that this number e, which is so important in mathematics, can never be known with complete accuracy. This may disturb you a bit, but it is not so strange as it first seems. The number π, with which you are well acquainted, is also a transcendental number. It's value to five decimal places is $\pi = 3.14159 \ldots$. It too cannot be written with complete accuracy, yet you have managed to work with π for many years without being particularly bothered by this elusive quality. From now on, you should consider e in a similar light.

Returning to our discussion of the function $E(x)$, we have just defined the number e:

$$E(1) = e$$

Can you compute any values of the function $E(x)$ for values of x different from $x = 1$? One approach might be to use a computer again and again to evaluate the sum in Equation (4.4) for various values of x. (Exercise Set 4.1 contains computer programs that can be used for such computations.) For instance, we might try

$$E(2) = 1 + \frac{2}{1!} + \frac{2^2}{2!} + \frac{2^3}{3!} + \frac{2^4}{4!} + \cdots$$

or

$$E(\sqrt{5}) = 1 + \frac{\sqrt{5}}{1!} + \frac{(\sqrt{5})}{2!} + \frac{(\sqrt{5})^3}{3!} + \frac{(\sqrt{5})^4}{4!} + \cdots$$

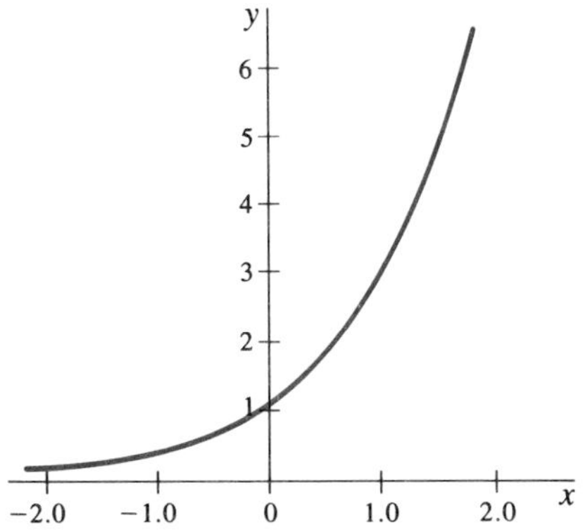

FIGURE 4.2
Graph of $y = E(x)$

In fact, such a procedure has been carried out, and every book of mathematical tables contains a table of values of the function $E(x)$. A typical table is reproduced in Table 4.1. We can use these values to sketch the graph of the function $y = E(x)$ (Figure 4.2). We observe that $E(x)$ is positive for any x and that its graph rises as we move along from left to right, with y tending to $+\infty$ as $x \to +\infty$, and to $0+$ as $x \to -\infty$.

Such information is useful, but one of the most important properties of the function $E(x)$ is one that we are not likely to discover by looking at its graph. We state this fact as a theorem.

TABLE 4.1

x	$E(x)$	$E(-x)$
0.00	1.0000	1.0000
0.05	1.0513	0.9512
0.10	1.1052	0.9048
0.15	1.1618	0.8607
0.20	1.2214	0.8187
0.25	1.2840	0.7788
0.50	1.6487	0.6065
0.75	2.1170	0.4724
1.0	2.7183	0.3679
1.5	4.4817	0.2231
2.0	7.3891	0.1353
2.5	12.182	0.0821
3.0	20.086	0.0498
3.5	33.115	0.0302
4.0	54.598	0.0183
4.5	90.017	0.0111
5.0	148.41	0.0067
5.5	244.69	0.0041
6.0	403.43	0.0025
6.5	665.14	0.0015
7.0	1096.6	0.0009
7.5	1808.0	0.0006
8.0	2981.0	0.0003
8.5	4914.8	0.0002
9.0	8103.1	0.0001
10.0	22,026	0.00005

THEOREM 4.1 For any real numbers a and b,

$$E(a+b) = E(a)E(b)$$

Unfortunately, we do not have enough mathematical know-how at this time to be able to give a rigorous proof of this theorem. Instead, we shall present what we hope is an intuitively convincing argument for its validity.

PROOF From Equation (4.4) we know that

$$E(\quad) = 1 + \frac{(\quad)}{1!} + \frac{(\quad)}{2!} + \frac{(\quad)}{3!} + \cdots$$

In particular,

$$E(a) = 1 + a + \frac{a^2}{2} + \frac{a^3}{6} + \cdots$$

$$E(b) = 1 + b + \frac{b^2}{2} + \frac{b^3}{6} + \cdots$$

and

$$E(a+b) = 1 + (a+b) + \frac{(a+b)^2}{2} + \frac{(a+b)^3}{6} + \cdots$$

Now let us assume that we can multiply the expressions for $E(a)$ and $E(b)$ together in the same manner in which we would multiply two polynomials.

$$\begin{aligned}
E(a)E(b) = {} & 1\left(1 + b + \frac{b^2}{2} + \frac{b^3}{6} + \cdots\right) \\
& + a\left(1 + b + \frac{b^2}{2} + \frac{b^3}{6} + \cdots\right) \\
& + \frac{a^2}{2}\left(1 + b + \frac{b^2}{2} + \frac{b^3}{6} + \cdots\right) \\
& + \frac{a^3}{6}\left(1 + b + \frac{b^2}{2} + \frac{b^3}{6} + \cdots\right) \\
& + \cdots \\
= {} & 1 + b + \frac{b^2}{2} + \frac{b^3}{6} + \cdots \\
& + a + ab + \frac{ab^2}{2} + \frac{ab^3}{6} + \cdots \\
& + \frac{a^2}{2} + \frac{a^2b}{2} + \frac{a^2b^2}{4} + \frac{a^2b^3}{12} + \cdots \\
& + \frac{a^3}{6} + \frac{a^3b}{6} + \frac{a^3b^2}{12} + \frac{a^3b^3}{36} + \cdots
\end{aligned}$$

which, when added vertically, yields

$$\begin{aligned}
& = 1 + (a + b) + \left(\frac{a^2}{2} + ab + \frac{b^2}{2}\right) + \left(\frac{a^3}{6} + \frac{a^2b}{2} + \frac{ab^2}{2} + \frac{b^3}{6}\right) + \cdots \\
& = 1 + (a + b) + \frac{a^2 + 2ab + b^2}{2} + \frac{a^3 + 3a^2b + 3ab^2 + b^3}{6} + \cdots \\
& = 1 + (a + b) + \frac{(a + b)^2}{2} + \frac{(a + b)^3}{6} + \cdots
\end{aligned}$$

But this is $E(a + b)$. So, for any numbers a and b,

$$E(a + b) = E(a)E(b)$$ ⊟

Let us see how this result can be used to give us additional information about the function $E(x)$. First of all, we already know that $E(1) = e$. Applying Theorem 4.1 with $a = b = 1$, we find

$$E(2) = E(1 + 1) = E(1)E(1) = e \cdot e = e^2$$

Next, consider

$$E(3) = E(2 + 1) = E(2)E(1) = e^2 \cdot e = e^3$$

You can see the trend. Inductively, we could show that for, any positive integer p,

$$E(p) = e^p$$

Now let us try to compute the function $E(x)$ for some negative values of x. We know that $E(0) = 1$, so

$$1 = E(0) = E[1 + (-1)] = E(1)E(-1) = eE(-1)$$

It follows that $E(-1) = \frac{1}{e} = e^{-1}$. In general,

$$1 = E(0) = E[p + (-p)] = E(p)E(-p) = e^p \cdot E(-p)$$

so that $E(-p) = \frac{1}{e^p} = e^{-p}$ for any integer p.

Can we evaluate the function $E(x)$ for fractional values of x by using the same idea? For instance, let $x = \frac{1}{2}$. Then

$$e = E(1) = E(\tfrac{1}{2} + \tfrac{1}{2}) = E(\tfrac{1}{2})E(\tfrac{1}{2}) = [E(\tfrac{1}{2})]^2$$

But if $[E(\frac{1}{2})]^2 = e$, then

$$E(\tfrac{1}{2}) = \sqrt{e} = e^{1/2}$$

where we have chosen the positive square root, because $E(x)$ is positive for all x.

How about $E(\frac{1}{3})$? You can probably guess that

$$E(\tfrac{1}{3}) = e^{1/3}$$

To show this, simply write

$$e = E(1) = E(\tfrac{1}{3} + \tfrac{2}{3}) = E(\tfrac{1}{3})E(\tfrac{2}{3})$$

But $E(\frac{2}{3}) = E(\frac{1}{3} + \frac{1}{3}) = E(\frac{1}{3})E(\frac{1}{3})$, so

$$e = E(\tfrac{1}{3})E(\tfrac{2}{3}) = E(\tfrac{1}{3})E(\tfrac{1}{3})E(\tfrac{1}{3}) = [E(\tfrac{1}{3})]^3$$

or, taking cube roots,

$$E(\tfrac{1}{3}) = \sqrt[3]{e} = e^{1/3}$$

In fact, we can use this same (monotonous) procedure to show inductively that

$$E\left(\frac{1}{q}\right) = e^{1/q} \qquad \text{for any positive integer } q$$

Thus, for example, $E(\frac{1}{7}) = e^{1/7}$, $E(\frac{1}{9}) = e^{1/9}$, and so on.

If we choose x to be an arbitrary nonnegative rational number, then $x = p/q$ for some integers p and q. In this case,

$$E\left(\frac{p}{q}\right) = E\underbrace{\left(\frac{1}{q} + \frac{1}{q} + \cdots + \frac{1}{q}\right)}_{p \text{ times}} = \underbrace{E\left(\frac{1}{q}\right)E\left(\frac{1}{q}\right)\cdots E\left(\frac{1}{q}\right)}_{p \text{ times}}$$

$$= \left[E\left(\frac{1}{q}\right)\right]^p = (e^{1/q})^p = e^{p/q}$$

Finally, it is easily checked that, for any rational number $r > 0$, $E(-r) = e^{-r}$.

We may summarize our findings in the following way: The function $E(x)$, given by Equation (4.4) is also given by

$$E(x) = e^x$$

for x any rational number. (Recall that a rational number is any real number, positive or negative, that can be written as the ratio of two integers.)

Are there any kinds of numbers x that we haven't considered? Yes, there are. They include $x = \sqrt{2}$, $x = \pi$, and any other irrational number. Does $E(\sqrt{2})$ equal $e^{\sqrt{2}}$ or not? (Come to think of it, what does $e^{\sqrt{2}}$ mean, anyway?) For such numbers we refer again to Equation (4.4). For instance, we would define

$$e^{\sqrt{2}} = E(\sqrt{2}) = 1 + \sqrt{2} + \frac{(\sqrt{2})^2}{2!} + \frac{(\sqrt{2})^3}{3!} + \cdots$$

and

$$e^{\pi} = 1 + \pi + \frac{\pi^2}{2!} + \frac{\pi^3}{3!} + \cdots$$

With this understanding we make the following definition.

DEFINITION 4.2 For any real number x,

$$e^x = E(x)$$

where $E(x)$ is that function for which $E'(x) = E(x)$ and $E(0) = 1$. □

From now on, we will drop the notation $E(x)$ in favor of the customary e^x. In summary, we have discovered that

(4.5)

(i) $e^x = 1 + \frac{x}{1!} + \frac{x^2}{2!} + \frac{x^3}{3!} + \cdots$

(ii) $e^{a+b} = e^a e^b$ for any a, b

(iii) e^x is positive for any choice of x

(iv) the graph of $y = e^x$ (shown in Figure 4.2) rises as x increases, with $y \to +\infty$ as $x \to +\infty$, and $y \to 0$ as $x \to -\infty$

(v) $\frac{d}{dx}(e^x) = e^x$

It is property (v) that led us to the discovery of function $y = e^x$ in the first place. This function is called the **exponential function**. Many pocket calculators have a key labeled e^x or exp that computes values of this function.

EXAMPLE 1 It is known that, when an amount of money A_0 is invested in an account that earns a rate of interest r compounded m times per year, then after t years the amount will have grown to

$$A = A_0\left(1 + \frac{r}{m}\right)^{mt}$$

For example, \$2500 invested at 9% compounded monthly will grow, after 10 years, to

$$A = \$2500\left(1 + \frac{0.09}{12}\right)^{120} \approx \$6{,}128.39$$

Continuous compounding of interest is the limit of compounding m times per year as $m \to \infty$. After t years of continuous compounding, it can be shown that A_0 will have grown to

$$A = A_0 e^{rt}$$

What would be the result if the \$2500 above had been invested for 10 years in an account that earns 9% compounded continuously?

SOLUTION $A = A_0 e^{rt} = 2500\, e^{0.09(10)} = 2500(2.4596) \approx \$6{,}149.01$

EXAMPLE 2 Find dy/dx when $y = xe^x$.

SOLUTION Observe that the function $y = xe^x$ is the product of the two functions $f(x) = x$ and $g(x) = e^x$. By the product rule for differentiation,

$$\frac{dy}{dx} = f(x)g'(x) + g(x)f'(x) = xe^x + e^x \cdot 1 = xe^x + e^x$$

where we have made use of the fact that $\frac{d}{dx}(e^x) = e^x$ [(v) of Properties (4.5)].

EXAMPLE 3 Find dy/dx when $y = e^{x+2}$.

SOLUTION Observe first that, in view of (ii) of Properties (4.5), we have

$$e^{x+2} = e^x e^2$$

Of course e^2 is just a constant. Therefore

$$\frac{dy}{dx} = e^2 e^x = e^{x+2}$$

Thus e^{x+2} is also a function that is equal to its own derivative.

PRACTICE EXERCISE Show that, for any choice of the number a, the function e^{x+a} is equal to its own derivative

$$\frac{d}{dx}(e^{x+a}) = e^{x+a}$$

EXAMPLE 4 Differentiate $g(x) = e^{5x}$.

SOLUTION We first observe that $e^{5x} = (e^x)^5$. If we set $u = e^x$, we have $(d/dx)\,u = e^x$ and $g(x) = u^5$. Therefore, by the generalized power rule,

Equation (2.15),

$$g'(x) = 5(e^x)^4 e^x = 5 \cdot e^{4x} \cdot e^x = 5e^{4x+x} = 5e^{5x}$$

$$\uparrow n \quad \uparrow u^{n-1} \quad \nwarrow \frac{du}{dx}$$

PRACTICE EXERCISE Show that, for any number k,

$$\frac{d}{dx}(e^{kx}) = ke^{kx}$$

(In particular, $d/dx\,(e^{-x}) = -e^{-x}$.)

EXAMPLE 5 Differentiate $\dfrac{e^{-x}}{(x+1)}$.

SOLUTION Put $f(x) = e^{-x}$ and $g(x) = x + 1$. Then $f' = -e^{-x}$ and $g' = 1$, so applying the quotient rule to the function $f(x)/g(x) = \dfrac{e^{-x}}{(x+1)}$ yields

$$\left(\frac{e^{-x}}{x+1}\right)' = \frac{(x+1)(-e^{-x}) - e^{-x}(1)}{(x+1)^2} = \frac{-xe^{-x} - 2e^{-x}}{(x+1)^2}$$

Section 4.1 EXERCISES

In each of Exercises 1–6 use a calculator to find the approximate value of $E(x)$ at the given value of x by taking the sum of the first seven terms of the infinite series for $E(x)$. Compare your result with the value given in Table 4.1.

1. $E(-1)$
2. $E(-2)$
3. $E(3)$
4. $E(-\frac{1}{2})$
5. $E(\frac{1}{3})$
6. $E(-\frac{1}{3})$
7. In the text we found that $E(1) = e$ is approximately 2.718. Check that your result in Exercise 1—namely, your approximation to $E(-1) = e^{-1}$—is approximately $\dfrac{1}{2.718}$.
8. Verify $(d/dx)(e^{kx}) = ke^{kx}$ directly from the series definition of the exponential function. That is, replace x by kx in every term of the series, then differentiate, and finally factor k out of every term.
9. Verify $(d/dx)\,(e^{x^2}) = 2xe^{x^2}$ directly from the series definition of the exponential function.
10. Verify that $E(\frac{1}{3}) = e^{1/3}$ and that $E(-\frac{1}{3}) = e^{-1/3}$, as approximated in Exercises 5 and 6, are indeed reciprocals of each other (approximately).

Differentiate each of the functions in Exercises 11–27.

11. $e^x + e^{-x}$
12. xe^{2x}
13. x^2e^{-x}
14. $e^x - e^{-x}$
15. $\dfrac{e^{x+2}}{x^2}$
16. $\dfrac{e^{3x-2}}{x-1}$
17. $\dfrac{e^x - e^{-x}}{e^x + e^{-x}}$
18. $\dfrac{x^2+1}{e^{2x}}$
19. e^{-x^2}
20. $e^xe^{x^2}$
21. e^{x+x^2}
22. $\dfrac{3}{e^x}$
23. $e^{\sqrt{x}}$
24. $e^{1/x}$
25. $(e^x)^4$
26. e^{4x}
27. e^{x^4}
28. Show that $y = ce^x$, for any constant c, is a function that is equal to its own derivative.
29. Sketch the graph of $y = e^{-x}$ (without choosing any values of x) by making use of the graph of $y = e^x$.

30. (a) Given any function $y = f(x)$, show that the graph of $y = f(x + a)$ is a translation of the graph of $y = f(x)$ by a units to the left.

(b) Use this fact to give a geometrical interpretation of the practice exercise preceding Example 4 where $f(x) = e^x$ and $f(x + a) = e^{x+a}$.

31. We know that $\frac{d}{dx}(e^x) = e^x$. Express this fact in terms of the limit of the Newton quotient and deduce that

$$\lim_{h \to 0} \frac{e^h - 1}{h} = 1$$

32. See Example 1. Continuous compounding is often used as an approximation to daily compounding. Suppose \$1000 is invested at 8% for 5 years. To what will this sum grow: (a) if compounded daily; (b) if compounded continuously?

33. Eleven percent compounded continuously for 1 year is equivalent to what rate of simple annual interest? (This is called the **effective yield**).

34. Mr. Smith estimates that it will cost \$20,000 for his son's first year in college, 10 years from now. In order to have \$20,000 in 10 years, how much must Mr. Smith put in an investment that earns 11.5% compounded continuously. (This amount is called the **present value** of the \$20,000.)

35. How much would one have to deposit now at 8% compounded continuously to have \$100,000 in 25 years?

36. Sketch the graph of $f(x) = xe^{-x}$. Be sure to find all local extrema, inflection points, and asymptotes. (*Hint*: Use the fact that $\lim_{x \to \infty} xe^{-x} = 0$ and $\lim_{x \to -\infty} xe^{-x} = -\infty$.)

IF YOU WORK WITH A COMPUTER (OPTIONAL)

37. Use either of the following computer programs to approximate: (a) e; (b) e^2; (c) e^{-1}; (d) $e^{1/3}$.
(*Note*: These programs can easily be modified to sum any series in which the nth term is simply expressed in terms of the $(n - 1)$th term, n, and x.) If your version of BASIC does not allow long variable names, make the appropriate modifications to the BASIC version of the program.

```
10  REM - Exponential Function (BASIC version)

20  INPUT "Compute e to what power";x

30  REM - stop when last term summed < epsilon
40  epsilon=.00001

50  REM - start with initial term 1
60  prevterm = 1
70  sum = 1
80  n=1

90  REM - compute nth term
100 term=prevterm*x/n
110 sum = sum + term
120 IF ABS(term)<epsilon THEN 160
130 prevterm = term
140 n = n + 1
150 GOTO 100
```

```
(* Exponential Fn. - Pascal version *)
program Expo (input, output);
  const
    epsilon = 0.00001;

  var
    prevterm, term, sum, x : real;
    n : integer;

begin
  write('Compute e to what power? ');
  readln(x);

  prevterm := 1;                (* initial term *)
  sum := 1;
  n := 1;

  repeat
    term := prevterm * x / n;   (* compute nth term *)
    sum := sum + term;
    prevterm := term;
    n := n + 1
  until abs(term) < epsilon;
```

(*Programs continue at top of next page*)

```
160 PRINT "e to the x approximately";sum
170 PRINT n;"terms were summed."
180 END
```

```
writeln('e to the x approximately ', sum : 1 : 6);
writeln(n : 1, ' terms were summed.')

end.
```

For Exercises 38 and 39 use either of the computer programs that follow Exercise 39.

38. The monthly payment for an n-year, $r\%$ mortgage of face amount (principal) A is given by the formula

$$\text{Monthly payment} = \frac{Ar}{1200}\left[\frac{1}{1-\left(\dfrac{1}{1+r/1200}\right)^{12n}}\right]$$

(Mortgages are figured on the basis of monthly compounding.) Find the monthly payment on:

(a) a \$12,000 10-year mortgage at 16%

(b) a \$50,000 25-year mortgage at 12.5%

(c) a \$50,000 25-year mortgage at 8%

39. Which has smaller monthly payments for the same principal: a 25-year mortgage at 12% or a 30-year mortgage at 14%?

```
10    REM - Monthly Payment Prog - BASIC version

20    INPUT "Amount of loan, no commas";A
30    INPUT "Interest rate (e.g., enter 9% as 9)"; r
40    INPUT "How many years";N

50    i = r/1200
60    w = 1/(1+i)
70    pmt = A*i / ( 1-w^(12*N) )

80    PRINT "Monthly payment: ";
90    PRINT USING "$$###.##";pmt

100   END
```

```
(* Monthly Payment Program - Pascal version *)
program Loan (input, output);
 var
  A, r, i, w, pmt, N : real;

begin
 write('Amount of loan, no commas)? ');
 readln(A);
 write('Interest rate (e.g., enter 9% as 9)? ');
 readln(r);
 write('How many years? ');
 readln(N);

 i := r / 1200;
 w := 1 / (1 + i);
 pmt := A * i / (1 - exp(12 * N * ln(w)));

 writeln('Monthly payment = $', pmt : 1 : 2)

end.
```

SECTION 4.2

The Chain Rule

In Section 2.7, we introduced the generalized power rule for differentiating a power of a differentiable function $u = u(x)$:

$$\frac{d}{dx}[u(x)]^n = n[u(x)]^{n-1}u'(x)$$

or, in abbreviated form,

$$\frac{d}{dx}u^n = nu^{n-1}\frac{du}{dx}$$

For example, if $g(x) = \sqrt{x + x^3} = (x + x^3)^{1/2}$, we saw (in Example 21 in Chapter 2) that

$$g'(x) = \frac{1}{2}(x + x^3)^{-1/2}(1 + 3x^2)$$

Let us reinterpret this last example. In the language of Chapter 1, g is really the composition of the function $f(u) = u^{1/2}$ and the function $u(x) = x + 3x^2$, as illustrated in Figure 4.3.

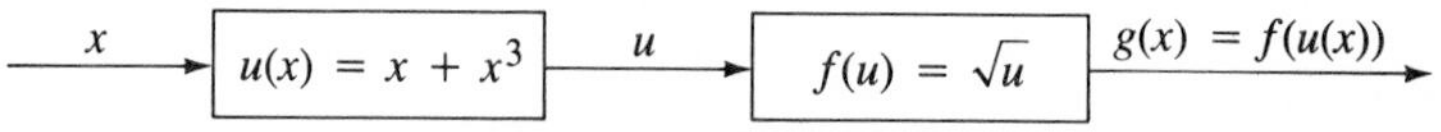

FIGURE 4.3

The derivative of $g(x)$ turned out to be the product of the derivative of f and the derivative of u:

$$g'(x) = f'(u)u'(x) = \frac{1}{2}u^{-1/2}(1 + 3x^2) = \frac{1}{2}(x + x^3)^{-1/2}(1 + 3x^2)$$

The next theorem, the chain rule for differentiation, establishes the generality of this situation. This rule will enlarge the class of functions we can differentiate more than any other single differentiation formula.

THEOREM 4.3 (The Chain Rule) Let $u(x)$ and $f(u)$ be two differentiable functions with derivatives $u'(x)$ and $f'(u)$, respectively. Consider the composite function $g = f \circ u$—that is,

$$g(x) = f(u(x))$$

Then $g(x)$ is differentiable and

(4.6) $$g'(x) = f'(u(x))u'(x)$$ ⊟

If we let $y = f(u)$, where $u = u(x)$, the chain rule can be expressed in Leibniz notation as

(4.7) $$\frac{dy}{dx} = \frac{dy}{du}\frac{du}{dx}$$

This form is more easily remembered, because it appears as though the du's on the right cancel. Of course dy/du and du/dx are not really fractions but indivisible symbols for derivatives.

EXAMPLE 6 Differentiate $e^{(x+x^2)}$.

SOLUTION Let $y = f(u) = e^u$, where $u = x + x^2$. Then

$$\frac{dy}{dx} = \frac{dy}{du}\frac{du}{dx} = e^u(1 + 2x) = e^{(x+x^2)}(1 + 2x)$$

Note that in the last step, we substituted $u = x + x^2$. That is, we evaluated $f'(u) = e^u$ at $u(x)$. The need for this step is evident in Equation (4.6).

We shall not give a proof of the chain rule but shall simply indicate why it is plausible. Suppose $y = f(u)$, where $u = u(x)$. It is important to remember that a derivative is a rate of change. Therefore, if we take any x in the domain of $f \circ u$ and x begins to vary, then u changes du/dx times as fast as x, and as a consequence, y changes dy/du times as fast as u. It seems reasonable, then, to say that y changes $\left(\frac{dy}{du}\frac{du}{dx}\right)$ times as fast as x. But this latter rate—the rate of change of y with respect to x—is dy/dx.

EXAMPLE 7 Scientists believe that a certain city's average carbon monoxide pollution, in parts per million (ppm), is given approximately by

$$C = 0.2p^{3/2} + 120$$

when the city's population is p thousand people. The population t years from now is estimated to be

$$p = 20 + 0.8t$$

thousands. Find the rate at which carbon monoxide pollution will be changing in 5 years.

SOLUTION When $t = 5$, $p = 24$. Therefore

$$\frac{dC}{dt} = \frac{dC}{dp}\frac{dp}{dt} = 0.3p^{1/2}(0.8) = 0.3\sqrt{24}(0.8) \approx 1.18 \text{ ppm/yr}$$

EXAMPLE 8 Let $z = 8 + 4y^3$ and $y = e^x + x^2$. Find $\frac{dz}{dx}$.

SOLUTION $\frac{dz}{dy} = 12y^2$ and $\frac{dy}{dx} = e^x + 2x$. Thus

$$\frac{dz}{dx} = \frac{dz}{dy}\frac{dy}{dx} = (12y^2)(e^x + 2x)$$
$$= 12(e^x + x^2)^2(e^x + 2x)$$

Section 4.2 EXERCISES

In each of Exercises 1–6 you are given two functions $u(x)$ and $f(u)$. Write the expression for the composite function $g(x) = f(u(x))$ and find $g'(x)$.

1. $u(x) = 2x - 3, \quad f(u) = u^5$

2. $u(x) = x^5, \quad f(u) = 2u - 3$

3. $u(x) = \frac{1}{2x - 3}, \quad f(u) = u^5$

4. $u(x) = x^5, \quad f(u) = \frac{1}{2u - 3}$

5. $u(x) = x^2 - 2x + 3, \quad f(u) = \sqrt{u}$

6. $u(x) = \sqrt{x}, \quad f(u) = u^2 - 2u + 3$

Each of the expressions in Exercises 7–10 can be viewed as a composite function $g(x) = f(u(x))$. Write expressions for $u(x)$ and $f(u)$ and find the derivative of the given function $g(x)$.

7. $(3x - 2)^{17}$

8. $\dfrac{1}{(5x + 7)^{88}}$

9. $\sqrt{3x^2 - x + 1}$

10. $e^{x^5 + x + 1}$

Use the chain rule to differentiate each of the functions in Exercises 11–25 (do not simplify).

11. $(5x + 7)^{88}$

12. $(x - 1)^{19}$

13. $(x^2 + x - 3)^{49}$

14. $\dfrac{1}{(x + 2)^9}$

15. $\dfrac{1}{(3x - 2)^{17}}$

16. $\sqrt{3x + 2}$

17. $\sqrt{5x^2 + 4x - 1}$

18. $(3x^2 - x + 1)^5$

19. $(3x^2 - x + 1)^{-1}$

20. $(3x^2 - x + 1)^{1/2}$

21. $\dfrac{(x^2 + x + 1)^3}{(e^x + e^{2x})^2}$

22. $\sqrt{(5x + 7)^3 + 1}$

23. $e^{\sqrt{x^5 + x + 1}}$

24. $\sqrt{e^{x^5 + x + 1}}$

25. $\dfrac{e^{-x^2}}{x^2 + 1}$

Differentiate each of the expressions given in Exercises 26–35. Specify the rules used and in what order they are used. (Do not simplify.)

26. $(3x^5 + x^4 - 2)^8(2x - 7)^9$

27. $(3x^5 + x^4 - 2)^8 + (2x - 7)^9$

28. $\dfrac{(3x^5 + x^4 - 2)^8}{(2x - 7)^9}$

29. $\dfrac{\sqrt{3x^5 + x^4 - 2}}{(2x - 7)^9}$

30. $\dfrac{(3x^5 + x^4 - 2)^9 + (2x - 7)^8}{(3x + 2)^6}$

31. $\sqrt{\dfrac{(x^2 + x + 1)^5 - 2}{(3x - 2)^7 + 1}}$

32. $(e^{2x} - e^{3x})^{75}$

33. $e^{3x^2 - 5x + 4}$

34. $\dfrac{x^2e^{-x^3}}{(e^x - e^{3x})^2}$

35. $\dfrac{x^2e^{-x^3}}{(e^x - e^{3x})^2}$

In Exercises 36–42, where m and n are positive integers, differentiate.

36. $(2x - 3)^n$

37. $(x^2 + x + 1)^n$

38. e^{nx}

39. e^{-x^n}

40. $(e^{mx} + e^{nx})^{m+n}$

41. $\dfrac{e^{mx} - e^{nx}}{e^{mx} + e^{nx}}$

42. $(e^{mx} + e^{-nx})(x^m - x^n)$

43. If a company spends x dollars per month on advertising, its profit P is given by

$$P = \frac{-x^2}{500} + 12x - 6000$$

The advertising budget is at present \$2000 per month and is increasing by \$100 each month. Find the rate of change of profit.

44. A company that manufactures x vacuum cleaners per week finds that its cost and revenue are given by

$$C = 6000 + 3x$$

$$R = 15x - \frac{x^2}{1600}$$

Production is presently 4000 vacuum cleaners and is increasing at the rate of 250 per week. Find the rate of increase of cost, revenue, and profit.

SECTION 4.3

Inverse Functions

Consider any collection of points in the x-y plane and then consider the new points obtained by interchanging the x- and y-coordinates of each of the original pairs. Thus, if our original collection contains the point (a, b), we reverse the coordinates and consider the new point (b, a).

EXAMPLE 9 Given the points (2, 8), (−1, 2), and (4, 3), each marked with a box in Figure 4.4, interchanging the coordinates yields the "new" points (8, 2), (2, −1), and (3, 4), each of which is marked with a dot. Under our coordinate-interchanging transformation, note that any points on the line $y = x$, such as (2, 2), (3, 3), and (π, π), remain unchanged and that these are the *only* points with this property.

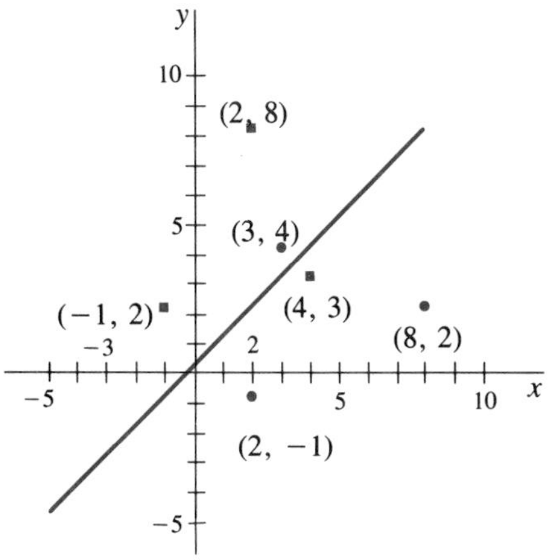

FIGURE 4.4

In fact, the line $y = x$ is what is called an **axis of symmetry**. If you can imagine the entire x-y plane as folded along the dashed line $y = x$ (see Figure 4.5), each point (a, b) will touch the corresponding point (b, a).

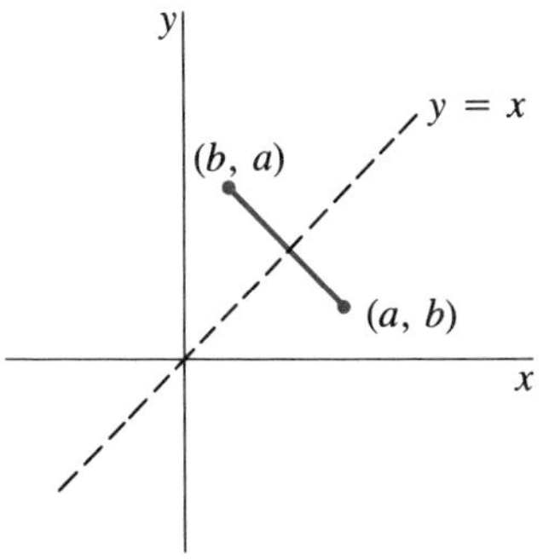

FIGURE 4.5

Now suppose that we have the graph of some function $y = f(x)$, as shown in Figure 4.6(a). We apply to each point $(a, f(a))$ on the graph our coordinate-interchanging transformation and then plot all the new pairs $(f(a), a)$ as shown in Figure 4.6(b). In this way we obtain a new graph. It is the mirror image of the original graph, using the axis $y = x$ as a mirror.

Let us see, for instance, what our coordinate-interchanging transformation will do to a (nonvertical) line l that contains the two points (x_1, y_1) and (x_2, y_2). It seems fairly clear that our transformation will send l to a new line l' with the points (y_1, x_1) and (y_2, x_2) on it. If the equation of the given line l is $y = mx + b$, its slope m is given by

$$m = \frac{y_2 - y_1}{x_2 - x_1}$$

The slope of l' is also easily determined. If we denote it by m', then

$$m' = \frac{x_2 - x_1}{y_2 - y_1} = \frac{1}{m}$$

Of course, if the line l is vertical, its slope m is undefined (this occurs when $x_1 = x_2$), and the slope m' of the line l' is then 0. On the other hand, if line l is horizontal, then $m = 0$ and m' is undefined.

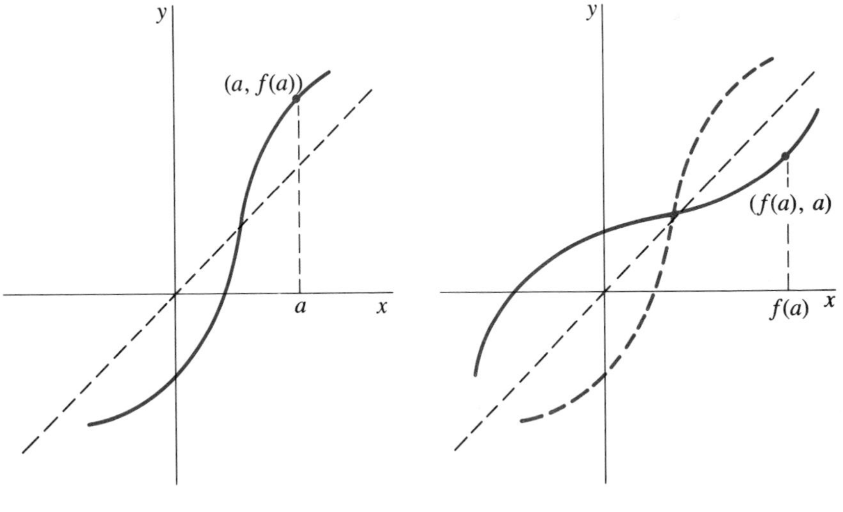

(a) $y = f(x)$

(b) $x = f(y)$

FIGURE 4.6

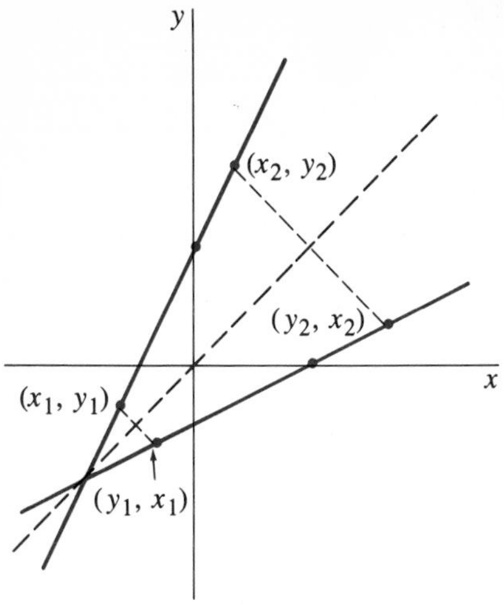

FIGURE 4.7

Let us now try to write down the equation of the line l'. We already know its slope is $1/m$, so it suffices to find a specific point on it. Our original line has equation $y = mx + b$, which means that it must cross the y-axis at the point $(0, b)$. Our transformed line must therefore pass through the point $(b, 0)$. Thus the equation of l' is

$$y - 0 = \frac{1}{m}(x - b)$$

or

$$y = \frac{1}{m}(x - b)$$

The reasoning that led us to this equation is fine, but there is a much quicker way to get the same result. Our transformation interchanges the x- and y-coordinates of every point. Let's simply take the equation of our line, $y = mx + b$, and interchange x and y in that equation, to get

$$x = my + b$$

If we now solve this for y, we find that

$$y = \frac{1}{m}(x - b)$$

which is the same equation for l' that we found before.

Note that, if $m = 0$, we cannot solve for y by dividing by m. In this case, our original equation is $y = b$. The transformed equation simply replaces y with x to yield $x = b$ (see Figure 4.8).

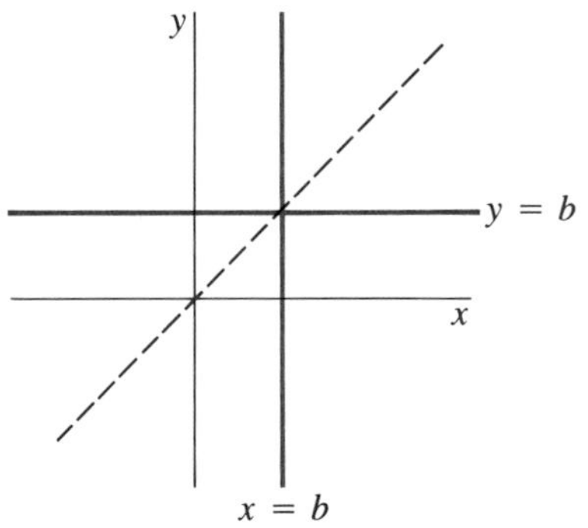

FIGURE 4.8
Interchanging transformation applied to a horizontal line yields a vertical line.

Let's see what our interchanging transformation does when applied to a graph other than a straight line.

EXAMPLE 10 What does our coordinate-interchanging transformation do to the curve $y = \sqrt[3]{x + 1}$?

SOLUTION Just as before, we can figure out what our new graph will look like by interchanging the roles of the x and y variables in the equation

$y = \sqrt[3]{x+1}$. Thus our new graph (Figure 4.9) will consist of all those points whose coordinates satisfy the equation

$$x = \sqrt[3]{y+1}.$$

When writing equations, we usually try to solve them for y as a function of x. To do this here, we cube both sides of the foregoing expression to get

$$x^3 = y + 1$$

or

$$y = x^3 - 1$$

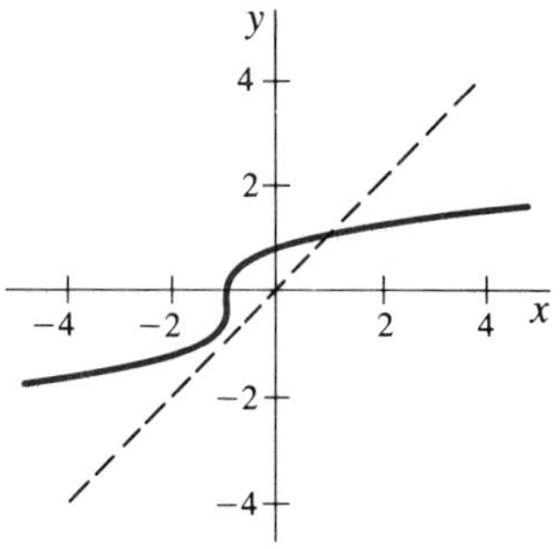

FIGURE 4.9
$y = \sqrt[3]{x+1}$

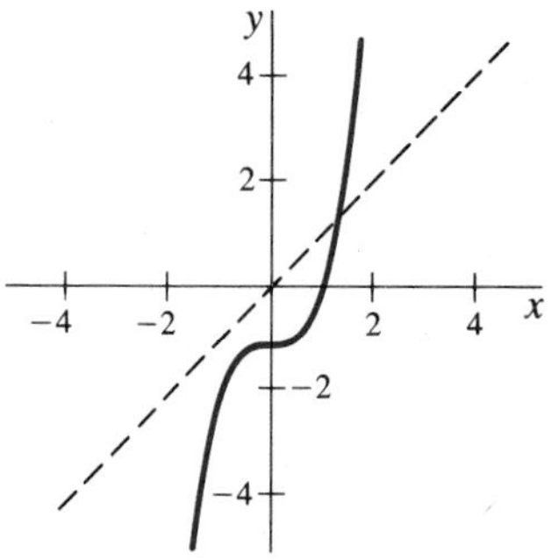

FIGURE 4.10
$y = x^3 - 1$

As it turns out, there is a very curious and important relationship between each function and the one into which it is transformed. To see what it is, consider again the straight line $y = mx + b$ and the transformed line

$$y = \frac{1}{m}(x - b)$$

If we set

$$f(x) = mx + b$$

and

$$g(x) = \frac{1}{m}(x - b)$$

we can compute the composite function $g \circ f$:

$$g(f(x)) = g(mx + b) = \frac{1}{m}[(mx + b) - b] = x$$

as well as the composite function $f \circ g$:

$$f(g(x)) = f\left[\frac{1}{m}(x - b)\right] = m\left[\frac{1}{m}(x - b) + b\right] = x$$

Similarly, if we consider the functions $f(x) = \sqrt[3]{x+1}$ and $g(x) = x^3 - 1$ of Example 10, then

$$g(f(x)) = g(\sqrt[3]{x+1}) = (\sqrt[3]{x+1})^3 - 1 = x$$

and

$$f(g(x)) = f(x^3 - 1) = \sqrt[3]{(x^3 - 1) + 1} = x$$

again! Let's formulate our discovery in the form of a theorem.

THEOREM 4.4 Given any function f, suppose g is a second function with the property that, whenever a point (a, b) lies on the graph of $y = f(x)$, the point (b, a) lies on the graph of $y = g(x)$. Then

(4.8)
$$g(f(x)) = x \quad \text{for all } x \text{ in the domain of } g \circ f, \text{ and}$$
$$f(g(x)) = x \quad \text{for all } x \text{ in the domain of } f \circ g$$ ⊟

DEFINITION 4.5 Functions f and g that satisfy Equations (4.8) are called **inverses** of each other. That is, g is called the inverse of f, and f is called the inverse of g. ⊟

Thus, if f is the inverse of g, it "un-does" whatever g does, and by the same token, g "un-does" whatever f does.

EXAMPLE 11 Find the inverse function for $y = \sqrt[5]{x^3 + 2}$.

SOLUTION Interchanging x and y yields

$$x = \sqrt[5]{y^3 + 2}$$

We now solve the resulting equation for y. Raising both sides to the 5th power, we have

$$x^5 = y^3 + 2$$

Thus

$$y^3 = x^5 - 2$$

and

$$y = \sqrt[3]{x^5 - 2}$$

EXAMPLE 12 Find the inverse of $y = 3x^9 + 3$.

SOLUTION Interchanging x and y yields

$$x = 3y^9 + 3$$

so that

$$y = \sqrt[9]{\frac{x - 3}{3}}$$

EXAMPLE 13 The demand function for a particular commodity is

$$p = \frac{23.50}{1 + 0.02x}$$

where p is the unit price and x is the number of units sold. Find x as a function of p.

SOLUTION Essentially, we are asked to invert the demand function. There is no need to interchange the names of the variables in this type of problem, we simply solve for x.

$$p(1 + 0.02x) = 23.50$$

$$p + 0.02px = 23.50$$

$$0.02px = 23.50 - p$$

$$x = \frac{50(23.50 - p)}{p}$$

Unfortunately, we encounter some difficulties when we try to apply the "nice" theory of inverses that we have developed to certain functions. We can best illustrate the problems that can arise by looking at another example.

EXAMPLE 14 What is the inverse of $y = x^2$?

SOLUTION The graph of the function $y = x^2$ is, of course, a parabola (see Figure 4.11). We have seen it many times before. If we apply our coordinate-interchanging transformation, we obtain

$$x = y^2$$

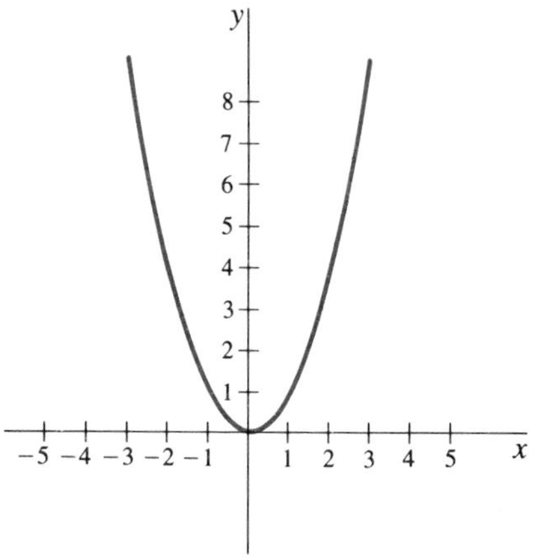

FIGURE 4.11
$y = x^2$

This is also a parabola (see Figure (4.12). It is symmetric with respect to the x-axis rather than the y-axis, and it opens to the right. It is the mirror image of the parabola that is the graph of the function $y = x^2$, using the line $y = x$ as a mirror. Unfortunately, $x = y^2$ is *not* the equation of a function, because its graph has the property that some vertical lines cross it in more than one point. This fact means that, if we try to solve the equation $x = y^2$ for y, we have

$$y = +\sqrt{x} \quad \text{and} \quad y = -\sqrt{x}$$

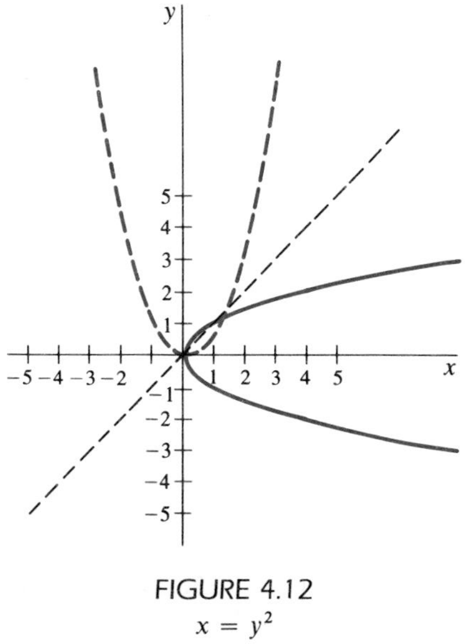

FIGURE 4.12
$x = y^2$

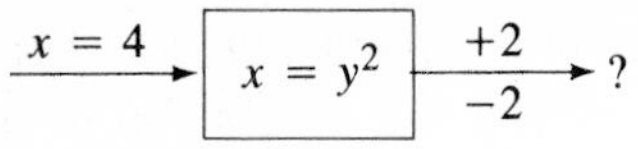

FIGURE 4.13

There are two choices, which represent the top and bottom halves of the graph in Figure 4.12, respectively. We cannot solve for y uniquely in terms of x.

In other words, the problem is that, in a function-machine picture of $x = y^2$ (where y is expressed in terms of x), there are two possible choices for output y. For instance, if the input is $x = 4$, the output y could be either $+2$ or -2 (see Figure 4.13). In order to avoid this difficulty, we restrict the domain of the function $f(x) = x^2$ in such a way that no horizontal line crosses its graph at more than one point. The easiest way to do this is to throw away the left half of the graph and consider only

$$f(x) = x^2 \qquad x \geq 0$$

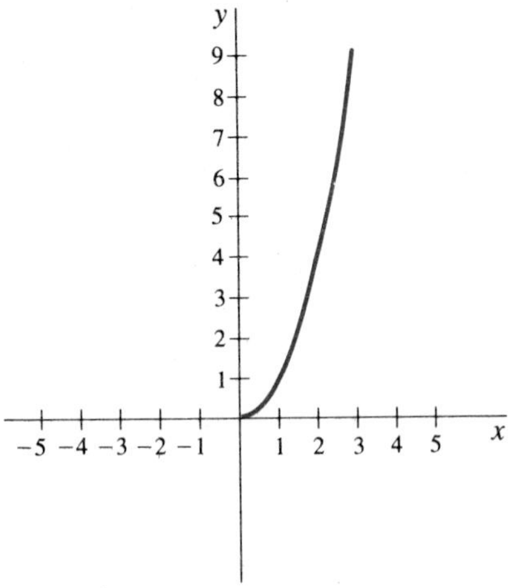

FIGURE 4.14
$f(x) = x^2 \qquad x \geq 0$

If $y = x^2$ for $x \geq 0$ (see Figure 4.15) interchanging x- and y-coordinates yields

$$x = y^2 \qquad y \geq 0$$

Now, because we always want $y \geq 0$, we can solve for y by taking the nonnegative square root of both sides to obtain

$$y = \sqrt{x}$$

Observe that this is a perfectly well defined function whenever $x \geq 0$. Moreover, if we set $g(x) = \sqrt{x}$ for $x \geq 0$ (see Figure 4.15), then

$$f(g(x)) = (f(\sqrt{x})) = (\sqrt{x})^2 = x \qquad \text{for } x \geq 0$$

and

$$g(f(x)) = g(x^2) = \sqrt{x^2} = x \qquad \text{for } x \geq 0$$

The functions f and g are inverses of one another on the restricted domain $x \geq 0$.

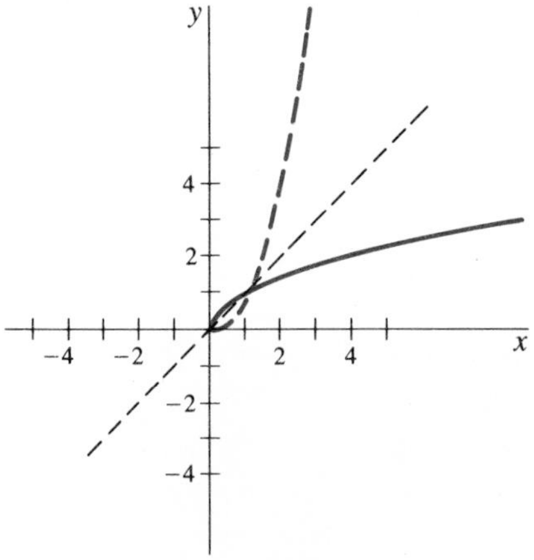

FIGURE 4.15
$g(x) = \sqrt{x}, x \geq 0$

In summary, any function that has the property that no horizontal line intersects its graph in more than one point has an inverse. Functions possessing an inverse are therefore either always increasing (the graph is rising) or always decreasing (the graph is falling) as we move to the right along the graph. For a function that is neither always increasing nor always decreasing on its entire domain (such as $y = x^2$) we can define a **local inverse** on some smaller interval where the curve is always increasing or always decreasing (see Figure 4.16).

In the next section we will apply our discussion of inverse functions to the exponential function $E(x) = e^x$ introduced in Section 4.1.

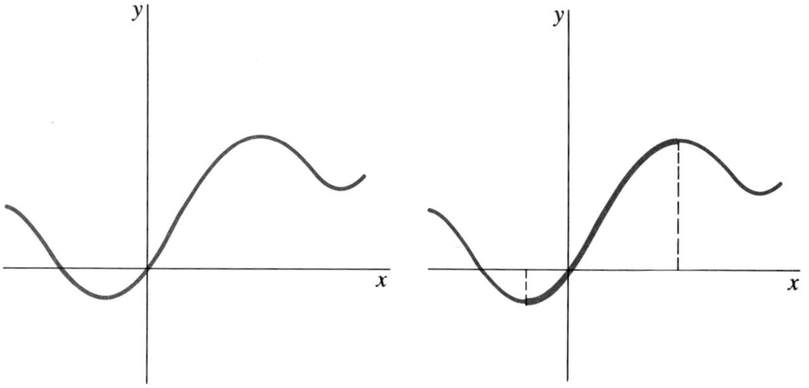

(a) Complete graph of $y = f(x)$

(b) Restriction of the graph of $y = f(x)$ to an interval on which it is increasing

FIGURE 4.16

Section 4.3 EXERCISES

In Exercises 1–6 plot the given point, and find the point that is symmetric to it with respect to the line $y = x$.

1. (2, 5) **2.** (3, 0) **3.** $(-2, -3)$

4. $(-2, -2)$ **5.** (0, 6) **6.** $(-\sqrt{2}, 0)$

In each of Exercises 7–14 a function $y = f(x)$ is given.

i. Find its inverse function in the form $y = g(x)$.

ii. Verify that f and g are indeed inverse functions of each other. That is, verify that $f(g(x)) = x$ and $g(f(x)) = x$ for all appropriate values of x.

iii. Sketch the two graphs on the same set of axes.

7. $y = 3x - 2$ **8.** $y = 5x$ **9.** $y = x + 1$

10. $y = \frac{x}{3} + 2$ **11.** $y = -x$ **12.** $y = -x + 3$

13. $y = x^3 + 2$ **14.** $y = x^{1/3} + 2$

15. Discuss the inverse of $y = 3$ (if there is one).

For each of the functions $y = f(x)$ in Exercises 16–34, find the inverse (if it exists) and state the domain of definition.

16. $y = (x^5 - 1)^{1/3}$ **17.** $y = x^{1/5} - 1$

18. $y = x^2 + 2$ **19.** $y = x^4$

20. $y = x^2 - 2x - 3$ **21.** $y = \frac{1}{x}$

22. $y = \frac{1}{x^2}$ **23.** $y = 8$

24. $y = x^2 + 2x + 1$ **25.** $y = \frac{x}{x + 1}$

26. $y = x^3 + 4$ **27.** $y = \sqrt{x - 2}$

28. $y = (x + 2)^3$ **29.** $y = (x - 1)^4$

30. $y = x^2 - 4$ **31.** $y = \frac{1}{x + 3}$

32. $y = (x + 3)^{1/2}$

33. $y = (x + 4)^{1/3}$

34. $y = (2 - x)^3$

35. The monthly demand x for a certain brand of portable radio is related to its price p by the following demand function. Express x as a function of p.

$$p = 24\left(1 - \frac{\sqrt{x}}{50}\right) \qquad 0 \le x \le 2500$$

36. The price p of a new video casette recorder is related to the monthly demand by the following demand function. Express x as a function of p.

$$p = 575 - \frac{(x + 25)^2}{2000} \qquad 0 \le x \le 1047$$

SECTION 4.4

Implicit Differentiation

Consider the point (4, 2) on the parabola $x = y^2$. In Figure 4.17 we have drawn the tangent line to the curve at that point. How can we compute the slope of that tangent line?

At first glance, this may strike you as a trivial problem. We have seen that, for any function $y = f(x)$ and a point (a, b) lying on its graph, the derivative

$$\left.\frac{dy}{dx}\right|_{x=a} = f'(a)$$

represents the slope of the tangent line at the given point.

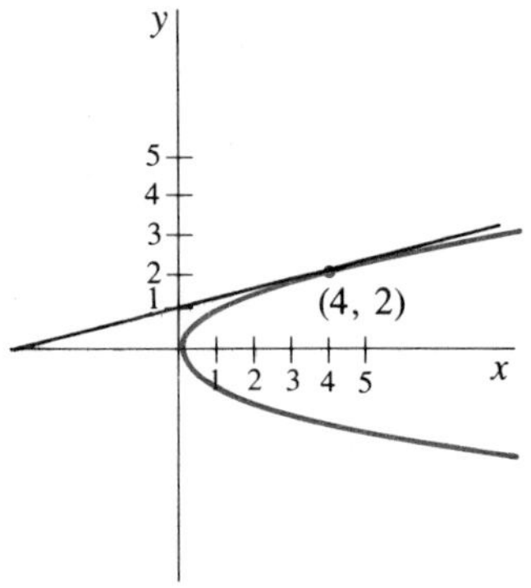

FIGURE 4.17
$x = y^2$

A more thoughtful reexamination of Figure 4.17 now reveals a hidden difficulty. The graph of the equation $x = y^2$ does not represent the graph of a function (because there are vertical lines that intersect the graph in more than one point.) Accordingly, we cannot compute a derivative; we do not have a function to differentiate. And yet our intuition tells us that the slope of the tangent line makes perfectly good sense. How can we resolve this apparent paradox?

There are two alternative solutions. First observe that, although the entire graph $x = y^2$ does not represent the graph of a function, its top half, described by $\sqrt{x} = y$, does (see Figure 4.18). Because the bottom portion does not play any role in determining the slope of the tangent line at the point (4, 2), we simply disregard it. The resulting function $y = \sqrt{x} = x^{1/2}$ is said to be defined *implicitly* by the equation $x = y^2$. We can now compute the derivative directly as follows:

$$\left.\frac{dy}{dx}\right|_{x=4} = \left.\frac{1}{2}x^{-1/2}\right|_{x=4} = \frac{1}{2}(4^{-1/2}) = \frac{1}{4}$$

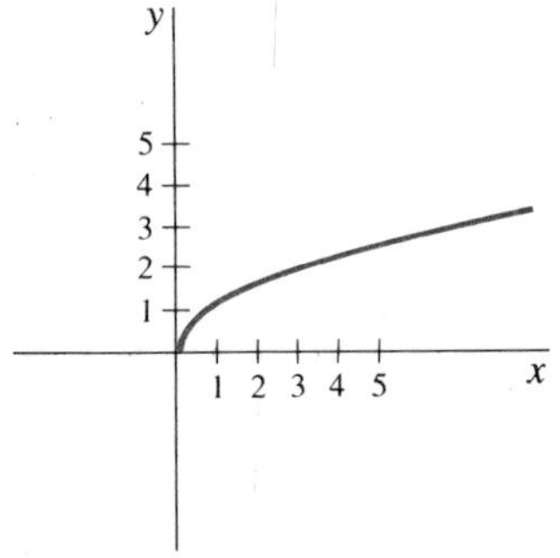

FIGURE 4.18
$y = \sqrt{x}$

A second approach, known as the **method of implicit differentiation**, does not require us to actually determine any implicitly defined functions. Instead we simply *imagine* that we have solved $x = y^2$ for y in terms of x near the point (4, 2) to obtain some function $y = u(x)$. We emphasize again that we do not actually know the function $u(x)$. We merely suppose that it could be found in principle. Thus, substituting $y = u(x)$ into $x = y^2$ yields

$$x = (u(x))^2$$

Next we differentiate this relation, utilizing the chain rule on the right-hand side

$$1 = 2(u(x))^1 u'(x)$$

Thus

$$u'(x) = \frac{1}{2u(x)}$$

We now replace $u(x)$ with y; our answer becomes

$$y' = \frac{1}{2y}$$

Finally, we evaluate our derivative at the point (4, 2),

$$y'|_{(4,2)} = \left.\frac{dy}{dx}\right|_{(4,2)} = \frac{1}{2(2)} = \frac{1}{4}$$

to obtain the same answer that we obtained before.

EXAMPLE 15 Find y' if $x = y^5 + y$.

SOLUTION In this case we don't know how to find y as a function of x. Let's just imagine, however, that y is given by some expression $u(x)$ (which we don't actually know). Then

$$x = (u(x))^5 + u(x)$$

and, upon differentiation,

$$1 = 5(u(x))^4 u'(x) + u'(x)$$

Thus

$$1 = (5u^4(x) + 1)u'(x)$$

or

$$u'(x) = \frac{1}{5u^4(x) + 1}$$

Replacing u by y and u' by y', we find that

$$y' = \frac{1}{5y^4 + 1}$$

EXAMPLE 16 Find the equation of the tangent line to the curve $x^2 - y^2 = 5$ at the point $(3, -2)$.

SOLUTION We do *not* try to solve the equation $x^2 - y^2 = 5$ for y as a function of x. Rather, we simply view y as a function of x, $y = u(x)$, where $u(x)$ is unspecified. Thus our equation becomes

$$x^2 - (u(x))^2 = 5$$

Differentiating both sides of this equation produces

$$2x - 2u(x)u'(x) = 0$$

or

$$u'(x) = \frac{x}{u(x)}$$

Letting $y = u(x)$ and $y' = u'$ yields

$$y' = \frac{x}{y}$$

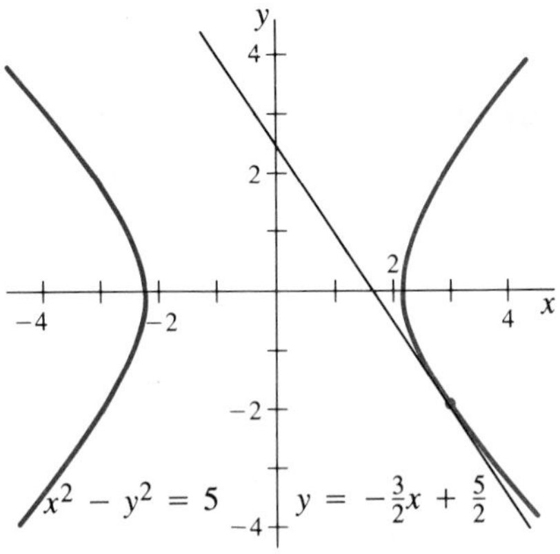

FIGURE 4.19

At the point $(3, -2)$, $y' = -3/2$. This is the slope of the tangent line at the given point. The equation of the tangent line is then

$$y - (-2) = -\frac{3}{2}(x - 3)$$

which can be simplified to

$$y = -\frac{3}{2}x + \frac{5}{2}$$

The graph of $x^2 - y^2 = 5$ and its tangent line at $(3, -2)$ are shown in Figure 4.19. Observe that this graph is not the graph of a function, because there are vertical lines that intersect the graph at more than one point. Nevertheless, at each point of the graph there is a clearly defined tangent line. Implicit differentiation permits us to determine the slope of that tangent line.

EXAMPLE 17 Find y' if $y = e^{x-y}$.

SOLUTION Viewing $y = u(x)$ for some unspecified function $u(x)$, we have $u(x) = e^{x-u(x)}$. Implicit differentiation yields

$$u'(x) = e^{x-u(x)} \cdot (1 - u'(x))$$

which, when we solve for $u'(x)$, gives

$$u'(x) = \frac{e^{x-u(x)}}{1 + e^{x-u(x)}}$$

Setting $y = u(x)$ and $y' = u'$ produces

$$y' = \frac{e^{x-y}}{1 + e^{x-y}}$$

You may be wondering why we bother to set $y = u(x)$, because at the end of each problem we replace each $u(x)$ by y anyway. To see why, suppose we are trying to differentiate a formula that contains the term y^2. Setting $y = u(x)$, we may then differentiate $[u(x)]^2$ by the chain (or power) rule to get

$$2u(x) \cdot u'(x)$$

If we now set $y = u(x)$ and $y' = u'$, this derivative becomes

$$2y \cdot y'$$

If we had asked you to differentiate y^2 directly (without first setting $y = u(x)$), you might have written simply $2y$, forgetting the y'. Our introduction of the function $u(x)$ is a device to help you differentiate correctly. Provided you understand this, you can avoid the introduction of $u(x)$ entirely. But you must remember to be sure to apply the chain rule to every term that contains a y.

EXAMPLE 18 Find $\frac{dy}{dx}$ at the point $(3, -3)$ if $4x^2 + 3xy - y^2 = 0$.

SOLUTION We will differentiate implicitly, thinking of y as some expression $u(x)$, although we will never actually write it that way. Thus

$$\underbrace{8x}_{\substack{\text{derivative}\\ \text{of } 4x^2}} + 3\underbrace{(xy' + y)}_{\substack{\text{derivative}\\ \text{of } xy\\ \text{by product rule}}} - \underbrace{2yy'}_{\substack{\text{derivative of}\\ y^2}} = 0$$

Setting $(x, y) = (3, -3)$ yields

$$24 + 3(3y' - 3) + 6y' = 0$$

which tells us that

$$\left.\frac{dy}{dx}\right|_{(3,-3)} = y'|_{(3,-3)} = -1$$

EXAMPLE 19 Find the derivative dy/dx when $x = y^3$.

SOLUTION This is a straightforward implicit differentiation problem:

$$1 = 3y^2y'$$

or

$$y' = \frac{1}{3y^2}$$

Note that our original problem $x = y^3$ is easily solved for y, so our answer becomes

$$y' = \frac{1}{3y^2} = 3\frac{1}{(x^{1/3})^2} = \frac{1}{3x^{2/3}} = \frac{1}{3}x^{-2/3}$$

which of course agrees with the result obtained by applying the power rule to $x^{1/3}$.

You may recall that, so far, we have actually *proved* the power rule only for nonzero integral exponents (we have *assumed* it for nonintegral exponents). The following example suggests how this rule's validity can be established for rational (fractional) exponents (see Exercise 47).

EXAMPLE 20 Use implicit differentiation and the power rule for positive integer exponents to find the derivative of $y = x^{5/4}$.

SOLUTION Raising each side to the 4th power gives

$$y^4 = (x^{5/4})^4 = x^5$$

Differentiating implicitly, we find that

$$4y^3y' = 5x^4$$

or

$$y' = \frac{5x^4}{4y^3}$$

Finally, substituting $y = x^{5/4}$, we obtain

$$y' = \frac{5x^4}{4(x^{5/4})^3} = \frac{5x^4}{4(x^{15/4})} = \frac{5}{4}x^{4-(15/4)} = \frac{5}{4}x^{(16/4-(15/4)} = \frac{5}{4}x^{1/4}$$

Thus $\frac{d}{dx}(x^{5/4}) = \frac{5}{4}x^{1/4}$, as expected.

We can perform a similar computation for any rational exponent, and in fact the power rule is valid for any real exponent.

THEOREM 4.6 (The Power Rule) Let r be any real number. Then, for $x > 0$, the function x^r is differentiable and its derivative is given by

$$\frac{d}{dx}(x^r) = rx^{r-1}$$

□

EXAMPLE 21 Differentiate $y = x^{-5/8}$.

SOLUTION By Theorem 4.6,

$$\frac{dy}{dx} = -\frac{5}{8}x^{(-5/8)-1} = -\frac{5}{8}x^{-13/8}$$

EXAMPLE 22 Differentiate $y = \sqrt{x^3 - 2x} = (x^3 - 2x)^{1/2}$.

SOLUTION An application of Theorem 4.6, combined with the chain rule, gives

$$y' = \frac{1}{2}(x^3 - 2x)^{-1/2}(3x^2 - 2)$$

Section 4.4 EXERCISES

In Exercises 1–10, differentiate.

1. $x^{2/3}$
2. $x^{7/5} - x^{-7/5}$
3. $6x^{5/11} - \frac{1}{2}x^{2/9}$
4. $x^{3\sqrt{2}} + 5x^{2\sqrt{3}}$

5. $(3x - 2)^{1/5}$

6. $(x^2 + 2x + 5)^{2/7}$

7. $\dfrac{(x^2 + 1)^{1/3}}{(x^3 - x + 1)^{2/5}}$

8. $\dfrac{x^{1/3} + x^{-1/3}}{x^{5/4}}$

9. $(e^{3x} - e^{x^2})^{2/3}$

10. $(x + 1)^{3/5}(x^2 + 2)^{2/3}$

In Exercises 11–13 find the derivative.

11. $y = x^{3/4}$ at $x = 5$

12. $y = x^{-1/2}$ at $x = 7$

13. $y = x^{\sqrt{3}}$ at $x = 8$

In Exercises 14–17 find the equation of the tangent line to the curve.

14. $y = x^{3/2}$ at $(1, 1)$

15. $y = x^{3/2}$ at $(9, 27)$

16. $y = x^{-2/3}$ at $(-8, 1/4)$

17. $y = x^{5/3}$ at $(-8, -32)$

In each of Exercises 18–27 find $y' = dy/dx$ in two ways: (i) via implicit differentiation and (ii) by solving for y explicitly and then differentiating. Your answers should be equivalent.

18. $3x + 2y = 5$

19. $2x^2 - y = 9$

20. $3xy - 7 = 0$

21. $2x - y^2 = 9$

22. $x^2 + y^2 = 9$

23. $5x^2 + 4x + y^2 = 18$

24. $4x^2 + 9y^2 = 36$

25. $y^3 - 3 = 2x^2 - 4x$

26. $\sqrt{x} + \sqrt{y} = 2$

27. $x^{1/3} + y^{1/3} = 1$

In Exercises 28–36 use implicit differentiation to find $y' = dy/dx$.

28. $y = e^{xy}$

29. $x^3 + 2y^2 - y = 8$

30. $y^3 = 2x + 5y$

31. $2x^3y - 3x^2y^2 + x - y = 7$

32. $3x = e^{2y}$

33. $xy + x^2 + y^2 - 2 = 0$

34. $xe^y = x^3 + e^x$

35. $xy = x^2 + y^2$

36. $xy^5 + x^2y^4 - 3x^5 = x + y$

In Exercises 37–42 find the equation of the tangent line to the curve.

37. $x^2 + y^2 = 25$ at the point $(3, 4)$

38. $(x - 2)^2 + (y + 1)^2 = 25$ at the point $(6, 2)$

39. $xy = 6$ at the point $(-2, -3)$

40. $x^2 + xy - y^2 = x$ at the point $(1, 1)$

41. $x^3y + 2x^2y^2 + x - y = 3$ at the point $(-1, 2)$

42. $y^3 + y = x$ at the point $(10, 2)$

In Exercises 43–45 use the technique of implicit differentiation to show that the given statement is true.

43. The tangent line to the circle $x^2 + y^2 = a^2$ at the point (x_0, y_0) has the equation $x_0x + y_0y = a^2$.

44. The equation of the tangent line to the ellipse

$$\frac{x^2}{a^2} + \frac{y^2}{b^2} = 1$$

at the point (x_0, y_0) is

$$\frac{x_0x}{a^2} + \frac{y_0y}{b^2} = 1$$

45. The equation of the tangent line to the hyperbola

$$\frac{x^2}{a^2} - \frac{y^2}{b^2} = 1$$

at the point (x_0, y_0) is

$$\frac{x_0x}{a^2} - \frac{y_0y}{b^2} = 1$$

46. Find the equation of the tangent line to the parabola $y^2 = 2ax$ at the point (x_0, y_0).

47. Let r be any rational number (so r is of form $r = p/q$, where p and q are integers). Prove that

$$\frac{d}{dx}(x^r) = rx^{r-1}$$

by using implicit differentiation, as in Example 20.

SECTION 4.5

Logarithmic Functions and Their Derivatives

One very important function that you encountered in high school that we have not mentioned up to this point is the logarithm. You probably remember looking up numbers in logarithm tables and determining mantissas and characteristics. The justification for studying logarithms was that they had some properties that permitted a remarkable simplification of certain algebraic expressions. Those properties are as follows:

(i) For any two positive numbers a and b,

$$\log ab = \log a + \log b$$

Thus logarithms "reduce multiplication to addition."

(ii) For any two positive numbers a and b,

$$\log(a/b) = \log a - \log b$$

Logarithms "reduce division to subtraction."

(iii) For any positive number a and any real number r,

$$\log(a^r) = r \log a$$

Logarithms "convert powers to multiplication."

These are certainly remarkable properties for a function to have!

In high school you worked primarily with logarithms to the base 10, which are often written $\log_{10}x$. These logarithms possess all three of the properties just outlined, and they satisfy the relation $\log_{10}10 = 1$. It will probably be helpful if we recall how logarithms to base 10 are defined.

DEFINITION 4.7 For any $x > 0$, $\log_{10}x$ is defined as that number y for which

$$10^y = x$$ ⊟

That is, $\log_{10}x$ is the power to which the base 10 must be raised to produce the number x. In particular,

(4.9)
$$\log_{10}10 = 1 \quad \text{because} \quad 10^1 = 10$$
$$\log_{10}1 = 0 \quad \text{because} \quad 10^0 = 1$$
$$\log_{10}0.01 = -2 \quad \text{because} \quad 10^{-2} = \frac{1}{10^2} = 0.01$$

and so on.

In calculus, and in fact in all mathematics courses from this point on, it is convenient to work not with logarithms to the base 10, but rather with logarithms to the base e. You recall that we discussed the number $e = 2.71828\ldots$ at some length in Section 4.1. Logarithms to the base e are called **natural logarithms** and are denoted by $\ln x$. By analogy with Definition 4.7, we have Definition 4.8.

DEFINITION 4.8 For any $x > 0$, $\ln x$ is defined as that number y for which

$$e^y = x$$ ⊟

That is, ln x is the power to which the base e must be raised to produce x. In other words, $y = \ln x$ and $x = e^y$ mean the same thing. Thus, for instance,

$$\ln 1 = 0, \qquad \text{because} \quad e^0 = 1,$$
$$\ln e = 1, \qquad \text{because} \quad e^1 = e,$$
$$\ln e^3 = 3, \qquad \text{because} \quad e^3 = e^3,$$
$$\ln e^{1/2} = \frac{1}{2} \qquad \text{because} \quad e^{1/2} = e^{1/2}$$

In fact, we must have

(4.10)
$$\ln e^x = x$$

because $e^x = e^x$. Also, if $y = \ln x$, then $x = e^y$, so that

(4.11)
$$e^{\ln x} = x \qquad x > 0$$

If we set $f(x) = e^x$ and $g(x) = \ln x$, then Equations (4.10) and (4.11) tell us that

$$g(f(x)) = x \qquad \text{for all } x$$

and
$$f(g(x)) = x \qquad \text{for all } x > 0$$

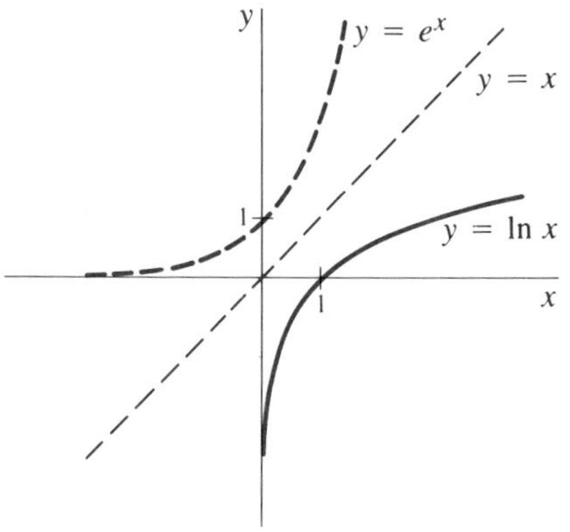

FIGURE 4.20
Graph of $y = \ln x$

In the language of Section 4.3, we have discovered that the exponential function $f(x) = e^x$ and the natural logarithm function $g(x) = \ln x$ are inverses of one another. This fact enables us immediately to obtain the graph of the function $g(x) = \ln x$. All we have to do is sketch $f(x) = e^x$ and then use the line $y = x$ as a mirror. The graph of $g(x) = \ln x$ is the reflection of $f(x) = e^x$ in this mirror (see Figure 4.20). We note that ln x is defined only when $x > 0$, and its graph rises as we move along it from left to right.

There is a great deal that we can now establish about the natural logarithm function because of the fact that $f(x) = e^x$ and $g(x) = \ln x$ are inverses. To start, let us prove that $g(x) = \ln x$ has the properties (4.9) that we usually associate with logarithms.

THEOREM 4.9 For any a and b positive and r any real number:

(i) $\ln ab = \ln a + \ln b$
(ii) $\ln(a/b) = \ln a - \ln b$
(iii) $\ln(a^r) = r \ln a$

PROOF

(i) Let $c = \ln a$ and $d = \ln b$. Then, from the definition of the logarithm,

$$e^c = a \quad \text{and} \quad e^d = b$$

Thus
$$ab = e^c e^d = e^{c+d}$$

That is, $c + d$ is the power to which e must be raised to produce ab. This is equivalent to saying

$$\ln (ab) = c + d = \ln a + \ln b$$

(ii) We again set $c = \ln a$ and $d = \ln b$, so that

$$e^c = a \quad \text{and} \quad e^d = b$$

Then $\frac{a}{b} = \frac{e^c}{e^d} = e^{c-d}$. Raising e to the $(c - d)$ power produces a/b, so we have

$$\ln\left(\frac{a}{b}\right) = c - d = \ln a - \ln b$$

(iii) This time we simply let $c = \ln a$, so that $e^c = a$. Then

$$a^r = (e^c)^r = e^{rc}$$

so that, by the definition of the natural logarithm,

$$\ln a^r = rc = r \ln a$$

□

EXAMPLE 23 If $\ln 2 = 0.6931$ (to four places), find:

(a) $\ln 4$ (b) $\ln 0.5$

SOLUTION

(a) We know that $4 = 2 \cdot 2$. Thus

$$\ln 4 = \ln(2 \cdot 2) = \ln 2 + \ln 2 = 0.6931 + 0.6931 = 1.3862$$

(b) Here we observe that $0.5 = \frac{1}{2} = 2^{-1}$. Thus

$$\ln 0.5 = \ln 2^{-1} = -1 \ln 2 = -0.6931$$

PRACTICE EXERCISE Given that $\ln 2 = 0.6931$ and that $\ln 3 = 1.0986$, find:

(a) $\ln 12$ (b) $\ln 1.5$ (c) $\ln\left(\frac{2}{9}\right)$ (d) $\frac{\ln 2}{\ln 9}$

[Your answers should be: (a) 2.4848; (b) 0.4055; (c) −1.5041; (d) 0.3155.]

Note that, in part (d), $\ln 2/\ln 9$ is *not* the same as $\ln\left(\frac{2}{9}\right)$, and Theorem 4.9 (ii) is not applicable.

As with every other function we have studied, we want to compute the derivative of the natural logarithm.

THEOREM 4.10 For any $x > 0$,

$$(\ln x)' = \frac{d}{dx}(\ln x) = \frac{1}{x}$$

PROOF Let $y = \ln x$. Then, from the definition of the natural logarithm, $x = e^y$. Differentiating this equation implicitly yields

$$1 = e^y \frac{dy}{dx} = x \frac{dy}{dx}$$

so that, for $y = \ln x$, we have

$$y' = \frac{dy}{dx} = \frac{1}{x}$$

Note how we have made use of the fact that the exponential function and the natural logarithm are inverses of one another. □

Combining Theorem 4.10 with the chain rule yields Corollary 4.11.

COROLLARY 4.11 If u is any differentiable function of x, then whenever $u(x) > 0$,

$$\frac{d}{dx}(\ln u) = \frac{1}{u}\frac{du}{dx}$$ □

EXAMPLE 24 Differentiate $y = \ln(x^2 + 8)$.

SOLUTION Applying Corollary 4.11 with $u = x^2 + 8$ yields

$$y' = \underbrace{\frac{1}{x^2 + 8}}_{\frac{1}{u}} \cdot \underbrace{2x}_{\frac{du}{dx}}$$

EXAMPLE 25 Find the derivative of $\ln(e^x + 1)$.

SOLUTION

$$\frac{d}{dx}[\ln(e^x + 1)] = \frac{1}{e^x + 1} \cdot (e^x + 1)' = \frac{e^x}{e^x + 1}$$

EXAMPLE 26 Differentiate $y = \ln \sqrt{\dfrac{x + 1}{x - 2}}$.

SOLUTION Rather than letting

$$u = \sqrt{\frac{x + 1}{x - 2}}$$

and applying Corollary 4.11 directly, let us first try to simplify the formula for y by making use of the special properties of logarithms given in Theorem 4.9. We have, by property (iii) of Theorem 4.9,

$$\ln\left(\frac{x + 1}{x - 2}\right)^{1/2} = \frac{1}{2}\ln\left(\frac{x + 1}{x - 2}\right)$$

By Property ii, this can be simplified to $\frac{1}{2}[\ln(x + 1) - \ln(x - 2)]$. Thus

$$y = \tfrac{1}{2}[\ln(x + 1) - \ln(x - 2)]$$

and so

$$\begin{aligned} y' &= \frac{1}{2}\left[\frac{1}{x + 1} \cdot 1 - \frac{1}{x - 2} \cdot 1\right] = \frac{1}{2}\left(\frac{1}{x + 1} - \frac{1}{x - 2}\right) \\ &= \frac{-3}{2(x + 1)(x - 2)} \end{aligned}$$

LOGARITHMIC DIFFERENTIATION The previous example suggests a method that is frequently useful when we are differentiating a function with a complicated formula. The method is probably best illustrated by working a few examples.

EXAMPLE 27 Find y' if $y = \dfrac{x^2(x - 3)^2}{(x^2 + 4)^5}$.

SOLUTION Your first inclination, if faced with this problem, would probably be to differentiate by using the quotient rule, treating the numerator as a product, and applying the power rule to each of the terms $(x - 3)^2$ and $(x^2 + 4)^5$. It certainly sounds complicated (and it is!), although it can be done. But let's try to simplify the task by taking the natural logarithm of both sides of the equation.

$$\begin{aligned} \ln y &= \ln\left[\frac{x^2(x - 3)^2}{(x^2 + 4)^5}\right] = \ln x^2 + \ln(x - 3)^2 - \ln(x^2 + 4)^5 \\ &= 2 \ln x + 2 \ln(x - 3) - 5 \ln(x^2 + 4) \end{aligned}$$

Now we have an equation that is not solved explicitly for y. But we can still differentiate this equation by using implicit differentiation. Keep in mind that y is a function of x. The derivative of $\ln y(x)$ is given by $\frac{1}{y(x)}\, y'(x)$, according to Corollary 4.11. Thus we have

$$\frac{1}{y} y' = 2\left(\frac{1}{x}\right) + 2\left(\frac{1}{x - 3}\right) - 5\left(\frac{1}{x^2 + 4}\right)(2x)$$

We can now multiply both sides of the resulting equation by y to yield

$$y' = \left[\frac{2}{x} + \frac{2}{x - 3} - \frac{10x}{x^2 + 4}\right] y = \left[\frac{2}{x} + \frac{2}{x - 3} - \frac{10x}{x^2 + 4}\right]\left[\frac{x^2(x - 3)^2}{(x^2 + 4)^5}\right]$$

If desired, the three terms in the bracket can be combined into a single term by combining them into fractions having a common denominator.

In summary, our idea for differentiating a complicated function is first to take the natural logarithm of both sides of the formula for the function and then to differentiate. This technique is known as **logarithmic differentiation**.

EXAMPLE 28 Find y' if $y = a^x$ for some positive constant a.

SOLUTION Let's take the natural logarithm of both sides, so, in accordance with property (iii) of Theorem 4.9, we have

$$\ln y = \ln a^x = x \ln a$$

Keeping in mind that $\ln a$ is just a constant, upon differentiating both sides with respect to x, we get

$$\frac{1}{y} y' = \ln a \quad \text{or} \quad y' = y \ln a$$

Because $y = a^x$, this is the same as $y' = a^x \ln a$

or
$$\frac{d}{dx}(a^x) = a^x \ln a$$

Note that our answer was *not* $y' = xa^{x-1}$, as many students are tempted to write. The formula

$$(x^n)' = nx^{n-1}$$

is valid for functions of the form

$$(\text{Variable base})^{\text{constant exponent}}$$

In Example 28, by contrast, we were faced with the problem of differentiating a function of the form

$$(\text{Constant base})^{\text{variable exponent}}$$

Another remark worth making is that, if we happen to choose $a = e$ in Example 28, our result tells us that

$$(e^x)' = e^x \ln e$$

But $\ln e = 1$, so

$$(e^x)' = e^x$$

This agrees with what we already know.

Let's try a problem in which we differentiate a function of the type

$$(\text{Variable base})^{\text{variable exponent}}$$

EXAMPLE 29 Differentiate $y = x^x$.

SOLUTION We take the natural logarithm of both sides.

$$\ln y = x \ln x$$

Differentiating implicitly with respect to x, we get

$$\frac{1}{y} \cdot y' = x\frac{1}{x} + \ln x = 1 + \ln x$$

so
$$y' = \frac{d}{dx}(x^x) = y(1 + \ln x) = x^x(1 + \ln x)$$

Section 4.5 EXERCISES

In Exercises 1–12 suppose ln 2, ln 3, and ln 5 are known. How would you find each of the following? First write your answers simply in terms of "ln 2, ln 3, and ln 5." Then use a calculator and the approximate values ln 2 = 0.6931, ln 3 = 1.0986, and ln 5 = 1.6094 to approximate your answers.

1. $\ln 8$ **2.** $\ln 6$ **3.** $\ln 25$

4. $\ln 27$ **5.** $\ln 2.5$ **6.** $\ln 0.125$

7. $\ln \frac{9}{25}$ **8.** $\ln 3\sqrt{5}$ **9.** $\ln \sqrt[3]{4}$

10. $\ln \sqrt{30}$ **11.** $\ln 0.1$ **12.** $\ln 360$

Rewrite (simplify) each of the expressions in Exercises 13–24.

13. $\ln(e^{2x})$ **14.** $e^{\ln x^2}$

15. $e^{\ln(1/x)}$ **16.** $\ln e^{1/x}$

17. $e^{-\ln x^3}$ **18.** $\ln e^{-x^3}$

19. $\ln\left(\frac{1}{e^{-x^2}}\right)$ **20.** $\ln(xe^x)$

21. $e^{\ln x^2 + \ln 3}$ **22.** $e^{\ln x - \ln y}$

23. $\ln(e^{-\ln x^2})$ **24.** $\ln e^{1/x} - \ln e^{x^2}$

Find the derivative in each of Exercises 25–41 (do not simplify).

25. $\ln(3x)$ **26.** $\ln(x^2 + 1)$

27. $x \ln x$ **28.** $x^2/\ln(3x)$

29. $x^2\sqrt{\ln(3x)}$ **30.** xe^{-x^2}

31. $\ln(x^2\sqrt{x-1})$ **32.** $\ln(\ln x)$

33. $(\ln ax)e^{bx}$ **34.** $\ln(xe^{-x})$

35. $xe^{1/x}$ **36.** $\ln(2 \ln x)$

37. $\ln \frac{e^x}{e^x + e^{-x}}$ **38.** $\frac{e^x - e^{-x}}{e^x \ln x}$

39. $\ln(\ln x^3)$ **40.** $\ln(\sqrt{x} + xe^{-x})$

41. $e^{(e^x)}$

In Exercises 42–50 find the tangent line to the given curve at the point indicated. (Leave your answer in terms of the logarithm and exponential functions.)

42. $y = \ln(x^2 + 3)$ at $x = -1$

43. $y = e^{-x}$ at $x = 2$

44. $y = \ln(1 - x)$ at $x = -1$

45. $y = xe^{-x}$ at $x = 0$

46. $y = e^{-1/x}$ at $x = 1$

47. $y = x \ln x$ at $x = e^2$

48. $y = x/\ln x$ at $x = 5$

49. $y = \frac{\ln x}{x}$ at $x = 1$

50. $y = \ln(x^2)$ at $x = e$

Differentiate each of the functions in Exercises 51–57. (In many cases you will want to use logarithmic differentiation.)

51. 10^x **52.** $\sqrt{2^x}$ **53.** $x^{\sqrt{2}}$

54. x^π **55.** $x^{\ln x}$ **56.** $x^{(x^x)}$

57. $(x^x)^x$

In Exercises 58–62 find the equation of the tangent line to the given curve at the given point.

58. $y = 2^x$ at $x = 1$ **59.** $y = 3^{-x}$ at $x = -1$

60. $y = x^x$ at $x = 2$ **61.** $y = \pi^x$ at $x = 1$

62. $y = x(2^x)$ at $x = -1$

Verify the results of the differentiations in Exercises 63–69 (*a* and *b* are constants).

63. $\frac{d}{dx}\left[\frac{1}{2a}\ln\frac{(x+a)}{(x-a)}\right] = \frac{1}{a^2 - x^2}$

64. $\frac{d}{dx}\left[\frac{x}{a} - \frac{b\ln(ax+b)}{a^2}\right] = \frac{x}{ax+b}$

65. $\frac{d}{dx}\left[\frac{1}{b}\ln\left(\frac{x}{ax+b}\right)\right] = \frac{1}{x(ax+b)}$

66. $\frac{d}{dx}\left[\ln(x + \sqrt{a^2 + x^2})\right] = \frac{1}{\sqrt{a^2 + x^2}}$

67. $\frac{d}{dx}\left[\ln(x + \sqrt{x^2 - a^2})\right] = \frac{1}{\sqrt{x^2 - a^2}}$

68. $\frac{d}{dx}\left(\frac{1}{a}\ln\frac{x}{a + \sqrt{a^2 + x^2}}\right) = \frac{1}{x\sqrt{a^2 + x^2}}$

69. $\frac{d}{dx}\left(\frac{\sqrt{a^2 + x^2}}{a^2 x}\right) = \frac{-1}{x^2\sqrt{a^2 + x^2}}$

70. Show that: (a) $\frac{d^2}{dx^2}(x \ln x) = \frac{1}{x}$ (b) $\frac{d^3}{dx^3}(x^2 \ln x) = \frac{2}{x}$

(c) $\frac{d^{n+1}}{dx^{n+1}}(x^n \ln x) = \frac{n!}{x}$ (inductively)

71. (a) Verify that, for fixed $a > 0$, $a^x = e^{x \ln a}$.

(b) Use this fact to prove directly (rather than by logarithmic differentiation) that $(d/dx)(a^x) = (a^x) \ln a$.

72. For any $n \geq 1$, show (inductively) that

$$\frac{d^n}{dx^n}(xe^x) = (x + n)e^x$$

73. Use the fact that $x^x = e^{x \ln x}$ (for $x > 0$) to show directly that

$$\frac{d}{dx}(x^x) = x^x(1 + \ln x)$$

Use logarithmic differentiation to find the derivative dy/dx in each of Exercises 74–77.

74. $y = x(x - 1)(x - 2)$

75. $y = x^2(x - 1)^2(x - 2)^2$

76. $y = (x - a_1)(x - a_2)(x - a_3) \ldots (x - a_n)$

77. $y = (x - a_1)^2(x - a_2)^2(x - a_3)^2 \cdots (x - a_n)^2$

78. Explain how you would differentiate a function of the form

$$(\text{Constant base})^{\text{constant exponent}}$$

SECTION 4.6

Exponential Growth and Decay: A First Look at Differential Equations

Any equation involving an unknown function and one or more of its derivatives is called a **differential equation**. Such equations are the natural language in which to express many of the general laws in such fields as physics, chemistry, biology, astronomy, sociology, and economics. In fact, it is safe to say that differential equations arise in just about every field in which mathematics is an important tool. Some examples of differential equations follow.

$$\frac{d}{dt}(mv) = F$$

This is **Newton's second law of motion**. Here m represents the mass of an object, v is its velocity, and F is an external force acting on the object.

$$L\frac{d^2Q}{dt^2} + R\frac{dQ}{dt} + \frac{1}{C}Q(t) = E(t)$$

This equation from electrical circuit theory describes the charge Q of a condenser in a circuit with capacitance C, resistance R, inductance L, and impressed voltage $E(t)$.

$$\frac{d^2x}{dt^2} + kx = 0$$

This equation describes the oscillation of a spring. The constant k is the "spring constant," a measure of the stiffness of the spring.

$$\frac{dP}{dt} = rP(K - P)$$

This is called the **logistic equation**. It describes the growth of a population P whose resources are limited. The constant r is the intrinsic growth rate of the species, and K denotes the carrying capacity of the environment (that is, its ability to sustain the population).

In general, given any function $y = f(x)$, its derivative dy/dx represents the rate of change of the variable y with respect to x; that is, it tells how a change in x produces a change in y. In almost every natural process, the variables and their rates of change are connected with one another by means of some underlying scientific principle describing that process. When we express this connection in mathematical terms, the result is often a differential equation. A few examples should help to clarify this idea.

EXAMPLE 30 Consider the simple differential equation

$$\frac{dy}{dx} = y$$

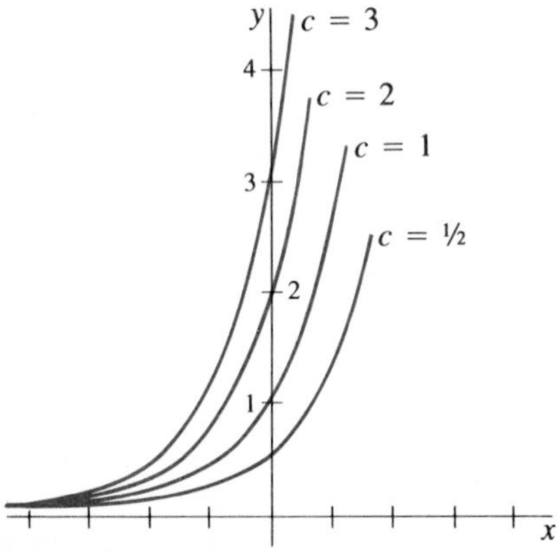

FIGURE 4.21
Graph of $y = ce^x$

We say that we have solved the equation when we have determined a function $y = y(x)$ that satisfies it. Such a function itself is called a **solution** of the differential equation. Here we are looking for a function whose derivative dy/dx is equal to the function itself. Do we know such a function? We certainly should! In fact, we devoted Section 4.1 to the discovery of a function with precisely this property—namely, the exponential function

$$y = y(x) = e^x$$

We can verify that it "works" by simply substituting it and its derivative $dy/dx = e^x$ into the differential equation $dy/dx = y$. But are there any other functions that might work? After a bit of thought, you might discover that $y = 2e^x$ also works, because its derivative is $dy/dx = 2e^x = y$. Similarly, $y = 5e^x$ has derivative $dy/dx = 5e^x = y$, so it too is a solution. In fact, for any choice of constant c,

$$y(x) = ce^x \tag{4.12}$$

satisfies our differential equation. It can be shown that functions of the form $y = ce^x$ are the only functions that satisfy the differential equation of Example 30. Because the formula $y = ce^x$ provides all solutions of this differential equation by appropriate choice of the arbitrary constant c, it is called the **general solution** of the differential equation. Figure 4.21 shows the graph of $y = ce^x$ for various choices of c.

Even if we know the form of the general solution, there are still a very large number of distinct individual solutions—in fact, there is a different solution for each different choice of c. How can we single out one of these solutions from all of the others for consideration? The answer is that we can't unless we have more information. Often this additional information is given by specifying what value the solution should have when $x = 0$. For example, we might require that

$$y(0) = 4 \tag{4.13}$$

In graphical terms, we are stipulating that our solution should cross the y-axis at $y = 4$. In this case, our function $y(x) = ce^x$ must satisfy

$$4 = y(0) = ce^0 = c \cdot 1 = c$$

Thus $c = 4$ and $y(x) = 4e^x$. Condition (4.13) is an example of an **initial condition**; it specifies the value that our solution initially takes on when $x = 0$. Observe that it enables us to select one solution from among all the solutions of the differential equation. The equation given in Example 30 and an initial condition such as Condition (4.13), taken together, constitute an **initial value problem**.

Another illustration of the notion of a differential equation is offered by a problem from Newtonian physics.

EXAMPLE 31 According to the law of gravity discovered by Sir Isaac Newton, an object falling to earth experiences a constant acceleration of -32 feet per second^2. (The minus sign arises because we measure distance from the ground upward, and the acceleration is downward towards the earth.) Because acceleration is by definition the derivative of velocity, we see that the **differential equation** describing this phenomenon is

(4.14)
$$\frac{dv}{dt} = -32$$

where $v(t)$ denotes the velocity of the object at time t.

Suppose that at time $t = 0$ we throw a ball directly upward with an **initial velocity** of 20 ft/sec. That is, our **initial condition** is

(4.15)
$$v(0) = 20$$

Let us now determine the velocity $v(t)$ for each time $t > 0$.

We begin by asking what functions $v(t)$ satisfy Equation (4.14). That is, what functions have derivative equal to -32? Because -32 is a constant, we are led to guess that a linear function is a candidate for $v(t)$. In fact it is easy to see that $v(t) = -32t$ will do, because

$$\frac{dv}{dt} = v'(t) = -32$$

More generally,

(4.16)
$$v(t) = -32t + c$$

will satisfy Equation (4.14) for any constant c, because the derivative of a constant is 0.

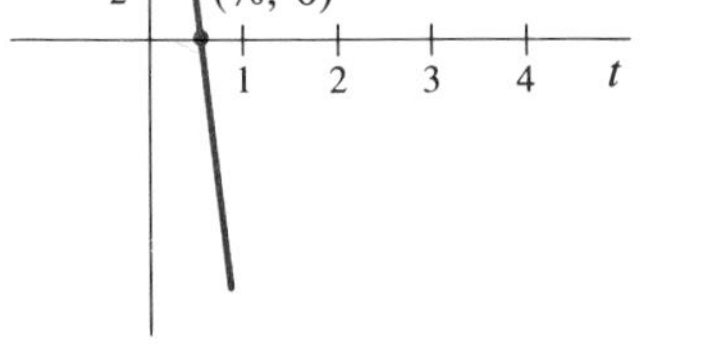

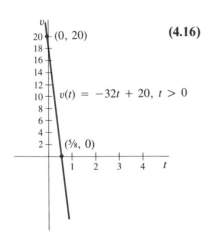

FIGURE 4.22

Expression (4.16) is the general solution of our differential equation. *Every* solution of the differential equation has the form $v(t) = -32t + c$ for some constant c.

Finally, we employ the initial condition $v(0) = 20$ in Equation (4.16) to determine the constant c:

$$20 = v(0) = -32 \cdot 0 + c$$

so $c = 20$. Thus

$$v(t) = -32t + 20 \qquad \text{for } t > 0$$

If we plot the linear function $v(t)$ as shown in Figure 4.22, we note that, for times t less than $\frac{5}{8}$, $v(t)$ is positive so that the ball is rising, whereas, for times t greater than $\frac{5}{8}$, $v(t)$ is negative and the ball is falling. Also, at $t = \frac{5}{8}$ the velocity is momentarily equal to 0.

EXAMPLE 32 Another example of a differential equation that is similar to the one in Example 30, but slightly more complicated, is

$$\frac{dy}{dx} = ky \tag{4.17}$$

for some fixed number $k \neq 0$. This equation expresses the idea that we wish to determine a function $y(x)$ whose derivative $\frac{dy}{dx}$ is equal to k times the function itself. Alternatively, we may interpret the equation as stating that the derivative of y is proportional to y itself, with constant of proportionality k. As before, we ask ourselves whether we know of such a function. And again we find that we do. If $y(x) = e^{kx}$, then $\frac{dy}{dx} = ke^{kx} = ky$. Thus $y = e^{kx}$ is a solution of Equation (4.17). By analogy with Example 30, we guess that

$$y(x) = ce^{kx} \tag{4.18}$$

is also a solution for every choice of constant c. This is easily verified because, for $y = ce^{kx}$ we have $\frac{dy}{dx} = kce^{kx}$, so that

$$\underbrace{kce^{kx}}_{\frac{dy}{dx}} = \underbrace{k}_{k}(\underbrace{ce^{kx}}_{y})$$

Because it contains the arbitrary constant c, $y(x) = ce^{kx}$ is called the general solution of Equation 4.17. Specifying an initial condition for our solution when $x = 0$ serves to determine the value of c. For instance, if we require that

$$y(0) = 2$$

then we must have

$$2 = y(0) = ce^{k \cdot 0} = ce^0 = c \cdot 1$$

so that $c = 2$ and

$$y(x) = 2e^{kx}$$

Of course, the names we give our variables are immaterial. If we write the differential equation as

$$\frac{dM}{dt} = kM \tag{4.19}$$

this is really the same equation as Equation (4.17). In Equation (4.19), $M = M(t)$ might be interpreted as representing the amount of a substance (or population) whose rate of change (with respect to time) is proportional (with proportionality constant k) to the amount present. The general solution of this equation was found to be

$$M(t) = ce^{kt} \tag{4.20}$$

Because we have $M(0) = ce^{k \cdot 0} = c$ when $t = 0$, the constant c represents the amount of substance M that we start with at time $t = 0$. This is our initial condition. If k is positive, M is growing, and our solution to Equation (4.20) exhibits what is called **exponential growth**; if k is negative, it exhibits **exponential decay**. The proportionality constant k is sometimes called the **rate of exponential growth** or **the rate of exponential decay**, according to whether k is positive or negative. The two cases are illustrated in Figure 4.23.

Let us now give an application of these ideas.

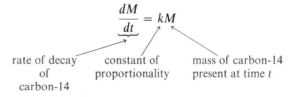

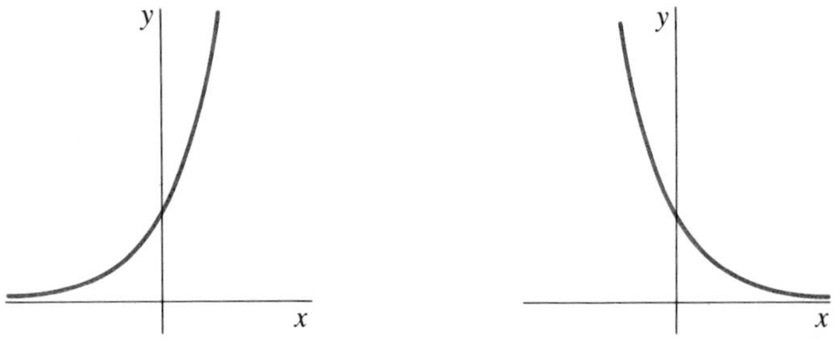

(a) Exponential growth: $M(t) = M(0)e^{kt}$ $\quad k > 0$

(b) Exponential decay: $M(t) = M(0)e^{kt}$ $\quad k < 0$

FIGURE 4.23

EXAMPLE 33 It is well known that radioactive elements, such as uranium and radium, decay into other elements and subatomic particles at a rate that is proportional to the amount of the radioactive substance present. The time required for half of a given quantity of this radioactive substance to decompose is known as its **half-life**. The half-life is independent of the amount of the element that we start with. For example, a certain radioactive isotope of carbon, carbon-14, is known to have a half-life of about 5600 years. This means that, no matter how much carbon-14 we start with, 5600 years later exactly half that amount will remain. After another 5600 years, exactly half of *that* amount (or a quarter of the original) will be left, and so on.

The fact that the rate of decay of a radioactive substance is proportional to the amount of the substance present can be expressed by means of a differential equation. Let

$$M = M(t) = \text{mass of carbon-14 present at time } t$$

Then dM/dt represents the rate of decay of the mass of carbon-14 with respect to time, and we have

$$\frac{dM}{dt} = kM$$

rate of decay of carbon-14 ($\frac{dM}{dt}$); constant of proportionality (k); mass of carbon-14 present at time t (M)

We know that the general solution of this equation is

$$M(t) = M(0)e^{kt}$$

In this situation, the number k is called the **rate constant**. Its value depends on the properties of carbon-14. If we assume that at time $t = 0$ we start with 100 grams of carbon-14, then $M(0) = 100$, so

$$M(t) = 100e^{kt}$$

We know that after 5600 years the mass of carbon-14 remaining will be only 50 grams. Thus, at $t = 5600$ years, we have $M = 50$, so

$$50 = 100e^{k(5600)}$$

or

$$e^{5600k} = \frac{50}{100} = 0.5$$

We can solve this equation for k by using the properties of the natural logarithm developed in the previous section. Taking the natural logarithm of both sides of the equation yields

$$\ln(e^{5600k}) = 5600k = \ln 0.5 = \ln \tfrac{1}{2} = -\ln 2$$

or

$$5600k \approx -0.6931$$

Thus we must have

$$k \approx -0.000124$$

Using this value in our formula for $M(t)$, we get

$$M(t) = 100e^{-0.000124t}$$

This formula enables us to compute the amount of carbon-14 that will be present at any time t. Observe that k is a negative number so that our solution exhibits exponential *decay*, as we would expect.

This simple idea forms the basis for an important technique used by archaeologists and geologists called **radiocarbon dating**. Carbon-14 is present at all times in the atmosphere as a result of certain interactions between cosmic ray neutrons and nitrogen. Animals and plants, while they are alive, are constantly breathing this substance and incorporating it into their tissues. The proportion of carbon-14 present in the animal or plant remains approximately constant throughout its life, but after death no more is absorbed and the carbon-14 starts to decay. Thus, if a human bone that is dug up contains only half as much radioactive carbon as that of a man living today, the person from whom the bone came must have lived about 5600 years ago. If only a quarter as much of the radioactive substance remains, the person lived about 5600 + 5600 = 11,200 years ago. This method has permitted scientists to date such famous historical artifacts as the Dead Sea Scrolls (see Exercise 60 in this section) and the prehistoric monuments at Stonehenge.

EXAMPLE 34 **Continuously Compounded Interest** Some savings accounts pay interest that is compounded continuously. This is another example of exponential growth, because the rate of growth of the amount of money in the account,

dA/dt, is proportional to the amount at any given time. That is,

$$\frac{dA}{dt} = kA$$

If we denote the initial amount by A_0, then, as we have seen before, the solution to this differential equation is $A(t) = A_0e^{kt}$, which we usually write with r in place of k to indicate that the growth constant k is the interest rate r, so

$$A(t) = A_0e^{rt}$$

Let's use this equation to solve the following problems.

(a) How much money would be in a savings account after 5 years if \$1000 were left to earn 6% interest compounded continuously?

(b) How much should new parents deposit in a savings account earning 8% in order to pay \$60,000 toward their child's college education 18 years from now?

(c) If these parents have only \$10,000 to deposit now, what interest rate will guarantee them \$60,000 in 18 years?

SOLUTION

(a) $A_0 = \$1000$, $r = 0.06$, and $t = 5$, so $A = 1000e^{(0.06)5} = 1000e^{0.3} \approx$ \$1350.

(b) $A(18) = \$60{,}000$, $r = 0.08$, and $t = 18$, so $\$60{,}000 = A_0e^{(0.08)18}$. Solving for A_0 yields

$$A_0 = \frac{\$60{,}000}{e^{(0.08)18}} \approx \$14{,}216$$

(c) $A(18) = \$60{,}000$, $A_0 = \$10{,}000$, and $t = 18$, so $60{,}000 = 10{,}000e^{r(18)}$. We solve this equation for r as follows:

$$\frac{60{,}000}{10{,}000} = 6 = e^{r(18)}$$

$$\ln 6 = \ln(e^{r(18)}) = 18r$$

$$r = \frac{\ln 6}{18} \approx 0.0995$$

The parents should look for an interest rate close to 10%.

EXAMPLE 35 A bacteria culture is known to grow at a rate proportional to the amount present. Initially 1000 strands of the bacteria are observed in the culture; after 3 hours there are 3000 strands. How many strands of bacteria will there be after 5 hours?

SOLUTION Let $M = M(t)$ be the number of strands of bacteria after t hours have elapsed. Then dM/dt is the rate of growth of M, and it must be proportional to M. That is, there is a constant $k > 0$ for which

$$\frac{dM}{dt} = kM$$

The general solution of this equation is

$$M(t) = M(0)e^{kt}$$

When $t = 0$ there are 1000 strands of bacteria, so

$$M(t) = 1000\, e^{kt}$$

We must try now to find the value of the constant k. Well, at time $t = 3$ there are 3000 strands of bacteria:

$$M(3) = 1000\, e^{3k} = 3000$$

or

$$e^{3k} = 3$$

In order to determine k from this equation, we simply take the natural logarithm of both sides. Because $\ln e^{3k} = 3k$, we have $3k = \ln 3$, so

$$k = \frac{\ln 3}{3} = \frac{1.0986}{3} = 0.3662$$

Substituting this value for k in our formula for M yields

$$M(t) = 1000\, e^{0.3662t}$$

Our problem will be solved if we can determine how many strands of bacteria there will be after 5 hours. This requires us to evaluate our formula for M when $t = 5$.

$$M(5) = 1000\, e^{0.3662(5)} = 1000\, e^{1.8310} = 6240$$

EXAMPLE 36 In a tank containing 50 gallons of water, 20 pounds of salt are dissolved. Stirring distributes the salt evenly thoughout the water. Fresh water is entering the tank through a hose at the top at the rate of 2 gallons per minute, and the well-stirred solution is leaving through an opening at the bottom at the same rate (see Figure 4.24).

(a) Find a formula for the amount of salt $M(t)$ in the tank at time t.

(b) How long will it take for there to be only 5 pounds of salt left?

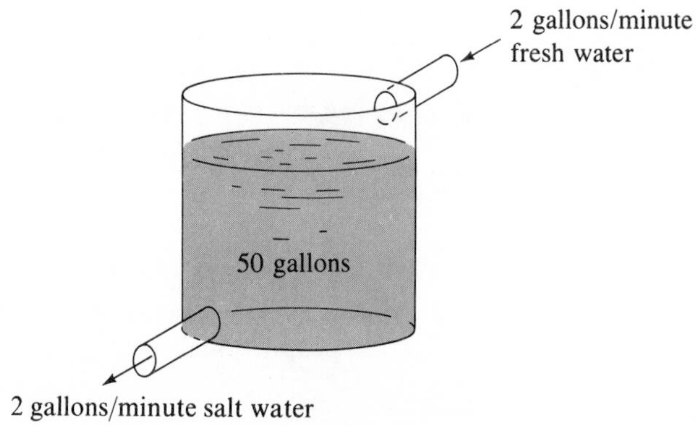

FIGURE 4.24

SOLUTION

(a) If we let $M(t)$ denote the amount of salt in the tank at time t, then the rate of change of M, $M'(t)$, must be the rate at which salt is leaving (no new salt is entering the tank).

In each minute, 2 gallons of solution leave out of a total of 50 gallons. Thus 2/50 of the amount of salt in the tank leaves each minute. Thus

$$\frac{dM}{dt} = -\frac{2}{50}M = -0.04M$$

This is a differential equation of the form of Equation (4.19) with $k = -0.04$, so its general solution is

$$M(t) = M(0)e^{-0.04t}$$

Initially, at $t = 0$, there are 20 pounds of salt in the tank, and so

$$M(t) = 20e^{-0.04t}$$

(b) We want to know when there will be only 5 pounds of salt remaining. That is, when will $M(t) = 5$? Thus

$$5 = 20e^{-0.04t}$$

or

$$0.25 = e^{-0.04t}$$

Taking the natural logarithm of both sides produces

$$\ln(0.25) = \ln(e^{-0.04t}) = -0.04t$$

Thus

$$t = \frac{\ln(0.25)}{-0.04} = \frac{-1.3863}{-0.04} = 34.66$$

There will be 5 pounds of salt left after approximately 35 minutes.

The preceding example illustrates the idea of **compartmental analysis**. If a substance is present in a system in several distinguishable locations and if it passes from one location to another at some particular rate, each location is called a *compartment* for the substance. Thus Example 36 involves the use of a single compartment. This idea is often useful for analyzing biological systems.

For instance, we may view the blood stream as a single compartment in which there is present a certain amount of radioactive iodine that is being used as a tracer. Some of the iodine will be absorbed by the thyroid gland, while some will be filtered by the kidneys into the urine. It seems reasonable to suppose that the amount absorbed by each organ will be proportional to the amount present in the blood stream. The thyroid and kidneys here are playing the same role as the opening at the bottom of the tank in Example 36. If $M(t)$ denotes the amount of iodine present in the blood at time t, then the rate of absorption of iodine by the thyroid is $k_1M(t)$ for some constant of proportionality k_1, and the rate of filtration by the kidneys is $k_2M(t)$ for some constant k_2. The total rate of decrease is thus

$$\frac{dM}{dt} = -k_1 M - k_2 M = (-k_1 - k_2)M$$

which is again an equation of the same type as Equation (4.19). Conclusions that can be drawn on the basis of this simple equation are in good agreement with experimental evidence.

EXAMPLE 37 **Newton's law of cooling** states that the surface temperature of an object changes at a rate that is proportional to the difference between the temperature of the object and that of its surrounding (called the **ambient temperature**). It provides a satisfactory description of facts that can be observed experimentally in many circumstances.

A metal bar at a temperature of 100°F is brought into a room with constant temperature of 40°. After 20 minutes the temperature of the bar is 80°. Find the temperature of the bar after 1 hour (60 minutes).

SOLUTION Let $T(t)$ = the surface temperature of the bar at time t, and let A denote the ambient temperature. Then $T(t) - A$ is the difference between the temperature of the bar and room temperature. Newton's law of cooling tells us that

$$\frac{dT}{dt} = k\left[T(t) - A\right] \tag{4.21}$$

for some constant of proportionality k. In this particular problem, $A = 40$. Now, if we let $M(t) = T(t) - 40$, then $dM/dt = dT/dt$, so Equation (4.21) becomes

$$\frac{dM}{dt} = kM$$

As we know, this has the solution

$$M(t) = M(0)e^{kt}$$

or, because $M(0) = T(0) - 40 = 100 - 40 = 60$ and $T(t) = 40 + M(t)$,

$$T(t) = 40 + 60e^{kt}$$

After 20 minutes ($t = 20$), $T = 80$. Thus

$$80 = 40 + 60e^{20k}$$

so $e^{20k} = \frac{2}{3}$. Taking the natural logarithm of both sides, we find that

$$k = \frac{1}{20}\ln\left(\frac{2}{3}\right) = -0.02027$$

which in turn implies that

$$T(t) = 40 + 60\,e^{-0.02027t}$$

After 60 minutes,

$$T(60) = 40 + 60\,e^{-0.02027(60)} \approx 57.8^\circ$$

EXAMPLE 38 A certain radioactive material decays at a rate proportional to the amount present. A block of this material that originally had a mass of 100 grams is observed, after 20 years, to have a mass of 80 grams. Find the half-life of the substance.

SOLUTION If $M(t)$ denotes the amount of radioactive material at time t, then, as in Example 33,

$$M(t) = 100\, e^{kt}$$

By observation, if $t = 20$ years, then $M = 80$ grams, so

$$80 = 100\, e^{20k}$$

Taking natural logarithms, we get

$$20k = \ln\left(\frac{80}{100}\right) = \ln(0.8)$$

and

$$k = \frac{\ln 0.8}{20} = \frac{-0.2231}{20} = -0.0112$$

Substituting this into our expression for M yields

$$M(t) = 100\, e^{-0.0112t}$$

You will recall that the half-life represents the length of time it takes for half of the substance to decay. Thus, after that length of time, only 50 grams will remain. Setting

$$50 = M(t) = 100\, e^{-0.0112t}$$

we find

$$e^{-0.0112t} = \frac{1}{2}$$

and so

$$-0.0112t = \ln \frac{1}{2}$$

Thus

$$t = \frac{\ln 1/2}{-0.0112} = \frac{-0.6931}{-0.0112} = 61.9 \text{ years}$$

SEMILOGARITHMIC COORDINATES (OPTIONAL) Imagine that we decide to study the growth of a population of squirrels. The number of squirrels at the end of each year* is given in Table 4.2.

* Data adapted from R. L. Smith, *Ecology and Field Biology*, 2nd ed. (New York: Harper & Row, 1974), p. 316.

TABLE 4.2

Year	0	1	2	3	4	5	6	7	8	9	10
Number of squirrels	30	38	52	69	87	110	141	180	230	293	375

From theoretical considerations, we suspect that the growth rate of the squirrel population is proportional to the number of squirrels. Phrased in another way, we expect that the population is growing exponentially. How can we confirm or reject this hypothesis on the basis of our data?

One approach would be to plot on a piece of graph paper the ordered pairs (t, N), where t represents the number of years elapsed and N the number of squirrels at the end of t years. If we connect these points together, the resulting curve should look like an exponential curve, with equation of the form

$$N = ce^{kt} \tag{4.22}$$

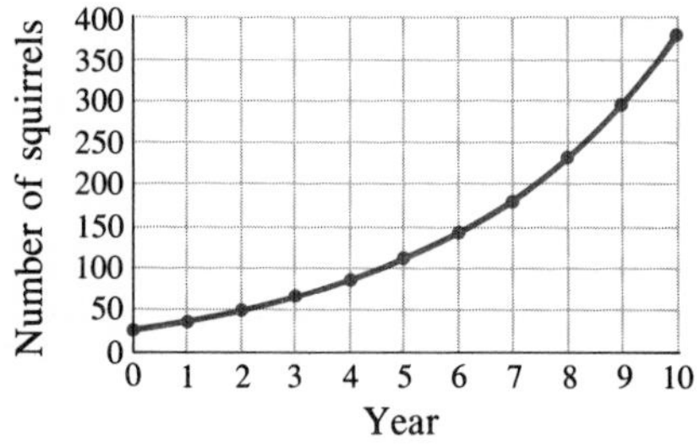

FIGURE 4.25

for appropriate choice of constant c and k. In Figure 4.25 we have begun the construction of just such a curve. It looks like it might be an exponential, but are we sure? Maybe it is part of a parabola. And how can we estimate the sizes of c and k? Can we solve these problems?

Let's suppose that Equation (4.22) holds and take the natural logarithm of both sides of the formula. Then, by the properties of logarithms,

$$\ln N = \ln ce^{kt} = \ln c + \ln(e^{kt}) = \ln c + kt$$

Let's set $y = \ln N$ and $b = \ln c$. Thus

$$y = kt + b$$

which is simply the equation of a straight line with slope k and y-intercept b. Let's construct a new table (Table 4.3) that contains all the information of Table 4.2 as well as the values of $y = \ln N$ (correct to one decimal place).

TABLE 4.3

Year	0	1	2	3	4	5	6	7	8	9	10
N	30	38	52	69	87	110	141	180	230	293	375
$y = \ln N$	3.4	3.6	4.0	4.2	4.5	4.7	4.9	5.2	5.4	5.7	5.9

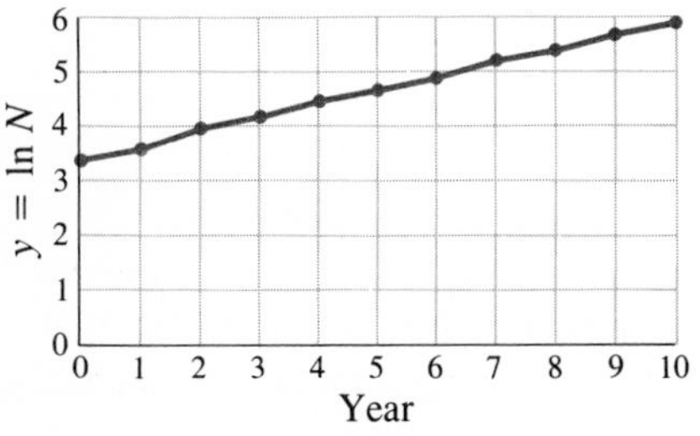

(a)
FIGURE 4.26

If we now construct a new graph with t- and y-axes, it follows that our squirrel population is growing exponentially if and only if the set of all points (t, y) lie on a straight line. In Figure 4.26(a) these points are plotted and connected. We see that the points do, in fact, appear to lie very nearly on a straight line. We also observe that this line crosses the y-axis when $b = 3.4$. Thus $\ln c = b = 3.4$ and $c = e^{3.4} = 30$. Moreover, if we choose two points on our line, we can compute its slope k. (Since all the points are not exactly collinear, the computed value of k will vary slightly with the choice of the two points. A more precise method of fitting a line to a set of points is discussed in Sec-

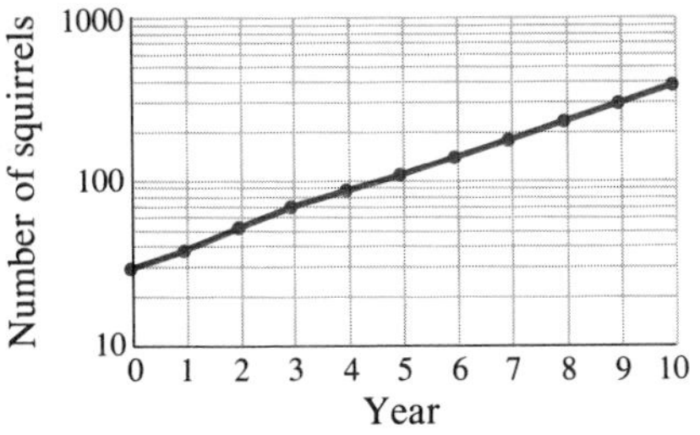

(b)
FIGURE 4.26

tion 8.4.) For example, if we choose (0, 3.4) and (10, 5.9), then

$$k = \frac{5.9 - 3.4}{10 - 0} = \frac{2.5}{10} = 0.25$$

Thus N is given approximately by the formula

$$N = 30\, e^{0.25t}$$

How nicely this trick of using logarithms solved our problems!

Because this idea is such a nice one, mathematicians have devised a way to make use of it without going to the trouble of computing $y = \ln N$, as we did in Table 4.3. This is done by using special graph paper to plot the points (t, N) directly: see Figure 4.26(b). We now describe how such graph paper is designed.

Suppose we take the y-axis of Figure 4.26(a) and replace each labeled y value by its corresponding N value. The correspondence between $y = \ln N$ and $N = e^y$ is illustrated in Table 4.4. We can imagine that every point on the vertical axis is labeled with its N value rather than its y value. What we have now is this: an "N-axis" on which each point N has distance $y = \ln N$ above the point originally marked $y = 0$. For example, the point $N = 30$ will be $\ln 30 \approx 3.40$ units above the point originally marked $y = 0$. Such a vertical axis is called a **logarithmic scale** (Figure 4.27).

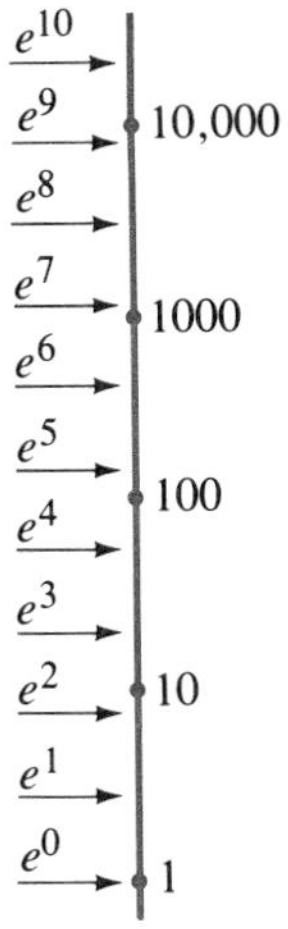

FIGURE 4.27
Vertical logarithmic scale

TABLE 4.4

y	N
0	$e^0 = 1$
1	$e^1 \approx 2.72$
2	$e^2 \approx 7.39$
3	$e^3 \approx 20.09$
4	$e^4 \approx 54.60$
5	$e^5 \approx 148.41$
6	$e^6 \approx 403.43$

The newly labeled N-axis looks strange at first: equal changes in N are not represented by equal vertical distances. Rather, the further up the scale we move, the greater the change in N that each unit of distance on the graph represents. Indeed, from Figure 4.27, it is evident that the powers of e are equally spaced. Graph paper on which one axis possesses the usual linear scale, while the other axis has a logarithmic scale, is called **semilogarithmic paper**. Usually, the points where N is a power of ten—these are equally spaced too—are the values of N that are marked out in advance on a logarithmic scale, rather than the powers of e [see Figures 4.26(b) and 4.27].

Let's solidify our understanding of these ideas by working an example.

EXAMPLE 39 Catalase is an enzyme that decomposes peroxide into water and oxygen. Its enzymatic activity decreases during exposure to sunlight in the presence of oxygen. Determine, from the data given in Table 4.5*, whether the decrease in

* Mitchell and Anderson, "Catalase Photo Inactivation," *Science*, 150, 74(1965).

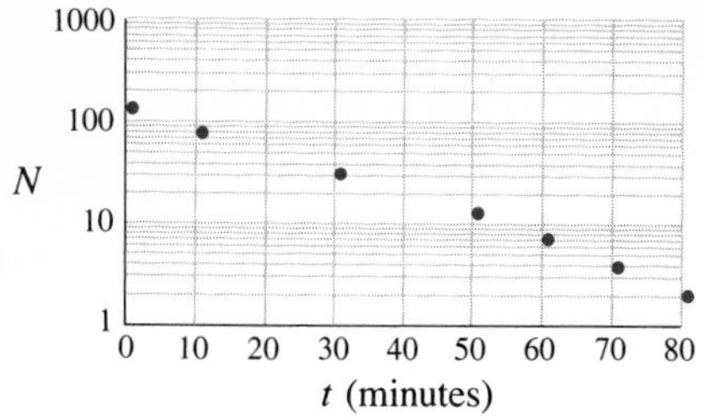

FIGURE 4.28

enzymatic activity obeys an exponential decay law $N = ce^{kt}$, where N represents the concentration of catalase (in μg/10 ml) as a function of the time t (in minutes). Can you also estimate c and k?

TABLE 4.5

t (minutes)	0	10	30	50	60	70	80
N (μg/ml)	121	74	30	12	6.7	3.7	2.0

SOLUTION We construct a plot of the pairs (t, N) on a piece of semi-logarithmic paper (see Figure 4.28). The points all lie on or near a straight line, so we conclude that the decay is exponential. To approximate c, we simply observe that the line crosses the N-axis at (0, 121) and so $c = 121$.

It is somewhat more complicated to find k. Recall that k is the slope of the straight line found by plotting the points $(t, \ln N)$, as we did in Figure 4.26(a). Thus, if we choose two points on our line—say (50, 12), and (0, 121)—then

$$k = \frac{\ln 121 - \ln 12}{0 - 50} = \frac{4.796 - 2.485}{-50} = -0.0462$$

It follows that (approximately)

$$N = 121\,e^{-0.0462t}$$

Section 4.6 EXERCISES

1. (a) Find the *general solution* of the differential equation $\frac{dy}{dx} = 2x - 3$.

(b) Find the *particular solution* corresponding to the initial condition $y(0) = 5$. That is, solve the initial value problem.

2. (a) Find the *general solution* of the differential equation $\frac{dy}{dx} = 3x^2 + 4x + 2$.

(b) Find the *particular solution* corresponding to the initial condition $y(0) = 9$.

3. (a) Find the *general solution* of the differential equation $\frac{dy}{dx} = \frac{1}{x^2} - x + 1$.

(b) Find the *particular solution* corresponding to the initial condition $y(-1) = 5$.

4. (a) Find the *general solution* of the differential equation $\frac{dy}{dx} = e^x$.

(b) Find the *particular solution* corresponding to the initial condition $y(0) = 1$.

5. (a) Find the *general solution* of the differential equation $\frac{dy}{dx} = -3x$.

(b) Find the *particular solution* corresponding to the initial condition $y(0) = 8$.

6. (a) Find the *general solution* of the differential equation $\frac{dy}{dx} = -3y$.

(b) Find the *particular solution* corresponding to the initial condition $y(0) = 8$.

7. Verify that $y = xe^x$ is a solution of the differential equation

$$\frac{dy}{dx} = e^x(x + 1)$$

8. Verify that $y = xe^{-x}$ is a solution of the differential equation

$$\frac{dy}{dx} = e^{-x} - y$$

9. Verify that $y = \frac{\ln x}{x}$ is a solution of the differential equation

$$\frac{dy}{dx} = \frac{1 - \ln x}{x^2}$$

10. Verify that $y = xe^x$ is a solution of the differential equation

$$\frac{d^2y}{dx^2} = y + 2e^x$$

11. Verify that $y = x \ln x - x$ is a solution of the differential equation

$$\frac{dy}{dx} = \ln x$$

12. Show that, for any positive integer r, $y = xe^x$ is a solution of the differential equation

$$\frac{d^ry}{dx^r} = y + re^x$$

13. Verify that $y = e^{x^2/2} + c$ (c any constant) is a solution of the differential equation $\frac{dy}{dx} = xy$.

14. Verify that $y = xe^x$ is a solution of the differential equation

$$\frac{d^2y}{dx^2} - \frac{dy}{dx} = \frac{y}{x}$$

15. Verify that $y = \sqrt{x^2 + 1}$ is a solution of

$$\frac{dy}{dx} = \frac{x}{\sqrt{x^2 + 1}}$$

16. Verify that both $y = \frac{1}{2}\left(\frac{x - 1}{x + 1}\right)$ and $y = \frac{-1}{x + 1}$ are solutions of

$$\frac{dy}{dx} = \frac{1}{(x + 1)^2}$$

17. Verify that $y = \sqrt{x^2 + 3}$ is a solution of $\frac{dy}{dx} = \frac{x}{y}$.

Find $f(x)$ under the conditions given in Exercises 18–21.

18. $f''(x) = 12, f'(1) = 3, f(1) = 4$

19. $f''(x) = 6x + 2, f'(1) = 1, f(1) = 3$

20. $f'''(x) = 6, f''(1) = 6, f'(1) = 4, f(1) = 0$

21. $f'(x) = \frac{1}{x}, f(e) = 3$

In Exercises 22–25 find the equation of motion $s = f(t)$ of an object (where s = distance, v = speed, and a = acceleration).

22. $a = -32$; when $t = 0$, we have $s = 0$ and $v = 64$

23. $a = 4t$; when $t = 0$, we have $s = 4$ and $v = 16$

24. $a = 6t - 1$; when $t = 0$, we have $s = 9$ and $v = 48$

25. $v = 6t^2 - t$; when $t = 2$, we have $s = 4$

26. A ball is thrown upward from the ground at a speed of 48 ft/sec.
    (a) Find the equation of motion $s = f(t)$. *Hint*: $s''(t) = -32 = v'(t)$.
    (b) How high is it after 1 sec? After 2 sec?
    (c) How high does it rise? When does it reach its highest point?
    (d) When does it hit the ground? With what speed?

27. A ball is dropped from a platform 60 ft high. Find its equation of motion. When does it hit the ground? With what speed?

28. A man is standing on the roof of a building 200 ft high. When does a ball hit the ground if the man: (a) drops it; (b) throws it downward at 12 ft/sec; (c) throws it upward at 12 ft/sec.

29. A ball is dropped from a height of 60 ft. It bounces off the ground with a speed $\frac{3}{4}$ the speed of its impact. How high does it go on the first bounce? On the second?

30. A projectile is thrown upward with initial speed v_0 from an initial position s_0. What is the equation of motion? What is the maximal height?

31. What constant acceleration will take a car from 0 to 60 mph (88 ft/sec) in 20 sec?

32. What constant acceleration is needed for a car starting from rest to travel 1000 ft in 10 sec?

33. A car moving at 88 ft/sec (60 mph) is subjected to a constant deceleration (negative acceleration) of 11 ft/sec^2. How far does it travel before coming to rest?

34. What constant deceleration is needed for a car moving at 60 mph to stop within 200 ft?

35. Suppose we have 1 gram of radioactive material and that its exponential decay rate constant is $k = -0.001$. Determine how much of this material there will be in (a) 100 years; (b)1000 years.

36. Suppose we have 0.01 grams of a radioactive material and its rate of exponential decay is $k = -0.001$. Determine how much there was (a) 1000 years ago; (b) 1 million years ago.

37. Accounts that compound interest a fixed number of times each year (such as quarterly, monthly, or daily) earn less interest per year than those that compound interest continuously. To see this, consider the interest earned on \$1000 for 2 years at 6%. Calculate the amount when interest is compounded continuously by using $A(t) = A_0 e^{rt}$. For noncontinuous compounding, use

$$A = A_0\left(1 + \frac{i}{m}\right)^{mn}$$

where i = interest rate (expressed as a decimal number), m = number of times per year that interest is compounded, and n = number of years. How much money will be in the account if interest is compounded: (a) quarterly, (b) monthly, (c) daily, (d) continuously?

38. Compound interest earns more money for an account than simple interest. For example, interest compounded continuously at 8% earns \$8.33 of interest in one year on a deposit of \$100. Here 8.33% is called the **effective annual interest rate**. Refer to Exercise 37 and find the effective annual interest rate when an annual rate of 7% is compounded: (a) quarterly, (b) monthly, (c) daily, (d) continuously.

39. While rummaging through your grandparents' attic, you discover an overdue library book from 1920. If the maximum fine then was \$2.00, but it would have earned 5% interest compounded continuously, how much would it cost to pay the fine and interest in 1986?

40. You have just inherited \$10,000 and would like to invest it to save for a \$20,000 down payment on a house or condominium. If you find an investment paying 9% compounded continuously, how long will you have to wait to earn the down payment?

41. (Sales Decline) Marketing studies show that, if all other market factors remain unchanged but advertising of a product is halted, sales of the unadvertised product will decline continuously at a rate proportional to the current sales at any time t. That is,

$$\frac{ds}{dt} = -kS$$

The solution to this differential equation is $S(t) = S_0 e^{-kt}$. Assume that advertised sales of big burgers are 5000 per week at a local store. After one week without advertising, sales have dropped off to 4000 big burgers per week.

(a) How many big burgers will be sold during the third week without advertising?

(b) When sales drop below 1000 per week, big burger will have to close. When will this occur if advertising does not resume?

42. Suppose $f(t) = ce^{kt}$, c and k constant. Show that, if $f(t_1)$ and $f(t_2)$ are known for any t_1, t_2, then we can determine the constant of proportionality k. In fact,

$$k = \frac{\ln f(t_2) - \ln f(t_1)}{t_2 - t_1}$$

In Exercises 43–45 consider a chemical reaction in which a substance decomposes at a rate proportional to the amount of the substance present.

43. If 10 lb of the substance becomes 5 lb in 6 hr, how long does it take before there is only 1 lb left?

44. If 10 lb of the substance reduces to 9 lb in 1 hr, how long does it take before there are only 5 lb left?

45. If 50 lb of our substance shrink to 15 lb in 5 hr, how long does it take for 95% of the substance to disappear? How long for 98% to disappear?

46. Suppose the constant of proportionality for a process of exponential decay is $k = -6$. When will the amount of material left be: (a) half the original amount, (b) one-third of the original amount; (c) one-tenth of the original amount?

47. Suppose 1% of a radioactive material decomposes in 15 years. What is the material's half-life? When will 99% of the material be gone?

48. Suppose the rate of increase of the population of the United States is proportional to the population already present. Say the population in 1950 was 200 million and the population in 1975 is 230 million.

(a) What will the population be in the year 2000?

(b) When will the population reach 300 million?

(c) What is the constant of proportionality k?

49. Suppose the population of Egypt grows at the rate of 1% per year. How long does it take for the population to double?

50. Assume that the population of the earth is now 3 billion and that it is expected to reach 8 billion in 100 years. At what rate must the population grow per year in order for this to occur (i.e., what is the growth constant k)?

51. A culture of bacteria grows at a rate proportional to the number of bacteria.

(a) If it takes 20 hours for 10^7 bacteria to double in number, how long does it take for 10^7 bacteria to become 10^8?

(b) If it takes 15 hours for the culture to double in size, how long does it take for the culture to triple in size?

52. A culture of bacteria grows at the rate of 2% per day. How long does it take for it to double?

53. A culture of bacteria doubles in size in 9 hours. What is the rate of exponential growth k?

(a) Suppose we are in a situation of exponential growth, with rate of exponential growth k; then $f(t) = ce^{kt}$. Show that the "doubling time" D [the time it takes for $f(t)$ to double] depends only on k (not on c or t). More precisely,

$$D = \frac{\ln 2}{k} \quad \text{or} \quad kD = \ln 2$$

(b) Show that this relationship carries over to exponential decay. That is, if H denotes the half-life, then

$$kH = \ln 2$$

(c) Show that $f(t)$ is given by

$$f(t) = f(0)\, 2^{t/D}$$

for exponential growth, and by

$$f(t) = f(0)\, 2^{-t/H}$$

for exponential decay.

54. Consider the general situation of exponential decay: ce^{kt}, $k < 0$. Determine when the amount of material is: (a) $\frac{1}{3}$ the original amount; (b) $\frac{1}{10}$ the original amount.

55. The temperature $T = T(t)$ of a cup of hot coffee decreases at a rate proportional to the difference between the temperature of the coffee and the constant temperature A of the surrounding air. Suppose the air temperature A is 70°F and it takes 10 minutes for the coffee to go from almost boiling—say, 200°F—to reasonably lukewarm—say, 100°F. Find the coffee temperature after: (a) 5 minutes; (b) 20 minutes; (c) 1 hour; (d) 5 hours.

56. Refer to Exercise 55. When does the temperature of the coffee reach (a) 140°F; (b) 120°F; (c) 85°F; (d) 65°F?

57. Refer to Exercises 55 and 56. Find a formula for $T - A$ for the general case of an initial difference in temperature c and constant of proportionality k.

58. A couple goes into a restaurant to chat over a cup of coffee. The man gets his coffee, adds cream at room temperature, waits 10 min, and then drinks. The woman, on the other hand, lets her coffee cool for 10 min, adds the same amount of cream at room temperature, and drinks immediately. Whose coffee is hotter? (You may wish to make additional assumptions.) Does it matter how much cream is added? Does it matter what the room temperature is? Does it matter what the initial temperature of the coffee is?

59. The atmospheric pressure P (in millimeters of mercury) at an altitude of x kilometers above sea level varies according to the law $dP/dx = kP$, where k is a constant. $P(0) = 760$ and $P(5) = 411$.

(a) Find $P(7)$.

(b) Find the altitude for which $P(x) = 100$.

60. The Dead Sea Scrolls are believed to have been written about 100 B.C. What percentage of their original carbon-14 content should remain today?

61. If \$2500 is invested in a certificate of deposit that earns 10% interest compounded continuously, how much will the certificate be worth in 18 months?

62. In how many years will the value of the certificate in Exercise 61 triple?

SECTION 4.7

Related Rates

There is a wide class of problems that arise in situations in which two (or more) rates of change are related to each other in some way. For example, the inflation rate is related to the prime interest rate, and a change in body weight is related to a change in food intake. From our knowledge of one of these rates, we attempt to deduce the other. This generally requires that we use the technique of implicit differentiation developed in Section 4.4.

Some examples should serve to illustrate both the nature of related-rates problems and their solution.

EXAMPLE 40 Suppose a man is standing atop a 10-ft ladder that is leaning against a wall with its base 4 ft from the bottom of the wall. Assume that the ladder starts to slip at time $t = 0$ and that its base slides outward at 2 ft/sec (see Figure 4.29). After 2 seconds, how fast is the top of the ladder moving?

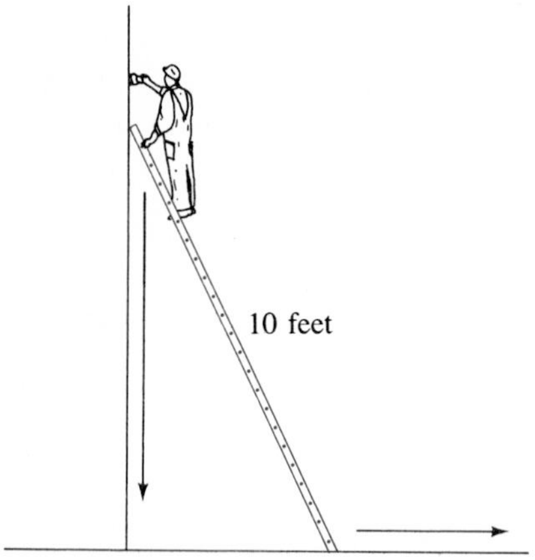

FIGURE 4.29

SOLUTION Let us begin our analysis by introducing some variables. Let's denote by $x = x(t)$ the distance of the base of the ladder from the wall at time t (see Figure 4.30). Also let us denote the distance from the top of the ladder to the ground by $y = y(t)$.

We are given that the base of the ladder moves outward at 2 ft/sec. Thus

$$\frac{dx}{dt} = 2$$

We are asked to find the rate at which $y(t)$ is decreasing. That is, we wish to find dy/dt. Well, how are the variables x and y related? From Figure 4.30 we see that

$$x^2 + y^2 = 100$$

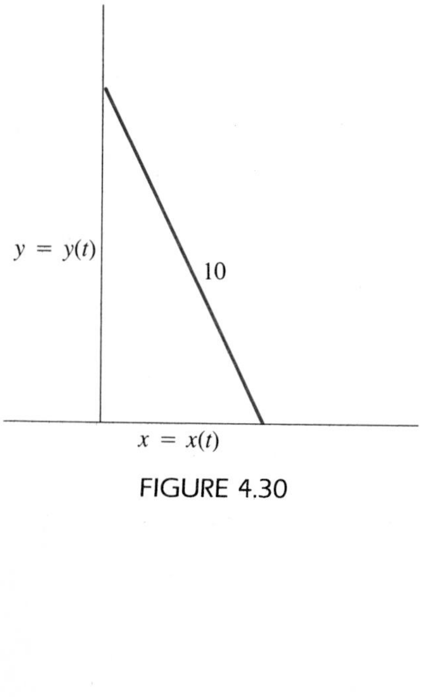

FIGURE 4.30

We call this a **static relation** between x and y; it holds at every time t. Differentiating the static relation implicitly with respect to t, we obtain

$$2x\frac{dx}{dt} + 2y\frac{dy}{dt} = 0$$

which can be solved for the desired rate of change of y:

$$\frac{dy}{dt} = -\frac{x}{y}\frac{dx}{dt}$$

Note that the height y varies in a fairly complicated manner. As the base of the ladder moves further from the wall, x increases while y decreases. This means that the top of the ladder is moving downward at an ever-increasing speed, even though the base of the ladder is moving at the constant speed of 2 ft/sec.

We are given that $x(0) = 4$ at $t = 0$. Thus, after 2 seconds, x will have increased an additional 4 ft, so $x(2) = 8$. Because

$$x^2 + y^2 = 100$$

we have, at time $t = 2$,

$$y = \sqrt{100 - x^2} = \sqrt{100 - (8)^2} = \sqrt{36} = 6$$

Combining these results with $dx/dt = 2$ yields

$$\left.\frac{dy}{dt}\right|_{t=2} = -\frac{2x(2)}{y(2)} = -\frac{2(8)}{6} = -8/3$$

The minus sign signifies that y is decreasing.

Let's summarize the basic ingredients of our method for solving related-rates problems in the following guide.

GUIDE FOR RELATED-RATES PROBLEMS

1 We draw a picture illustrating the statement of the problem.

2 We label, or name, with variables $x, y, z, \ldots$ all those quantities whose rates of change either are given to us or we are asked to find. Keep in mind that such phrases as *feet per second*, *miles per hour*, and *cubic inches per minute*, represent rates of change. A question that begins "How fast . . ." is asking us to find the rate of change of some quantity.

3 After naming the appropriate variables, we write down symbolically what we are given and what we are asked to find about the derivatives of these variables. Thus, for instance, in Example 40 we had $dx/dt = 2$, $dy/dt = ?$

4 We examine carefully our sketch from step 1 as labeled in step 2 and then try to find a static equation relating the variables whose derivatives we are either given or must find. Note that the static equation we seek will *not* involve the time t explicitly. Generally, such equations come directly from plane geometry (as in the use of the Pythagorean theorem in Example 40), from solid geometry, or from some underlying physical principle.

5 Once we have found the static equation, we differentiate both sides of it implicitly with respect to time t. In so doing we obtain an equation relating our variables and their derivatives, or rates of change. From this point, we make use of routine algebraic techniques to find the desired rate of change.

Let's consider some more examples.

EXAMPLE 41 An inverted conical reservoir with a base of radius 20 ft and a height of 100 ft is to be filled with water. Suppose that water is poured in at a rate of 28 cubic feet per second (ft^3/sec). How fast is the height of the water level (as measured from the bottom) rising at the point when the height is 56 ft?

SOLUTION In Figure 4.31 we have drawn a sketch of our reservoir. We see that we are given the rate at which the volume of water in the reservoir is increasing and are asked to find the rate of change of the height of the water. Thus, letting $V = V(t)$ denote the volume of water at time t, and letting $h = h(t)$ denote the height of the water level at time t, we have

$$\frac{dV}{dt} = 28 \text{ ft}^3/\text{sec} \qquad \frac{dh}{dt} = ?$$

We next need a static equation relating V to h. In this case, such an equation follows from solid geometry. In fact, the formula for the volume of a right circular cone of height h and radius of base r is given by

$$V = \frac{1}{3}\pi r^2 h \tag{4.23}$$

Equation (4.23) is certainly an equation involving V and h. The problem is that it also involves the radius r! So it is not yet the static equation we need. How can we eliminate r from Equation 4.23? We reason as follows:

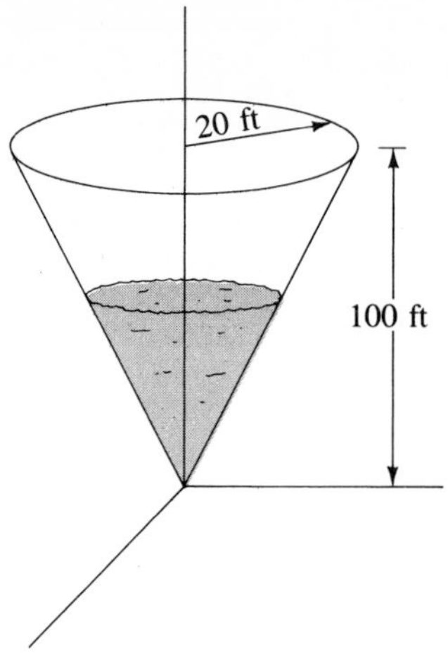

FIGURE 4.31

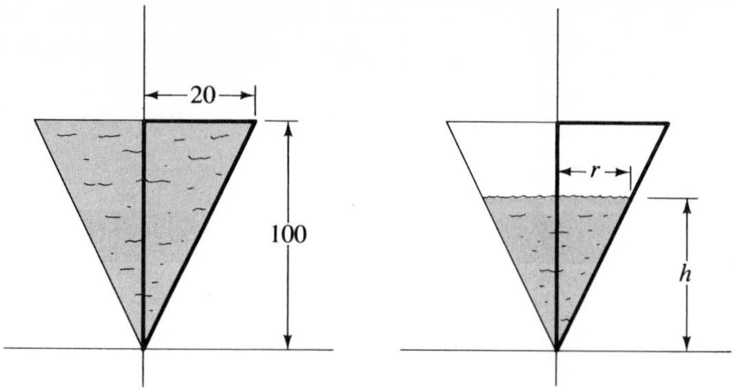

(a) Full reservoir

(b) Partially filled reservoir

FIGURE 4.32

If the reservoir is full, the height of the water level is 100 ft and its radius is 20 ft. If the reservoir is not full, it has some height h and some radius r (see Figure 4.32). But the two darkened triangles in Figure 4.32 are similar, so corresponding sides must be proportional. Thus

$$\frac{20}{100} = \frac{r}{h}$$

or

$$r = \frac{h}{5}$$

Plugging this into Equation 4.23 gives

$$V = \frac{1}{3}\pi\left(\frac{h}{5}\right)^2 h = \frac{\pi h^3}{75}$$

This is our static equation relating V and h. Next we differentiate both sides implicitly with respect to t.

$$\frac{dV}{dt} = \frac{3\pi}{75}h^2\frac{dh}{dt} = \frac{\pi h^2}{25}\frac{dh}{dt}$$

We know that $\frac{dV}{dt} = 28$. Thus

$$28 = \frac{\pi h^2}{25}\frac{dh}{dt}$$

or

$$\frac{dh}{dt} = \frac{700}{\pi h^2}$$

We were specifically asked to find the rate of change of the height when $h = 56$ ft. Substituting $h = 56$ finally gives

$$\frac{dh}{dt} = \frac{700}{\pi(56)^2} = \frac{25}{4\pi(28)} \approx 0.071 \text{ ft/sec}$$

EXAMPLE 42 Consider a single-celled spherical organism. Suppose that the volume of the cell is increasing at a rate of $1 \times 10^{-17}\ \text{cm}^3/\text{day}$. Find a formula for the rate of increase of the surface area of the cell.

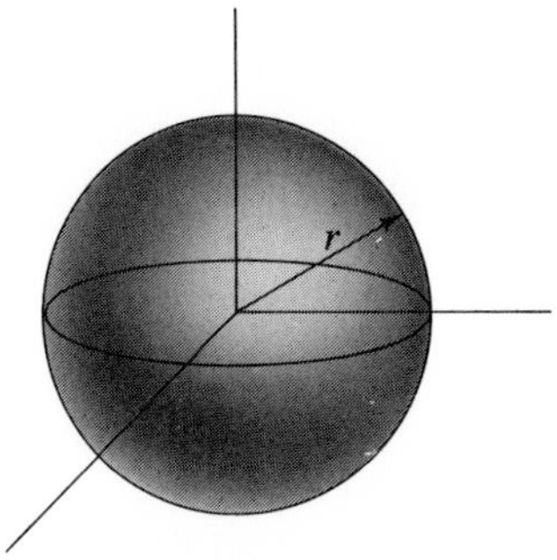

FIGURE 4.33

SOLUTION In Figure 4.33 we show a spherical cell. Let $V = V(t)$ denote the volume of the cell at time t, and let $S = S(t)$ denote its surface area. Our problem then tells us that

$$\frac{dV}{dt} = 1 \times 10^{-17}\ \text{cm}^3/\text{day}$$

and asks us to find $\frac{dS}{dt}$.

Again, we need a static equation relating S and V. Well, we do know formulas for the surface area and volume of a sphere, respectively, in terms of its radius:

$$S = 4\pi r^2$$

$$V = \frac{4}{3}\pi r^3$$

We seek to eliminate r and get an equation involving only S and V. From the second equation we see that

$$r^3 = \frac{3}{4\pi} V$$

so we take cube roots of both sides.

$$r = \left(\frac{3}{4\pi} V\right)^{1/3}$$

Substituting this expression into the formula for S gives

$$S = 4\pi\left[\left(\frac{3}{4\pi} V\right)^{1/3}\right]^2 = 4\pi\left(\frac{3}{4\pi}\right)^{2/3} V^{2/3}$$

For simplicity, let's write $\lambda = 4\pi\left(\frac{3}{4\pi}\right)^{2/3} \approx 4.836$, so that

$$S = \lambda V^{2/3}$$

This is the equation we need. Differentiating both sides implicitly with respect to t yields

$$\frac{dS}{dt} = \frac{2\lambda}{3} V^{-1/3}\frac{dV}{dt} = \frac{2\lambda}{3V^{1/3}}\frac{dV}{dt}$$

We are given $\frac{dV}{dt} = 1 \times 10^{-17}\ \text{cm}^3/\text{day}$, so

$$\frac{dS}{dt} = \frac{2\lambda}{3V^{1/3}}(1 \times 10^{-17}) \approx \frac{3.224 \times 10^{-17}}{V^{1/3}}$$

This is our desired formula.

Observe in Example 42 that the larger the volume V, the smaller the rate of increase of the surface area. This rather simple statement has some very profound biological implications. Certain biological functions, such as the exchange of oxygen and carbon monoxide through the cell wall, have rates that are proportional to the surface area of the cell. On the other hand, the metabolic capacity of a cell is proportional to its volume. It follows that, if the volume of the cell gets too large, $\frac{dS}{dt}$ will become small, the chemical exchange capability of the cell will be insufficient relative to its metabolic capacity, and the cell will die.

Similarly, in multicellular organisms some bodily functions, such as kidney filtration, depend on the surface area of the animal's volume. Such considerations imply a limit to the ultimate size of the organism.

For a more thorough discussion of such ideas (for example, why a 20-ft kangaroo couldn't jump 4 times higher than a 5-ft kangaroo), you are referred to the fascinating monograph by J. Maynard Smith, *Mathematical Ideas in Biology* (London–New York: Cambridge University Press, 1968).

Section 4.7 EXERCISES

1. The length of each side of a square is increasing at the rate of 7 in./sec. At what rate is the area of the square increasing when the side is 50 in. long?

2. How is Exercise 1 affected when the word *increasing* is changed to *decreasing*?

3. The radius of a circle is decreasing at the rate of 3 in. per minute. How fast is the area decreasing when the radius is 25 in.?

4. The radius of a circle is increasing at the rate of 3 in. per minute. How fast is the area increasing when the area is 100 in.^{2}?

5. The area of a circle is increasing at the rate of 7 in.^{2} per hour. At what rate is the radius increasing when the radius is 20 in.?

6. The area of a circle is decreasing at the rate of 4 in.^{2}/sec. At what rate is the radius decreasing when the area is 200 in.^{2}?

7. A 13-ft ladder is leaning against a wall. The foot of the ladder is slipping (away from the wall) at the rate of 3 in./sec. How fast is the top of the ladder sliding (down along the wall) when the foot of the ladder is 5 ft from the wall?

8. Two ships start out from the same point at noon. The first sails due north at 30 mph and the second sails due east at 40 mph. Find the rate at which the ships are separating (that is, the rate at which the distance between them is changing) at 3:30 P.M.

9. Do Exercise 8 if the second ship starts out at 2:00 P.M.

10. A conical reservoir, with its vertex down, has a height of 20 ft, and the diameter of its top is 12 ft. Water is pouring out at the vertex at the rate of 5 ft^3 per minute. How fast is the water level falling when: (a) the water is 10 ft deep; (b) the volume of water is 100 ft^3?

11. Sand is poured on a pile at the rate of 2 ft^3 per minute, and the sandpile always maintains a conical shape with the radius of the base being twice the height. At what rate is the height increasing when the height is 3 ft?

12. The radius of a circle is changing at the rate of 3 in./sec. When is the area changing at the rate of 15 in.^{2}/sec?

13. The area of a circle is changing at the rate of 10 in.^{2}/sec. When is the radius changing at the rate of 3 in./sec?

14. Discuss what happens in Exercises 3, 4, 5, 6 when the word *area* is replaced by *circumference*.

15. A large snowball is melting (where it has no chance at all!) at the rate of 12 ft^3/sec. How fast is the radius shrinking when the radius is 100 ft? $\left(V = \frac{4}{3}\pi r^3\right)$

16. You are blowing up a spherical balloon at a constant rate of 2 $in.^3$/sec. At what rate is the radius increasing when the volume is 100 $in.^3$?

17. Suppose it is known that a snowball melts (and a raindrop grows in volume) at a rate proportional to its surface area. Show that its radius changes at a constant rate (surface area $= 4\pi r^2$).

18. At noon, a ship that is 100 miles north of a fixed point X steams due south at a rate of 25 mph. At 2:00 P.M. another ship, which is 100 miles west of point X steams due east at a rate of 40 mph. Determine how fast the distance between the ships is changing (increasing or decreasing) at:

(a) 3:00 P.M. (b) 4:00 P.M. (c) 5:00 P.M. (d) 6:00 P.M.

19. A kite is 250 ft above the ground and is being blown horizontally and away from the person flying it at the rate of 4 ft/sec. At what rate is the string being let out when there are 600 ft of string already out?

20. Point A moves along the line $y = -x$ at the rate of α ft/sec while point B moves along the x-axis at the rate of β ft/sec. How fast is the distance between them changing when A is at $(a, -a)$ and B is at $(b, 0)$? Discuss the various cases.

21. A particle is moving along the circle $x^2 + y^2 = 25$. Determine when (that is, where) the rate of change of the y-coordinate is:

(a) equal to the rate of change of the x-coordinate;

(b) the negative of the rate of change of the x-coordinate;

(c) equal to 0;

(d) twice the rate of change of the x-coordinate.

22. A particle moves along the parabola $y = x^2 - 4x + 5$. Determine at which point on the curve the rate of change of the y-coordinate is:

(a) equal to the rate of change of the x-coordinate;

(b) twice the rate of change of the x-coordinate.

23. A man 6 ft tall walks away from a street lamp 15 ft high at the rate of 4 ft/sec. When he is 12 ft from the lamppost, how fast is:

(a) the length of his shadow changing?

(b) the tip of his shadow moving?

Does it matter how far he is from the lamppost?

24. The demand equation for a particular product is

$$p = \frac{8000}{x + 60}$$

Suppose the price p is now \$4 and is going up at \$0.50 per week. How fast is the demand x changing?

25. In Exercise 24 how fast is the revenue changing?

Key Mathematical Concepts and Tools

- The chain rule
- Exponential function
- Natural logarithm function
- Inverse functions
- Implicit differentiation
- Differential equations
- Related rates

Applications Covered in This Chapter

- Compound interest
- Exponential growth and decay
- Radioactive decay and radiocarbon dating
- Compartmental analysis
- Newton's law of cooling
- Graphing using semilogarithmic paper

FOUR Warm-Up Test

1. Find $\frac{dy}{dx}$ for each of the following.

(a) $y = \frac{2x}{(x^3 + 5)^4}$ (b) $y = xe^{3x^2}$

(c) $y = x^2 \ln(x + 1)$ (d) $y = e^{3e^{2x+1}}$

2. Boston is suffering from a shortage of tomatoes. If p is the price of tomatoes (in dollars per pound), and x is the supply (in tons), x and p are related by

$$p + xp = 3$$

Presently the price is \$1 per pound, and our supply of tomatoes is 2 tons. But the supply is decreasing at a rate of 0.1 ton per day. How fast is the price increasing (in dollars per day)?

3. Use logarithmic differentiation to compute y' for:

(a)

$$y = \frac{(x + 1)(x^2 - 6x)^{1/3}}{\sqrt{2x - 3}}$$

(b)

$$y = \frac{(x^3 - 4)^{1/5}}{(3x + 7)^{3/2}}$$

4. Find an equation for the tangent line to the curve $y = e^{x^2} + 4$ at the point (0, 5).

5. (a) An investor places \$35,000 in an account earning 13% compounded continuously. To how much money will this account accumulate in 10 years?

(b) To what rate of simple interest is 13% continuously compounded equivalent?

6. (a) Given two differentiable functions $y = f(u)$ and $u = u(x)$, write down the chain rule in Leibniz notation.

(b) Use the results of part (a) to differentiate $e^{(7-2x+x^3)}$.

7. Use implicit differentiation to find $\frac{dy}{dx}$, given that:

(a) $x^3 - 4xy + y^3 = 0$

(b) $x + ye^{xy} = 7$

(c) $e^x \ln(xy + 3) = 4$

8. Radioactive carbon-11 has a half-life of 20.5 min. How long does it take for a 200-gram sample of C^{11} to decay to a weight of only 40 g?

9. If x is the number of cars in Boston (in tens of thousands) and y is the carbon monoxide level in the air (in parts per million), then, according to one pollution model, x and y are related by

$$x^2 - 2xy - y = 3$$

Presently the number of cars is 50,000 and the carbon monoxide level is 2 parts per million. The number of cars is increasing at 5000 per year. How fast is the carbon monoxide level increasing?

10. A certain town had a population of 10,000 in 1962 and a population of 15,000 in 1975. Assume that the population grows exponentially.

(a) Estimate the town's population in 1990.

(b) In what year did its population reach 18,000?

FOUR Final Exam

1. Differentiate each of the following.

(a) $e^x \ln(x + 2)$ (b) $(3 + 4e^{2x})^5$

(c) xe^{-x^2} (d) $\frac{x + e^x}{x - e^x}$

2. Use implicit differentiation to find an equation for the tangent line to the ellipse $4x^2 + 5y^2 = 1$ at the point $(-\frac{1}{3}, \frac{1}{3})$.

3. A certain bank advertises that funds placed on deposit will double in 8 years if not drawn on.

   (a) Assuming that interest is being compounded continuously, find the interest rate being offered.

   (b) If \$1250 is deposited today, to how much will it accumulate at the end of 4 years?

4. Use logarithmic differentiation to compute the derivative of:

   (a) x^{x^2} (b) $\dfrac{x(x^2 + 3)^{1/3}}{(x - 1)^5}$

5. A colony of bacteria initially consists of 1200 bacteria. In 7 days the colony has grown to 1400 bacteria. Assume exponential growth.

   (a) Find the doubling time of this bacteria.

   (b) How fast will the bacteria be growing at the end of 10 days?

6. A man on a dock is pulling a rope that is fastened to the bow of a small boat. If the man's hands are 12 ft higher than the point where the rope is attached to the boat, and if he is pulling in the rope at the rate of 3 ft/sec, how fast is the boat approaching the dock when there are still 20 ft of rope out?

7. Find $\dfrac{dy}{dx}$ in each of the following.

   (a) $x^2 + 4xy - y^3 = 6$

   (b) $x \ln(x + y) = 3x$

   (c) $x = y^x$ (*Hint*: Take logs first)

8. Sketch the graph of the function $y = (x + 1)e^{-(x+2)}$, labeling all relevant information such as maxima, minima, and points of inflection, if any.

9. The demand equation for a certain product is

$$px + 75p = 10{,}000$$

   where p is the price in dollars and x the demand. The price is currently \$6.30, but it has been falling lately at \$0.35 per week.

   (a) How fast is the demand changing?

   (b) How fast is the revenue changing?

10. Radioactive carbon-14 has a half-life of 5750 years. The percentage of carbon-14 present in the remains of plants and animals can be used to determine their age. How old is an animal bone that has lost 32% of its carbon-14?

FIVE Integration

Outline

In Context

The earlier chapters of this book considered various aspects of *differential calculus*. This study involved the computation and application of the *derivative*, which measures rates of change.

In this chapter we begin the study of the second basic concept of elementary calculus, the *integral*. *Integral calculus* treats problems of summation, or aggregation. These two major branches of calculus are intimately related via the *fundamental theorem of calculus*, which we discuss in Section 5.3. The starting point for our study is a detailed examination of area.

SECTION 5.1

The Notion of Area

We wish to consider the problem of determining the area of an arbitrary region A in the plane. See, for example, the region in Figure 5.1. As a prelude to tackling this problem, we look first at some regions with more familiar shapes. For example, if A is a rectangle of length l and width w (see Figure 5.2), then the area of $A = lw$. Similarly, if A is a triangle with base b and height (altitude) h, (see Figure 5.3), then the area of $A = \frac{1}{2}bh$.

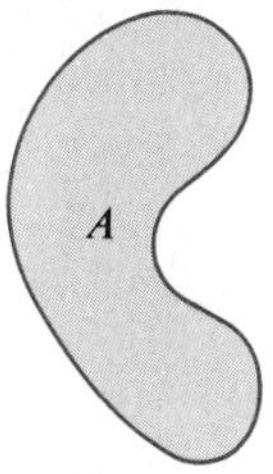

FIGURE 5.1

These formulas may be used to determine the areas of somewhat more complicated regions—namely, regions that can be decomposed into a collection of rectangles and triangles. To illustrate this procedure, suppose we wish to determine the area of a regular polygon of n sides, each of length s (see Figure 5.4). Such a region in the plane may be decomposed into n congruent isosceles triangles. If we denote one of these triangles by T, then

$$\text{area of } T = \frac{1}{2}sa$$

where a represents the height of triangle T (a is sometimes called the **apothem** of the regular polygon). Because there are n triangles, each congruent to T, we have

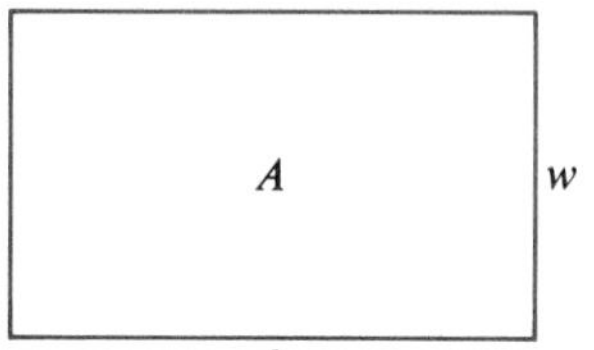

FIGURE 5.2

$$\text{Area of } A = (n)\left(\frac{1}{2}sa\right) = \frac{1}{2}(a)(ns)$$

But

$$ns = \underbrace{s + s + \cdots + s}_{n \text{ times}}$$

is precisely the sum of the lengths of all the sides of the polygon—that is, ns is the **perimeter** of A. Denoting the perimeter by p, we have

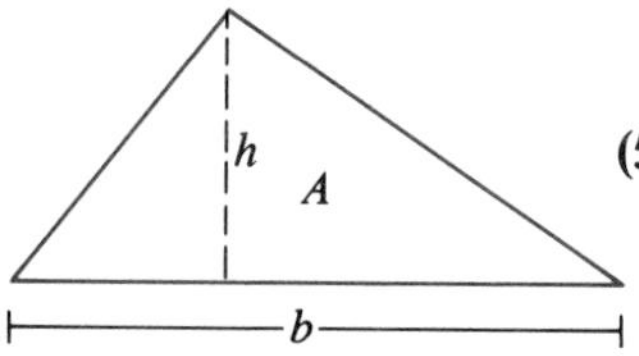

FIGURE 5.3

(5.1)
$$\text{Area of } A = \frac{1}{2}ap$$

What should we do when given a region that cannot be decomposed into rectangles and triangles? For example, suppose we are given a circle of radius r. Can we determine its area? (We are taking the point of view that the familiar formula for the area, πr^2, is not available to us.) To attack this problem, let us inscribe a regular polygon of n sides inside the given circle. Figure 5.5 shows three circles of radius r with, respectively, an inscribed square, an inscribed octagon (8 sides), and an inscribed icosagon (20 sides). From these figures it appears that, as the number of sides of the polygon increases, the area of the polygon gives a better and better approximation to the area of the circle. (Of course, the area of the polygon is always smaller than the area of the circle.) What happens to Equation (5.1) for the area of the polygon as n, the number of sides, increases? As n increases, it appears that the perimeter of the polygon approaches the circumference of the circle. For the purposes of this discussion, let us assume that we know the circumference of the circle—$2\pi r$. (This is equivalent to the standard definition of π as the ratio of the circumference to the radius in any circle.) In symbols,

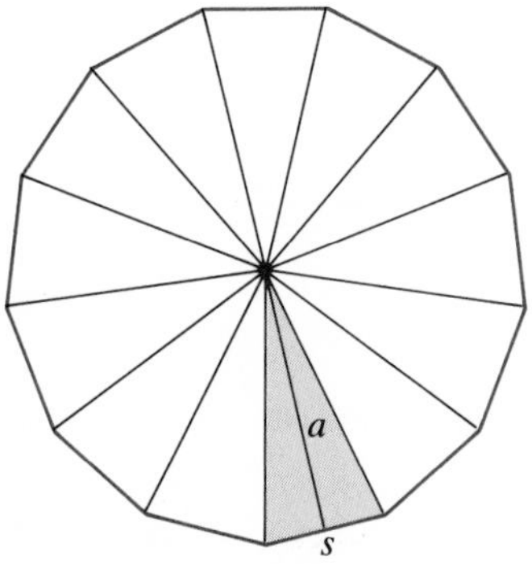

FIGURE 5.4

$$p \to 2\pi r \quad \text{as} \quad n \to \infty$$

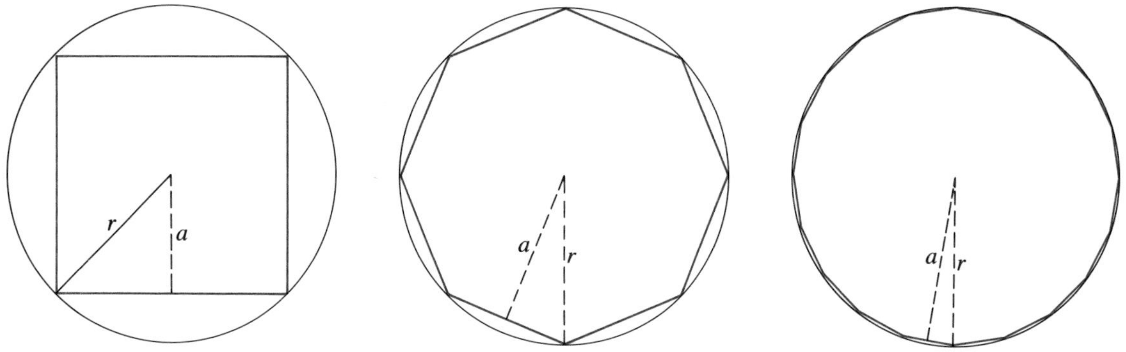

FIGURE 5.5

Also, as n increases, the length of the apothem of the polygon clearly approaches the radius r of the circle:

$$a \to r \quad \text{as} \quad n \to \infty$$

Thus, as n increases, we have

$$\left(\frac{1}{2}\right) ap \to \frac{1}{2}(r)(2\pi r) = \pi r^2$$

Our result is, of course, the "correct" formula for the area of the circle.

What happens if, instead of using inscribed regular polygons to approximate the area of the circle, we use *circumscribed* regular polygons for this purpose? Figure 5.6 provides sketches of a circumscribed square, octagon, and icosagon, respectively. It appears that, as the number of sides of the polygon increases, the area of the polygon approaches the area of the circle—although now each approximating area is slightly larger than the area of the circle. What happens as n, the number of sides of the regular polygon, increases? First of all, as the pictures indicate, the apothem a is always equal to r:

$$a = r$$

Furthermore, the perimeter p of the polygon approaches the circumference of

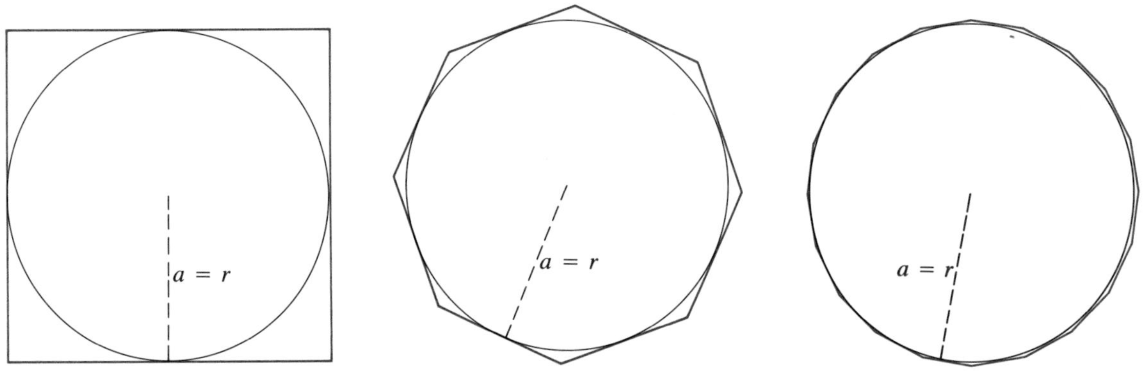

FIGURE 5.6

the circle, which we have already assumed to be $2\pi r$. Thus

$$p \to 2\pi r \quad \text{as} \quad n \to \infty$$

As $n \to \infty$, we have, using Equation (5.1),

$$\frac{1}{2}ap \to \frac{1}{2}(r)(2\pi r) = \pi r^2$$

so, the area of the circle is πr^2, as before.

Observe that, in the cases of both inscribed polygons and circumscribed polygons, we were led to use a limit process to solve our problem. We were able to arrive at the precise area of the circle only by allowing the number of sides of the approximating polygon to become arbitrarily large.

Suppose now that our region A is irregularly shaped, as in Figure 5.1, for example. How can we determine its area? The preceding discussion suggests that we should try to construct a region B that approximates the region A and such that the area of region B is easy to compute. Then, by constructing better and better approximations to A, we hope to be able to determine the area of A as a "limiting value." Moreover, it apparently will not matter whether we approximate A by regions that are slightly smaller or slightly larger than A itself.

Our next concern, therefore, is how to choose a region B that reasonably approximates the region A and at the same time is of such a form that its area is easy to compute. Such a choice was easy to make in the case of a circle, but for an irregularly shaped region A, it is not clear how to choose an approximating region B. One approach is to proceed according to the following scheme. Imagine that any region—in particular, the region of Figure 5.1—is drawn on a piece of graph paper. Suppose that both the horizontal and the vertical lines of the graph paper are spaced $\frac{1}{4}$ inch apart. Thus each box on the graph paper has area $\frac{1}{16}$ square inch. Choose the region B to consist of all those boxes (squares) that are contained in A or touch A at more than one point. The region B is the shaded region in Figure 5.7, and its area is easy to compute—one simply counts the number of boxes it contains and then multiplies by $\frac{1}{16}$.

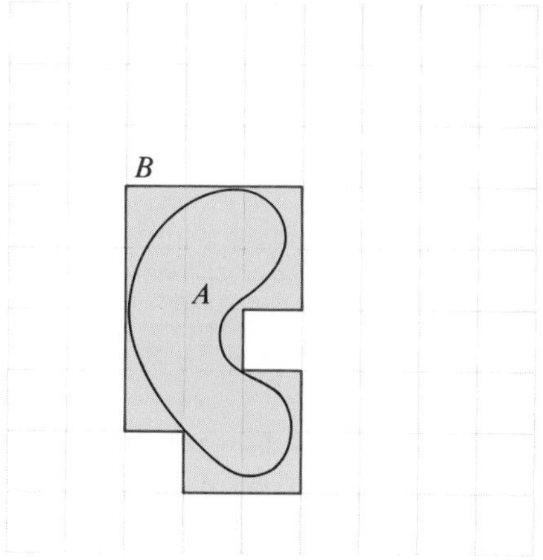

FIGURE 5.7

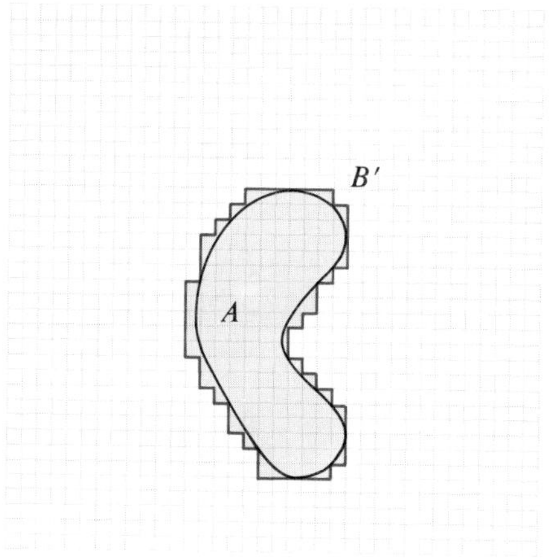

FIGURE 5.8

In Figure 5.7, B has area

$$(13)\left(\frac{1}{16}\text{ in}^2\right) \approx 0.81\text{ in}^2$$

Note that region B is somewhat larger than region A. Hence we know that the area of A is somewhat smaller than 0.81 square inches.

What happens if, instead of placing the region A on graph paper whose lines are spaced $\frac{1}{4}$ inch apart, we use graph paper with $\frac{1}{16}$-inch spacing? In this situation, following the same procedure as above, we construct a new approximating region B' (see Figure 5.8). By counting boxes carefully, we see that B' consists of 149 squares, each of area $1/256\text{ in}^2$. Thus B' has area

$$(149)\left(\frac{1}{256}\text{ in}^2\right) \approx 0.58\text{ in}^2$$

Note that the new approximating region, B', yields a smaller answer than did region B. Of course, region B' itself is still slightly larger than A, so it is clear that

$$\text{Area } A < \text{area } B' < \text{area } B$$

In particular, the area of A is somewhat smaller than 0.58 square inches. Naturally we expect our approximation to become more and more accurate as we use graph paper with smaller and smaller boxes.

We then see that, for an arbitrarily shaped region, the notion of area is intimately connected with that of the limit of approximations. We begin a more formal study of this approach in the next section.

SECTION 5.2

Area Under a Curve—The Definite Integral

In this section we will focus our attention primarily on the problem of computing the area of regions with a special character, such as that shown in Figure 5.9. Such a region is bounded above by a curve $y = f(x)$, $f(x) \geq 0$; below by the x-axis ($y = 0$); on the left by the vertical line $x = a$; and on the

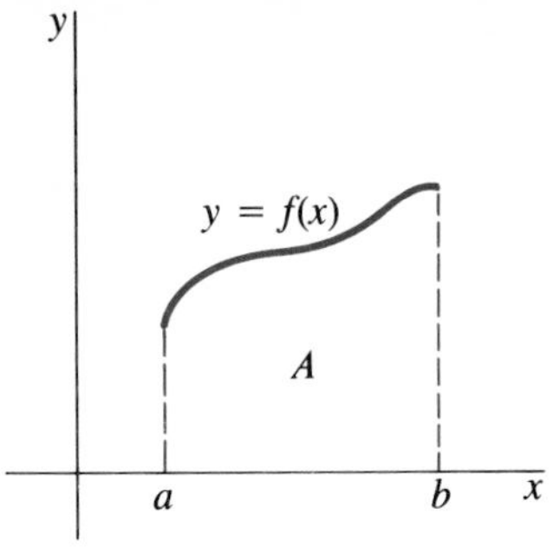

FIGURE 5.9

right by the vertical line $x = b$. In fact, consideration of such a region does not result in any great loss of generality, because complicated regions (such as that in Figure 5.1) can often be decomposed without great difficulty into smaller regions shaped like that in Figure 5.9 (see Figure 5.10).

The advantage of considering a region of the type pictured in Figure 5.9 is that the region is specified completely by a knowledge of the values of a and b and by the formula $y = f(x)$. This will enable us to compute its area analytically, without relying on our (questionable) ability to draw the region on an appropriate piece of graph paper.

At this point it is useful to introduce a symbolic notation for the area of the region shown in Figure 5.9.

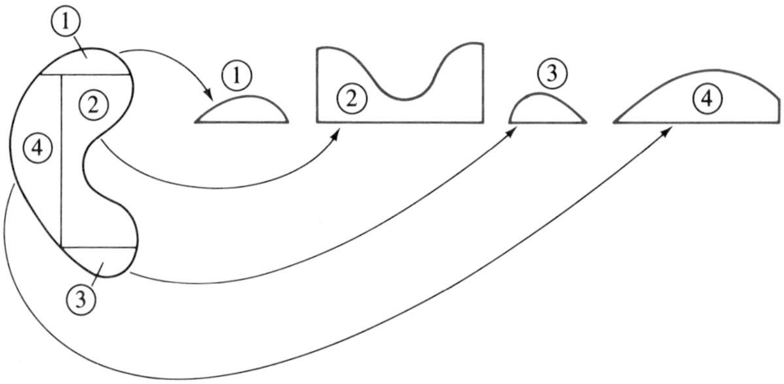

FIGURE 5.10

NOTATION 5.1 Assume $f(x) \geq 0$ for $a \leq x \leq b$. The area of the region bounded above by the graph of f, below by the x-axis, at the left by the line $x = a$, and at the right by the line $x = b$ is denoted

$$\int_a^b f(x)\,dx$$

(if this area exists) and is called the **definite integral** of f from a to b. ⊟

The foregoing is a *tentative* definition of the definite integral that will be generalized later. For now, $\int_a^b f(x)\,dx$ is to be viewed as just some specific number that represents an area. Motivation for this rather strange notation will be given in Section 5.3. Let us try to compute this number for a few choices of the function $y = f(x)$.

EXAMPLE 1 Compute $\int_0^2 (x + 1)\,dx$.

SOLUTION According to Notation 5.1, the symbol $\int_0^2 (x + 1)\,dx$ can be interpreted as the area of the region bounded above by the curve $y = x + 1$, below by the x-axis, to the left by the line $x = 0$, and to the right by the line $x = 2$. This region, which is pictured in Figure 5.11, is a trapezoid. You may recall the formula for the area of a trapezoid from high school geometry. It is

$$\text{Area} = \frac{1}{2}h(b_1 + b_2) \tag{5.2}$$

where b_1 and b_2 denote the lengths of the two parallel bases, and h represents the distance between these parallel bases. In our case the trapezoid is on its side. Rotating it clockwise by 90° should make it seem more familiar to you (see Figure 5.12). Its area is

$$\text{Area} = \left(\frac{1}{2}\right) \cdot 2(1 + 3) = 4$$

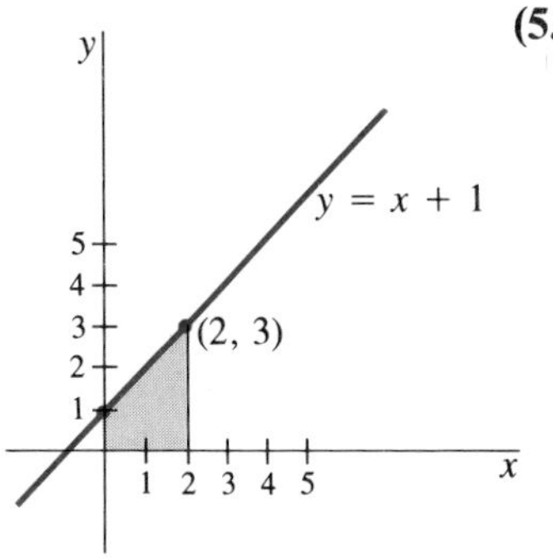

FIGURE 5.11

Thus we write

$$\int_0^2 (x + 1)\, dx = 4$$

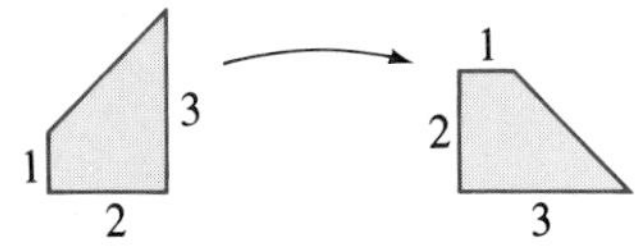

FIGURE 5.12

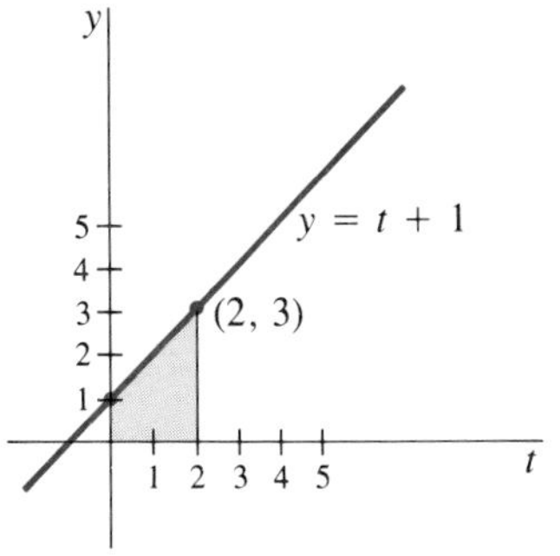

FIGURE 5.13

In the previous example, the variable x in the expression $\int_0^2 (x + 1)\, dx$ plays no significant role. It is called a **dummy variable**. We might just as well have written $\int_0^2 (t + 1)\, dt$ or $\int_0^2 (z + 1)\, dz$. To see this, observe that $\int_0^2 (t + 1)\, dt$ represents the area of the region pictured in Figure 5.13, and obviously this is the same as the region of Figure 5.11 with the horizontal axis relabeled as the t-axis instead of the x-axis. In general, the same remarks imply that $\int_a^b f(x)\, dx$ is the same as $\int_a^b f(t)\, dt$ or $\int_a^b f(u)\, du$, for example.

EXAMPLE 2 Evaluate $\int_{-1}^{1} \sqrt{1 - x^2}\, dx$.

SOLUTION In this case the region under consideration is that bounded above by $y = \sqrt{1 - x^2}$, below by the x-axis, to the left by $x = -1$, and to the right by $x = 1$. You may not immediately recognize the curve

$$y = \sqrt{1 - x^2}$$

But if we square both sides, we obtain

$$y^2 = 1 - x^2$$

or

$$x^2 + y^2 = 1$$

which we recognize as the unit circle. Thus the curve $y = \sqrt{1 - x^2}$ is in fact just the upper half of the unit circle. The region in question is shown in Figure 5.14. We know that the area of a circle of radius r is πr^2. Thus, for a circle of radius 1, the area is π. Therefore our semicircular region must have area $\pi/2$. In symbolic notation,

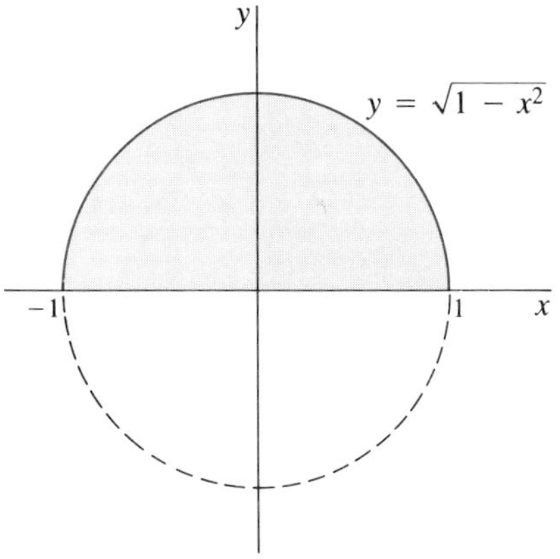

FIGURE 5.14

$$\int_{-1}^{+1} \sqrt{1 - x^2}\, dx = \frac{\pi}{2}$$

PRACTICE EXERCISE Draw the appropriate region and compute

(a) $\int_1^3 2\,dx$ (b) $\int_0^4 3z\,dz$

[Your answers should be 4 and 24.]

Note that, in each of these examples and exercises, the regions determined by the curves $y = f(x)$ had a special form: a trapezoid, a triangle, a rectangle, or a portion of a circle. This enabled us to quickly evaluate $\int_a^b f(x)\,dx$ by using area formulas with which we were already familiar. Obviously other functions $y = f(x)$ can yield regions that have a shape for which we have no formula readily available.

EXAMPLE 3 Let us now try to estimate the area

$$A = \int_0^4 3x^2\,dx$$

That is, we wish to estimate the area of the region bounded above by the curve

$$y = 3x^2$$

below by the x-axis, and on the right by the vertical line $x = 4$ (as shown in Figure 5.15). In other words, as it is commonly phrased, our problem is to find the area under the curve $y = 3x^2$ from $x = 0$ to $x = 4$.

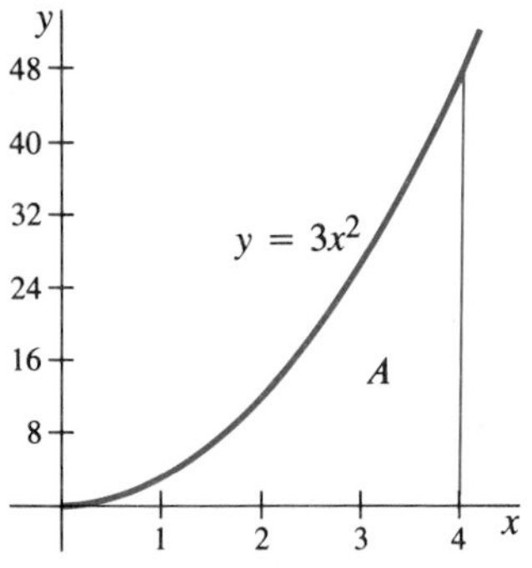

FIGURE 5.15

SOLUTION From a computational standpoint, it is convenient to approximate A by a region made up of rectangles, constructed as follows. We divide the interval [0, 4] into a number of equal subintervals. For each of these subintervals, we construct a rectangle whose base is the subinterval and whose height is the y-value at the right-hand end point of the subinterval. We sum the areas of all these rectangles to obtain an approximation to A. This process is illustrated in Figure 5.16, where the number of subintervals used is 4 and each rectangle has width 1.

The heights of the four rectangles are the values of $y = 3x^2$ at the points $x = 1, 2, 3$, and 4. These are given in Table 5.1. The sum of the area of the rectangles is therefore

$$\text{Area }(R_1) + \text{area }(R_2) + \text{area }(R_3) + \text{area }(R_4)$$
$$= (3 \times 1) + (12 \times 1) + (27 \times 1) + (48) \times 1 = 90$$

TABLE 5.1

x	$y = 3x^2$
1	3
2	12
3	27
4	48

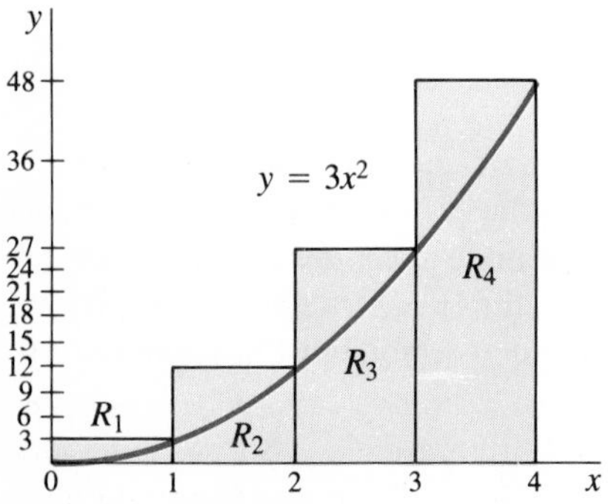

FIGURE 5.16

It is evident from Figure 5.16 that this is a rather poor estimate of the area $A = \int_0^4 3x^2\,dx$. A better estimate can be obtained by increasing the number n of approximating rectangles. In Figure 5.17, $n = 8$ and each rectangle has width $\frac{1}{2}$. The heights (and their sum) are given in Table 5.2. The total area of the rectangles $R_1, R_2, \ldots, R_8$ can be obtained by multiplying each height by the width $\frac{1}{2}$ and then summing or, equivalently, by summing the heights first and then multiplying by $\frac{1}{2}$. Thus our second estimate of the area A is

$$A \approx \text{area}(R_1) + \text{area}(R_2) + \cdots + \text{area}(R_8) = \frac{1}{2}(153) = 76.5$$

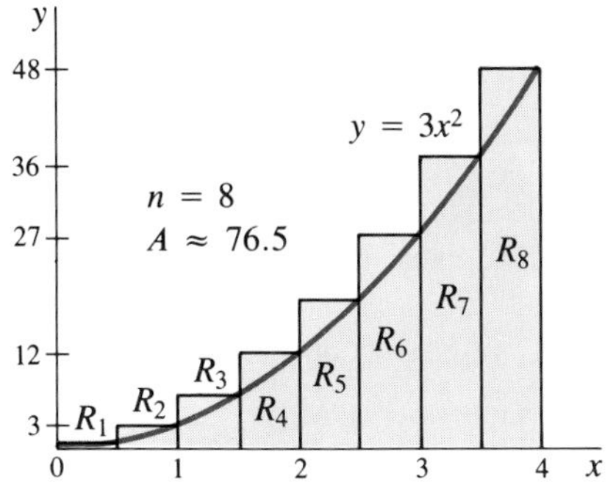

FIGURE 5.17

TABLE 5.2

x	$y = 3x^2$
0.5	0.75
1.0	3.00
1.5	6.75
2.0	12.00
2.5	18.75
3.0	27.00
3.5	36.75
4.0	48.00
	Sum = 153.00

PRACTICE EXERCISE Approximate the area A by means of $n = 16$ rectangles of width $\frac{1}{4}$. It is understood that the heights are to be chosen as before.

[You should construct a table similar to Tables 5.1 and 5.2. Your answer should be 70.125. See Figure 5.18.]

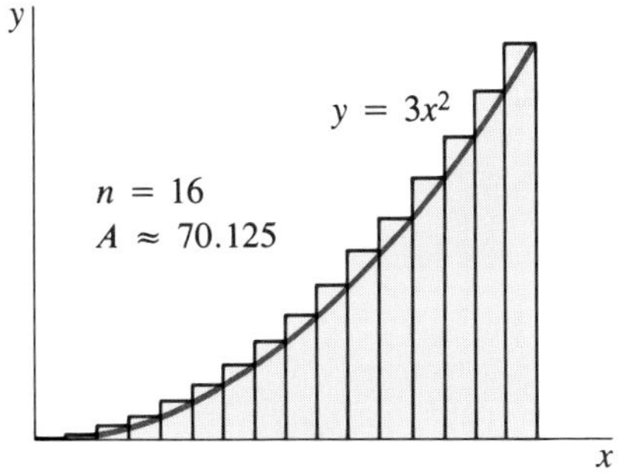

FIGURE 5.18

Figures 5.19(a) and (b) illustrate the approximation of A by $n = 32$ and $n = 64$ rectangles. The important point here is that, as n becomes larger and larger, the region composed of the n rectangles becomes a better and better approximation of the region below the curve. Consequently the (exact) area beneath the graph is the **limit** of our approximation as $n \to \infty$.

Obviously we can continue to approximate A by rectangles of smaller and smaller width—although the computations require lots of patience and it becomes all but impossible for us to draw a reasonable picture. For example, if we use 128 rectangles of width $\frac{1}{32}$, the approximate area will be 64.75; 1024

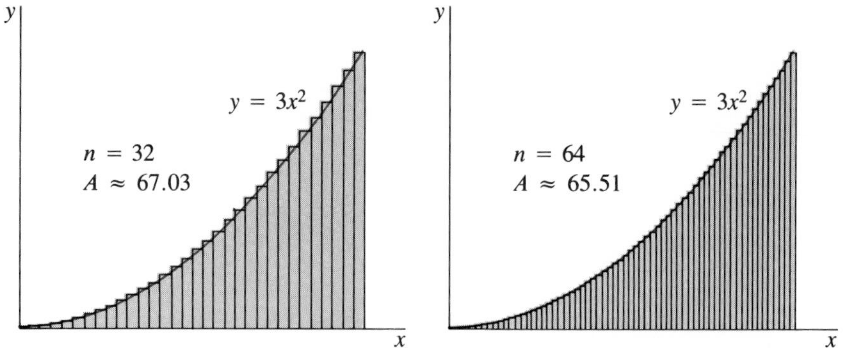

FIGURE 5.19

rectangles of width $\frac{1}{256}$ yield approximate area 64.09; and 4096 rectangles of width $\frac{1}{1024}$ yield approximate area 64.02. (Such computations are readily carried out by means of a computer. See the exercises at the end of this section.)

As we have refined our approximation, we have generated the numbers

$$90, 76.5, 70.125, 67.03, 65.51, 64.75, 64.09, 64.02$$

These numbers seem to be approaching 64, so we might guess that

$$A = \int_0^4 3x^2\,dx = 64 \qquad \text{(guess)}$$

Of course, we cannot be certain that this is correct. It would be equally valid to guess that the sequence of approximations is approaching, say, 64.0146. What we need—and will develop shortly—is an easy method for determining the limit of these rectangular approximations *exactly*. (Actually it is possible to derive a formula for the sum of n rectangles—as an expression involving n—and to then take its limit, algebraically, as $n \to \infty$. However, this technique is feasible only for the simplest of functions, so we shall not examine it in this book.)

Another comment we might make about this process concerns the heights of the rectangles. For each subinterval, we choose as the height of its rectangle the y-value at the right-hand end point. This was an arbitrary choice. We could have used the y-value at the left-hand end point, as in Figure 5.20(a);

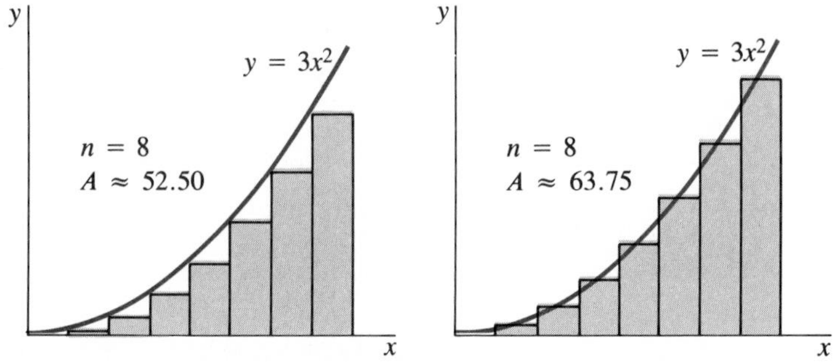

FIGURE 5.20

(a) Using left end points (b) Using midpoints

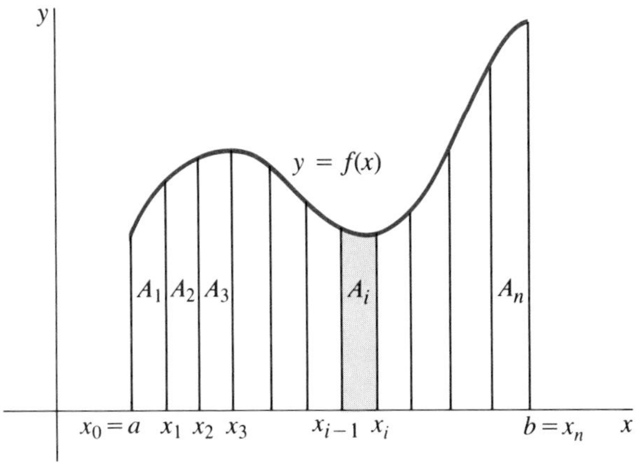

FIGURE 5.21

at the midpoint, as in Figure 5.20(b); or even at a random point in each interval. In any case—for a continuous function—the larger n is, the better the approximation.

Before considering further examples, we should formalize our procedure so as to obtain a careful definition of $\int_a^b f(x)\,dx$ based on the idea of approximating rectangles. Consider the area corresponding to $\int_a^b f(x)\,dx$ as pictured in Figure 5.21. As a first step, let's divide our region into n thin vertical strips, each one having the same width. Because the entire region has width $b - a$, each strip has width $(b - a)/n$. It is customary to denote the latter by Δx, where the Greek capital delta, Δ, means "increment" or "change in." Thus the subintervals have width

$$\Delta x = \frac{b - a}{n}$$

Let's denote the area of the first vertical strip by A_1, the area of the second by A_2, that of the third by A_3, and so on. All in all there are n strips, and the total area A is

$$A = A_1 + A_2 + \cdots + A_n$$

Denote the intermediate points on the x-axis between a and b by x_1, x_2, $x_3, \ldots, x_{n-1}$. (See Figure 5.21.) Observe that

$$x_1 = a + \Delta x$$

$$x_2 = x_1 + \Delta x = (a + \Delta x) + \Delta x = a + 2\,\Delta x$$

$$x_3 = x_2 + \Delta x = (a + 2\,\Delta x) + \Delta x = a + 3\,\Delta x$$

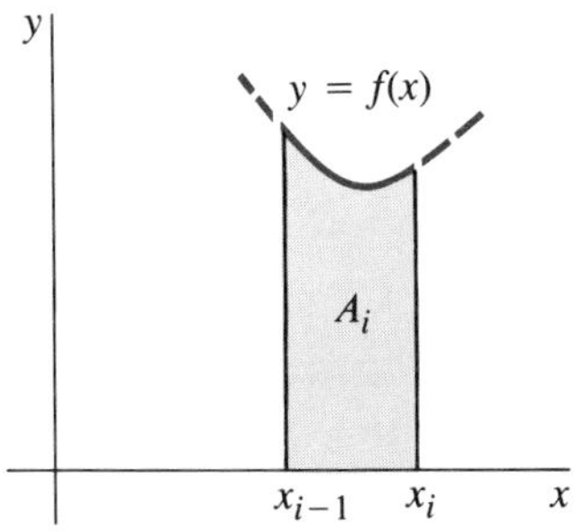

FIGURE 5.22

and that, in general,

$$x_i = a + i\,\Delta x \qquad i = 0, 1, 2, \ldots, n$$

Note that $x_0 = a$ and $x_n = a + n\,\Delta x = a + (b - a) = b$.

Now let us focus our attention for a moment on a single strip. We will consider the ith strip, which, of course, has area denoted by A_i. A magnified version of it appears in Figure 5.22. Let us examine the graph of $y = f(x)$ between x_{i-1} and x_i, and let us select *any* point z_i between x_{i-1} and x_i.

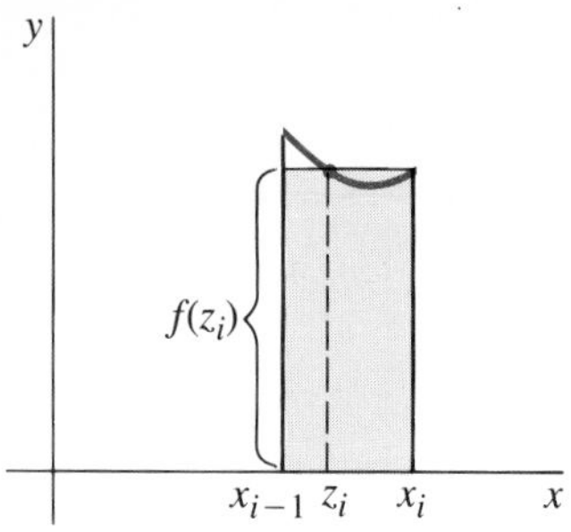

FIGURE 5.23

If we now draw a rectangle of width $\Delta x = (b - a)/n$ and height $f(z_i)$ (See Figure 5.23), it will have area given by

$$f(z_i)\,\Delta x$$

and this should approximately equal the area A_i. We write $A_i \approx f(z_i)\,\Delta x$. In precisely the same way, we can construct a rectangle corresponding to each of the regions $A_1, A_2, \ldots, A_n$. Each rectangle will have width Δx and height equal to some value of the function $y = f(x)$ on the interval. When we are finished, we will have n rectangles whose total area approximates the area of our original region. (This situation is illustrated in Figure 5.24.) Thus

$$A = A_1 + A_2 + \cdots + A_n \approx f(z_1)\,\Delta x + f(z_2)\,\Delta x + \cdots + f(z_n)\,\Delta x$$

We emphasize again that what we have done is construct an approximating region, composed entirely of rectangles, that provides an estimate for the area A.

This is perhaps a good time to digress for a moment to introduce convenient short-hand notation. The Greek letter Σ (sigma) in mathematics is used to denote a sum. It simply means "add them up." A typical expression involving **sigma notation** is

$$\sum_{i=3}^{8} i^2$$

This means to take the formula i^2 and set $i = 3$ in the formula. Next set $i = 4$, $i = 5$, and so on, until we reach the value $i = 8$. Having done this, we take all the numbers we have obtained and we "add them up." Thus

$$\sum_{i=3}^{8} i^2 = 3^2 + 4^2 + 5^2 + 6^2 + 7^2 + 8^2 = 199$$

The value $i = 3$ tells us what number to start with in the formula i^2. The number $i = 8$ tells us when to stop. The i itself is called the **index of summation.**

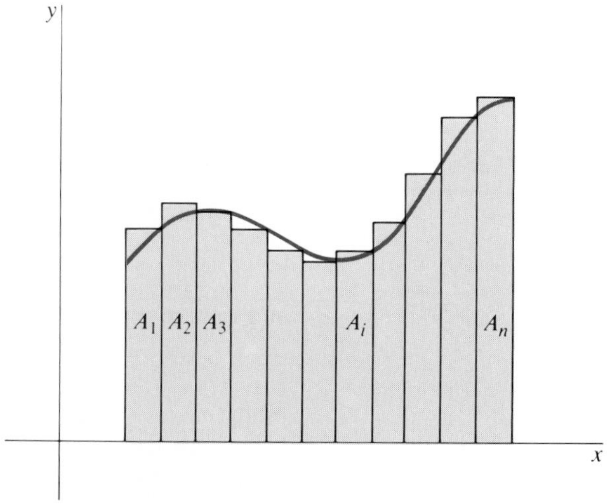

FIGURE 5.24

EXAMPLE 4 Write the expression $x_1 + x_2 + x_3 + x_4 + x_5$ in sigma notation.

SOLUTION Here we have an expression that is the sum of five terms. The first term starts with $i = 1$, and the last one has $i = 5$. Therefore

$$x_1 + x_2 + x_3 + x_4 + x_5 = \sum_{i=1}^{5} x_i$$

PRACTICE EXERCISE (a) Show that $\sum_{i=2}^{6} (i)(i - 1) = 70$.

(b) Write the expression $f(x_1) + f(x_2) + \cdots + f(x_{12})$ in sigma notation.

[Your answer for part (b) should be $\sum_{i=1}^{12} f(x_i)$.]

Using our new notation we can write

$$f(z_1)\,\Delta x + f(z_2)\,\Delta x + \cdots + f(z_n)\,\Delta x = \sum_{i=1}^{n} f(z_i)\,\Delta x$$

(Convince yourself that these are really the same expression.) Therefore

$$A \approx \sum_{i=1}^{n} f(z_i)\,\Delta x \qquad (5.3)$$

Sums of the form of Equation 5.3 are known as **Riemann sums**, after Georg F. B. Riemann (1826–1866), the eminent German mathematician who helped develop the modern theory of integration.

If, in each subinterval, we choose z_i to be the right endpoint—that is, $z_i = x_i$—then Equation (5.3) will be called the *right-hand rule*. If we choose z_i to be the left end point—that is, $z_i = x_{i-1}$—then Equation (5.3) will be called the *left-hand rule*.

What should we expect will happen if we choose n very large? Because each rectangle has width $\frac{b - a}{n}$, this is equivalent to requiring that we approximate A by very, very thin rectangles. On the basis of our previous discussion and the figures, we expect that, for any "nicely" shaped region, our sum will get closer to the true area A. This is indeed the case. It can be shown that, if $f(x)$ is any nonnegative continuous function on $[a, b]$, then the sums $\sum_{i=1}^{n} f(z_i)\,\Delta x$, where the z_i may be chosen arbitrarily in the interval from x_{i-1} to x_i, always have a unique limit as $n \to \infty$. This limit is the area under the curve, which we have denoted $\int_a^b f(x)\,dx$.

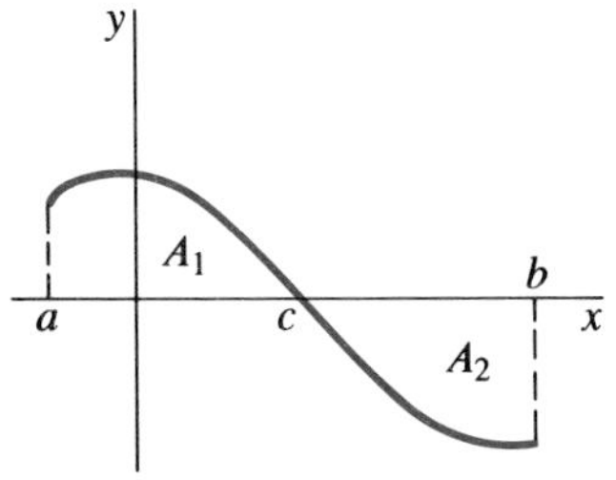

FIGURE 5.25

Up to now we have assumed that $f(x) \geq 0$ for all x in $[a, b]$. We now wish to drop that assumption and require only that f be continuous on the interval $[a, b]$. Thus f may now drop below the x-axis, as shown in Figure 5.25. However, it is still true that the Riemann sums for such a function approach a unique limit as $n \to \infty$, and we now formally define this limit to be the definite integral of f over $[a, b]$.

DEFINITION 5.2 Let f be continuous on the interval $[a, b]$. Then the **definite integral** of f from a to b, denoted $\int_a^b f(x)\,dx$, is defined to be the limit approached by the Riemann sums of f as $n \to \infty$:

$$\int_a^b f(x)\,dx = \lim_{n\to\infty} \sum_{i=1}^{n} f(z_i)\,\Delta x_i$$

The endpoints a and b are sometimes called the **limits of integration**. (This usage of the term *limit* has nothing to do with the concept of limit used previously in the book.) The function being integrated is called the **integrand**. ⊟

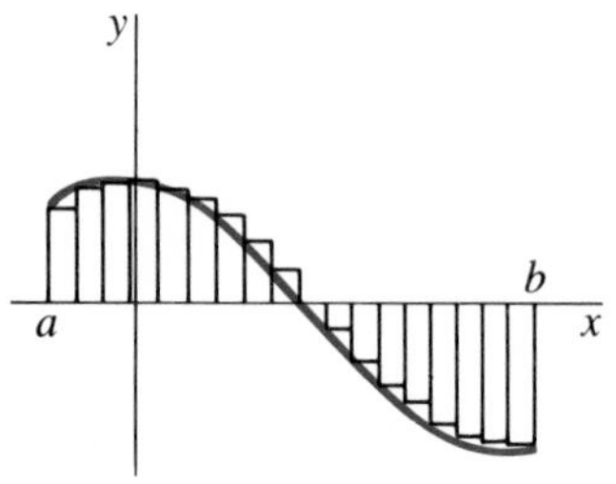

FIGURE 5.26

The integral sign, $\int$, is actually an elongated S, reminding us of the word *sum*. To the early pioneers of calculus, the integral was viewed as a sum of infinitely many rectangles of infinitesimal width dx. This is how the notation $\int_a^b f(x)\,dx$ arose historically. To indicate passage to the limit, Σ is rewritten $\int$, and Δx is rewritten dx.

Actually, in the case of a function that is sometimes negative, the interpretation of the definite integral as area has to be applied with caution, because some of the terms $f(z_i)\,\Delta x$ in a Riemann sum will be negative. In other words, in Figure 5.26 the areas of those rectangles that are below the x-axis are subtracted, rather than added, in the Riemann sum. When we pass to the limit as $n \to \infty$, we find that the definite integral gives us the area above the x-axis *minus* the area below:

$$\int_a^b f(x)\,dx = A_1 - A_2$$

If we want to find the sum $A_1 + A_2$ (in Figure 5.25), we must compute separately $\int_a^c f(x)\,dx$ and $|\int_c^b f(x)\,dx|$ and add them together. We shall consider this type of problem a little later. As we shall learn soon, the definite integral has many applications in economics and the sciences apart from area. We are using area only to motivate the concept and to make its definition easy to visualize.

EXAMPLE 5 Using the right-hand rule and $n = 4$, estimate

$$\int_0^1 \frac{1}{1 + x^2}\,dx$$

SOLUTION For the right-hand rule, $z_i = x_i$, so

(5.4) $$\int_a^b f(x)\,dx \approx \sum_{i=1}^{n} f(x_i)\,\Delta x = \left(\sum_{i=1}^{n} f(x_i)\right)\left(\frac{b - a}{n}\right)$$

We have $a = 0$, $b = 1$, $n = 4$ and $f(x) = \dfrac{1}{1 + x^2}$. Then

$$\Delta x = \frac{b - a}{n} = \frac{1 - 0}{4} = 0.25$$

and the x_i's and their y-values are given in Table 5.3. Substituting these values into Equation 5.4, we find

$$\int_0^1 \frac{1}{1 + x^2}\, dx \approx \left(\sum_{i=1}^{4} f(x_i)\right)\left(\frac{b - a}{n}\right)$$

$$= [0.94 + 0.84 + 0.64 + 0.50]0.25 = 0.73$$

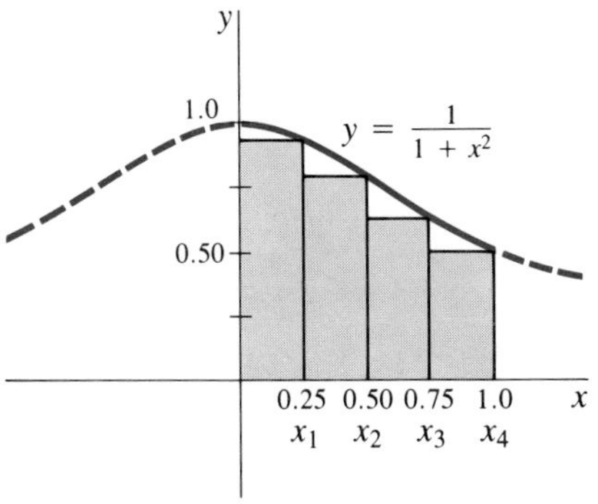

FIGURE 5.27

TABLE 5.3

i	$x_i = a + i\,\Delta x$	$f(x_i) = \dfrac{1}{1 + (x_i)^2}$
1	0.25	0.94
2	0.50	0.80
3	0.75	0.64
4	1.00	0.50

If you examine the sketch of $y = \dfrac{1}{1 + x^2}$ in Figure 5.27, you will see that, for this particular function, the right-hand rule amounted to approximating each subregion by a slightly smaller rectangle. Thus we expect that our estimate is too small. We will show later that the precise value of $\int_0^1 \dfrac{1}{1 + x^2}\, dx$ is $\pi/4$, which is roughly 0.78.

For the left-hand rule, $z_i = x_{i-1}$ for each i, so

$$\int_a^b f(x)\, dx \approx \sum_{i=1}^{n} f(x_{i-1})\, \Delta x = \left(\sum_{i=1}^{n} f(x_{i-1})\right)\left(\frac{b - a}{n}\right)$$

PRACTICE EXERCISE Use the left-hand rule with $n = 4$ to estimate $\int_0^1 \dfrac{1}{1 + x^2}\, dx$.

[Your answer should be roughly 0.85. The left-hand rule in this case is equivalent to approximating each subregion in the sketch by a slightly larger rectangle (Figure 5.28).]

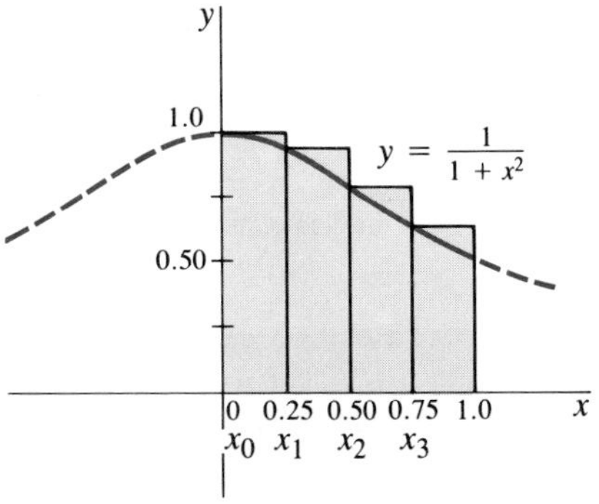

FIGURE 5.28

Section 5.2 EXERCISES

Interpret the integral in each of Exercises 1–10 as an area. Which ones can you evaluate exactly? For those you can evaluate exactly, what is the value?

1. $\int_0^4 5\,dx$
2. $\int_1^4 x\,dx$
3. $\int_0^4 2x\,dx$
4. $\int_1^3 x^2\,dx$
5. $\int_0^3 x^2\,dx$
6. $\int_1^2 \frac{1}{x}\,dx$
7. $\int_{-1}^{+1} e^x\,dx$
8. $\int_0^1 (5 - 3x)\,dx$
9. $\int_0^2 (2x + 3)\,dx$
10. $\int_0^1 \sqrt{1 - x^2}\,dx$

Approximate the definite integrals (areas) in Exercises 11–19 by the right-hand rule, using $n = 4$ and $n = 8$.

11. $\int_1^5 x\,dx$
12. $\int_0^4 x\,dx$
13. $\int_1^3 x^2\,dx$
14. $\int_0^4 (x^2 + 1)\,dx$
15. $\int_0^4 x^3\,dx$
16. $\int_1^5 \frac{1}{x}\,dx$
17. $\int_1^5 \frac{3}{x^2}\,dx$
18. $\int_{-1}^3 (2x + 3)\,dx$
19. $\int_0^4 e^x\,dx$

In Exercises 20–23, use definite integral notation to represent the area of the region described.

20. Bounded by the x-axis, the line $y = 3x + 3$, the y-axis, and the vertical line $x = 3$.
21. Enclosed by the curve $y = x^4$, the x-axis, and the vertical lines $x = 3$ and $x = 5$.
22. Enclosed by the curve $y = 1/x$, the x-axis, and the vertical lines $x = 1$ and $x = 4$.
23. Between the curve $y = 4 - x^2$ and the x-axis.

Write out the meaning of the summations in Exercises 24–34.

24. $\sum_{i=1}^{10} i$
25. $\sum_{i=1}^{20} i$
26. $\sum_{i=1}^{n} i$
27. $\sum_{i=1}^{15} i^2$
28. $\sum_{i=2}^{n} i(i - 1)$
29. $\sum_{i=1}^{n} \frac{1}{i}$
30. $\sum_{i=1}^{n} (i^3 + i)$
31. $\sum_{n=1}^{8} n \ln n$
32. $\sum_{i=1}^{n} a_i b_i$
33. $\sum_{i=1}^{n} f(x_i)$
34. $\sum_{1}^{n} x_i\, f(x_i)$

Write each of the sums in Exercises 35–38 in summation notation.

35. $1 + \frac{1}{2} + \frac{1}{3} + \frac{1}{4} + \cdots + \frac{1}{17}$
36. $1^2 + 2^2 + 3^2 + \cdots + n^2$
37. $\frac{1}{1\cdot 2} + \frac{1}{2\cdot 3} + \frac{1}{3\cdot 4} + \cdots + \frac{1}{25\cdot 26}$
38. $1 - \frac{1}{2} + \frac{1}{3} - \frac{1}{4} + \frac{1}{5} - \cdots + \frac{1}{99}$

(*Hint*: $(-1)^n = 1$ if n is even; $(-1)^n = -1$ if n is odd.)

In Exercises 39–48 (i) use the right-hand rule with $n = 4$ to estimate each of the definite integrals; (ii) do the same for the left-hand rule with $n = 4$; (iii) do the same for the right-hand rule with $n = 8$; and (iv) do the same for the left-hand rule with $n = 8$.

39. $\int_0^4 5\,dx$
40. $\int_1^5 x\,dx$
41. $\int_1^5 (2x + 3)\,dx$
42. $\int_0^4 x\,dx$
43. $\int_0^2 (2x + 3)\,dx$
44. $\int_1^5 x^2\,dx$
45. $\int_0^2 (x^2 + 1)\,dx$
46. $\int_{-2}^2 (x^3 + 3x + 5)\,dx$
47. $\int_0^3 (6 - x)\,dx$
48. $\int_0^2 (4 - x^2)\,dx$

In Exercises 49–56 use the right-hand rule and the left-hand rule—with $n = 10$—to estimate the area under the curve.

49. $y = 2x + 3$ from $x = 0$ to $x = 5$
50. $y = x^2$ from $x = 1$ to $x = 6$
51. $y = \frac{1}{x}$ from $x = 1$ to $x = 2$

52. $y = \dfrac{1}{1 + x}$ from $x = 0$ to $x = 5$

53. $y = x^3$ from $x = 0$ to $x = 3$

54. $y = x^4$ from $x = 0$ to $x = 10$

55. $y = e^x$ from $x = 0$ to 1

56. $y = \ln x$ from $x = 1$ to 2

(Obviously, tables or a calculator should be used.)

Use the right-hand rule and the left-hand rule with $n = 4$ to estimate the integrals in Exercises 57–60.*

57. $\displaystyle\int_0^{\pi} \sin x \, dx$

58. $\displaystyle\int_{-\pi/2}^{\pi/2} \cos x \, dx$

59. $\displaystyle\int_0^{4} e^x \, dx$

60. $\displaystyle\int_e^{5e} \ln x \, dx$

IF YOU WORK WITH A COMPUTER (OPTIONAL)

In Exercises 61–64, use either of the computer programs that follow to approximate the indicated Riemann sums by the right-hand rule. (For each problem, modify the function definition in the program.)

61. $\displaystyle\int_0^4 3x^2 \, dx$ using $n = 100$

62. $\displaystyle\int_0^1 \frac{1}{1 + x^2} \, dx$ using $n = 32$

63. $\displaystyle\int_{-1}^1 e^{-x^2} \, dx$ using $n = 64$

64. $\displaystyle\int_{-2}^2 (x^3 + 3x + 5) \, dx$ using $n = 128$

65. Modify the program you used so that it computes Riemann sums by the left-hand rule. Then repeat Exercises 61–64.

```
10   REM - Riemann Sums - BASIC version

20   DEF FNf(x) = 3*x*x
30   INPUT "End points a,b";a,b
40   INPUT "How many rectangles";n
50   d = (b - a)/n
60   x=a

70   FOR i = 1 TO n
80     x = x + d
90     sum = sum + FNf(x)
100  NEXT i
110  sum = sum * d
```

```
(* Riemann Sums - Pascal version *)

program Integral (input, output);

  var
    a, b, d, x, sum : real;
    i, n : integer;

  function f (x : real) : real;
  begin
    f := 3 * x * x
  end;

begin
  write('Value of a? ');
  readln(a);
  write('Value of b? ');
  readln(b);
  write('How many rectangles? ');
  readln(n);
  d := (b - a) / n;
  x := a;
```

(*Programs continue at top of next page*)

* The trigonometric functions sin and cos are treated in Chapter 9.

```
120 PRINT "Integral approximately"; sum

130 END
```

```
for i := 1 to n do
  begin
    x := x + d;
    sum := sum + f(x)
  end;
sum := sum * d;

writeln('Integral approximately ', sum : 1 : 4)

end.
```

SECTION 5.3

The Fundamental Theorem of Calculus

In Section 5.2 we used numerical methods for estimating the value of a definite integral $\int_a^b f(x)\,dx$. For most frequently encountered functions, however, it is possible to compute quickly the exact value of the integral. This is accomplished by exploiting a fundamental relationship that exists between the concepts of area and derivative.

DEFINITION 5.3 Given a function f, any differentiable function F for which $F'(x) = f(x)$ for all x in the domain of f is called an **antiderivative** of f. ⊟

EXAMPLE 6 Find all the antiderivatives of $f(x) = 2x$.

SOLUTION Keep in mind that we are trying to do the *reverse* of differentiation. We are looking for a function F whose derivative will equal f. It is not hard to see that

$$F(x) = x^2$$

is an answer, for

$$F'(x) = 2x = f(x)$$

Thus $F(x) = x^2$ is one antiderivative of f. Are there any others? It is clear that $F(x) = x^2 + 1$ will also work, as will $F(x) = x^2 + 2$, or $F(x) = x^2 - \pi$. In fact, if c is any constant, then $F(x) = x^2 + c$ is an antiderivative of $f(x) = 2x$ also.

As this discussion suggests, the following theorem guarantees that, as soon as we find one antiderivative of a function f, we immediately know all others.

THEOREM 5.4 Suppose F is an antiderivative of f on the interval I. That is, $F'(x) = f(x)$ for all x in I. If G is any other antiderivative of f on I, then

$$G(x) = F(x) + c \qquad x \text{ in } I$$

for some constant c.

PROOF By assumption, $G'(x) = f(x) = F'(x)$. Hence the function $H(x) = G(x) - F(x)$ has derivative

$$H'(x) = G'(x) - F'(x) = f(x) - f(x) = 0$$

for every x in I. Now, a function whose derivative is 0 throughout an interval must be a constant: $H(x) = c$. Hence $G(x) - F(x) = c$ or $G(x) = F(x) + c$, as claimed. ⊟

PRACTICE EXERCISE Find all the antiderivatives of (a) $4x^3$, (b) x^2, (c) $x^{1/2}$, (d) $1/x$, and (e) e^x.

[Your answers should be (a) $x^4 + c$; (b) $\dfrac{x^3}{3} + c$; (c) $\dfrac{2}{3}x^{3/2} + c$; (d) $\ln x + c$; and (e) $e^x + c$.]

The method of finding these answers is basically to make an "educated guess" based on a mixture of the basic differentiation formulas we have learned and adjusting constants if need be. The term $+c$ appearing in each answer is included to give the form of the most general antiderivative. It is called the **constant of integration**.

Note, too, that antidifferentiation problems (problems in which we are asked to find antiderivatives) have a very nice feature: We can always tell at the end of the problem whether we've done it correctly! All we have to do is differentiate our answer and see if it works. For instance, our answer in part (c) of the foregoing practice exercise was $F(x) = \frac{2}{3}x^{3/2} + c$. Then, differentiation yields

$$F'(x) = \frac{3}{2}\left(\frac{2}{3}\right)x^{1/2} = x^{1/2} = f(x)$$

as desired.

Note: In the following discussion, we shall assume $f(x) \geq 0$ for all x in $[a, b]$, so that the definite integrals involved can be interpreted as areas. However, this is only a convenience, and the proofs could easily be modified for more general f.

The following surprising result provides a fundamental link between the definite integral and antidifferentiation.

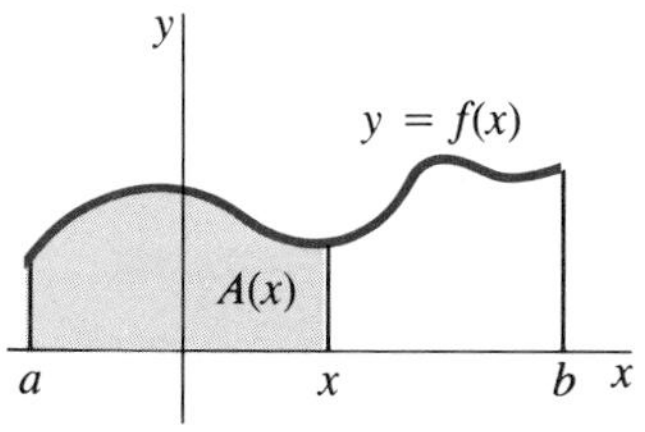

FIGURE 5.29

THEOREM 5.5 For any number x, $a \leq x \leq b$, denote by $A(x)$ the area under the graph of the continuous function $y = f(x)$ from a to x (Figure 5.29).

Then A is an antiderivative of f. That is,

$$A'(x) = f(x)$$ ⊟

Note, that if we choose $x = b$, then $A(b)$ is the total area we wish to find. If we choose $x = a$, then $A(a)$ is the area under the curve from $x = a$ to $x = a$, which is clearly 0:

(5.5) $$A(a) = 0$$

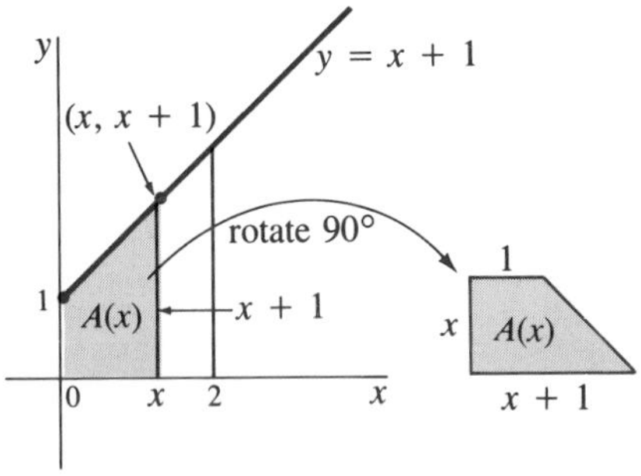

FIGURE 5.30

Before proving this theorem, let us try it out on a familiar example, the region below the graph of $y = x + 1$ between 0 and 2. If $0 \leq x \leq 2$, we see from Figure 5.30 that $A(x)$ is the area of a trapezoid with height x, lower base $x + 1$, and upper base 1. Thus, from Equation (5.2)

$$A(x) = \frac{x}{2}\left[(x + 1) + 1\right] = \frac{1}{2}(x^2 + 2x) = \frac{x^2}{2} + x$$

Differentiating, we find that $A'(x) = x + 1$, or that $A(x)$ is an antiderivative of $x + 1$, just as Theorem 5.5 promises.

As a further example, suppose we wish to find the area beneath the graph of $y = x^2$ from $x = 1$ to $x = 4$. In Figure 5.31 we have sketched $A(x)$, for $1 \leq x \leq 4$. According to Theorem 5.5, $A(x)$ is an antiderivative of x^2. Therefore

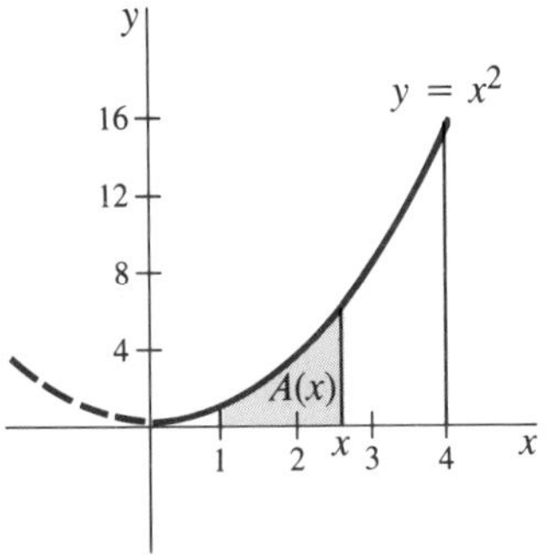

FIGURE 5.31

$$A(x) = \frac{x^3}{3} + c$$

where c is some constant. If we knew the value of this constant, we could quickly find the total area under $y = x^2$ from 1 to 4 just by substituting $x = 4$ into the formula for $A(x)$. As you can see, we are quite close to solving the area problem without the tedious process of summing approximating rectangles and trying to calculate the limit.

Note: If f is not assumed to be nonnegative, then in the statement of Theorem 5.5, $A(x)$ would be replaced by

$$\int_a^x f(t)\,dt$$

and the theorem's conclusion would read

(5.6) $$\frac{d}{dx}\left(\int_a^x f(t)\,dt\right) = f(x)$$

(Because x is used as one of the endpoints of the interval over which we are integrating, a different variable, such as t, must be used for the variable of integration.)

PROOF OF THEOREM 5.5 $A(x)$ is not given by a formula, so we must resort to the definition of the derivative as a limit and show that

$$A'(x) = \lim_{h\to 0} \frac{A(x + h) - A(x)}{h} = f(x)$$

If $a \le x < x + h \le b$, then

$$A(x + h) = \text{the area under the curve from } a \text{ to } x + h$$

$$A(x) = \text{the area under the curve from } a \text{ to } x$$

Taking the difference, we get

$$A(x + h) - A(x) = \text{the area under the curve between } x \text{ and } x + h$$

Now, if h is small, $A(x + h) - A(x)$ is the area of a thin strip that is approximately a rectangle of height $f(x)$ and base h, and

$$A(x + h) - A(x) \approx hf(x)$$

with the accuracy of this approximation improving as $h \to 0$ [compare parts (a) and (b) of Figure 5.32]. Dividing by h, we have

$$\frac{A(x + h) - A(x)}{h} \approx f(x)$$

Taking the limit of both sides as $h \to 0$, we obtain the **equality**

(5.7)
$$A'(x) = \lim_{h \to 0} \frac{A(x + h) - A(x)}{h} = f(x)$$

(You may notice that in the above discussion, we assumed $h > 0$, and then let h approach zero. Actually, Equation (5.7) holds even if we let $h < 0$, as long as h approaches 0.) ⊟

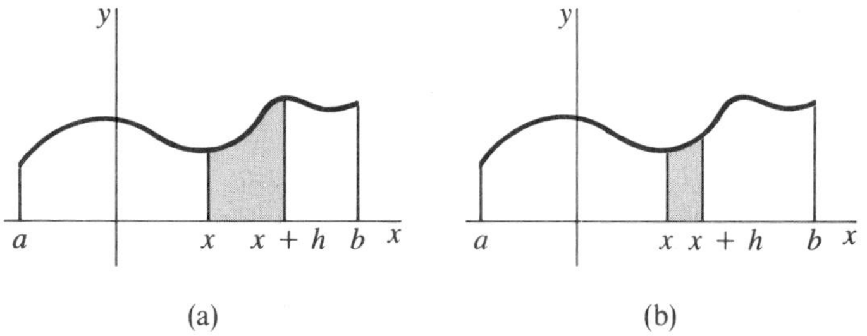

FIGURE 5.32

We are now ready to state the principal result of this chapter.

THEOREM 5.6 (The Fundamental Theorem of Calculus) Let f be a continuous function on $[a, b]$. Then, if F is any antiderivative of f (that is, if $F'(x) = f(x)$ for all x in $[a, b]$),

(5.8)
$$\int_a^b f(x)\,dx = F(b) - F(a)$$ ⊟

Theorem 5.6 (the fundamental theorem of calculus) is a prescription for finding the definite integral of f from a to b without summing up areas of approximating rectangles. In addition, the equals sign in Theorem 5.6 tells us that our result will be exact, not merely an approximation. What's the catch? Well, really there isn't one. We just have to be able to find any antiderivative of f in order to write down our answer. In Chapter 7 we shall learn how to find the antiderivative of a great many functions. (Still certain functions, such as $f(x) = e^{-x^2/2}$, do not have an antiderivative that we can write down, and for such integrals as $\int_0^3 e^{-x^2/2}\,dx$, the only route we can take is that of numerical integration, or approximation. We will study more powerful techniques for numerical integration in Chapter 7.

Before we give a proof of the fundamental theorem, let's see how it can be applied to a few examples.

EXAMPLE 7 Use the fundamental theorem of calculus to compute $\int_0^2 (x + 1)\,dx$.

SOLUTION In order to evaluate $\int_0^2 (x + 1)\,dx$, we should first find some antiderivative of $f(x) = x + 1$. $F(x) = \dfrac{x^2}{2} + x$ will work, because its derivative is given by

$$F'(x) = \frac{2x}{2} + 1 = x + 1 = f(x)$$

Now, having chosen $F(x) = \dfrac{x^2}{2} + x$, we see that, according to Theorem 5.6,

$$\int_0^2 (x + 1)\,dx = F(2) - F(0)$$

Now, $F(2) = \dfrac{(2)^2}{2} + 2 = 4$, and $F(0) = \dfrac{(0)^2}{2} + 0 = 0$. We conclude that

$$\int_0^2 (x + 1)\,dx = 4 - 0 = 4$$

This answer agrees with the one we found in Example 1, as it must. ⊟

NOTATION 5.7 It is customary to denote the expression $F(b) - F(a)$ by $F(x)|_a^b$. Thus, if F is an antiderivative of f, the fundamental theorem says that

$$\int_a^b f(x)\,dx = F(x)\Big|_a^b$$ ⊟

Perhaps surprisingly, it does not matter which antiderivative of f we use. To see this, recall that, if G is any other antiderivative of f, then $G(x) = F(x) + c$ for some constant c. But then

$$G(x)\Big|_a^b = G(b) - G(a) = [F(b) + c] - [F(a) + c] = F(x)\Big|_a^b$$

Consequently, we usually choose the simplest antiderivative we can find. That is, we take $c = 0$.

EXAMPLE 8 Find the area of the region bounded above by the parabola $y = 4 - x^2$ and below by the x-axis.

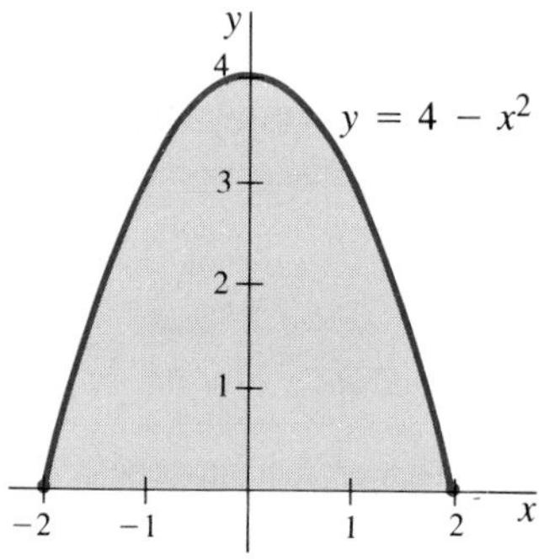

FIGURE 5.33

SOLUTION The region we are interested in is shown shaded in Figure 5.33. The parabola crosses the x-axis at $x = -2$ and at $x = +2$ so that we are concerned with computing

$$\int_{-2}^{2} (4 - x^2)\,dx$$

Our first step is to try to determine an antiderivative F for the integrand $f(x) = 4 - x^2$. This time

$$F(x) = 4x - \frac{x^3}{3}$$

will do, because

$$F'(x) = 4 - \frac{3x^2}{3} = 4 - x^2 = f(x)$$

Next we apply the fundamental theorem, which tells us that

$$\int_{-2}^{2} (4 - x^2)\,dx = \left(4x - \frac{x^3}{3}\right)\Bigg|_{-2}^{2} = \left(8 - \frac{8}{3}\right) - \left(-8 + \frac{8}{3}\right)$$

$$= \frac{16}{3} - \left(\frac{-16}{3}\right) = \frac{32}{3}$$

PROOF OF THE FUNDAMENTAL THEOREM Theorem 5.5 tells us that A is one antiderivative of f. If F is any other antiderivative of f, then A and F must differ by a constant, as we proved earlier. Therefore

(5.9) $$A(x) = F(x) + c$$

for some constant c.

How can we determine the value of c? We have previously noted that $A(a) = 0$ because it represents a region with no area. If we now set $x = a$, we find

$$0 = A(a) = F(a) + c$$

or

$$c = -F(a)$$

We can now use this value of c in Equation (5.9) to get

$$A(x) = F(x) - F(a)$$

Letting $x = b$ produces

$$A(b) = F(b) - F(a)$$

and because, by definition, $A(b) = \int_a^b f(x)\,dx$, we have finally shown that

$$\int_a^b f(x)\,dx = F(b) - F(a)$$

for any antiderivative F of f. ◻

EXAMPLE 9 Find the total area bounded by the x-axis and the graph $y = 1 - x^3$ between $x = 0$ and $x = 2$ (see Figure 5.34).

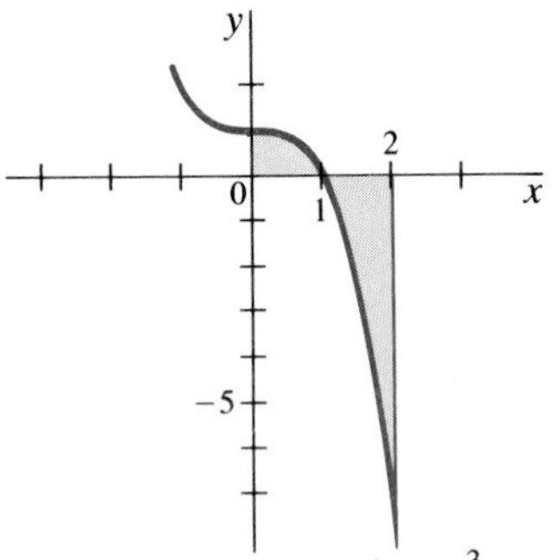

$y = 1 - x^3$

FIGURE 5.34

SOLUTION We must find a function $F(x)$ whose derivative $F'(x)$ equals $1 - x^3$. Such a function will have to be a fourth-degree polynomial in order to produce a third-degree polynomial upon differentiation. A little thought reveals that $F(x) = x - \dfrac{x^4}{4}$ will do nicely. Then

$$\int_0^2 (1 - x^3)\,dx = \left(x - \frac{x^4}{4}\right)\Big|_0^2 = 2 - \frac{16}{4} = -2$$

Something has obviously gone awry, because an area cannot be negative! What we have overlooked is that the graph of $y = 1 - x^3$ is partly below the x-axis. As we explained in connection with Figure 5.26, the definite integral counts the area of the portion of the region below the x-axis *negatively*. To obtain the actual total area, we must compute separate integrals from 0 to 1 and from 1 to 2 and add their absolute values. Thus

$$\int_0^1 (1 - x^3)\,dx = \left(x - \frac{x^4}{4}\right)\Big|_0^1 = 3/4$$

$$\int_1^2 (1 - x^3)\,dx = \left(x - \frac{x^4}{4}\right)\Big|_1^2 = (-2) - \left(\frac{3}{4}\right) = -\frac{11}{4}$$

Accordingly, the total area sought is

$$A = \frac{3}{4} + \left|-\frac{11}{4}\right| = \frac{7}{2}$$

EXAMPLE 10 Use the fundamental theorem to compute

$$\int_0^{\ln 2} e^{3x}\,dx$$

SOLUTION We seek a function $F(x)$ with $F'(x) = e^{3x}$. You may check that $F(x) = (1/3)^{e^{3x}}$ will do. Thus

$$\begin{aligned}\int_0^{\ln 2} e^{3x}\,dx &= \left[\frac{1}{3}e^{3x}\right]_0^{\ln 2}\\ &= \frac{1}{3}e^{3\ln 2} - \frac{1}{3}e^0\\ &= \frac{1}{3}(e^{\ln 2})^3 - \frac{1}{3}e^0\\ &= \frac{1}{3}(2^3) - \frac{1}{3}(1) = \frac{7}{3} \qquad (\text{because } e^{\ln 2} = 2)\end{aligned}$$

Section 5.3 EXERCISES

Find all antiderivatives for the functions $f(x)$ given in Exercises 1–14.

1. x^5 **2.** $\frac{1}{x^5}$ **3.** $x^3 + \frac{x}{2} + 1$

4. $\frac{1}{2x^3}$ **5.** $\sqrt{x}$ **6.** $x^{7/5} + x^{3/2}$

7. $\frac{1}{\sqrt{x}}$ **8.** $\frac{1}{x-1}$ **9.** $\frac{1}{3x}$

10. $2.5/x$ **11.** xe^{x^2} **12.** $ax + b$

13. $ax^2 + bx + c$ **14.** $e^x + e^{-x}$

What does each of Exercises 15–19 represent as an area?

15. $\int_1^5 (2x + 3)\, dx$ **16.** $\int_2^{16} \frac{1}{x}\, dx$

17. $\int_0^3 (x^3 + 1)\, dx$ **18.** $\int_2^{10} \frac{1}{5x}\, dx$

19. $\int_0^5 (3x^2 + 8x + 1)\, dx$

Draw a sketch and find the area under the curve for each of Exercises 20–26.

20. $y = 3x + 5$ between $x = 1$ and $x = 6$

21. $y = 6 - x - x^2$ between $x = -3$ and $x = 2$

22. $y = x^3$ between $x = 0$ and $x = 1$

23. $y = x^4$ between $x = -2$ and $x = 2$

24. $y = \frac{1}{x}$ between $x = 2$ and $x = 8$

25. $y = e^{-x}$ between $x = 0$ and $x = 3$

26. $y = \sqrt{x + 1}$ between $x = 0$ and $x = 3$

27. If you examine the proof of the fundamental theorem of calculus, you will observe that it deals with the situation where $h > 0$. What happens when $h < 0$?

SECTION 5.4

The Indefinite Integral and the Method of Substitution

We have seen that the problem of evaluating a definite integral

$$\int_a^b f(x)\, dx$$

is easily solved, provided we know an antiderivative F of the function f. For this reason the problem usually focused on in calculus books is not that of evaluating specific definite integrals like

$$\int_2^3 \ln x\, dx \quad \text{or} \quad \int_0^4 \sqrt{4 - x}\, dx \quad \text{or} \quad \int_0^1 \frac{1}{1 + x^2}\, dx$$

but rather that of finding the antiderivative of functions like $\ln x$, $\sqrt{4 - x}$, and $\frac{1}{1 + x^2}$. Once these antiderivatives are known, the fundamental theorem of calculus quickly leads to the solution of the appropriate integrals above. Moreover, as we shall see, antidifferentiation has many applications in its own right, apart from evaluating definite integrals.

It is customary to use the notation

$$\int f(x)\, dx$$

without limits a and b to denote the general problem of finding all the antiderivatives of the function f. The symbol $\int f(x)\,dx$ is also called the **indefinite integral** of the function f. Thus the statement "Evaluate $\int 2x\,dx$" means to find all functions F whose derivative $F'(x) = 2x$. As we have seen, all such functions have the form $F(x) = x^2 + c$ for some constant c, so we write

$$\int 2x\,dx = x^2 + c$$

In order to evaluate indefinite integrals in some simple cases, we shall now deduce a few antidifferentiation rules based on our repertoire of differentiation formulas. First, from the rules for differentiation of sums (Theorem 2.6) and of constant multiples (Corollary 2.9), we can immediately infer that

(5.10)
$$\int \left[f(x) + g(x) \right] dx = \int f(x)\,dx + \int g(x)\,dx$$

(5.11)
$$\int kf(x)\,dx = k \int f(x)\,dx$$

for k a constant.

The first of these rules (which can be generalized to a sum of any number of terms) tells us that we can antidifferentiate a sum term by term, as in

$$\int (2x + 3x^2)\,dx = \int 2x\,dx + \int 3x^2\,dx = x^2 + x^3 + c$$

(Note that we do not write down a separate constant of integration after each term on the right, but rather collect them into a single constant c.)
Equation (5.11) permits us to "pull" constants out from inside an integral, as in

$$\int 5e^x\,dx = 5 \int e^x\,dx = 5e^x + c$$

(because e^x is its own derivative).

Next let us consider the antiderivative of x^n, where n is a number other than -1. A multiple of x^{n+1} would yield a multiple of x^n when differentiated, so we would expect that

$$\int x^n\,dx = ax^{n+1} + c$$

for some constant a. Because we require that

$$\frac{d}{dx}\left(ax^{n+1}\right) = a(n + 1)x^n = x^n$$

it is clear that $a = \dfrac{1}{n + 1}$. Consequently, we have

(5.12)
$$\int x^n\,dx = \frac{x^{n+1}}{n + 1} + c \qquad n \neq -1$$

(This would make no sense if $n = -1$. Why?) Together with Equations (5.10) and (5.11), this rule enables us to antidifferentiate any sum of multiples of powers of x. As special cases, we have the rules

(5.13)
$$\int kx^n\,dx = k\frac{x^{n+1}}{n+1} + c \qquad n \neq -1$$

(5.14)
$$\int k\,dx = kx + c$$

where (5.14) is the case of (5.13) in which $n = 0$.

Among the infinitely many antiderivatives of a given function, at most one can pass through a specified point. Thus, as the next example illustrates, being given a single point on the antiderivative's graph determines a unique value for the constant of integration.

EXAMPLE 11 A company's marginal cost at a production level of x units is

$$C'(x) = 0.0015x^2 - 2x + 250$$

Find the cost function $C(x)$ if the cost to manufacture a thousand units is \$250,000.

SOLUTION We must antidifferentiate, because we wish to obtain $C(x)$ from its derivative. Using our rules, we find that $C(x)$ must have the form

$$C(x) = 0.0005x^3 - x^2 + 250x + c$$

where c is a constant to be determined. $C(1000) = 250{,}000$, so we can obtain an equation solvable for c by substituting $x = 1000$, $C = 250{,}000$ in the foregoing equation:

$$250{,}000 = 0.0005(1000)^3 - 1000^2 + 250(1000) + c.$$

This yields $c = 500{,}000$, so

$$C(x) = 0.0005x^3 - x^2 + 250x + 500{,}000$$

PRACTICE EXERCISE Find a function satisfying (a) $f''(x) = -4/x^3$, (b) $f'(1) = 6$, and (c) $f(2) = 7$.

[Your answer should be $f(x) = 4x - 2/x$.]

EXAMPLE 12 Evaluate $\int (1 + x)^2\,dx$.

SOLUTION To apply the rules at our disposal thus far, we must square $(1 + x)^2$ as $1 + 2x + x^2$, and then

$$\int (1 + 2x + x^2)\,dx = x + x^2 + \frac{1}{3}x^3 + c$$

In a little while, we will derive an alternative method for the preceding example, which will make the evaluation of, say, $\int (1 + x)^{20}\,dx$ just as easy.

The following rules are easily established by differentiating their right-hand sides.

(5.15)
$$\int e^{kx}\,dx = \frac{1}{k}e^{kx} + c \qquad k \text{ constant} \neq 0$$

(5.16)
$$\int \frac{1}{x}\,dx = \ln x + c \qquad x > 0$$

The second of these is valid for any interval where $x > 0$, because the natural logarithm is defined only for positive x. However, we can give a more general formula, valid for all nonzero x:

(5.17)
$$\int \frac{1}{x}\,dx = \ln|x| + c \qquad x \neq 0$$

To see this, differentiate the right side. By the chain rule,

(5.18)
$$\frac{d}{dx}\ln|x| = \frac{1}{|x|}\frac{d}{dx}|x|$$

Now,

$$\text{for } x > 0, \quad |x| = x \text{ and } \frac{d}{dx}|x| = \frac{d}{dx}(x) = 1$$

whereas

$$\text{for } x < 0, \quad |x| = -x \text{ and } \frac{d}{dx}|x| = \frac{d}{dx}(-x) = -1$$

In either case, Equation (5.18) becomes

$$\frac{d}{dx}\ln|x| = \frac{1}{x}$$

which establishes Equation (5.17).

EXAMPLE 13 Evaluate $\int \left(e^{-x/2} + \frac{1}{2x}\right) dx$.

SOLUTION Applying Equation (5.15) with $k = -1/2$ and Equation (5.17), we have

$$\int \left(e^{-x/2} + \frac{1}{2x}\right) dx = \int e^{-x/2}\,dx + \frac{1}{2}\int \frac{dx}{x}$$
$$= -2e^{-x/2} + \frac{1}{2}\ln|x| + c$$

In order to work out many of the applications of integration in the next chapter, we shall need one important technique that follows from the chain rule. Later, in Chapter 7, we shall develop further techniques.

Suppose that F is an antiderivative of f, so that $F' = f$ or, equivalently,

(5.19) $$\int f(x)\,dx = F(x) + c$$

Recall the chain rule (Section 4.2), which says that, if g is a differentiable function, then

$$\frac{d}{dx}F(g(x)) = F'(g(x))g'(x) = f(g(x))g'(x)$$

The latter, stated in equivalent integral form, is

(5.20) $$\int f(g(x))g'(x)\,dx = F(g(x)) + c$$

We shall make this somewhat complicated formula simpler through a notational device. We introduce a new variable u by letting $u = g(x)$. Then $\frac{du}{dx} = g'(x)$, and Equation (5.20) becomes

(5.21) $$\int f(u)\frac{du}{dx}\,dx = F(u) + c$$

EXAMPLE 14 Evaluate $\int 2x(x^2 + 4)^5\,dx$.

SOLUTION Note that $2x$ is the derivative of $x^2 + 4$. Hence, if we let $u = x^2 + 4$, $\frac{du}{dx} = 2x$, the integral can be rewritten

$$\int 2x(x^2 + 4)^5\,dx = \int u^5\frac{du}{dx}\,dx$$

This has the form of the left side of Equation (5.21), provided that we let $f(u) = u^5$. Then we may use $F(u) = u^6/6$ so that

$$\int u^5\frac{du}{dx}\,dx = \frac{u^6}{6} + c = \frac{(x^2 + 4)^6}{6} + c$$

(The last step is necessary because the original question asked for the antiderivatives of a function of x.) Therefore we have

$$\int 2x(x^2 + 4)^5\,dx = \frac{(x^2 + 4)^6}{6} + c$$

which can be verified by differentiating the right side.

As this example illustrates, the key is to find something (u) in the integrand whose derivative is present in the integrand too.

The term $\frac{du}{dx}\,dx$ in Equation (5.21) is customarily abbreviated du, as though $\frac{du}{dx}$ were a fraction and the dx's canceled. Of course, $\frac{du}{dx}$ is *not* a fraction, and we are only introducing a notational device that allows us to rewrite Equation (5.21) as

(5.22)
$$\int f(u)\,du = F(u) + c$$

where u is a function of x and $du = \frac{du}{dx}\,dx$. This now looks just like an equivalent of Equation (5.19), but there is this important difference: In Equation (5.19) x is an independent variable; in Equation (5.22) u is a function of the independent variable.

EXAMPLE 15 Evaluate $\int x^2\sqrt{x^3 - 4}\,dx$.

SOLUTION If the integrand contained $3x^2$ instead of x^2, it would be easy to find something in the integrand $(x^3 - 4)$ whose derivative is also present. Here we are off by a factor of 3, but being off by a constant factor will never be a problem, as we shall see.

Accordingly, let $u = x^3 - 4$ so that $\frac{du}{dx} = 3x^2$, or (using the abbreviation discussed above), $du = 3x^2\,dx$. Now our integral contains $x^2\,dx$ rather than $3x^2\,dx$, and $x^2\,dx$ is just $\frac{1}{3}\,du$.

We can now re-express our integral in terms of u and du, as follows:

$$\int x^2\sqrt{x^3 - 4}\,dx = \int u^{1/2}\left(\frac{1}{3}\,du\right) = \frac{1}{3}\int u^{1/2}\,du$$
$$= \frac{1}{3}\cdot\frac{2}{3}u^{3/2} + c = \frac{2}{9}(x^3 - 4)^{3/2} + c$$

Because we introduce, or "substitute," a new variable u into the integral, the foregoing method is called **change of variable** or the **substitution method**.

Here is a summary of the steps involved in this technique.

SUBSTITUTION METHOD—STEP BY STEP

1 Look within the integrand for a function whose derivative—or a constant multiple of whose derivative—is also present. Call this function u. (If none exists, try another method.)

2 Compute $du = \frac{du}{dx}\,dx$.

3 Re-express the integral in terms of u and du. (No x or dx can remain.)

4 Evaluate the resulting integral.

5 Convert your answer into an equivalent expression involving only x, by substituting $u = u(x)$.

EXAMPLE 16 Evaluate $\int_0^2 \frac{x\,dx}{\sqrt{2x^2+1}}$.

SOLUTION This time we have a definite integral. Because the derivative of $2x^2 + 1$ is a multiple of x and we have x in the integrand, we let $u = 2x^2 + 1$. Then $du = 4x\,dx$, so $x\,dx = du/4$. Therefore

$$\int \frac{x\,dx}{\sqrt{2x^2+1}} = \frac{1}{4}\int u^{-1/2}\,du = \frac{1}{4}(2u^{1/2}) + c = \frac{1}{2}\sqrt{2x^2+1} + c$$

and
$$\frac{1}{2}\left(\sqrt{2x^2+1}\right)\Big|_0^2 = \frac{1}{2}(3-1) = 1$$

Note that, after antidifferentiation with respect to u, we converted our answer back to a function of x and then substituted the values $x = 0$ and $x = 2$. It would be incorrect to use the x-values 0 and 2 as limits of integration in the integral with respect to u.

Alternatively, we can avoid converting back to a function of x by making the u integral into a definite integral. However, we must be sure to use the values of u that correspond to $x = 0$ and $x = 2$. When $x = 0$, $u = 2x^2 + 1 = 1$; and when $x = 2$, $u = 9$. Thus

$$\frac{1}{4}\int_1^9 u^{-1/2}\,du = \frac{1}{2}\left(\sqrt{u}\,\Big|_1^9\right) = \frac{1}{2}(3-1) = 1$$

as before.

In the next chapter we shall sample the great variety of applications of both the definite and the indefinite integral. In Chapter 7 we shall learn more advanced techniques of antidifferentiation.

Section 5.4 EXERCISES

Find all antiderivatives of each of the functions in Exercises 1–23.

1. $4x - 7$

2. $\frac{7x+4}{5}$

3. $5x^2 + 8x$

4. $\frac{5x^2+8x}{3}$

5. $\frac{5x^2+8x}{3x}$

6. $\frac{2}{x^3}$

7. $x^{17} + \frac{1}{x^{17}}$

8. $x^{1/3} + 3x^{5/2} - x^{2/3}$

9. $(1+x)^{3/2}$

10. $(2-3x)^{2/7}$

11. $(2-3x)^{15}$

12. $\sqrt{4x-7}$

13. $(2-3x)^{-5}$

14. $\frac{1}{\sqrt{5-2x}}$

15. $e^{-x} + x^2$

16. e^{-3x}

17. $x(1+x^2)$

18. $x\sqrt{1+x^2}$

19. $x(4+3x^2)^{10}$

20. $\frac{1}{x+1}$

21. $\frac{1}{3x+2}$

22. $\frac{x}{\sqrt{1+x^2}}$

23. $\frac{x}{1+x^2}$

Evaluate the indefinite integrals in Exercises 24–37.

24. $\int (3x + 8)\, dx$

25. $\int \frac{(8x - 3)\, dx}{4}$

26. $\int \frac{(4x^2 - 5x + 3)\, dx}{2}$

27. $\int \frac{(4x^2 - 5x + 3)\, dx}{x}$

28. $\int \left(x + \frac{1}{x^2}\right) dx$

29. $\int (3x^{2/5} - 5x^{3/2})\, dx$

30. $\int (3 + 4x)^{3/2}\, dx$

31. $\int e^{5x}\, dx$

32. $\int \frac{3\, dx}{x}$

33. $\int \frac{5\, dx}{x - 2}$

34. $\int \frac{x}{x^2 - 1}\, dx$

35. $\int xe^{-x^2}\, dx$

36. $\int \frac{6x - 5}{3x^2 - 5x + 2}\, dx$

37. $\int (2x + 1)\sqrt{x^2 + x + 1}\, dx$

Verify, by differentiation, the validity of each of the equations in Exercises 38–41.

38. $\int \ln x\, dx = x \ln x - x + c$

39. $\int xe^x\, dx = xe^x - e^x + c$

40. $\int \frac{dx}{a^2 - x^2} = \frac{1}{2a} \ln\left|\frac{x + a}{x - a}\right| + c$ $\quad$ (a constant)

41. $\int \frac{\sqrt{a^2 - x^2}}{x}\, dx = \sqrt{a^2 - x^2} - a \ln\left|\frac{a + \sqrt{a^2 - x^2}}{x}\right| + c$ $\quad$ (a constant)

In Exercises 42–46 find a function with the given properties.

42. $f'(x) = x^2 + x + 1, \quad f(0) = 0$

43. $f'(x) = 4x^3 - x^2 + 5x + 2, \quad f(-1) = 3$

44. $f'(x) = \frac{1}{x + 1}, \quad f(1) = 2 \ln 2$

45. $f'(x) = \sqrt{x + 1}, \quad f(3) = -4$

46. $f'(x) = \frac{2}{\sqrt[3]{x}}, \quad f(8) = -1$

Evaluate the integrals in Exercises 47–82.

47. $\int \frac{e^{\ln x}}{x}\, dx$

48. $\int \frac{1 + e^x}{e^x}\, dx$

49. $\int xe^{x^2}\, dx$

50. $\int \frac{(\ln x)^2}{x}\, dx$

51. $\int \frac{x^2\, dx}{\sqrt{4 + 5x^3}}$

52. $\int x^2(5 + x^3)\, dx$

53. $\int \frac{dx}{x\sqrt{x}}$

54. $\int \frac{dx}{\sqrt{x}(1 + \sqrt{x})}$

55. $\int \frac{dx}{3x - 5}$

56. $\int x\sqrt{1 - 4x^2}\, dx$

57. $\int e^{-5x}\, dx$

58. $\int 3^x\, dx$

59. $\int (1 + x)^{20}\, dx$

60. $\int \frac{e^x - e^{-x}}{e^x + e^{-x}}\, dx$

61. $\int \frac{3x^2 - 1}{(x^3 - x + 7)^{2/3}}\, dx$

62. $\int \frac{dx}{x \ln x}$

63. $\int_0^6 e^{x/3}\, dx$

64. $\int_e^{e^3} \frac{dx}{x \ln x}$

65. $\int_0^1 xe^{-x^2}\, dx$

66. $\int_0^1 xe^{(x^2+1)}\, dx$

67. $\int (e^x + e^{-x})^2\, dx$

68. $\int_0^1 \frac{x}{4 - x^2}\, dx$

69. $\int_0^{1/2} \frac{x}{4x^2 + 1}\, dx$

70. $\int_0^2 x\sqrt{3x^2 + 4}\, dx$

71. $\int \frac{\ln x^5}{x}\, dx$

72. $\int 8(x^3 - 4x^2)^3(3x^2 - 8x)\, dx$

73. $\int \frac{6x - 15x^4}{x^2 - x^5}\, dx$

74. $\int e^{(4t+t^3)}(9t^2 + 12)\, dt$

75. $\int \frac{(x + 1)\, dx}{\sqrt[3]{x^2 + 2x + 2}}$

76. $\int x^3 e^{x^4}\, dx$

77. $\int \frac{1 + (1/x)^5}{x^2}\, dx$

78. $\int \frac{\sqrt{\ln x}}{x}\, dx$

79. $\int \frac{dx}{x(\ln x)^2}$

80. $\int \frac{\ln \sqrt{x}}{x}\, dx$

81. $\int e^x \sqrt{e^x}\, dx$

82. $\int x^2 \sqrt[5]{3 - x^3}\, dx$

83. A company's marginal revenue when x units are sold is

$$R'(x) = 1000 + 0.02x - 3x^2$$

Find the revenue function, $R(x)$. (*Note*: $R = 0$ when $x = 0$.)

84. Suppose the rate at which the value of a certain computer is depreciating t years after purchase is given by

$$V'(t) = 100t - 800 \qquad 0 \le t \le 5$$

Find $V(t)$ if the machine is purchased new (at $t = 0$) for \$4000.

85. In t hr of operation, a distillery produces $W(t)$ gal of whiskey. Suppose the rate of production at time t is

$$W'(t) = 50 + 2t - \frac{t^2}{10}$$

Find the formula for $W(t)$.

86. A bag of bagels is placed in a freezer where the temperature is 20°F. After t min the average temperature of the bagels is decreasing at the rate

$$\frac{dT}{dt} = -4.5e^{-0.09t}$$

Find the temperature T as a function of t, if $T(10) = 40$°F.

Key Mathematical Concepts and Tools

Area
Riemann sum
Right-hand rule
Left-hand rule
Fundamental theorem of calculus
Definite integral
Indefinite integral and the antiderivative
Substitution method

Applications Covered in This Chapter

Area under a curve
Deriving the cost function given marginal cost

FIVE Warm-Up Test

1. Use the right-hand rule with $n = 3$ to estimate the value of:

(a) $\int_1^4 (x - x^2)\, dx$ (b) $\int_0^1 \frac{1}{(x + 1)^3}\, dx$

2. Evaluate the following summations.

(a) $\sum_{i=3}^{i=7} (i + i^2)$ (b) $\sum_{i=0}^{i=4} (1 + (-1)^i)$

(c) $\sum_{i=1}^{4} \frac{i - 1}{i}$

3. Evaluate the following definite integrals exactly by interpreting each integral as the area of a simple geometrical figure. Do not use antidifferentiation.

(a) $\int_{-1}^{+2} (2x + 3)\, dx$ (b) $\int_3^7 4\, dx$

(c) $\int_{-8}^{8} \sqrt{64 - x^2}\, dx$

4. Find, by integration, the area bounded above by the curve $y = -32 + 12x - x^2$ and below by the x-axis.

5. Evaluate the following indefinite integrals.

(a) $\int xe^{-x^2}\, dx$ (b) $\int \frac{x^2}{4 + x^3}\, dx$

(c) $\int \left(\frac{5}{x^2} + 6 + 3x\right) dx$

6. Find all antiderivatives for:

(a) $\dfrac{3}{x^{1/3}} - x^{5/2} + \dfrac{2}{x}$ (b) $\dfrac{x}{(4 - x^2)^{3/2}}$

(c) 6

7. Find the unique function $f(x)$ that satisfies all of the following conditions.

(i) $f'''(x) = 0$ for all x

(ii) $f(1) = 0$, $f(-1) = 0$

(iii) $f'(2) = 8$

8. The marginal cost of a certain commodity is given by

$$3x + 2\sqrt{x + 1}$$

and the fixed cost is \$1500. Find the cost function $C(x)$.

9. (a) Write down a definite integral that represents the area under the curve $y = \dfrac{1}{x}$ between $x = 1$ and $x = 3$.

(b) Estimate the value of ln 3 by using the left-hand rule with $n = 4$ on the definite integral of part (a).

10. A baseball is thrown vertically upward with an initial velocity of 96 ft/sec from a height of 7 ft. How high above the ground will the ball be after 2 sec? (*Hint*: The acceleration due to gravity is independent of time and is equal to -32 ft/sec^2.)

FIVE Final Exam

1. Evaluate the following definite integrals exactly by interpreting each integral as the area of a simple geometric figure. Do not use antidifferentiation.

(a) $\int_2^5 7\,dx$ (b) $\int_1^6 x\,dx$

(c) $\int_0^5 (5 - x)\,dx$ (d) $\int_{-6}^{+6} \sqrt{36 - x^2}\,dx$

2. Evaluate the following summations.

(a) $\sum_{i=1}^{7} (i - 3)$ (b) $\sum_{i=2}^{9} i^2$

(c) $\sum_{i=3}^{5} \dfrac{i}{i + 1}$ (d) $\sum_{i=1}^{6} (-1)^i$

3. Use the left-hand rule with $n = 4$ to estimate the value of:

(a) $\int_0^2 (x^2 - 1)\,dx$ (b) $\int_2^3 \dfrac{1}{x^2}\,dx$

4. Evaluate the following indefinite integrals.

(a) $\int \left(\dfrac{1}{x} + 1\right) dx$ (b) $\int e^{x+3}\,dx$

(c) $\int x(5 + 4x^2)^8\,dx$

5. (a) Write down the definite integral for the area of the region bounded above by the curve $y = 20 - x - x^2$ and below by the x-axis.

(b) Evaluate this integral.

6. Find all antiderivatives for:

(a) $\dfrac{1}{2x + 3}$ (b) $\dfrac{x}{9 + x^2}$

(c) $e^{3x} - \dfrac{2}{\sqrt{x}}$

7. Find the area bounded above by $y = e^{3x}$, below by the x-axis, on the left by the line $x = 0$, and on the right by the line $x = \ln 3$.

8. (a) Write down a definite integral that represents the area ($= \pi/4$) of that portion of the unit circle within the first quadrant.

(b) Use the right-hand rule with $n = 5$ to obtain a crude approximation to $\pi/4$ from the definite integral you found in part (a).

9. Explicitly find the function $f(x)$ that satisfies:

(a) $f'(x) = x^3 - 4x + 7$, $f(1) = 2$

(b) $f'(x) = xe^{x^2}$, $f(0) = 4$

10. Find the unique function $f(x)$ that satisfies all of the following conditions.

(i) $f'''(x) = 30$ for all x

(ii) $f(0) = -4$

(iii) $f'(0) = 2$

(iv) $f''(1) = 24$

SIX

Applications of Integration

Outline

In Context

In Chapter 5 we studied in some detail the area A of the plane region below the graph of an equation $y = f(x)$ and above some interval (a, b). We sliced our region into thin subregions, each of width $x = (b - a)/n$. Then we used thin rectangles R_k to approximate each of the smaller regions:

$$A \approx R_1 + R_2 + \cdots + R_n = \sum_{k=1}^{n} R_k.$$

Each rectangle, in turn, had area $R_k = f(z_k)\,\Delta x$, where z_k was an arbitrary point in the kth subinterval $[x_{k-1}, x_k]$, so that

$$A \approx \sum_{k=1}^{n} f(z_k)\,\Delta x$$

(see Figure 6.1). This is called a *Reimann sum.* As $n \to \infty$, our approximation became more and more accurate, so that in the limit it approached the true area:

$$A = \lim_{n\to\infty} \sum_{k=1}^{n} f(z_k)\,\Delta x$$

We summarized the entire process by writing

$$\lim_{n\to\infty} \sum_{k=1}^{n} f(z_k)\,\Delta x = \int_a^b f(x)\,dx$$

Frequently, in applications, we wish to compute some quantity other than area. Often we break this quantity down into many smaller units whose sum is the desired quantity. Then we approximate each small piece by something more easily computed. This procedure results in a Riemann sum just like the foregoing (though it is no longer interpreted as an area). The limit of this sum as the number of approximating pieces becomes very large is a definite integral. In the next few sections, we will exploit this idea to solve a variety of problems.

SECTION 6.1

The Integral as an Average

Suppose we set up a thermometer at a certain location between 12 noon and 12 midnight. We know from experience that the temperature readings will vary throughout the day. How can we determine the average temperature during this 12-hour period?

One approach to this problem might be to take measurements at 1 P.M., 2 P.M., 3 P.M., . . . , 12 P.M. and then sum these measurements and divide by 12. For example, if these measurements happened to be as shown in Table 6.1, the average temperature would be

$$\frac{62 + 64 + 67 + 68 + 67 + 66 + 66 + 64 + 61 + 59 + 57 + 56}{12} = \frac{757}{12} = 63.08^\circ$$

TABLE 6.1

Time	1 P.M.	2 P.M.	3 P.M.	4 P.M.	5 P.M.	6 P.M.	7 P.M.	8 P.M.	9 P.M.	10 P.M.	11 P.M.	12 P.M.
Temperature	62°	64°	67°	68°	67°	66°	66°	64°	61°	59°	57°	56°

Now we know that this answer is only an approximate one, because we have taken only hourly measurements. It is conceivable (though admittedly unlikely) that after 4 P.M., when the temperature was 68°, a sudden thunderstorm passed by, dropping the temperature quickly to 58° at 4:30 P.M. and then allowing it to move back up to 67° at 5 P.M. If this were the case, our estimated average temperature of 63.08° would be inaccurate.

One way to improve our answer is to take more frequent temperature readings, say every half hour between 12 noon and 12 midnight (at 12:30 P.M., 1 P.M., 1:30 P.M., . . . , 12 P.M.). Then we would have 24 measurements in all, and we would take their average. Again we will come up with an answer that is only approximate, but it is likely to be more accurate than our previous answer. By this time you can probably guess what's coming next.

Suppose we have a device that can continuously record the temperature between time $t = a$ and $t = b$. If we denote temperature by T, a typical plot of the temperature versus time might be as shown in Figure 6.2. Imagine that this curve is the graph of $T = f(t)$ for some appropriate function f.

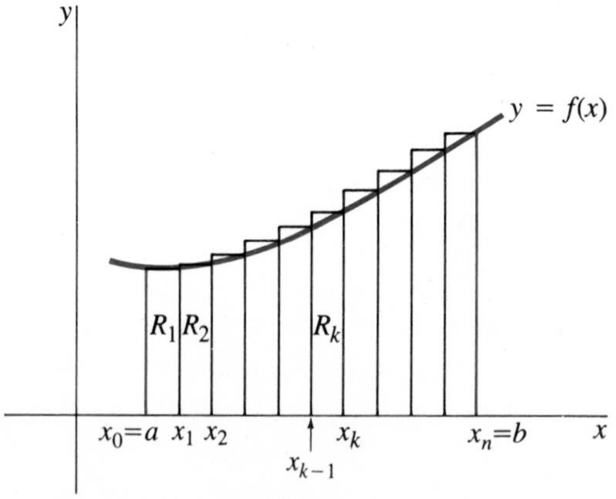

FIGURE 6.1

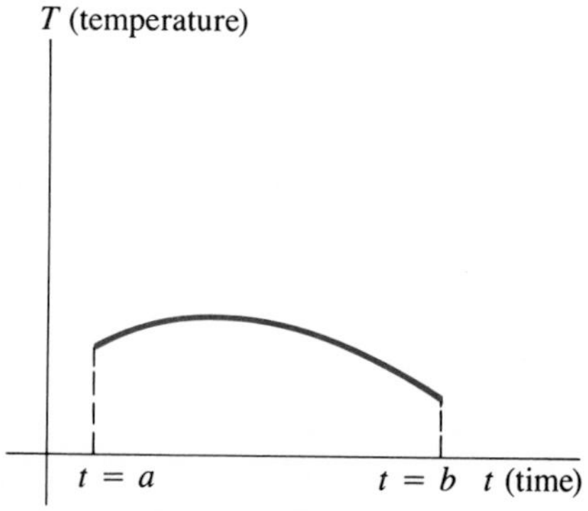

FIGURE 6.2

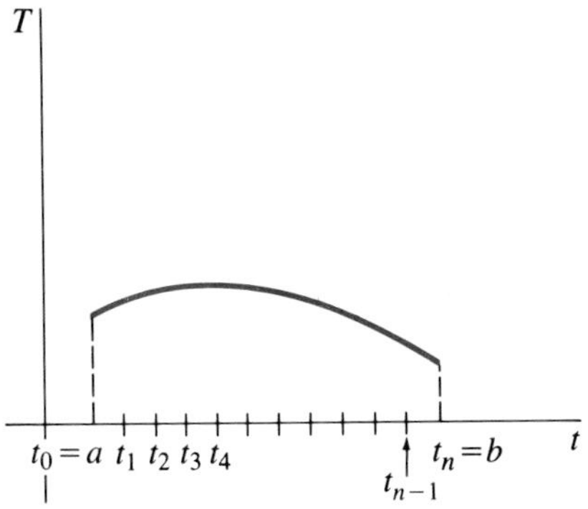

FIGURE 6.3

Next let us subdivide the time interval between $t = a$ and $t = b$ into n equally spaced subintervals, each of length

$$\Delta t = \frac{b - a}{n}$$

Denote the points of subdivision by $t_0 = a, t_1, t_2, \ldots, t_n = b$ (see Figure 6.3). In this case our temperature readings throughout the time interval $[a, b]$ are

$$f(t_1), f(t_2), f(t_3), \ldots, f(t_n)$$

Our estimate of the average temperature is

$$\frac{f(t_1) + f(t_2) + f(t_3) + \cdots + f(t_n)}{n} = \sum_{k=1}^{n} \frac{f(t_k)}{n} = \frac{1}{n} \sum_{k=1}^{n} f(t_k)$$

If we recall that $\Delta t = \dfrac{b - a}{n}$, then $\dfrac{\Delta t}{b - a} = \dfrac{1}{n}$. Our average temperature estimate now becomes

$$\frac{1}{n} \sum_{k=1}^{n} f(t_k) = \frac{\Delta t}{b - a} \sum_{k=1}^{n} f(t_k)$$

or, equivalently,

$$\left(\frac{1}{b - a}\right) \sum_{k=1}^{n} f(t_k)\, \Delta t$$

We can improve our approximation by taking more and more equally spaced temperature readings (that is, we let $n \to \infty$). Because we are then taking the limit of a Riemann sum, our average temperature between time $t = a$ and $t = b$ is given by

$$\frac{1}{b - a} \int_a^b f(t)\, dt$$

In the general case where $y = f(x)$ is a function that does not necessarily represent the temperature, our discussion suggests that we make the following definition.

DEFINITION 6.1 For any function f, defined and continuous on $a \le x \le b$, we define the **average value** (or **mean value**) of f on $[a, b]$ to be

$$\frac{1}{b - a} \int_a^b f(x)\, dx$$ ⊟

EXAMPLE 1 Find the average value of $f(x) = x^2 - x$ on the interval $[1, 3]$.

SOLUTION According to the definition, the average is

$$\frac{1}{b - a} \int_a^b f(x)\, dx = \frac{1}{3 - 1} \int_1^3 (x^2 - x)\, dx = \frac{1}{2} \int_1^3 (x^2 - x)\, dx$$

$$= \frac{1}{2}\left(\frac{x^3}{3} - \frac{x^2}{2}\right)\Bigg|_1^3 = \frac{1}{2}\left[\left(9 - \frac{9}{2}\right) - \left(\frac{1}{3} - \frac{1}{2}\right)\right] = \frac{7}{3}$$

EXAMPLE 2 A drug is injected into the bloodstream. Gradually, the drug is eliminated from the bloodstream by the kidneys. The fraction of the drug remaining at any time t is described by the function

$$f(t) = e^{-0.29t}$$

where t is measured in hours. Find the average fraction of the drug present in the bloodstream during the first hour.

SOLUTION We simply apply the definition with $f(t) = e^{-0.29t}$, $a = 0$, and $b = 1$. Thus

$$\frac{1}{b-a}\int_a^b f(t)\,dt = \frac{1}{1-0}\int_0^1 e^{-0.29t}\,dt = \int_0^1 e^{-0.29t}\,dt$$

$$= -\frac{1}{0.29}e^{-0.29t}\Big|_0^1 = \frac{1}{0.29}[1 - e^{-0.29}]$$

$$\approx \frac{1}{0.29}[1 - 0.748] \approx 0.87$$

In other words, an average of 87% of the drug will be present in the bloodstream during the first hour.

EXAMPLE 3 An object is moving in such a way that its velocity at each instant in time is described by the equation

$$v(t) = 12 + 2t - t^2$$

Find its average velocity between $t = 0$ and $t = 3$.

SOLUTION

$$\frac{1}{b-a}\int_a^b v(t)\,dt = \frac{1}{3-0}\int_0^1 (12 + 2t - t^2)\,dt$$

$$= \frac{1}{3}\left(12t + t^2 - \frac{t^3}{3}\right)\Big|_0^3 = \frac{1}{3}(36 + 9 - 9) = 12$$

Before concluding this section, it will be worthwhile to give a geometric interpretation of Definition 6.1. Suppose $f(x) \geq 0$ on $[a, b]$. Then $\int_a^b f(x)\,dx$ represents the area under the curve $y = f(x)$ between $x = a$ and $x = b$. Now if we let $\bar{f}$ denote the average value of f on $[a, b]$, then, according to Definition 6.1,

$$\bar{f} = \frac{1}{b-a}\int_a^b f(x)\,dx$$

or

$$(\bar{f})(b-a) = \int_a^b f(x)\,dx$$

Thus $\bar{f}$ can be thought of as the height of a rectangle whose base has length $(b - a)$ and whose area is exactly the same as the area under the curve $y = f(x)$ between $x = a$ and $x = b$ (see Figure 6.4).

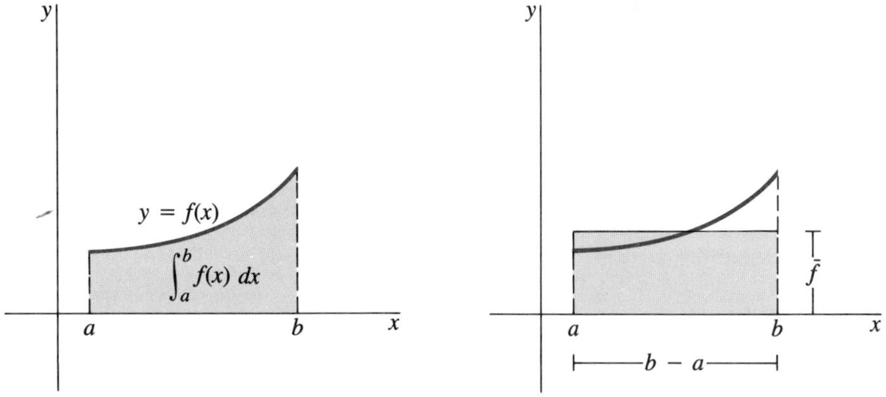

(a) Area beneath the graph (b) Rectangle with the same area

FIGURE 6.4

Section 6.1 EXERCISES

In Exercises 1–7 find the average of $\bar{f}$ of the function f as given.

1. $f(x) = 1 + x^2, \quad 0 \le x \le 2$
2. $f(x) = x^2 + e^x, \quad 0 \le x \le 1$
3. $f(z) = \dfrac{1}{\sqrt{z^2 + 6z + 9}}, \quad 2 \le z \le 4$
4. $f(x) = \dfrac{x}{x^2 + 1}, \quad 0 \le x \le 1$
5. $y = e^t, \quad 0 \le t \le 3$
6. $y = \dfrac{3x^2 + 4x}{(x^3 + 2x^2 + 9)^{1/2}}, \quad 1 \le x \le 2$
7. $y = xe^{x^2}, \quad 0 \le x \le 2$
8. Consider the problem of drug application, as in Example 2, but with the following modifications: Find the drug dose (in grains) required to assure an average level of 0.5 grain in the period from 2 to 5 hr after administration of the drug.
9. Again referring to Example 2, assume that an initial dose of 3 grains is administered at $t = 0$ and that a second 3-grain dose is administered 8 hr later. Find the average level in the bloodstream during the second 8-hr period.
10. In the study of body burden (the level at which potentially harmful pollutants are present in the organs and blood of the body), the following differential equation has been proposed.*

$$\frac{dz}{dt} = \alpha\left(\frac{dw}{dt} - \frac{dx}{dt}\right) - cz$$

In this equation $z(t)$ represents the amount of the pollutant per unit volume of blood at time t, $w(t)$ represents the total intake of the pollutant in the time interval between 0 and t, and $x(t)$ represents the total amount of excretion of the pollutant in the time interval between 0 and t; α and c are constants. With the simplifying assumption that $w(t) - x(t)$, the excess of intake over excretion, remains constant, we have $(dw/dt) - (dx/dt) = 0$ and the differential equation reduces to $dz/dt = -cz$. Thus $z = z_0 e^{-ct}$ for $t \ge 0$, where z_0 represents the amount of

* J. S. Rustage, "Mathematical Models in Medicine," International Journal of Mathematical Education in Science and Technology, Vol. 2, (1971): 193–203.

pollutant per unit volume in the blood at time 0. Write the expression for the average amount of pollutant in the blood per unit volume (a) between times $t = 5$ and $t = 10$; (b) between times $t = T$ and $t = 2T$. (c) What happens to your answer to part (b) as $T \to \infty$?

11. Over a 10-hr period in a certain location, measurements of barometric pressure varied according to the formula

$$B(t) = 30 + t^3 - 3t \qquad 0 \le t \le 10$$

Find the average barometric pressure over this period.

12. A large corporation records its daily profits each day for one year. It finds that the profit $P(t)$ on day t, $1 \le t \le 365$, is given approximately by

$$P(t) = \$12{,}000 + \frac{t}{100}(365 - t)$$

Develop a quick way for this company to approximate its average daily profit without computing

$$\frac{P(1) + P(2) + \cdots + P(365)}{365}$$

13. The cost function for mining x pounds of a certain ore is given by

$$C(x) = 1800 + (10{,}000 - \sqrt{x})^{2/3} \qquad 0 \le x \le 100 \text{ lb}$$

Find the average marginal cost in mining between 60 and 70 lb.

14. We have seen that the amount of money in a savings account earning continuously compounded interest is given by $A(t) = A_0e^{rt}$, where r is the annual interest rate as a decimal, A_0 is the initial principal, and t is the time in years. Find the average amount in the account if \$5000 is invested at 7%: (a) over the first 10 years; (b) during the second 10 years.

15. We saw in Example 33 in Chapter 4 that the amount of a radioactive substance present at time t satisfies the equation $P(t) = P_0e^{kt}$, where k is negative. For carbon-14, we found $k \approx -0.000124$. If the initial amount of carbon-14 was 30 g, what was the average amount of carbon-14 over the next 1000 years?

16. If a 3000-bushel storage bin takes 10 min to empty, and the volume of grain left in the bin at any time is given by $V(t) = 30(10 - t)^2$ bushels, what is the average amount of grain in the bin while it is emptying?

SECTION 6.2

Area Between Curves and Its Applications

Let us consider two curves $y = f(x)$ and $y = g(x)$ defined for $a \le x \le b$. Assume that, for every such x, we have $f(x) \ge g(x)$. Thus the graph of $y = f(x)$ lies above that of $y = g(x)$, as shown in Figure 6.5. In this section we shall develop a simple expression for the area A enclosed between these two curves for $a \le x \le b$, and we shall describe a number of business and economic applications.

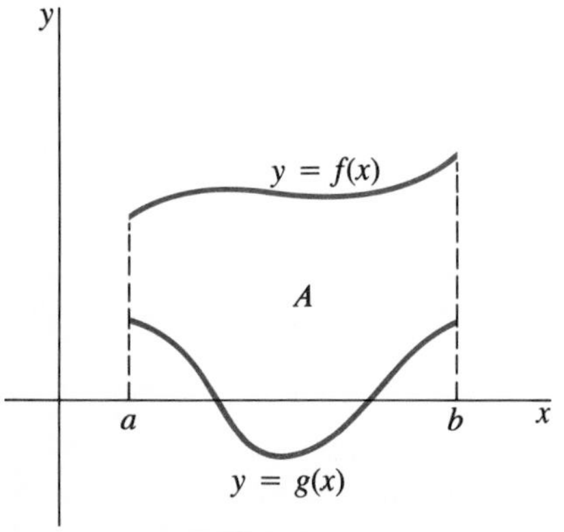

FIGURE 6.5

By analogy with our work in Chapter 5, we begin by slicing our region into thinner subregions, each having width

$$\Delta x = \frac{b - a}{n}$$

In a familiar manner, we use thin rectangles R_k to approximate each of our smaller regions (see Figure 6.6):

$$A \approx R_1 + R_2 + \cdots + R_n = \sum_{k=1}^{n} R_k$$

Note that, for each k, we have used the values of f and g at the right-hand end point x_k of the kth subinterval to determine the height of the rectangle R_k. Thus its upper edge is at height $f(x_k)$ and its lower edge is at $g(x_k)$. Accordingly,

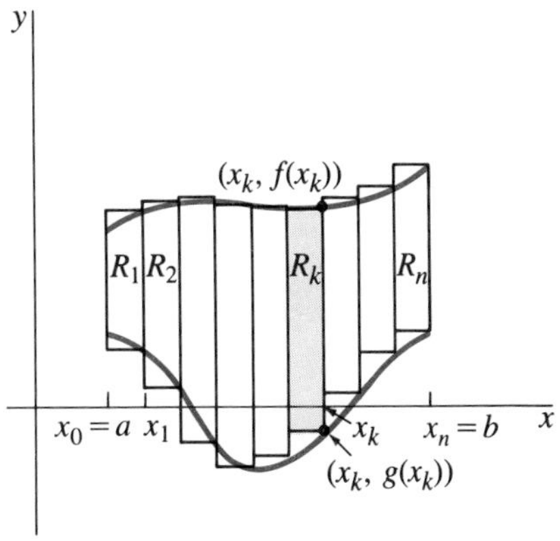

FIGURE 6.6

the rectangle R_k has area

$$R_k = \underbrace{[f(x_k) - g(x_k)]}_{\text{height}} \underbrace{\Delta x}_{\text{width}}$$

Our approximation to A now takes the form

(6.1) $$A \approx \sum_{k=1}^{n} [f(x_k) - g(x_k)]\,\Delta x$$

Notice that this is a Riemann sum analogous to those obtained in our discussion of the area under a single curve. If we now take more and more, but thinner and thinner rectangles, we obtain in the limit

(6.2) $$A = \int_a^b [f(x) - g(x)]\,dx$$

Note that the right-hand side of Equation (6.1) differs from the right-hand side of Equation (6.2) only in that the Δx is replaced by dx and the $\sum_{k=1}^{n}$ is replaced by $\int_a^b$. This is a good intuitive way to pass to the limit in such an approximation procedure.

REMARK In the event that both $f(x) > 0$ and $g(x) > 0$ for $a \leq x \leq b$, there is a simple geometric interpretation of Equation (6.1) that makes it very easy to remember. There are three areas that we can consider in the present situation:

1 The area A, lying between the curves $y = f(x)$ and $y = g(x)$. See Figure 6.7(a).

2 The area B, lying between the x-axis and the curve $y = g(x)$. See Figure 6.7(b). From Chapter 5 we know that B is given by

$$B = \int_a^b g(x)\,dx$$

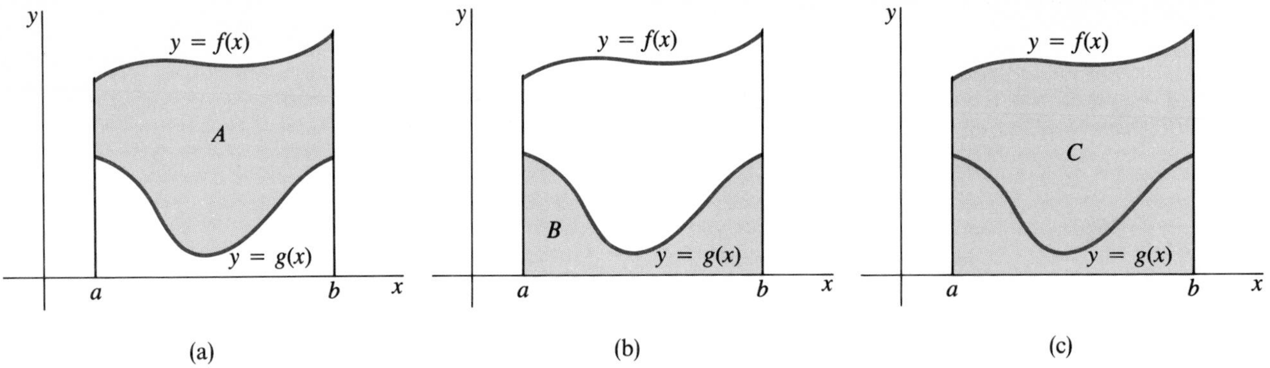

FIGURE 6.7

3 The area C, lying between the x-axis and the curve $y = f(x)$. See Figure 6.7(c). From Chapter 5, we know that

$$C = \int_a^b f(x)\,dx$$

Now it is evident from Figure 6.7 that

$$A = C - B$$

Making use of the formulas we have for B and C yields

$$A = \int_a^b f(x)\,dx - \int_a^b g(x)\,dx = \int_a^b [f(x) - g(x)]\,dx$$

which agrees with Equation (6.2). ⊟

EXAMPLE 4 Find the area enclosed between the two curves $y = x^2$ and $y = x^3$ for $0 \le x \le 1$.

SOLUTION Figure 6.8 shows that the curve $y = x^3$ lies below $y = x^2$ for $0 \le x \le 1$. Thus we apply Equation (6.2) with $f(x) = x^2$ and $g(x) = x^3$.

$$A = \int_0^1 [x^2 - x^3]\,dx = \left(\frac{x^3}{3} - \frac{x^4}{4}\right)\Bigg|_0^1 = \frac{1}{12}$$

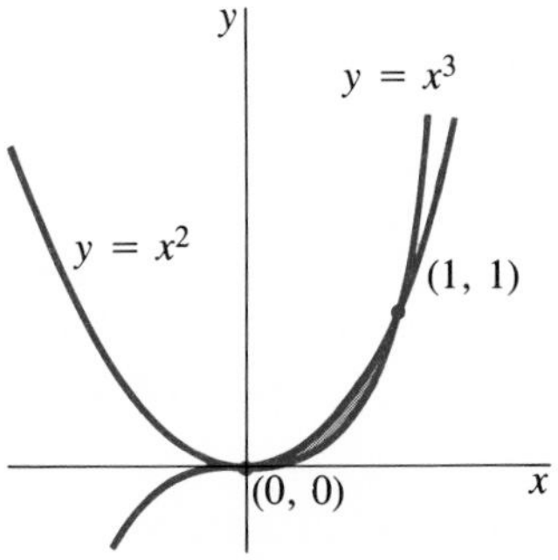

FIGURE 6.8

EXAMPLE 5 Find the area between the curves $y = 4x$ and $y = x^3 + x^2 - 2x$.

SOLUTION As a preliminary step, let us determine where the straight line $y = 4x$ intersects the cubic $y = x^3 + x^2 - 2x$. This will help us draw a sketch. We set

$$x^3 + x^2 - 2x = 4x$$

so that

$$x^3 + x^2 - 6x = 0$$

which factors to yield

$$(x)(x - 2)(x + 3) = 0$$

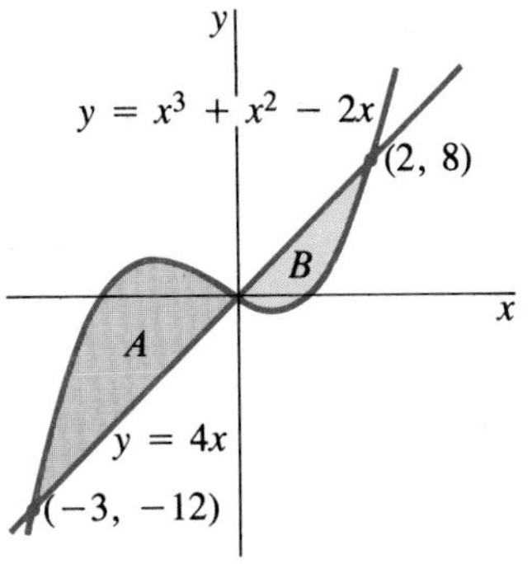

FIGURE 6.9

The curves intersect at $x = 0$, $x = 2$, and $x = -3$ (see Figure 6.9).

Here we have a situation we have not encountered before. If we examine the region A, where $-3 \le x \le 0$, we see that the cubic lies above the straight line. For the region B, however, where $0 \le x \le 2$, the straight line lies above the cubic. Because Equation (6.2) requires that $y = f(x)$ lie above $y = g(x)$, we cannot apply this formula simultaneously to both regions A and B. Accordingly, we must treat these regions separately. We have

$$A = \int_{-3}^{0} [(x^3 + x^2 - 2x) - 4x]\,dx = \int_{-3}^{0} (x^3 + x^2 - 6x)\,dx$$

$$= \left(\frac{x^4}{4} + \frac{x^3}{3} - 3x^2\right)\Bigg|_{-3}^{0} = \frac{63}{4}$$

$$B = \int_{0}^{2} [4x - (x^3 + x^2 - 2x)]\,dx = \int_{0}^{2} (6x - x^3 - x^2)\,dx$$

$$= \left(3x^2 - \frac{x^3}{3} - \frac{x^4}{4}\right)\Bigg|_{0}^{2} = \frac{16}{3}$$

The total area between the two curves is simply the sum of these two areas.

$$A + B = \frac{63}{4} + \frac{16}{3} = \frac{253}{12}$$

EXAMPLE 6 A manufacturing company finds that, if it produces 100 items, its profit is zero. (This is the break-even point. The company will lose money if production is smaller than 100 items and will make money if production exceeds 100.) It estimates that its marginal revenue function is

$$\frac{dR}{dx} = 200 + 2x - 0.003x^2$$

and its marginal cost is given by

$$\frac{dC}{dx} = 150 + x$$

Find the profit when the company produces 120 items.

SOLUTION We begin by recalling that

$$\text{Profit} = \text{revenue} - \text{cost}$$

$$P(x) = R(x) - C(x)$$

so

$$\text{Marginal profit} = \text{marginal revenue} - \text{marginal cost}$$

$$\frac{dP}{dx} = \frac{dR}{dx} - \frac{dC}{dx}$$

We are interested in finding $P(120)$. From the fundamental theorem of calculus,

$$\int_{100}^{120} \left(\frac{dP}{dx}\right) dx = P(120) - P(100)$$

But we are given that the profit for producing 100 items is zero.

$$P(100) = 0$$

Thus

$$P(120) = \int_{100}^{120} \left(\frac{dP}{dx}\right) dx = \int_{100}^{120} \left(\frac{dR}{dx} - \frac{dC}{dx}\right) dx \qquad \textbf{(6.3)}$$

$$= \int_{100}^{120} [(200 + 2x - 0.003x^2) - (150 + x)]\, dx$$

$$= \int_{100}^{120} (50 + x - 0.003x^2)\, dx = \left(50x + \frac{x^2}{2} - 0.001x^3\right)\Bigg|_{100}^{120}$$

$$= (6000 + 7200 - 1728) - (5000 + 5000 - 1000)$$

$$= 2472$$

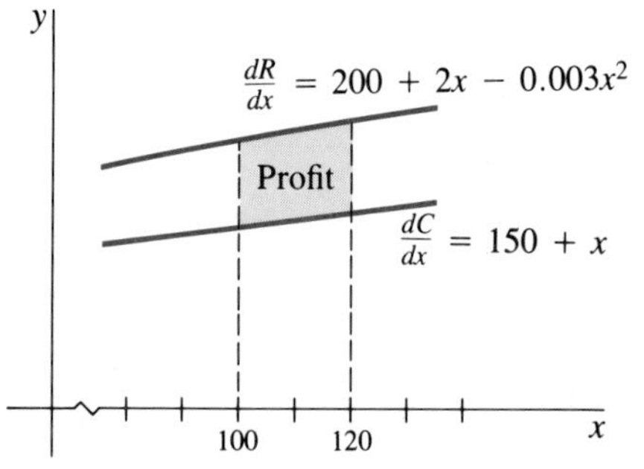

FIGURE 6.10

Note that our total profit, $P(120)$, is given in Equation (6.3) as

$$\int_{100}^{120} \left(\frac{dR}{dx} - \frac{dC}{dx}\right) dx$$

This is exactly the area enclosed between the marginal revenue function and the marginal cost function (see Figure 6.10).

CONSUMERS' SURPLUS AND PRODUCERS' SURPLUS Let $p = D(x)$ be the demand curve for a certain product, where p is the unit price and x is the number of units consumers are willing to purchase at price p. Let $p = S(x)$ be the supply curve for the same product. This relates the number of units x that producers are willing to sell at the price p. Typical examples of these curves are illustrated in Figure 6.11.

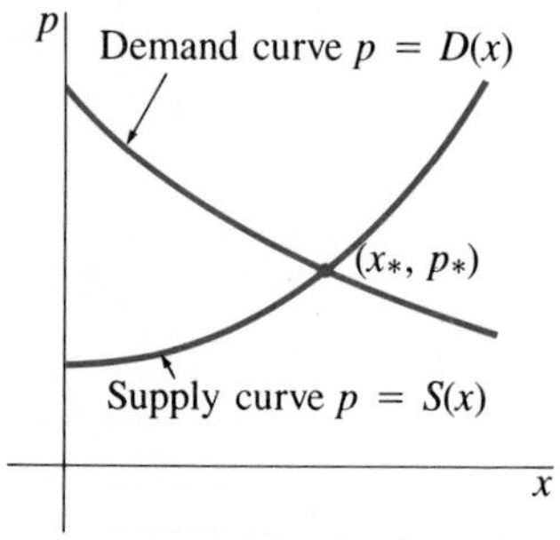

FIGURE 6.11

In a competitive market situation, the demand x_* and the price p_* at which consumer willingness to buy equals supplier willingness to sell is called **market equilibrium**. The point (x_*, p_*) is the intersection of the demand curve and the supply curve.

For example, if the demand and supply equations are

$$p = D(x) = -3x + 18$$

$$p = S(x) = x^2/4 + 2$$

the market equilibrium can be found by equating these two functions and solving for the equilibrium demand x_*.

$$-3x + 18 = x^2/4 + 2$$

$$-12x + 72 = x^2 + 8$$

$$0 = x^2 + 12x - 64 = (x - 4)(x + 16)$$

Because the demand cannot be negative, we discard the negative root, so $x_* = 4$. Substituting this into either of the two original equations gives $p_* = 6$.

Even though, in a competitive market, all consumers need only pay the market (equilibrium) price, many of them are willing to pay more. In our example, if the price were \$3, demand would be $x = 5$ units (found by setting $p = 3$ in the demand equation). The amount of money saved by consumers who would have been willing to pay more, but didn't have to (because of free-market conditions), is called the consumers' surplus.

To find the consumers' surplus, we calculate the maximum that consumers would pay if they had to and subtract what they actually pay, namely $p_* x_*$. Let us imagine that a monopolist is totally in control of the market, makes available to consumers only a small amount, Δx, of the product at a time, and sells it for the highest price obtainable. Thus we partition the interval $[0, x_*]$ into a number n of equal subintervals, each of width $\Delta x = x_*/n$. Now, when the monopolist makes available the first Δx units of the product, the price is $D(x_1)$, and these units yield revenue $D(x_1)\,\Delta x$. Under free-market conditions, the Δx units would yield revenue $p_*\,\Delta x$, so the consumers who purchase them would save the difference, $D(x_1)\,\Delta x - p_*\,\Delta x$, or

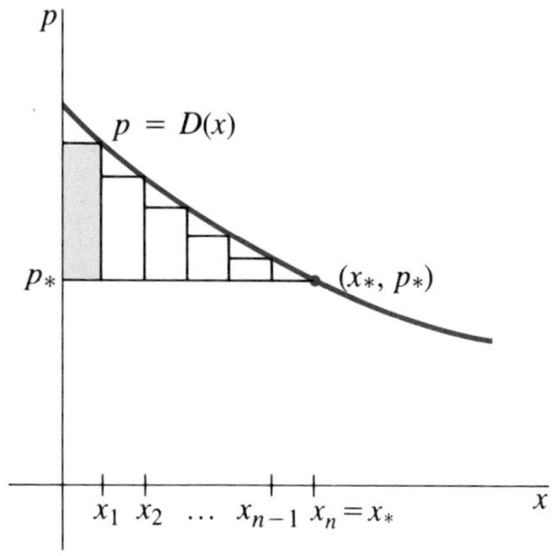

FIGURE 6.12

$$[D(x_1) - p_*]\,\Delta x$$

Note that this is the area of the first (shaded) rectangle in Figure 6.12.

After the first Δx units are purchased, the monopolist makes available another Δx units, which are purchased by the consumers who [were not willing to pay $D(x_1)$, but] are willing to pay $D(x_2)$. The revenue is $D(x_2)\,\Delta x$, but under free-market conditions it would be only $p_*\,\Delta x$, so these consumers would save

$$[D(x_2) - p_*]\,\Delta x$$

the area of the second rectangle in Figure 6.12.

Continuing in this fashion for each of the remaining subintervals and summing, we see that the total amount saved by all consumers (if they pay only the equilibrium price) is

$$\sum_{k=1}^{n} (D(x_k) - p_*)\,\Delta x$$

Of course, the monopolist will increase his revenue by choosing Δx to be small so that he makes available a tiny quantity of the product at a time. For this reason the **consumers' surplus** (CS) is defined to be

$$\lim_{n\to\infty} \sum_{k=1}^{n} (D(x_k) - p_*)\,\Delta x$$

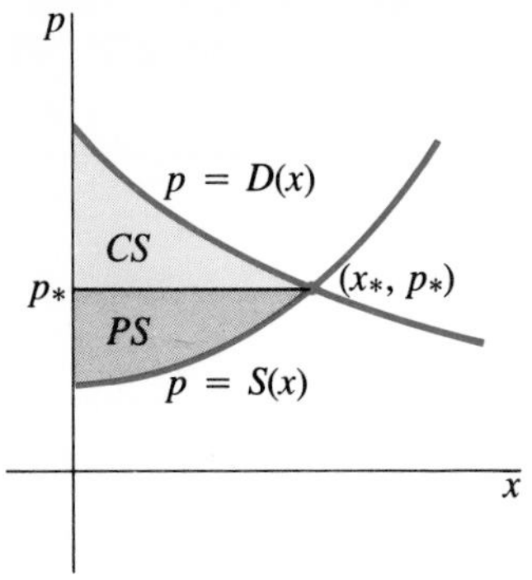

FIGURE 6.13

which we recognize as the definition of a definite integral. Therefore

$$\text{CS} = \int_0^{x_*} (D(x) - p_*)\, dx \tag{6.4}$$

The consumers' surplus is thus the area between the curves $p = D(x)$ and $p = p_*$ from 0 to x_* (Figure 6.13).

Now, just as consumers benefit from not having to pay more than the market price, the producers benefit because, as the supply curve indicates, some would be willing to sell at a lower price. The aggregate benefit to producers from a competitive market is called the **producers' surplus** (PS) and is defined to be the area between the supply curve $p = S(x)$ and the line $p = p_*$. Thus

$$\text{PS} = \int_0^{x_*} (p_* - S(x))\, dx \tag{6.5}$$

EXAMPLE 7 Compute the consumers' surplus and the producers' surplus for the demand and supply curves

$$p = D(x) = -3x + 18$$

$$p = S(x) = x^2/4 + 2$$

SOLUTION Earlier we found that $(x_*, p_*) = (4, 6)$. From Equation (6.4) we know that the consumers' surplus is the integral

$$\begin{aligned} \text{CS} &= \int_0^4 [(-3x + 18) - 6]\, dx = \int_0^4 (-3x + 12)\, dx \\ &= \left(\frac{-3x^2}{2} + 12x\right)\Bigg|_0^4 \\ &= \$24 \end{aligned}$$

And Equation (6.5) reveals that the producers' surplus is

$$\begin{aligned} \text{PS} &= \int_0^4 \left[6 - \left(\frac{x^2}{4} + 2\right)\right] dx = \int_0^4 \left(4 - \frac{x^2}{4}\right) dx \\ &= \left(4x - \frac{x^3}{12}\right)\Bigg|_0^4 = \$32/3 \end{aligned}$$

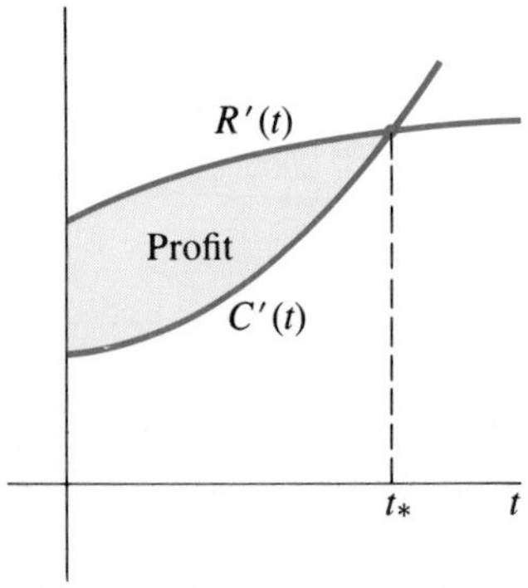

FIGURE 6.14

MAXIMIZING PROFIT OVER TIME Suppose we are in a situation where revenue, cost, and profit are functions of time, so that $R(t)$, $C(t)$, and $P(t)$ are *cumulative* amounts accrued from time 0 to time t. We will be given the derivatives $R'(t)$ and $C'(t)$, which are the time rates at which revenue and cost are accumulating, in units such as thousands of dollars per month. Typically the revenue rate exceeds the cost rate at the outset (so that a profit is generated), but after a while the cost rate rises and overtakes the revenue rate. Therefore the optimal time to terminate the undertaking is that time t_* at which $R'(t_*) = C'(t_*)$ (see Figure 6.14). After that time, $P'(t) = R'(t) - C'(t)$ is negative.

EXAMPLE 8 Suppose that

$$R'(t) = 15 + 0.3t$$
$$C'(t) = 5 + 0.5t$$

where both are in thousands of dollars per month. When should this undertaking be terminated, and what will be the total profit up to that time?

SOLUTION Equating these two functions, we find that their graphs intersect at $t_* = 50$ (see Figure 6.15), so

$$P'(t) = R'(t) - C'(t) > 0 \quad \text{for} \quad 0 \le t < 50$$

After time $t_* = 50$, P' becomes negative and our accumulated profit begins to decline. Therefore the optimal time to terminate is at 50 months. The profit accumulated up to that time (because $P(0) = 0$) is

$$\begin{aligned} P(50) = P(50) - P(0) &= \int_0^{50} P'(t)\,dt \\ &= \int_0^{50} [R'(t) - C'(t)]\,dt \\ &= \int_0^{50} [(15 + 0.3t) - (5 + 0.5t)]\,dt = \int_0^{50} (10 - 0.2t)\,dt \\ &= (10t - 0.1t^2)\Big|_0^{50} = \$250 \end{aligned}$$

Note that this is the area between the graphs of R' and C' from 0 to 50.

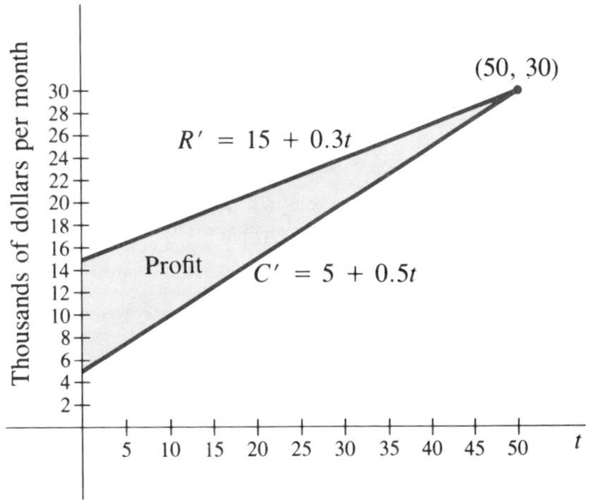

FIGURE 6.15

EXAMPLE 9 Suppose that, in Example 8, the revenue and costs are attributable to a new piece of equipment, which, after t months, has a salvage value of

$$S(t) = -\frac{9}{4}t + 135$$

thousands of dollars. Under this additional assumption, when should the equipment be sold?

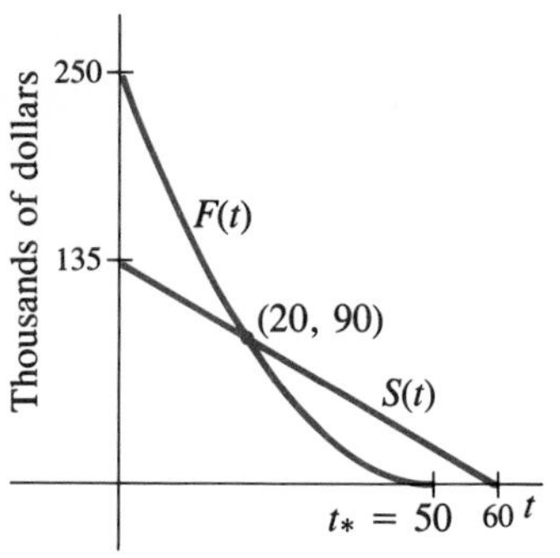

FIGURE 6.16

SOLUTION The equipment should be sold at such time as the future profit that can be earned from it equals the salvage value. To illustrate this, let $F(t)$ be the profit (ignoring the salvage value after the 50 months) generated by the equipment from time t to time $t_* = 50$. That is,

$$F(t) = \int_t^{50} P'(x)\,dx = \int_t^{50} (10 - 0.2x)\,dx = 250 - 10t + 0.1t^2$$

(In Example 8 we computed $F(0) = 250$.) $F(t)$ and $S(t)$ are graphed together in Figure 6.16. We can find the intersection of these two curves by equating and solving for t.

$$250 - 10t + 0.1t^2 = -\frac{9}{4}t + 135$$

Multiplying by 40 and collecting terms, we get

$$4t^2 - 310t + 4600 = 0$$

or

$$(4t - 230)(t - 20) = 0$$

Only the root $t = 20$ is between 0 and t_*.

As we can see from Figure 6.16, at 20 months the potential future profit from the equipment is the same as the salvage value from selling the equipment at that time. Prior to this time, the potential future profit is greater than the salvage value; after it, the salvage value is larger. Therefore 20 months is the optimal time to dispose of the equipment.

(*Note*: We have not taken into account here the fact that d dollars of salvage value received at a particular time t could be invested as a lump sum to earn interest, whereas d dollars of future profits, distributed bit by bit from time t to time t_*, are not all available at once for investment purposes and so have a lesser "present value" at time t. Had this been taken into consideration, the optimal time for equipment disposal would have been slightly earlier than 20 months. The concept of present value is discussed in Section 6.4.)

INCOME DISTRIBUTION—THE LORENZ CURVE The statistician Max Lorenz introduced a graphical device, now known as the Lorenz curve, for depicting distribution of income among a given population of income recipients. In Figure 6.17 the x-axis indicates cumulative percentages (as decimals) of income recipients (or of families) ranked from poorest to richest. For example, $x = 0.30$ corresponds to the poorest 30% of families. The y-axis indicates cumulative share of income.

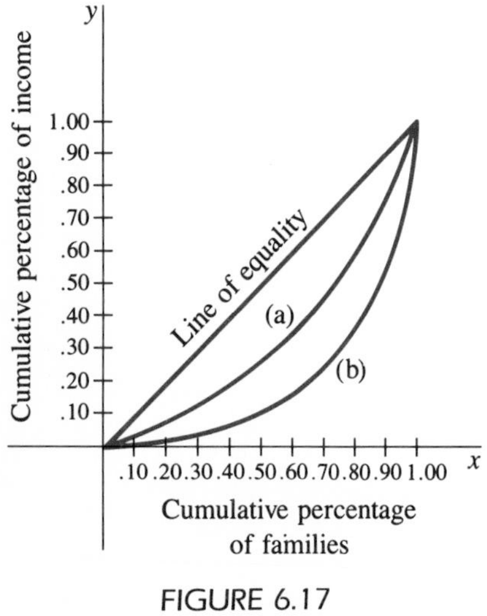

FIGURE 6.17

In Figure 6.17 curve (a) is typical of Lorenz curves for modern industrialized societies such as the United States or Great Britain. Here we see that the bottom 60% of families earn about 35% of the income. Curve (b) is typical of less developed countries. Here the bottom 60% of families earn only about 15% of the income.

The "line of equality," $y = x$, represents perfect equality of distribution, in which all families receive the same share of the total income. In general, the lower a Lorenz curve dips below the line of equality, the greater the inequality of income distribution. One measure of this inequality is the area between the Lorenz curve and the line of equality. The ratio of this area to the total area below the line of equality is called the **concentration ratio** (CR) or the **coefficient of inequality**. Thus, if $y = f(x)$ is our Lorenz curve, the concentration ratio is

(6.6)
$$\text{CR} = \frac{\int_0^1 (x - f(x))\,dx}{\int_0^1 x\,dx} = \frac{\left.\frac{x^2}{2}\right|_0^1 - \int_0^1 f(x)\,dx}{\left.\frac{x^2}{2}\right|_0^1} = 1 - 2\int_0^1 f(x)\,dx$$

(In the last step, we multiplied numerator and denominator by 2.)

EXAMPLE 10 The income distribution in a certain developed country is approximated by the following Lorenz curve:

$$f(x) = 0.3x + 0.15x^2 + 0.55x^3 \qquad 0 \le x \le 1$$

(a) Find $f(0.30)$ and interpret the result.

(b) What percentage of income is possessed by the richest 20% of income recipients?

(c) Find the concentration ratio.

SOLUTION Computation with a pocket calculator will be facilitated by rewriting $f(x)$ as

$$f(x) = [(0.55x + 0.15)x + 0.3]x$$

(a) We find that $f(0.30) \approx 0.12$, which means that the poorest 30% of income recipients earn only about 12% of the income.

(b) Because $f(0.80) \approx 0.62$, the bottom 80% of recipients earn about 62% of income. Consequently, the top 20% earn about 38% of the income.

(c)
$$\begin{aligned} \text{CR} &= 1 - 2\int_0^1 (0.3x + 0.15x^2 + 0.55x^3)\,dx \\ &= 1 - 2[0.15x^2 + 0.05x^3 + 0.1375x^4]\Big|_0^1 \\ &= 0.325 \end{aligned}$$

Section 6.2 EXERCISES

In Exercises 1–16 sketch and find the area of the region bounded by the given curves.

1. $y = x^2 - 4x + 10$ and $y = x + 6$

2. $y = x^3 + 2$ and $y = x^3 - x^2 + 3$

3. $y = x^2 - 2x$ and $x = y$

4. $x = y^2 - 1$ and $x - y = 1$

5. $y^3 = x^2 - 2x + 1$ and $3y = x + 3$

6. $y = \dfrac{4}{(x+2)^2}$ and $3x + y = 1$ for $x \geq -1$

7. $y = x^2 + 2x$ and $y = x^3 + 3x^2 + 3x$

8. $x + y = 8$, the x-axis, the y-axis

9. $x + y = -8$, the x-axis, the y-axis

10. $y = 6 - x^2$ and $y = x$

11. $y = x^2 - 1$ and $y + 1 = x^3$

12. $y = x^2$ and $y = 3x$

13. $y = x^4$ and $y = 32 - x^4$

14. $y = e^x$, $x = 0$, $y = 1$ and $x = 4$

15. $y = 2e^{-x}$, $y = 0$, $x = 0$, and $x = 1$

16. $y = 4 - x$, $y = x$ and $y = 4$

17. Write the equation of the parabola passing through $(-a, 0)$, $(a, 0)$, and $(0, b)$, where $a > 0$ and $b > 0$. Show that the area enclosed between this parabola and the x-axis is given by $\frac{2}{3}$ (the length of its base) $\times$ (its height).

18. Find the area of the triangle formed by the x-axis and the two lines $y = x + 5$ and $y = -4x + 3$.

19. Find the area of the triangle with vertices (3, 2), (1, −4), and (0, 6).

20. Use integration to find the area of a pizza slice, given that the whole pizza is 10 inches in diameter and that the pizza has been divided into 8 slices. Specifically, find the area enclosed by $x^2 + y^2 = 25$, the line $y = x$, and the x-axis.

21. An importer finds that, when she imports and sells 100 transistor components, her profit is $1250. She knows her marginal revenue function for this item to be

$$\frac{dR}{dx} = 180 + 3x - 0.02x^2$$

and her marginal cost is given by

$$\frac{dC}{dx} = 85 + 0.21x^2$$

What is the profit (or loss) when she imports and sells 220 transistor components?

22. The marginal revenue function in producing a certain item is given by

$$\frac{dR}{dx} = -0.028x^2 + 3x + 32$$

and $\dfrac{dC}{dx}$, the marginal cost function, is given by

$$\frac{dC}{dx} = 3x + 4$$

How many items should this company produce in order to maximize its profit? What will this maximum profit be, assuming that it is known that the overhead in not producing *any* items is $200?

In Exercises 23–26 find the market equilibrium point (x_*, p_*), the consumers' surplus, and the producers' surplus.

23. $p = D(x) = 90 - 2x, \quad p = S(x) = 4x$

24. $p = D(x) = 60 - 4x - x^2, \quad p = S(x) = 2x + 33$

25. $p = D(x) = \dfrac{12 - 4x}{x + 1}, \quad p = S(x) = x + 3$

26. $p = D(x) = 500 - x^2, \quad p = 200 + 5x$

27. The price p, in cents, when x units are demanded is

$$p = \frac{500}{1 + x}$$

Find the consumers' surplus if market demand is 49 units.

28. Suppose the revenue rate for a certain project is a constant $50 thousand per month, and the cost rate is $(20 + 3t) thousand per month t months after the project is begun. When should the project be terminated and what will the (cumulative) profit be up to that time?

29. Assume that the rate, in hundreds of thousands of dollars per year, at which a certain oil field is bringing in revenue t years after drilling began is

$$R'(t) = 14 - 2\sqrt{t}$$

and that the cost rate (in the same units) is

$$C'(t) = 2 + \sqrt{t}$$

When will further drilling cease to be profitable, and what will be the total profit realized up to that time?

30. An advertising campaign costs two thousand dollars per week. Revenue attributable to the campaign is earned at the rate of $R'(t) = (10 - \sqrt{t})$ thousand dollars per week.

(a) Find the optimal time to terminate the campaign.

(b) How much profit will have been earned by that time?

31. The income distribution in a certain developing country is modeled by the following Lorenz curve:

$$f(x) = 0.8x^3 + 0.2x \qquad 0 \le x \le 1$$

(a) What percentage of national income is earned by the poorer 50% of families?

(b) Compute the concentration ratio.

32. Compute the concentration ratio for the Lorenz curve

$$f(x) = 0.24x^2 + 0.76x$$

and compare it with that in Exercise 31. In which country is income distributed more nearly evenly?

SECTION 6.3

Volumes (Optional)

In this section we shall be concerned with the problem of computing the volumes of certain solids. As usual, our point of view, is to consider the integral as a sum of infinitely many "infinitesimally" thin pieces. Our discussion should be fairly intuitive.

We begin by considering the problem of finding the volume generated by rotating a plane region about a straight line not passing through the region's interior. A volume generated in this way is called a **volume of revolution**. The simplest example of this is to rotate about the x-axis the region between the x-axis and the curve $y = f(x)$ from $x = a$ to $x = b$, where f is nonnegative on the interval $[a, b]$ (see Figure 6.18). Let us denote this volume by V.

Imagine now that we take a large knife and slice V into thin slices perpendicular to the x-axis, much as you would slice a loaf of bread (see Figure 6.19). If we denote the volumes of the individual slices by $V_1, V_2, \ldots, V_n$, then

$$V = V_1 + V_2 + \ldots + V_n = \sum_{k=1}^{n} V_k$$

Let us look at the typical slice, V_k, whose right-hand side is located at a distance x_k from the origin. See Figure 6.19(b). What does this typical slice look like? First of all, each slice has a circular cross section (as shown in Figure 6.20)

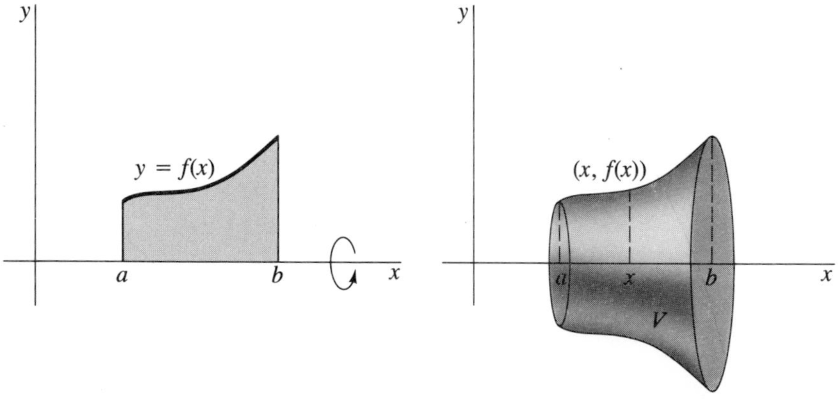

FIGURE 6.18

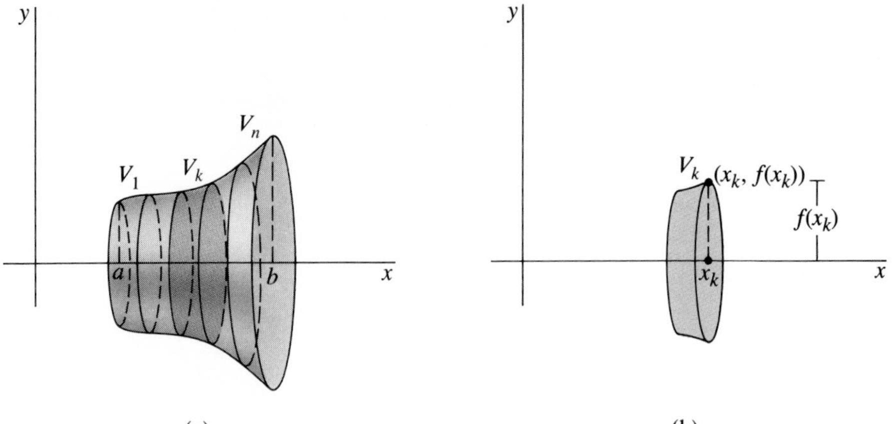

(a) (b)

FIGURE 6.19

whose radius depends on where the slice is made. For a slice made at x_k, the radius of the resulting circle is just $f(x_k)$ units.

The subvolume V_k itself looks something like a thin coin, or disc, except that it does not have a straight edge. Let us take the thickness of each disc to be

$$\Delta x = \frac{b - a}{n}$$

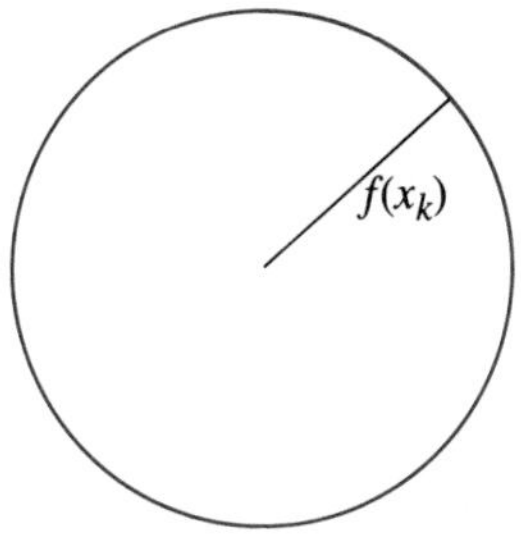

FIGURE 6.20

We will approximate the volume V_k by the volume of an actual disc, which is simply a very thin cylinder (see Figure 6.21). The volume of a cylinder is given by the formula

$$\underbrace{(\pi r^2)}_{\text{cross-sectional area}} \cdot \underbrace{h}_{\text{cylinder height}}$$

In the case of our approximating cylinder (disc), $r = f(x_k)$ and $h = \Delta x$ (our disc is standing on edge, and so its "height" is the width Δx). Thus

$$V_k \approx \pi(f(x_k))^2 \, \Delta x$$

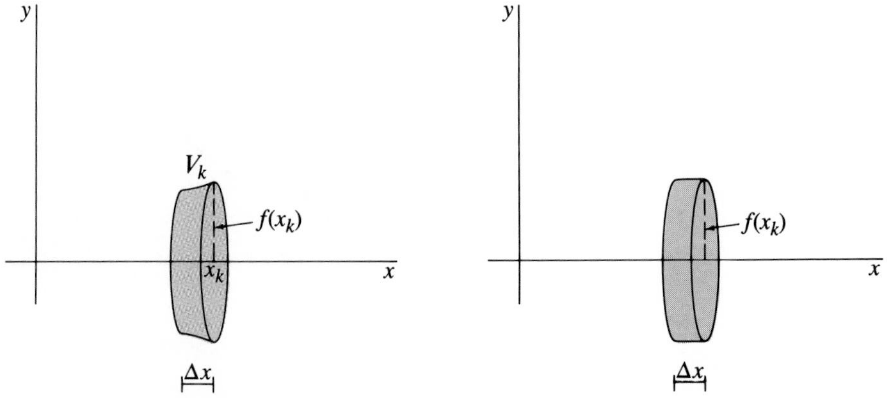

(a) Typical slice V_k (b) Approximating disc

FIGURE 6.21

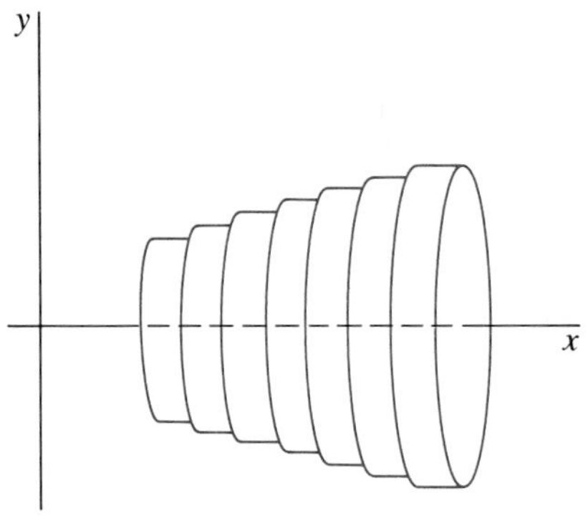

FIGURE 6.22

The entire region V can be approximated by stacking discs of appropriate size one next to the other (see Figure 6.22).

Because $V = \sum_{k=1}^{n} V_k$, we now have

(6.7) $$V \approx \sum_{k=1}^{n} \pi[f(x_k)]^2 \, \Delta x$$

Notice that this is a Riemann sum analogous to those obtained in our discussion of areas and averages. If we now take more and more, but thinner and thinner slices, we achieve better and better accuracy, and so we obtain in the limit

(6.8) $$V = \int_a^b \pi[f(x)]^2 \, dx$$

This is the volume of revolution generated by rotating the region under $y = f(x)$ between $x = a$ and $x = b$ about the x-axis.

EXAMPLE 11 Find the volume generated by rotating the region beneath the curve $y = x^2 + 1$ between $x = 1$ and $x = 4$ about the x-axis. A typical slice is shown shaded in Figure 6.23.

SOLUTION The desired volume looks like a "solid lampshade." We simply have to apply Equation (6.8) with $f(x) = x^2 + 1$, $a = 1$, and $b = 4$. Thus

$$\begin{aligned} V &= \int_1^4 \pi(x^2 + 1)^2 \, dx = \int_1^4 \pi(x^4 + 2x + 1) \, dx \\ &= \pi\left[\frac{x^5}{5} + \frac{2x^3}{3} + x\right]_1^4 \\ &= \pi\left[\frac{1024}{5} + \frac{128}{3} + 4\right] - \pi\left[\frac{1}{5} + \frac{2}{3} + 1\right] \\ &= \frac{1248}{5}\pi \end{aligned}$$

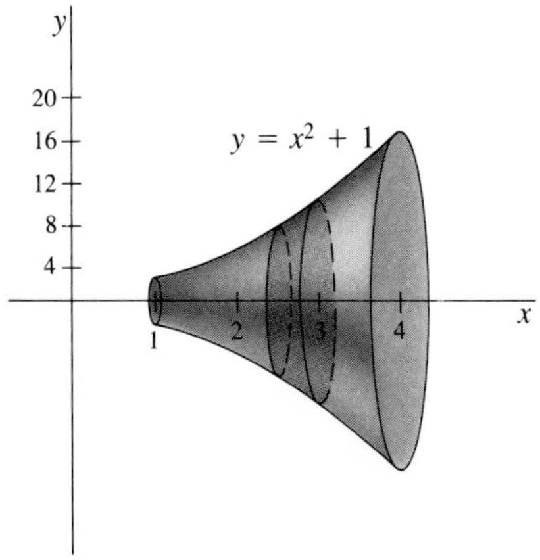

FIGURE 6.23

EXAMPLE 12 Use the disc method to derive the formula for the volume of a sphere of radius r.

SOLUTION A sphere of radius r can be obtained by rotating a semicircle around the x-axis. This situation is depicted in Figure 6.24.

The equation of a semicircle is just $y = \sqrt{r^2 - x^2}$, $-r \le x \le r$. If we apply Equation (6.8) with

$$f(x) = \sqrt{r^2 - x^2} \qquad a = -r \qquad b = r$$

then

$$\begin{aligned} V &= \int_{-r}^{r} \pi(\sqrt{r^2 - x^2})^2\,dx = \int_{-r}^{r} \pi(r^2 - x^2)\,dx \\ &= \pi\left(r^2x - \frac{x^3}{3}\right)\Bigg|_{-r}^{r} = \pi\left(r^3 - \frac{r^3}{3}\right) - \pi\left(-r^3 + \frac{r^3}{3}\right) \\ &= \frac{4}{3}\pi r^3 \end{aligned}$$

This, of course, is the well-known formula.

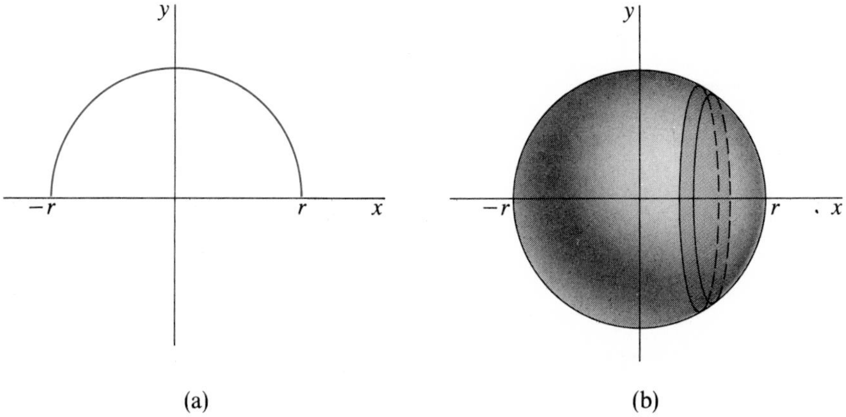

FIGURE 6.24

EXAMPLE 13 Find the volume generated by rotating about the x-axis the region between the curves $y = \sqrt{x}$ and $y = x^3$.

SOLUTION The region between the curves $y = \sqrt{x}$ and $y = x^3$ is shown in Figure 6.25. Note that $y = \sqrt{x}$ lies above $y = x^3$ for $0 \le x \le 1$. In this situation the solid obtained by rotating this region about the x-axis has a hollow inside. This volume of revolution cannot be determined by a simple application of Equation (6.8) because of the presence of two functions, $f(x) = \sqrt{x}$ and $g(x) = x^3$. (A typical slice of volume, as shown in Figure 6.26, is a thin disc with a hole in it—a Lifesaver candy shape, or a washer.)

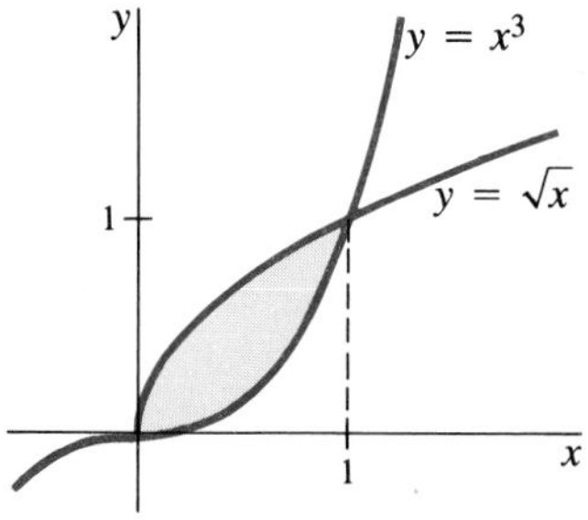

FIGURE 6.25

Now, if our problem had been simply to find the volume generated by rotating $f(x) = \sqrt{x}$, $0 \le x \le 1$, about the x-axis, we could have applied Equation (6.8) to obtain its volume V_1 (see Figure 6.27) as follows:

$$V_1 = \int_0^1 \pi[f(x)]^2\,dx = \int_0^1 \pi(\sqrt{x})^2\,dx$$
$$= \int_0^1 \pi x\,dx$$
$$= \frac{\pi x^2}{2}\bigg|_0^1 = \frac{\pi}{2}$$

Similarly, if we had been given only $g(x) = x^3$, $0 \le x \le 1$, we could have

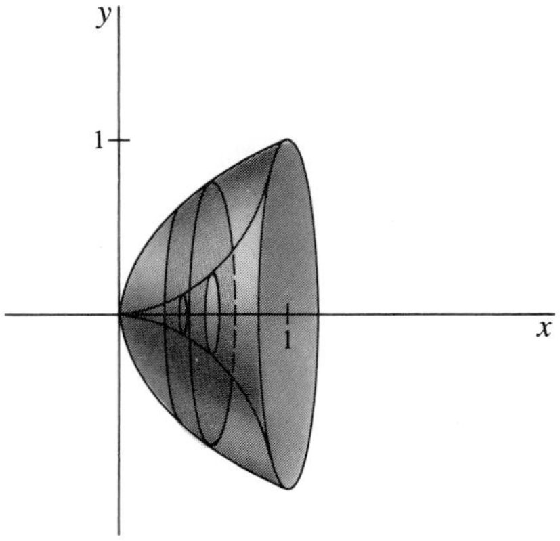

FIGURE 6.26

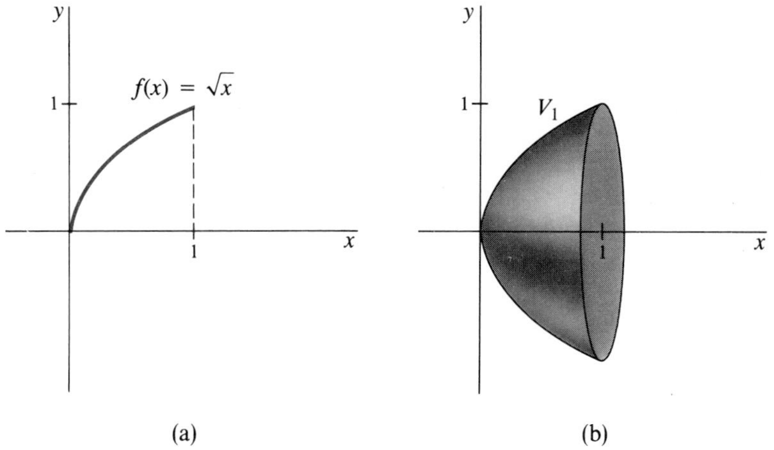

FIGURE 6.27

computed its volume V_2 (see Figure 6.28) as follows:

$$V_2 = \int_0^1 \pi[g(x)]^2\,dx = \int_0^1 \pi(x^3)^2\,dx = \int_0^1 \pi x^6\,dx$$
$$= \pi\frac{x^7}{7}\bigg|_0^1 = \frac{\pi}{7}$$

If we now go back and compare Figures 6.27 and 6.28 with Figure 6.26, we can see that the desired volume V is simply the volume V_1 with V_2 removed:

(6.9)
$$V = V_1 - V_2 = \int_0^1 \pi(\sqrt{x})^2\,dx - \int_0^1 \pi(x^3)^2\,dx$$
$$= \frac{\pi}{2} - \frac{\pi}{7} = \frac{5\pi}{14}$$

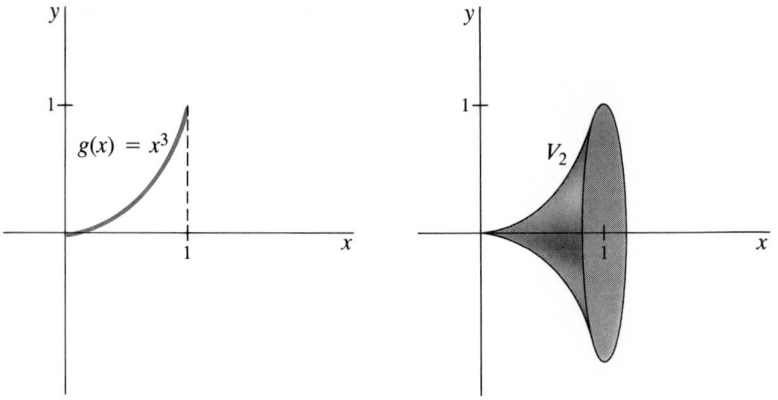

(a) FIGURE 6.28 (b)

In general, suppose we consider a plane region bounded above by $y = f(x)$, below by $y = g(x)$ ($g(x) \geq 0$), to the left by $x = a$, and to the right by $x = b$. See Figure 6.29(a). If this region is rotated about the x-axis, it generates a volume of revolution like that shown in Figure 6.29(b). By analogy with Equation (6.9), its volume V is the volume of the solid generated by rotating the outer curve, minus the volume of the hole generated by the inner curve. Thus

(6.10)
$$V = \int_a^b \pi[f(x)]^2\,dx - \int_a^b \pi[g(x)]^2\,dx$$
$$= \int_a^b \pi([f(x)]^2 - [g(x)]^2)\,dx$$

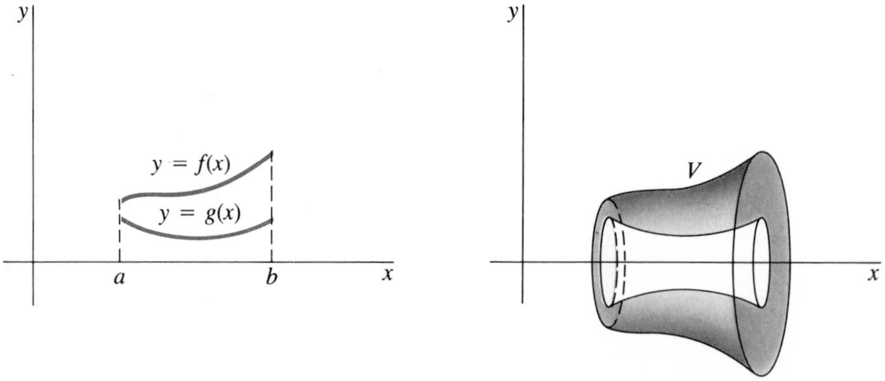

(a) (b)

FIGURE 6.29

EXAMPLE 14 Find the volume generated by rotating about the x-axis the region between the curves $y = x + 2$ and $y = x^2$ for $0 \leq x \leq 2$.

SOLUTION The given plane region and the resulting volume of revolution are shown in Figure 6.30. For $0 \leq x \leq 2$, we have $f(x) = x + 2$ above

$g(x) = x^2$. Applying Equation (6.10) yields

$$V = \int_0^2 \pi\,[(x+2)^2 - (x^2)^2]\,dx = \int_0^2 \pi(x^2 + 4x + 4 - x^4)\,dx$$

$$= \pi\left(\frac{x^3}{3} + 2x^2 + 4x - \frac{x^5}{5}\right)\Big|_0^2 = \frac{184\pi}{15}$$

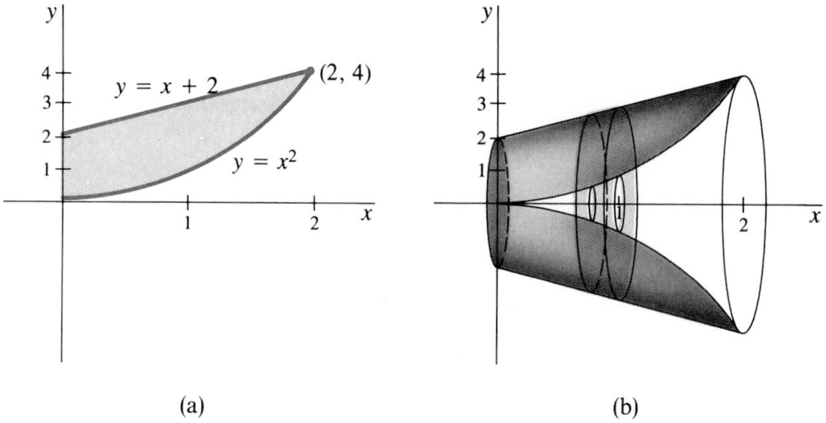

FIGURE 6.30

Section 6.3 EXERCISES

In Exercises 1–12 use the disc method to find the volume generated when the region bounded by the given curves is revolved about the x-axis.

1. $y = x^2 + 2x$, x-axis, $x = 2$
2. $y = x^{2/3}$, x-axis, $x = 1$, and $x = 8$
3. $y = e^{2x} + 3$, $y = 4 - e^x$, $x = 0$, and $x = \ln 2$
4. $y = \dfrac{1}{x}\sqrt{\ln x}$, x-axis, $x = 1$, and $x = e^2$
5. $y = \dfrac{\sqrt{2x+4}}{\sqrt{x^2+4x+5}}$, x-axis, $x = -2$, and $x = -1$
6. $y = \left(x + \dfrac{1}{x}\right)$, $y = \dfrac{1}{x}$, $x = 1$, and $x = 5$
7. $y = x^2$, $y = 0$, $x = 1$
8. $y = x^{1/2}$, x-axis, $x = 9$
9. $y = 1/x$, x-axis, $x = \frac{1}{2}$, $x = 1$
10. $y = x^2$, $y = 1$, $x = 0$
11. $y = x^2$, $x = y^2$ (*Hint*: First find where these curves intersect.)
12. $y = x^2$, $y = 8 - x^2$
13. An apple (which is assumed to be spherical) of radius $\sqrt{5}$ cm has its core removed. The cylindrical core has radius 1 cm. What is the volume of the cored apple? (*Hint*: Rotating the shaded region shown in the figure about the x-axis yields half the answer.)

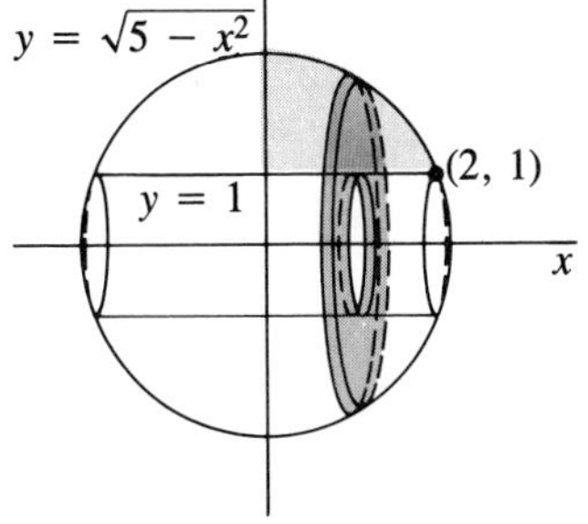

FIGURE FOR EXERCISE 13

14. Show that the volume of a circular cone of height h and base radius r is $V = \frac{1}{3}\pi r^2 h$ by computing the volume generated by rotating a right triangle about one of its legs, as shown in the figure.

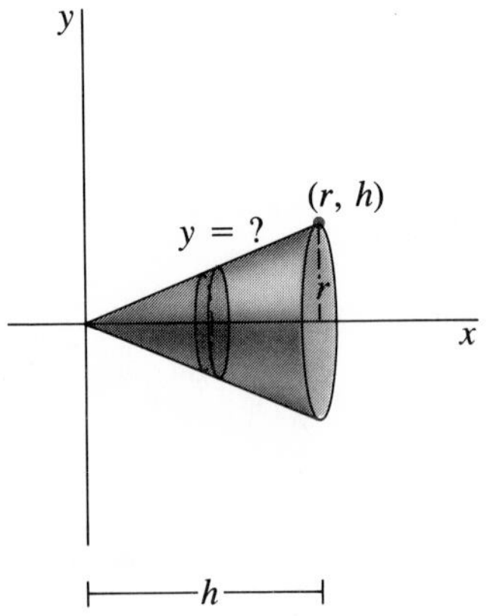

FIGURE FOR EXERCISE 14

15. An equation of an ellipse with major axis a and minor axis b is given by $\dfrac{x^2}{a^2} + \dfrac{y^2}{b^2} = 1$, where $a > b$. What is the volume of the ellipsoid generated by rotating this ellipse about the x-axis?

16. What is the volume of the solid cap chopped off a solid sphere of radius 3 ft by a plane 2 ft from the sphere's center at its closest point?

SECTION 6.4

Other Applications: Net Change, Present Value, and Velocity and Acceleration

In this section we consider a variety of additional applications of both the definite and the indefinite integral.

NET CHANGE Suppose $Q(t)$ is the amount of some commodity available at time t. If we know the formula for $Q'(t)$, the rate at which Q is changing, we can compute the net change in Q from $t = a$ to $t = b$ as the definite integral

$$\int_a^b Q'(t) = Q(t)\Big|_a^b = Q(b) - Q(a)$$

EXAMPLE 15 The salvage value of a piece of machinery is changing at the rate

$$V'(t) = 400(t - 5) \qquad 0 \le t \le 5$$

dollars per year, t years after purchase. Find the change in salvage value during the third year (that is, between $t = 2$ and $t = 3$).

SOLUTION The change during the third year is $V(3) - V(2)$, the difference between the value at the end and the value at the beginning of that year. This is

$$\begin{aligned} V(3) - V(2) &= \int_2^3 V'(t)\,dt = \int_2^3 400(t - 5)\,dt \\ &= 200(t - 5)^2\Big|_2^3 = 200[(-2)^2 - (-3)^2] \\ &= -1000 \end{aligned}$$

Thus the salvage value decreases (because of the minus sign) by a thousand dollars.

As the next example illustrates, the independent variable need not be time.

EXAMPLE 16 A company's marginal cost is

$$C'(x) = \frac{x}{x^2 + 1} \qquad x \geq 0$$

Find the increase in cost due to raising production from $x = 3$ to $x = 7$. (Here x is in hundreds of units and C is in tens of thousands of dollars.)

SOLUTION

$$C(7) - C(3) = \int_3^7 C'(x)\, dx = \int_3^7 \frac{x}{x^2 + 1}\, dx$$

Letting $u = x^2 + 1$, $du = 2x\, dx$, we have

$$\int_3^7 \frac{x}{x^2 + 1}\, dx = \frac{1}{2}\int_{10}^{50} \frac{du}{u} = \frac{1}{2}(\ln 50 - \ln 10)$$

$$= \frac{1}{2}\ln 5 \approx 0.8047$$

or approximately \$8047.

PRESENT VALUE Suppose we wish to invest a single lump sum V and want to have it grow to a desired amount A after t years. If our money earns interest at rate r, compounded continuously, then (see Example 34 in Chapter 4) $A = Ve^{rt}$. Accordingly (just solve for V), the amount we must invest now, to have A in t years, is

$$V = Ae^{-rt}$$

This is called the **present value** of A in t years. For example, the present value of \$1000 in 10 years, assuming continuous compounding at 9%, is

$$V = 1000e^{-0.09(10)} \approx \$406.57$$

Now suppose that we are able to get in on an investment that will earn money continuously from $t = t_1$ to $t = t_2$ years and that, at any time t, is earning at the rate of $A(t)$ dollars per year. The total amount of money earned between $t = t_1$ and $t = t_2$ is given by the integral

$$\int_{t_1}^{t_2} A(t)\, dt$$

This is just another example of computing a net change from a rate. This is the limit of Riemann sums, $\Sigma A(t_k)\, \Delta t$, in which each term approximates the earnings during a small time interval Δt that contains t_k.

However, if we are interested in the present value of such an investment, we must multiply each term in the Riemann sum by e^{-rt_k} (again assuming continuous compounding at rate r). From this it follows that the present value of the investment is

(6.11)
$$V = \int_{t_1}^{t_2} A(t)e^{-rt}\,dt$$

EXAMPLE 17 An inventor estimates that her latest invention will bring her royalties at the rate of

$$1200(1 + e^{0.15t})$$

dollars per year t years from now, for $0 \leq t \leq 10$. What is the most you would be willing to pay for exclusive rights to this invention, if the prevailing interest rate is 8% compounded continuously?

SOLUTION You should pay no more than the present value of the total predicted royalties, which, by Equation (6.11), is

$$\begin{aligned} V &= \int_0^{10} 1200(1 + e^{0.15t})e^{-0.08t}\,dt \\ &= 1200\int_0^{10} (e^{-0.08t} + e^{+0.07t})\,dt \\ &= 1200\left(\frac{e^{-0.08t}}{-0.08} + \frac{e^{+0.07t}}{+0.07}\right)\Bigg|_0^{10} \\ &\approx 1200(23.151 - 1.786) \\ &\approx \$25{,}639 \end{aligned}$$

VELOCITY AND ACCELERATION Suppose an object is moving in a straight line, and its distance s from some reference point is given by $s = s(t)$. Then, as we saw in Chapter 2, the object's **velocity** v is the rate of change of s and, accordingly, is the derivative

$$v(t) = \frac{ds}{dt}$$

Similarly, the object's **acceleration** a is the rate of change of v and is given by

$$a(t) = \frac{dv}{dt} = \frac{d^2s}{dt^2}$$

Proceeding in the opposite direction, if we are given the acceleration, we can integrate (antidifferentiate) to get the velocity and then integrate once more to get the distance. At each step we can determine the constant of integration, provided we are given additional information.

EXAMPLE 18 As was known to Galileo (1564–1642), under the influence of gravity all bodies fall with the same acceleration (assuming air resistance is negligible). Near the

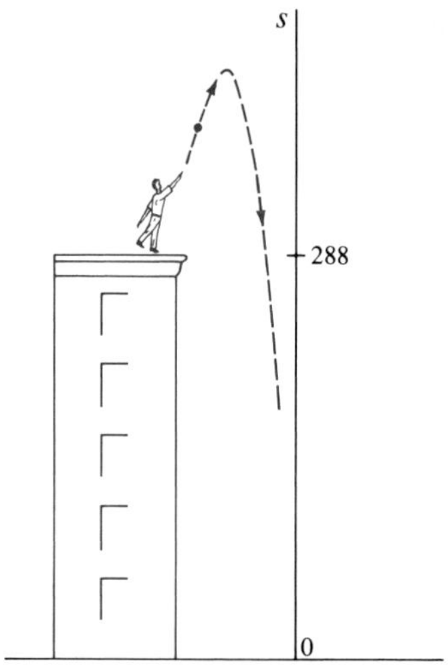

FIGURE 6.31

earth's surface, this acceleration has constant magnitude of approximately 32 feet per second per second, which is usually abbreviated 32 ft/sec^2.

Suppose a man standing on the edge of a tall building's roof throws a ball upward with an initial velocity of 48 feet per second (see Figure 6.31). Assume that the roof is 288 feet above street level.

(a) What is the maximal height reached by the ball?
(b) When does the ball reach the street?
(c) How fast is it moving when it lands?

SOLUTION

(a) Let s denote distance (or height) above the ground, so that $s = 0$ at street level and $s = 288$ at the top of the building. When the ball is rising, s increases, so $v = ds/dt > 0$. When the ball is falling, s decreases, so $v = ds/dt < 0$. At the exact instant when the ball is at its highest point, $v = 0$.

Throughout the ball's entire flight, v decreases (from positive through 0 to negative). Consequently $a = dv/dt$ is negative, so

(6.12) $$a(t) = -32$$

is the equation we want to antidifferentiate. We also have the **initial conditions**

(6.13) $$v(0) = 48$$

(6.14) $$s(0) = 288$$

From Equation (6.12) we obtain

$$v(t) = \int a(t)\, dt = -32t + c$$

where c is a constant to be determined.

Setting $t = 0$ and using Equation (6.13) yields

$$48 = v(0) = -32(0) + c = c$$

and so $c = 48$. Consequently,

(6.15) $$v(t) = -32t + 48$$

Integrating again, we have

$$s(t) = \int v(t)\, dt = \int (-32t + 48)\, dt = -16t^2 + 48t + d$$

where d is another constant. Again setting $t = 0$, but this time using Equation (6.14), we get

$$288 = s(0) = -16(0)^2 + 48(0) + d = d$$

hence $d = 288$. Therefore

(6.16) $$s(t) = -16t^2 + 48t + 288$$

Now, as we noted earlier, the ball achieves its maximal height when $v(t) = 0$ or, by Equation (6.15), when

$$-32t + 48 = 0$$

which yields $t = 48/32 = 3/2$. Its height at this moment is

$$s(3/2) = -16\left(\frac{3}{2}\right)^2 + 48\left(\frac{3}{2}\right) + 288$$
$$= 324 \text{ feet}$$

(b) We have only to observe that the ball lands when $s(t) = 0$. Thus we solve

$$s(t) = -16t^2 + 48t + 288 = 0$$

Dividing through by -16 yields

$$t^2 - 3t - 18 = (t - 6)(t + 3) = 0$$

or $t = 6$ (the negative root is extraneous). Therefore the ball strikes the ground at $t = 6$ seconds. Its speed then is

$$v(6) = -32(6) + 48$$
$$= -144 \text{ feet per second}$$

(negative because the motion is downward).

EXAMPLE 19 What initial velocity would be required to throw a stone from the ground to a height of 50 feet?

SOLUTION As in the previous example, we begin by integrating $a(t) = -32$ to obtain

$$v(t) = -32t + c$$

When $t = 0$, v is the initial velocity we are trying to find. Call it v_0. Thus

(6.17) $$v(t) = -32t + v_0$$

Integrating a second time, we have

$$s(t) = -16t^2 + v_0t + d$$

where $d = s(0) = 0$, the initial height. Hence

(6.18) $$s(t) = -16t^2 + v_0t$$

Now, at the instant that $v = 0$ (at the stone's highest point), s is required to be 50. Imposing this condition on Equation (6.17) and Equation (6.18) gives us the two equations

$$0 = -32t + v_0$$

$$50 = -16t^2 + v_0t$$

From the first, $t = v_0/32$. Substituting this into the second equation produces

$$\begin{aligned} 50 &= -16\left(\frac{v_0}{32}\right)^2 + v_0\left(\frac{v_0}{32}\right) \\ &= \left(\frac{v_0^2}{32}\right)\left(\frac{-16}{32} + 1\right) \\ &= \frac{v_0^2}{64} \end{aligned}$$

Consequently, $v_0 = \sqrt{50 \cdot 64} \approx 56.57$ feet per second.

Section 6.4 EXERCISES

1. A company's marginal revenue and marginal cost (in hundreds of dollars per unit per day) are given by

$$R'(x) = 15 - 3x$$
$$C'(x) = 2$$

(a) If production is raised from 3 units per day to 5 units per day, how much will the company's revenue and costs change?

(b) How much will profits change?

2. A cake, initially at 60°F, is placed inside a freezer. After t hr the temperature T of the cake is dropping at the rate

$$T'(t) = -\frac{60}{(1 + t)^2}$$

degrees per hour.

(a) How much does the temperature of the cake drop in the first hour?

(b) How much does the temperature of the cake drop in the second hour?

3. Redo Exercise 2 under the assumption that the rate of temperature change is given instead by

$$T'(t) = -40e^{-0.8t}$$

4. A company's marginal revenue when x units are produced is

$$R'(x) = \frac{40}{\sqrt{x + 1}} + 6$$

(a) How much revenue is earned from the first 3 units?

(b) How much revenue is earned from the next 3 units (the fourth through the sixth)?

5. An advertising campaign costs a certain clothing designer

$$C'(t) = 2000(1 + e^{-0.02t})$$

dollars per day t days after the campaign is begun.

(a) Sketch the graph of C'.

(b) Find the cost of the first 30 days of this campaign. (Ignore any fixed start-up costs: Consider $C(0)$ to be zero.)

6. An annuity will pay \$100,000 at maturity in 20 years. If money earns 8% compounded continuously, how much should you be willing to pay now to purchase the annuity contract?

7. A novelist believes that his new book will be a hit and earn him royalties of at least \$5000 per year for the next 5 years. If money earns 10% compounded continuously, what would be a reasonable lump-sum payment to offer the novelist in exchange for the royalties for the next 5 years? Assume royalties are paid at the end of each year.

8. A coal mine will yield estimated revenue at the rate of

$$R'(t) = 100 + 900e^{-0.3t} \qquad 0 \le t \le 20$$

thousands of dollars per year, t years from now. Assuming that this estimate is reliable and that investments are available that will earn 10% compounded continuously, how much is this coal mine worth to a prospective buyer?

9. If a brick were dropped from the top of the Bunker Hill Monument (221 ft high), how long would it take to reach the ground?

10. Find the constant acceleration a required to make a car go from rest to 60 mph (88 ft/sec) in 352 ft?

11. An object is tossed vertically upward from the ground with an initial velocity of 96 ft/sec. What is the maximum height reached by the object?

12. An athlete can jump 4 ft off the ground. What initial velocity is needed to jump that high?

13. The acceleration due to gravity near the surface of the moon is only about one-sixth that for earth: $a \approx -\frac{16}{3}$ ft/sec^2. Suppose that the athlete in Exercise 12 were transported to the moon and jumped up with the same initial velocity. How high would she jump?

Key Mathematical Concepts and Tools

Integral as an average

Area between two curves

Volume of a solid of revolution

Applications Covered in This Chapter

Average temperature

Deriving profit at a given production level

Consumers' surplus

Producers' surplus

Maximizing profit over time

Income distribution—the Lorenz curve

Net change in a quantity over an interval

Present value of money

Velocity and acceleration

SIX Warm-Up Test

1. Find the area bounded by the two curves $y = 9 - x^2$ and $y = x^2 - 4x - 21$.

2. A new drug is being tested. After injection it is gradually eliminated from the bloodstream via the kidneys. It is found that, t hours after injection, the fraction of the original dose remaining is given by

$$f(t) = e^{-0.41t}$$

(a) Find the average fraction of the drug in the bloodstream during the third full hour after injection.

(b) Find the initial dose (in grams) required in order to assure an average level of 0.7 gram in the period from $t = 5$ to $t = 8$ hr.

3. Find the average value $\bar{f}$ of the function f.

(a) $f(x) = \dfrac{x^2}{x^3 + 4}, \quad 1 \le x \le 4$

(b) $f(x) = 2x^2e^{-x^3} + \dfrac{4}{x}, \quad 2 \le x \le e$

4. Let $p = D(x) = -4x + 39$ and $p = S(x) = \dfrac{x^2}{2} + 15$ be the demand and supply functions for a given product.

(a) Find the market equilibrium (x_*, p_*).

(b) Compute the consumers' surplus (CS) and the producers' surplus (PS).

5. A mail-order business aimed at making a quick profit is begun. It is estimated that, as a function of time t in months, the time rate of change of revenue is given by $R'(t) = 10 + 0.3t$, and the time rate of change of cost is given by $C'(t) = 7 + 0.4t$. Here revenue and cost are measured in thousands of dollars.

(a) After how many months would you suggest that the business close down?

(b) What would be the total profit obtained over the life of this business?

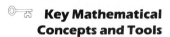

6. Find the volume generated by rotating the region beneath the curve $y = 2e^x + 3$ between $x = 0$ and $x = 1$ about the x-axis.

7. The salvage value of a certain computer system is changing at the rate

$$V'(t) = -3200(7 + 6t - t^2) \qquad 0 \le t \le 7$$

dollars per year, t years after purchase. How much does the salvage value drop in the first 3 years? In the second 3 years?

8. What minimum initial velocity v_0 is required to throw a ball from a height of 6 ft to a height of 40 ft?

9. A company maintains cash reserves (in thousands of dollars) of

$$C(x) = 1 + 12x - x^2 \qquad 0 \le x \le 12$$

where x is the number of months after the first of the year. What is the average cash reserve for the first quarter (that is, the first 3 months of the year)?

10. Find the volume generated when the region between the curves $y = x^2$ and $y = 4x - x^2$ is revolved about the x-axis.

SIX Final Exam

1. Find the average value $\bar{f}$ of the function f.
   (a) $f(x) = (x + 2)(x^2 + 4x)^{1/2}$, $1 \le x \le 3$
   (b) $f(x) = \dfrac{x^2 + 3x - 4}{x}$, $2 \le x \le 5$

2. A train moves on a straight track in such a way that its acceleration after t seconds is given by $a(t) = 3t^3 + 5t$ ft/sec^2. Its velocity at time $t = 1$ is 22 ft/sec.
   (a) What is the train's velocity at time $t = 5$?
   (b) What is its average velocity between $t = 0$ and $t = 5$?

3. A colony of bacteria is known to double its size every 10 days. It has been found that the average count between the beginning of the 15th and the end of the 18th day is 12,000. Find the initial size of the colony.

4. Find the total area bounded by the two curves $y = x^3 - 6x^2 + 8x$ and $y = x^2 - 4x$.

5. A sphere of radius 5 is cut by a plane at a distance 3 from its center (at closest point). Find the volume of the solid "cap" sliced off.

6. Let $p = D(x) = -9x + 105$ and $p = S(x) = x^2/5 + 55$ be the demand function and the supply function for a given product, where p is in dollars.
   (a) Find the market equilibrium (x_*, p_*)
   (b) Compute the consumers' surplus (CS) and the producers' surplus (PS).

7. After t hr of operation a coal mine is producing coal at the rate of $40 + 2t - t^2/5$ tons of coal/hr. Find the average rate of coal production during the first 3 hr of operation of the mine.

8. An annuity will pay $85,000 at maturity in 15 years. If money earns 12% compounded continuously, what is the present value of this annuity?

9. A political campaign incurs costs at the rate of

$$C'(t) = 2200(2 + e^{-0.18t})$$

dollars per day t days after its initiation. The campaign lasts 30 days.
   (a) Find the total cost if you are told that the initial start-up cost is 10,000.
   (b) What would be the additional cost if the campaign were extended from 30 to 60 days?

10. Use calculus to find the area of the triangle with vertices (0, 4) (0, 9), and (1, 7).

SEVEN

More on Integration: Techniques, Differential Equations, and Modeling

Outline

In Context

We have relied successfully on the integration formulas shown as Equations (7.1) to (7.6) to handle the variety of applications considered in the preceding chapter. However, many types of integrals that arise in practice require more advanced techniques. In this chapter we shall study a few of these, as well as learn how to antidifferentiate with the aid of integral tables (Appendix B).

We shall also continue our study of differential equations and see how they can be used to construct mathematical models for solving physical, biological, sociological, and economic problems.

(7.1) $$\int [f(x) + g(x)]\, dx = \int f(x)\, dx + \int g(x)\, dx$$

(7.2) $$\int kf(x) dx = k \int f(x)\, dx$$

(7.3) $$\int kx^n\, dx = k\,\frac{x^{n+1}}{n+1} + c \qquad n \neq -1$$

(7.4) $$\int e^{kx}\, dx = \frac{1}{k} e^{kx} + c \qquad k \neq 0$$

(7.5) $$\int \frac{1}{x}\, dx = \ln|x| + c \qquad x \neq 0$$

(7.6) $$\int f(g(x))g'(x)\, dx = F(g(x)) + c \qquad \text{where } F' = f$$

SECTION 7.1

Integration by Parts

Integration by parts is one of the most powerful and versatile of the integration techniques. Like the method of substitution (Section 5.4), it is based on the idea of replacing a given integral that we do not know how to evaluate with a second one that is more easily computed.

The basis for the method is the product rule for differentiation:

$$f(x)g'(x) + f'(x)g(x) = [f(x)g(x)]'$$

This may also be written as

$$f(x)g'(x) = [f(x)g(x)]' - f'(x)g(x)$$

Let us take the antiderivative of both sides of this expression.

(7.7)
$$\int f(x)g'(x)\,dx = \int [f(x)g(x)]'\,dx - \int f'(x)g(x)\,dx$$

Now the symbol $\int [f(x)g(x)]'\,dx$ asks us to find an antiderivative of the function $[f(x)g(x)]'$. Obviously, $f(x)g(x)$ is such an antiderivative, because its derivative is $[f(x)g(x)]'$. Making use of this fact in Equation (7.7), we obtain

(7.8)
$$\int f(x)g'(x)\,dx = f(x)g(x) - \int f'(x)g(x)\,dx$$

This is called the **integration-by-parts formula**. It will frequently enable us to evaluate antiderivatives when all other methods fail. Let us see how we can apply this formula.

EXAMPLE 1 Evaluate $\int xe^{2x}\,dx$.

SOLUTION If this problem appeared in a section other than this one you might be tempted to try various substitutions, such as $u = e^{2x}$. The result is a still more complicated integral.

Because nothing else seems to work, we will try to use the integration-by-parts formula, Equation (7.8). This formula is applicable to an expression of the form

$$\int f(x)g'(x)\,dx$$

In the case of the integral $\int xe^{2x}\,dx$, we must decide which term we should denote by $f(x)$ and which by $g'(x)$. There are two obvious possibilities. We might try

$$f(x) = x \qquad g'(x) = e^{2x}$$

or we might set

$$f(x) = e^{2x} \qquad g'(x) = x$$

Let's try them both in turn to see what happens.

Possibility 1: $f(x) = x,\ g'(x) = e^{2x}$
In order to apply Equation (7.8) we will have to know $f(x)$, $f'(x)$, $g(x)$, and

$g'(x)$. By assumption, $f(x) = x$ and $g'(x) = e^{2x}$. This information is summarized by the following simple scheme.

$f(x) = x$	$g(x) = ?$
$f'(x) = ?$	$g'(x) = e^{2x}$

Obviously, if $f(x) = x$, then $f'(x) = 1$. On the other hand, if $g'(x) = e^{2x}$, then $g(x)$ is a function whose derivative is e^{2x}; $g(x) = e^{2x}/2$ will do, by Equation (7.4). We can now complete our table.

$f(x) = x$	$g(x) = e^{2x}/2$
$f'(x) = 1$	$g'(x) = e^{2x}$

We now need only make the appropriate substitutions into the integration-by-parts formula:

$$\int \underbrace{x}_{f(x)}\underbrace{e^{2x}}_{g'(x)}\,dx = \underbrace{x}_{f(x)}\underbrace{\frac{e^{2x}}{2}}_{g(x)} - \int \underbrace{(1)}_{f'(x)}\underbrace{\frac{e^{2x}}{2}}_{g(x)}\,dx$$

Now

$$\int (1)\left(\frac{e^{2x}}{2}\right)dx = \frac{1}{2}\int e^{2x}\,dx = \frac{e^{2x}}{4} + c$$

by Equation (7.4), so

$$\int xe^{2x}\,dx = \frac{xe^{2x}}{2} - \frac{e^{2x}}{4} - c$$

Of course the constant $-c$ can be replaced by $+c$, because each represents an arbitrary constant of integration.

You should try to differentiate our answer to check that it works.

Possibility 2: $f(x) = e^{2x}$, $g'(x) = x$
Once again we construct a table with our choices of $f(x)$ and $g'(x)$.

$f(x) = e^{2x}$	$g(x) = ?$
$f'(x) = ?$	$g'(x) = x$

The table is easily completed by differentiating $f(x) = e^{2x}$ to get $f'(x) = 2e^{2x}$ and by antidifferentiating $g'(x) = x$ to obtain $g(x) = x^2/2$.

$f(x) = e^{2x}$	$g(x) = x^2/2$
$f'(x) = 2e^{2x}$	$g'(x) = x$

Now

$$\int \underbrace{e^{2x}}_{f(x)}\,\underbrace{x}_{g'(x)}\,dx = \underbrace{e^{2x}}_{f(x)}\underbrace{\left(\frac{x^2}{2}\right)}_{g(x)} - \int \underbrace{2e^{2x}}_{f'(x)}\underbrace{\left(\frac{x^2}{2}\right)}_{g(x)}\,dx$$

We have replaced the problem of evaluating $\int xe^{2x}\,dx$ with that of evaluating $\int x^2e^{2x}\,dx$. This replacement problem, unfortunately, looks worse than the original. It appears that, for this example, Possibility 2 for the choice of $f(x)$ and $g'(x)$ is not a very good one.

Equation (7.8) is more easily remembered when rewritten in Leibniz notation. Let $u = f(x)$ and $v = g(x)$. Then $du = f'(x)\,dx$ and $dv = g'(x)\,dx$, and the integration-by-parts formula becomes

(7.9)
$$\int u\,dv = uv - \int v\,du$$

In evaluating $\int xe^{2x}\,dx$, we let

$$u = x \qquad dv = e^{2x}\,dx$$

and found $du = dx$, $v = e^{2x}/2$.

The previous example points out the primary difficulty that most students have when trying to apply the integration-by-parts formula. At an early stage, you must decide which function to select as u and which (together with dx) to call dv. (Remember, dv always includes dx.) There are a few cases where the choice is immaterial, but most of the time one choice affords a substantial reduction in the complexity of the integral (as in Possibility 1 above), whereas the other makes matters more difficult (Possibility 2). Although it is impossible to give a rule of thumb that holds in all cases, here is a rough guide that is often helpful.

INTEGRATION-BY-PARTS HINT

1 Try letting dv be the "most sophisticated" portion of the integrand that you know how to antidifferentiate—together with dx.

2 Let u be the remainder of the integrand. This should be a function whose derivative is not inordinately complex.

Of course, the use of the phrase *most sophisticated* is not very precise, but our intuition is sufficient here to serve us quite nicely. For instance, most people would agree that e^{2x} is "more sophisticated" than x.

EXAMPLE 2 Evaluate $\int x \ln x\,dx$.

SOLUTION Here we are again faced with an integrand that is a product of two functions, x and $\ln x$. The function $\ln x$ is "more sophisticated" than x, *but we do not know its antiderivative.* Thus, according to the foregoing hint, $\ln x\,dx$ is not a good choice for dv. Therefore, we make the following selections.

$$u = \ln x \qquad dv = x\,dx$$

so that

$$du = \frac{dx}{x} \qquad v = \frac{x^2}{2}$$

Consequently,

$$\int \underset{\uparrow}{x} \underset{\underset{u}{\uparrow}}{\ln x} \underset{\uparrow}{dx} = \underset{\underset{v}{\uparrow}}{\frac{x^2}{2}} \underset{\underset{u}{\uparrow}}{\ln x} - \int \underset{\underset{v}{\uparrow}}{\frac{x^2}{2}} \underset{\underset{du}{\uparrow}}{\frac{dx}{x}}$$

($x\,dx$ = dv)

$$= \frac{x^2}{2} \ln x - \frac{1}{2}\int x\,dx$$

$$= \frac{x^2}{2} \ln x - \frac{x^2}{4} + c$$

EXAMPLE 3 Evaluate $\int \ln x\,dx$.

SOLUTION This example seems quite difficult. It does not fall within any of the categories for which we have a substitution available, and it does not seem that the integration-by-parts formula is applicable because the integrand consists of the single function ln x, rather than a product. However, we can resort to the somewhat artificial device of rewriting our problem as

$$\int (1)(\ln x)\,dx$$

Now the integrand *is* a product of two functions, 1 and ln x. The function ln x is the "more sophisticated" but we do not know its antiderivative. (In fact, the very problem we are trying to solve is to find the antiderivative of ln x.) Accordingly, we use

$$u = \ln x \qquad dv = 1\,dx = dx$$

$$du = \frac{dx}{x} \qquad v = x$$

so

$$\int \underset{\underset{u}{\uparrow}}{\ln x}\, \underset{\underset{dv}{\uparrow}}{dx} = \underset{\underset{v}{\uparrow}}{x}\, \underset{\underset{u}{\uparrow}}{\ln x} - \int \underset{\underset{v}{\uparrow}}{x}\, \underset{\underset{du}{\uparrow}}{\frac{dx}{x}}$$

$$= x \ln x - \int dx$$

$$= x \ln x - x + c$$

PRACTICE EXERCISE Evaluate $\int \sqrt{x} \ln x\,dx$.

[Your answer should be equivalent to $\frac{2}{3}x^{3/2}(\ln x - \frac{2}{3}) + c$.]

EXAMPLE 4 Evaluate $\int x^2 e^{2x}\,dx$.

SOLUTION Following the integration-by-parts hint, we select

$$u = x^2 \qquad dv = e^{2x}\,dx$$

$$du = 2x\,dx \qquad v = \frac{1}{2}e^{2x}$$

and obtain

$$\int \underset{u}{\underset{\uparrow}{x^2}}\,\underset{dv}{\underset{\uparrow}{e^{2x}}\,dx} = \underset{u}{\underset{\uparrow}{x^2}}\,\underset{v}{\underset{\uparrow}{\frac{e^{2x}}{2}}} - \int \underset{v}{\underset{\uparrow}{\left(\frac{1}{2}e^{2x}\right)}}\,\underset{du}{\underset{\uparrow}{2x\,dx}}$$

$$= \frac{x^2 e^{2x}}{2} - \int xe^{2x}\,dx$$

The last term, $\int xe^{2x}\,dx$, is not immediately integrable, but it can be evaluated by using the integration-by-parts formula again. In fact, we did this in Example 1, where we found that

$$\int xe^{2x}\,dx = \frac{xe^{2x}}{2} - \frac{e^{2x}}{4} - c$$

By substituting this into the foregoing expression, we obtain

$$\int x^2 e^{2x}\,dx = \frac{x^2 e^{2x}}{2} - \left[\frac{xe^{2x}}{2} - \frac{e^{2x}}{4} - c\right]$$

$$= \frac{x^2 e^{2x}}{2} - \frac{xe^{2x}}{2} + \frac{e^{2x}}{4} + c$$

This example shows that it is occasionally necessary to integrate by parts several times until an answer is found. Each successive application of the integration-by-parts formula results in a slightly simpler antidifferentiation problem.

EXAMPLE 5 Compute $\int x\sqrt{1 - 2x}\,dx$.

SOLUTION The integrand is the product of the functions x and $\sqrt{1 - 2x} = (1 - 2x)^{1/2}$. Both terms are easily integrated, but the term $(1 - 2x)^{1/2}$ is clearly the "most sophisticated" of the two. Thus we set

$$u = x \qquad dv = (1 - 2x)^{1/2}\,dx$$

$$du = dx \qquad v = -\frac{1}{3}(1 - 2x)^{3/2}$$

Then

(7.10) $$\int x\sqrt{1 - 2x}\,dx = -\frac{x}{3}(1 - 2x)^{3/2} + \frac{1}{3}\int (1 - 2x)^{3/2}\,dx$$

The integral on the right can be evaluated by the method of substitution. Let $w = (1 - 2x)$ so that $dw = -2\,dx$. We then have $dx = -\frac{1}{2}\,dw$ and

(7.11) $$\int (1 - 2x)^{3/2}\,dx = -\frac{1}{2}\int w^{3/2}\,dw = -\frac{1}{5}w^{5/2} = -\frac{1}{5}(1 - 2x)^{5/2}$$

[We omitted "$+c$" here, because we shall not need it until the very end when the evaluation of Equation (7.10) is complete.] Finally, substituting Equation (7.11) into Equation (7.10), we reach

$$\int x\sqrt{1 - 2x}\,dx = -\frac{x}{3}(1 - 2x)^{3/2} - \frac{1}{15}(1 - 2x)^{5/2} + c$$

We close with an exercise showing that the selection of u and dv can occasionally require some imagination in breaking up the integrand.

PRACTICE EXERCISE Show that

$$\int 2x^3\sqrt{x^2 + 1}\,dx = \frac{2}{3}x^2(x^2 + 1)^{3/2} - \frac{4}{15}(x^2 + 1)^{5/2} + c$$

by setting $u = x^2$ and $dv = 2x(x^2 + 1)^{1/2}\,dx$. Then try to solve this problem by making the more obvious choices for u and dv.

Section 7.1 EXERCISES

Evaluate the integrals in Exercises 1–26.

1. $\int xe^x\,dx$

2. $\int x^2e^{-x}\,dx$

3. $\int x^3e^x\,dx$

4. $\int \frac{\ln x}{x}\,dx$

5. $\int (\ln x)^2\,dx$

6. $\int (\ln x)^3\,dx$

7. $\int x^2 \ln x\,dx$

8. $\int \sqrt{x}\ln x\,dx$

9. $\int xe^{-x^2}\,dx$

10. $\int \frac{e^x\,dx}{1 + e^x}$

11. $\int e^x\sqrt{1 + e^x}\,dx$

12. $\int \frac{e^x\,dx}{(1 + e^x)^{3/2}}$

13. $\int \frac{e^x}{\sqrt{1 + e^x}}\,dx$

14. $\int \frac{1 + e^{2x}}{e^x}\,dx$

15. $\int \frac{1 + e^x}{e^{2x}}\,dx$

16. $\int \frac{e^{2x}}{1 + e^x}\,dx$

17. $\int x^2(e^x - 1)\,dx$

18. $\int x^3\sqrt{3x^2 + 1}\,dx$

19. $\int x^{-3}\ln x\,dx$

20. $\int x^2e^{3x}\,dx$

21. $\int x^3\sqrt{4 - x^2}\,dx$

22. $\int \frac{\ln x}{x^2}\,dx$

23. $\int \ln\sqrt{x}\,dx$

24. $\int_0^3 xe^{x/3}\,dx$

25. $\int_1^5 \ln x\,dx$

26. $\int_0^2 4x(x + 2)^{-3}\,dx$

27. (a) Prove the recursion formula

$$\int x^n e^x \, dx = x^n e^x - n \int x^{n-1} e^x \, dx$$

(b) Use this formula repeatedly to find $\int x^5 e^x \, dx$ and $\int x^6 e^x \, dx$.

28. (a) Prove the recursion formula

$$\int x^m (\ln x)^n \, dx = \frac{x^{m+1}(\ln x)^n}{m+1} - \frac{n}{m+1} \times \int x^m (\ln x)^{n-1} \, dx \qquad m, n \neq -1$$

(b) Use this formula to find $\int x^4 (\ln x)^3$.

SECTION 7.2

Integration of Rational Functions (Partial Fractions)

In this section we shall develop a systematic technique for finding the antiderivatives of certain *rational functions*; that is, functions that are the ratio of two polynomials, such as $\frac{x^2 - x}{x + 2}, \frac{x - 1}{x^3 - 3x + 1}, \frac{3x + 2}{x^4 + 7}$. We will limit our discussion initially to those rational functions whose denominators are polynomials of degree one, i.e., linear functions.

As always, we proceed by working some examples.

RATIONAL FUNCTIONS WITH FIRST-DEGREE DENOMINATOR

EXAMPLE 6 Evaluate

$$\int \frac{1}{x - 2} \, dx$$

SOLUTION We try the substitution

$$u = x - 2 \qquad du = dx$$

Then

$$\int \frac{1}{x - 2} \, dx = \int \frac{1}{u} \, du = \ln|u| + c$$

which, in terms of x, is

$$\int \frac{1}{x - 2} \, dx = \ln|x - 2| + c$$

PRACTICE EXERCISE Evaluate $\int \frac{1}{x + 4} \, dx$.

[Your answer should be $\ln|x + 4| + c$.]

EXAMPLE 7 Evaluate $\int \frac{x}{x - 2} \, dx$.

SOLUTION In this case the substitution

$$u = x - 2 \qquad du = dx$$

gives

$$\int \frac{x}{x-2}\,dx = \int \frac{u+2}{u}\,du = \int \left(1 + \frac{2}{u}\right) du$$
$$= u + 2\ln|u| + c$$

We therefore have

$$\int \frac{x}{x-2}\,dx = x - 2 + 2\ln|x-2| + c$$

There is an alternative method for obtaining this integral. If we carry out the long division indicated in the rational function $\frac{x}{x-2}$, as follows,

$$\begin{array}{r} 1 \\ x-2\,\overline{)\,x} \\ x-2 \\ \hline \textcircled{2} \end{array}$$

Result of multiplying $1 \cdot (x-2)$ — Subtract — Remainder

we find that

$$\frac{x}{x-2} = 1 + \frac{2}{x-2}$$

Therefore

$$\int \frac{x}{x-2}\,dx = \int \left(1 + \frac{2}{x-2}\right) dx$$
$$= x + 2\ln|x-2| + c$$

Note that our two answers differ by the additive constant -2, which will disappear upon differentiation. These are equivalent answers.

EXAMPLE 8 Compute $\int \frac{x^2+2}{x+4}\,dx$.

SOLUTION We once again try long division.

$$\begin{array}{r} x \quad - 4 \\ x+4\,\overline{)\,x^2 \qquad\quad + 2} \\ x^2 + 4x + 2 \\ \hline -4x + 2 \\ -4x - 16 \\ \hline \textcircled{18} \end{array}$$

Bring down $+2$

Product of x and $x + 4$

Product of -4 and $x + 4$

Remainder

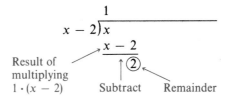

This time $\dfrac{x^2 + 2}{x + 4} = x - 4 + \dfrac{18}{x + 4}$, so that

$$\int \frac{x^2 + 2}{x + 4}\,dx = \int \left(x - 4 + \frac{18}{x + 4}\right) dx$$
$$= \frac{x^2}{2} - 4x + 18 \ln|x + 4| + c$$

What we did in Examples 7 and 8 was to convert an **improper rational function** (one whose numerator has degree greater than or equal to that of its denominator) into the sum of a polynomial and a **proper rational function** (one whose numerator has degree strictly less than that of its denominator). Every improper rational function can be so converted by long division.

PRACTICE EXERCISE Show, in general, that

$$\int \frac{dx}{ax + b} = \frac{1}{a} \ln|ax + b| + c \qquad \text{if } a \neq 0$$

Next we consider the integration of rational functions whose denominators factor into two or more linear functions. In order to motivate the discussion to follow, let us consider the problem of combining into a single fraction the difference

$$\frac{1}{x - 2} - \frac{1}{x + 3}$$

A least common denominator is $(x - 2)(x + 3)$, so

$$\frac{1}{x - 2} - \frac{1}{x + 3} = \frac{1(x + 3) - 1(x - 2)}{(x - 2)(x + 3)} = \frac{5}{x^2 + x - 6} \tag{7.12}$$

It follows that, if we are faced with the problem of evaluating

$$\int \frac{5}{x^2 + x - 6}\,dx$$

then, from Equation (7.12), this is exactly the same as evaluating

$$\int \left(\frac{1}{x - 2} - \frac{1}{x + 3}\right) dx = \int \frac{1}{x - 2}\,dx - \int \frac{1}{x + 3}\,dx$$

But that is easy! By the formula of the foregoing practice exercise,

$$\int \frac{5}{x^2 + x - 6}\,dx = \int \frac{1}{x - 2}\,dx - \int \frac{1}{x + 3}\,dx$$
$$= \ln|x - 2| - \ln|x + 3| + c$$

Basically, we took advantage of the fact that the rational function $\dfrac{5}{(x - 2)(x + 3)}$ can be decomposed into the sum of two terms, each of which has a first-degree denominator. Once we knew the decomposition, the integration itself was fairly simple.

As another illustration of this idea, suppose we try to evaluate

$$\int \frac{x^2 + 1}{(x - 1)(x + 2)(x - 3)}\, dx$$

This looks fairly difficult. But if we were to tell you that

(7.13)
$$\frac{x^2 + 1}{(x - 1)(x + 2)(x - 3)} = \frac{-1/3}{x - 1} + \frac{1/3}{x + 2} + \frac{1}{x - 3}$$

you could make use of this decomposition to write

$$\begin{aligned}\int \frac{x^2 + 1}{(x - 1)(x + 2)(x - 3)}\, dx &= -\frac{1}{3}\int \frac{1}{x - 1}\, dx + \frac{1}{3}\int \frac{1}{x + 2}\, dx + \int \frac{1}{x - 3}\, dx \\ &= -\frac{1}{3}\ln|x - 1| + \frac{1}{3}\ln|x + 2| + \ln|x - 3| + c\end{aligned}$$

Of course, this solution is not entirely satisfactory because it depends on our being able to find the decomposition given in Equation (7.13), just as our previous example depended on Equation (7.12). But if we ignore for a moment the difficulties involved in finding such decompositions, we can see that the approach of splitting a rational function into the sum of terms with lower-degree denominators certainly seems to simplify integration.

In general, it can be shown that any polynomial can, *in principle*, be factored into the product of first-degree terms (terms of the form $ax + b$) and irreducible (nonfactorable) quadratic terms ($ax^2 + bx + c$). For instance,

$$x^3 - 2x^2 - 5x + 6 = (x - 1)(x + 2)(x - 3)$$

$$x^3 - 2x^2 + x - 2 = (x - 2)(x^2 + 1)$$

$$x^5 - x^4 + 6x^3 - 6x^2 + 9x - 9 = (x - 1)(x^2 + 3)(x^2 + 3)$$

In practice it is sometimes very difficult to factor a polynomial. (Try to factor $x^3 - 3x^2 + 3$, for example.) However, the polynomials we will encounter in this chapter will factor without extraordinary difficulty. In fact, we will deal only with polynomials that factor completely as a product of linear (first-degree) factors.

We try to decompose our rational function into a sum of fractions each having one of the factors of the denominator as its denominator. (Each fraction will itself be a rational function of the type we learned to integrate in our earlier examples.) In some sense we are trying to do the reverse of what we do when we add fractions by forming a least common denominator. The technique by which this decomposition is accomplished is generally called the **method of partial fractions**. Other books occasionally refer to it as the **method of undetermined coefficients**.

DENOMINATOR FACTORABLE INTO DISTINCT LINEAR FACTORS Let us see how the method of partial fractions works.

EXAMPLE 9 Evaluate $\int \frac{1}{x^2 - 1}\, dx$.

SOLUTION The denominator is easily factored.

$$x^2 - 1 = (x - 1)(x + 1)$$

This suggests that we should try to write

$$\frac{1}{x^2 - 1} = \frac{A}{x - 1} + \frac{B}{x + 1}$$

for suitable constants A and B. How can we figure out what the values of A and B should be? We multiply both sides of this equation by the least common denominator $(x - 1)(x + 1)$ to obtain

$$1 = A(x + 1) + B(x - 1) = (A + B)x + (A - B)$$

This can be an identity only if the coefficient of x on the right, $A + B$, equals the coefficient of x on the left, 0; and if the constant term on the right, $A - B$, equals the constant term on the left, 1. Therefore, we require

$$A + B = 0 \longleftarrow \text{equating coefficients of } x$$

$$A - B = 1 \longleftarrow \text{equating constant terms}$$

These simultaneous equations are readily solved to yield

$$A = \frac{1}{2} \qquad B = -\frac{1}{2}$$

We can now substitute these values into one original equation to obtain

$$\frac{1}{x^2 - 1} = \frac{\frac{1}{2}}{x - 1} + \frac{-\frac{1}{2}}{x + 1}$$

so

$$\int \frac{1}{x^2 - 1}\,dx = \int \frac{\frac{1}{2}}{x - 1}\,dx + \int \frac{-\frac{1}{2}}{x + 1}\,dx$$

$$= \frac{1}{2}\ln|x - 1| - \frac{1}{2}\ln|x + 1| + c$$

EXAMPLE 10 Evaluate $\displaystyle\int \frac{x}{x^2 - 3x + 2}\,dx$.

SOLUTION The polynomial $x^2 - 3x + 2$ factors into $(x - 2)(x - 1)$, so we try to write

$$\frac{x}{x^2 - 3x + 2} = \frac{A}{x - 2} + \frac{B}{x - 1}$$

If we multiply both sides by $(x - 2)(x - 1)$, we obtain

$$x = A(x - 1) + B(x - 2)$$

or

$$x = (A + B)x + (-A - 2B)$$

This, in turn, implies that

$$A + B = 1 \longleftarrow \text{equating coefficients of } x$$

$$-A - 2B = 0 \longleftarrow \text{equating constant terms}$$

We solve these equations simultaneously to find

$$A = 2 \qquad B = -1$$

Therefore $$\int \frac{x}{x^2 - 3x + 2}\,dx = \int \frac{2}{x - 2}\,dx + \int \frac{-1}{x - 1}\,dx$$
$$= 2\ln|x - 2| - \ln|x - 1| + c$$

In general, suppose we have a proper rational function $R(x)$. Then $R(x)$ has the form

$$R(x) = \frac{p(x)}{q(x)}$$

where p and q are polynomials and the degree of p is less than the degree of q. If q factors into n *distinct* linear factors, then $R(x)$ has the form

$$R(x) = \frac{p(x)}{(a_1x + b_1)(a_2x + b_2)\cdots(a_nx + b_n)}$$

In this case it is possible to find n numbers $A_1, A_2, \ldots, A_n$ such that

$$R(x) = \frac{p(x)}{(a_1x + b_1)(a_2x + b_2)\cdots(a_nx + b_n)}$$
$$= \frac{A_1}{a_1x + b_1} + \frac{A_2}{a_2x + b_2} + \cdots + \frac{A_n}{a_nx + b_n}$$

This is called the **partial-fractions decomposition** of $R(x)$.

EXAMPLE 11 Evaluate $\displaystyle\int \frac{x^2 + 1}{(x - 1)(x + 2)(x + 3)}\,dx.$

SOLUTION This time the denominator is already factored and we need only determine A, B, and C to satisfy

$$\frac{x^2 + 1}{(x - 1)(x + 2)(x - 3)} = \frac{A}{x - 1} + \frac{B}{x + 2} + \frac{C}{x - 3}$$

Multiplying both sides by $(x - 1)(x + 2)(x - 3)$ yields

(7.14) $$x^2 + 1 = A(x + 2)(x - 3) + B(x - 1)(x - 3) + C(x - 1)(x + 2)$$

Now, as before, we could multiply out the right side, equate coefficients of like powers of x on both sides, and solve the resulting system of three simultaneous equations in three unknowns. However, there is an alternative, quick method for determining A, B, and C.

If Equation (7.14) is to be an identity, it must hold for all x, in particular for $x = 1$, $x = -2$, and $x = 3$ (the roots of the denominator in Example 11).

Setting $x = 1$ on both sides gives

$$2 = A(3)(-2)$$

because the last two summands on the right vanish at $x = 1$. Consequently $2 = -6A$ and $A = -\frac{1}{3}$.

Similarly, setting $x = -2$ in Equation (7.14) yields $5 = B(-3)(-5)$, hence $B = 1/3$. Finally, setting $x = 3$ in Equation (7.14) gives $10 = C(2)(5)$, so that $C = 1$. Therefore

$$\frac{x^2 + 1}{(x - 1)(x + 2)(x - 3)} = \frac{-\frac{1}{3}}{x - 1} + \frac{\frac{1}{3}}{x + 2} + \frac{1}{x - 3}$$

and

$$\begin{aligned}\int \frac{x^2 + 1}{(x - 1)(x + 2)(x - 3)}\,dx &= -\frac{1}{3}\int \frac{1}{x - 1}\,dx + \frac{1}{3}\int \frac{1}{x + 2}\,dx + \int \frac{1}{x - 3}\,dx \\ &= -\frac{1}{3}\ln|x - 1| + \frac{1}{3}\ln|x + 2| + \ln|x - 3| + c\end{aligned}$$

DENOMINATOR FACTORABLE INTO LINEAR FACTORS (NOT ALL DISTINCT) Consider the problem of combining into a single fraction the sum

$$\frac{1}{(x - 1)^2} + \frac{2}{x + 3}$$

The least common denominator is $(x - 1)^2(x + 3)$, so

(7.15) $$\frac{1}{(x - 1)^2} + \frac{2}{x + 3} = \frac{1(x + 3) + 2(x - 1)^2}{(x - 1)^2(x + 3)} = \frac{2x^2 - 3x + 5}{(x - 1)^2(x + 3)}$$

Alternatively, consider the problem of adding

$$\frac{1}{x - 1} + \frac{1}{(x - 1)^2} + \frac{2}{x + 3}$$

Again, the least common denominator is $(x - 1)^2(x + 3)$, and

(7.16) $$\begin{aligned}\frac{1}{x - 1} + \frac{1}{(x - 1)^2} + \frac{2}{x + 3} &= \frac{1(x - 1)(x + 3) + 1(x + 3) + 2(x - 1)^2}{(x - 1)^2(x + 3)} \\ &= \frac{3x^2 - x + 2}{(x - 1)^2(x + 3)}\end{aligned}$$

The two rational functions

$$\frac{2x^2 - 3x + 5}{(x - 1)^2(x + 3)} \quad \text{and} \quad \frac{3x^2 - x + 2}{(x - 1)^2(x + 3)}$$

certainly look similar, yet one of them arose from a sum of two terms [with denominators $(x - 1)^2$ and $(x + 3)$, respectively], whereas the other is the sum of three terms [with denominators $(x - 1)$, $(x - 1)^2$, and $(x + 3)$]. How are we to know, when we are trying to decompose a function like those in Equation (7.15) or Equation (7.16), which terms to try in our partial-fractions decomposition? The answer is that we allow for all possibilities that give rise to the same least common denominator. Thus, if the denominator of the rational function $R(x)$ contains the linear factor $ax + b$ repeated k times, our partial-fractions decomposition should include *all* the terms.

$$\frac{A_1}{ax + b} + \frac{A_2}{(ax + b)^2} + \cdots + \frac{A_k}{(ax + b)^k}$$

For instance, if we wish to decompose *any* rational function whose denominator is $(x - 1)^2(x + 3)$, we should try

$$\frac{p(x)}{(x - 1)^2(x + 3)} = \frac{A}{x - 1} + \frac{B}{(x - 1)^2} + \frac{C}{x + 3}$$

If $p(x) = 2x^2 - 3x + 5$, as it did in Equation (7.15), then the coefficient A will turn out to be 0. If, on the other hand, $p(x) = 3x^2 - x + 2$, as in Equation (7.16), then A will turn out to be 1. Unfortunately, we can't tell in advance which terms are required, so we must allow for them all.

EXAMPLE 12 Evaluate $\displaystyle\int \frac{1}{x^2(x + 1)}\,dx$.

SOLUTION The denominator of our integral is already factored, and it contains the repeated factor x. Accordingly, we try

$$\frac{1}{x^2(x + 1)} = \frac{A}{x} + \frac{B}{x^2} + \frac{C}{x + 1}$$

Multiplying both sides by $x^2(x + 1)$, we obtain

(7.17) $$1 = Ax(x + 1) + B(x + 1) + Cx^2$$

If we set $x = 0$, we get $1 = B$. If we set $x = -1$, we get $1 = C$. We have now used all the roots of the denominator of our integrand. Therefore we choose some other convenient (small) x, such as $x = 1$. Setting $x = 1$ in Equation (7.17) and remembering that $B = 1, C = 1$, we have

$$\begin{aligned} 1 &= A(1)(2) + B(2) + C(1) \\ &= 2A + 2 + 1 = 2A + 3 \end{aligned}$$

from which we obtain $A = -1$. Therefore

$$\begin{aligned} \int \frac{1}{x^2(x + 1)}\,dx &= \int \frac{-1}{x}\,dx + \int \frac{1}{x^2}\,dx + \int \frac{1}{x + 1}\,dx \\ &= -\ln|x| - \frac{1}{x} + \ln|x + 1| + c \end{aligned}$$

We could also do this problem by rewriting Equation (7.17) as

$$1 = (A + C)x^2 + (A + B)x + B$$

and equating like coefficients to obtain the system

$$A + C = 0$$
$$A + B = 0$$
$$B = 1$$

which yields $A = -1$, $B = 1$, $C = 1$, just as before.

EXAMPLE 13 Evaluate $\int \frac{x}{(x-1)^3}\,dx$.

SOLUTION This denominator contains the linear factor $(x - 1)$ repeated three times. We set

$$\frac{x}{(x-1)^3} = \frac{A}{x-1} + \frac{B}{(x-1)^2} + \frac{C}{(x-1)^3}$$

Multiplying both sides by $(x - 1)^3$, we obtain

$$\begin{aligned} x &= A(x-1)^2 + B(x-1) + C \\ &= Ax^2 + (-2A + B)x + (A - B + C) \end{aligned}$$

Upon equating like powers of x, we find

$$A = 0$$
$$-2A + B = 1$$
$$A - B + C = 0$$

Thus $$A = 0,\ B = 1,\ C = 1$$

Finally,

$$\begin{aligned} \int \frac{x}{(x-1)^3}\,dx &= \int \frac{1}{(x-1)^2}\,dx + \int \frac{1}{(x-1)^3}\,dx \\ &= -(x-1)^{-1} - \frac{(x-1)^{-2}}{2} + c \end{aligned}$$

DENOMINATOR CONTAINING IRREDUCIBLE QUADRATIC FACTORS If the denominator of the rational function $R(x)$ contains irreducible quadratic terms in its factorization, and none of these terms is repeated, then corresponding to each factor of the form $ax^2 + bx + c$, the partial-fractions decomposition should contain a term of form

$$\frac{Ax + B}{ax^2 + bx + c}$$

For example,

$$\frac{2x^2 + 7x + 13}{(x^2+1)(x+4)} = \frac{x+3}{x^2+1} + \frac{1}{x+4}$$

and the term $\frac{x+3}{x^2+1}$ cannot be decomposed any further, because $x^2 + 1$ does *not* factor into linear factors.

Rational functions with denominators that are irreducible quadratics can be handled via what are called trigonometric substitutions, but we will not treat such substitutions in this book.

Section 7.2 EXERCISES

Evaluate the following integrals.

1. $\displaystyle\int \frac{3\,dx}{x-7}$

2. $\displaystyle\int \frac{5\,dx}{3x-4}$

3. $\displaystyle\int \frac{x}{x+3}\,dx$

4. $\displaystyle\int \frac{3x-4}{5x}\,dx$

5. $\displaystyle\int \frac{x^2-x-6}{x-3}\,dx$

6. $\displaystyle\int \frac{x+3}{x-3}\,dx$

7. $\displaystyle\int \frac{2x+1}{2x-1}\,dx$

8. $\displaystyle\int \frac{x^2+3x+5}{x+2}\,dx$

9. $\displaystyle\int \frac{x^2+1}{x+1}\,dx$

10. $\displaystyle\int \frac{x+1}{x^2-1}\,dx$

11. $\displaystyle\int \frac{x^3+1}{x+1}\,dx$

12. $\displaystyle\int \frac{x^3-1}{x+1}\,dx$

13. $\displaystyle\int \frac{x^3+x^2+x+1}{x+2}\,dx$

14. $\displaystyle\int \frac{x^3+6x^2+5x-6}{x+2}\,dx$

15. $\displaystyle\int \frac{3x^3-14x^2+9x-4}{4-x}\,dx$

16. $\displaystyle\int \frac{x+4}{2-x}\,dx$

17. $\displaystyle\int \frac{x+6}{3-x}\,dx$

18. $\displaystyle\int \frac{8x^3+1}{2x+1}\,dx$

19. $\displaystyle\int \frac{x^3-1}{x-1}\,dx$

20. $\displaystyle\int \frac{6}{3x-4}\,dx$

21. $\displaystyle\int \frac{5\,dx}{(x-4)(x+3)}$

22. $\displaystyle\int \frac{x\,dx}{(x-4)(x+3)}$

23. $\displaystyle\int \frac{x^2+1}{(x-4)(x+3)}\,dx$

24. $\displaystyle\int \frac{2}{x(x-3)}\,dx$

25. $\displaystyle\int \frac{x+1}{x(x-3)}\,dx$

26. $\displaystyle\int \frac{x^2+1}{x(x-3)}\,dx$

27. $\displaystyle\int \frac{dx}{x(x-3)^2}$

28. $\displaystyle\int \frac{dx}{x^2(x-3)}$

29. $\displaystyle\int \frac{x^2+x+1}{x^2(x-3)}\,dx$

30. $\displaystyle\int \frac{x+1}{x^2(x-3)^2}\,dx$

31. $\displaystyle\int \frac{dx}{x^2-4}$

32. $\displaystyle\int \frac{x+1}{x^2-5x-6}\,dx$

33. $\displaystyle\int \frac{x^2+x+1}{x(x+1)(x-2)}\,dx$

34. $\displaystyle\int \frac{dx}{(x-2)^3}$

35. $\displaystyle\int \frac{x^2+1}{(x-2)^3}\,dx$

36. $\displaystyle\int \frac{x^2+4}{x^2-4}\,dx$

37. $\displaystyle\int \frac{x+2}{x^2-3x-4}\,dx$

38. $\displaystyle\int \frac{x^2-1}{(x-1)(x+2)(x+3)}\,dx$

39. $\displaystyle\int \frac{x^2+1}{x(x-1)(x+2)}\,dx$

40. $\displaystyle\int \frac{x-2}{(x^2-4)(x^2-9)}\,dx$

41. $\displaystyle\int \frac{x^3+3}{x^2-x-2}\,dx$

42. $\displaystyle\int \frac{x-2}{x^2-x}\,dx$

43. $\displaystyle\int \frac{x^2}{x^3-1}\,dx$

44. $\displaystyle\int \frac{x^2+x+1}{x^3(x-1)^2}\,dx$

45. $\displaystyle\int \frac{x^2\,dx}{x^2+3x+2}$

46. $\displaystyle\int \frac{dx}{1-x^2}$

47. $\displaystyle\int \frac{x^3\,dx}{x^2+3x+2}$

48. $\displaystyle\int \frac{x}{x^2+4x-5}\,dx$

49. $\displaystyle\int \frac{x^3+4x^2+3x+2}{x^2+3x+2}\,dx$

50. $\displaystyle\int \frac{x+1}{x^2+4x-5}\,dx$

51. $\displaystyle\int \frac{x^3+2x+1}{(x^2+x-2)(x^2-2x-3)}\,dx$

52. $\displaystyle\int \frac{dx}{x(x+1)^2}$

53. $\displaystyle\int \frac{x^2+3}{(x+2)(x-1)^2}\,dx$

54. $\displaystyle\int \frac{x + 3}{2x^3 - 8x}\,dx$

55. $\displaystyle\int \frac{x^3\,dx}{x^2 - 2x + 1}$

56. $\displaystyle\int \frac{2x^2 - 3x}{(x^2 - 2x - 8)(x + 1)}\,dx$

57. $\displaystyle\int \frac{3x + 2}{(x + 1)(x^2 + 2x - 8)}\,dx$

58. $\displaystyle\int \frac{x^2 + 5}{(x - 2)(x^2 + x - 12)}$

59. $\displaystyle\int \frac{x^2 + 2x + 3}{(x - 3)(x^2 + 3x + 2)}\,dx$

60. $\displaystyle\int \frac{x^2 + 4x + 2}{(x + 2)(x^2 + 2x - 3)}\,dx$

SECTION 7.3

Integration Using Tables

In addition to the methods of integration by substitution, parts, and partial fractions, there are many other techniques covered in more advanced courses. Our goal in this text is not to make you an expert on antidifferentiation, but rather to help you acquire enough facility to handle many of the integrals that commonly arise in practical applications.

Appendix B contains a list of integration formulas that can be used as a supplement to the methods you have learned so far. A few examples will illustrate how such a table can be used.

EXAMPLE 14 Evaluate $\displaystyle\int \frac{x\,dx}{(3 + 2x)^2}$.

SOLUTION Except for the fact that the variable is denoted x rather than u, this has the same form as the left side of Equation 4 in Appendix B:

$$\int \frac{u\,du}{(a + bu)^2} = \frac{1}{b^2}\left(\frac{a}{a + bu} + \ln|a + bu|\right) + c$$

If we set $a = 3$, $b = 2$ in this formula and change u to x, we have

$$\int \frac{x\,dx}{(3 + 2x)^2} = \frac{1}{4}\left(\frac{3}{3 + 2x} + \ln|3 + 2x|\right) + c$$

PRACTICE EXERCISE Evaluate $\displaystyle\int_0^4 \frac{dt}{\sqrt{t^2 + 9}}$.

[Your answer, obtained via Equation 30 in the table, should be $\ln 9 - \ln 3 = \ln 3$.]

Often a substitution must first be made to convert the given integral into a form that fits some table entry, as in the next example.

EXAMPLE 15 Evaluate $\displaystyle\int \frac{\sqrt{16 - 49x^2}}{x}\,dx$.

SOLUTION The only formula in the table that resembles this is Equation 38:

$$\int \frac{\sqrt{a^2 - u^2}}{u}\,du = \sqrt{a^2 - u^2} - a \ln\left|\frac{a + \sqrt{a^2 - u^2}}{u}\right| + c$$

In order to make 16 correspond to a^2, and $49x^2$ to u^2, we set

$$a = 4 \qquad u = 7x$$

Then $du = 7dx$ and $dx = \frac{1}{7}\,du$. We therefore have

$$\int \frac{\sqrt{16 - 49x^2}}{x}\,dx = \frac{1}{7}\int \frac{\sqrt{a^2 - u^2}}{u/7}\,du = \int \frac{\sqrt{a^2 - u^2}}{u}\,du$$

$$= \sqrt{a^2 - u^2} - a \ln\left|\frac{a + \sqrt{a^2 - u^2}}{u}\right| + c$$

$$= \sqrt{16 - 49x^2} - 4 \ln\left|\frac{4 + \sqrt{16 - 49x^2}}{7x}\right| + c$$

EXAMPLE 16 Evaluate $\int \frac{e^x}{(1 - e^{2x})^{3/2}}\,dx$.

SOLUTION This does not resemble anything in the integral table, so we might attempt a substitution. If we try $u = e^{2x}$ or $u = 1 - e^{2x}$, we will be quickly stymied, because the integral contains $e^x\,dx$ rather than $e^{2x}\,dx$.

Suppose we try $u = e^x$. This gives $du = e^x\,dx$ and $1 - e^{2x} = 1 - u^2$. Then

$$\int \frac{e^x}{(1 - e^{2x})^{3/2}}\,dx = \int \frac{du}{(1 - u^2)^{3/2}}$$

which can be evaluated via Equation 39 in Appendix B with $a = 1$.

$$\int \frac{e^x}{(1 - e^{2x})^{3/2}}\,dx = \int \frac{du}{(1 - u^2)^{3/2}}$$

$$= \frac{u}{\sqrt{1 - u^2}} + c = \frac{e^x}{\sqrt{1 - e^{2x}}} + c$$

A number of formulas in the table are used to reduce a given integral to a simpler one. Such formulas are called **reduction formulas**. Here is a typical example.

EXAMPLE 17 Evaluate $\int (\ln 3x)^2\,dx$.

SOLUTION To fit this to Equation 48, we set

$$u = 3x, \quad x = u/3, \quad dx = du/3, \quad n = 2$$

to obtain

$$\int (\ln 3x)^2\,dx = \frac{1}{3}\int (\ln u)^2\,du = \frac{1}{3}\left[u(\ln u)^2 - 2\int \ln u\,du\right] \tag{7.18}$$

Now, by Equation 45

$$\int \ln u\, du = u \ln u - u$$

Substituting this into Equation (7.18), we get

$$\int (\ln 3x)^2\, dx = \frac{1}{3}[u(\ln u)^2 - 2u \ln u + 2u] + c$$
$$= x(\ln 3x)^2 - 2x \ln(3x) + 2x + c$$

(Note that we always add $+c$ only in the final step.)

PRACTICE EXERCISE Compute $\int_0^1 x^3 e^x\, dx$.

[Your answer, obtained by using Equation 42 twice and then Equation 41, should be $6 - 2e$.]

One other technique we might mention is **completing the square**, which can often be used on integrals involving quadratic functions.

EXAMPLE 18 Evaluate $\displaystyle\int \frac{dx}{x^2 - 2x - 3}$.

SOLUTION We complete $x^2 - 2x$ in the denominator to a perfect square by adding and subtracting 1 (the square of half of the coefficient of x).

$$x^2 - 2x - 3 = (x^2 - 2x \underline{+ 1}) \underline{- 1} - 3 = (x - 1)^2 - 4$$

Then, using the substitution $u = x - 1$ and Equation 24, we obtain

$$\int \frac{dx}{(x-1)^2 - 4} = \int \frac{du}{u^2 - 4} = \frac{1}{4}\ln\left|\frac{u-2}{u+2}\right| + c = \frac{1}{4}\ln\left|\frac{x-3}{x+1}\right| + c$$

PRACTICE EXERCISE Evaluate $\int \sqrt{x^2 - 8x}\, dx$.

[Your answer should be $\frac{1}{2}[(x - 4)\sqrt{x^2 - 8x} - 16 \ln|x - 4 + \sqrt{x^2 - 8x}|] + c$.]

Section 7.3 EXERCISES

Evaluate the indefinite integrals in Exercises 1–10.

1. $\displaystyle\int \frac{dx}{x(2 + x)^2}$

2. $\displaystyle\int \frac{dx}{(1 - 9x^2)^{3/2}}$

3. $\displaystyle\int \frac{\sqrt{4t^2 + 1}}{t}\, dt$

4. $\displaystyle\int y^2 e^{-y}\, dy$

5. $\displaystyle\int \frac{\sqrt{9 - x^2}}{x}\, dx$

6. $\displaystyle\int \sqrt{16t^2 + 25}\, dt$

7. $\displaystyle\int \frac{dz}{z(4 + z)}$

8. $\displaystyle\int \frac{ds}{s^2\sqrt{2 + s}}$

9. $\displaystyle\int \frac{x^2\,dx}{\sqrt{7 + x}}$

10. $\displaystyle\int \frac{x^2 + 6x + 5}{(x + 3)^2}\,dx$

11. Redo Example 18 by the method of partial fractions.

Evaluate the definite integrals in Exercises 12–15.

12. $\displaystyle\int_0^1 \frac{x}{\sqrt{x + 1}}\,dx$

13. $\displaystyle\int_1^e (\ln x)^2\,dx$

14. $\displaystyle\int_4^{11} \frac{dx}{x^2 - 9}$

15. $\displaystyle\int_0^{\sqrt{3}/4} \frac{dx}{(1 - 4x^2)^{3/2}}$

16. The rate at which a new experimental drug is absorbed into the bloodstream is estimated to be $te^{-0.19t}$ micrograms per hr, t hr after injection of a standardized dose. How much is absorbed during the first 5 hr?

17. The demand x for a certain product is related to its price by the equation

$$p = \frac{600}{\sqrt{125 + x^2}}$$

Find the consumers' surplus if the market equilibrium demand is $x = 10$ units.

18. Find the area beneath the graph of $f(x) = x/(x + 4)^2$, from $x = 0$ to $x = 4$.

19. Find the average value of $f(x) = xe^{-x}$, for $0 \le x \le 2$.

SECTION 7.4

Separable Differential Equations

Earlier in the text we encountered the notion of a **differential equation**. One example is

$$\frac{dy}{dt} = ky \tag{7.19}$$

which describes exponential growth (if $k > 0$) or decay (if $k < 0$), and which we studied in Section 4.6.

In general, a differential equation is just an equation involving an unknown function (denoted here by y) and one or more of its derivatives. The number of such equations is endless, and, as we have pointed out, differential equations have application to nearly every field in which mathematics is used. There is a vast body of literature that systematically explores the various techniques for determining the solution of differential equations. To study them in detail would require years. We will content ourselves with an examination of one particular class of differential equation that includes Equation (7.19) as a special case. The ideas involved are best motivated by considering a simple example.

EXAMPLE 19 Solve the differential equation

$$\frac{dy}{dt} = \frac{t}{y} \tag{7.20}$$

SOLUTION We have already seen that it is sometimes useful to treat the derivative dy/dt algebraically as though it is really a fraction whose numerator is dy and whose denominator is dt. If we now cross-multiply in Equation (7.20), we get

$$y\,dy = t\,dt$$

Observe that the left-hand side of this equation contains all the terms that have the variable y in them, whereas the right-hand side contains all the t's. The variables are said to be *separated.*

Our next step is to antidifferentiate (integrate) both sides of our equation. Symbolically,

$$\int y\,dy = \int t\,dt$$

Thus

$$\frac{y^2}{2} = \frac{t^2}{2} + C_1$$

or

$$y^2 = t^2 + 2C_1$$

where C_1 is a constant of integration. Because $2C_1$ represents *any* constant, we simply relabel it c. Solving for y, we obtain the general solution

$$y = \sqrt{t^2 + c} \qquad \text{provided } y \geq 0$$

The constant c is determined by specifying an initial condition.

We can easily verify that this relation does, in fact, satisfy Equation (7.20) by differentiating it and substituting the result into the equation

$$\underbrace{\frac{1}{2}(t^2 + c)^{-1/2}(2t)}_{\substack{\uparrow \\ dy/dt}} = \underbrace{\frac{t}{\sqrt{t^2 + c}}}_{t/y}$$

More generally, we wish to consider equations that have the form

(7.21) $$\frac{dy}{dt} = f(y)g(t)$$

This can be rewritten as

$$\frac{1}{f(y)}\,dy = g(t)\,dt$$

The process is called **separation of variables**. Now, provided we are able to antidifferentiate both sides of this relation,

$$\int \frac{1}{f(y)}\,dy = \int g(t)\,dt$$

we will have obtained the desired solution.

EXAMPLE 20 Solve Equation (7.19)

$$\frac{dy}{dt} = ky \qquad y > 0$$

via separation of variables (k is a nonzero constant).

SOLUTION We can rewrite our equation in the form

$$\frac{1}{y}\,dy = k\,dt$$

Here the variables have been separated. Then

$$\int \frac{1}{y}\,dy = \int k\,dt$$

so $$\ln y = kt + C_1$$

We may solve for y as a function of t by exponentiating both sides of the equation.

$$e^{\ln y} = y = e^{kt + C_1} = e^{C_1}e^{kt}$$

Because C_1 is a constant, e^{C_1} is another constant that we can denote by c, and our solution is

$$y = ce^{kt}$$

This is, of course, the same answer we found in Section 4.6. There, however, we had to guess the answer. Here we obtained it in a systematic fashion.

EXAMPLE 21 Find the solution of the differential equation

$$\frac{dy}{dt} = \sqrt{ty}$$

for which $y = 1$ when $t = 0$ (that is, $y(0) = 1$).

SOLUTION We separate the variables, so that

$$\frac{1}{\sqrt{y}}\,dy = \sqrt{t}\,dt$$

Thus

$$\int \frac{1}{\sqrt{y}}\,dy = \int \sqrt{t}\,dt$$

so $$2y^{1/2} = \frac{2}{3}t^{3/2} + c$$

Now, when $t = 0$, we want $y = 1$:

$$2(1)^{1/2} = \frac{2}{3}(0)^{3/2} + c$$

Thus $$c = 2$$

and our solution takes the form

$$2y^{1/2} = \frac{2}{3}t^{3/2} + 2$$

This can be solved for y in terms of t.

$$y = \left(\frac{1}{3}t^{3/2} + 1\right)^2$$

Note how the initial condition $y(0) = 1$ served to determine the value of the constant of integration c.

EXAMPLE 22 Solve the differential equation

$$\frac{dy}{dt} = k(1 - y)y$$

together with the initial condition $y = 0.1$ when $t = 0$.

SOLUTION We easily separate the variables to write

$$\frac{1}{(1 - y)y}\,dy = k\,dt$$

and so

(7.22) $$\int \frac{1}{(1 - y)y}\,dy = \int k\,dt = kt + c$$

We are now faced with the problem of finding the antiderivative

$$\int \frac{1}{(1 - y)y}\,dy$$

Because the integrand has a denominator that is factored into linear factors, you may recognize this as a problem that can be handled by the method of partial fractions. We set

$$\frac{1}{(1 - y)y} = \frac{a}{1 - y} + \frac{b}{y} = \frac{ay + b(1 - y)}{(1 - y)y} = \frac{(a - b)y + b}{(1 - y)y}$$

These expressions will be equal provided

$$a - b = 0$$

$$b = 1$$

Thus $$a = b = 1$$

and
$$\int \frac{1}{(1 - y)y}\,dy = \int \frac{1}{1 - y}\,dy + \int \frac{1}{y}\,dy$$
$$= -\ln(1 - y) + \ln y$$
$$= \ln\left(\frac{y}{1 - y}\right)$$

If we substitute this expression into Equation (7.22), we get

$$\ln\left(\frac{y}{1 - y}\right) = kt + c$$

We can evaluate the constant of integration by utilizing our initial condition $y = 0.1$ when $t = 0$.

$$\ln\left(\frac{0.1}{1 - 0.1}\right) = k \cdot 0 + c = c$$

or

$$c = \ln\left(\frac{1}{9}\right)$$

Thus

$$\ln\left(\frac{y}{1 - y}\right) = kt + \ln\left(\frac{1}{9}\right)$$

It remains only for us to solve for y as a function of t. We exponentiate both sides of the equation to obtain

$$e^{\ln[y/(1-y)]} = \frac{y}{1 - y} = e^{kt + \ln(1/9)} = e^{\ln(1/9)}e^{kt} = \frac{1}{9}e^{kt}$$

Thus

$$\frac{y}{1 - y} = \frac{1}{9}e^{kt}$$

or

$$9y = (1 - y)e^{kt} = e^{kt} - e^{kt}y$$

Putting all terms containing y on the left-hand side yields

$$e^{kt}y + 9y = e^{kt}$$

$$y = \frac{e^{kt}}{e^{kt} + 9}$$

Finally, we can multiply both numerator and denominator by e^{-kt} to get

$$y = \frac{1}{1 + 9e^{-kt}}$$

PRACTICE EXERCISE Use the method demonstrated in Example 22 to solve the slightly more general differential equation

$$\frac{dy}{dt} = k(A - y)y$$

together with the initial condition $y(0) = 0.1A$, where both A and k are constants ($A > 0$, $k \neq 0$).

$\left[\text{Your answer should be } y = \frac{A}{1 + 9e^{-kAt}}.\right]$

EXAMPLE 23 Solve the differential equation

$$\frac{dy}{dt} = k(A - y)$$

subject to the initial condition $y(0) = 0$.

SOLUTION We write

$$\frac{1}{A-y}\,dy = k\,dt$$

$$\int \frac{1}{A-y}\,dy = \int k\,dt$$

$$-\ln(A-y) = kt + c.$$

From the initial condition $y(0) = 0$, we see that

$$-\ln(A-0) = k \cdot 0 + c = c$$

so our constant of integration is $c = -\ln A$. We have

$$-\ln(A-y) = kt - \ln A$$

If we multiply the entire equation by -1, we find

$$\ln(A-y) = -kt + \ln A$$

Taking exponentials yields

$$e^{\ln(A-y)} = A - y = e^{-kt+\ln A} = e^{-kt}e^{\ln A} = Ae^{-kt}$$

Finally,

$$A - y = Ae^{-kt}$$

and so

$$y = A - Ae^{-kt} = A(1 - e^{-kt})$$

Let us now look at some specific applications of the technique of separation of variables.

EXAMPLE 24 As we have seen, a body falling subject to gravitational influence has the differential equation

$$\frac{dv}{dt} = -32\ \text{ft/sec}^2$$

satisfied by its velocity. Strictly speaking, this differential equation holds only in the absence of opposing air resistance. Suppose now that air resistance has been found to be proportional to the square of the velocity; that is, it is equal to bv^2 for some constant b. For the sake of definiteness, let us set b equal to 8. Thus the differential equation for the velocity of our falling body is now

(7.23) $$\frac{dv}{dt} = -32 + 8v^2$$

Unlike the situation in which air resistance is neglected whatever the initial velocity v_0 specified, Equation (7.23) possesses a solution $v(t)$ that has a finite *limiting value* as $t \to +\infty$. This value is called the **free-fall velocity** and is independent of v_0. Let us now find the free-fall velocity corresponding to Equation (7.23).

SOLUTION We assume an initial condition $v(0) = v_0$, with $-2 < v_0 < 0$ (v is negative for downward motion). Then

$$\frac{dv}{dt} = -8(4 - v^2) = -8(2 + v)(2 - v)$$

so by separation of variables,

$$\int \frac{dv}{(2 + v)(2 - v)} = -\int 8\,dt = -8t + c$$

Now, by using partial fractions, we find that

$$\frac{1}{(2 + v)(2 - v)} = \frac{\frac{1}{4}}{2 + v} + \frac{\frac{1}{4}}{2 - v}$$

so

$$\begin{aligned}\int \frac{dv}{(2 + v)(2 - v)} &= \frac{1}{4}\int \frac{dv}{2 + v} + \frac{1}{4}\int \frac{dv}{2 - v} \\ &= \frac{1}{4}\ln(2 + v) - \frac{1}{4}\ln(2 - v) + c \\ &= \frac{1}{4}\ln\left(\frac{2 + v}{2 - v}\right) + c\end{aligned}$$

Thus $\frac{1}{4}\ln\left(\frac{2 + v}{2 - v}\right) = -8t + c$

So that, upon exponentiation,

$$\frac{2 + v}{2 - v} = ke^{-32t} \qquad \text{for a constant } k$$

Solving for $v(t)$ now yields

$$(2 + v) = 2ke^{-32t} - vke^{-32t}$$

or

$$v(1 + ke^{-32t}) = 2(ke^{-32t} - 1)$$

and

$$v(t) = \frac{2(ke^{-32t} - 1)}{1 + ke^{-32t}} \qquad \text{for } t \geq 0$$

The value of k depends on the initial velocity $v(0) = v_0$. Because we are interested only in finding the free-fall velocity, we take the limit as $t \to +\infty$ in the foregoing expression for $v(t)$. Noting that $e^{-32t} \to 0$ as $t \to +\infty$, we obtain

$$\lim_{t \to +\infty} v(t) = -2$$

Thus the free-fall velocity is equal to -2 feet per second. (The minus sign, as usual, signifies downward motion.)

A second type of application, but one whose solution is also amenable to the separation-of-variables technique, is the following.

EXAMPLE 25 A second-order chemical equation is one in which one molecule of each of two different substances A and B interact to produce one molecule of a new substance X.

A chemist would write the "equation"

$$\mathrm{A} + \mathrm{B} \to \mathrm{X}$$

Letting a and b represent the initial concentration (in moles) of substances A and B, respectively, it can be shown that, if $x(t)$ represents the concentration of X at time t, then the rate of change of $x(t)$ will be jointly proportional to the remaining concentrations of A and B at time t. That is,

$$\frac{dx}{dt} = k(a - x)(b - x)$$

for some positive constant k.

For simplicity, let us take $k = 1$. If we now assume that the reaction begins with 5 moles of substance A, 7 moles of substance B, and 0 moles of substance X, let us find the concentration of $x(t)$ at each subsequent time $t > 0$.

SOLUTION With the values given, our underlying differential equation is

$$\frac{dx}{dt} = (5 - x)(7 - x) \tag{7.24}$$

and the initial condition is

$$x(0) = 0$$

Using separation of variables on Equation (7.24), we obtain

$$\int \frac{dx}{(5 - x)(7 - x)} = \int dt = t + c$$

Now, using partial-fraction decomposition, we obtain

$$\frac{1}{(5 - x)(7 - x)} = \frac{+\frac{1}{2}}{5 - x} + \frac{-\frac{1}{2}}{7 - x}$$

so

$$\begin{aligned}\int \frac{dx}{(5 - x)(7 - x)} &= \frac{1}{2}\int \frac{dx}{5 - x} - \frac{1}{2}\int \frac{dx}{7 - x}, \\ &= -\frac{1}{2}\ln(5 - x) + \frac{1}{2}\ln(7 - x) \\ &= \frac{1}{2}\ln\left(\frac{7 - x}{5 - x}\right)\end{aligned}$$

Thus

$$\frac{1}{2}\ln\left(\frac{7 - x}{5 - x}\right) = t + c$$

from which (upon doubling and exponentiating and relabeling e^{2c} as m) we obtain

$$\frac{7 - x}{5 - x} = me^{2t}, \qquad \text{for } t \geq 0$$

Solving for $x(t)$, we find

$$7 - x = 5me^{2t} - xme^{2t}$$

so

$$x(me^{2t} - 1) = 5me^{2t} - 7$$

and thus

$$x(t) = \frac{5me^{2t} - 7}{me^{2t} - 1} \qquad \text{for } t \geq 0$$

Setting $t = 0$ and noting that $x(0) = 0$ gives

$$0 = \frac{5m - 7}{m - 1}$$

which determines that $m = \frac{7}{5}$, and so, explicitly,

(7.25)
$$x(t) = \frac{7e^{2t} - 7}{\frac{7}{5}e^{2t} - 1} = 35\left(\frac{e^{2t} - 1}{7e^{2t} - 5}\right) \qquad \text{for } t \geq 0$$

We may now find the limiting concentration of $x(t)$ by letting $t \to +\infty$ in Equation (7.25).

We have

(7.26)
$$\begin{aligned} \lim_{t \to +\infty} x(t) &= \lim_{t \to +\infty} 35\left(\frac{e^{2t} - 1}{7e^{2t} - 5}\right) \\ &= \lim_{t \to +\infty} 35\left(\frac{1 - e^{-2t}}{7 - 5e^{-2t}}\right) \\ &= \frac{35}{7} \\ &= 5 \end{aligned}$$

so that the concentration of $x(t)$ tends toward 5 moles. This is really no surprise to a chemist. The reaction will continue until all of the substance of lesser initial concentration (in this case A) disappears. *Mathematically*, this happens at time $t = +\infty$, but for the chemist's purposes the reaction may be "over" quite rapidly because of the negative exponents in Equation (7.26).

In the problems at the end of this section you will encounter a number of other important physical and biological situations in which differential equations arise. Depending on the underlying process, their interpretations may differ, but their mathematical solutions will always be based on the technique of separation of variables.

Section 7.4 EXERCISES

In Exercises 1–15 use the separation-of-variables technique to solve the differential equation with the initial conditions given. You need not solve explicitly for y in terms of x.

1. $\dfrac{dy}{dt} = t^2y^2 + y^2, \quad y(0) = \dfrac{1}{2}$

2. $\dfrac{dy}{dx} = \dfrac{e^y}{e^{2y} + 1}, \quad y(0) = \ln 2$

3. $\dfrac{dy}{dt} = \dfrac{(y^2 - 1)t}{y}, \quad y(0) = \dfrac{\sqrt{3}}{2}$

4. $\dfrac{dy}{dt} = ye^{-t} - 5e^{-t}, \quad y(0) = \pi$

5. $\dfrac{dy}{dx} = (y - 4)^2, \quad y(0) = 1$

6. $(x^2 - 9)\dfrac{dy}{dx} = y^2 - 4, \quad y(0) = 0$

7. $\left(\dfrac{e^x}{e^{y^2}}\right)\dfrac{dy}{dx} = \dfrac{1}{y}, \quad y(0) = -1$

8. $\dfrac{dy}{dt} = \dfrac{3t + 4}{y}, \quad y(0) = 4$

9. $\dfrac{dy}{dt} = \dfrac{(5t + 1)y}{3}, \quad y(0) = 7$

10. $\dfrac{dy}{dt} = \dfrac{1 + 3t}{4y^2t}, \quad y(1) = 0$

11. $\dfrac{dy}{dt} = \dfrac{3t - 4}{\ln y}, \quad y(0) = 1$

12. $\dfrac{dy}{dt} = \dfrac{1 - t^2}{e^y}, \quad y(0) = 1$

13. $\dfrac{dy}{dt} = \dfrac{e^t + 2}{\ln y}, \quad y(0) = e$

14. $\dfrac{dy}{dt} = \dfrac{t + 4}{(1 + y)^2}, \quad y(0) = -1$

15. $\dfrac{dy}{dt} = (t^3 - t)(1 + y)^2, \quad y(0) = 3$

16. In the study of blood plasma it is found that, for a suitable, constant a, the number of charged ions in solution at time t, $N(t)$, satisfies the differential equation

$$\frac{dN}{dt} = a^2 - N^2$$

Write a formula for $N(t)$ in terms of $N(0) = N_0$, the number of charged ions at time $t = 0$.

17. According to Newton's law of cooling, the rate of change of temperature of a body in a surrounding medium is directly proportional to the difference between the body's temperature and that of the surroundings. That is, $dT/dt = k(T - T_s)$, where T_s is the surrounding temperature. A cup of coffee at 160° is placed in a room where the temperature is 70°. After 5 min, the temperature of the coffee is 130°.
   (a) Find the temperature after another 10 min.
   (b) Find the time at which the temperature of the coffee will be 80°.

18. In a chemical reaction where two constituents A and B combine to produce a third constituent X, the rate of formation of X, dx/dt, is jointly proportional to $a - x$ and $b - x$, where a and b are the initial concentrations of A and B, respectively, and x is the concentration of X at time t. That is,

$$\frac{dx}{dt} = k(a - x)(b - x)$$

for some suitable rate constant k. Taking $k = 0.15$, $a = 10$, $b = 8$, and $x(0) = 0$, find the concentration of X produced by time $t = 3$.

19. The rate of change of the population of a certain species of birds is directly proportional to the amount present at any given time but with the proportionality factor $k = e^{-0.2t}$. That is, this factor varies with time in an exponentially decaying manner. Thus,

$$\frac{dp}{dt} = e^{-0.2t}p$$

where t is measured in years. At time $t = 0$ the population of this species is 1200 individuals.
   (a) Find the population after 5 years.
   (b) Find the time it takes for the population to grow to 50,000 individuals.

20. A ship is churning along at a speed of 40 mph when its propeller breaks. If the resistance of the water is proportional (at any instant) to the speed of the ship, how far will the ship coast in 2 min if in $\frac{1}{2}$ mi its speed is reduced to 25 mph?

21. In relativity theory, the mass of a moving object changes with its velocity v. Denoting this mass by $M(v)$, it can be

verified experimentally that

$$\frac{dM}{dv} = M \cdot \frac{v}{c^2 - v^2}$$

where c is the speed of light.
A certain particle has rest mass M_0.

(a) What is the formula for its mass when it is moving at velocity v?

(b) How fast must the particle be moving in order for its mass to be increased by a factor of 2?

22. A variable resistor of x ohms is connected in parallel with a resistor of 10 ohms. The overall resistance of the system, denoted by $R(x)$, varies with x according to the differential equation

$$\frac{dR}{dx} = 1 - \frac{R}{5} + \frac{R^2}{100}$$

When $x = 10$ ohms, $R(10) = 5$ ohms.

(a) Find the equation for $R(x)$ as a function of x.

(b) Find the total resistance R of the system when x is 30 ohms.

(c) What happens to $R(x)$ as $x \to \infty$? As $x \to 0$?

In Exercises 23–28 find the most general solution of the given differential equation.

23. $P\dfrac{dV}{dP} + V = 0$ (Find V as a function of P.)

24. $x\dfrac{dx}{dt} + t\sqrt{1 - x^2} = 0$

25. $\dfrac{dr}{d\theta} = (1 + r^2)\,\theta e^{2\theta}$

26. $\dfrac{dx}{dy} = x \ln x$

27. $\dfrac{dx}{dy} = -\dfrac{9}{4}\dfrac{y}{x}$

28. $\dfrac{dx}{dy} = -\dfrac{9}{4}\dfrac{x}{y}$

SECTION 7.5

Mathematical Modeling

One of the most important uses of the calculus in such fields as biology, sociology, and economics is in the construction of **mathematical models**. Roughly speaking, this is a process in which we attempt to formulate equations whose solution reflects in a mathematical way the significant features of some underlying (physical, sociological, or economic) problem. Generally, our assumptions about the underlying problem are so oversimplified that our first models are not complex enough to incorporate all the important aspects of our problem. The manner in which the actual behavior differs from the resulting mathematical behavior can often suggest ways in which the model should be modified. Eventually, if we obtain a good enough model, it can be useful for predicting future behavior.

In order to illustrate these ideas, we will formulate some mathematical models for two problems arising in different fields. The first is the sociological problem of *social diffusion*. That is, we wish to study the spread in a population of a bit of information, a fad, or a belief or the adoption of some technological innovation. Such problems all have the characteristic that, for each individual in our population, we are able to decide whether he or she does or does not have the property being studied. The second class of problems arises in biology in the study of population growth.

As we shall see, these two problems yield almost identical mathematical equations. This is a common occurrence in applied mathematics. In fact, numerous other problems that we could study give rise to essentially the same models: the spread of an epidemic in a population, the rate at which certain chemical reactions occur, the decay of radioactive substances, and absorption of x-rays by the body.

We will restrict ourselves to problems in which our assumptions yield a single differential equation. In each case the differential equation will be solvable by the method of separation of variables that we studied in Section 7.4.

EXPONENTIAL GROWTH Let us first consider a simple problem in social diffusion. In order to formulate a model, it is necessary for us to identify clearly any underlying assumptions. Simply stated, we will assume the following: *The rate of spread of the property* (which might be a belief, a piece of information, or even a style of clothing) *is proportional to the number of people who already have the property.*

Note that our assumption is simply the statement of a hypothesis about the nature of a phenomenon. The extent to which this hypothesis is correct or not is a question not for the mathematician, whose concern is to investigate the mathematical consequences of the assumption, but for the investigator who is concerned with the property under consideration.

In order to clarify our assumption, let us be a little more specific about our problem. Imagine that the mayor of a large city goes on television and announces his resignation and that N_0 people see his announcement live. Suppose that each person who has heard the news will tell k people about it during each successive hour. Thus in the first hour we expect kN_0 people to hear about the statement by word of mouth. More generally, if we let $N(t) =$ number of people who have heard the news by time t, then $kN(t)$ represents the rate at which the story will be spreading throughout the city. But we may also measure the rate of spread of the story by dN/dt. This results in the differential equation

(7.27) $$\frac{dN}{dt} = kN$$

subject to the initial condition

(7.28) $$N(0) = N_0$$

That is, at the time $t = 0$, N_0 people—or only those who were watching their televisions at that time—know the story. The equations (7.27) and (7.28) represent what is called a **diffusion model**.

The differential equation, Equation (7.27), is one that we have previously encountered on several occasions. We know, therefore, that the solution of it, subject to the additional restriction of Equation (7.28), is given by

(7.29) $$N(t) = N_0 e^{kt}$$

Thus our assumption resulted in a model that produced **exponential growth**. The graph of $N(t)$ is shown in Figure 7.1. Note that we do not include any portion of the graph for $t < 0$, because the news had not yet been announced at such times.

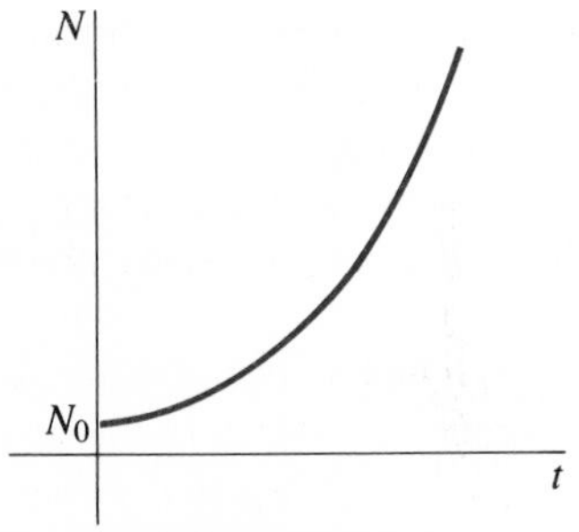

FIGURE 7.1
$N(t) = N_0 e^{kt}$

Of course, we all recognize that it is impossible for the spread of the news really to be exponential. After all, the exponential function gets arbitrarily large as t does, whereas in reality there are only a fixed number of people in the city. It is important to realize, however, that our mathematics was in no way responsible for this. Exponential growth is a logical consequence of our assumption, and we will have to reexamine that assumption if we are to construct a better model.

Before doing that, let us shift our attention to the biological problem of population growth. Suppose that, in a large population, *the birth rate of each individual per unit time is b, whereas the mortality (or death) rate of each individual per unit time is m, where b and m are constants* ($b > m$).

Under this assumption, $k = b - m$ represents the net addition to the population of each individual per unit time. For example, we might be studying the growth, in a culture, of single-celled organisms that reproduce by fission. Perhaps we expect each organism to divide and reproduce once every 2 hours, and the average life expectancy of any one particular cell is 4 hours. Then $b = \frac{1}{2}$ per hour, and $m = \frac{1}{4}$ per hour. Imagine that, at time $t = 0$, we place N_0 cells in the culture. Then, during the first hour, we expect the population to increase by kN_0 individuals. More generally, if we let $N(t) =$ number of individuals in the population at time t, then dN/dt, the rate of growth of the population, must satisfy

$$\frac{dN}{dt} = kN$$

$$N(0) = N_0$$

which is precisely the same model as the one obtained in Equations (7.27) and (7.28) for the diffusion problem.

Although exponential growth does describe very closely the population size of certain single-celled organisms, at least for a while, in actuality we do not expect the size of our population to increase indefinitely. Eventually, due to physical limitations of space or diminishing food supply, we expect the growth rate to slow down. Again it is necessary for us to question the validity of our assumption.

THE LOGISTIC CURVE Let us return to the problem of the diffusion, in a large city, of the news of the mayor's resignation. For definiteness, we suppose that the total population of the city is A and that $N_0 = \frac{1}{10}A = 0.1A$. That is, 10% of the population learns of the news directly from the mayor's announcement on television. As before, $N(t)$ will denote the number of people at time t who have heard the news.

In our previous model we assumed that each person who heard the news would tell k people about it during each successive hour. In reality, this does not seem plausible. Initially, when only a small fraction of the population knows of the event, each person who has heard the news will probably tell it to many people. At later times (when almost everyone will know about the resignation), it will be difficult to find *anyone* to tell who does not already know.

There are many ways to express the relationship described in the preceding paragraph. The simplest is to replace the constant k with a linear expression of the form

$$k = l[A - N(t)]. \tag{7.30}$$

Here l is a constant and $(A - N(t))$ represents the number of individuals in the population at time t who have not yet heard of the mayor's resignation. Observe that, if almost everyone knows, then $N(t)$ is almost A, so that $k = l[A - N(t)]$ is small. If $N(t)$ is small, on the other hand, then k is large and the news will be spread more rapidly by each individual.

We may now replace k in Equation (7.27) with its counterpart in Equation (7.30) to obtain

$$\frac{dN}{dt} = l[A - N(t)]N(t) \tag{7.31}$$

subject to the initial condition

$$N(0) = 0.1A \tag{7.32}$$

We may summarize the assumptions that resulted in this model as follows: *The rate of spread of the property is proportional to the number of people who have the property and also to the number of people who do not have it.*

Equation (7.31) is one that we encountered in the practice exercise following Example 22. There we used separation of variables to obtain the solution

$$N(t) = \frac{A}{1 + 9e^{-Alt}} \tag{7.33}$$

Having solved the initial value problem, Equations (7.31) and (7.32), the question arises whether this solution $N(t)$ is really a true representation of the phenomenon we are studying. What does its graph look like? In order to answer these questions, let us employ our old curve-sketching techniques. By rewriting Equation (7.33) in the form

$$N(t) = A(1 + 9e^{-Alt})^{-1}$$

we can readily compute the derivative

$$N'(t) = \frac{9A^2le^{-Alt}}{(1 + 9e^{-Alt})^2}$$

From the fact that the exponential function is always positive, we can see that both the numerator and denominator of $N'(t)$ are always positive. Thus $N(t)$ is increasing on $[0, \infty)$. We already know that $N(0) = 0.1A$, which is the y-intercept of the curve. By employing the quotient rule, we can compute $N''(t)$.

$$N''(t) = \frac{(1 + 9e^{-Alt})^2(-9A^3l^2)e^{-Alt} - 9A^2le^{-Alt}[2(1 + 9e^{-Alt})(-9Ale^{-Alt})]}{(1 + 9e^{-Alt})^4}$$

$$= \frac{9A^3l^2(-e^{-Alt} + 9e^{-2Alt})}{(1 + 9e^{-Alt})^3}$$

Although this computation was somewhat involved, from it we can see that

$$N'' = 0 \quad \text{when} \quad -e^{-Alt} + 9e^{-2Alt} = 0$$

Multiplying both sides of this equation by e^{+2Alt}, we find

$$-e^{Alt} + 9 = 0$$

or

$$e^{Alt} = 9$$

Taking the natural logarithm of both sides, we find that

$$t = \frac{\ln 9}{Al} \tag{7.34}$$

In a similar manner, we find that $N'' > 0$ if

$$-e^{-Alt} + 9e^{-2Alt} > 0$$

which happens when

$$0 < t < \frac{\ln 9}{Al}$$

Thus N is concave up on $\left(0, \frac{\ln 9}{Al}\right)$. In exactly the same way, we find that N is concave down on $\left(\frac{\ln 9}{Al}, \infty\right)$. The point $t = \frac{\ln 9}{Al}$ of Equation (7.34) is therefore a point of inflection. From Equation (7.33) we see that $N = A/2$ at this point.

Finally we examine $N(t)$ for asymptotes. There are no vertical asymptotes because the denominator of N never vanishes. However, because

$$\lim_{t\to\infty} e^{-Alt} = \lim_{t\to\infty} \frac{1}{e^{Alt}} = 0$$

we have

$$\lim_{t\to\infty} N(t) = \lim_{t\to\infty} \frac{A}{1 + 9e^{-Alt}} = A$$

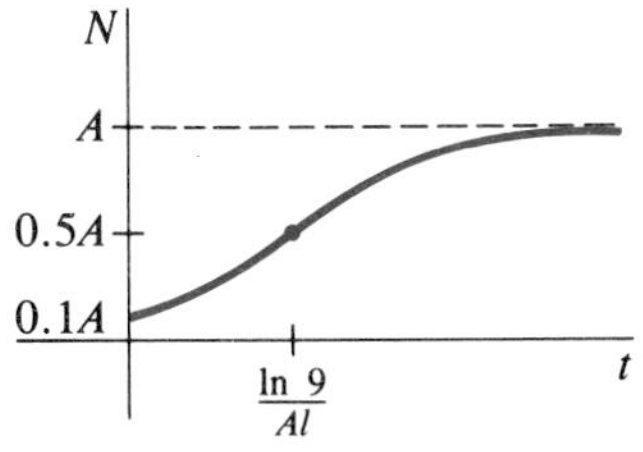

FIGURE 7.2
The logistic curve

Thus N has a horizontal asymptote at $N = A$. Putting these characteristics together, we obtain the curve shown in Figure 7.2, which is known as the **logistic curve**. The horizontal asymptote at A indicates that, if we just wait long enough, nearly everyone will have heard the news of the mayor's resignation.

We now turn our attention once again to the problem of population growth. Let us examine in more detail the plausibility of our earlier assumption. As we indicated before, if the resources of the community are limited, it appears likely that, as the population increases, so will the death rate. For example, there might not be enough food to go around. The simplest way to express such a relationship mathematically is to assume that the mortality rate m is proportional to the size of the population N. In summary, we assume that, in a large population, *the birth rate of each individual per unit time is b (a constant), whereas the mortality rate of each individual per unit time is proportional to the present size of the population.*

If we let c denote the proportionality constant of the preceding sentence, we have assumed that

$$m = cN(t)$$

Thus, if we put

$$k = b - m = b - cN(t)$$

then k represents the net addition to the population of each individual per unit time. Our differential equation now becomes

$$\frac{dN}{dt} = kN(t) = [b - cN(t)]N(t) \tag{7.35}$$

This may be rewritten as

$$\frac{dN}{dt} = c\left[\frac{b}{c} - N(t)\right]N(t)$$

If we now let $c = l$ and $b/c = A$, we find that this is exactly Equation (7.31), which we know has the logistic curve as its solution. Note that Equation (7.35) was arrived at by a very much different line of reasoning from the one we employed in the social diffusion problem. Nevertheless, the resulting mathematical models are the same.

You may be interested to know that there is a very large body of scientific literature giving examples of the close correspondence between growth of laboratory populations and the theoretical logistic curve shown in Figure 7.2.* As one example, you might consult E. P. Odum, *Fundamentals of Ecology*, 3d ed. (Philadelphia: Saunders, 1971) for data on the growth of yeast in a culture.

The logistic curve also arises in many other applications that we have chosen not to discuss in detail. For instance, it accurately models autocatalytic chemical reactions. These are reactions in which one or more of the reaction products catalyze further reactions. It also describes the rate of the spread of certain diseases.

CONSTANT SOURCE DIFFUSION As a final example of the modeling of a problem in social diffusion, we consider a situation that has no natural analog in any problem of biological population growth. In order to motivate the model, consider the problem of a company that is trying to market a new detergent, "Cleano." The company decides to do all of its advertising on television. For the sake of the example, we may suppose that, if a commercial is run during prime time, it reaches 10% of the population. Unlike the problem of the mayor's resignation, which many people are likely to talk about, it seems unlikely that news of Cleano's introduction will be spread by word of mouth. It seems much more likely that people will learn of it only through commercials.

---

* The exact shape of the graph depends on the initial value, $N(0)$. For example, if $N(0) > A/2$, there is no inflection point.

Now the next time the commercial is run in prime time, it will again reach 10% of the population. However, some of these people will already have heard about Cleano from the preceding commercial. Thus the ad will inform 10% of those people who have not heard of it previously. Let us now state our assumptions more precisely.

In a city of population A, we assume that diffusion of a piece of information spreads from a constant source, independent of the number of people who know the information. We further assume that *the number of people who hear of the news is proportional* (*with proportionality constant k*) *to the number who do not yet know it.*

(It is worthwhile at this stage to remark that our motivating example about the introduction of Cleano does not quite fit into the framework of the preceding assumption, because TV commercials are not run continuously—even though it seems that they are—but rather occur at fixed time intervals. We will sidestep this issue by saying simply that we are trying to formulate a model based on this assumption and that our example was intended to convince you of the plausibility of such an assumption.)

Having written down our assumptions, we must now translate them into an equation. As before, let $N(t)$ = number of people at time t who know the information being spread. Then $A - N(t)$ denotes the number of people who do not yet know. Our assumptions state that the rate of spread of the information, dN/dt, is proportional to $A - N(t)$. That is,

(7.36)
$$\frac{dN}{dt} = k[A - N(t)]$$

A reasonable initial condition is

(7.37)
$$N(0) = 0$$

(Thus, at time $t = 0$, before we start advertising, no one knows about Cleano.)

We have already solved Equation (7.36) with the initial condition given in Equation (7.37) by separation of variables (see Example 23). For convenience, we recall the solution:

$$N(t) = A - Ae^{-kt} = A(1 - e^{-kt})$$

What does the graph of this equation look like? We again employ our curve-sketching techniques of Chapter 3. Note first that

$$N'(t) = Ake^{-kt}$$

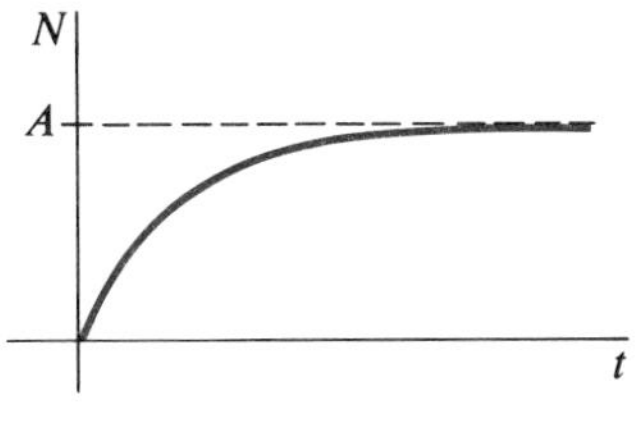

FIGURE 7.3
$N(t) = A(1 - e^{-kt})$

This function is always positive, so $N(t)$ is increasing on $[0, \infty)$. In addition,

$$N'' = -k^2Ae^{-kt}$$

is always negative, which tells us that N is concave down on $[0, \infty)$. We know that $N(0) = 0$ from Equation (7.37). Finally,

$$\lim_{t\to\infty} N(t) = \lim_{t\to\infty} A(1 - e^{-kt})$$
$$= \lim_{t\to\infty} A\left(1 - \frac{1}{e^{kt}}\right) = A$$

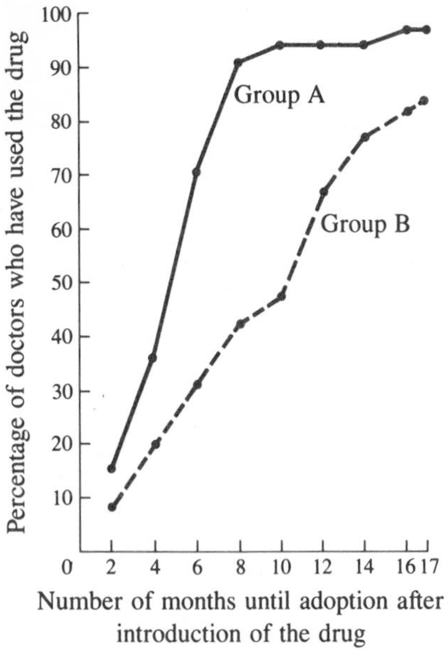

FIGURE 7.4

so that the curve has a horizontal asymptote at $N = A$. This says that the advertising will "saturate" the population as time goes on. Figure 7.3 shows the graph of $N(t)$. It may be remarked that certain chemical reactions that are supplied with a constant external heat source also result in equations of the same form as Equations (7.36) and (7.37).

Are the mathematical models we have developed in this chapter simply interesting academic exercises, or can they really help us gain some insight into the problems they are supposed to describe? The answer is that they can, and the final example we will now discuss should help convince you of this fact. The problem we will examine, as well as the data we will use, is taken from an article by James Coleman, Elihu Katz, and Herbert Menzel, entitled "The Diffusion of on Innovation among Physicians," which appeared in the journal *Sociometry*, Vol. 20 (1957), pages 253–270.

This paper studied the rate at which a new drug was adopted for use by a group of physicians practicing in the Midwest. The decision to adopt a new drug or not is generally made on the basis of a number of different factors. In examining the data concerning the adoption of one particular drug, it was discovered that one factor that affected the decision to adopt was how "socially integrated" the doctor was within the medical community. That is, did he tend to speak frequently with his colleagues about medical problems and techniques, or did he tend to work alone? Of course, the classification of a doctor as being "socially integrated" or "isolated" is not always easy to make, but the investigators made the distinction on the basis of interviews in which each doctor was asked a series of sociometric questions.

Let Group A consist of the "integrated" physicians and Group B consist of the "isolated" physicians. Figure 7.4 depicts the adoption rates found for the two groups. The vertical axis represents the cumulative percentage of doctors who have adopted the drug, and the horizontal axis represents the time, in months, until adoption.

The investigators sought to explain the difference in shape between the two curves plotted in Figure 7.4. One possible explanation is that the integrated doctors are more apt to learn of a new drug through conversation with their colleagues (a logistic growth model), whereas isolated physicians would rely chiefly on journal advertisements, drug salesmen, and other media (a constant source diffusion model).

This is not the only explanation possible, however. A second possible explanation that was considered is that the doctors in the two groups differ by nature in their receptivity to new developments in medicine, and that this difference in personality accounts for the difference in the graphs. Here is how the researchers chose between these alternatives.

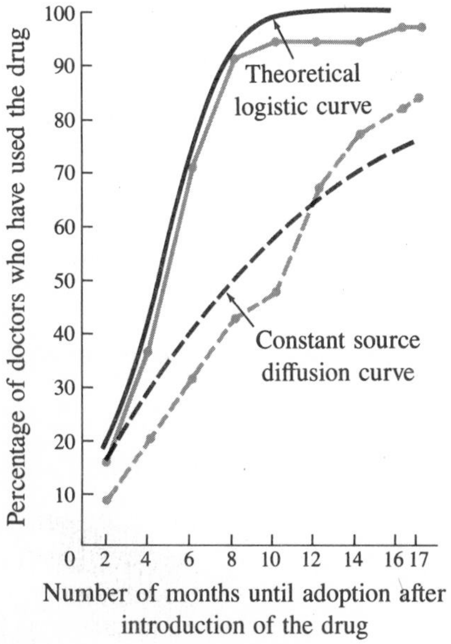

FIGURE 7.5

Figure 7.5 shows (for appropriate choices of the constants) theoretical curves that fit the data well in the view of the researchers. The curve for Group A is well approximated by a logistic curve (like that in Figure 7.2). The curve for Group B is well approximated by a curve representing constant source diffusion (like that in Figure 7.3). This indicated to the investigators that the first explanation is more likely: doctors in Group A spread news of the new drug by word of mouth, while those in Group B rely almost exclusively on advertising. Had the second explanation been operative, both curves would have been logistic with different values of the parameters.

Section 7.5 EXERCISES

1. A 200-gal water tank is full, but unfortunately there are 40 lb of salt dissolved in it. Suppose that pure water is pumped into the tank at the rate of 3 gal/min and that the tank water (in which, it is assumed, the salt is uniformly distributed) is pumped out at the same rate.
   (a) Find an expression for the amount of salt in the tank at time t.
   (b) How much salt is left after $\frac{1}{2}$ hr?
   (c) When will there be 15 lb of salt left?
2. Do Exercise 1 if pure water is pumped in at the rate of 4 gal/min and water from the tank is pumped out at the rate of 3 gal/min. (Assume here that the tank has infinite capacity.)
3. Do Exercise 1 if pure water is pumped in at the rate of 3 gal/min, and the tank water is pumped out at the rate of 4 gal/min.
4. Let $Q = Q(t)$ denote the amount of a radioactive substance present at time t, and let $Q_0 = Q(0)$ denote the amount present at $t = 0$. Let H, called the **half-life**, denote the time it takes for half of the substance to decay. If H equals 756 years, how long does it take before only 25% of the substance remains? When does $\frac{1}{8}$ of the original substance remain?
5. (See Exercise 4.) Show that the half-life H is independent of Q_0, the amount present at the start.
6. As you know the amount of radioactive substance present (with notation as in Exercise 4) is given by $Q(t) = Q_0e^{-kt}$ for some positive constant k. Show that, if the half-life H is known, we can find Q via the formula
$$Q(t) = Q_0 2^{-t/H}$$
7. Start with 200 g of radioactive material. After 3 years there are 50 g left. How much is left after 6 years? After 10 years? When is there only 5% of the original material left?
8. At time $t = 12$ hours we have 75 g of radioactive material, and at time $t = 15$ hours we have 72 g. How much was there at the start (at $t = 0$)?
9. A water tank contains 200 gal of water in which 20 lb of salt have been dissolved. Water containing $\frac{1}{2}$ lb of salt per gal is pumped into the tank at the rate of 4 gal/min, and the well-stirred mixture is pumped out of the tank at the rate of 4 gal/min. (Compare Example 36 in Chapter 4.)
   (a) Find an expression for the amount of salt in the tank at time t.
   (b) How much salt will there be in 1 hr?
   (c) When will there be 50 lb of salt in solution?
10. During the 1849 gold rush the population of a California gold-mining town grew at a rate proportional to the current population. If the population in 1849 was 1000 and in 1851 it was 10,000, what was the population in 1854? In 1857?
11. (See Exercise 10.) At the same time, the population in a prairie town decreased at a rate proportional to the population. If the population was 20,000 in 1849 and 16,000 in 1851, what was the population in 1854? In 1857? When could this town be called a ghost town?

For Exercises 12–14 the rate of increase in amount of an investment earning continuously compounded interest is proportional to the amount.

12. If you invest \$200 at 5%, how much will you have after 5 years?
13. If your mother owed a fine of \$1 for a library book due 30 years ago and the interest compounded continuously at 5% must also be paid, what would she have to pay now?
14. How much should be deposited in a savings account earning 9% compounded continuously so that a newborn child may withdraw \$80,000 for college at age 18?

SECTION 7.6

Numerical Integration

In Section 5.2 we found that we could approximate the area under the graph of a continuous nonnegative function $f(x)$ as a sum of rectangles, using either the right-hand or left-hand rule. Actually, both of these rules usually require a large number of rectangles to achieve more than two or three significant digits of accuracy. This in itself is not a tremendous problem for today's high-speed digital computers, but the accumulation of round-off errors due to the large number of computations is.

In this section we will look at two alternative methods for approximating a definite integral. Both of these generally give better accuracy with fewer subintervals. Under certain conditions, we shall be able to tell just how accurate these approximations are for a specific function.

You may be wondering why we need numerical integration techniques at all, inasmuch as the fundamental theorem reduces the evaluation of a definite integral to finding an antiderivative. That is true, but for many functions, finding an antiderivative may be exceedingly difficult. And for others (including the function e^{-x^2}, which plays a vital role in statistics), there are no known antiderivatives expressible in terms of any of the functions studied in this text.

THE TRAPEZOIDAL RULE Suppose we wish to estimate $\int_a^b f(x)\,dx$, where f is continuous and nonnegative on the interval $[a, b]$. We partition $[a, b]$ into n equal subintervals of length $\Delta x = (b - a)/n$ by the points

$$a = x_0, x_1, x_2, x_3, \ldots, x_n = b$$

where $x_i - x_{i-1} = \Delta x$ for $i = 1, 2, \ldots, n$. For each i we shall denote $f(x_i)$ by y_i; see Figure 7.6(a). The idea behind the trapezoidal rule is to abandon our reliance on approximating rectangles and instead to make use of trapezoids.

To approximate the area A_i of the typical subregion shown in Figure 7.6(b), we draw the line segment connecting the points (x_{i-1}, y_{i-1}) and (x_i, y_i) and form a trapezoid (Figure 7.7), whose area by Equation (5.4), is

$$A_i = (y_{i-1} + y_i)\,\Delta x/2$$

If we now construct such a trapezoid for each i from 1 to n and sum their areas, we have

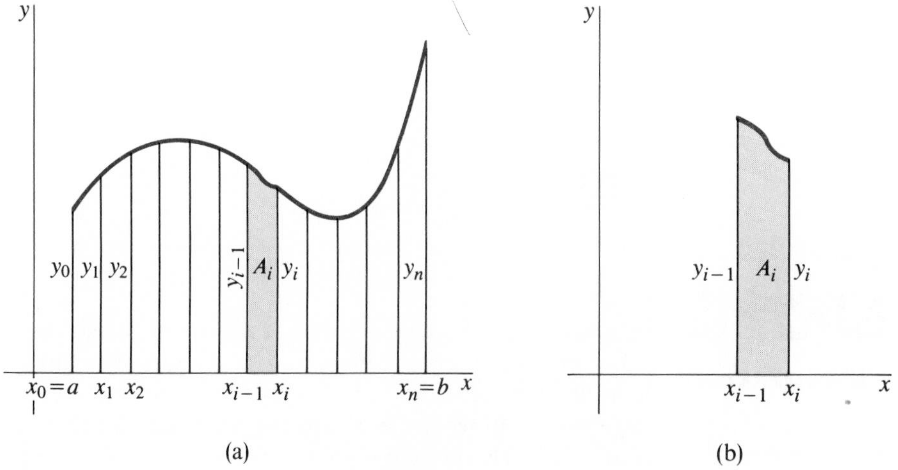

FIGURE 7.6

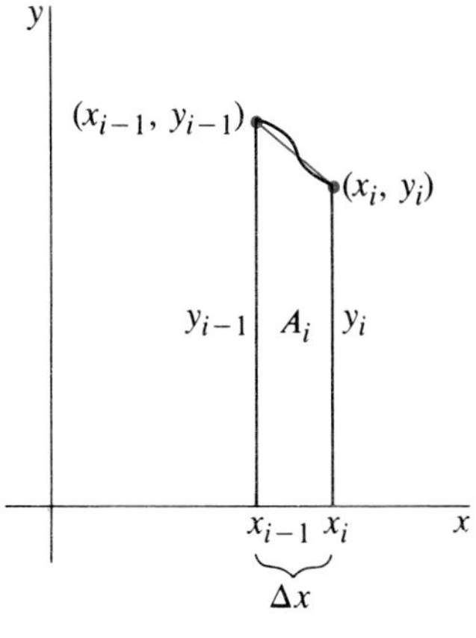

FIGURE 7.7

$$A_1 + A_2 + \cdots + A_n \approx \underbrace{(y_0 + y_1)\,\Delta x/2}_{\text{1st trapezoid}} + \underbrace{(y_1 + y_2)\,\Delta x/2}_{\text{2nd trapezoid}} + \underbrace{(y_2 + y_3)\,\Delta x/2}_{\text{3rd trapezoid}} + \cdots + \underbrace{(y_{n-1} + y_n)\,\Delta x/2}_{n\text{th trapezoid}}$$

$$= \frac{\Delta x}{2}[(y_0 + y_1) + (y_1 + y_2) + (y_2 + y_3) + \cdots + (y_{n-1} + y_n)]$$

$$= \frac{\Delta x}{2}[y_0 + 2y_1 + 2y_2 + 2y_3 + \cdots + 2y_{n-1} + y_n]$$

$$= \Delta x\left[\frac{y_0}{2} + y_1 + y_2 + y_3 + \cdots + y_{n-1} + \frac{y_n}{2}\right]$$

Note that, in the last expression, each y except the first and last appears with coefficient 1. Although we have assumed up to this point that f is nonnegative, this assumption is not necessary: The trapezoidal rule can be used to approximate an integral of any continuous function. We can now sum up our discussion.

THE TRAPEZOIDAL RULE Let f be a continuous function on the interval $[a, b]$, and let

$$a = x_0, x_1, x_2, x_3, \ldots, x_n = b$$

be a partition of $[a, b]$ into n equal subintervals of width $\Delta x = (b - a)/n$. Let $y_i = f(x_i)$ for $i = 0, 1, 2, \ldots, n$. Then

(7.38) $$\int_a^b f(x)\,dx \approx \Delta x\left[\frac{y_0}{2} + y_1 + y_2 + y_3 + \cdots + y_{n-1} + \frac{y_n}{2}\right]$$ ⊟

Figure 7.8 illustrates the use of this rule for $f(x) = 3x^2$ on $[0, 4]$ with $n = 4$. Note how the polygonal line formed by the trapezoid tops hugs the graph. (See the practice exercise that follows.) Let's try a few examples.

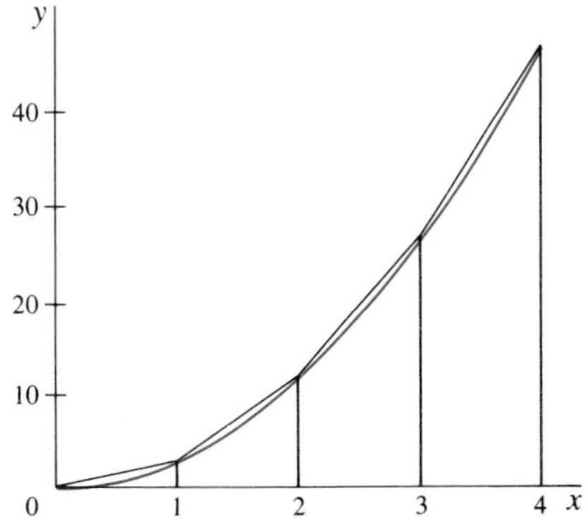

$y = 3x^2$
FIGURE 7.8

EXAMPLE 26 Use the Trapezoidal Rule with $n = 8$ to approximate $\int_0^4 3x^2\,dx$.

SOLUTION Here $\Delta x = (b - a)/n = (4 - 0)/8 = 0.5$. The values of x_i and $y_i = f(x_i)$ are given in Table 7.1.

TABLE 7.1

i	x_i	$y_i = 3x_i^2$
0	0	0.00
1	0.5	0.75
2	1.0	3.00
3	1.5	6.75
4	2.0	12.00
5	2.5	18.75
6	3.0	27.00
7	3.5	36.75
8	4.0	48.00

Substituting these values into Equation (7.38) yields

$$\int_0^4 3x^2\,dx \approx 0.5\left[\frac{1}{2}(0) + 0.75 + 3.00 + 6.75 + 12.00 + 18.75 + 27.00 + 36.75 + \frac{1}{2}(48)\right] = 64.5$$

This is within 1% of the exact area, which is 64. Yet the amount of computation is roughly the same as was required by the right-hand and left-hand rules. (Compare Table 7.1 with Table 5.2.) We have increased our accuracy without substantially increasing the computation required.

PRACTICE EXERCISE Use the trapezoidal rule with $n = 4$ to approximate $\int_0^4 3x^2\,dx$ (Figure 7.8).

[Your answer should be 66.]

There is an alternative method for deriving the trapezoidal rule. In Section 5.2 we used both the left-hand and right-hand rules to approximate several definite integrals. You may have noticed that, for those functions for which the right-hand rule produced an approximation that was too large, the left-hand rule produced one that was too small, and vice versa. Accordingly, one might be led to use the average of the left-hand rule and the right-hand rule as a numerical approximation technique.

Left-hand rule: $$\int_a^b f(x)\,dx \approx \left(\frac{b-a}{n}\right)[f(x_0) + f(x_1) + \cdots + f(x_{n-1})]$$

Right-hand rule: $$\int_a^b f(x)\,dx \approx \left(\frac{b-a}{n}\right)[f(x_1) + f(x_2) + \cdots + f(x_n)]$$

$$\text{Average} = \frac{\text{left-hand rule} + \text{right-hand rule}}{2}$$
$$= \left(\frac{b-a}{n}\right)\left[\frac{f(x_0)}{2} + f(x_1) + \cdots + f(x_{n-1}) + \frac{f(x_n)}{2}\right]$$

and this formula is exactly the trapezoidal rule given in Equation (7.38).

PRACTICE EXERCISE Use the trapezoidal rule with $n = 4$ to approximate $\int_0^1 \frac{1}{1+x^2}\,dx$.

[Your answer should be ≈ 0.783. The exact value of this integral is $\pi/4 \approx 0.7854$.]

As a final example of the trapezoidal rule, let's consider the problem of estimating $\int_a^b e^{-x^2/2}\,dx$ for various values of a and b. Because there is no known antiderivative for $e^{-x^2/2}$ in terms of elementary functions, there is no other path to follow than numerical integration for evaluating the definite integral of $f(x) = e^{-x^2/2}$!

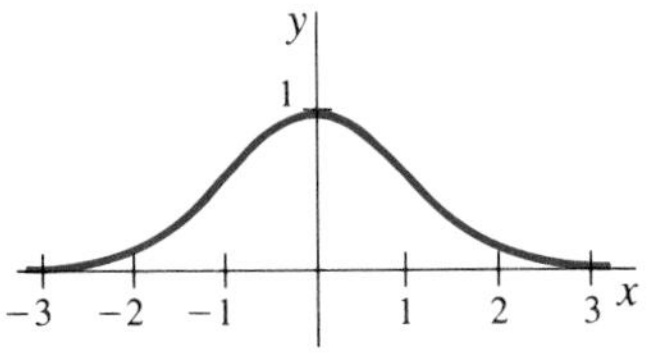

FIGURE 7.9
$y = e^{-x^2/2}$

The particular function $e^{-x^2/2}$ is graphed in Figure 7.9. (The graph was sketched by using our curve-sketching techniques from Chapter 3.) It is a very important function in statistics. Essentially, it is the famous bell-shaped curve you refer to when you ask whether your instructor "marks on a curve." More precisely, the curve

$$y = \frac{1}{\sqrt{2\pi}} e^{-x^2/2}$$

(obtained from $y = e^{-x^2/2}$ simply by multiplying by a constant $1/\sqrt{2\pi}$) is called the **density function for the standard normal distribution**. It has applications in every field where statistics are used. It is sufficiently important, in fact, that there are hand-books that tabulate

$$\int_a^b \frac{1}{\sqrt{2\pi}} e^{-x^2/2}\,dx$$

for various values of a and b. These numbers represent certain probabilities whose significance we will discuss in Chapter 10. For now, let us content ourselves with using the trapezoidal rule to approximate the values of $\int_a^b e^{-x^2/2}\,dx$ for a few choices of a and b.

EXAMPLE 27 Estimate $\int_0^3 e^{-x^2/2}\,dx$ by using the trapezoidal rule with $n = 6$.

SOLUTION In this example,

$$\Delta x = \frac{b-a}{n} = \frac{3-0}{6} = 0.5$$

In Table 7.2 we have tabulated the values of $e^{-x^2/2}$ (to 2 decimal places) for $x_0 = 0$, $x_1 = 0.5$, $x_2 = 1$, $x_3 = 1.5$, and so on. These values are found by consulting a table of values of the exponential function or by using a calculator.

TABLE 7.2

i	x_i	$y_i = e^{-x_i^2/2}$
0	0	1.00
1	0.5	0.88
2	1	0.61
3	1.5	0.32
4	2.0	0.14
5	2.5	0.04
6	3.0	0.01

From Equation (7.38) we have

$$\int_0^3 e^{-x^2/2}\,dx \approx 0.5\left[\frac{1}{2}(1.00) + 0.88 + 0.61 + 0.32 + 0.14 + 0.04 + \frac{1}{2}(0.01)\right]$$
$$= 1.25$$

(This is correct to three significant digits.)

EXAMPLE 28 Estimate $\int_{-3}^{3} e^{-x^2/2}\,dx$.

SOLUTION A glance at the graph of $y = e^{-x^2/2}$ in Figure 7.9 reveals that it is symmetric about the y-axis. That means, for instance, that

$$\int_{-3}^{0} e^{-x^2/2}\,dx = \int_0^3 e^{-x^2/2}\,dx$$

From our calculation in the previous example, we now have

$$\int_{-3}^{3} e^{-x^2/2}\,dx = \int_{-3}^{0} e^{-x^2/2}\,dx + \int_0^3 e^{-x^2/2}\,dx$$
$$\approx 1.25 + 1.25 = 2.50$$

PRACTICE EXERCISE

(a) Use the trapezoidal rule with $n = 2$, together with Table 7.2, to estimate $\int_0^1 e^{-x^2/2}\,dx$.

(b) Use the trapezoidal rule with $n = 4$ to estimate $\int_0^2 e^{-x^2/2}\,dx$.

(c) By symmetry, approximate $\int_{-1}^{1} e^{-x^2/2}\,dx$ and $\int_{-2}^{2} e^{-x^2/2}\,dx$.

[Your answers should be (a) 0.84; (b) 1.19; (c) 1.68; and (d) 2.38.]

SIMPSON'S RULE In using the trapezoidal rule, we were replacing the graph of our function f by a polygonal curve—that is, by a curve made up of straight line segments (the tops of the trapezoids). In Simpson's rule, the portion of the graph over a pair of adjacent subintervals is approximated by a parabolic arc. The derivation of this rule is more complex than that of the trapezoidal rule, so we shall simply state the rule

SIMPSON'S RULE Let f be a continuous function on the interval $[a, b]$, and let

$$a = x_0, x_1, x_2, x_3, \ldots, x_n = b$$

be a partition of $[a, b]$ into n equal subintervals of width $\Delta x = (b - a)/n$, where n is even. Let $y_i = f(x_i)$ for $i = 0, 1, 2, \ldots, n$. Then

(7.39)
$$\int_a^b f(x)\,dx \approx \frac{\Delta x}{3}[y_0 + 4y_1 + 2y_2 + 4y_3 + 2y_4 + \cdots + 2y_{n-2} + 4y_{n-1} + y_n]$$

Except for the first and last terms, the coefficients alternate 4, 2, 4, 2, 4, 2, ... Let us try this rule on a few familiar examples.

EXAMPLE 29 Use Simpson's rule with $n = 8$ to approximate $\int_0^4 3x^2\,dx$.

SOLUTION As in Example 26, $n = 8$ and $\Delta x = 0.5$. Using the values from Table 7.1, we have

$$\int_0^4 3x^2\,dx \approx (0.5/3)[0.00 + 4(0.75) + 2(3.00) + 4(6.75) + 2(12.00) + 4(18.75) + 2(27.00) + 4(36.75) + 48.00] = 64.00$$

which is exact (as was to be expected because $3x^2$ is a quadratic).

Simpson's rule, which involves parabolic approximations, is exact for any even n for any quadratic function and, as it turns out, for any cubic as well.

EXAMPLE 30 Use Simpson's rule with $n = 4$ to approximate $\int_0^3 e^{-x^2/2}\,dx$.

SOLUTION $\Delta x = 3/4 = 0.75$. The required y-values are given in Table 7.3

TABLE 7.3

i	x_i	$y_i = e^{-x_i^2/2}$
0	0.00	1.0000
1	0.75	0.7548
2	1.50	0.3247
3	2.25	0.0796
4	3.00	0.0111

Substituting these values into Equation (7.39) yields

$$\int_0^4 e^{-x^2/2}\,dx \approx (0.75/3)[1.000 + 4(0.7548) + 2(0.3247) + 4(0.0796) + 0.0111] \approx 1.250$$

which is correct to four significant digits.

HOW ACCURATE ARE THE RULES IN GENERAL? If we know enough about the function f and its derivatives, it is possible to calculate upper bounds on the errors involved in using these rules to approximate $\int_a^b f(x)\,dx$. Specifically, we have the following two theorems, whose proofs can be found in *Interface: Calculus and the Computer*, David A. Smith, 2nd ed., Philadelphia: Saunders College Publishing, 1984.

THEOREM 7.1 Let f be a function with a continuous second derivative f'' on the interval $[a, b]$, and suppose that K is a number satisfying

$$|f''(x)| \le K \qquad \text{for } a \le x \le b$$

Then the error E_n in approximating $\int_a^b f(x)\,dx$ by the trapezoidal rule with n intervals satisfies

$$|E_n| \le \frac{K(b-a)^3}{12n^2}$$ ⊟

THEOREM 7.2 Let f be a function with a continuous fourth derivative $f^{(4)}$ on the interval $[a, b]$, and suppose that K is a number satisfying

$$|f^{(4)}(x)| \le K \qquad \text{for } a \le x \le b$$

Then the error E_n in approximating $\int_a^b f(x)\,dx$ by Simpson's rule with n (even) intervals satisfies

$$|E_n| \le \frac{K(b-a)^5}{180n^4}$$ ⊟

How do we use these theorems? Well, suppose we want to compute an approximation to $\int_a^b f(x)$ that is within ϵ of the exact integral, where ϵ is typically some small number, such as 0.0001 or 0.00001. If we are using the trapezoidal rule, then, by Theorem 7.1, we need only choose n large enough to make

$$\frac{K(b-a)^3}{12n^2} \le \epsilon$$

which is equivalent to choosing

$$n \ge \sqrt{\frac{K(b-a)^3}{12\epsilon}}$$

EXAMPLE 31 What value of n will ensure that the trapezoidal rule approximation of $\int_0^2 x^4\,dx$ will be within 0.0001 of the exact value?

SOLUTION Let $f(x) = x^4$. Then $f'(x) = 4x^3$ and $f''(x) = 12x^2$. For $0 \le x \le 2$, $|f''(x)| \le 48$; hence we may take $K = 48$ as an upper bound for $|f''(x)|$ on $[0, 2]$. The desired degree of accuracy will be achieved if we choose n

to be an integer satisfying

$$n \geq \sqrt{\frac{48(2^3)}{12(0.0001)}} \approx 565.7$$

Accordingly (round up!), we need to use at least 566 subintervals.

Similar reasoning applies to Simpson's rule. To achieve accuracy ϵ, we need only choose n large enough to make

$$\frac{K(b-a)^5}{180n^4} \leq \epsilon$$

which is equivalent to choosing

$$n \geq \left[\frac{K(b-a)^5}{180\epsilon}\right]^{1/4}$$

PRACTICE EXERCISE What value of n will ensure that the Simpson's rule approximation of $\int_0^2 x^4\,dx$ will be within 0.0001 of the exact value?

[Here $f^{(4)} = 24$, a constant. Your answer should be $n \geq 16$ subintervals (even!).]

The following exercises include computer programs in both BASIC and Pascal that can be used to carry out the computations of the trapezoidal rule and Simpson's rule.

Section 7.6 EXERCISES

Approximate each of the definite integrals in Exercises 1–18 by using the trapezoidal rule with (a) $n = 4$; (b) $n = 8$.

1. $\int_1^5 x\,dx$
2. $\int_1^5 (2x + 3)\,dx$
3. $\int_0^2 x^2\,dx$
4. $\int_0^4 (x^2 + 1)\,dx$
5. $\int_1^2 \frac{1}{x}\,dx$
6. $\int_1^3 \frac{1}{x^2}\,dx$
7. $\int_0^1 \frac{1}{1+x}\,dx$
8. $\int_1^5 x^3\,dx$
9. $\int_0^4 x^4\,dx$
10. $\int_{-1}^1 \frac{1}{1+x^4}\,dx$
11. $\int_0^3 e^{-x^2/2}\,dx$
12. $\int_0^4 \sqrt{x}\,dx$
13. $\int_0^1 e^x\,dx$
14. $\int_0^1 e^{-x}\,dx$
15. $\int_1^2 \left(x + \frac{4}{x}\right)dx$
16. $\int_1^5 \ln x\,dx$
17. $\int_0^8 xe^x\,dx$
18. $\int_2^6 \frac{x}{1+x^2}\,dx$

Approximate each of the definite integrals in Exercises 19–26 by using Simpson's rule with (a) $n = 4$ and (b) $n = 8$.

19. $\int_0^4 x\,dx$
20. $\int_0^2 (2x + 3)\,dx$

21. $\int_0^2 (6 - x^2)\,dx$

22. $\int_0^4 (x^2 + x + 1)\,dx$

23. $\int_1^5 \frac{1}{x^2}\,dx$

24. $\int_0^1 \frac{dx}{1 + x^2}$

25. $\int_2^6 \frac{e^x}{x}$

26. $\int_1^2 e^{-x^2/2}\,dx$

27. What value of n will ensure that the trapezoidal rule appproximation of $\int_0^1 x^5\,dx$ will be within 0.01 of the exact value?

28. Repeat Exercise 27 for Simpson's rule.

29. For estimating $\int_1^3 \left(x + \frac{2}{x}\right) dx$ with Simpson's rule:

(a) Show that $K = 48$ is an upper bound (in fact, the smallest upper bound) for $|f^{(4)}(x)|$ on [1, 3].

(b) Determine a value of n that ensures that the Simpson's rule estimate is within 0.0001 of the exact value.

IF YOU WORK WITH A COMPUTER (OPTIONAL)

Use either of the following trapezoidal rule computer programs to do Exercises 30–35. Use $n = 50$.

30. $\int_0^2 \sqrt{x^3 + 2}\,dx$

31. $\int_0^3 e^{-x^2/2}\,dx$

32. $\int_1^5 x \ln x\,dx$

33. $\int_1^3 \left(x + \frac{2}{x}\right) dx$

34. $\int_0^1 xe^{2x}\,dx$

35. $\int_0^1 (1 + x^2)^{3/2}\,dx$

```
10   REM  Trapezoidal rule - BASIC version

20   REM - Define integrand
30   DEF FNf(x) = 3*x*x

40   INPUT "End points of interval [a,b]"; a,b
50   INPUT "How many subintervals"; N

60   deltax = (b-a)/N
70   trapsum=0
80   x=a

90   REM - sum up y1, y2, . . . ,yN-1
100  FOR i=1 TO N-1
110      x = x + deltax
120      trapsum = trapsum + FNf(x)
130  NEXT i

140  REM  Add in y0/2 + yN/2
150  trapsum = trapsum + (FNf(a)+FNf(b))/2
```

```
(* Trapezoidal Rule - Pascal Version *)
program TrapRule (input, output);

 var
  a, b, x, deltax, trapsum : real;
  N, i : integer;

 function f (x : real) : real;
 begin
  f := 3 * x * x
 end;

begin
 write('Value of a? ');
 readln(a);
 write('Value of b? ');
 readln(b);
 write('How many subintervals? ');
 readln(N);

 deltax := (b - a) / N;
 trapsum := 0;
 x := a;
```

(Programs continue at top of next page)

```
160  REM Multiply by deltax
160  trapsum = trapsum * deltax

170  PRINT "Trapezoidal approximation ="; trapsum

180  END
```

```
(* sum up y1, y2, ..., yN-1 *)
 for i := 1 to N - 1 do
  begin
   x := x + deltax;
   trapsum := trapsum + f(x)
  end;

(* add in y0/2 + yN/2; then multiply by deltax *)
 trapsum := trapsum + (f(a) + f(b)) / 2;
 trapsum := trapsum * deltax;
 writeln('Trap. rule sum is ', trapsum : 1 : 4)

end.
```

Use either of the following Simpson's rule computer programs to do Exercises 36–41. Use $n = 20$.

36. $\displaystyle\int_0^2 \frac{dx}{\sqrt{x^4 + 2}}$

37. $\displaystyle\int_{-2}^2 e^{-x^2/2}\, dx$

38. $\displaystyle\int_0^2 xe^{-x}\, dx$

39. $\displaystyle\int_0^1 \sqrt{x^3 + 2x + 1}\, dx$

40. $\displaystyle\int_{-1}^1 2\sqrt{1 - x^2}\, dx$

41. $\displaystyle\int_0^1 \frac{4}{1 + x^2}\, dx$

42. Modify the program you used for Exercises 36-41 so that it uses an initial value of, say $n = 4$, to compute the Simpson's rule approximation. Then have it repeatedly double n and compute new approximations until two successive approximations differ in absolute value by less than ϵ. Use this modified program to repeat:

(a) Exercise 36 with $\epsilon = 0.00005$

(b) Exercise 40 with $\epsilon = 0.00001$

(c) Exercise 41 with $\epsilon = 0.000005$

```
10    REM - Simpson's rule - BASIC version

20    DEF FNf(x) = SQR(1 - x*x)

30    INPUT "End points of interval [a,b]";a,b
40    INPUT "Number of subintervals (EVEN)";N

50    deltax=(b-a)/N
60    sum = 0

70    REM add y2 + y4 + y6 + .... + yN-2
80    FOR i = 2 TO N-2 STEP 2
90      sum = sum + FNf(a + i*deltax)
100   NEXT i

110   REM and then double
120   sum = sum * 2

130   REM add4y1 + 4y3 + .... + 4yN-1
140   FOR i = 1 TO N-1 STEP 2
150     sum = sum + 4*FNf(a + i*deltax)
160   NEXT i
```

```
(* Simpson's rule - Pascal Version *)
program Simpson (input, output);

 var
  a, b, deltax, sum : real;
  N, i : integer;

 function f (x : real) : real;
 begin
  f := 4 * sqrt(1 - x * x)
 end;

begin
 write('Value of a? ');
 readln(a);
 write('Value of b? ');
 readln(b);
 write('How many subintervals (EVEN)? ');
 readln(N);

 deltax := (b - a) / N;
 sum := 0;
```

(*Programs continue at top of next page*)

```
170 REM add y0 + yN
180 sum = sum + FNf(a) + FNf(b)

190 REM Simpson's rule sum
200 sum = sum*deltax/3

210 PRINT "The estimate of the integral is"; sum
220 END
```

```
(* add y2 + y4 + y6 + . . . + yN-2; then double *)
  i := 2;
  while i <= N - 2 do
    begin
      sum := sum + f(a + i * deltax);
      i := i + 2
    end;
  sum := sum * 2;

(* add in 4y1 + 4y3 + . . . + 4yN-1 *)
  i := 1;
  while i <= N - 1 do
    begin
      sum := sum + 4 * f(a + i * deltax);
      i := i + 2
    end;

(* add in y0 + yn *)
  sum := sum + f(a) + f(b);
  sum := sum * deltax / 3;
  writeln('Simpson''s rule approx. is ', sum : 1 : 4)

end.
```

SECTION 7.7

Improper Integrals (Optional)

We saw in Section 5.3 how the value of a definite integral $\int_a^b f(x)\,dx$ can be computed from knowledge of an antiderivative $F(x)$ of $f(x)$. More precisely, the fundamental theorem of calculus tells us that

$$\int_a^b f(x)\,dx = F(b) - F(a)$$

In our previous discussion a and b were ordinary real numbers. We never considered the situation where a or b (or possibly both) is infinite. This section is devoted to the study of such situations, which arise quite frequently in probability theory and statistics (Chapter 10).

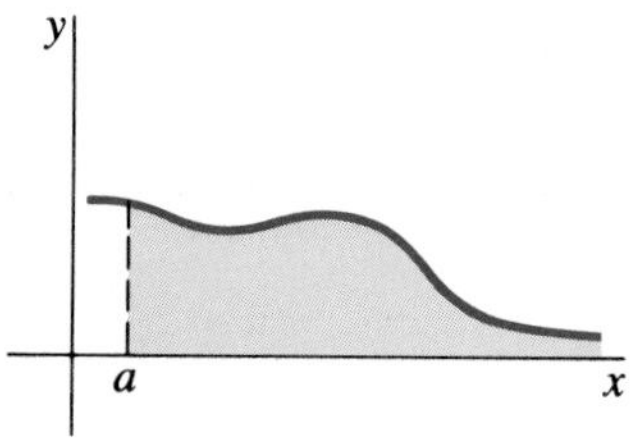

FIGURE 7.10
$y = f(x)$

For purposes of illustration, let us suppose that a is an ordinary real number and $b = \infty$. Suppose further that $f(x)$ is a continuous nonnegative function (by which we mean that $f(x) \geq 0$ for all x in the interval $[a, \infty)$). What meaning shall we give to the following expression?

$$\int_a^\infty f(x)\,dx$$

In order to be consistent with our previous interpretation of $\int_a^b f(x)\,dx$ for ordinary real numbers a and b (that is, "finite" numbers) as the area under the curve, there is only one possible meaning we can give—namely, that $\int_a^\infty f(x)\,dx$ must stand for the area under the curve. In other words, $\int_a^\infty f(x)\,dx$ represents the area of the region bounded above by the curve $y = f(x)$, below by the x-axis, and on the left by the vertical line $x = a$, and unbounded on the right. Thus $\int_a^\infty f(x)\,dx$ is the area of the shaded region shown in Figure 7.10. However, this does not settle the question completely; we need to look more carefully at the area under the curve from a to ∞.

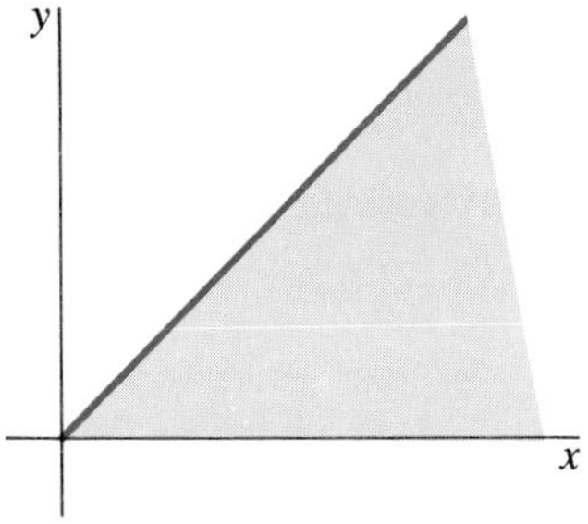

FIGURE 7.11

$y = x$

Consider, for example, the nonnegative continuous function $f(x) = x$. Then $\int_0^\infty x\,dx$, which represents the area under the curve $y = x$ from 0 to ∞ (as shown in Figure 7.11), must clearly be infinite—just look at the figure!

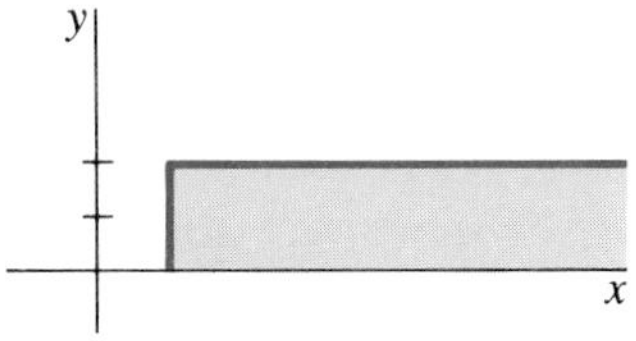

FIGURE 7.12

$y = 2$

In similar fashion, if we take $f(x) = 2$, then $\int_1^\infty 2\,dx$ represents the area under the curve (that is, the horizontal line) $y = 2$ from 1 to ∞. This area, as shown in Figure 7.12, may be viewed as a "rectangle" extending infinitely in one direction—so the area is surely infinite.

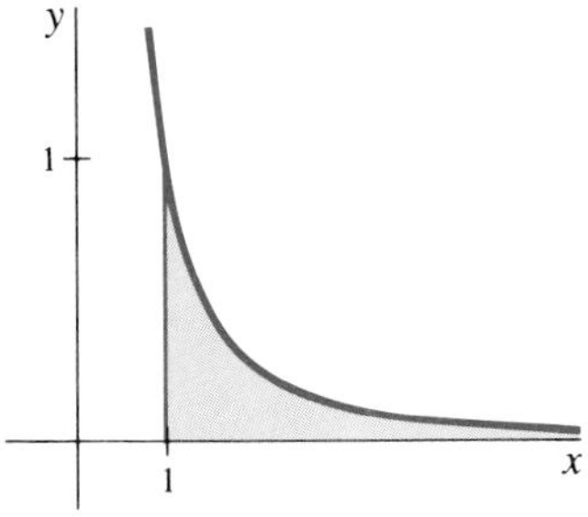

FIGURE 7.13

$$y = \frac{1}{x^2}$$

On the other hand, suppose we consider the nonnegative continuous function $f(x) = 1/x^2$. Then, as shown in Figure 7.13, $\int_1^\infty (1/x^2)\,dx$ represents the area under the curve $y = 1/x^2$ from 1 to ∞. Now, it is far from clear whether this area is finite or infinite; more detailed scrutiny of the area is required. How should we go about trying to compute this area?

Take a number b (b should be thought of as positive and large) and consider the definite integral $\int_1^b (1/x^2)\,dx$. According to our work in Section 5.3, $\int_1^b (1/x^2)\,dx$ is a perfectly good number, and we can compute it via the fundamental theorem of calculus. Of course, $\int_1^b (1/x^2)\,dx$ represents the area under the curve $y = 1/x^2$ from $x = 1$ to $x = b$ (see Figure 7.14), and this area should be viewed as an approximation to the area under the curve $y = 1/x^2$ from $x = 1$ to ∞, for b large. In other words, $\int_1^b (1/x^2)\,dx$ serves as an approximation to $\int_1^\infty (1/x^2)\,dx$. The error we make in using this approximation is clearly $\int_b^\infty (1/x^2)\,dx$, the area under the curve to the right of $x = b$. The crucial observation for us here is that, as b gets larger and larger, the definite integral $\int_1^b (1/x^2)\,dx$ keeps getting larger and larger and provides a better and better approximation to $\int_1^\infty (1/x^2)\,dx$. These remarks lead us to the following definition.

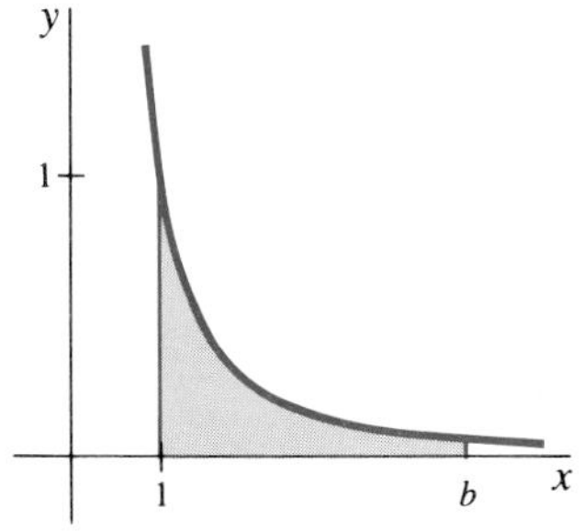

FIGURE 7.14

$$y = \frac{1}{x^2}$$

DEFINITION 7.3 Suppose $f(x)$ is a nonnegative continuous function on the interval $[a, \infty)$. We define $\int_a^\infty f(x)\,dx$, which is called an **improper integral**, by

(7.40)
$$\int_a^\infty f(x)\,dx = \lim_{b\to\infty} \int_a^b f(x)\,dx$$

If, as b gets larger and larger, the definite integral $\int_a^b f(x)\,dx$ becomes arbitrarily large (that is, if $\int_a^b f(x)\,dx \to \infty$), we say that the improper integral $\int_a^\infty f(x)\,dx$ **diverges**, and we write

$$\int_a^\infty f(x)\,dx = \infty$$

which signifies that the area under the curve $y = f(x)$ to the right of $x = a$

is infinite. If, on the other hand, as b gets larger and larger, the definite integral $\int_a^b f(x)\,dx$ becomes arbitrarily close to some number A (that is, if $\int_a^b f(x)\,dx \to A$, or—what is the same—if $\lim_{b\to\infty} \int_a^b f(x)\,dx = A$), we say that the improper integral $\int_a^\infty f(x)\,dx$ **converges** to A. We then write

$$\int_a^\infty f(x)\,dx = A$$

which signifies that A is the area under the curve $y = f(x)$ to the right of $x = a$. ⊟

Let us turn to some illustrations of how this definition works.

EXAMPLE 32 Evaluate the improper integral

$$\int_1^\infty \frac{1}{x^2}\,dx$$

SOLUTION This is the example considered, but not settled, earlier. For any number $b > 1$, we have

$$\int_1^b \frac{1}{x^2}\,dx = \frac{-1}{x}\bigg|_1^b = \frac{-1}{b} - (-1) = 1 - \frac{1}{b}$$

To evaluate $\int_1^\infty (1/x^2)\,dx$ we must find the limit of $\int_1^b (1/x^2)\,dx$ as $b \to 0$. But

$$\lim_{b\to\infty}\left(1 - \frac{1}{b}\right) = 1$$

Therefore

$$\int_1^\infty \frac{1}{x^2}\,dx = 1$$

Thus the area under the curve $y = 1/x^2$ to the right of $x = 1$ is finite—in fact, this area is 1.

EXAMPLE 33 Evaluate the improper integral

$$\int_2^\infty \frac{1}{x+4}\,dx$$

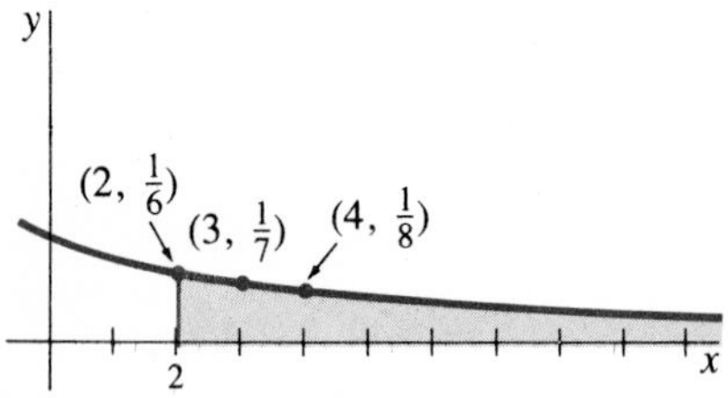

FIGURE 7.15

$$y = \frac{1}{x+4}$$

SOLUTION Here we want to find the area under the hyperbola $y = 1/(x + 4)$ to the right of $x = 2$, as indicated in Figure 7.15. To do this, we must examine

$$\lim_{b\to\infty} \int_2^b \frac{1}{x+4}\,dx$$

Now, for any $b > 2$, we have

$$\int_2^b \frac{1}{x+4}\,dx = \ln|x+4|\bigg|_2^b = \ln(b+4) - \ln 6$$

As $b \to \infty$, $\ln 6$ stays fixed but $\ln(b + 4) \to \infty$ (because $\ln x \to \infty$ as $x \to \infty$). Therefore,

$$\lim_{b\to\infty} \int_2^b \frac{1}{x + 4}\,dx = \infty$$

and the improper integral

$$\int_2^\infty \frac{1}{x + 4}\,dx$$

diverges. The area under the curve, as sketched in Figure 7.15, is infinite.

EXAMPLE 34 Evaluate the improper integral

$$\int_0^\infty xe^{-x}\,dx$$

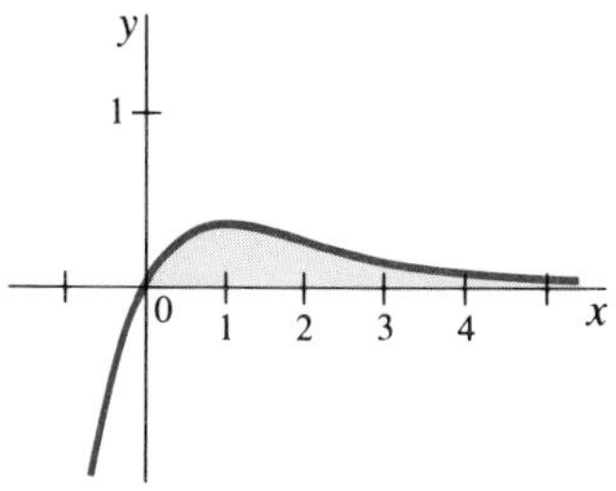

FIGURE 7.16
$y = xe^{-x}$

SOLUTION First let us draw a sketch. If we use the graphing techniques from Chapter 3 (see also Exercise 36 of Section 4.1), we find that the area to be evaluated is the shaded region shown in Figure 7.16.

Next we need to compute the definite integral

$$\int_0^b xe^{-x}\,dx$$

for any $b > 0$. Now, we can easily show—via integration by parts—that

$$\int xe^{-x}\,dx = -xe^{-x} - e^{-x} + c$$

Consequently,

$$\begin{aligned}\int_0^b xe^{-x}\,dx &= (-xe^{-x} - e^{-x})\Big|_0^b \\ &= (-be^{-b} - e^{-b}) - (-1) \\ &= 1 - be^{-b} - e^{-b}\end{aligned}$$

Thus we must examine

$$\lim_{b\to\infty} \int_0^b xe^{-x}\,dx = \lim_{b\to\infty} (1 - be^{-b} - e^{-b})$$

From our earlier work, we know that $\lim_{x\to\infty} e^x = \infty$, so $\lim_{x\to\infty} e^{-x} = \lim_{x\to\infty} \frac{1}{e^x} = 0$ and also that $\lim_{x\to\infty} xe^{-x} = 0$. Therefore,

$$\lim_{b\to\infty} \int_0^b xe^{-x}\,dx = \lim_{b\to\infty} (1 - be^{-b} - e^{-b}) = 1$$

This shows that the improper integral converges to 1:

$$\int_0^\infty xe^{-x}\,dx = 1$$

EXAMPLE 35 Evaluate the improper integral

$$\int_{-\infty}^{1} e^{-x}\, dx$$

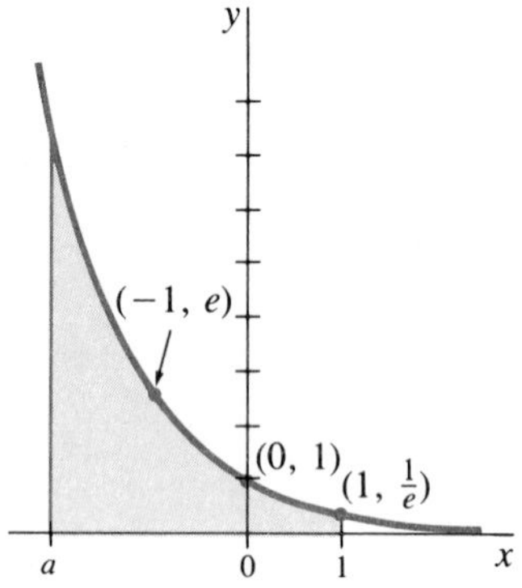

FIGURE 7.17
$y = e^{-x}$

SOLUTION This integral does not quite fall under Definition 7.3 of the improper integral, but it is clear how the definition should be extended to take care of it. Thus $\int_{-\infty}^{1} e^{-x}\, dx$ represents the area under the curve $y = e^{-x}$ to the left of the vertical line $x = 1$, and it is defined by

$$\int_{-\infty}^{1} e^{-x}\, dx = \lim_{a \to -\infty} \int_{a}^{1} e^{-x}\, dx$$

It is obvious from Figure 7.17 that $\int_{-\infty}^{1} e^{-x}\, dx$ diverges, but let us show this in a formal way. For any $a < 0$, we have

$$\int_{a}^{1} e^{-x}\, dx = -e^{-x}\Big|_{a}^{1} = -e^{-1} - (-e^{-a}) = e^{-a} - \frac{1}{e}$$

thus

$$\lim_{a \to -\infty} \int_{a}^{1} e^{-x}\, dx = \lim_{a \to -\infty} \left(e^{-a} - \frac{1}{e}\right) = \infty$$

so

$$\int_{-\infty}^{1} e^{-x}\, dx = \infty$$

The integral diverges.

Now, suppose the function $f(x)$ is continuous on $(-\infty, +\infty)$. What meaning is to be assigned to the following improper integral?

$$\int_{-\infty}^{\infty} f(x)\, dx$$

The answer is that, as a matter of convention, this integral is defined to be

$$\int_{-\infty}^{\infty} f(x)\, dx = \int_{-\infty}^{0} f(x) + \int_{0}^{\infty} f(x)\, dx \tag{7.41}$$

and the integral is said to *converge* when both integrals on the right side converge; if one (or both) of the integrals on the right side diverges, then the integral is said to *diverge*.

REMARK When trying to evaluate an improper integral $\int_{-\infty}^{\infty} f(x)\, dx$, we are to use Definition (7.3). Do *not* try the formulation

$$\int_{-\infty}^{+\infty} f(x)\, dx = \lim_{b \to \infty} \int_{-b}^{+b} f(x)\, dx$$

It may yield the wrong answer. ⊟

Section 7.7 EXERCISES

In Exercises 1–10 evaluate the indicated integrals.

1. $\int_0^{\infty} e^{-5x}\,dx$

2. $\int_1^{\infty} \ln x\,dx$

3. $\int_0^{\infty} \frac{2x}{1+x^2}\,dx$

4. $\int_6^{\infty} \frac{dx}{x^2-4x-5}$

5. $\int_{-\infty}^{-4} \frac{1}{x^2-x-12}\,dx$

6. $\int_{-\infty}^{1} e^{3x}\,dx$

7. $\int_1^{\infty} \frac{dx}{x(x+5)}$

8. $\int_2^{\infty} \frac{dx}{x(x-1)}$

9. $\int_2^{\infty} e^{-x}\,dx$

10. $\int_0^{\infty} \frac{dx}{(x+1)^3}$

In Exercises 11–19 find the area of the region bounded by the indicated curves.

11. $y = xe^{-x}$, x-axis, to the right of $x = 1$

12. $y = \frac{4}{x^2}$, x-axis, to the left of $x = -2$

13. $y = \frac{-x}{(x^2+1)^2}$, x-axis, to the left of $x = -1$

14. $y = (\ln e^{x^3})^{-1}$, x-axis, between $x = 1$ and $x = \infty$

15. $y = \frac{(\ln x)^2}{x}$, x-axis, to the right of $x = 1$

16. $y = \frac{x}{\sqrt{1+x^2}}$, x-axis, between $x = 0$ and $x = \infty$

17. $y = \frac{x}{\sqrt{4+x^2}}$, x-axis, between $x = -\infty$ and $x = 0$

18. $y = \frac{1}{x \ln x}$, x-axis, between $x = 2$ and $x = \infty$

19. $y = x^5 e^{-x}$, x-axis, to the right of $x = 0$

Key Mathematical Concepts and Tools

Integration by parts
Integration of rational functions by partial fractions
Integral tables
Separable differential equations
Separation of variables
Mathematical modeling
Numerical integration
Trapezoidal rule
Simpson's rule
Improper integrals

Applications Covered in This Chapter

Free-fall velocity
Second-order chemical reaction
Exponential spread of a rumor
Logistic growth
Constant source diffusion
Determining the mechanism by which use of a new drug is adopted
Area under the standard normal distribution

SEVEN Warm-Up Test

1. Evaluate the following integrals, using integration by parts if necessary.

(a) $\int (x+2)e^{3x}\,dx$ (b) $\int \frac{e^{-x}}{\sqrt{4+3e^{-x}}}\,dx$

(c) $\int x \ln[(2x+1)^2]\,dx$

2. (a) Describe as precisely as you can the trapezoidal rule for approximating $\int_a^b f(x)\,dx$. (Explicitly state the approximation formula.)

(b) Use the trapezoidal rule with $n = 6$ to approximate

$$\int_0^1 \sqrt{2+x^2}\,dx$$

3. Use the method of partial fractions to compute the following integrals.

(a) $\int \frac{x^2+2}{4-x}\,dx$ (b) $\int \frac{1-2x}{(x-2)(x-3)}\,dx$

(c) $\int \frac{x+3}{x^2+6x}\,dx$

4. Evaluate the following integrals by using, if necessary, a combination of the methods you have learned.

(a) $\int \frac{e^{3x}+4e^x}{e^{2x}-4}\,dx$ (b) $\int e^x \ln(e^x+2)\,dx$

5. Use the separation-of-variables technique to solve the following differential equations.

(a) $\frac{dy}{dx} = ye^{-x} + 3e^{-x}, \quad y(0) = -2$

(b) $\frac{dy}{dt} = -\frac{2}{5}\frac{(6-y)t}{y}, \quad y(5) = 5$

6. Use the table of integrals (Appendix B), if necessary, to evaluate:

(a) $\int \frac{dx}{(x+2)^2\sqrt{x+3}}$ (b) $\int \frac{dx}{(9+4x^2)^{3/2}}$

7. In a certain chemical reaction a substance C interacts with itself only to produce a new substance X according to the chemical "equation"

$$C + C \to X$$

Assume the law governing this interaction is

$$\frac{dx}{dt} = k(c - x)$$

where k is a rate constant, c is the initial concentration of substance C and $x(t)$ is the concentration of the product X at time t. Assuming that $k = 2$ and $c = 5$ moles, find a formula for $x(t)$ at each time $t > 0$, given that $x(0) = 0$.

8. Find the area of the unbounded region bounded above by the curve $y = (x+3)e^{-x}$, below by the x-axis, and on the left by the line $x = 0$.

9. (a) Use integration by parts to evaluate

$$\int x^3(3x^2-2)^{1/2}\,dx$$

(*Hint*: Rather than the "obvious" choice, set $u = x^2$, etc.)

(b) Redo part (a) via the substitution method.

10. A radioactive isotope decays with a half-life of 11 days. Suppose 6 grams remain after 20 days.

(a) How many grams will remain in another 20 days?

(b) What percentage of the initial amount disappears every 5 days?

SEVEN Final Exam

1. Use (repeated) integration by parts to evaluate

$$\int e^{2x}(2+5x+6x^2)\,dx$$

2. (a) Describe Simpson's rule as precisely as you can, and explicitly write down the formula for approximating $\int_a^b f(x)\,dx$ for a given even value of n.

(b) Exhibit the use of Simpson's rule with $n = 4$ to approximate

$$\int_1^2 \ln(5 - x^2)\,dx$$

3. Evaluate the following integrals using integration by parts:

(a) $\displaystyle\int \frac{x}{\sqrt{5 + x}}\,dx$ (b) $\displaystyle\int (\ln x)^2\,dx$

(c) $\displaystyle\int x^3 e^{x^2}\,dx$ (*Hint*: Let $u = x^2$, etc.)

4. Use the method of partial fractions to evaluate:

(a) $\displaystyle\int \frac{(x + 1)\,dx}{9 - 6x + x^2}$ (b) $\displaystyle\int \frac{4x - 12}{(x - 7)(x + 1)}\,dx$

(c) $\displaystyle\int \frac{x^3}{x^2 - x}\,dx$

5. Use the separation-of-variables technique to solve the following differential equations.

(a) $\displaystyle\frac{dy}{dx} = -\frac{y^2 + 4y + 4}{x}, \quad y(1) = 0$

(b) $\displaystyle\frac{dy}{dt} = \frac{(y - 1)t}{y^2(t + 1)}, \quad y(0) = 2$

[In part (b) you need not solve for y as an explicit function of t.]

6. Use the table of integrals, if necessary, to evaluate:

(a) $\displaystyle\int \frac{x^2 + x}{\sqrt{1 + 2x}}\,dx$ (b) $\displaystyle\int \frac{dx}{x^2\sqrt{25 - 9x^2}}$

7. A certain population of bacteria follows a logistic growth pattern with rate constant $l = 0.2$. That is,

$$\frac{dP}{dt} = 0.2(A - P)P$$

Here $P(t)$ represents the bacteria count at time t measured in hours. At present ($t = 0$), the bacteria count is 100 and the present rate of bacteria growth is $P'(0) = 8000$/hr. What will the count be 10 hr from now?

8. Evaluate the following improper integrals.

(a) $\displaystyle\int_2^\infty \frac{x^3}{(1 + x^4)^2}\,dx$ (b) $\displaystyle\int_{-\infty}^{-2} \frac{dx}{x^2 + x}$

9. Evaluate the following integrals, using, if necessary, a combination of the methods that you have learned.

(a) $\displaystyle\int \frac{(\ln x)^2}{(\ln x - 5)x}\,dx$ (b) $\displaystyle\int \frac{3u^2 + 2u + 1}{\sqrt{2 - u}}\,du$

10. (a) What is the present value of \$70,000 10 years from now, if the interest rate remains steady at 12% compounded continuously over this period?

(b) Redo part (a) if an interest rate of 14% prevails for the first 5 years and an interest rate of 12% for the second 5 years, both compounded continuously.

EIGHT Multivariable Calculus

Outline

In Context

Up to now we have been studying the calculus of functions of a single variable only. However, most things in life depend on a number of variables. For example, the consumer price index (CPI) is a function of the present prices of a number of goods and services that cost \$100 in 1967 (an arbitrarily chosen reference year). This chapter provides a brief introduction to the calculus of functions of several variables and its applications, including applied optimization and the method of least squares, an important tool in statistical analysis.

SECTION 8.1

Functions of Several Variables

In the previous chapters we studied the calculus of functions of a single independent variable. We introduced the notion of a function machine to represent the function pictorially, as shown in Figure 8.1. This machine accepts x as its input and yields $y = f(x)$ as output.

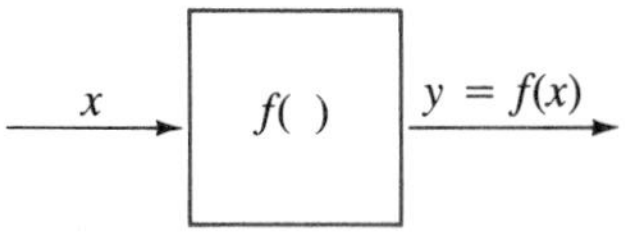

FIGURE 8.1
Function of a single variable

It is easy to find situations in which a single variable is insufficient to allow a mathematical description of a real-life situation. For example, the area of a rectangle, A, is given by

$$A = xy$$

where x represents the length of the rectangle and y its height (see Figure 8.2). In function notation we can write

$$A = f(x, y) = xy$$

Here the area is a function of two independent variables, x and y. A sketch of a function machine for a function of two variables is shown in Figure 8.3. It accepts x and y as distinct inputs and produces a single output, z. Some typical examples of functions of two variables are

$$f(x, y) = x^2 - 2xy + y^2$$

$$g(x, y) = e^{xy}$$

$$h(x, y) = \frac{xy}{x^2 + y^2}$$

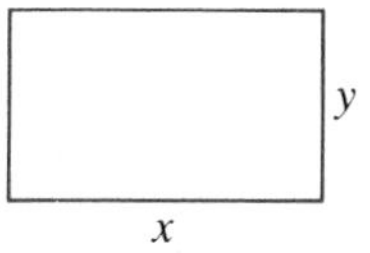

FIGURE 8.2
Area of a rectangle

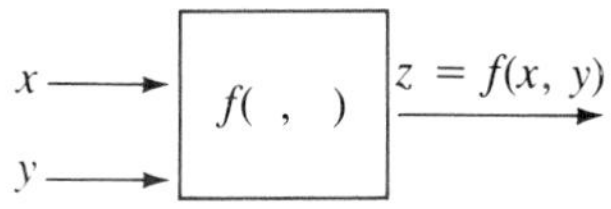

FIGURE 8.3
Function of two variables

EXAMPLE 1 Consider the weather map of the continental United States shown in Figure 8.4. Let x and y denote the longitude and latitude, respectively, of a

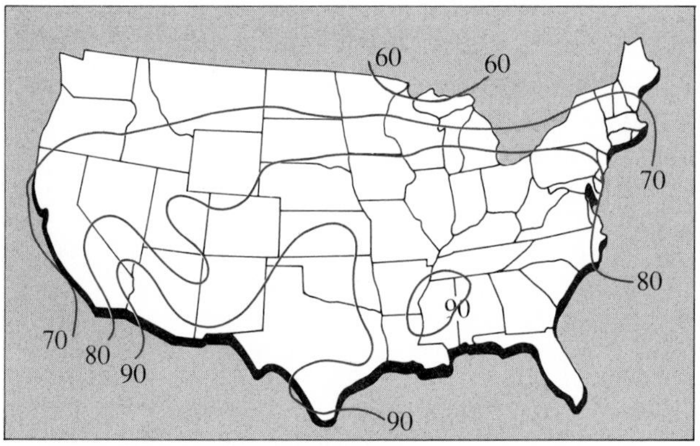

FIGURE 8.4

point on the map, and let $z = f(x, y)$ represent the temperature on the ground at that point at 12 noon. In the figure, curves, called **isotherms**, are drawn through points that have the same z value.

EXAMPLE 2 Let $z = f(x, y) = x^2 + 6xy$. Compute: (a) $f(2, 3)$; (b) $f(-1, 2)$; (c) $f(2, -1)$.

SOLUTION

(a) The number $f(2, 3)$ represents the value of the function f when $x = 2$ and $y = 3$. Thus $f(2, 3) = 2^2 + 6(2)(3) = 40$.

(b) Similarly, we compute $f(-1, 2) = (-1)^2 + 6(-1)(2) = -11$.

(c) Finally, $f(2, -1) = (2)^2 + 6(2)(-1) = -8$. Note that $f(-1, 2) \neq f(2, -1)$. Thus, in general, interchanging the values of x and y produces a different result.

PRACTICE EXERCISE Let $z = f(x, y) = x^3 + x \ln y$. Evaluate: (a) $f(2, 1)$; (b) $f(-3, e)$; (c) $f(0, 4)$.

[Your answers should be (a) 8, (b) -30, (c) 0.]

EXAMPLE 3 In economics, **capital goods** are defined as goods produced by the economic system for further production of other goods and services. An example might be a word processor used in an office to enhance the productivity of a secretary. Associated with a capital good is its **capitalized value** V. This is a measure of the value of the future earning power of that asset. The formula for V is

$$V = \frac{N}{r}$$

where $N =$ (perpetual) annual income, and $r =$ interest rate expressed as a decimal (for example, $5\% = 0.05$). Note that $V = V(N, r)$ is a function of two variables. Imagine that if we buy a word processor we can use it to type term papers for other students, and suppose further that we estimate we can earn \$8000 per year by doing this. (In this example, we ignore the value of our labor.) Suppose that interest rates are expected to average 12% per year for the foreseeable future. Then the capitalized value of the word processor is

$$V = \frac{8000}{0.12} = \$66{,}666.67$$

The lower the interest rate, the higher the capitalized value. This is because the value of an annual income of \$8000 is greater when alternative investments have a lower yield.

PRACTICE EXERCISE What is the capitalized value of an investment yielding income of \$95 per year if interest rates are 14%?

[Your answer should be \$678.57.]

The previous examples illustrate the concept of a function of two variables. It is an easy matter to formulate functions of three or more variables.

EXAMPLE 4 The volume of a rectangular box, denoted by V, is given by

$$V = xyz$$

where x, y, z are the length, width, and height of the box, respectively (see Figure 8.5). Thus, for instance,

$$V(3, 2, 4) = 3 \cdot 2 \cdot 4 = 24$$

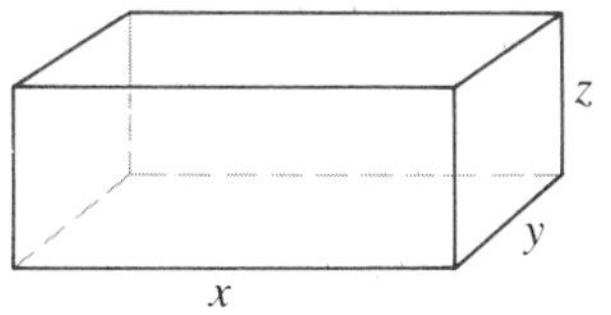

FIGURE 8.5

EXAMPLE 5 Consider again the weather map shown in Figure 8.4. Imagine a function g that has three independent variables: t, x, and y. The t denotes the time of day, whereas x and y represent the longitude and latitude of a point on the map. The value $g(t, x, y)$ is the temperature at time t at the point having longitude x and latitude y.

When a function depends on more than three independent variables, we usually label them simply $x_1, x_2, \ldots, x_n$. For instance,

$$f(x_1, x_2, x_3, x_4) = (x_1 + x_2)e^{x_3 x_4}$$

is a function of four variables. In economics, researchers frequently encounter functions with literally hundreds of input variables. For convenience, we will restrict ourselves almost exclusively to the study of functions of two variables.

The graph of a function of two variables is, in principle, drawn in a manner analogous to drawing the graph of a function of a single variable. Recall that, for a function $y = f(x)$, the graph requires two axes: one for the independent variable x and one for the dependent variable y. Associated with each value of x (in the function's domain) is the number $y = f(x)$, and the graph consists of all the ordered pairs $(x, f(x))$.

A function $z = f(x, y)$ requires three axes: two for the independent variables x and y and one for the dependent variable z. A point on the graph of this function consists of an ordered triple

$$(x, y, f(x, y))$$

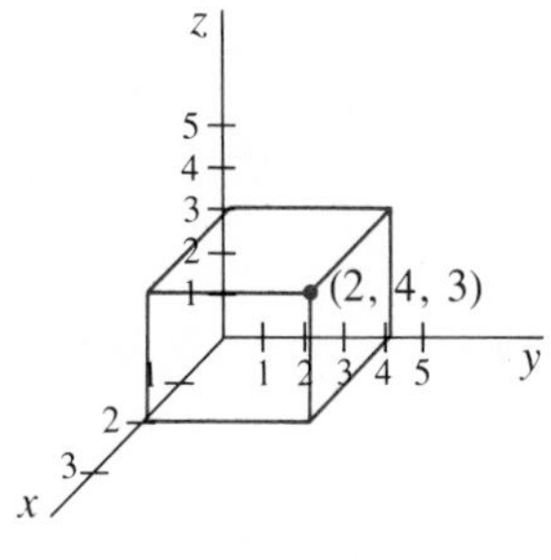

FIGURE 8.6 Three-dimensional coordinate system

A three-dimensional coordinate system is shown in Figure 8.6. We have shown

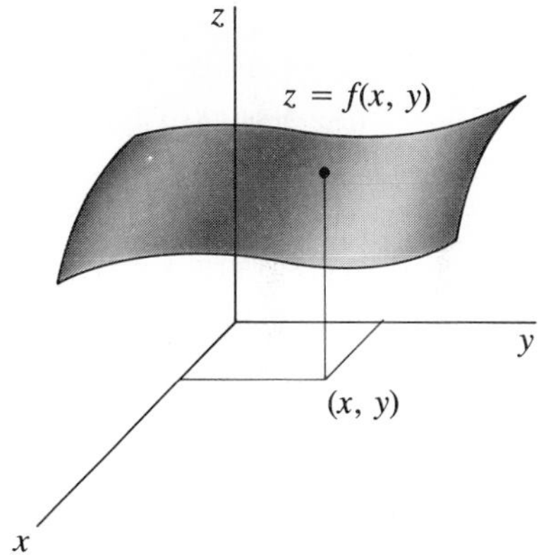

FIGURE 8.7

the point (2, 4, 3). Of course, the textbook page itself is two-dimensional, so it is impossible for us really to plot a point such as (2, 4, 3). Ideally the x-, y-, and z-axes of Figure 8.6 should be mutually perpendicular—an impossibility on a two-dimensional printed page.

The graph of $z = f(x, y)$ is obtained by selecting values of x and y such that (x, y) is in the domain of f and then computing the corresponding value of z. The resulting points together form a **surface**. Figure 8.7 shows a typical graph of a function of three variables. It looks something like the roof of a tent. The x- and y-coordinates provide the location of a point on the floor; the z-value is the height of the tent above that point (or the depth below, if z is negative).

Figures 8.8(a) and (b) show the graphs of the functions $f(x, y) = x^2 + y^2$ and $f(x, y) = y^2 - x^2$. For most functions it is relatively easy to compute z for given values of x and y. For example, if

$$z = f(x, y) = x^2 + 6xy$$

then

$$f(2, 3) = 2^2 + 6(2)(3) = 40$$

The difficulty in graphing such a function is not algebraic. It requires considerable artistic talent, however, to represent and describe the graph on a two-dimensional page. For this reason we will for the most part avoid drawing three-dimensional graphs.

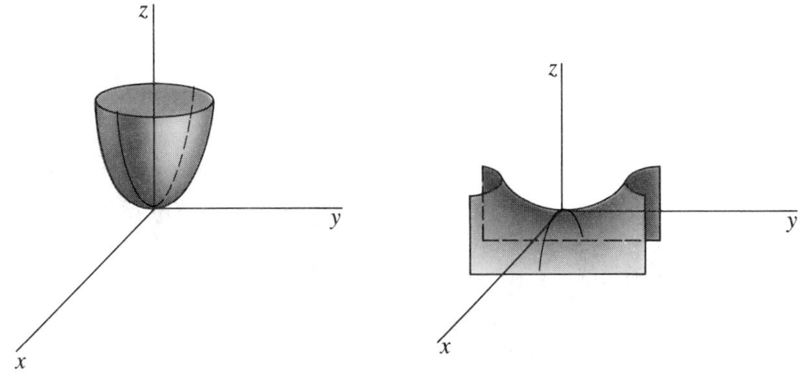

(a) $z = x^2 + y^2$

(b) $z = y^2 - x^2$

FIGURE 8.8

Section 8.1 EXERCISES

1. Let $z = f(x, y) = 4x^2y + 3x$. Compute $f(1, 2)$, $f(-1, 3)$, and $f(0, 1)$.

2. Consider the function

$$z = h(x, y) = \frac{x}{x^2 + y^2}$$

Compute $h(1, 2)$, $h(0, 3)$, and $h(9, -6)$. For which (x, y) pairs is this function not defined?

In Exercises 3–10 compute $f(3, -3)$, $f(2, 0)$, and $f(1, 4)$ for the given functions. If the value is not defined, say so.

3. $f(x, y) = xye^{2x}$

4. $f(x, y) = xy + 8x + 3 \ln x$

5. $f(x, y) = y \ln x + e^y$

6. $f(x, y) = x\sqrt{y}$

7. $f(x, y) = x^2 \ln y + 4$

8. $f(x, y) = \dfrac{x + y}{x - y}$

9. $f(x, y) = \dfrac{x - y}{x + y}$

10. $f(x, y) = x^2 + y^2$

11. For the function $z = f(x, y) = x^2y + x^3 + 5$, compute

$$f(3, 4) \cdot \frac{f(-4, 2)}{f(0, 0)}$$

12. The volume of a right circular cylinder is known to be given by the product of the area of its base by its height. Let r = radius of base and h = height.
   (a) Write down the formula for $V(r, h)$.
   (b) Compute $V(2, 3)$, $V(1, 3)$, and $V(2, 5)$.
   (c) Compute $V(0, 4)$ and $V(3, 0)$. How would you interpret your answers?

13. Consider the function $z = f(x, y) = xy^3 + x \ln y + 4$.
   (a) Evaluate $f(2, 1)$, $f(2, e)$, and $f(1, e^2)$
   (b) Fixing y at the value $y = e^3$, evaluate $f(1, e^3)$, $f(2, e^3)$, $f(3, e^3)$, and $f(4, e^3)$.
   (c) Plot z as a function of x for $z = f(x, e^3)$, where x varies from $-\infty$ to $+\infty$. Describe what you obtain.

14. For the function $z = f(x, y) = x^3 + ye^x$, compute (a) $f(1, 0)$; (b) $f(2, 1)$; (c) $f(\ln 4, 2)$.

15. Compute the capitalized value, at a prevailing interest rate of 11%, of:
   (a) A telephone answering system capable of producing \$15,000 per year in sales.
   (b) An addressograph machine capable of yielding \$9000 in sales per year.
   (c) Redo parts (a) and (b) if the interest rate is 8% rather than 11%.
   (d) Redo parts (a) and (b) if the interest rate is 14% rather than 11%.

16. The surface area of a hollow cylinder of base radius r and height h is given by the formula $S(r, h) = 2\pi rh$.
   (a) Write down a formula for the total surface area $S(r, h)$ of a tin can with closed top and bottom.
   (b) Write down a formula $S(r, h)$ for the surface area of a tin can with an open top.

17. A company has a current investment with capitalized value V. At an interest rate of r, it wishes to compare alternative investments.
   (a) Write down an expression for $Y(V, r)$ = the yield per year of an alternative investment that is required to match the capitalized value V of the current investment.
   (b) If the current investment has capitalized value $V = 20{,}000$ and the prevailing interest rate is 12%, what yield is required per year to make an alternate investment more advantageous than the present one?

SECTION 8.2

Partial Derivatives

In Chapter 2 we introduced the notion of the derivative of a function of a single variable. Chapter 3 presented some applications of the derivative. The remainder of this chapter is devoted to developing the corresponding concepts for a function of several variables. In particular, it focuses on the function $z = f(x, y)$ of two independent variables.

Let the variables x and y represent the longitude and latitude respectively, of a point in the continental United States, and let $z = f(x, y)$ represent the altitude of the land at this location, measured in feet above or below sea level. Figure 8.9 shows a portion of a map prepared by the Commonwealth of Massachusetts Department of Public Works of the area surrounding the town of Hopkinton, the starting point of the Boston Marathon. Note, for instance, that the Center School lies at approximate longitude 92° 18′ and latitude 78° 7′ and is about 450 feet above sea level.

Next envision a highway running directly from west to east across the country. That is, imagine a road of fixed latitude but varying longitude. Mathematically this highway is equivalent to fixing the y-coordinate of each

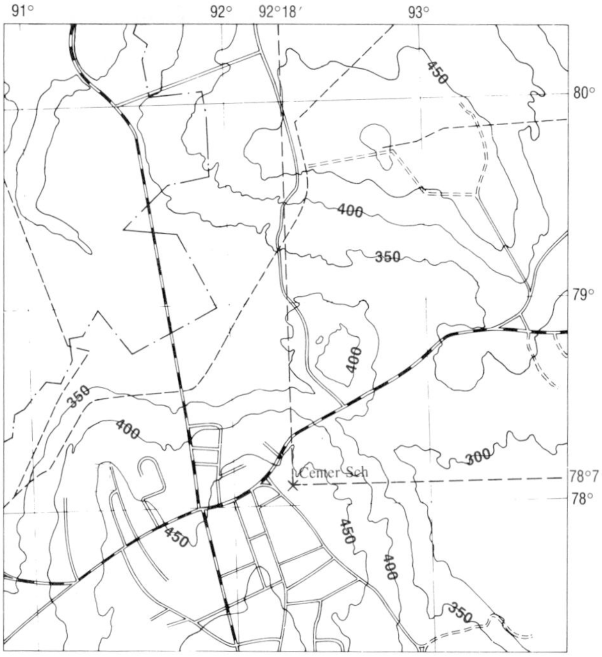

FIGURE 8.9
A typical map indicating elevation

point at the value y_0 (the latitude of the highway) and leaving the x-coordinate free to vary. Thus, on the highway, our function takes the form

$$z = f(x, y_0)$$

It is a function of x alone, because the y-value has been fixed.

Driving along this hypothetical west-to-east highway, a motorist will experience changes in elevation as the land rises and falls (see Figure 8.10). What rate of change of elevation does our motorist experience at a location (x_0, y_0)? In accordance with our earlier work, it must simply be the derivative of the function z with respect to x at the point where x takes the value x_0. This will tell us the steepness of the highway at (x_0, y_0) while traveling west to east. This slope is called the **partial derivative** of $z = f(x, y)$ with respect to x at the point (x_0, y_0).

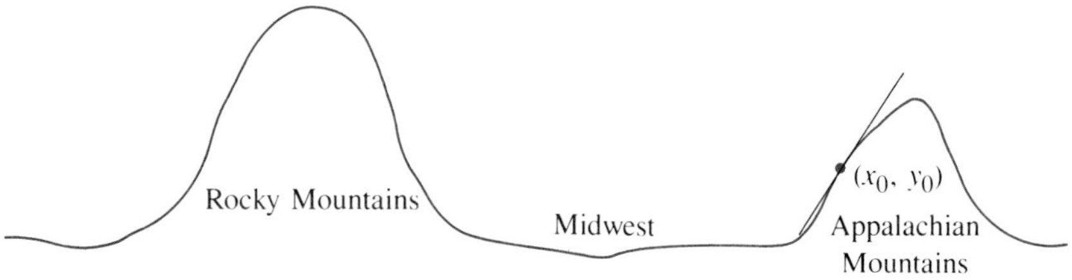

FIGURE 8.10
Land elevation on hypothetical west-to-east highway

DEFINITION 8.1 The **partial derivative of $z = f(x, y)$ with respect to x** is the derivative of $z = f(x, y)$ viewed as a function of x alone. The variable y is treated as a constant. It is denoted by the following equivalent notations:

$$f_x, \quad \frac{\partial f}{\partial x}, \qquad z_x, \quad \frac{\partial z}{\partial x}$$

Similarly, we write f_y, $\frac{\partial f}{\partial y}$, z_y, or $\frac{\partial z}{\partial y}$ to represent the **partial derivative of $z = f(x, y)$ with respect to y.** This corresponds to differentiating with respect to y while keeping x fixed. ⊟

You can interpret the value of $\frac{\partial f}{\partial y}$ at a point (x_0, y_0) as the slope, or rate of change of elevation, encountered by a motorist at the point (x_0, y_0) while driving along a south-to-north highway having fixed longitude $x = x_0$.

EXAMPLE 6 Let $z = f(x, y) = 6x^3y + y^4$. Compute: (a) $\partial f/\partial x$, or $\partial z/\partial x$; (b) $\partial f/\partial y$, or $\partial z/\partial y$.

SOLUTION

(a) In computing $\partial f/\partial x$ we must view y as a constant and differentiate with respect to x. As an aid to doing this, let us momentarily replace y by a constant a so that we have

$$z = f(x, a) = 6x^3a + a^4$$

The derivative with respect to x is now $18x^2a + 0$. If we now replace a by y, we have

$$\frac{\partial f}{\partial x} = 18x^2y$$

(b) In order to take the partial derivative with respect to y, we view x as a constant. If we set $x = a$, then

$$z = f(a, y) = 6a^3y + y^4$$

so that the derivative with respect to y would be $6a^3 + 4y^3$. Now, replacing a by x gives

$$\frac{\partial z}{\partial y} = 6x^3 + 4y^3$$

Note that in part (a) of the previous example the act of replacing the variable y with a constant, a, was merely a device to remind us to treat y as a constant. This step is unnecessary in practice once you are familiar with the techniques of partial differentiation.

The notations $\frac{\partial f}{\partial x}(x_0, y_0)$ and $\frac{\partial f}{\partial y}(x_0, y_0)$ represent the partial derivatives of f with respect to x and y, respectively, evaluated at the point (x_0, y_0).

EXAMPLE 7 For $z = f(x, y) = xe^{xy}$, evaluate:

(a) $\dfrac{\partial f}{\partial x}(1, 2)$ (b) $\dfrac{\partial f}{\partial y}(3, -2)$

SOLUTION

(a) To compute $\dfrac{\partial f}{\partial x}$ we view y as a constant. Recall that the derivative of xe^{ax} is determined from the product rule to be

$$x(ae^{ax}) + e^{ax}$$

Replacing a with y yields

$$\frac{\partial f}{\partial x} = xye^{xy} + e^{xy}$$

Finally, setting $x = 1$, $y = 2$ yields

$$\frac{\partial f}{\partial x}(1, 2) = (1)(2)e^{(1)(2)} + e^{(1)(2)} = 3e^2$$

(b) Similarly, $\dfrac{\partial f}{\partial y} = x^2e^{xy}$, so

$$\frac{\partial f}{\partial y}(3, -2) = (3)^2e^{(3)(-2)} = 9e^{-6}$$

As we noted earlier, a function may have any number of independent variables.

EXAMPLE 8 Let $W = f(x, y, z) = x^2 + y \ln z + xy$. Compute: (a) W_x; (b) W_y; (c) W_z.

SOLUTION

(a) $W_x = 2x + y$

(b) $W_y = \ln z + x$

(c) $W_z = y/z$

In Chapter 2 we discussed the notion of the second derivative $\dfrac{d^2f}{dx^2}$ of a function $y = f(x)$ of a single independent variable x. Now consider a function $z = f(x, y)$ of two independent variables x and y. There are four possible second partial derivatives to consider.

1 The derivative of f with respect to x and then with respect to x again. This is written

$$z_{xx} \quad \text{or} \quad \frac{\partial^2 z}{\partial x^2}$$

2 The derivative of f with respect to x and then with respect to y. This is written

$$z_{xy} \quad \text{or} \quad \frac{\partial^2 z}{\partial y\, \partial x}$$

Note that the order of the variables is reversed in the two different notations—something that is a frequent source of confusion.

3 The derivative of f first with respect to y and then with respect to x. This is written

$$z_{yx} \quad \text{or} \quad \frac{\partial^2 z}{\partial x\, \partial y}$$

4 The derivative of f with respect to y and then with respect to y again. This is written

$$z_{yy} \quad \text{or} \quad \frac{\partial^2 z}{\partial y^2}$$

Sometimes the letter f is used in place of z in these notations. The derivatives in (2) and (3) above are called **mixed partials**. Let us compute the four second-order partial derivatives for a specific function.

EXAMPLE 9 Let $z = f(x, y) = x^2 + 5x^2y + y^4$. Compute:

(a) $\dfrac{\partial^2 z}{\partial x^2}$ (b) $\dfrac{\partial^2 z}{\partial y \partial x}$ (c) $\dfrac{\partial^2 z}{\partial x \partial y}$ (d) $\dfrac{\partial^2 z}{\partial y^2}$

SOLUTION We first compute

$$\frac{\partial z}{\partial x} = 2x + 10xy$$

and

$$\frac{\partial z}{\partial y} = 5x^2 + 4y^3$$

Then we have

(a) $$\frac{\partial^2 z}{\partial x^2} = \frac{\partial}{\partial x}\left(\frac{\partial z}{\partial x}\right) = \frac{\partial}{\partial x}(2x + 10xy) = 2 + 10y$$

(b) $$\frac{\partial^2 z}{\partial y \partial x} = \frac{\partial}{\partial y}\left(\frac{\partial z}{\partial x}\right) = \frac{\partial}{\partial y}(2x + 10xy) = 10x$$

(c) $$\frac{\partial^2 z}{\partial x \partial y} = \frac{\partial}{\partial x}\left(\frac{\partial z}{\partial y}\right) = \frac{\partial}{\partial x}(5x^2 + 4y^3) = 10x$$

(d) $$\frac{\partial^2 z}{\partial y^2} = \frac{\partial}{\partial y}\left(\frac{\partial z}{\partial y}\right) = \frac{\partial}{\partial y}(5x^2 + 4y^3) = 12y^2$$

It is interesting to note that, in Example 9, the two mixed partial derivatives $\frac{\partial^2 z}{\partial x \partial y}$ and $\frac{\partial^2 z}{\partial y \partial x}$ were both equal to $10x$. This was not coincidental. In general the mixed partials are equal, provided they are continuous. We shall omit the proof of this fact.

EXAMPLE 10 In economics, the theory of production helps one analyze how much output of a particular good one can get from certain prescribed amounts of inputs, such as labor, machines, capital, or raw materials. There is a maximum obtainable amount of product for a given amount of inputs; this amount depends on the current state of technology. The formula relating inputs to outputs is called a **production function**.

One type of production function that occurs frequently is of the form

$$z = \beta L^{\alpha} K^{1-\alpha}$$

where $0 < \alpha < 1$ and α and β are constants. Here L represents the amount of labor and K the amount of capital used to produce the quantity z of a particular good. This is called a **Cobb–Douglas production function**.

It has been estimated that the production function for the United States in 1919 was (in appropriate units)

$$z = L^{0.76} K^{0.24} *$$

Find the marginal productivities of z.

SOLUTION The marginal productivity with respect to labor is simply the partial derivative of production with respect to labor:

$$\frac{\partial z}{\partial L} = 0.76 L^{-0.24} K^{0.24}$$

Similarly, the marginal productivity with respect to capital is

$$\frac{\partial z}{\partial K} = 0.24 L^{0.76} K^{-0.76}$$

Section 8.2 EXERCISES

In Exercises 1–10 find, for each given function, (a) $\partial f/\partial x$ and (b) $\partial f/\partial y$.

1. $f(x, y) = x \ln y + xy$
2. $f(x, y) = xe^y + 4x$
3. $f(x, y) = \frac{x + y}{y^2}$
4. $f(x, y) = x^2 y + e^{xy}$
5. $f(x, y) = \sqrt{x^2 + 10y}$
6. $f(x, y) = \frac{x - y}{x + y}$
7. $f(x, y) = e^{x+y} x^2 y$
8. $f(x, y) = e^{x^2} y$
9. $f(x, y) = y \ln x + x \ln y$
10. $f(x, y) = (x^2 + 5y)e^y$

* Paul H. Douglas and Grace T. Gunn, "The Production Function for American Manufacturing in 1919," *American Economic Review*, 31(1941), pp. 67–80.

In Exercises 11–20 take the functions of Exercises 1–10 consecutively, and compute the indicated values.

11. $f_x(1, 1)$, $f_y(3, 2)$, and $f_{xy}(4, 2)$

12. $f_x(3, 0)$ and $f_y(-1, 2)$

13. $\dfrac{\partial f}{\partial x}(3, 1)$ and $\dfrac{\partial^2 f}{\partial x^2}(1, 1)$

14. $f_{xx}(2, 2)$

15. $f_{yy}(4, 2)$

16. $\dfrac{\partial^2 f}{\partial y^2}(1, 2)$

17. $f(-1, 1)$

18. $f_x(2, 1) \cdot f_y(-2, 1)$

19. $\dfrac{f(2, 3)}{f_x(2, 3)}$

20. $[f(1, 2)f(-1, 2)]^2$

21. For the function $w = g(x, y) = \dfrac{x^2y + x + 1}{y}$, compute:

(a) w_x

(b) w_y

(c) w_{xx}

(d) w_{yy}

22. For $z = g(x, y) = xe^xy + y^2$, compute:

(a) $\dfrac{\partial g}{\partial x}$

(b) $\dfrac{\partial g}{\partial y}$

(c) g_{xy}

(d) g_{xx}

23. For the Cobb–Douglas production function of Example 10, with $\alpha = 0.80$ and $\beta = 2$ we have

$$z = f(L, K) = 2L^{0.8}K^{0.2}$$

Find:

(a) The marginal productivity with respect to labor.

(b) The marginal productivity with respect to capital.

24. Repeat Exercise 23 with $\alpha = 0.65$ and $\beta = 1.8$.

25. Repeat Exercise 23 with $\alpha = 0.45$ and $\beta = 1$.

26. A factory operation is found to follow a productivity function of the Cobb–Douglas type with $\beta = 200$ and $\alpha = 0.5$. L represents the amount of labor and K the amount of capital.

(a) Write down the productivity function $f(L, K)$.

(b) Find the marginal productivity of labor when $L = 105$ and $K = 92$.

(c) Find the marginal productivity of capital when $L = 84$ and $K = 100$.

27. Repeat Exercise 26 when $\beta = 100$ and $\alpha = 0.6$.

28. A certain small country is found to have a productivity function (not of the Cobb–Douglas type) $f(L, K) = (L^2 + 8K^2)^{3/2}$.

(a) Find the marginal productivity of labor when $L = 82$ and $K = 7$.

(b) Find the marginal productivity of capital when $L = 9$ and $K = 17$.

29. In a study by Douglas and Daly,* the production function of Canada was estimated to be of the Cobb–Douglas type with $\beta = 1$ and $\alpha = 0.42$.

(a) Find the marginal productivity function with respect to labor L.

(b) Find the marginal productivity function with respect to capital K.

30. The production function of Iowa farms in 1942† was estimated to be

$$z = x_1^{0.299}x_2^{0.256}x_3^{0.062}x_4^{0.200}x_5^{0.025}x_6^{0.158}$$

where z is the product, x_1 = land, x_2 = labor, x_3 = improvement, x_4 = liquid assets, x_5 = working assets, and x_6 = cash operating expenses. Find the marginal productivities.

31. Show that, for a Cobb–Douglas production function, if labor and capital are both multiplied by the same constant c, production is multiplied by c. In other words, show that $f(cL, cK) = cf(L, K)$.

32. Let $f(L, K) = \beta L^\alpha K^{1-\alpha}$, a Cobb–Douglas production function. Show that

$$L\frac{\partial f}{\partial L} + K\frac{\partial f}{\partial K} = f$$

33. For a certain suburb, let p_1 be the cost of commuting to work by bus, and let p_2 be the cost of commuting by cab. Let $f(p_1, p_2)$ be the number of people commuting to work by bus, and let $g(p_1, p_2)$ be the number commuting by cab. What would you expect to be true of the signs of $\partial f/\partial p_1$, $\partial f/\partial p_2$, $\partial g/\partial p_1$, and $\partial g/\partial p_2$?

34. The monthly payment for an n-year, \$$P$ mortgage at

* "The production Function for Canadian Manufacturers," *Journal of the American Statistical Assoc.*, XXXIX(1943), pp. 178–186.

† Gerhard Tintner, "A Note on the Derivation of Production Functions from Farm Records," *Econometrica*, 12(1944), pp. 26–34.

annual interest rate r (expressed as a decimal) is

$$f(P, r, n) = \frac{Pr/12}{\left[1 - \left(1 + \frac{r}{12}\right)^{-12n}\right]}$$

Compute the monthly payment when:

(a) $P = \$80{,}000$, $r = 0.14$, $n = 25$

(b) $P = \$100{,}000$, $r = 0.09$, $n = 30$

35. A company manufactures and sells tape recorders and radios. If recorders are priced at p each and radios at q, then the demand x for recorders and the demand y for radios are related to these prices by the following **joint demand functions**:

$$p = 150 - 5x + 2y$$

$$q = 240 + x - 6y$$

Suppose that fixed costs are \$400 and the unit costs for recorders and radios are \$40 and \$20, respectively. Find expressions for the revenue $R(x, y)$, the cost $C(x, y)$, and the profit $P(x, y)$.

36. An electronics outlet sells two competing grades of video tapes: standard, at \$$p$ each, and hi-grade, at \$$q$ each. The weekly demands x and y for the standard and hi-grade tapes are, respectively,

$$x = 450 - 24p + 13q$$

$$y = 200 + 20p - 18q$$

Find $\partial x/\partial p$, $\partial x/\partial q$, $\partial y/\partial p$, and $\partial y/\partial q$ and explain their significance. Are their signs reasonable? Why?

SECTION 8.3

Extrema of Functions of Two Variables

In Chapter 3 we studied the question of locating maxima and minima for a function of a single variable. In this section we extend that discussion to functions of several variables. For simplicity, we will restrict our discussion to a function $z = f(x, y)$ of two independent variables.

By analogy with Definition 3.2 we define a maximum of $z = f(x, y)$ as follows:

DEFINITION 8.2 The function $z = f(x, y)$ is said to have a **maximum** at the point (a, b) if $f(a, b) \geq f(x, y)$ for every (x, y) in the domain of f. The value $z = f(a, b)$ is called the **maximum value** of f.

Similarly, a **minimum** is said to occur at $(x, y) = (a, b)$ if $f(a, b) \leq f(x, y)$ for every point (x, y) in the domain of f. The value $z = f(a, b)$ in this case is called the **minimum value** of f. ⊟

Note that it is possible for the maximum (or minimum) value to be attained at more than one point.

In Section 8.2 we discussed a function $z = f(x, y)$ in which x and y represent the longitude and latitude, respectively, of a point in the continental United States, and $z = f(x, y)$ is the altitude of the land at this location, measured in feet above or below sea level. With this imagery in mind, we can view the location of a maximum as the site of the highest mountain peak in the United States (Mt. Whitney, in California, at 14,494 feet). The minimum is in Death Valley, California, 282 feet below sea level. As stated, our definition refers to the very highest peak and deepest valley. Of course, there are many other minor peaks and valleys.

DEFINITION 8.3 The function $z = f(x, y)$ is said to have a **local maximum** at the point (a, b) if $f(a, b) \geq f(x, y)$ for all points (x, y) near* the point (a, b).

Similarly, a **local minimum** occurs at a point (a, b) if $f(a, b) \leq f(x, y)$ for all (x, y) near (a, b). ⊟

Thus, for instance, the location of Mt. Rainier is a local maximum (the highest peak in the state of Washington), despite the fact that it is not the highest point in the continental United States.

Our previous study showed that, for functions of a single variable, the derivative provides an efficient way to locate potential maxima and minima. Specifically, for a function $y = f(x)$ we computed the derivative, $dy/dx = f'(x)$, and then set it equal to 0. This determined the critical points—those points at which the slope of the graph was 0. Let us generalize this procedure to functions of two variables.

FIRST DERIVATIVE TEST FOR FUNCTIONS OF TWO VARIABLES

If the differentiable function $z = f(x, y)$ attains either a local maximum or a local minimum at the point (a, b), then (assuming the partial derivatives exist), both $\dfrac{\partial f}{\partial x}(a, b)$ and $\dfrac{\partial f}{\partial y}(a, b)$ must vanish. Thus

$$\frac{\partial f}{\partial x}(a, b) = 0$$

$$\frac{\partial f}{\partial y}(a, b) = 0$$

Before we try to apply this test, let us see why it is valid.

Once again, consider the function $z = f(x, y)$ representing the altitude function at the location having longitude x and latitude y. Let us assume that the partial derivatives of f exist at every (x, y) in its domain. Next let (a, b) denote the coordinates of a local maximum of f. Thus (a, b) represents the location of a mountain peak. Now try to visualize driving west to east on a highway passing through the point (a, b), as shown in Figure 8.11. The y-value will remain constant at latitude $y = b$, while the elevation of the car will be described by $z = f(x, b)$ as x varies. The value $x = a$ on the highway, corresponding to the peak (a, b), represents a local maximum along our west-to-east highway, so the derivative of $z = f(x, b)$ with respect to x must be 0 at $x = a$. This is simply a restatement of the fact that

$$\frac{\partial f}{\partial x}(a, b) = 0$$

* By *near*, we mean "within some circular region centered at (a, b)."

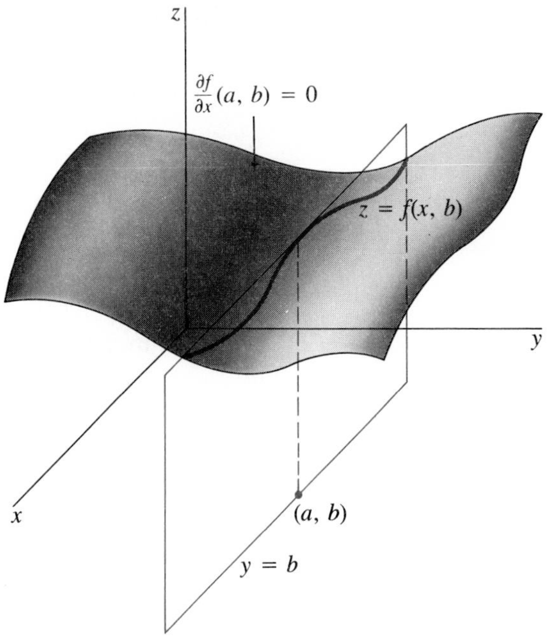

FIGURE 8.11
Hypothetical west-to-east highway with $y = b$

The condition that $\frac{\partial f}{\partial y}(a, b) = 0$ can be interpreted similarly by imagining a car traveling on a south–north route with fixed longitude $x = a$.

With the aid of the first-derivative test, we can now locate potential local maxima and minima.

EXAMPLE 11 Find the minimum of the function $z = 9x^2 - 6xy + 2y^2 - 6y + 11$.

SOLUTION The graph of this function is shown in Figure 8.12. It is evident from the picture that the function has a unique minimum. In order to determine the minimum, the first derivative test tells us to compute

$$\frac{\partial z}{\partial x} = 18x - 6y$$

and

$$\frac{\partial z}{\partial y} = -6x + 4y - 6$$

and then set these partial derivatives equal to 0. This produces the pair of simultaneous equations

$$18x - 6y = 0$$
$$-6x + 4y - 6 = 0$$

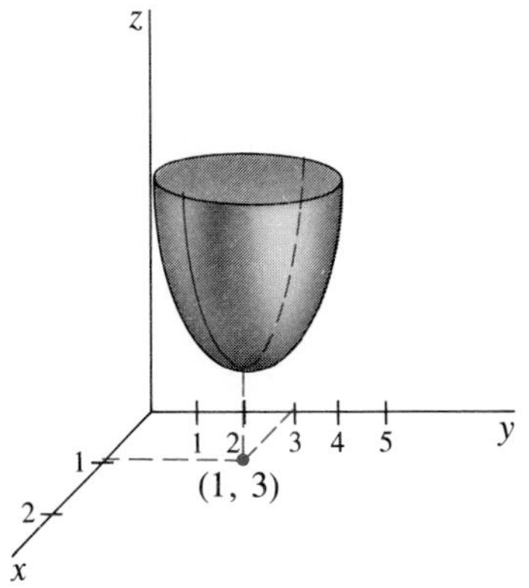

FIGURE 8.12
$z = 9x^2 - 6xy + 2y^2 - 6y + 11$

Multiplying the second equation by 3 and adding the resulting equations produces

$$\begin{aligned} 18x - 6y &= 0 \\ -18x + 12y &= 18 \\ \hline 6y &= 18 \\ y &= 3 \end{aligned}$$

Substituting this value into the first equation gives

$$18x - 6(3) = 0$$

so that $x = 1$. Thus the minimum point must be (1, 3). The value of $f(x, y)$ at this minimum point is

$$f(1, 3) = 9(1)^2 - 6(1)(3) + 2(3)^2 - 6(3) + 11 = 2$$

EXAMPLE 12 The distance from a point (x, y, z) in three-dimensional space to the origin (0, 0, 0) is given by the distance formula

$$d = \sqrt{x^2 + y^2 + z^2}$$

Let us attempt to find the point on the plane

$$z = x - y + 2$$

that is closest to the origin.

SOLUTION Figure 8.13 shows the graph of the plane $z = x - y + 2$. Because any point on the graph must satisfy its equation, the distance from a point on the plane to the origin is given by

$$d = \sqrt{x^2 + y^2 + (x - y + 2)^2}$$

Thus our problem is to minimize this function of two variables.

Before beginning our computations, we can simplify our task by observing that the minimum of d must occur at the same point as the minimum of the same expression without the square root sign (although the minimum value is different). So we shall minimize

$$f(x, y) = x^2 + y^2 + (x - y + 2)^2$$

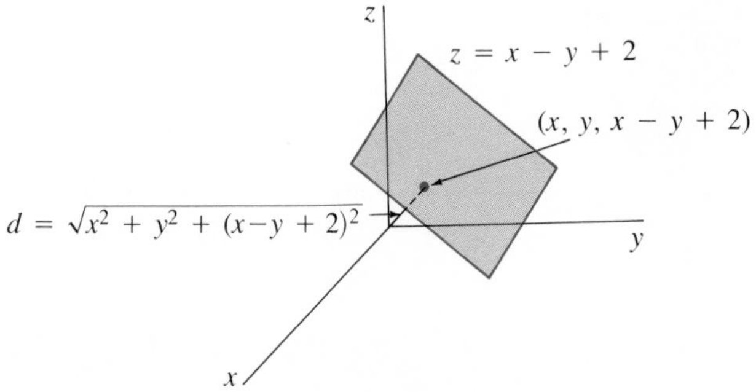

FIGURE 8.13

We compute

$$\frac{\partial f}{\partial x} = 2x + 2(x - y + 2)$$

$$\frac{\partial f}{\partial y} = 2y - 2(x - y + 2)$$

Next we set $\frac{\partial f}{\partial x} = \frac{\partial f}{\partial y} = 0$, to find

$$4x - 2y + 4 = 0$$

$$-2x + 4y - 4 = 0$$

We solve these equations simultaneously (multiply the second equation by 2 and add the equations) to obtain $x = -\frac{2}{3}, y = \frac{2}{3}$. Thus $(-\frac{2}{3}, \frac{2}{3})$ is the point on the plane $z = x - y + 2$ that is closest to the origin. The distance is found by substituting these values into the expression for d:

$$d = \sqrt{(-\tfrac{2}{3})^2 + (\tfrac{2}{3})^2 + (-\tfrac{2}{3} - \tfrac{2}{3} + 2)^2} = \sqrt{\tfrac{4}{3}}$$

EXAMPLE 13 A manufacturer produces two different items. The demand functions for the two items are found to be

$$p_1(x) = 28 - 3x$$

and

$$p_2(y) = 22 - 2y$$

where x and y are the quantities of the two items, and p_1 and p_2 are the corresponding prices. The joint cost function is given as

$$C(x, y) = x^2 + 3y^2 + 4xy$$

Find the production levels that will maximize the manufacturer's profits.

SOLUTION We know that

$$\text{Profit} = \text{revenue} - \text{cost}$$

The total revenue consists of the revenue from the sale of the two commodities, which we denote by $R_1(x)$ and $R_2(y)$, and these are given by

$$R_1(x) = xp_1(x) = 28x - 3x^2$$

$$R_2(y) = yp_2(y) = 22y - 2y^2$$

The profit function is now

$$f(x, y) = (28x - 3x^2) + (22y - 2y^2) - (x^2 + 3y^2 + 4xy)$$

We determine the maximum of this function by applying the first derivative test:

$$\frac{\partial f}{\partial x} = 28 - 6x - 2x - 4y$$

$$\frac{\partial f}{\partial y} = 22 - 4y - 6y - 4x$$

Thus we get

$$8x + 4y = 28$$
$$4x + 10y = 22$$

so that

$$x = 3 \quad \text{and} \quad y = 1$$

The maximal profit is

$$f(3, 1) = 57 + 20 - 24 = 53$$

EXAMPLE 14 A rectangular box is to have a volume of 50 cubic inches. Find the dimensions of the box that will have the smallest surface area.

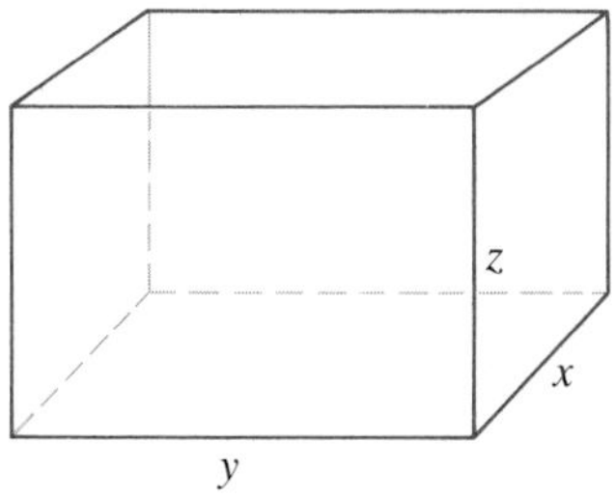

FIGURE 8.14

SOLUTION Let x, y, and z denote the length, width, and height of the box, respectively (see Figure 8.14). We have been told that

$$xyz = 50$$

so

$$z = \frac{50}{xy}$$

The surface area of the box is the sum of the areas of its six faces. Thus

$$\text{Surface area} = 2xy + 2xz + 2yz$$

Because $z = 50/xy$, we can substitute into the expression for the surface area to obtain a function of only two variables.

$$\begin{aligned}\text{Surface area} = f(x, y) &= 2xy + 2x(50/xy) + 2y(50/xy)\\ &= 2xy + (100/y) + (100/x)\end{aligned}$$

Our goal is to minimize $f(x, y)$. Employing the first derivative test, we compute

$$\frac{\partial f}{\partial x} = 2y - \frac{100}{x^2}$$

$$\frac{\partial f}{\partial y} = 2x - \frac{100}{y^2}$$

Setting these equal to 0 gives

$$y = \frac{50}{x^2} \quad \text{and} \quad x = \frac{50}{y^2}$$

so

$$y = \frac{50}{x^2} = \frac{50}{\left(\frac{50}{y^2}\right)^2} = \frac{y^4}{50}$$

or

$$y^4 = 50y$$

Now one solution of this equation is $y = 0$, which we can immediately discard because xyz must equal 50. Thus, because $y \neq 0$, we can divide through by y to obtain

$$y^3 = 50$$

or

$$y = \sqrt[3]{50}$$

The equation $x = 50/y^2$ now gives $x = \sqrt[3]{50}$ also. Finally, because $z = 50/xy$, we find that $z = \sqrt[3]{50}$ too.

We conclude that, of all rectangular boxes with a fixed volume of 50 cubic inches, the one having the minimal surface area must be a cube.

In the preceding examples we knew in advance from physical or geometrical intuition that the function we were considering had a unique maximal or minimal value. The first-derivative test then enabled us to find this extremum quickly.

More generally, however, if we start with a function $f(x, y)$ there may be a number of local extrema that will be found by applying the first-derivative test.

Any point (a, b) for which

(8.1) $$\frac{\partial f}{\partial x}(a, b) = 0 \quad \text{and} \quad \frac{\partial f}{\partial y}(a, b) = 0$$

is called a **critical point**. A critical point may be a local maximum or a local minimum, or it may be a saddle point. A **saddle point** is the multivariable analog of a critical point that is also a point of inflection. At a saddle point the graph is concave downward in some directions and concave upward in others. A typical saddle point is shown in Figure 8.15. The choice of terminology is obvious from the picture. In order to characterize a critical point as a local maximum, a local minimum, or a saddle point, we can use the second derivative test for maxima and minima. This test is similar in principle to the second derivative test for functions of a single variable that we studied in Section 3.3 though it is more complicated to apply. We simply state the test without proof and show how it is applied.

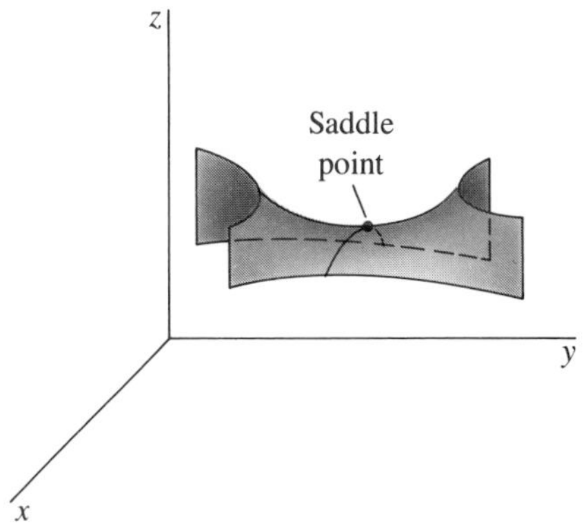

FIGURE 8.15
Saddle point for $f(x, y)$

SECOND DERIVATIVE TEST FOR MAXIMA AND MINIMA:

Suppose $f(x, y)$ is a function of two variables having continuous first-order and second-order partial derivatives near (a, b), and let

$$D(x, y) = \left(\frac{\partial^2 f}{\partial x^2}\right)\left(\frac{\partial^2 f}{\partial y^2}\right) - \left(\frac{\partial^2 f}{\partial x\, \partial y}\right)^2$$

Suppose further that (a, b) is a critical point for f. That is,

$$\frac{\partial f}{\partial x}(a, b) = \frac{\partial f}{\partial y}(a, b) = 0$$

Then:

(i) If $D(a, b) > 0$ and $\dfrac{\partial^2 f}{\partial x^2}(a, b) > 0$, f has a local minimum at (a, b).

(ii) If $D(a, b) > 0$ and $\dfrac{\partial^2 f}{\partial x^2}(a, b) < 0$, f has a local maximum at (a, b).

(iii) If $D(a, b) < 0$, f has a saddle point at (a, b).

(iv) If $D(a, b) = 0$, the second derivative test is inconclusive.

Let us demonstrate the use of the second derivative test with some examples.

EXAMPLE 15 Find all critical points of the function $f(x, y) = x^2 + y^2 - 4xy + 10x - 14y + 18$, and classify them using the second derivative test.

SOLUTION We first determine the critical points. We compute

$$\frac{\partial f}{\partial x} = 2x - 4y + 10$$

$$\frac{\partial f}{\partial y} = 2y - 4x - 14$$

and set these equal to 0.

$$2x - 4y + 10 = 0$$

$$-4x + 2y - 14 = 0$$

We solve simultaneously by doubling the first equation and adding it to the second.

$$\begin{array}{r} 4x - 8y + 20 = 0 \\ -4x + 2y - 14 = 0 \\ \hline -6y + 6 = 0 \\ y = 1 \end{array}$$

Substituting into either of the equations yields

$$x = -3$$

Thus $(-3, 1)$ is the only critical point. In order to determine whether this represents a local maximum, a local minimum, or a saddle point, we compute the second partial derivatives.

$$\frac{\partial^2 f}{\partial x^2} = 2 \qquad \frac{\partial^2 f}{\partial y^2} = 2 \qquad \frac{\partial^2 f}{\partial x\,\partial y} = -4$$

Therefore

$$D(-3, 1) = \left(\frac{\partial^2 f}{\partial x^2}\right)\left(\frac{\partial^2 f}{\partial y^2}\right) - \left(\frac{\partial^2 f}{\partial x\,\partial y}\right)^2 = (2)(2) - (-4)^2 = -12$$

By part (iii) of the second derivative test, $(-3, 1)$ is a saddle point.

EXAMPLE 16 Find and classify all critical points of the function $f(x, y) = x^3 + y^2 - 3xy - 6x + 5y + 1$.

SOLUTION $\dfrac{\partial f}{\partial x} = 3x^2 - 3y - 6.$

$$\frac{\partial f}{\partial y} = 2y - 3x + 5$$

Setting these equal to 0 gives

$$3x^2 - 3y - 6 = 0$$
$$2y - 3x + 5 = 0$$

The first equation can be solved for y in terms of x.

$$3y = 3x^2 - 6$$
$$y = x^2 - 2$$

We can then substitute this expression into the second equation.

$$2(x^2 - 2) - 3x + 5 = 0$$
$$2x^2 - 3x + 1 = 0$$

This factors to give $(2x - 1)(x - 1) = 0$, which has two roots,

$$x = \tfrac{1}{2}, \quad \text{and} \quad x = 1$$

From the equation $y = x^2 - 2$ we find that, when $x = \frac{1}{2}$, $y = -\frac{7}{4}$ and that, when $x = 1$, $y = -1$. Thus $(\frac{1}{2}, -\frac{7}{4})$ and $(1, -1)$ are the critical points.

To determine the nature of each of these critical points we compute

$$\frac{\partial^2 f}{\partial x^2} = 6x \qquad \frac{\partial^2 f}{\partial y^2} = 2 \qquad \frac{\partial^2 f}{\partial x\,\partial y} = -3$$

For the critical point $(\frac{1}{2}, -\frac{7}{4})$, we have

$$D(\tfrac{1}{2}, -\tfrac{7}{4}) = \left(\frac{\partial^2 f}{\partial x^2}(\tfrac{1}{2}, -\tfrac{7}{4})\right)\left(\frac{\partial^2 f}{\partial y^2}(\tfrac{1}{2}, -\tfrac{7}{4})\right) - \left(\frac{\partial^2 f}{\partial x\,\partial y}(\tfrac{1}{2}, -\tfrac{7}{4})\right)^2$$
$$= [(6)(\tfrac{1}{2})]2 - [-3]^2 = -3$$

From part (iii) of the second derivative test, $(\frac{1}{2}, -\frac{7}{4})$ is a saddle point.

We similarly test the point $(1, -1)$.

$$D(1, -1) = [(6)(1)][2] - [-3]^2 = 3$$

Since $\frac{\partial^2 f}{\partial x^2}(1, -1) = 6 > 0$, the point $(1, -1)$ is a local minimum according to part (i) of the second derivative test.

Section 8.3 EXERCISES

In Exercises 1–10 use the first-derivative test to find all critical points of the given function.

1. $f(x, y) = x^2 - 4x + y^2 + 6y + 12$
2. $f(x, y) = x^2 - 4x - y^2 - 4y + 15$
3. $f(x, y) = x^2 - 8xy + 3y^2 + 4x + 2y + 12$
4. $f(x, y) = \frac{x^3}{3} - x + \frac{y^2}{2} + 2y - 4$
5. $f(x, y) = \frac{x^3}{3} - x - \frac{y^2}{2} + 2y + 10$
6. $f(x, y) = \frac{x^4}{4} + \frac{2x^3}{3} + \frac{x^2}{2} + \frac{y^3}{3} - 9y + 1$
7. $g(x, y) = xy + e^y$
8. $g(x, y) = e^{xy} + 2x$
9. $f(x, y) = -x^2 + 2x - y^2 - 4y - 7$
10. $f(x, y) = x^{0.23}y^{0.77}$ for $x > 0, y > 0$

11–19. Apply the second derivative test to the critical points you found in Exercises 1–9 so as to determine which critical points give maxima, which give minima, and which are saddle points. If the test fails, say so.

20. An open-topped rectangular box is to have a volume of 100 in.^{3}. Find the dimensions of the box that will require the least amount of material to make (have the smallest surface area).

21. Suppose we turn Exercise 20 around a bit and specify that an open-topped rectangular box is to have surface area 10 in.^{2}. Does there exist a box of maximal volume with this restriction? If so, what are its dimensions? What about a box of minimal volume?

22. A manufacturer produces two different varieties of toothpaste. The demand functions for these two items are

$$p_1(x) = 56 - 3x$$
$$p_2(y) = 64 - 4y$$

where x and y are quantities of the two items, and p_1 and p_2 are the corresponding prices. The **joint cost function** is given by

$$C(x, y) = 2x^2 + 4y^2 + 8xy$$

Find the production levels that will maximize the manufacturer's profit.

23. Redo Exercise 22 for:

$$p_1(x) = 60 - 4x$$
$$p_2(y) = 56 - 3y$$

with $C(x, y)$ as in Exercise 22.

24. A company sells three products (call them I, II, and III) at \$6, \$12, and \$8 per unit, respectively. The cost of producing x units of product I, y units of product II, and z units of product III is

$$C(x, y, z) = 2x^2 + 3y^2 + 4z^2$$

Find the value of x, y, and z that will maximize the company's profit.

25. Suppose the demand functions for the products A and B that a company produces are constant. That is,

$$p_1(x) = p_1 \quad \text{and} \quad p_2(x) = p_2 \qquad \text{(constants)}$$

Let the cost function be $C(x, y)$. Show that the maximum profit will occur at production level (x, y), where

$$\frac{\partial C}{\partial x}(x, y) = p_1 \quad \text{and} \quad \frac{\partial C}{\partial y}(x, y) = p_2$$

In Exercises 26–29 find the optimal production level for p_1, p_2, and $C(x, y)$ as given. Also find the maximal profit.

26. $p_1 = \$10$ per unit $\quad p_2 = \$16$ per unit
$C(x, y) = x^2 + 4y^2$

27. $p_1 = \$8$ per unit $\quad p_2 = \$3$ per unit
$C(x, y) = xy$

28. $p_1 = \$23$ per unit $\quad p_2 = \$32$ per unit
$C(x, y) = x^2 + 2y^2 + 3xy$.

29. $p_1 = \$4$ per unit $\quad p_2 = \$14$ per unit
$C(x, y) = x^2 + 6y^2 + 2xy$

30. Postal regulations stipulate that the length plus the girth of a rectangular package cannot exceed 72 in. Find the dimensions of the package of largest volume that can be sent. (In the figure, z = length; girth = perimeter of cross section = $2x + 2y$.)

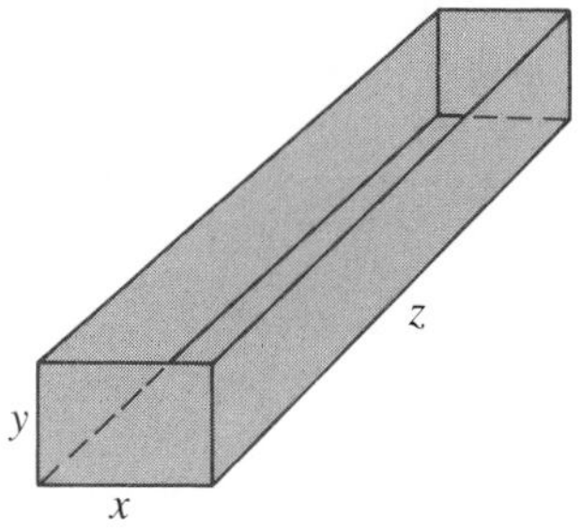

FIGURE FOR EXERCISE 30

31. Find three positive numbers whose sum is 30 and whose sum of squares is a minimum.

32. A manufacturing company estimates that its profit is given by

$$P(x, y) = -4x^2 + 4xy - 2y^2 + 16x - 8y + 23$$

where x is the amount spent on research and development, and y is the amount spent on advertising (P, x, and y are measured in thousands of dollars). How much should be budgeted for research and development and for advertising to maximize profit?

33. Redo Exercise 20 if the box is to have a vertical partition running across the middle.

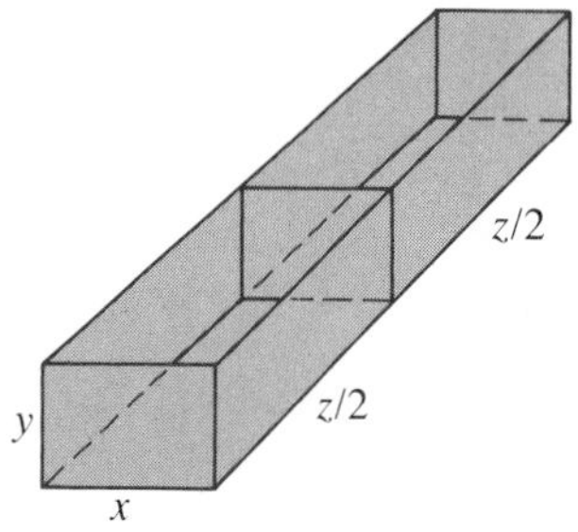

FIGURE FOR EXERCISE 33

SECTION 8.4

The Method of Least Squares

In Examples 28 and 29 in Chapter 3, we discussed the following type of problem. Given a collection of data points $(x_1, y_1), (x_2, y_2), \ldots, (x_n, y_n)$ which, from theoretical considerations, we believed should lie on a straight line through the origin, $y = mx$, we used optimization to select that value of m for which $y = mx$ best fits the given data.

In the more general problem, we try to find the equation of a line $y = mx + b$ that best fits a given set of data. (Our previous work corresponds to the special case $b = 0$.) Suppose that we have n data points, which we denote by

$$P_1(x_1, y_1), P_2(x_2, y_2), \ldots, P_n(x_n, y_n)$$

and let $y = mx + b$ be an arbitrary straight line. Let

$$(x_1, \tilde{y}_1), (x_2, \tilde{y}_2), \ldots, (x_n, \tilde{y}_n)$$

denote the points on the line $y = mx + b$ with the same x-coordinates as $P_1, P_2, \ldots, P_n$ (see Figure 8.16.)

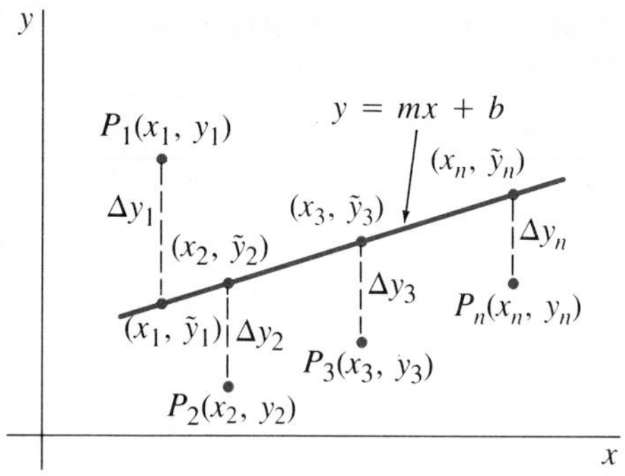

FIGURE 8.16

As before, the vertical deviations of the data points from the line are denoted by $\Delta y_1, \Delta y_2, \ldots, \Delta y_n$ and are given by

$$\Delta y_i = y_i - \tilde{y}_i \qquad i = 1, 2, \ldots, n$$

When the deviations are very small, the line fits the data very well; large deviations correspond to a poor fit. The criterion that is generally accepted to define the line of best approximation is to select the values of m and b that result in the smallest value for the sum of the squares of the deviations. That is, we seek to minimize:

(8.2)
$$E = (\Delta y_1)^2 + (\Delta y_2)^2 + \cdots + (\Delta y_n)^2$$

Note that if, all our data points lie on the line $y = mx + b$, then $\Delta y_i = 0$, $i = 1, 2, \ldots, n$, so $E = 0$ in this case. In general, however, $E > 0$. Expression (8.2) contains two variables, m and b. We obtain its minimum by applying the first-derivative test for a function of several variables.

Let us illustrate this by means of an example.

EXAMPLE 17 A computer manufacturer specializes in the production of small business computers. It currently has sales representatives operating in five sales territories and has collected the following data for the previous year.

TABLE 8.1

Sales territory number	*Population in 10,000's* x_i	*Sales in \$10,000's* y_i
1	2	6
2	5	12
3	3	8
4	7	12
5	2	8

(a) Determine the equation of the straight line that best approximates the data given in Table 8.1.

(b) Management is planning to hire a new sales representative to cover a sixth territory that has a population of 60,000 people. What should be the sales quota for the new territory?

SOLUTION

(a) We begin with the general equation of a line, $y = mx + b$. Using this equation, we next compute the y-values $\tilde{y}_1, \tilde{y}_2, \ldots, \tilde{y}_5$ for points on the line having the same x-coordinates as the given data. For instance, when $x_1 = 2$, then $\tilde{y}_1 = m(2) + b = 2m + b$. The deviations are simply the difference between the observed data values and the theoretical ones. We have tabulated these calculations in Table 8.2.

TABLE 8.2

Sales territory number	x_i	y_i	$\tilde{y}_i = mx_i + b$	$\Delta y_i = y_i - \tilde{y}_i$
1	2	6	$2m + b$	$6 - (2m + b)$
2	5	12	$5m + b$	$12 - (5m + b)$
3	3	8	$3m + b$	$8 - (3m + b)$
4	7	12	$7m + b$	$12 - (7m + b)$
5	2	8	$2m + b$	$8 - (2m + b)$

The sum of the squared deviations, which measures the accuracy of fit, is now given by

$$E = (6 - 2m - b)^2 + (12 - 5m - b)^2 + (8 - 3m - b)^2 + (12 - 7m - b)^2 + (8 - 2m - b)^2$$

We can determine the unique minimum value of E by computing $\partial E/\partial m$ and $\partial E/\partial b$ and then setting these partial derivatives equal to 0:

$$\begin{aligned}\frac{\partial E}{\partial m} &= 2(6 - 2m - b)(-2) + 2(12 - 5m - b)(-5) + 2(8 - 3m - b)(-3) \\ &\quad + 2(12 - 7m - b)(-7) + 2(8 - 2m - b)(-2) \\ &= 182m + 38b - 392\end{aligned}$$

$$\begin{aligned}\frac{\partial E}{\partial b} &= 2(6 - 2m - b)(-1) + 2(12 - 5m - b)(-1) + 2(8 - 3m - b)(-1) \\ &\quad + 2(12 - 7m - b)(-1) + 2(8 - 2m - b)(-1) \\ &= 38m + 10b - 92\end{aligned}$$

Thus we must solve the simultaneous equations

$$182m + 38b - 392 = 0$$

$$38m + 10b - 92 = 0$$

And we will leave it to you to find that

$$m = 1.128 \quad \text{and} \quad b = 4.915$$

Thus the line of best fit is

$$y = 1.128x + 4.915$$

(b) Once we have determined the line of best fit (also called the **regression line**), we can use it to make predictions. A population of 60,000 corresponds to an x-value of 6. Thus, if we want to predict the sales level in the new sales territory, we simply compute

$$y = 1.128(6) + 4.915 = 11.683$$

In other words, we can anticipate a sales level of \$116,830.00. Figure 8.17 shows the data from Table 8.1, together with the regression line.

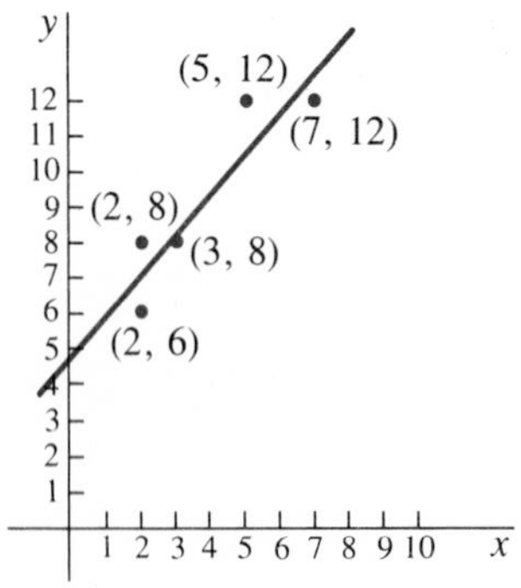

FIGURE 8.17
$y = 1.128x + 4.915$

The technique employed in the previous example is called the **method of least squares**. It also called **simple linear regression**. As noted in Example 28 in Chapter 3, this is an important statistical technique that is useful in numerous applications. Let us try to repeat the procedure just developed using more general notation.

Suppose we are given n data points

$$P_1(x_1, y_1), P_2(x_2, y_2), \ldots, P_n(x_n, y_n)$$

Let us try to find the regression line determined by these points. We construct Table 8.3 analogous to Table 8.2. The sum of the squared deviations is

$$E = [y_1 - (mx_1 + b)]^2 + [y_2 - (mx_2 + b)]^2 + \cdots + [y_n - (mx_n + b)]^2$$

To minimize E we first compute $\dfrac{\partial E}{\partial m}$ and $\dfrac{\partial E}{\partial b}$:

$$\begin{aligned}\frac{\partial E}{\partial m} &= 2[y_1 - mx_1 - b][-x_1] + 2[y_2 - mx_2 - b][-x_2] \\ &\quad + \cdots + 2[y_n - mx_n - b][-x_n] \\ &= 2[m(x_1^2 + x_2^2 + \cdots + x_n^2) - (x_1y_1 + x_2y_2 + \cdots + x_ny_n) \\ &\quad + b(x_1 + x_2 + \cdots + x_n)]\end{aligned}$$

$$\begin{aligned}\frac{\partial E}{\partial b} &= 2[y_1 - mx_1 - b][-1] + 2[y_2 - mx_2 - b][-1] \\ &\quad + \cdots + 2[y_n - mx_n - b][-1] \\ &= 2[-(y_1 + y_2 + \cdots + y_n) + m(x_1 + x_2 + \cdots + x_n) + nb]\end{aligned}$$

Next we set the resulting partial derivatives equal to 0 and divide both equations by 2. We obtain

(8.3) $$\begin{aligned} m(x_1^2 + x_2^2 + \cdots + x_n^2) + b(x_1 + x_2 + \cdots + x_n) &= (x_1y_1 + x_2y_2 + \cdots + x_ny_n) \\ m(x_1 + x_2 + \cdots + x_n) + nb &= y_1 + y_2 + \cdots + y_n\end{aligned}$$

We can use these equations to compute the regression line quite quickly by using a hand calculator, as we illustrate in the next example.

TABLE 8.3

x_i	y_i	$\tilde{y}_i = mx_i + b$	$\Delta y_i = y_i - y_i$
x_1	y_1	$mx_1 + b$	$y_1 - (mx_1 + b)$
x_2	y_2	$mx_2 + b$	$y_2 - (mx_2 + b)$
$\vdots$	$\vdots$	$\vdots$	$\vdots$
x_n	y_n	$mx_n + b$	$y_n - (mx_n + b)$

EXAMPLE 18 The earnings per share of American Telephone & Telegraph Company for the decade 1971–1980 are shown in the accompanying table.

(a) Compute the regression line for these data.

Year	1971	1972	1973	1974	1975	1976	1977	1978	1979	1980
Earnings per share	3.91	4.34	4.98	5.27	5.13	6.05	6.97	7.74	8.04	8.19

(b) Predict the earnings per share for 1981 on the basis of your calculations.

SOLUTION

(a) Let x_i denote the ith year. That is, $x_1 = 1$ corresponds to 1971; $x_2 = 2$ corresponds to 1972, and so on. Let y_i denote the earnings per share in the ith year. We construct Table 8.4 as our table of values.

TABLE 8.4

x_i	y_i	x_i^2	$x_i y_i$
1	3.91	1	3.91
2	4.34	4	8.68
3	4.98	9	14.94
4	5.27	16	21.08
5	5.13	25	25.65
6	6.05	36	36.30
7	6.97	49	48.79
8	7.74	64	61.92
9	8.04	81	72.36
10	8.19	100	81.90
Sum = 55	60.62	385	375.53

From this table, we can fill in the appropriate values into Equations (8.3).

$$385m + 55b = 375.53$$

$$55m + 10b = 60.62$$

The two linear equations must be solved simultaneously. We can do this by multiplying the first equation by 2 and the second by 11 and then subtracting the resulting equations.

$$\begin{aligned} 770m + 110b &= 751.06 \\ 605m + 110b &= 666.82 \\ \hline 165m \qquad\quad &= 84.24 \end{aligned}$$

$$m = 0.5105$$

To find the value of b, we use the second equation.

$$55(0.5105) + 10b = 60.62$$

$$b = 3.254$$

Thus the regression line is

$$y = 0.5105 + 3.254$$

(b) In order to predict the earnings of American Telephone & Telegraph in 1981, we simply set $x = 11$. Then

$$y = (0.5105)(11) + 3.254 \approx \$8.87$$

The actual earnings in that year were \$8.55 per share.

Actually the computational work can be simplified a bit. If we use the standard notation

$$\sum_{i=1}^{n} x_i = x_1 + x_2 + \cdots + x_n$$

to denote a sum, we can abbreviate Equations (8.3) in the unknowns m and b as follows:

$$m(\sum x_i^2) + b(\sum x_i) = \sum x_i y_i$$
$$m(\sum x_i) + bn = \sum y_i$$

where i runs from 1 to n in each sum. The solution to this system is

(8.4)

$$m = \frac{n \sum x_i y_i - (\sum x_i)(\sum y_i)}{n \sum x_i^2 - (\sum x_i)^2}$$
$$b = \frac{1}{n}(\sum y_i - m \sum x_i)$$

For example, in Example 18 these give

$$m = \frac{10(375.53) - 55(60.62)}{10(385) - (55)^2} = 0.5105$$

$$b = \frac{1}{10}[60.62 - 0.5105(55)] = 3.254$$

the same results as before.

PRACTICE EXERCISE Consider the following data for the consumption of wheat in the United States.

Year	*Consumption (millions of bushels)*
1910	36
1915	43
1920	34
1925	41
1930	47

(a) Fit the data with the best linear approximation, using the method of least squares.

(b) Extrapolate the data to predict wheat consumption in 1940. (*Hint*: Let 1910 correspond to $x_1 = 0$, 1915 to $x_1 = 1$, and so on.)

[Your answers should be (a) $y = 2x + 36.2$ and (b) 48.2 (millions of bushels).]

Section 8.4 EXERCISES

In Exercises 1–10

(a) Find the "best" least-squares straight line to fit the data.

(b) Compute the value of the error E corresponding to this "best" fit.

1.

x	y
1	7.1
2	9.8
3	12.9

2.

x	y
0	−2.1
−1	−7.2
3	13.1

3.

x	y
0	6
1	−1
2	8
3.4	2

4.

x	y
2	1
3	0
4	−5

5.

x	y
1	3
2	5
3	7
4	9

6.

x	y
1	1
2	2
3	−8
4	4

7.

x	y
0	1
2	2

8.

x	y
1	8
0	0.1
2	15.7

9.

x	y
1	1
2	4
3	9

10.

x	y
−2	3
−1	2.8
0	3.1
+1	2.9
+2	3.2

11. For the data of Exercise 1, extrapolate the value of y when $x = 4$.

12. For the data of Exercise 2, estimate the value of y when $x = 2$.

13. Five women go on a 3-month reducing diet. Their initial and final weights are given in the following table.

Initial weight	*Final weight*
150	124
160	126
170	137
180	140
190	142

(a) Find the best least-squares line fitting initial weight to final weight at the end of this reducing diet.

(b) Use your line to predict the final weight of a woman entering this plane with an initial weight of 175 pounds.

14. The earnings per share of Procter & Gamble from 1975 to 1981 are given in the accompanying table.

(a) Compute the regression line for these data, taking 1975 as year 1, etc.

Year	1975	1976	1977	1978	1979	1980	1981
Earnings	2.03	2.43	2.80	3.10	3.50	3.89	4.04

(b) Estimate the earnings per share for 1982 on the basis of your calculations.

15. Repeat Exercise 14 for Squibb Corporation.

Year	1975	1976	1977	1978	1979	1980	1981
Earnings	2.18	2.40	2.50	2.60	2.71	2.65	2.23

16. A new computer game cartridge reaches the market on January 1. Sales in four consecutive months are indicated in the following table, where month 1 = January, month 2 = February, and so on. Estimate the number of units to be sold in June—that is, month 6.

Month	*Number of units sold*
1	750
2	840
3	918
4	1005

IF YOU WORK WITH A COMPUTER (OPTIONAL)

Use either of the following computer programs to do Exercises 17–22. If your version of BASIC either (a) does not allow long variable names or (b) considers only the first two characters of a variable name significant, make the appropriate modifications to the BASIC version of the program.

17. Redo Example 18.

18. Redo the practice exercise on page 362.

19. The price range of shares of the Uplift Company, Inc., during the years 1980 to 1986 is given in the following table.

	High	*Low*
1980	20 1/2	18 3/8
1981	22 5/8	19 1/2
1982	27 7/8	23 1/8
1983	28 1/2	25 3/8
1984	35 1/8	29 5/8
1985	39 1/2	32

(a) Find the best least-squares line to fit the yearly high prices of this stock. (*Hint*: Take 1980 as year 0, 1981 as year 1, and so on.)

(b) Use this line to predict the approximate high price for 1988.

20. Refer to Exercise 19. Find the best least-squares line to fit the yearly low prices of the Uplift Company's stock.

21. Use the line you found in Exercise 20 to predict the lowest price for this stock in 1990.

22. Redo Exercise 14.

```
10   REM - Method of Least Squares - BASIC version

20   INPUT "How many points will you input"; n
30     sumx = 0                       'initialize sums
40     sumy = 0
50     sumx2 = 0
60     sumxy = 0

70   FOR i=1 TO n
80     INPUT "x, y"; x,y              'input next point
90     sumx = sumx + x
100    sumy = sumy + y
110    sumx2 = sumx2 + x*x
120    sumxy = sumxy + x*y
130  NEXT i

140  m = (n*sumxy - sumx*sumy) / (n*sumx2 - sumx*sumx)
150  b = (sumy - m*sumx) / n

160  PRINT "Least squares fit is  y ="; m; "x +"; b
170  END
```

```
(* Method of Least Squares - Pascal version *)
program LeastSquares (input, output);

 var
  n, i : integer;
  sumx, sumy, sumx2, sumxy, x, y, m, b : real;

begin
 write('How many points will you input? ');
 readln(n);
 sumx := 0;
 sumy := 0;
 sumx2 := 0;
 sumxy := 0;

 for i := 1 to n do
  begin
   write('x y (separated by a space)? ');
   readln(x, y);
   sumx := sumx + x;
   sumy := sumy + y;
   sumx2 := sumx2 + x * x;
   sumxy := sumxy + x * y
  end;

 m := (n * sumxy - sumx * sumy) / (n * sumx2 - sqr(sumx));
 b := (sumy - m * sumx) / n;
 write('Least squares fit is y = ', m : 1 : 4);
 writeln('x + ', b : 1 : 4)
end.
```

SECTION 8.5

Constrained Optimization—Lagrange Multipliers

In optimization problems, we are frequently asked to maximize or minimize an objective function while simultaneously restricting the permissible values of the independent variables. These restrictions are called **constraints**. It is easy to think of optimization problems in everyday life in which constraints arise from practical considerations: Automobile manufacturers try to design cars to be as safe as possible while keeping the cost of the car within a specified range. Airlines attempt to schedule their planes to maximize profits, subject to limitations on the number and type of planes they have available.

Let us consider a typical constrained optimization problem—one that we considered earlier in Chapter 3, Example 15.

A homeowner wishes to plant a rectangular vegetable garden. The garden will be located along the edge of a straight brook that runs through the property. He buys 50 feet of fencing to enclose his garden along the three sides that do not border the brook. What should be the dimensions of the garden if its area is to be maximized?

In mathematical terms, our goal was to maximize the objective function

$$A = xy$$

where x and y are the length and width of the garden, respectively, subject to the constraint

$$x + 2y = 50$$

(See Figure 3.51). Our solution consisted of solving the constraint equation for x in terms of y:

$$x = 50 - 2y$$

and then substituting this expression into the objective function:

$$A = xy = (50 - 2y)y$$

The resulting function of a single variable was easily maximized.

This method can be very cumbersome if the constraint equation is complicated. In this case it can be quite difficult to solve for one variable in terms of the other. Even worse, some constraint equations, such as

$$x + y - e^{xy} = 0$$

simply have no solution for y as a function of x, or vice versa.

The mathematician Joseph Lagrange (1736–1813) devised an alternative technique for solving constrained optimization problems that avoids this difficulty. It is reasonably straightforward, though the proof of its validity is somewhat complicated and so we have omitted it. The technique is called the **method of Lagrange multipliers**.

Consider the problem of optimizing the objective function $f(x, y)$ subject to the constraint equation $g(x, y) = 0$. We begin by introducing an auxiliary variable into the problem. This variable is usually denoted by the Greek letter λ (lambda) and is called the **Lagrange multiplier**. We next form a function of three variables

(8.5)
$$F(x, y, \lambda) = f(x, y) + \lambda g(x, y)$$

Then, according to Lagrange's method, at any local maximum or minimum of $f(x, y)$ subject to the constraint $g(x, y) = 0$, the following must be satisfied:

(8.6)
$$\frac{\partial F}{\partial x} = 0 \qquad \frac{\partial F}{\partial y} = 0 \qquad \frac{\partial F}{\partial \lambda} = 0$$

Equivalently,

$$\frac{\partial f}{\partial x} + \lambda \frac{\partial g}{\partial x} = 0 \qquad \frac{\partial f}{\partial y} + \lambda \frac{\partial g}{\partial y} = 0 \qquad g(x, y) = 0$$

In other words, any maximal or minimal value of $f(x, y)$ subject to the constraint $g(x, y) = 0$ must be a critical point of the unconstrained function of three variables defined by Equation (8.5).

Let us apply this method to the problem of designing a vegetable garden. Our objective function there was

$$f(x, y) = xy$$

The constraint equation, $x + 2y = 50$, must be written in the form $g(x, y) = 0$. Thus,

$$g(x, y) = x + 2y - 50 = 0$$

According to the foregoing discussion, we form the new function of three variables

$$F(x, y, \lambda) = xy + \lambda(x + 2y - 50)$$

We then compute

$$\frac{\partial F}{\partial x} = y + \lambda = 0$$

$$\frac{\partial F}{\partial y} = x + 2\lambda = 0$$

$$\frac{\partial F}{\partial \lambda} = x + 2y - 50 = 0$$

The result is a system of three equations in three unknowns. The first two are easily solved for λ.

$$\lambda = -y$$

$$\lambda = -\tfrac{1}{2}x$$

We can equate these two expressions to obtain

$$-\tfrac{1}{2}x = -y$$

or

$$x = 2y$$

This can be substituted into the third equation (the constraint equation) to give

$$2y + 2y - 50 = 0$$

$$y = 12.5$$

$$x = 2y = 25$$

which agrees with our previous solution.

Let's apply this method to some additional examples.

EXAMPLE 19 Find the point on the line $x - 2y + 3 = 0$ that is nearest the origin by using the method of Lagrange multipliers.

SOLUTION This problem is similar to Example 12. As in that problem, we observe that a point on the line will minimize the distance $d = \sqrt{x^2 + y^2}$ if it minimizes the square of the distance:

$$f(x, y) = x^2 + y^2$$

Our problem can now be rephrased as follows: Minimize the objective function $f(x, y) = x^2 + y^2$, subject to the constraint that the point (x, y) should lie on the line $g(x, y) = x - 2y + 3 = 0$. According to Equation 8.5, we write

$$F(x, y, \lambda) = x^2 + y^2 + \lambda(x - 2y + 3)$$

Then

$$\frac{\partial f}{\partial x} = 2x + \lambda = 0$$

$$\frac{\partial f}{\partial y} = 2y - 2\lambda = 0$$

$$\frac{\partial f}{\partial \lambda} = x - 2y + 3 = 0$$

We solve the first two equations for λ.

$$\lambda = -2x$$

$$\lambda = y$$

Then we set these values of λ equal to each other:

$$y = -2x$$

and substitute this value into the third equation.

$$x - 2(-2x) + 3 = 0$$

$$5x + 3 = 0$$

$$x = -\tfrac{3}{5}$$

$$y = -2x = \tfrac{6}{5}$$

You will note from the previous two examples that, although we have solved for λ in the course of our solution, the variable λ was an artificial one that we ourselves introduced. Accordingly, our final answer need not mention the value of λ that we have found, because the original statement of the problem did not require it.

This method is applicable also to problems with more than two independent variables. We can illustrate this by looking, once again, at Example 14.

EXAMPLE 20 Find the dimensions of a rectangular box of volume 50 cubic inches that has minimal surface area.

SOLUTION We want to minimize

$$A(x, y, z) = 2xy + 2xz + 2yz$$

subject to the constraint

$$xyz = 50$$

We can rewrite this constraint in the form $g(x, y, z) = 0$ by setting $g(x, y, z) = xyz - 50$. We now define

$$F(x, y, z, \lambda) = 2xy + 2xz + 2yz + \lambda(xyz - 50)$$

and compute

$$\frac{\partial F}{\partial x} = 2y + 2z + \lambda yz = 0$$

$$\frac{\partial F}{\partial y} = 2x + 2z + \lambda xz = 0$$

$$\frac{\partial F}{\partial z} = 2x + 2y + \lambda xy = 0$$

$$\frac{\partial F}{\partial \lambda} = xyz - 50 = 0$$

This is a system of four (nonlinear) equations in four unknowns, and it is not easy to see how to proceed at this point. One way is to multiply the first equation by x.

$$2xy + 2xz + \lambda xyz = 0$$

The last equation tells us that $xyz = 50$, so, by substitution,

$$2xy + 2xz + 50\lambda = 0$$

or

(8.7) $$xy + xz + 25\lambda = 0$$

Similarly, multiplying the second equation by y gives

$$2xy + 2yz + \lambda xyz = 0$$

Upon substituting $xyz = 50$, we get

(8.8) $$xy + yz + 25\lambda = 0$$

Finally, we multiply the third equation by z, which yields

$$2xz + 2yz + \lambda xyz = 0$$

and

(8.9) $$xz + yz + 25\lambda = 0$$

We can equate Equations (8.7) and (8.8) to produce

$$xy + xz + 25\lambda = xy + yz + 25\lambda$$

Subtracting common terms results in

$$xz = yz$$

$$x = y$$

Similarly, equating Equations (8.7) and (8.9) gives

$$xy + xz + 25\lambda = xz + yz + 25\lambda$$

$$xy = yz$$

$$x = z$$

Thus $x = y = z$. Because $xyz = 50$, we must have $x^3 = 50$, and

$$x = y = z = \sqrt[3]{50}$$

Of course, this agrees with our previous solution.

EXAMPLE 21 In Example 10 we introduced the concept of a production function. The production function for a particular product is given by

$$z = 36L^{1/3}K^{2/3}$$

where L and K are units of labor and capital, respectively. Suppose that each unit of labor costs \$20 and that each unit of capital costs \$30. Assume that \$60,000 altogether is available to spend on production. How many units of labor and how many of capital should be utilized to maximize production?

SOLUTION The cost of L units of labor costing \$20 each is $20L$, and the corresponding cost of K units of capital is $30K$. The total cost is constrained by the budget of \$60,000, and it seems reasonable to use all of the money available. Thus

$$20L + 30K = 60{,}000$$

or

$$2L + 3K = 6000$$

We can rewrite this in the form

$$g(L, K) = 2L + 3K - 6000 = 0$$

so that we can apply the Lagrange multiplier technique. We define

$$F(L, K, \lambda) = 36L^{1/3}K^{2/3} + \lambda(2L + 3K - 6000)$$

Equations (8.6) now become

$$\frac{\partial F}{\partial L} = 12L^{-2/3}K^{2/3} + 2\lambda = 0$$

$$\frac{\partial F}{\partial K} = 24L^{1/3}K^{-1/3} + 3\lambda = 0$$

$$\frac{\partial F}{\partial \lambda} = 2L + 3K - 6000 = 0$$

The first two equations can be solved for λ to give

$$\lambda = \frac{-12}{2} L^{-2/3}K^{2/3} = \frac{-6K^{2/3}}{L^{2/3}}$$

$$\lambda = \frac{-24}{3} L^{1/3}K^{-1/3} = \frac{-8L^{1/3}}{K^{1/3}}$$

These can be equated.

$$\frac{-6K^{2/3}}{L^{2/3}} = \frac{-8L^{1/3}}{K^{1/3}}$$

Upon multiplying both sides by $-L^{2/3}K^{1/3}$, we find that

$$6K = 8L$$

or

$$K = \tfrac{4}{3}L$$

Finally, we substitute this into the third equation.

$$2L + 3(\tfrac{4}{3}L) - 6000 = 0$$

$$6L - 6000 = 0$$

$$L = 1000$$

$$K = \tfrac{4}{3}L = 1333.33$$

Economics offers numerous problems to which the method of Lagrange multipliers is applicable, because every business wishes to operate optimally within practical limitations. As another illustration, consider the operation of a firm that is producing some good in a competitive environment. Imagine that there are two factors in the production process, which we denote by X and Y. Let

$$z = f(x, y)$$

represent the production function. That is, the firm can produce z units of output by utilizing x units of X and y units of Y. (For instance, in our previous example the two factors were labor and capital.) Suppose that each unit of the finished product can be sold at a price of p_z and that each unit of the inputs X and Y costs p_x and p_y, respectively. (The prices p_x, p_y, and p_z are being considered constants here.) How can the firm be operated most profitably?

The profit is given by the revenue taken in from selling the finished product, less the cost of materials used in production. The revenue produced by selling z units at p_z dollars each is simply

$$zp_z$$

Similarly, the costs of purchasing x and y units of factors X and Y are xp_x and xp_y. The total profit T is therefore given by

$$T = T(x, y, z) = zp_z - (xp_x + yp_y)$$

Our goal is now to maximize T, subject to the production function

$$z = f(x, y)$$

This latter equation can be rewritten in the form

$$g(x, y, z) = z - f(x, y) \\ = 0$$

We have now phrased our problem in such a way that we can apply the method of Lagrange multipliers. We form the function of four variables

$$F(x, y, z, \lambda) = zp_z - xp_x - yp_y + \lambda(z - f(x, y))$$

Thus maximal profitability will occur when

(8.10)

$$\frac{\partial F}{\partial x} = -p_x - \lambda \frac{\partial f}{\partial x} = 0$$

$$\frac{\partial F}{\partial y} = -p_y - \lambda \frac{\partial f}{\partial y} = 0$$

$$\frac{\partial F}{\partial z} = p_z + \lambda = 0$$

$$\frac{\partial F}{\partial \lambda} = z - f(x, y) = 0$$

Observe that the first two equations can be solved for λ to give

$$\lambda = -\frac{p_x}{\frac{\partial f}{\partial x}}$$

$$\lambda = -\frac{p_y}{\frac{\partial f}{\partial y}}$$

If we equate these two expressions for λ, we obtain

$$-\frac{p_x}{\frac{\partial f}{\partial x}} = -\frac{p_y}{\frac{\partial f}{\partial y}}$$

or

$$\frac{\frac{\partial f}{\partial x}}{\frac{\partial f}{\partial y}} = \frac{p_x}{p_y}$$

That is, at the most profitable production level, the ratio of the marginal productivities is equal to the price ratio. This is a well-known result in economics.

We conclude by applying this discussion to a specific example.

EXAMPLE 22 The production function for a firm is

$$z = 10.5 - \frac{1}{x^2} - \frac{1}{y^2}$$

The price at which the firm can sell each unit of output is $p_z = \$4$, and the inputs are priced at $p_x = \$27$ and $p_y = \$1$. Find the amounts of the factors x and y that should be utilized to maximize profit.

SOLUTION The profit function T is given by

$$T(x, y, z) = 4z - 27x - y$$

The production function constraint is

$$g(x, y, z) = z + \frac{1}{x^2} + \frac{1}{y^2} - 10.5 = 0$$

We therefore form

$$F(x, y, z, \lambda) = 4z - 27x - y + \lambda\left(z + \frac{1}{x^2} + \frac{1}{y^2} - 10.5\right)$$

and compute

$$\frac{\partial F}{\partial x} = -27 - \frac{2\lambda}{x^3} = 0$$

$$\frac{\partial F}{\partial y} = -1 - \frac{2\lambda}{y^3} = 0$$

$$\frac{\partial F}{\partial z} = 4 + \lambda = 0$$

$$\frac{\partial F}{\partial \lambda} = z + \frac{1}{x^2} + \frac{1}{y^2} - 10.5 = 0$$

These equations are readily solved. The third equation shows that

$$\lambda = -4$$

Substituting this value into the first equation yields

$$-27 - \frac{2(-4)}{x^3} = 0$$

$$x^3 = \frac{8}{27}$$

$$x = \frac{2}{3}$$

Similarly, substitution of $\lambda = -4$ into the second equation gives

$$-1 - \frac{2(-4)}{y^3} = 0$$

$$y^3 = 8$$

$$y = 2$$

Finally, the optimal production level is given by the last equation.

$$z + \frac{1}{(\frac{2}{3})^2} + \frac{1}{(2)^2} - 10.5 = 0$$

$$z = 10.5 - \frac{9}{4} - \frac{1}{4}$$

$$= 8$$

Thus our manufacturer should produce 8 units of output, utilizing $\frac{2}{3}$ unit of X and 2 units of Y.

PRACTICE EXERCISE The production function for a firm is

$$z = 5 - \frac{1}{\sqrt{x}} - \frac{1}{\sqrt{y}}$$

The prices for the inputs are $p_x = 1$ and $p_y = 8$, and the output price is $p_z = 2$. Find the optimal production schedule.

[Your answer should be $x = 1, y = \frac{1}{4}, z = 2$.]

Section 8.5 EXERCISES

Solve all of the exercises by the method of Lagrange multipliers.

1. Minimize $x^2 + y^2 + 5$ subject to the constraint $x + 2y = 5$.

2. Find the point on the straight line $y = 5x + 6$ that is closest to the origin.

3. Find the point on the straight line $y = 5x + 6$ that is closest to the point $(-1, 3)$.

4. Maximize $x^2 - 3y^2$ subject to the constraint $y = -x + 2$.

5. Maximize $3x^2 - y^2$ subject to the constraint $y = -2x - 3$.

6. Find three positive numbers x, y, and z with $x + 2y + 2z = 30$ and their product xyz as large as possible.

7. A 20-in. string is to be cut once and the two pieces used to form the perimeter of a circle and of a square. What should the lengths of the two cut pieces be in order to minimize the two enclosed areas?

8. In a microcomputer manufacturing plant, the production function for a certain circuit chip is found to satisfy the Cobb–Douglas function

$$z = 80L^{1/5}K^{4/5}$$

where L and K are units of labor and capital, respectively. Suppose each unit of labor costs \$3000 and each unit of capital costs \$2200. Assume further that a total of \$1,000,000 is available to spend on production. How many units of L and of K should be utilized to maximize production?

9. Look at Exercise 8 again. Without knowing the total amount available for production costs—only that the amount is finite—can you find the ratio of K to L (that is, K/L) in any optimal production setting?

10. Redo Exercise 8 for the Cobb–Douglas function

$$z = 75L^{2/7}K^{5/7}$$

11. Redo Exercise 9 for the Cobb–Douglas function

$$z = 75L^{2/7}K^{5/7}$$

12. Find the dimensions of the rectangle of maximal area that can be inscribed in the ellipse

$$\frac{x^2}{a^2} + \frac{y^2}{b^2} = 1$$

(Your answer will be in terms of a and b.)

13. Find the point on the unit circle $x^2 + y^2 = 1$ that is closest to the point (5, 8).

14. The production function for a certain firm is

$$z = 85 - \frac{1}{x^2} - \frac{2}{y^2}$$

The firm sells each unit of output at a price $p_z = \$28$, and the inputs are priced at $p_x = \$9$ and $p_y = \$3$, respectively. Find the number of units of the factors x and y that should be utilized to maximize profit.

For Exercises 15–18 consider the production function

$$z = 8 - \frac{2}{\sqrt{x}} - \frac{4}{y^{3/2}}$$

Given that the input and output prices are as listed, find the production schedule that maximizes profit in each case.

15. $p_x = 2, p_y = 1, p_z = 3$

16. $p_x = 4, p_y = 2, p_z = 5$

17. $p_x = 3, p_y = 3, p_z = 3$

18. $p_x = 2, p_y = 4, p_z = 6$

19. Find the point on the plane $2x + 4y - z = 38$ that is closest to the origin.

20. A rancher wishes to fence off two adjacent rectangular corrals using a total of 600 yd of fencing. What should the overall dimensions of the double corral be to maximize the total area?

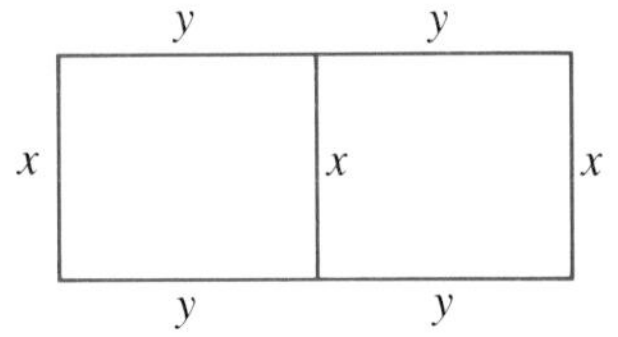

FIGURE FOR EXERCISE 20

21. Suppose the Post Office accepts only packages whose length plus girth (perimeter of cross section) does not exceed 72 in. What is the maximum volume of a package that will be accepted? (See Exercise 30 of Section 8.3.)

SECTION 8.6

Double Integration and Its Applications

VOLUMES FROM CROSS SECTIONS Just as we used area to motivate the single-variable integral, we can use the concept of volume to motivate the double integral. We considered volumes of revolution in (the optional) Section 6.3. Our present discussion, which is independent of that section, will treat more general volumes.

Suppose we wish to find the volume of a rather irregularly shaped loaf of bread. We set up an x-axis roughly parallel to the longest dimension of the loaf, as shown in Figure 8.18. Suppose this loaf extends from $x = a$ to $x = b$. Imagine now that we take a large knife and cut the loaf into n thin slices of width $\Delta x = (b - a)/n$ perpendicular to the x-axis (Figure 8.19a). If we denote the volume of the individual slices by $V_1, V_2, \ldots, V_n$, then the volume V of the whole loaf is the sum

$$V = V_1 + V_2 + \cdots + V_n = \sum_{k=1}^{n} V_k$$

Let's look at a typical slice, V_k, whose right face is at distance x_k from the origin (Figure 8.19b). If we let $A(x)$ denote the area of a slice (or cross section)

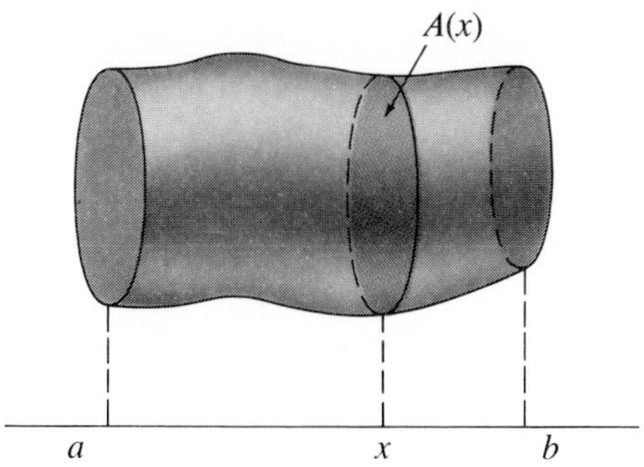

FIGURE 8.18

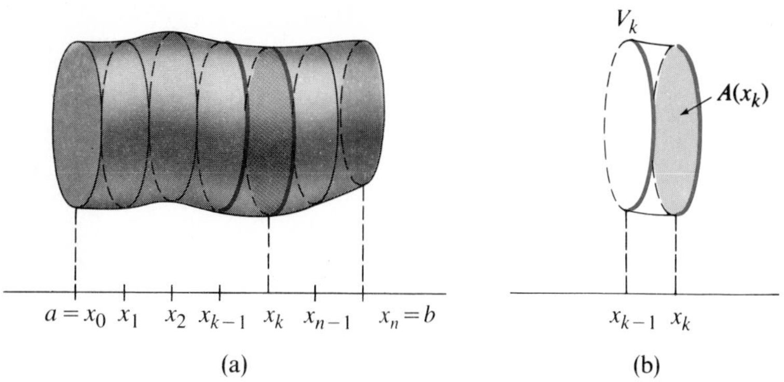

FIGURE 8.19

of the loaf made by a knife slice at x, then the right face of the kth slice has area $A(x_k)$.

Now the cross-sectional area, $A(x)$, varies from one end of the loaf to the other. However, if (as we shall assume) A is a continuous function, then when n is sufficiently large (so that Δx is small), the variation in $A(x)$ from $x = x_{k-1}$ to $x = x_k$ will not be appreciable. Accordingly, the volume of the kth slice will be approximately $V_k \approx A(x_k)\,\Delta x$. The volume of the entire loaf is then approximated as

$$V \approx \sum_{k=1}^{n} A(x_k)\,\Delta x$$

and the accuracy of the approximation improves as n becomes larger. We recognize the sum on the right as a Riemann sum. Passing to the limit as n goes to infinity, we obtain the result

(8.11)
$$V = \int_a^b A(x)\,dx$$

Thus the area of the loaf of bread—and in fact of any solid with continuously varying cross-sectional area $A(x)$—is obtained by integrating $A(x)$. In the particular case studied in Section 6.3, $A(x)$ was $\pi f(x)^2$, the area of the circular cross section at x, which had radius $f(x)$.

EXAMPLE 23 Find the volume of a pyramid whose base is a square of side b and whose altitude is h.

SOLUTION As shown in Figure 8.20(a), we set up an x-axis along the line containing the pyramid's altitude drawn from the vertex to the base, with the base being the cross section at $x = 0$. (Although the axis is vertical rather than horizontal as in Figure 8.18, the principle is the same.) The cross section at height x above the base is a square, whose side we denote by $s = s(x)$. In particular, $s(0) = b$ and $s(h) = 0$.

The shaded right triangle in Figure 8.20(a) is redrawn in Figure 8.20(b). By similar triangles,

$$\frac{\frac{s}{2}}{h - x} = \frac{\frac{b}{2}}{h}$$

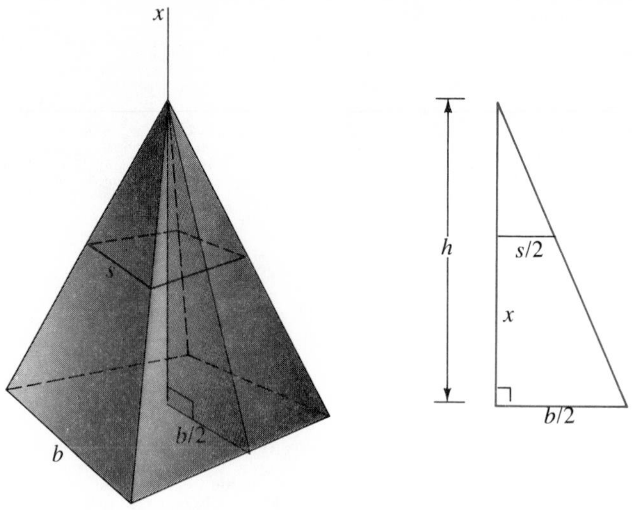

(a) (b)

FIGURE 8.20

so

$$s = \frac{b}{h}(h - x)$$

Substituting $A(x) = s^2 = \dfrac{b^2}{h^2}(h - x)^2$ into Equation (8.11), we compute the pyramid's volume as

$$V = \frac{b^2}{h^2}\int_0^h (h - x)^2\,dx = \frac{b^2}{h^2}\left[-\frac{(h - x)^3}{3}\right]\Bigg|_0^h$$
$$= \frac{b^2}{h^2}\left[0 - \left(-\frac{h^3}{3}\right)\right] = \frac{1}{3}b^2h$$

which is $\frac{1}{3}$ the area of the base times the height.

PRACTICE EXERCISE

One of the most surprising accomplishments of the ancient Egyptians was their discovery of the formula for the volume of a truncated square pyramid (see Figure 8.21). (An example of the use of the formula appears in the *Moscow Mathematical Papyrus*, written about 1850 B.C.*)

If the lower and upper bases are squares of sides b and a, respectively, and if the altitude (distance between the bases) is h, find an expression for the volume as a function of a, b, and h. (*Hint*: If you let $s(x)$ denote the side of the square cross section at height x above the lower base, then $s(0) = b$ and $s(h) = a$. Use these values and the fact that $s(x)$ is a linear function of x to deduce that $s(x) = b + (a - b)x/h$.)

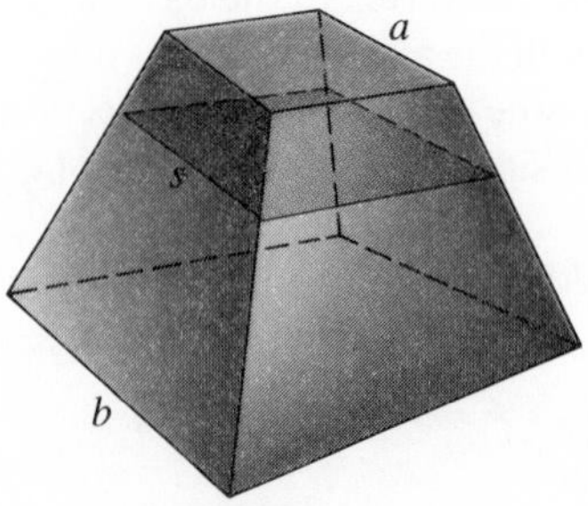

FIGURE 8.21

* For some thoughts on how the Egyptians might have discovered this formula, see Richard L. Faber, *Foundations of Euclidean and Non-Euclidean Geometry* (New York: Marcel Dekker, 1983), pp. 34–35.

$\left[\text{Your answer should be } V = \dfrac{h}{3}\left(\dfrac{b^3 - a^3}{b - a}\right) = \dfrac{h}{3}(a^2 + ab + b^2).\right]$

ITERATED INTEGRALS Let us suppose now that the solid whose volume we wish to find is the region below the graph of some nonnegative, continuous function $f(x, y)$ and above the rectangle R in the xy-plane specified by the inequalities $a \leq x \leq b$ and $c \leq y \leq d$. The cross section made by a slicing plane at some x between a and b is shaded in Figure 8.22. Because we will compute the volume by means of Equation (8.11), we need an expression for $A(x)$, the area of this shaded slice.

But this is the area below a curve: The curve lies in a vertical plane parallel to the yz-plane and is in fact the curve

$$z = f(x, y) \qquad c \leq y \leq d \quad x \text{ held fixed}$$

Consequently $A(x)$ is given by the integral

$$A(x) = \int_c^d f(x, y)\, dy$$

which is evaluated just like any other definite integral, provided we remember to treat x as a constant (the resulting evaluation will be an expression involving x). Accordingly the volume V is given by

(8.12) $$V = \int_a^b A(x)\, dx = \int_a^b \left[\int_c^d f(x, y)\, dy\right] dx$$

This latter expression is called an **iterated integral**. Here are some examples of how such integrals are computed.

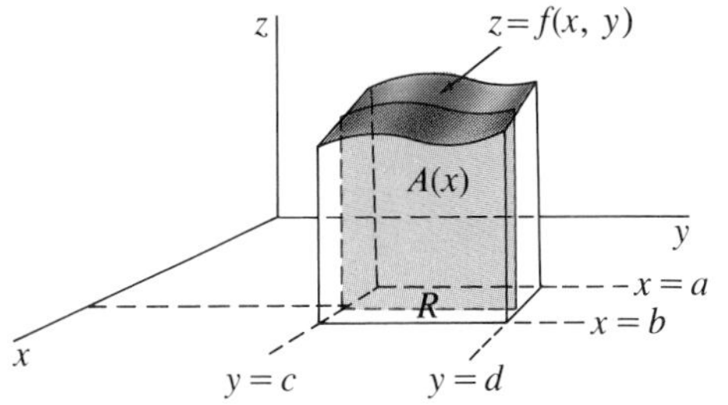

FIGURE 8.22

EXAMPLE 24 Find the volume of the solid in the first quadrant bounded above by the plane $z = 48 - 3x - 8y$ and below by the rectangle specified by $0 \leq x \leq 4$ and $0 \leq y \leq 3$.

SOLUTION The cross section at x (shown shaded in Figure 8.23) has area

$$A(x) = \int_0^3 (48 - 3x - 8y)\, dy = (48y - 3xy - 4y^2)\Big|_0^3$$
$$= 144 - 9x - 36 = 108 - 9x$$

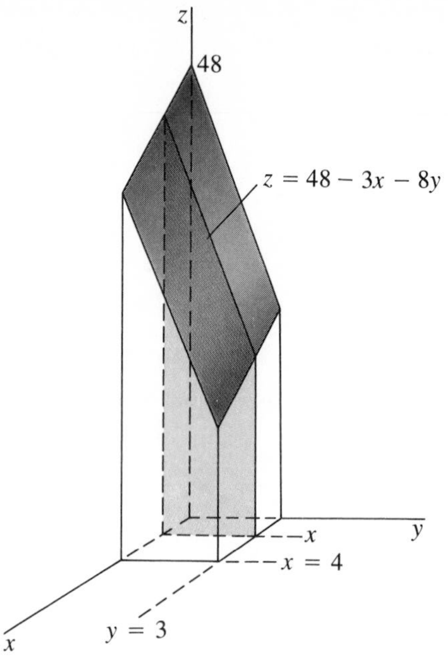

FIGURE 8.23

Note that the antiderivative of $-3x$ with respect to y is $-3xy$ because x is held constant. The volume of the solid is therefore

$$\begin{aligned} V &= \int_0^4 A(x)\,dx = \int_0^4 \left[\int_0^3 (48 - 3x - 8y)\,dy\right] dx \\ &= \int_0^4 (108 - 9x)\,dx \\ &= (108x - 9x^2/2)\Big|_0^4 \\ &= 432 - 72 = 360 \end{aligned}$$

To shorten the notation, it is customary to omit the brackets in an iterated integral. Equation (8.11) is therefore usually written as

$$\begin{aligned} V &= \int_a^b A(x)\,dx \\ &= \int_a^b \int_c^d f(x, y)\,dy\,dx \end{aligned}$$

The inner integration (with respect to y here) is always performed first.

EXAMPLE 25 Find the volume of the solid beneath the graph of the paraboloid $z = 100 - 4x^2 - y^2$, for $0 \leq x \leq 3$ and $0 \leq y \leq 6$ (see Figure 8.24).

SOLUTION Integrating the cross section, we have

$$\begin{aligned} V &= \int_0^3 \left[\int_0^6 (100 - 4x^2 - y^2)\,dy \right] dx \\ &= \int_0^3 \left[\left(100y - 4x^2y - \frac{y^3}{3}\right)\Big|_0^6 \right] dx \\ &= \int_0^3 (528 - 24x^2)\,dx \\ &= (528x - 8x^3)\Big|_0^3 \\ &= 1368 \end{aligned}$$

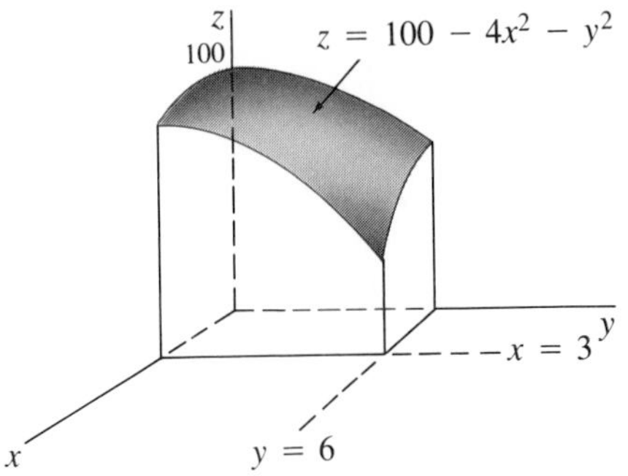

FIGURE 8.24

Up to now, we have been using slices perpendicular to the x-axis and obtaining the volume as the integral of the cross-sectional area $A(x)$. We can also use slices perpendicular to the y-axis. If $A(y)$ denotes the area of the solid's slice at y, then

$$A(y) = \int_a^b f(x, y)\,dx$$

and the volume of the solid shown in Figure 8.22 can be computed by the formula

$$\begin{aligned} V &= \int_c^d A(y)\,dy \\ &= \int_c^d \int_a^b f(x, y)\,dx\,dy \end{aligned}$$

As always, the inner integration (with respect to x here) is performed first.

PRACTICE EXERCISE Rework Example 25 by integrating in the other order.

$\left[\text{Your answer should be } V = \int_0^6 \int_0^3 (100 - 4x^2 - y^2)\,dx\,dy = 1368.\right]$

THE DOUBLE INTEGRAL Volumes can be approached also through a limit process (that is, Riemann sums), as was area in Chapter 5. Suppose f is a nonnegative continuous function defined on the rectangle R: $a \leq x \leq b$, $c \leq y \leq d$. We partition R into smaller subrectangles by equally spaced vertical lines and equally spaced horizontal lines. Let n be the number of these subrectangles, and denote their width and height by Δx and Δy (see Figure 8.25).

Number the subrectangles 1 through n in some fashion (it doesn't matter how). For each k between 1 and n, choose an arbitrary point (x_k, y_k) in the kth subrectangle. Consider the rectangular box or column whose base is the kth subrectangle and whose height is $f(x_k, y_k)$ (see Figure 8.26). This column, whose volume is $f(x_k, y_k)\, \Delta x\, \Delta y$, approximates the portion of our solid above the kth subrectangle.

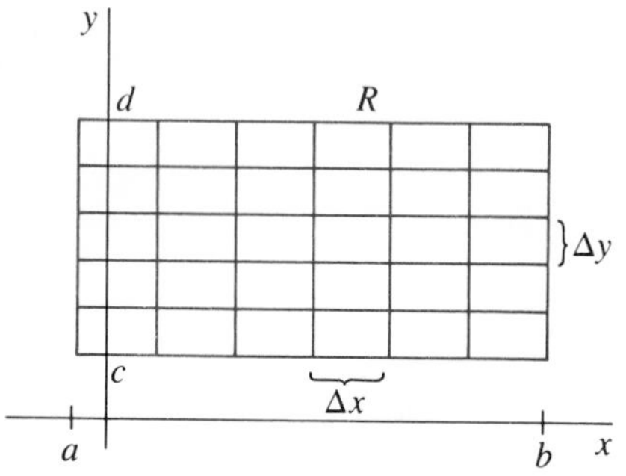

FIGURE 8.25

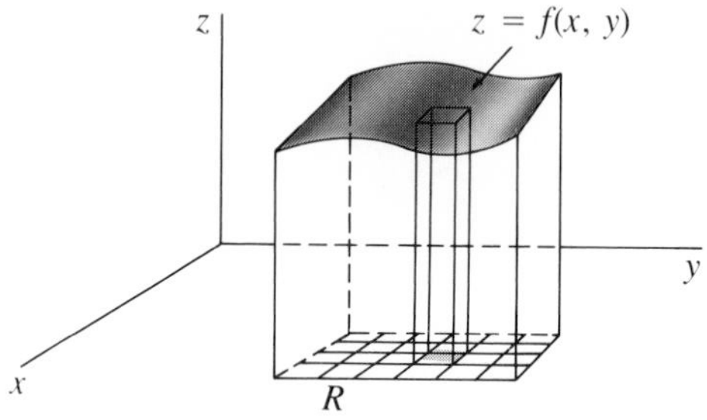

FIGURE 8.26

Consequently the total volume V of our solid is approximated by the sum

$$V \approx \sum_{k=1}^{n} f(x_k, y_k)\, \Delta x\, \Delta y$$

The exact volume is the limit approached by this sum as n goes to infinity and the dimensions of the individual subrectangles go to 0. This limit exists for any continuous function and is called the **double integral** of f over R and denoted

$$\iint_R f(x, y)\, dA$$

As with single-variable integrals, the function being integrated, $f(x, y)$, is called the **integrand**. It is known that, if f is continuous, the double integral is equal in value to the iterated integrals discussed previously.

THEOREM 8.4 If $f(x, y)$ is a continuous function defined on the rectangle R: $a \leq x \leq b$, $c \leq y \leq d$, then

$$\iint_R f(x, y)\, dA = \int_a^b \int_c^d f(x, y)\, dy\, dx = \int_c^d \int_a^b f(x, y)\, dx\, dy \qquad \boxminus$$

The earlier developers of the calculus thought of the symbol dA as representing an infinitesimal element of area in the form of a rectangle of sides

dx and dy. The term $f(x, y)\,dA$ was thought of as the volume of an infinitesimally thin rectangular column, and the double integral was the sum of the volumes of all the columns resulting from a partition of R. In practice, we evaluate a double integral by computing either of the corresponding iterated integrals, choosing whichever order of integration seems more easily carried out. From now on we shall use the terms **double integral** and **iterated integral** interchangeably.

INTEGRALS OVER NONRECTANGULAR REGIONS Up to now, the region R over which we have integrated has been a rectangle in the xy-plane. It is possible to carry out an iterated integral over more general regions. In particular, we will consider double integrals over regions of either of the two types depicted in Figure 8.27. For type A regions, we evaluate the double integral as follows:

$$\iint_R f(x, y)\,dA = \int_a^b \int_{g(x)}^{h(x)} f(x, y)\,dy\,dx$$

whereas for type B regions, we use

$$\iint_R f(x, y)\,dA = \int_{g(y)}^{h(y)} f(x, y)\,dx\,dy$$

Note that in both cases the limits of integration in the inner integral are functions of the variable of integration for the outer integral. In most cases, it will not be necessary to make a three-dimensional sketch. All that is needed is to sketch the region R and decide whether it is of type A or type B (or both) and, accordingly, which order of integration is most appropriate. Here are a few examples to illustrate the procedure.

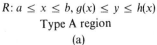

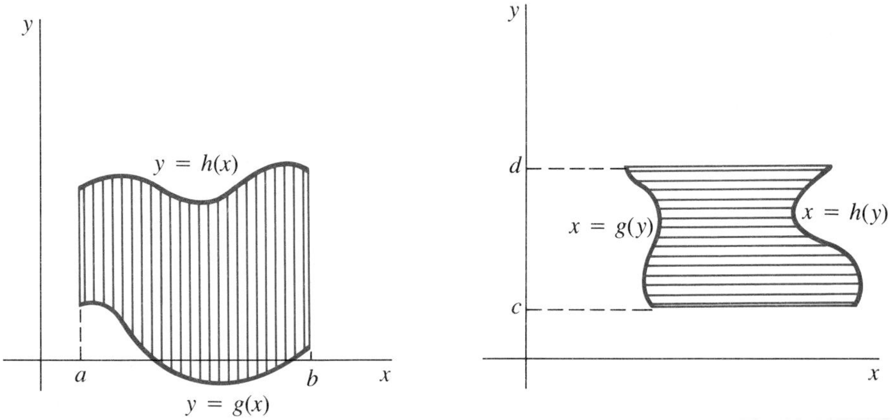

$R: a \le x \le b,\ g(x) \le y \le h(x)$
Type A region
(a)

$R: c \le y \le d,\ g(y) \le x \le h(y)$
Type B region
(b)

FIGURE 8.27

EXAMPLE 26 Evaluate the double integral

$$\iint_R (2x + y)\, dA$$

for R = the region enclosed by the curves $y = 2x$ and $y = x^2$.

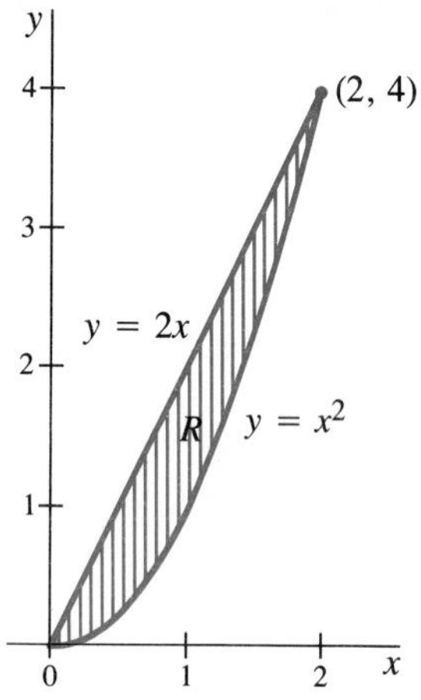

FIGURE 8.28

SOLUTION The region R, which is sketched in Figure 8.28, is of type A, with x varying from 0 to 2. Accordingly,

$$\begin{aligned}\iint_R (2x + y)\, dA &= \int_0^2 \int_{x^2}^{2x} (2x + y)\, dy\, dx \\ &= \int_0^2 \left(2xy + \frac{y^2}{2}\right)\Bigg|_{x^2}^{2x} dx \\ &= \int_0^2 \left[(4x^2 + 2x^2) - \left(2x^3 + \frac{x^4}{2}\right)\right] dx \\ &= \int_0^2 \left(6x^2 - 2x^3 - \frac{x^4}{2}\right) dx \\ &= \left(2x^3 - \frac{x^4}{2} - \frac{x^5}{10}\right)\Bigg|_0^2 = \frac{48}{10} = 4.8\end{aligned}$$

PRACTICE EXERCISE Let R be the region in the first quadrant bounded by the y-axis and the lines $y = x$ and $y = 2 - x$. Evaluate

$$\iint_R xy\, dA$$

[Your answer should be $\frac{1}{3}$.]

EXAMPLE 27 Let R be the region bounded by the graphs of $y = \sqrt{x}$ and $y = -x/2 + 4$ and the x-axis (see Figure 8.29). Compute

$$\iint_R y\, dA$$

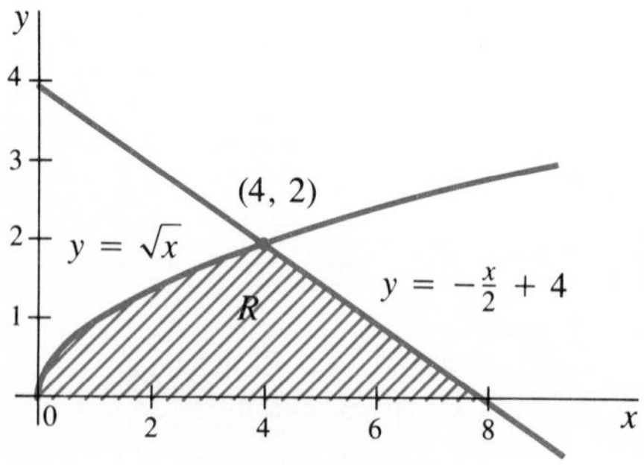

FIGURE 8.29

SOLUTION If we compare Figure 8.29 with Figure 8.27, we see that R is not of type A because it does not have a unique "top" curve. From $x = 0$ to $x = 4$, the region extends vertically from the x-axis to the parabola $y = \sqrt{x}$, whereas from $x = 4$ to $x = 8$, it extends vertically from the x-axis to the line $y = -x/2 + 4$.

On the other hand, we can view R as made up of two type A regions R_1 and R_2, by drawing the dividing line $x = 4$ (see Figure 8.30). The desired

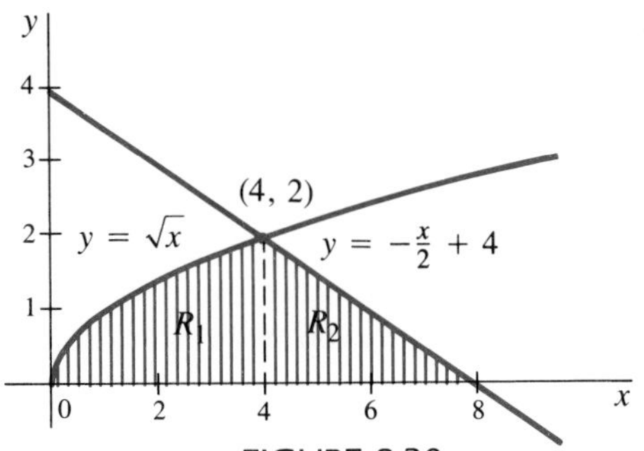

FIGURE 8.30

double integral can then be computed as the sum

$$\iint_R y\,dA = \iint_{R_1} y\,dA + \iint_{R_2} y\,dA$$

$$= \int_0^4 \int_0^{\sqrt{x}} y\,dy\,dx + \int_4^8 \int_0^{-\frac{x}{2}+4} y\,dy\,dx$$

$$= \int_0^4 \left(\frac{y^2}{2}\Big|_0^{\sqrt{x}}\right) dx + \int_4^8 \left(\frac{y^2}{2}\Big|_0^{\frac{8-x}{2}}\right) dx$$

$$= \int_0^4 \frac{x}{2}\,dx + \frac{1}{8}\int_4^8 (8-x)^2\,dx$$

$$= \frac{x^2}{4}\Big|_0^4 - \frac{1}{24}(8-x)^3\Big|_4^8$$

$$= 4 - \frac{1}{24}(0) + \frac{64}{24} = \frac{20}{3}$$

Alternatively, if we invert the equations of the boundary curves $y = \sqrt{x}$ and $y = -x/2 + 4$ (that is, solve them for x as functions of y), we can interpret R as a region of type B, as shown in Figure 8.31. We can then more easily compute the integral as follows:

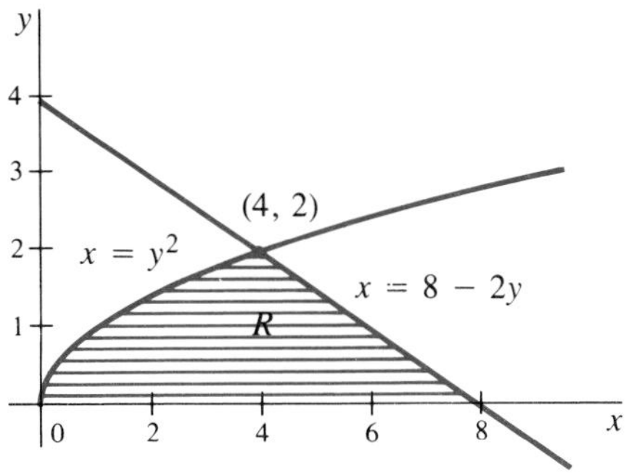

FIGURE 8.31

$$\iint_R y\,dA = \int_0^2 \int_{y^2}^{8-2y} y\,dx\,dy = \int_0^2 \left(yx\Big|_{y^2}^{8-2y}\right) dy$$

$$= \int_0^2 (8y - 2y^2 - y^3)\,dy$$

$$= \left(4y^2 - \frac{2}{3}y^3 - \frac{y^4}{4}\right)\Big|_0^2$$

$$= 16 - \frac{16}{3} - 4 = \frac{20}{3}$$

which is the same answer as before (as of course it must be).

The preceding example illustrates how the shape of the region often determines the more efficient choice for the order of integration. The next example shows how the function being integrated can sometimes determine that choice.

EXAMPLE 28 Evaluate

$$\iint_R e^{y^2}\,dA$$

for R the region bounded by the graphs of $y = 2x$, $y = 2$, and $x = 0$.

SOLUTION The region R is shown in Figure 8.32. If we view it as a region of

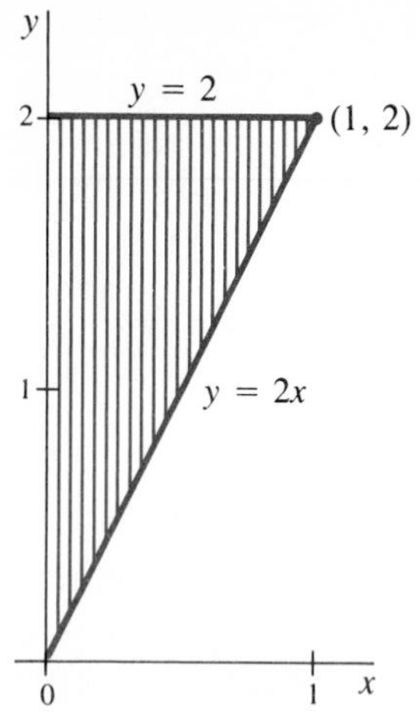

FIGURE 8.32

type A, we can write

$$\iint_R e^{y^2}\,dA = \int_0^1 \int_{2x}^2 e^{y^2}\,dy\,dx$$

However, because there is no known antiderivative for e^{y^2} in terms of elementary functions, we are unable to proceed further. Accordingly, we view R as a type B region by rewriting the equations for its boundary curves, as indicated in Figure 8.33. Then we have

$$\begin{aligned}\iint_R e^{y^2}\,dA &= \int_0^2 \int_0^{y/2} e^{y^2}\,dx\,dy \\ &= \int_0^2 \left(xe^{y^2}\Big|_0^{y/2} \right) dy \\ &= \int_0^2 \frac{y}{2} e^{y^2}\,dy \\ &= \frac{1}{4} e^{y^2}\Big|_0^2 = \frac{1}{4}(e^4 - 1)\end{aligned}$$

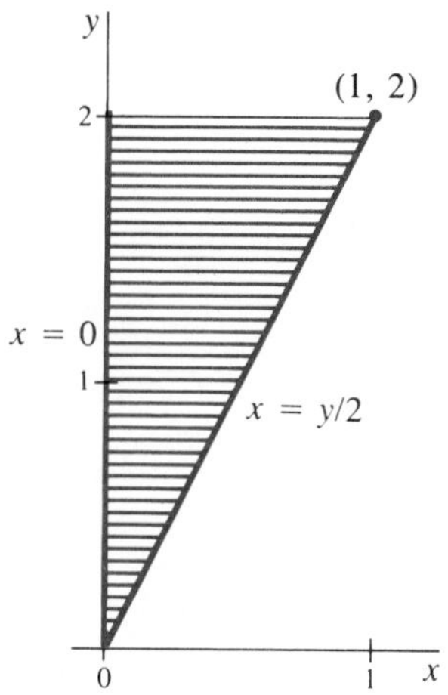

FIGURE 8.33

APPLICATIONS OF DOUBLE INTEGRALS So far, our principal interpretation of the double integral has been as a volume, at least for nonnegative integrands. However, there is a variety of more practical uses for double integrals, of which we shall be able to consider only a few in this text.

Average Value In Section 6.1 we learned that the single-variable integral could be used to compute the average value of a continuous function $f(x)$ defined on some interval $[a, b]$. The average value, denoted $\bar{f}$ was defined by

$$\bar{f} = \frac{1}{b - a} \int_a^b f(x)\,dx$$

By analogy, if $f(x, y)$ is defined and continuous on some region R in the xy-plane, then the **average value** of f, denoted $\bar{f}$, is defined by

$$\bar{f} = \frac{1}{\text{Area of } R} \iint_R f(x, y)\, dA$$

(The area of R is the analog of the length of the interval $[a, b]$.)

EXAMPLE 29 The number of units of a certain product produced by a factory in a given month is given by the following Cobb–Douglas production function (see Example 10):

$$f(L, K) = 200\, L^{0.2}\, K^{0.8}$$

where L is the number of units of labor (say, worker-days), and K is the number of units of capital (say, thousands of dollars). L and K vary within the limits

$$20 \leq L \leq 30 \qquad 40 \leq K \leq 50$$

Find the average number of units produced in a month.

SOLUTION Our region R is the rectangle (in what we might call the LK-plane) determined by these inequalities. Accordingly,

$$\text{Area of } R = (30 - 20)(50 - 40) = 100$$

We will integrate f over R and then divide by 100.

$$\begin{aligned} \int_{40}^{50} \int_{20}^{30} 200\, L^{0.2}\, K^{0.8}\, dL\, dK &= \frac{200}{1.2} \int_{40}^{50} \left(L^{1.2} \Big|_{20}^{30} \right) K^{0.8}\, dK \\ &= \frac{200}{1.2(1.8)} \left(L^{1.2} \Big|_{20}^{30} \right) \left(K^{1.8} \Big|_{40}^{50} \right) \\ &\approx 799{,}055 \end{aligned}$$

Now, dividing by the area of R, we have

$$\bar{f} \approx \frac{799{,}055}{100} \approx 7991 \text{ units}$$

PRACTICE EXERCISE Find the average value of the function $f(x, y) = x^2 + 6xy$, on the rectangle R: $0 \leq x \leq 2$, $1 \leq y \leq 3$. Don't forget to divide the integral by the area of R!

[Your answer should be $\frac{40}{3}$.]

Population Densities Suppose we are interested in finding the total population in a certain urban center. We can set up a coordinate system with the origin at center city, with the directions of the axes conveniently chosen, and with miles for the units on both axes. Suppose that, on the basis of demographic studies, we have a formula giving the **population density**, $f(x, y)$, in individuals per square mile, at each point (x, y). Then, if R is the region occupied by the city, its total population P is obtained by integrating the density function over R:

$$\text{Total population} = \iint_R f(x, y)\, dA$$

Intuitively, we may think of dA as an infinitesimal subregion of the city located at (x, y). Multiplying the density $f(x, y)$ (in individuals per square mile) by the area of dA (in square miles) gives the number of individuals $f(x, y)\, dA$ in the subregion. Evaluating the integral (which is really a limit of Riemann sums) "sums up" the population contributions from all the subregions dA that make up R.

EXAMPLE 30 Suppose that a certain city has a population of 2000 people per square mile at its center (the origin of our coordinate system) but that the density falls off with increasing distance from the center according to the rule

$$f(x, y) = 2000 - 12x^2 - 12y^2$$

Compute the total population in the region

$$R\colon -3 \le x \le 3, -2 \le y \le 2$$

SOLUTION The total population can be obtained by evaluating the integral

$$\int_{-3}^{3} \int_{-2}^{2} (2000 - 12x^2 - 12y^2)\, dy\, dx$$

We can simplify the computations somewhat by exploiting the symmetry of the density function $f(x, y)$. Because of the squared terms, $f(x, y) = f(-x, y)$ and $f(x, y) = f(x, -y)$. This implies that the foregoing iterated integral is equal in value to four times the integral over the portion of R that lies in the first quadrant. Consequently the population is

$$\begin{aligned} 4\int_0^3 \int_0^2 (2000 - 12x^2 - 12y^2)\, dy\, dx &= 4\int_0^3 (2000y - 12x^2y - 4y^3)\Big|_0^2\, dx \\ &= 4\int_0^3 (4000 - 24x^2 - 32)\, dx \\ &= 4(4000x - 8x^3 - 32x)\Big|_0^3 \\ &= 4(12{,}000 - 216 - 96) \\ &= 46{,}752 \end{aligned}$$

The concept of integrating a density to obtain a whole occurs in many fields. An electrical engineer might integrate the "charge density" on a condensor plate to compute the total electrical charge. A theoretical physicist or astronomer might integrate the density of matter to obtain the total mass within a certain volume of space (this would be what is known as a **triple integral**). A biologist might integrate the density of bacteria in a culture medium to find the total number of bacteria.

EXAMPLE 31 Suppose that a triangular metal plate carries an electrical charge, the density of which varies from point to point because of the presence of a magnetic field. Assume that the plate is in the shape of the region R bounded by the x- and y-axes and the line $x + y = 1$ and that the charge density (in suitable units) is

$$f(x, y) = 2000\, e^{-(x+y)}$$

Find the total charge on the plate.

SOLUTION We simply integrate the density f over the region R (see Figure 8.34).

$$\begin{aligned}
2000 \int_0^1 \int_0^{1-x} e^{-(x+y)}\, dy\, dx &= 2000 \int_0^1 \left(-e^{-(x+y)} \Big|_0^{1-x} \right) dx \\
&= 2000 \int_0^1 (-e^{-1} + e^{-x})\, dx \\
&= 2000 \left(-\frac{x}{e} - e^{-x} \right) \Big|_0^1 \\
&= 2000 \left[\left(-\frac{1}{e} - \frac{1}{e} \right) - (-1) \right] \\
&= 2000 \left(1 - \frac{2}{e} \right) \approx 528.5
\end{aligned}$$

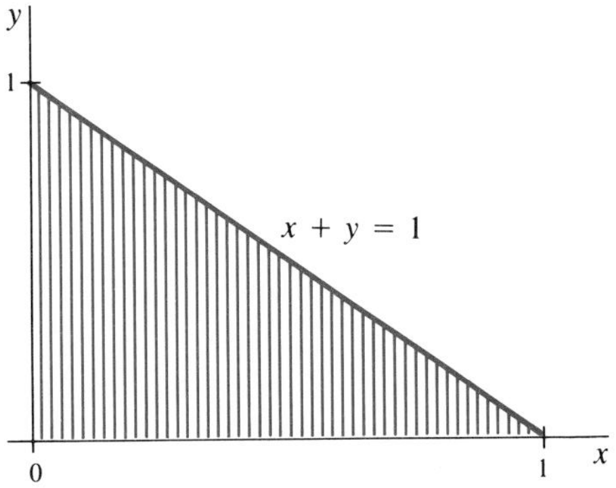

FIGURE 8.34

Section 8.6 EXERCISES

In Exercises 1–6 evaluate the iterated integrals.

1. $\int_0^2 \int_x^{x^2} 3\, dy\, dx$

2. $\int_0^1 \int_0^{3x} xy^2\, dy\, dx$

3. $\int_1^3 \int_0^{\sqrt{1+y^2}} x\, dx\, dy$

4. $\int_0^2 \int_1^{e^y} \frac{y}{x}\, dx\, dy$

5. $\int_0^1 \int_0^{\sqrt{x}} \frac{y}{\sqrt{x^2 + 4}}\, dy\, dx$

6. $\int_0^2 \int_0^x \frac{1}{1 + x^2}\, dy\, dx$

In Exercises 7–12 set up and evaluate an iterated integral (or sum of integrals) of the given function $f(x, y)$ over the region R.

7. $f(x, y) = x/\sqrt{x^2 + y^2}$; R: the triangle formed by the lines $x = 0$, $x = y$, and $y = 2$.

8. $f(x, y) = x$; R: the semicircle bounded by the x-axis and the graph of $y = \sqrt{1 - x^2}$.

9. $f(x, y) = e^{-(x+y)}$; $R: 0 \le x \le 1, 0 \le y \le 3$.

10. $f(x,y) = x^2y$; R: the triangle formed by the lines $y = x$, $y = 2 - x$, and $y = 0$.

11. $f(x, y) = \dfrac{1}{x + y + 1}$; R: the triangle formed by the lines $y = x$, $y = 2 - x$, and $y = 0$.

12. $f(x, y) = x + y$; R: the region bounded by the y-axis and the right half of the circle $x^2 + y^2 = 4$.

In Exercises 13–16 for each iterated integral, write an equivalent iterated integral (or sum of integrals) with the order of integration reversed. Then evaluate both to verify that their values are equal.

13. $\displaystyle\int_0^2 \int_0^{2-x} 2xy\,dy\,dx$

14. $\displaystyle\int_0^4 \int_0^{\sqrt{x}} (x + y)\,dy\,dx$

15. $\displaystyle\int_0^1 \int_{e^x}^{e} y\,dy\,dx$

16. $\displaystyle\int_0^1 \int_x^{2-x^2} x\,dy\,dx$

In Exercises 17–19 set up and evaluate an integral (or sum of integrals) for the given function $f(x, y)$ over the given region.

17. $f(x, y) = 3x$; R: the triangle formed by the lines $x = 0$, $x = y$, and $y = 2$.

18. $f(x, y) = x$; R: the semicircle bounded by the x-axis and the graph of $y = \sqrt{4 - x^2}$.

19. $f(x, y) = y$; R: the triangle formed by the lines $y = x$, $x + y = 2$, and $y = 0$.

In Exercises 20–23 find the volume of the solid beneath the graph of $f(x, y)$ and above the region R in the xy-plane.

20. $f(x, y) = x + y$; $R: 0 \le x \le 2, 1 \le y \le 3$.

21. $f(x, y) = y^2$; $R: 2 \le x \le 4, 0 \le y \le 3$.

22. $f(x, y) = 12 - 3x - 2y$; R: the triangular region bounded by the x-axis, the y-axis, and the line $3x + 2y = 12$.

23. $f(x, y) = e^{y/x}$; R: the trapezoid bounded by the lines $y = x$, $y = 0$, $x = 1$, and $x = 2$.

In Exercises 24–27 find the average value of the given function $f(x, y)$ over the given region R.

24. $f(x, y) = e^{-(x+y)}$; $R: 0 \le x \le 2; 0 \le y \le 1$.

25. $f(x, y) = x - y$; R: the triangle formed by the lines $y = 2x$, $x = 2$, and $y = 0$.

26. $f(x, y) = y$; R: the triangle formed by the lines $y = x$, $x + y = 2$, and $y = 0$.

27. $f(x, y) = \dfrac{x}{1 + y^2}$; R: the region bounded by the graphs of $x = 0$, $y = x^2$, and $y = 9$.

28. A microscopic slide is in the shape of the square $-1 \le x \le 1$, $-1 \le y \le 1$ (where x and y are measured in centimeters). The density of bacteria at (x, y) is

$$f(x, y) = 200 - 3x^2 - 3y^2$$

bacteria per square centimeter. Find the total number of bacteria on the slide.

Key Mathematical Concepts and Tools

Function of several variables

Three-dimensional coordinate systems

Partial derivatives

Extrema for a function of two variables (local and absolute maxima and minima)

First derivative test

Critical point

Saddle point

Second derivative test

Optimization

Constrained optimization and Lagrange multipliers

Double integral

Applications Covered in This Chapter

Temperature as a function of longitude and latitude
Value of a capital asset as a function of income and interest rates
Cobb–Douglas production functions
Minimizing distance from a point to a surface
Maximizing profit for a producer of two items
Constructing the box of minimal surface area for a given volume
Method of least squares—simple linear regression
Maximizing area of a garden of given perimeter
Maximizing production
Volume
Average value
Population density

EIGHT Warm-Up Test

1. Let $g(x, y) = \frac{x^2}{y} e^{-3x}$. Compute:

   (a) $\frac{\partial g}{\partial x}$ (b) $\frac{\partial g}{\partial y}$

   (c) $\frac{\partial^2 g}{\partial x\, \partial y}(-1, -2)$

2. A machine tool operation follows a Cobb–Douglas production function with $\alpha = 0.7$, $\beta = 2$. That is,

$$z = f(L, K) = 2L^{0.7} K^{0.3}$$

   (a) Find the marginal productivity of labor L.

   (b) Find the marginal productivity of capital K.

3. Find all critical points of the function

$$f(x, y) = x^3 + y^3 - 9xy + 27$$

   Determine whether they give maxima or minima, or are saddle points.

4. An oil company produces two grades of gasoline. The demand functions are $p_1(x) = 64 - 4x$ for the higher grade and $p_2(y) = 77 - 3y$ for the lower grade, where x and y are quantities in millions of gallons of the two grades, and p_1 and p_2 the corresponding prices in dollars. The joint cost function is

$$C(x, y) = 3x^2 + 8y^2 + 11xy$$

   Find the production levels that will maximize the oil company's profit.

5. Using the method of least squares, find the "best" straight-line fit to the following data:

x	y
0	−2.8
0.1	2.1
−1	−7.8
2	7.2

6. Consider the following data on earnings per share. (*Hint*: Take 1984 as year 0, 1985 as year 1, and so on.)

Year	1984	1985	1986
Earnings	2.15	2.23	2.47

   (a) Compute the regression line for these data.

   (b) Predict the 1990 earnings per share.

7. Use the technique of Lagrange multipliers to find the point on the straight line $y = 7x - 1$ that is closest to the origin.

8. For each of the following iterated integrals, (i) evaluate it as written and (ii) rewrite and evaluate an equivalent

iterated integral (or sum of integrals) with the order of integration reversed.

(a) $\int_0^2 \int_{x^3}^8 y^{-1/3}\, dy\, dx$ (b) $\int_0^1 \int_x^{2-x} 3\, dy\, dx$

(c) $\int_0^1 \int_{e^y}^e \frac{1}{x}\, dx\, dy$

9. Consider the production function

$$z = 8 - \frac{2}{x^{3/2}} - \frac{1}{y^{1/2}}$$

Find the optimal production schedule, given that the input and output prices are $p_x = 4$, $p_y = 3$, and $p_z = 8$

10. Find the dimensions of the rectangle of maximal area that can be inscribed in the upper half of the circle of radius 4. (Assume that one side of the rectangle must lie on the base (diameter) of this semicircle.)

EIGHT Final Exam

1. Find all critical points of the function $x^3 - 24xy + 8y^2 + 24 - 40y + 5$. Determine whether they give maxima or minima, or are saddle points.

2. Use Lagrange multipliers to find the values of x and y that maximize the function $f(x, y) = 5x + 3y$ subject to the constraint $x^2 + 3y^2 = 112$.

3. Using the method of least squares, find the best straight-line fit to the following data:

x	y
0.1	6
0.2	8
0.3	10.3
0.4	14.2

4. Find the point on the circle $x^2 + y^2 = 36$ that is:
   (a) Closest to the point $(-1, 1)$.
   (b) Furthest from the point $(-1, 1)$.

5. Find the optimal production level for p_x, p_y, and $C(x, y)$ as given. Also find the maximum profit.
   (a) $p_x = \$12$ per unit, $p_y = \$18$ per unit, $C(x, y) = 3x^2 + 5y^2$
   (b) $p_x = \$6$ per unit, $p_y = \$14$ per unit, $C(x, y) = 2x^2 + 8y^2 + 4xy$

6. An oil drum made of steel is to hold 100 liters of oil. It is to be in the shape of a right circular cylinder closed at the top and the bottom. Find the dimensions (radius of base and height) that will minimize the amount of steel used in the manufacture of this drum.

7. Let $f(x, y) = \dfrac{x}{x^2 + y} + 4$. Compute:
   (a) $\dfrac{\partial f}{\partial x}$ (b) $\dfrac{\partial f}{\partial y}$
   (c) $f_{xx}(-2, 2)$ (d) $f_{yy}(1, 3)$

8. The following monthly sales data have been tabulated for a large appliance manufacturer.

Month	1 = Jan	2 = Feb	3 = March	4 = April
Sales	1250	1300	1370	1430

   Sales figures represent numbers of units sold.
   (a) Compute the regression line for these data.
   (b) Predict the sales figures for June and for August.

9. In each of the following problems, sketch the region R. Then set up and evaluate a single iterated integral of f over R.
   (a) $f(x, y) = 4xy$; R: the triangle formed by the lines $y = x/2$, $y = 0$, and $x = 2$.
   (b) $f(x, y) = \sqrt{1 - y^2}$; R: the triangle formed by the lines $x = 2y$, $x = 0$, and $y = 1$.
   (c) $f(x, y) = y$; R: the region in the first quadrant bounded by the graphs of $y = 1/x$, $y = x$, and $y = \frac{1}{2}$.

10. The production function for a certain industry is $f(L, K) = 3L^{0.4}K^{0.6}$, where K and L are the number of units of labor and of capital used. At a labor cost of \$76 per unit and a capital cost of \$83 per unit, find the amount of labor and capital that should be used in order to minimize cost, given that 870 units are to be produced.

NINE

The Trigonometric Functions

Outline

In Context

In this chapter we introduce an important class of functions that has been conspicuously absent in the text up to this point. These are the trigonometric functions. Unlike the other functions we have studied, the trigonometric functions are *periodic*; that is, they have graphs that repeat themselves over and over. This makes them especially useful in applications to natural phenomena with repetitive behavior, such as heartbeat, population or business cycles, and acoustics. We have chosen to collect all the information about these functions into a single, self-contained chapter.

SECTION 9.1

The Trigonometric Functions: Basics

In this section we introduce the trigonometric functions. Many of you are familiar with most, and possibly all, of the basic facts about the trigonometric functions. For those who have no previous experience with trigonometry, you will find in this section (along with its problems) a complete but concise development—starting from scratch—of all the essentials of trigonometry.

Let us begin by introducing the sine and cosine functions. These functions are essential for the study of phenomena in which a pattern is repeated over and over again. Such cycles occur frequently in biological rhythms; menstruation, breathing, and seasonal variation in birth rates are good examples.

As a starting point, consider the unit circle. Recall that this is the circle of radius 1 centered at the origin and that its equation is

$$x^2 + y^2 = 1$$

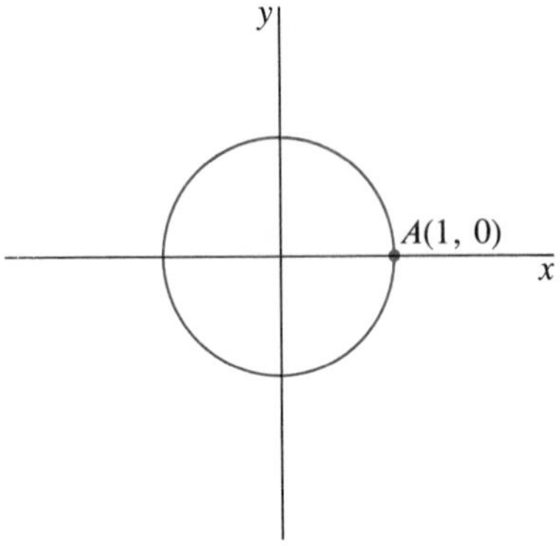

FIGURE 9.1

The point (1, 0) satisfies this equation, so it lies on the unit circle. Denote this point by A, as shown in Figure 9.1. Imagine that we have placed a nail at point A. Choose any real number $\alpha \geq 0$ and cut a piece of string of length α, as shown in Figure 9.2. Now attach one end of the string to the nail at point A, and then "wrap" the string around the unit circle in the counterclockwise direction as far as it will go, as shown in Figure 9.3. The string will end at some point P on the unit circle and, of course, the coordinates of P will satisfy the equation $x^2 + y^2 = 1$.

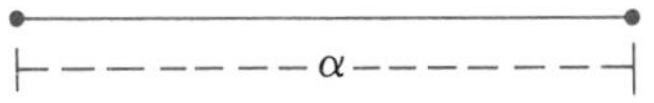

FIGURE 9.2

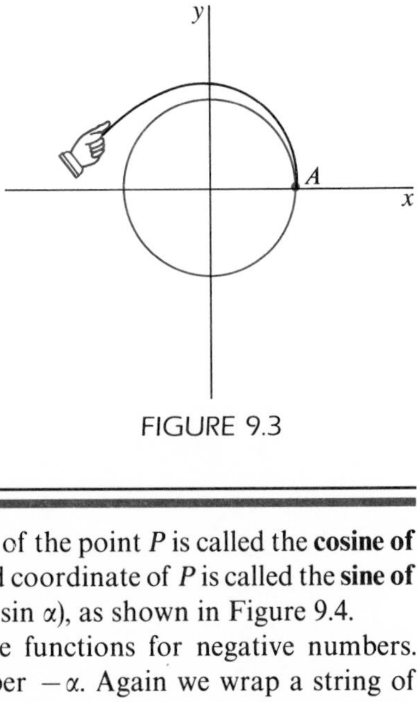

FIGURE 9.3

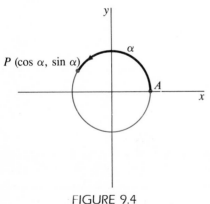

FIGURE 9.4

DEFINITION 9.1 The first coordinate of the point P is called the **cosine of** $\boldsymbol{\alpha}$ and denoted by $\cos \alpha$, whereas the second coordinate of P is called the **sine of** $\boldsymbol{\alpha}$ and denoted by $\sin \alpha$. Thus $P = (\cos \alpha, \sin \alpha)$, as shown in Figure 9.4.

We also define the sine and cosine functions for negative numbers. Suppose we are given the negative number $-\alpha$. Again we wrap a string of length α around the unit circle, but this time we do so in the clockwise direction. As before, let P denote the point where the string ends. By analogy with the first part of this definition the first coordinate of P is called the **cosine of** $\boldsymbol{-\alpha}$ and denoted by $\cos(-\alpha)$, whereas the second coordinate of P is called the **sine of $-\alpha$** and denoted by $\sin(-\alpha)$, as shown in Figure 9.5. Thus the sine and cosine functions are defined for all real numbers. ⊟

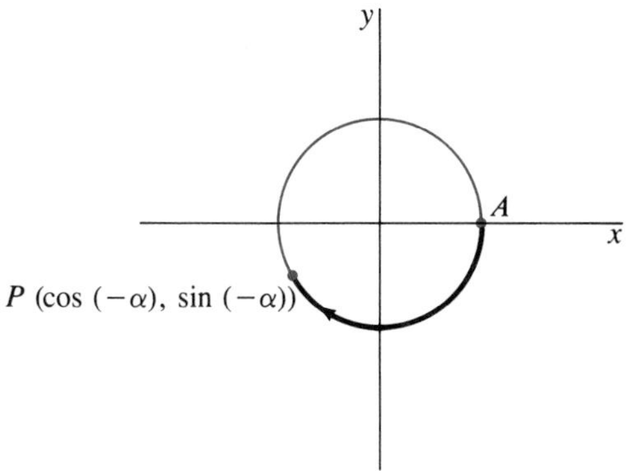

FIGURE 9.5

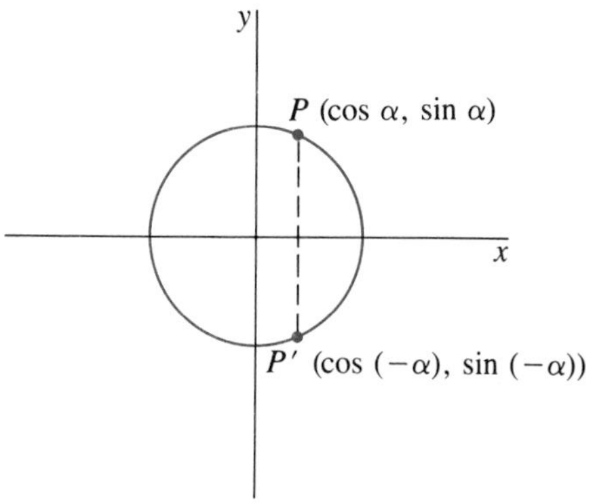

FIGURE 9.6

Two useful relationships follow immediately from this definition. First, for $\alpha \geq 0$, consider the points $P = (\cos\alpha, \sin\alpha)$ and $P' = (\cos(-\alpha), \sin(-\alpha))$ on the unit circle. From the symmetry of the circle and the way points P and P' are reached, it is easy to see that P and P' have the same x-coordinate, whereas their y-coordinates are negatives of each other (Figure 9.6). Therefore,

(9.1) $$\cos(-\alpha) = \cos\alpha \qquad \sin(-\alpha) = -\sin\alpha$$

Second, for every real number α (positive, negative, or zero) the point $P = (\cos\alpha, \sin\alpha)$ is on the unit circle and hence satisfies the equation $x^2 + y^2 = 1$. Consequently,

(9.2) $$\cos^2\alpha + \sin^2\alpha = 1 \qquad \text{for any number } \alpha$$

Here we are using the standard notation $\cos^2\alpha$ for $(\cos\alpha)^2$ and $\sin^2\alpha$ for $(\sin\alpha)^2$.

PERIODICITY OF SINE AND COSINE According to the definition, we know—at least in principle—how to find $\cos\alpha$ and $\sin\alpha$ for any real number α. Let us see how we can use our scheme to find $\cos\alpha$ and $\sin\alpha$ when $\alpha = 0$. We must place a nail at $A = (1, 0)$ and then, starting from that point, wrap a string of length 0 in the counterclockwise direction around the unit circle as far as it will go. Its end point is, by Definition 9.1, $(\cos 0, \sin 0)$. But, because the length of the string is 0, the end point is the starting point $(1, 0)$, as shown in Figure 9.7. Therefore we have

(9.3) $$\cos 0 = 1 \qquad \sin 0 = 0$$

Let us apply the same method to determine $\cos 2\pi$ and $\sin 2\pi$. All we have to do is figure out how far a string of length 2π will reach around the unit circle. Fortunately this is easy, because the circumference of a circle is 2π times its radius. Thus the unit circle (whose radius is 1) has circumference 2π. So a string of length 2π will wrap around exactly once and end up where it

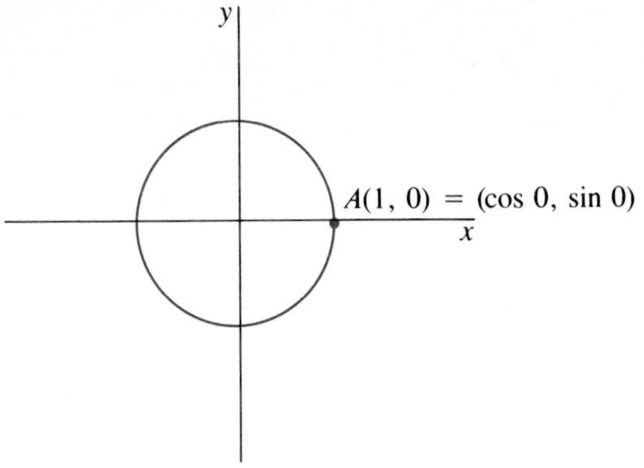

FIGURE 9.7

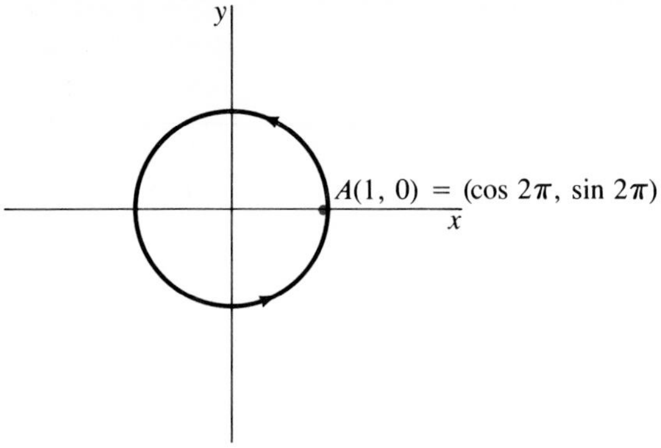

FIGURE 9.8

started—at the point (1, 0), as shown in Figure 9.8. Consequently we have

(9.4) $$\cos 2\pi = 1 \qquad \sin 2\pi = 0$$

Now, for any real number α, let us consider the point determined by a string of length α—namely, $(\cos \alpha, \sin \alpha)$. By the same reasoning that we employed above, we see that a string of length $\alpha + 2\pi$, whose end point is then $(\cos(\alpha + 2\pi), \sin(\alpha + 2\pi))$ has exactly the same end point as a string of length α. This means that

(9.5) $$\cos(\alpha + 2\pi) = \cos \alpha \qquad \sin(\alpha + 2\pi) = \sin \alpha$$

for every real number α. In other words, the sine and cosine functions "repeat themselves" every 2π units. This property is called **periodicity**, and we say that the sine and cosine functions are **periodic**, with period 2π.

ANGLES AND RADIAN MEASURE Let α be any real number, and suppose we have located the point $P(\cos \alpha, \sin \alpha)$ on the unit circle. Recall what this means: A string of length α that is attached at one end to the point $A(1, 0)$ will wrap around the circle until it reaches point P, as shown in Figure 9.9. Alternatively, it means that the arc AP has length α. Denoting the center of the unit circle (which is the origin) by O, let us draw the radii OA and

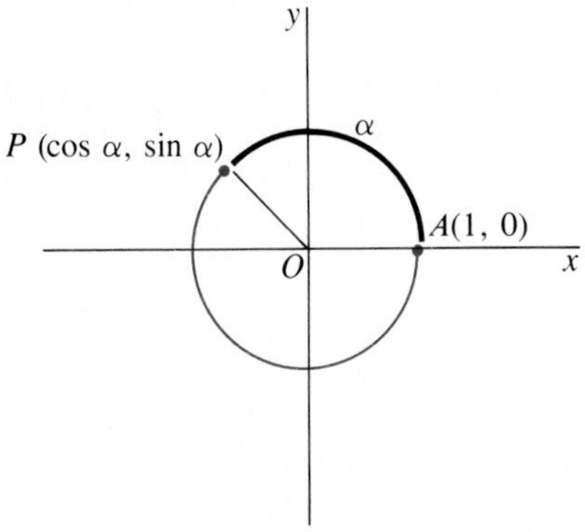

FIGURE 9.9

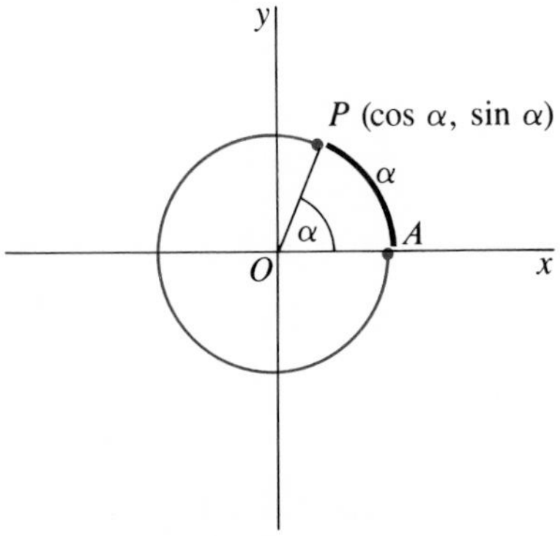

FIGURE 9.10

OP (Figure 9.9). The lines OA and OP form an angle (denoted $< AOP$) that is said to be in **standard position** (because side OA is horizontal). The point O is called the **vertex** of the angle; OA is called the **initial side** of the angle, and OP is called the **terminal side**. Now refer to Figure 9.10.

DEFINITION 9.2 The measure of the angle AOP is said to be **α radians** when the arc AP that it cuts off on the unit circle has length α. The **cosine of an angle** of α radians is defined to be $\cos \alpha$, and the **sine of an angle** of α radians is defined to be $\sin \alpha$. ⊟

Thus the cosine and sine functions that were defined previously for lengths have now been defined also for angles measured in radians. For example, when we write $\sin \pi/3$, we may think of this as the sine of an angle of $\pi/3$ radians.

Of course, you are accustomed to measuring angles in degrees, but this should cause no serious difficulty. Just as we frequently measure length in feet, inches, yards, or miles and can then convert easily from one of these units of measurement to another, the same applies for angles. All we need do is find the relationship between radians and degrees. Because the circumference of the unit circle is 2π, the angle at the center that measures once around the circle is 2π radians. On the other hand, as is well known, this angle is 360 degrees—so we have the fundamental relation

(9.6) $$2\pi \text{ radians} = 360 \text{ degrees}$$

Now consider an arbitrary angle. If we take the ratio of the measure of this angle to the measure of the angle for once around the circle, this proportion comes out the same no matter what unit is used for measuring angles. Consequently we have

$$\frac{\text{radians}}{2\pi} = \frac{\text{degrees}}{360}$$

In other words, given an angle that measures d degrees or r radians, d and r are related by the formula

(9.7) $$\frac{r}{2\pi} = \frac{d}{360}$$

so it is easy to convert from radians to degrees, and vice versa.

EXAMPLE 1 To convert $\pi/6$ radians to degrees, set $r = \pi/6$ in Equation (9.7). This yields

$$\frac{d}{360} = \frac{\pi/6}{2\pi} = \frac{1}{12}$$

so $d = 30°$ (° denotes degrees). Thus an angle of $\pi/6$ radians is the same as one of 30°.

EXAMPLE 2 To convert 135° to radians, put $d = 135$ in Equation (9.7). This yields

$$\frac{r}{2\pi} = \frac{135}{360} = \frac{3}{8}$$

so that $r = 3\pi/4$ radians.

For reasons that are not readily apparent here, it is much more natural in the study of calculus to measure angles in radians than in degrees—and we shall do so henceforth. To make sure that you have full control over the relationship between these two units of measurement and to assist you in making conversions, you should review the entries given in Table 9.1. The fact that 1 radian (which is $180/\pi$ degrees) is approximately 57° is of no particular interest to us.

TABLE 9.1

Radians	0	$\pi/6$	$\pi/4$	$\pi/3$	$\pi/2$	$2\pi/3$	$3\pi/4$	$5\pi/6$	π	$5\pi/4$	$3\pi/2$	$7\pi/4$	2π
Degrees	0	30	45	60	90	120	135	150	180	225	270	315	360

COSINE AND SINE FOR SPECIAL NUMBERS There are certain special values of α for which, because of the geometric properties of the unit circle, it is easy to find $\cos\alpha$ and $\sin\alpha$. We have already done so for $\alpha = 0$ in Equation (9.3) and for $\alpha = 2\pi$ in Equation (9.4). Now let us reinforce the ideas introduced thus far by computing $\cos\alpha$ and $\sin\alpha$ for a few special values of α.

A string of length π will reach halfway around the unit circle, because the circumference of this circle is 2π. It follows (as indicated in Figure 9.11) that the point with coordinates $(\cos\pi, \sin\pi)$, being the halfway point, is precisely the point $(-1, 0)$. Consequently

(9.8) $$\cos\pi = -1 \qquad \sin\pi = 0$$

As another example, consider the case $\alpha = \pi/2$. A string of length $\pi/2$ ends at the point $(\cos\pi/2, \sin\pi/2)$, which is one quarter of the way around the circle. But (as indicated in Figure 9.12) this is clearly the point (0, 1).

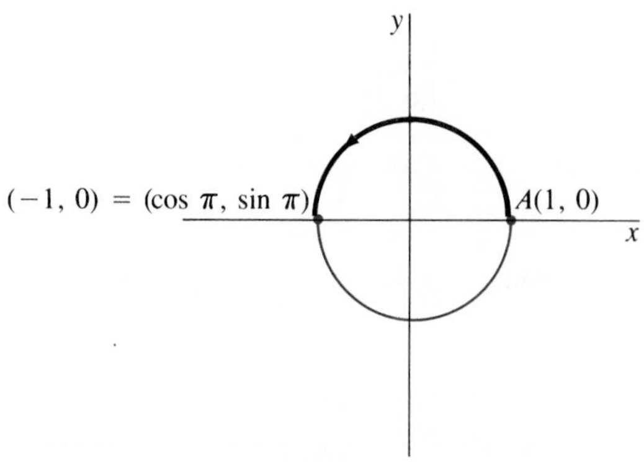

FIGURE 9.11

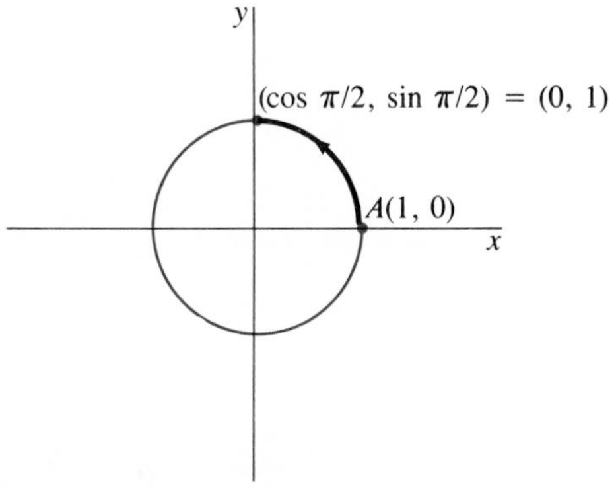

FIGURE 9.12

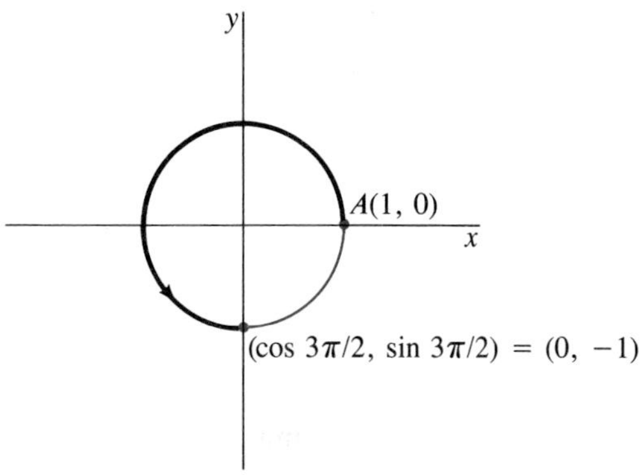

FIGURE 9.13

Consequently

(9.9) $\cos \pi/2 = 0 \qquad \sin \pi/2 = 1$

In similar fashion, for the case $\alpha = 3\pi/2$ (see Figure 9.13) we have

(9.10) $\cos 3\pi/2 = 0 \qquad \sin 3\pi/2 = -1$

Let us turn our attention to the problem of finding the cosine and sine for values of α that do not produce a point on one of the coordinate axes. There are certain values of α for which this is not hard to do.

EXAMPLE 3 We determine the cosine and sine for $\alpha = \pi/4$. A string of length $\pi/4$ ends at the point $P(\cos \pi/4, \sin \pi/4)$ on the unit circle, and, because $\pi/4$ is one-eighth of the circumference of the circle, this point must lie midway on the circle between the points (1, 0) and (0, 1), as shown in Figure 9.14. By symmetry, the coordinates of this midway point must be equal (in other words, the midway point lies on the line $y = x$), so

$$\cos \pi/4 = \sin \pi/4$$

In addition, because the point P is on the unit circle, its coordinates satisfy the equation $x^2 + y^2 = 1$. Hence

$$\cos^2(\pi/4) + \sin^2(\pi/4) = 1$$

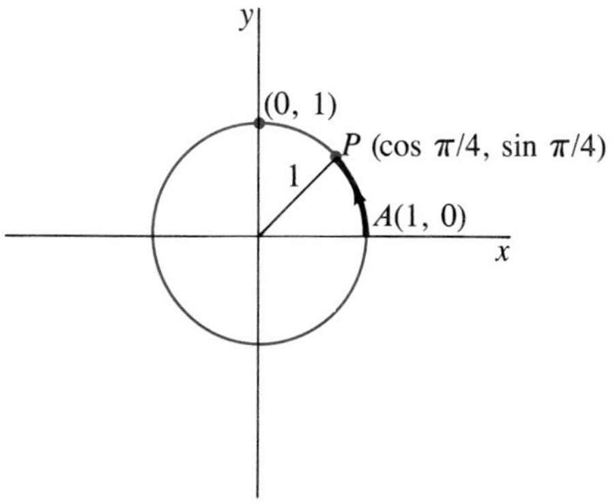

FIGURE 9.14

From these two relations, we have

$$\sin^2(\pi/4) + \sin^2(\pi/4) = 1$$

or $$2\sin^2(\pi/4) = 1$$

Therefore $$\sin^2(\pi/4) = 1/2$$

and $$\sin \pi/4 = \frac{1}{\sqrt{2}} = \frac{\sqrt{2}}{2}$$

(Note that we have taken the positive square root for $\sin \pi/4$. It has to be this way because our point $P(\cos \pi/4, \sin \pi/4)$ lies in the first quadrant, and $\sin \pi/4$ must therefore be positive.) We have shown then, that

(9.11) $$\cos \pi/4 = \sin \pi/4 = \sqrt{2}/2$$

Now let us turn to $\alpha = 3\pi/4$. Because $3\pi/4 = \pi/2 + \pi/4$ and a quarter of the circle has length $\pi/2$, it follows that a string of length $3\pi/4$ ends at the point $(\cos 3\pi/4, \sin 3\pi/4)$, which is midway on the unit circle between the points $(0, 1)$ and $(-1, 0)$. Moreover, as you can easily convince yourself, this point is symmetric, with respect to the y-axis, to the point $(\cos \pi/4, \sin \pi/4)$ (Figure 9.15). The symmetry of these points with respect to the y-axis tells us that

$$\cos 3\pi/4 = -\cos \pi/4 \quad \text{and} \quad \sin 3\pi/4 = \sin \pi/4$$

So, by virtue of Equation (9.11),

(9.12) $$\cos 3\pi/4 = -\sqrt{2}/2 \qquad \sin 3\pi/4 = \sqrt{2}/2$$

We leave it to you to show, by further use of symmetry, that

(9.13)
$$\cos 5\pi/4 = -\sqrt{2}/2 \qquad \sin 5\pi/4 = -\sqrt{2}/2$$
$$\cos 7\pi/4 = \sqrt{2}/2 \qquad \sin 7\pi/4 = -\sqrt{2}/2$$

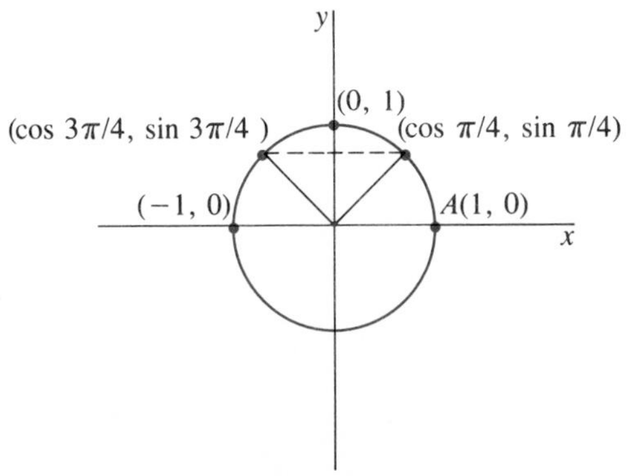

FIGURE 9.15

EXAMPLE 4 Let us try to determine cosine and sine for $\alpha = \pi/3$. A string of length $\pi/3$ ends at the point $P(\cos \pi/3, \sin \pi/3)$ on the unit circle, and, because $\pi/3$ is one-sixth of the circumference of the circle, this point must lie one-third of the way along the circle from the point (1, 0) to the point (−1, 0). Figure 9.16 (in which we also plot the points for $\alpha = 2\pi/3, \pi, 4\pi/3$, and $5\pi/3$) illustrates this. (Note that the inscribed figure is a regular hexagon.) Now, the triangle POA is isosceles. Because an angle of $\pi/3$ radians is 60° (so angle POA is 60°) and the sum of the angles of a triangle is 180°, it follows that angles OPA and PAO are both 60°. Thus triangle POA is equilateral. If we drop a perpendicular from P to the x-axis and call its foot Q, then Q bisects the segment $\overline{OA}$ and hence $\overline{OQ} = \frac{1}{2}$ (Figure 9.17). Applying the Pythagorean theorem, we have

$$\overline{OP}^2 = \overline{OQ}^2 + \overline{PQ}^2$$

or $1^2 = (1/2)^2 + \overline{PQ}^2$. It follows that

$$\overline{PQ} = \frac{\sqrt{3}}{2}$$

Now, according to the way coordinates are assigned to points in the plane,

$$\overline{OQ} = x\text{-coordinate of } P \quad \text{and} \quad \overline{PQ} = y\text{-coordinate of } P$$

Consequently

(9.14) $$\cos \pi/3 = 1/2 \qquad \sin \pi/3 = \sqrt{3}/2$$

We may also proceed one step further. Because the point ($\cos 2\pi/3$, $\sin 2\pi/3$) is clearly symmetric, with respect to the y-axis, to the point ($\cos \pi/3$, $\sin \pi/3$)—see Figure 9.16—we have

$$\cos 2\pi/3 = -\cos \pi/3 \quad \text{and} \quad \sin 2\pi/3 = \sin \pi/3$$

Therefore

(9.15) $$\cos 2\pi/3 = -1/2 \qquad \sin 2\pi/3 = \sqrt{3}/2$$

It is left to you, as an exercise, to show (by making use of symmetry) that

(9.16) $$\cos 4\pi/3 = -1/2 \qquad \sin 4\pi/3 = -\sqrt{3}/2$$
$$\cos 5\pi/3 = 1/2 \qquad \sin 5\pi/3 = -\sqrt{3}/2$$

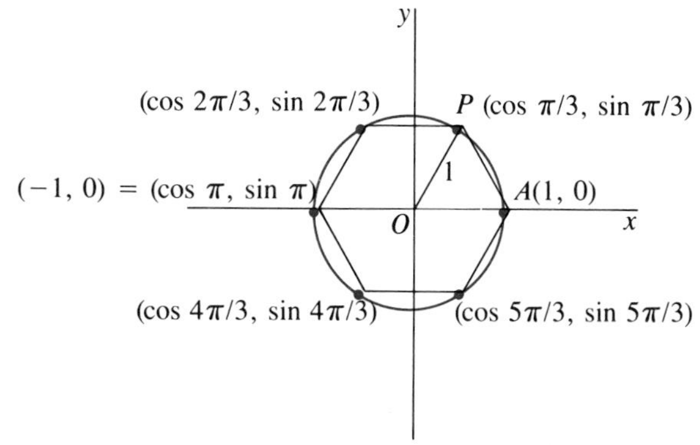

FIGURE 9.16

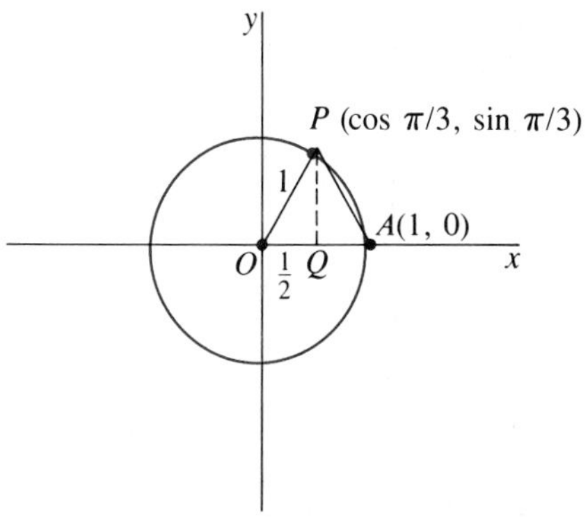

FIGURE 9.17

PRACTICE EXERCISE Exploit the special geometry of the circle to show that

$$\cos \pi/6 = \sqrt{3}/2 \qquad \sin \pi/6 = 1/2$$
$$\cos 5\pi/6 = -\sqrt{3}/2 \qquad \sin 5\pi/6 = 1/2$$
$$\cos 7\pi/6 = -\sqrt{3}/2 \qquad \sin 7\pi/6 = -1/2$$
$$\cos 11\pi/6 = \sqrt{3}/2 \qquad \sin 11\pi/6 = -1/2$$

More specifically, prove the formulas for $\pi/6$ by using Figure 9.18 and noting that angle POP' is $\pi/3$ (or 60°), that triangle POP' is equilateral, that the x-axis is the perpendicular bisector of the segment PP', and so on. Then apply symmetry techniques to find cosine and sine for $5\pi/6$, $7\pi/6$, and $11\pi/6$.

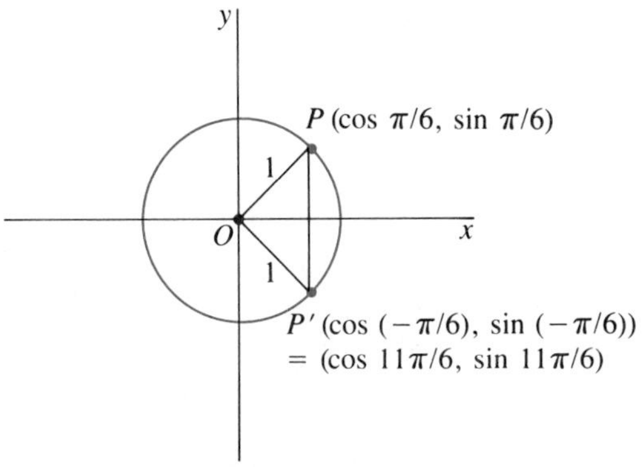

FIGURE 9.18

For ready reference let us collect all the values of cosine and sine that we have determined in Table 9.2.

For most values of α that one might choose, such as $\alpha = 1$, $\alpha = \frac{5}{4}$, or $\alpha = \sqrt{2}$, there is no convenient way to use the geometry of the circle, as we have done, to find $\cos \alpha$ and $\sin \alpha$. Fortunately, however, there are other approaches

TABLE 9.2

Angle in radians	0	$\pi/6$	$\pi/4$	$\pi/3$	$\pi/2$	$2\pi/3$	$3\pi/4$	$5\pi/6$	π	$7\pi/6$	$5\pi/4$	$4\pi/3$	$3\pi/2$	$5\pi/3$	$7\pi/4$	$11\pi/6$
Angle in degrees	0	30	45	60	90	120	135	150	180	210	225	240	270	300	315	330
cosine	1	$\sqrt{3}/2$	$\sqrt{2}/2$	$\frac{1}{2}$	0	$-1/2$	$-\sqrt{2}/2$	$-\sqrt{3}/2$	-1	$-\sqrt{3}/2$	$-\sqrt{2}/2$	$-1/2$	0	$1/2$	$\sqrt{2}/2$	$\sqrt{3}/2$
sine	0	$1/2$	$\sqrt{2}/2$	$\sqrt{3}/2$	1	$\sqrt{3}/2$	$\sqrt{2}/2$	$1/2$	0	$-1/2$	$-\sqrt{2}/2$	$-\sqrt{3}/2$	-1	$-\sqrt{3}/2$	$-\sqrt{2}/2$	$-1/2$

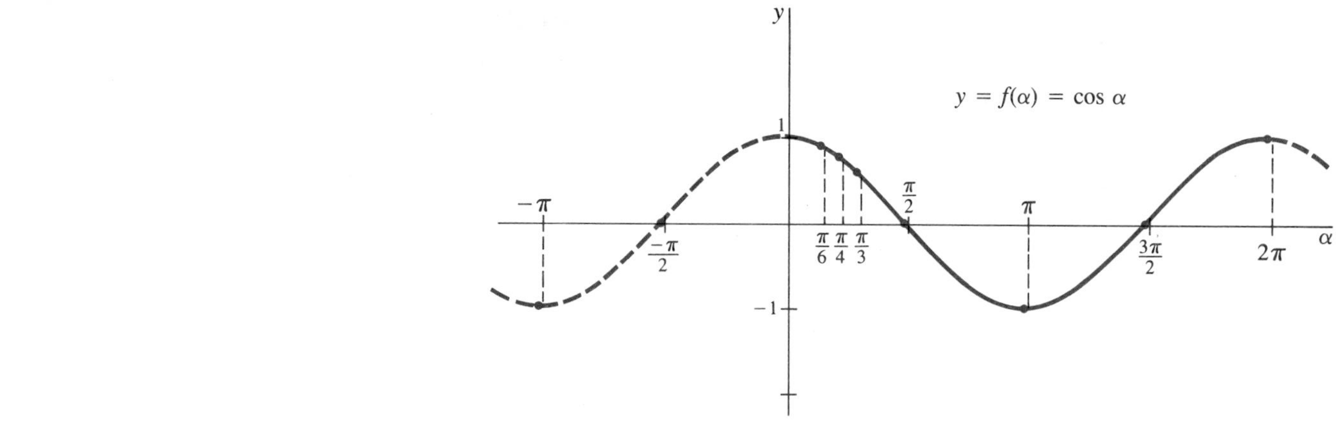

FIGURE 9.19

(of a rather advanced nature) for computing cos α and sin α in general. These approaches are the basis for standard trigonometric tables.

Let us turn to the graphs of the cosine and sine functions. Figure 9.19 is a sketch of $y = f(\alpha) = \cos \alpha$, using the points $(\alpha, \cos \alpha)$ from Table 9.2. Note that, once we have drawn the graph on the interval from 0 to 2π (by which we mean for all values of α between 0 and 2π), all we have to do is repeat the picture, over and over again, because of the periodicity of the cosine function [see Equation (9.5)]. Figure 9.20 is a sketch of the graph $y = f(\alpha) = \sin \alpha$, using the points $(\alpha, \sin \alpha)$ from Table 9.2.

We may note that the domain of both the cosine function and the sine function is the set of all real numbers—because we can wrap *any* length of string around the unit circle, and wrapping in the clockwise direction is associated with negative α's. The range of both the cosine and sine functions is the set of all real numbers y that satisfy $-1 \leq y \leq 1$. This is clear as soon as one examines the x- and y-coordinates of the point $P(\cos \alpha, \sin \alpha)$ as it traces its path around the unit circle. In particular, neither the cosine nor the sine function can ever take on any value greater than 1 or less than -1, because the point $P(\cos, \alpha, \sin \alpha)$ always satisfies the equation $\cos^2\alpha + \sin^2\alpha = 1$ and the square of any real number is always ≥ 0.

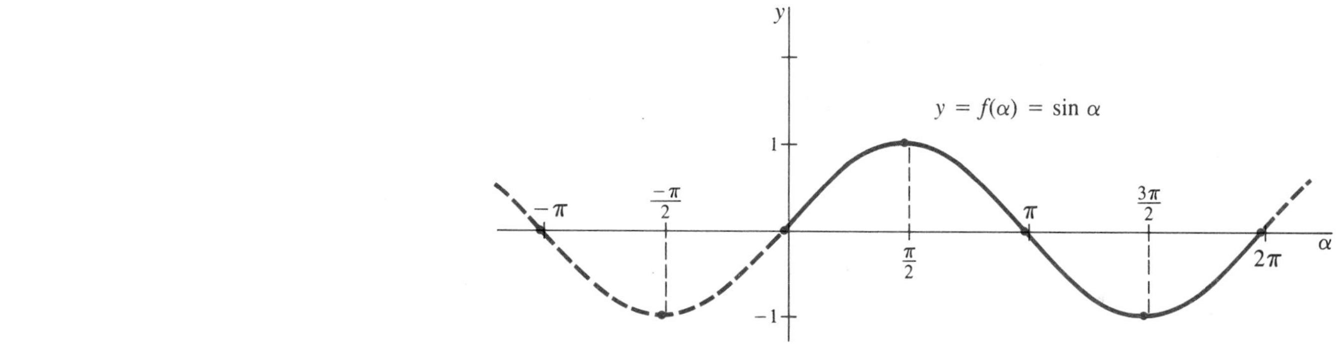

FIGURE 9.20

You may observe from Figures 9.19 and 9.20 that the graphs of the cosine and sine functions are essentially the same. More precisely, if we shift the graph of the cosine function to the right by $\pi/2$, the result is the graph of the sine function.

There are four additional standard trigonometric functions: **tangent** (tan), **cotangent** (cot), **secant** (sec), and **cosecant** (csc). They are defined as follows:

$$\tan \alpha = \frac{\sin \alpha}{\cos \alpha}$$

$$\cot \alpha = \frac{1}{\tan \alpha} = \frac{\cos \alpha}{\sin \alpha}$$

$$\sec \alpha = \frac{1}{\cos \alpha}$$

$$\csc \alpha = \frac{1}{\sin \alpha}$$

Because division by 0 is meaningless, it is understood that these definitions apply only when the denominator is not equal to 0. If the denominator is 0 for some α, the function is said to be **undefined** for that α. For example, $\tan \pi/2$ is undefined, as are $\cot 0$, $\sec 3\pi/2$, $\csc \pi$, and so on.

PRACTICE EXERCISE (a) Prove the following identities:

$$\sec^2\alpha = 1 + \tan^2\alpha$$

$$\csc^2\alpha = 1 + \cot^2\alpha$$

(b) Prove the following formulas:

$$\tan(-\alpha) = -\tan \alpha \qquad \cot(-\alpha) = -\cot \alpha$$

$$\sec(-\alpha) = \sec \alpha \qquad \csc(-\alpha) = -\csc \alpha$$

TABLE 9.3

Angle in radians	0	$\pi/6$	$\pi/4$	$\pi/3$	$\pi/2$	$2\pi/3$	$3\pi/4$	$5\pi/6$	π	$7\pi/6$	$5\pi/4$	$4\pi/3$	$3\pi/2$	$5\pi/3$	$7\pi/4$	$11\pi/6$
Angle in degrees	0	30	45	60	90	120	135	150	180	210	225	240	270	300	315	330
tangent	0	$\sqrt{3}/3$	1	$\sqrt{3}$	und	$-\sqrt{3}$	-1	$-\sqrt{3}/3$	0	$-\sqrt{3}/3$	1	$\sqrt{3}$	und	$-\sqrt{3}$	-1	$-\sqrt{3}/3$
cotangent	und	$\sqrt{3}$	1	$\sqrt{3}/3$	0	$-\sqrt{3}/3$	-1	$-\sqrt{3}$	und	$\sqrt{3}$	1	$\sqrt{3}/3$	0	$-\sqrt{3}/3$	-1	$-\sqrt{3}$
secant	1	$2\sqrt{3}/3$	$\sqrt{2}$	2	und	-2	$-\sqrt{2}$	$-2\sqrt{3}/3$	-1	$-2\sqrt{3}/3$	$-\sqrt{2}$	-2	und	2	$\sqrt{2}$	$2\sqrt{3}/3$
cosecant	und	2	$\sqrt{2}$	$2\sqrt{3}/3$	1	$2\sqrt{3}/3$	$\sqrt{2}$	2	und	-2	$-\sqrt{2}$	$-2\sqrt{3}/3$	-1	$-2\sqrt{3}/3$	$-\sqrt{2}$	-2

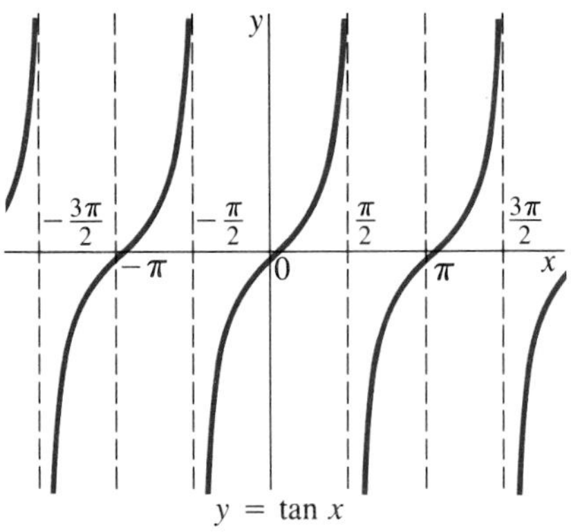

FIGURE 9.21

(c) Show that the additional trigonometric functions are periodic, of period 2π. That is, show that

$$\tan(\alpha + 2\pi) = \tan\alpha \qquad \cot(\alpha + 2\pi) = \cot\alpha$$
$$\sec(\alpha + 2\pi) = \sec\alpha \qquad \csc(\alpha + 2\pi) = \csc\alpha$$

(d) Verify the entries in Table 9.3 of values of tan, cot, sec, and csc (und means "undefined"). This table complements Table 9.2.

(e) Verify that the graph of the tangent function looks as shown in Figure 9.21 and additionally that $\tan(\alpha + \pi) = \tan\alpha$.

(f) Find the domain and the range for the tan, cot, sec, and csc functions.

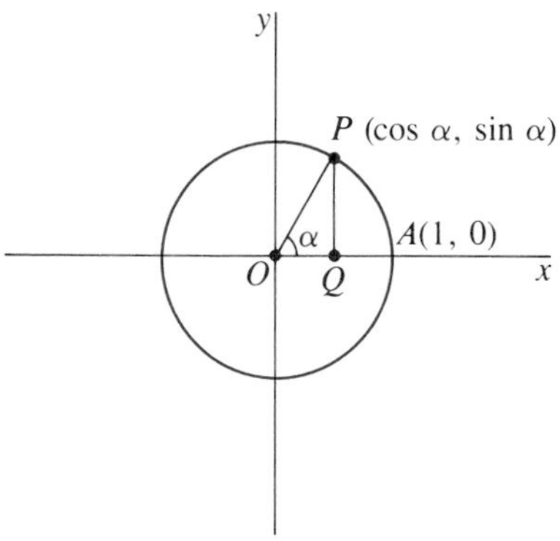

FIGURE 9.22

THE RIGHT TRIANGLE In high school you probably learned about trigonometric functions, but there these functions were defined with reference to right triangles. Our purpose here is to relate our current definitions of the trigonometric functions to the definitions you may have seen previously.

To begin with, we consider an angle of α radians $(0 < \alpha < \pi/2)$ and place it in standard position (see Definition 9.2). The vertex of an angle is at the origin, $O = (0, 0)$. The initial side meets the unit circle at $A(1, 0)$, and the terminal side meets the unit circle at $P(\cos\alpha, \sin\alpha)$. Let us drop a perpendicular from the point P to the x-axis and denote the foot of this perpendicular by Q, as shown in Figure 9.22. The coordinates of Q are clearly $(\cos\alpha, 0)$. What are the lengths of the sides of the right triangle OQP? Obviously, $\overline{OP} = 1$ because it is a radius of the unit circle. Also, from the way coordinates are assigned to a point in the plane (or by a simple application of the distance formula), we have

$$\overline{OQ} = \cos\alpha \quad \text{and} \quad \overline{PQ} = \sin\alpha$$

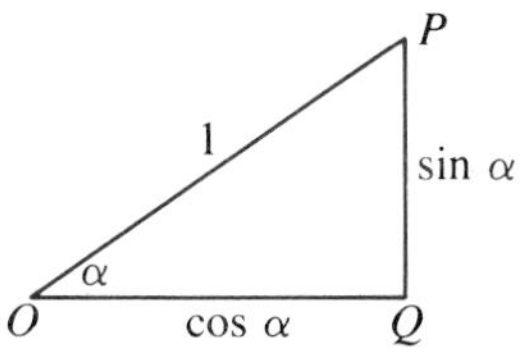

FIGURE 9.23

Thus we are dealing with the right triangle depicted in Figure 9.23.

Now let us consider an arbitrary right triangle $O'Q'P'$, in which angle $P'O'Q'$ is α radians, as shown in Figure 9.24. As is customary, we have labeled the sides of triangle $O'Q'P'$ as follows: The hypotenuse $O'P'$ is denoted by hyp; the side $P'Q'$, which is opposite the angle $P'O'Q'$, is denoted by opp; and the side $O'Q'$, which is adjacent to the angle $P'O'Q'$, is denoted by adj.

The right triangles OQP and $O'Q'P'$ are similar (which means that all three pairs of corresponding angles are equal). From plane geometry, we know that corresponding sides of similar triangles are proportional. For instance,

$$\frac{\overline{OQ}}{\overline{OP}} = \frac{\overline{O'Q'}}{\overline{O'P'}}$$

and

$$\frac{\overline{PQ}}{\overline{OP}} = \frac{\overline{P'Q'}}{\overline{O'P'}}$$

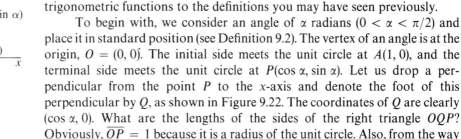

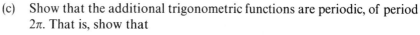

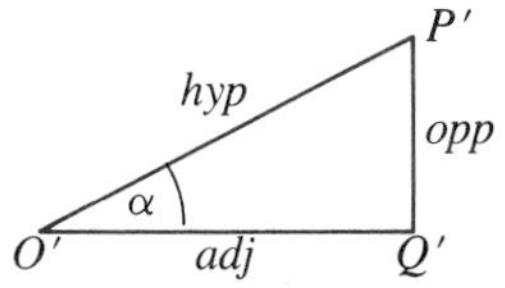

FIGURE 9.24

which yields $\cos\alpha/1 = \text{adj}/\text{hyp}$ and $\sin\alpha/1 = \text{opp}/\text{hyp}$. We have proved that, in any right triangle, the cosine and sine (as defined in this section) of α are given by

(9.17) $$\cos\alpha = \text{adj/hyp} \qquad \sin\alpha = \text{opp/hyp}$$

From these relations, it follows immediately that

(9.18) $$\tan\alpha = \text{opp/adj}$$

PRACTICE EXERCISE You have just seen that, for any angle α in a right triangle, the definitions of sin, cos, and tan introduced in this section turn out to be the same as the familiar high school definitions. Of course, the same statement clearly applies for the remaining trigonometric functions: cot, sec, and csc.

Consider an arbitrary angle α placed in standard position. Choose any point on the terminal side, call it $P(x, y)$, and denote its distance from the origin by r so that

$$r = \sqrt{x^2 + y^2}$$

A typical sketch of this arrangement appears in Figure 9.25. Use similar triangles (be careful of signs) to show that

(9.19) $$\cos\alpha = x/r \qquad \sin\alpha = y/r \qquad \tan\alpha = y/x$$

These formulas are often used in high school as the definitions of the trigonometric functions for angles α greater than $\pi/2$. (Of course, they are also valid for angles α between 0 and $\pi/2$.) Thus we see that our current definitions of the trigonometric functions are really the same as the old high school definitions.

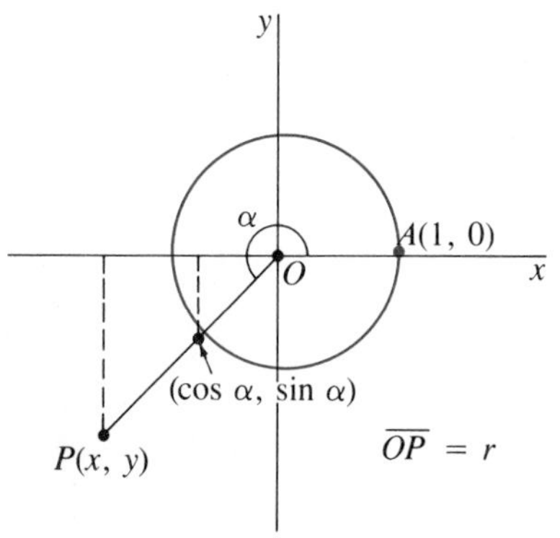

FIGURE 9.25

PRACTICE EXERCISE Consider an arbitrary triangle. Denote its vertices by A, B, and C and the lengths of the sides opposite these vertices by a, b, and c, respectively. No harm will come if we also use the symbol for a vertex to denote the angle at that vertex.

We can always place triangle ABC in a standard position as follows. Point C is at the origin. Point B is on the x-axis to the right of C, so its coordinates are $(a, 0)$. Point A is above the x-axis, so it falls in either the first quadrant, as shown in Figure 9.26(a), or the second quadrant, as shown in

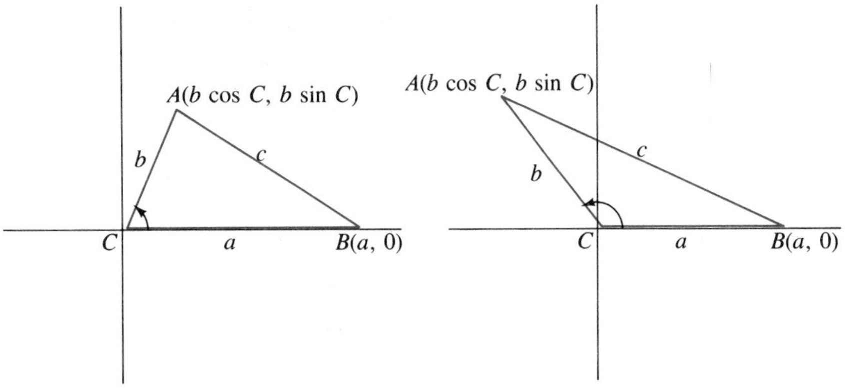

FIGURE 9.26

Figure 9.26(b). Making use of Equation (9.19), we find that the coordinates of A are $(b \cos C, b \sin C)$.

By applying the distance formula between A and B show that

(9.20) $$c^2 = a^2 + b^2 - 2ab \cos C$$

Equation (9.20) is known as the **law of cosines**. It enables us to compute one side of a triangle if we know the other two sides and the angle between them. (Of course, the roles of a, b, and c are interchangeable, so we also have $b^2 = a^2 + c^2 - 2ac \cos B$ and $a^2 = b^2 + c^2 - 2bc \cos A$.) Observe that, if ABC is a right triangle with the right angle at C, Equation (9.20) becomes

$$c^2 = a^2 + b^2$$

For this reason, the law of cosines should be viewed as a generalization of the Pythagorean theorem.

SOME USEFUL TRIGONOMETRIC IDENTITIES Once we have defined the trigonometric functions, there is a literally endless collection of formulas and relationships involving them that we could derive. Fortunately we have no need for most of these formulas. However, there are a few that we will use later in this chapter. The starting point for developing these relationships is

(9.21) $$\cos(\alpha - \beta) = \cos \alpha \cos \beta + \sin \alpha \sin \beta$$

We outline the derivation of Equation (9.21) in the problems at the end of this section. If we replace β by $-\beta$ in Equation (9.21) and make use of Equation (9.1), we obtain

(9.22) $$\cos(\alpha + \beta) = \cos \alpha \cos \beta - \sin \alpha \sin \beta$$

One of the elementary consequences of Equation (9.21) is obtained by setting $\alpha = \pi/2$:

$$\begin{aligned}\cos(\pi/2 - \beta) &= \cos \pi/2 \cos \beta + \sin \pi/2 \sin \beta \\ &= (0)(\cos \beta) + (1)(\sin \beta) = \sin \beta\end{aligned}$$

Thus, for any angle β,

(9.23) $$\cos(\pi/2 - \beta) = \sin \beta$$

Now, if we replace β here by $\pi/2 - \beta$, the result is

$$\cos[\pi/2 - (\pi/2 - \beta)] = \sin(\pi/2 - \beta)$$

Thus, for any angle β,

(9.24) $$\sin(\pi/2 - \beta) = \cos \beta$$

To illustrate the last two formulas, let us set $\beta = \pi/3$. Then

$$\pi/2 - \beta = \pi/2 - \pi/3 = \pi/6$$

so Equation (9.23) becomes

$$\cos \pi/6 = \sin \pi/3$$

A glance at Table 9.22 shows that these are, in fact, both equal to $\sqrt{3}/2$. Similarly, Equation (9.24) yields

$$\sin \pi/6 = \cos \pi/3$$

Both are equal to $1/2$.

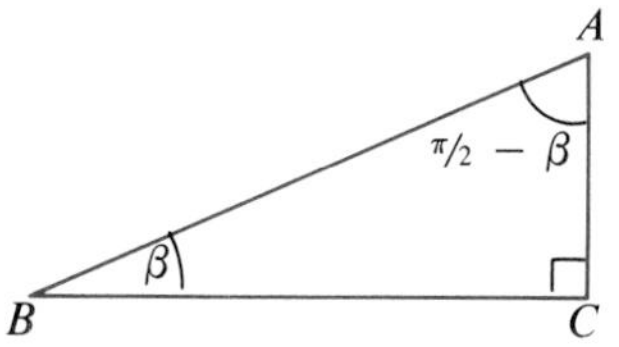

FIGURE 9.27

REMARK If the angle β satisfies $0 < \beta < \pi/2$, it may be placed in a right triangle as shown in Figure 9.27.

The angle at A is clearly $\pi/2 - \beta$, the "complement" of β. According to the definitions of sine and cosine in a right triangle, Equation (9.17), we have

$$\cos(\pi/2 - \beta) = \frac{\overline{AC}}{\overline{AB}} = \sin \beta$$

$$\sin(\pi/2 - \beta) = \frac{\overline{BC}}{\overline{AB}} = \cos \beta$$

Thus Equations (9.23) and (9.24) should be viewed as generalizations of the well-known facts that in a right triangle the sine of an angle equals the cosine of its complement and, equivalently, the cosine of an angle equals the sine of its complement. (Two positive angles are said to be **complementary** when their sum is $\pi/2$, and **supplementary** when their sum is π.) ⊟

Finally, we derive the two last consequences of the formula for $\cos(\alpha - \beta)$.

$$\begin{aligned}\sin(\alpha + \beta) &= \cos[\pi/2 - (\alpha + \beta)] = \cos[(\pi/2 - \alpha) - \beta] \\ &= \cos(\pi/2 - \alpha)\cos \beta + \sin(\pi/2 - \alpha)\sin \beta \\ &= \sin \alpha \cos \beta + \cos \alpha \sin \beta\end{aligned}$$

Thus, for any α and β,

(9.25) $$\sin(\alpha + \beta) = \sin \alpha \cos \beta + \cos \alpha \sin \beta$$

If we replace β with $-\beta$, we find that

$$\sin(\alpha - \beta) = \sin\alpha\cos\beta - \cos\alpha\sin\beta \tag{9.26}$$

Section 9.1 EXERCISES

In Exercises 1–16 find the values without using calculators or tables.

1. $\tan 240°$	**2.** $\cos 5\pi/4$	**3.** $\sin(-\pi/6)$
4. $\csc 30°$	**5.** $\csc 180°$	**6.** $\cos(-\pi/3)$
7. $\sec 7\pi/6$	**8.** $\sec 315°$	**9.** $\cot 120°$
10. $\tan 3\pi/4$	**11.** $\tan(-\pi/4)$	**12.** $\cot 225°$
13. $\sin 2\pi/3$	**14.** $\sin 90°$	**15.** $\sec(-\pi/2)$
16. $\cos 300°$		

In Exercises 17–25 use a right triangle and the information given to determine the desired value.

17. $\sin\alpha = 3/4$	$\cos\alpha = ?$
18. $\cot\alpha = 12/5$	$\sec\alpha = ?$
19. $\cos\alpha = 5/13$	$\tan\alpha = ?$
20. $\sec\alpha = 5/3$	$\sin\alpha = ?$
21. $\tan\alpha = 8/15$	$\sec\alpha = ?$
22. $\sin\alpha = 4/5$	$\sec\alpha = ?$
23. $\csc\alpha = 17/8$	$\tan\alpha = ?$
24. $\cos\alpha = 4/7$	$\sin\alpha = ?$
25. $\sec\alpha = 13/12$	$\csc\alpha = ?$

Convert the measures given in Exercises 26–34 from radians to degrees or from degrees to radians. Do not use tables.

26. $5\pi/4$	**27.** $120°$	**28.** $3\pi/20$
29. $7\pi/6$	**30.** $-\pi/6$	**31.** $270°$
32. $210°$	**33.** $315°$	**34.** $-\pi/3$

In Exercises 35–38 use the sum and difference identities (9.21), (9.22), (9.25) and (9.26) for sine and cosine to calculate the values.

35. $\sin 15°$	**36.** $\cos 105°$
37. $\cos \pi/12$	**38.** $\sin 5\pi/12$

Sketch the graphs in Exercises 39–46. Recall that $y = f(x - a)$ is a horizontal translation of the graph $y = f(x)$ and that $y = f(x) + b$ is a vertical translation of the graph of $y = f(x)$.

39. $y = \sin(x - \pi/6)$

40. $y = \cos(x + \pi/3)$

41. $y = \tan(x + \pi/4)$

42. $y = \sin x + 1$

43. $y = \csc(x - \pi/2)$

44. $y = \cos x - 1$

45. $y = \cos(x + \pi/4) + 3$

46. $y = \sin(x - \pi/3) - 2$

47. Prove Equation (9.21). That is, prove that

$$\cos(\alpha - \beta) = \cos\alpha\cos\beta + \sin\alpha\sin\beta$$

by applying the law of cosines, Equation (9.20), to triangle OAB in the accompanying figure, where angle $AOB = \alpha - \beta$.

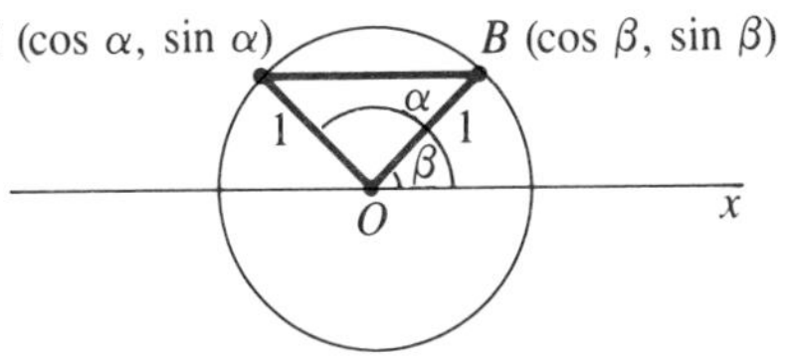

FIGURE FOR EXERCISE 47

SECTION 9.2

Derivatives of the Trigonometric Functions

In this section we shall find the derivatives of the basic trigonometric functions. We shall use radian measure throughout. Let us begin our discussion with the sine function.

In order to determine the derivative of the function $f(x) = \sin x$, we must examine the difference quotient

$$\frac{f(x+h) - f(x)}{h}$$

and see what happens to it as h tends toward 0. We may think of our function in the form

$$f(\quad) = \sin(\quad)$$

so that

$$f(x+h) = \sin(x+h)$$

Our difference quotient becomes

$$\frac{\sin(x+h) - \sin x}{h} \tag{9.27}$$

What happens to this expression as h tends toward 0? It is not easy to decide, because this expression is not so "nice" as some of the expressions we obtained in computing derivatives in Section 2.4. In these examples the denominator, h, could be divided into the numerator to simplify the difference quotient. In fact, this "dividing away" of the denominator h seemed crucial to our development, because the next step in each case was to let h tend toward 0!

But in the difference quotient facing us here, there seems to be no easy way to eliminate the h in the denominator. Because of this difficulty, we must take a more involved approach. We proceed by expanding $\sin(x + h)$ according to the formula for the sine of the sum of two angles, Equation (9.25).

$$\sin(x+h) = \sin x \cos h + \cos x \sin h$$

Thus Equation (9.27) becomes

$$\frac{\sin x \cos h + \cos x \sin h - \sin x}{h}$$

which in turn may be rewritten as

$$\sin x\left(\frac{\cos h - 1}{h}\right) + \cos x\left(\frac{\sin h}{h}\right) \tag{9.28}$$

If we set $u(h) = \dfrac{\cos h - 1}{h}$ and $v(h) = \dfrac{\sin h}{h}$, we need to decide whether $\lim_{h\to 0} u(h)$ and $\lim_{h\to 0} v(h)$ exist and, if so, to evaluate these limits. In order to help us do this, we have reproduced, in Table 9.4, a portion of a standard table of four-decimal-place values of the trigonometric functions. For simplicity, we look first at the function $v(h) = \dfrac{\sin h}{h}$ for values of h near 0 and tending toward it.

TABLE 9.4

h (in radians)	$\sin h$	$\cos h$	$\tan h$
0.00	0.0000	1.0000	
0.01	0.0100	1.0000	
0.02			
0.03			
0.04			
0.05	0.0500	0.9988	
0.06			
0.07			
0.08			
0.09			
0.10	0.0998	0.9950	
0.11			
0.12			
0.13			
0.14			
0.15			
0.16			
0.17			
0.18			
0.19			
0.20	0.1987	0.9801	

We see that

$$\text{If } h = 0.20, \quad v(h) = \frac{\sin 0.20}{0.20} \approx \frac{0.1987}{0.20} = 0.9935$$

$$\text{If } h = 0.10, \quad v(h) = \frac{\sin 0.10}{0.10} \approx \frac{0.0998}{0.10} = 0.9980$$

$$\text{If } h = 0.05, \quad v(h) = \frac{\sin 0.05}{0.05} \approx \frac{0.0500}{0.05} = 1.0000$$

$$\text{If } h = 0.01, \quad v(h) = \frac{\sin 0.01}{0.01} \approx \frac{0.0100}{0.01} = 1.0000$$

where the sine values are accurate to four decimal places. It appears that as h tends toward 0, the numbers $\dfrac{\sin h}{h}$ are tending toward 1. We guess, therefore, that

(9.29) $$\lim_{h \to 0} \frac{\sin h}{h} = 1$$

What about $\lim_{h \to 0} u(h)$? Well, as in the case of the function $v(h)$, let us choose values of h tending toward 0.

$$\text{If } h = 0.20, \quad v(h) = \frac{\cos 0.20 - 1}{0.20} \approx \frac{0.9801 - 1}{0.20} = \frac{-0.0199}{0.20} = -0.0995$$

$$\text{If } h = 0.10, \quad v(h) = \frac{\cos 0.10 - 1}{0.10} \approx \frac{0.9950 - 1}{0.10} = \frac{-0.0050}{0.10} = -0.0500$$

$$\text{If } h = 0.05, \quad v(h) = \frac{\cos 0.05 - 1}{0.05} \approx \frac{0.9988 - 1}{0.05} = \frac{-0.0012}{0.05} = -0.0240$$

$$\text{If } h = 0.01, \quad v(h) = \frac{\cos 0.01 - 1}{0.01} \approx \frac{1.0000 - 1}{0.01} = 0$$

where, again, our values are accurate to four decimal places. Now, as h tends toward 0, the numbers $\frac{\cos h - 1}{h}$ tend toward 0. Therefore, we conjecture that

(9.30)
$$\lim_{h \to 0} \frac{\cos h - 1}{h} = 0$$

We will soon present a careful geometric proof that the limits shown in Equations (9.29) and (9.30) are correct. For now, let us accept them and return to the difference quotient that is Equation (9.28). If we take the limit of this as h tends toward 0, then making use of Equation (9.30) and the fact that $\sin x$ does not vary as h varies, we find that the term

$$\sin x \left(\frac{\cos h - 1}{h} \right)$$

must approach 0. Similarly, making use of Equation (9.29) and the fact that $\cos x$ does not vary as h varies, we find that

$$\cos x \left(\frac{\sin h}{h} \right)$$

approaches $(\cos x)(1) = \cos x$. Combining these facts suggests that, if $f(x) = \sin x$, then $\lim_{h \to 0} \frac{f(x + h) - f(x)}{h} = 0 + \cos x = \cos x$. We summarize this discussion in Theorem 9.3.

THEOREM 9.3 If $f(x) = \sin x$, then f is differentiable and

$$\frac{d}{dx} \sin x = \cos x \quad \text{or equivalently} \quad \int \cos x \, dx = \sin x + c$$ □

COROLLARY 9.4 If u is a differentiable function, then $f(x) = \sin u(x)$ is differentiable and $f'(x) = \cos u(x) \cdot u'(x)$.

PROOF This is simply the chain rule applied to the composition of $(f \circ u)(x) = \sin u(x)$. □

EXAMPLE 5 Differentiate $y = x^2 \sin x$.

SOLUTION All we have to do is apply the product rule with $u(x) = x^2$ and $v(x) = \sin x$. Thus $f'(x) = u(x)v'(x) + u'(x)v(x) = x^2 \cos x + 2x \sin x$.

EXAMPLE 6 Find $\dfrac{dy}{dx}$ if $y = \sin 3x$.

SOLUTION By Corollary 9.4, $\dfrac{dy}{dx} = (\cos 3x)(3) = 3 \cos 3x$.

Let us now give a careful proof of the two limits, Equations (9.29) and (9.30).

THEOREM 9.5 $\displaystyle\lim_{h \to 0} \frac{\sin h}{h} = 1.$

PROOF Let us consider values of h for which

$$0 < h < \pi/2$$

We turn our attention to Figure 9.28, in which the circle is centered at O, with radius 1. Thus

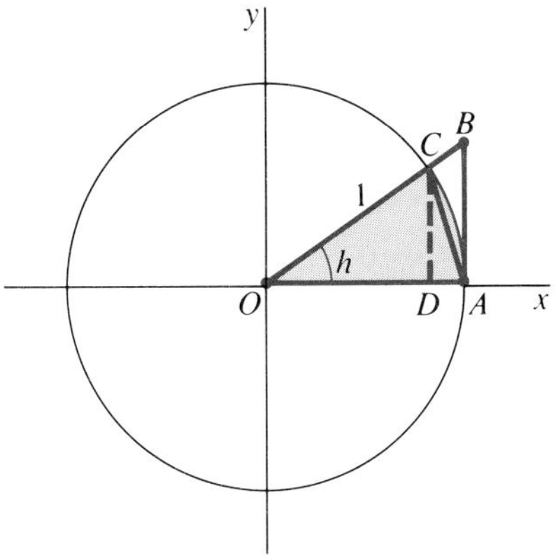

FIGURE 9.28

$$\overline{OC} = \overline{OA} = 1$$

because they are both radii of the circle. We are supposing that angle BOA is an angle of h radians. If we look at triangle OCD, we see that

$$\sin h = \frac{\overline{CD}}{\overline{OC}} = \frac{\overline{CD}}{1} = \overline{CD}$$

In a similar manner, a glance at right triangle BOA reveals that

$$\tan h = \frac{\overline{AB}}{\overline{OA}} = \frac{\overline{AB}}{1} = \overline{AB}$$

Let us put these facts aside for a moment, to consider some areas in Figure 9.28. It is clear that

$$\text{Area } \Delta OAC < \text{area sector } OAC < \text{area } \Delta OAB, \tag{9.31}$$

because each figure is contained in the next. (*Note*: "sector OAC" is the shaded "slice of pie".) Recall that the area of a triangle is given by one-half the product of its base and its altitude. Therefore

$$\text{Area } \Delta OAC = \tfrac{1}{2}(\overline{OA})(\overline{CD}) = \tfrac{1}{2}(1)(\sin h) = \tfrac{1}{2}\sin h$$

and

$$\text{Area } \Delta OAB = \tfrac{1}{2}(\overline{AB})(\overline{OA}) = \tfrac{1}{2}(\tan h)(1) = \tfrac{1}{2}\tan h$$

The area of a sector is given by $(\pi r^2)\left(\dfrac{h}{2\pi}\right) = \tfrac{1}{2}r^2 h$, where r is the radius of the circle. In this case, because $r = 1$,

$$\text{Area sector } OAC = \tfrac{1}{2} h$$

Substituting these values into Equation (9.31) yields

$$\tfrac{1}{2} \sin h < \tfrac{1}{2} h < \tfrac{1}{2} \tan h$$

Multiplying by 2, we find that

$$\sin h < h < \tan h$$

Now, because $0 < h < \pi/2$ by assumption, $\sin h$ is positive so that we may divide by it and preserve the direction of the inequality. This gives

$$1 < \frac{h}{\sin h} < \frac{\tan h}{\sin h} = \frac{1}{\cos h}$$

Let us flip these numbers upside down; that has the effect of turning the inequalities around. (To see this, simply note that $2 < 3$, but $\frac{1}{2} > \frac{1}{3}$.) Thus

$$1 > \frac{\sin h}{h} > \cos h.$$

As h tends toward 0, the number 1 stays fixed. Also, as h tends toward 0, $\cos h$ tends toward 1. We see that the number $\sin h/h$ is "sandwiched" between two quantities that are both tending toward 1. Well, then, we certainly must have *it* tending to 1 also. This proves what we want in the situation where $h > 0$.

Finally, suppose that $-\pi/2 < h < 0$. Then $-h$ is a positive number, and

$$\frac{\sin(-h)}{-h} = \frac{-\sin h}{-h} = \frac{\sin h}{h}$$

It follows that we obtain the same limit regardless of whether h tends toward 0 as a positive or a negative number. ⊟

THEOREM 9.6 $\displaystyle\lim_{h \to 0} \frac{\cos h - 1}{h} = 0$

PROOF We can establish this limit easily by making use of a bit of algebra.

$$\frac{\cos h - 1}{h} = \frac{(\cos h - 1)}{h} \cdot \frac{\cos h + 1}{\cos h + 1} = \frac{\cos^2 h - 1}{h(\cos h + 1)}$$

↑ We are multiplying by the same number in the numerator and in the denominator.

Now we know that for any h, $\cos^2 h + \sin^2 h = 1$. Therefore

$$\frac{\cos h - 1}{h} = \frac{-\sin^2 h}{h(\cos h + 1)} = (\sin h)\left(\frac{\sin h}{h}\right)\left(\frac{-1}{\cos h + 1}\right)$$

and

$$\lim_{h\to 0} \frac{\cos h - 1}{h} = \left(\lim_{h\to 0} \sin h\right)\left(\lim_{h\to 0} \frac{\sin h}{h}\right)\left(\lim_{h\to 0} \frac{-1}{\cos h + 1}\right)$$

As h tends toward 0, we know that $\lim_{h\to 0} \frac{\sin h}{h} = 1$. We also know that $\cos h$ tends toward 1 as h tends toward 0, so that $\lim_{h\to 0} \frac{-1}{\cos h + 1} = -\frac{1}{2}$. But as h tends toward 0; $\sin h$ also tends toward 0, so that $\lim_{h\to 0} \sin h = 0$. Combining these limits, we find that

$$\lim_{h\to 0} \frac{\cos h - 1}{h} = (0)(1)(-\tfrac{1}{2}) = 0$$ □

THEOREM 9.7 The cosine function $f(x) = \cos x$ is differentiable, and its derivative is $f'(x) = -\sin x$. That is,

$$\frac{d}{dx}(\cos x) = -\sin x \quad \text{or equivalently} \quad \int \sin x\, dx = -\cos x + c$$

PROOF According to Equation (9.24), we may write

$$y = \cos x = \sin(\pi/2 - x)$$

We can now apply Corollary 9.4 with $u(x) = \pi/2 - x$ to obtain

$$y' = [\cos(\pi/2 - x)][-1] = -\cos(\pi/2 - x) = -\sin x$$ □

By the chain rule, we now have

COROLLARY 9.8 If u is a differentiable function, then $f(x) = \cos u(x)$ is differentiable and $f'(x) = -\sin u(x) \cdot u'(x)$. □

EXAMPLE 7 Differentiate $f(x) = \cos(\ln x)$.

SOLUTION By Corollary 9.8,

$$f'(x) = -\sin(\ln x) \cdot \frac{d}{dx}(\ln x) = -\sin(\ln x) \cdot 1/x$$

THEOREM 9.9 The tangent function $f(x) = \tan x$ is differentiable and

$$\frac{d}{dx}(\tan x) = \sec^2 x \quad \text{or equivalently} \quad \int \sec^2 x\, dx = \tan x + c$$

PROOF The key to this result is the observation that

$$\tan x = \sin x/\cos x$$

By the quotient rule,

$$\frac{d}{dx}(\tan x) = \frac{d}{dx}\left(\frac{\sin x}{\cos x}\right) = \frac{(\cos x)(\cos x) - \sin x(-\sin x)}{(\cos x)^2}$$

$$= \frac{\cos^2 x + \sin^2 x}{\cos^2 x} = \frac{1}{\cos^2 x} = \sec^2 x \qquad \square$$

PRACTICE EXERCISE Establish the rules for the differentiation of the remaining trigonometric functions.

(a) $\frac{d}{dx}(\cot x) = -\csc^2 x$

(b) $\frac{d}{dx}(\sec x) = \sec x \tan x$

(c) $\frac{d}{dx}(\csc x) = -\csc x \cot x$

(*Hint*: Use the quotient rule.)

EXAMPLE 8 Find the derivative of $\tan\left(\frac{2x+3}{5x-7}\right)$.

SOLUTION By the chain rule,

$$\frac{d}{dx}\left[\tan\left(\frac{2x+3}{5x-7}\right)\right] = \sec^2\left(\frac{2x+3}{5x-7}\right)\cdot\left[\frac{(5x-7)(2)-(2x+3)(5)}{(5x-7)^2}\right]$$

$$= \frac{-29}{(5x-7)^2}\sec^2\left(\frac{2x+3}{5x-7}\right)$$

EXAMPLE 9 Find $\frac{dy}{dx}$ if $y = \sec^3 2x = (\sec 2x)^3$.

SOLUTION $\frac{dy}{dx} = 3(\sec 2x)^2 \cdot (\sec 2x \tan 2x) \cdot 2$

$$= 6 \sec^3 2x \tan 2x$$

EXAMPLE 10 A beacon in a lighthouse 1 mile from a straight shoreline rotates at the rate of 3 revolutions per minute. How fast is the ray of light moving along the shore when it passes a point 2 miles from the lighthouse?

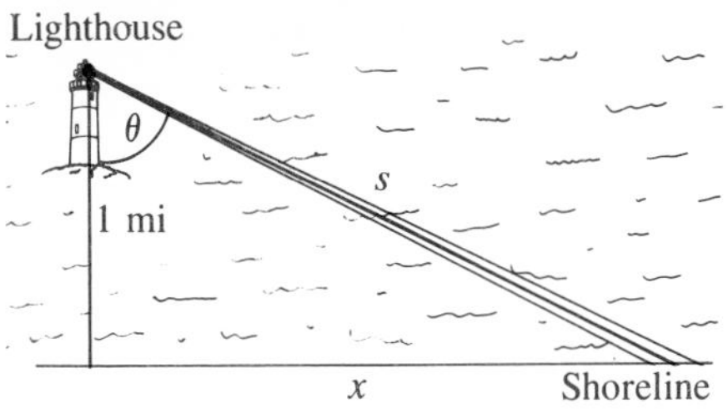

FIGURE 9.29

SOLUTION Refer to Figure 9.29. We have been given the fact that $\frac{d\theta}{dt} = 3(2\pi)$ radians/minute. We are asked to compute $\frac{dx}{dt}$ when $s = 2$. In order to solve this problem we need some relationship between the variables. One natural choice is $\tan\theta = x/1$. Consequently, if we differentiate the equation with respect to the time t, we obtain

$$\frac{dx}{dt} = \sec^2\theta\,\frac{d\theta}{dt}$$

At the instant when $s = 2$, we have $x = \sqrt{s^2 - 1} = \sqrt{3}$ and $\sec\theta = 2$. Thus

$$\left.\frac{dx}{dt}\right|_{s=2} = (2)^2(6\pi) = 24\pi$$

This problem is an example of a *related-rates* problem (see Section 4.7). We were asked to relate the rate of movement of the light along the shore to the rate of rotation of the beacon in the lighthouse.

Section 9.2 EXERCISES

Use the derivatives of the trigonometric functions derived in this section and the derivative rules from Chapters 2 and 4 to determine the derivatives of the functions given in Exercises 1–30.

1. $x^3\cos x$
2. $x^4\sin x$
3. $\sin^3 x$
4. $e^x\sec x$
5. $\frac{\cos x}{x^2}$
6. $\sec(e^{3x})$
7. $e^x\tan x$
8. $\sec^3 2x$
9. $\cos^2 x$
10. $e^{\tan x}$
11. $e^{\sin x}$
12. $\sin^3 4x$
13. $\csc^4 3x$
14. $\frac{x^3}{\tan x}$
15. $\sqrt{\tan x}$
16. $\ln(\cos x)$
17. $\tan^5 4x$
18. $\ln|\cot x - \csc x|$
19. $\cot\sqrt{x}$
20. $e^{\cos x}$
21. $\ln(\sin x)$
22. $\cos e^x$
23. $\tan(1/x)$
24. $\sqrt{\sin 3x}$
25. $\sin e^x$
26. $\ln(\sec x)$
27. $\ln|\csc x - \cot x|$
28. $\csc(e^{\sqrt{x}})$
29. $\sec(x^4 + \sqrt{x})$
30. $\tan^4\sqrt[3]{x}$

Use the limits discussed in this section, $\lim_{h\to 0}\frac{\sin h}{h} = 1$ and $\lim_{h\to 0}\frac{\cos h - 1}{h} = 0$, to find the limits in Exercises 31–34.

31. $\lim_{x\to 0}\frac{\tan x}{x}$
32. $\lim_{x\to 0}\frac{\sin^2 x}{x}$
33. $\lim_{x\to 0}\frac{\sin 2x}{x}$
34. $\lim_{x\to 0} x\cot 2x$
35. $\lim_{x\to 0}\frac{\sin^2(x/2)}{x^2}$

Use the trigonometric identities from Section 9.1 and your knowledge of the derivatives of trigonometric functions to solve the following related-rates problems.

36. Two ships, A and B, are sailing away from the point O along routes such that the angle $AOB = 120°$. How fast is the distance between them changing if at a certain instant $OA = 8$ km, $OB = 6$ km, ship A sails at 20 km/hr, and ship B sails at 30 km/hr?

37. An airplane climbing at an angle of 45° passes directly over a ground radar station at an altitude of 2 mi. A later reading shows that the distance from the radar station to the plane is 7 mi and is increasing at 9 mi/min. What is the speed of the plane in miles per hour?

38. Two ships, *Sunstar* and *Seabreeze*, leave an island at the same time. *Sunstar* travels due east at a rate of 15 mi/hr. *Seabreeze* travels in a straight course making an angle of 60° with the path of *Sunstar* at a rate of 20 mi/hr. How fast are they separating at the end of 2 hr?

SECTION 9.3

Integrals Involving Trigonometric Functions

Strictly speaking, the examples with which we begin this section are just further illustrations of the use of the substitution method of Section 5.4. They present no new ideas. However, because of the many identities involving the six basic trigonometric functions, antidifferentiation of trigonometric expressions offers a great variety of problems—and so appears more complex.

In the spirit of our development, we would like to instill in you a feeling for how to tackle various types of antidifferentiation problems rather than trying to develop a comprehensive list of indefinite integrals. For this reason we will limit our textual development almost exclusively to problems involving $\sin x$, $\cos x$, $\sec x$, and $\tan x$. You can generally solve problems involving $\csc x$ and $\cot x$ by considering analogous problems involving $\sec x$ and $\tan x$, respectively, and the former *will* appear in the problem sets.

EXAMPLE 11 Evaluate $\int (1 + \sin x)^{1/2} \cos x\, dx$.

SOLUTION If we try the substitution

$$u = 1 + \sin x$$

then

$$du = \cos x\, dx$$

which is a part of our integrand. Thus

$$\int (1 + \sin x)^{1/2} \cos x\, dx \quad \xrightarrow{\text{becomes}} \quad \int u^{1/2}\, du = \frac{2}{3} u^{3/2} + c$$

Converting back to x's via the equation $u = 1 + \sin x$ yields

$$\int (1 + \sin x)^{1/2} \cos x\, dx = \frac{2}{3}(1 + \sin x)^{3/2} + c$$

EXAMPLE 12 Evaluate $\displaystyle\int \frac{\sec^2 x}{(4 + \tan x)^2}\, dx$.

SOLUTION Here we try the substitution

$$u = 4 + \tan x$$

so that

$$du = \sec^2 x\, dx$$

Note, again, that $\sec^2 x\, dx$ is part of our integrand. Thus

$$\int \frac{\sec^2 x}{(4 + \tan x)^2}\, dx \xrightarrow{\text{becomes}} \int u^{-2}\, du = -\frac{1}{u} + c$$

which, in terms of x, gives

$$\int \frac{\sec^2 x}{(4 + \tan x)^2}\, dx = -\frac{1}{4 + \tan x} + c$$

EXAMPLE 13 Evaluate $\int \tan x\, dx$.

SOLUTION We wish to find a function whose derivative is $\tan x$. This problem is not nearly so straightforward as

$$\int \sin x\, dx = -\cos x + c \quad \text{or} \quad \int \cos x\, dx = \sin x + c$$

Let us try a substitution. The only available choice seems to be

$$u = \tan x$$

But then

$$du = \sec^2 x\, dx$$

which is not a part of our integrand. Is there an alternative approach we might try?

Suppose we recall that

$$\tan x = \frac{\sin x}{\cos x}$$

and consider $\int \frac{\sin x}{\cos x}\, dx$. Now, we try the substitution

$$u = \cos x$$

Then

$$du = -\sin x\, dx$$

and

$$\int \frac{\sin x}{\cos x}\, dx \xrightarrow{\text{becomes}} \int -\frac{1}{u}\, du$$

This is an antiderivative we have encountered before:

$$\int -\frac{1}{u}\, du = -\ln|u| + c$$

In terms of x,

$$\int \tan x\, dx = \int \frac{\sin x}{\cos x}\, dx = -\ln|\cos x| + c$$

Because for any number $a > 0$, $-\ln a = \ln\left(\frac{1}{a}\right)$, this answer is often written as

$$\int \tan x\, dx = -\ln|\cos x| + c = \ln|\cos x|^{-1} + c$$

$$= \ln\left|\frac{1}{\cos x}\right| + c = \ln|\sec x| + c$$

PRACTICE EXERCISE We have indicated that integrals involving cot x closely resemble those involving tan x. As an illustration of this idea, show that

$$\int \cot x\, dx = \ln|\sin x| + c = -\ln|\csc x| + c$$

EXAMPLE 14 Evaluate $\int x \sin x^2\, dx$.

SOLUTION Here, for the first time, the argument of our trigonometric function is not simply x but a function of x, namely x^2. We try to set

$$u = x^2$$

Then $$du = 2x\, dx$$

We have an x present in our integrand, rather than a $2x$, so we write

$$\frac{1}{2}\, du = x\, dx$$

Then $$\int x \sin x^2\, dx \xrightarrow{\text{becomes}} \int \tfrac{1}{2} \sin u\, du = -\tfrac{1}{2} \cos u + c$$

Converting back to x's, we find

$$\int x \sin x^2\, dx = -\tfrac{1}{2} \cos (x^2) + c$$

EXAMPLE 15 Evaluate $\int \tan^2 x\, dx$.

SOLUTION What should we try to set equal to u? Let us try to think of

$$\int \tan^2 x\, dx \quad \text{as} \quad \int (\tan x)(\tan x)\, dx$$

and set $$u = \tan x$$

Then $$du = \sec^2 x\, dx$$

which is not present in our integrand. It seems we may have tried the wrong substitution for u. Let us instead try

$$u = \tan^2 x$$

Now $$du = 2 \tan x \sec^2 x\, dx$$

which does not seem any better. Is there anything we can do?

You may remember that, for any number x,

$$1 + \tan^2 x = \sec^2 x$$

This can be rewritten as $\tan^2 x = \sec^2 x - 1$, so that

$$\int \tan^2 x \, dx = \int (\sec^2 x - 1) \, dx$$

Now we know that $\sec^2 x$ is the derivative of $\tan x$, whereas 1 is the derivative of x, so

$$\int \tan^2 x \, dx = \int (\sec^2 x - 1) \, dx = \tan x - x + c$$

for arbitrary constant c.

Our solution to this problem may be a little unsettling to you. After all, we did not end up using our substitution method at all, but seem to have jumped directly to the answer after replacing $\tan^2 x$ by $\sec^2 x - 1$. The point is that, in doing integration problems, we can take advantage of any valid formula we can think of that will help. It is even a good idea to *guess* at the answer if we can.

INTEGRATING POWERS OF SIN *x* AND COS *x*

EXAMPLE 16 Evaluate:

(a) $\int (1 - \sin^2 x) \cos x \, dx$ (b) $\int \cos^3 x \, dx$

SOLUTION

(a) Letting $u = \sin x$, we have

$$du = \cos x \, dx$$

Conveniently, $\cos x$ is present in our integrand. Thus

$$\int (1 - \sin^2 x) \cos x \, dx \xrightarrow{\text{becomes}} \int (1 - u^2) \, du = u - \frac{u^3}{3} + c$$

and we conclude that

$$\int (1 - \sin^2 x) \cos x \, dx = \sin x - \frac{\sin^3 x}{3} + c$$

(b) The obvious substitution to try here is

$$u = \cos x$$
$$du = -\sin x \, dx$$

But because the integrand, $\cos^3 x$, contains no sine function, it does not appear that this substitution will work.

Why is it that the integral of part (a) was so easy and this one is so hard? Well, in part (a), once we chose $u = \sin x$, its derivative $\frac{du}{dx} = \cos x$ was present and was just what we needed to convert the integrand from x's to u's. With this in mind, let us try to rewrite our integral so that it too will contain the term $\cos x$.

Making use of the fact that $\sin^2 x + \cos^2 x = 1$, we find

$$\int \cos^3 x\, dx = \int \cos^2 x \cos x\, dx = \int (1 - \sin^2 x) \cos x\, dx$$

This is *precisely* the same integral we considered in part (a), so

$$\int \cos^3 x\, dx = \sin x - \frac{\sin^3 x}{3} + c$$

EXAMPLE 17 Compute $\int \sin^5 2x\, dx$.

SOLUTION By analogy with the previous problem, we write

$$\begin{aligned} \int \sin^5 2x\, dx &= \int (\sin^4 2x)(\sin 2x)\, dx \\ &= \int (\sin^2 2x)^2 (\sin 2x)\, dx \\ &= \int (1 - \cos^2 2x)^2 \sin 2x\, dx \end{aligned}$$

Now we try to let $u = \cos 2x$. Then

$$du = -2 \sin 2x\, dx$$

and so

$$-\frac{1}{2}\, du = \sin 2x\, dx$$

Therefore

$$\begin{aligned} \int \sin^5 2x\, dx = \int (1 - \cos^2 2x)^2 \sin 2x\, dx &\xrightarrow{\text{becomes}} \int -\frac{1}{2}(1 - u^2)^2\, du \\ &= \int -\frac{1}{2}(1 - 2u^2 + u^4)\, du \\ &= -\frac{1}{2}\left(u - \frac{2u^3}{3} + \frac{u^5}{5}\right) + c \end{aligned}$$

Upon restoration of the original variable x, we find

$$\int \sin^5 2x\, dx = -\frac{1}{2}\left(\cos 2x - \frac{2(\cos 2x)^3}{3} + \frac{(\cos 2x)^5}{5}\right) + c$$

We remark that the method used in part (b) of Example 16 and in Example 17 can, in principle, be used to find the antiderivative of any product of sines and cosines (of the same argument), provided one of the terms is raised to an odd power; for example,

$$\int \cos^{85} x \, dx \quad \text{or} \quad \int \sin^5 x \cos^{12} x \, dx \quad \text{or} \quad \int \sin^2 2x \cos^{17} 2x \, dx$$

One further example should help to solidify these ideas.

EXAMPLE 18 Evaluate $\int \sin^2 3x \cos^3 3x \, dx$.

SOLUTION

$$\begin{aligned}\int \sin^2 3x \cos^3 3x \, dx &= \int \sin^2 3x \cos^2 3x \cos 3x \, dx \\ &= \int \sin^2 3x(1 - \sin^2 3x) \cos 3x \, dx \\ &= \int (\sin^2 3x - \sin^4 3x) \cos 3x \, dx\end{aligned}$$

Setting $u = \sin 3x$

we find $du = 3 \cos 3x \, dx$

$$\frac{1}{3} du = \cos 3x \, dx$$

Now $\int (\sin^2 3x - \sin^4 3x) \cos 3x \, dx \xrightarrow{\text{becomes}} \int \frac{1}{3}(u^2 - u^4) \, du$

$$= \frac{1}{3}\left(\frac{u^3}{3} - \frac{u^5}{5}\right) + c$$

so

$$\int \sin^2 3x \cos^3 3x \, dx = \frac{1}{3}\left(\frac{\sin^3 3x}{3} - \frac{\sin^5 3x}{5}\right) + c$$

PRACTICE EXERCISE Evaluate $\int (\cos x)^{1/2} \sin^5 x \, dx$.

[Your answer should be $-\frac{2}{3}(\cos x)^{3/2} + \frac{4}{7}(\cos x)^{7/2} - \frac{2}{11}(\cos x)^{11/2} + c$]

Up until this point we have considered only problems that involve an odd power of either the sine or the cosine. What about problems containing only even powers of these functions? For example, we might consider $\int \sin^2 x \, dx$. We leave it for you to convince yourself that the "obvious" choice of substitution, $u = \sin x$, only complicates matters. For such problems, we must resort to an alternative technique based on two additional trigonometric identities:

(9.32) $$\sin^2 x = \frac{1 - \cos 2x}{2}$$

(9.33) $$\cos^2 x = \frac{1 + \cos 2x}{2}$$

Let us look at a few examples in which we use these formulas.

EXAMPLE 19 Compute $\int \sin^2 x\, dx$.

SOLUTION

$$\int \sin^2 x\, dx = \int \frac{1 - \cos 2x}{2}\, dx$$

$$= \int \left(\frac{1}{2} - \frac{1}{2}\cos 2x\right) dx = \frac{1}{2}x - \frac{1}{4}\sin 2x + c$$

PRACTICE EXERCISE Show that $\int \cos^2 x\, dx = \frac{1}{2}x + \frac{1}{4}\sin 2x + c$.

EXAMPLE 20 Evaluate $\int \cos^4 2x\, dx$.

SOLUTION We write $\cos^4(2x) = (\cos^2 2x)^2$. Because our integrand involves only even powers of $\cos x$, we make use of Equation (9.33), with x replaced by $2x$, to obtain

$$\cos^4(2x) = \left(\frac{1 + \cos 4x}{2}\right)^2 = \frac{1}{4} + \frac{\cos 4x}{2} + \frac{\cos^2 4x}{4}$$

It follows that

$$\int \cos^4 2x\, dx = \int \frac{1}{4}\, dx + \int \frac{\cos 4x}{2}\, dx + \int \frac{\cos^2 4x}{4}\, dx$$

$$= \frac{1}{4}x + \frac{1}{8}\sin 4x + \frac{1}{4}\int \cos^2 4x\, dx$$

Now we are left with the problem of integrating $\cos^2 4x$, which is again an integral involving only even powers of x, so we fall back once again on Equation (9.33), with x replaced by $4x$.

$$\cos^2 4x = \frac{1 + \cos 8x}{2}$$

Therefore

$$\frac{1}{4}\int \cos^2 4x\, dx = \frac{1}{4}\int \frac{1 + \cos 8x}{2}\, dx$$

$$= \frac{1}{4}\left[\int \frac{1}{2}\, dx + \int \frac{\cos 8x}{2}\, dx\right] = \frac{1}{4}\left[\frac{1}{2}x + \frac{1}{16}\sin 8x\right] + c$$

$$= \frac{1}{8}x + \frac{1}{64}\sin 8x + c$$

Finally we combine this with the partial answer we obtained above, to get

$$\int \cos^4 2x\, dx = \tfrac{1}{4}x + \tfrac{1}{8}\sin 4x + (\tfrac{1}{8}x + \tfrac{1}{64}\sin 8x) + c$$

$$= \tfrac{3}{8}x + \tfrac{1}{8}\sin 4x + \tfrac{1}{64}\sin 8x + c$$

PRACTICE EXERCISE Use the technique we just developed to evaluate $\int \sin^2 x \cos^2 x \, dx$.

[Your answer should be $\frac{1}{8}x - \frac{1}{32} \sin 4x + c$]

INTEGRATING POWERS OF TAN x AND SEC x In the earlier part of this section, we integrated various combinations of sin x and cos x. We based our solutions primarily on the basic relationships

$$\frac{d}{dx}(\sin x) = \cos x \qquad \frac{d}{dx}(\cos x) = -\sin x$$

$$\sin^2 x + \cos^2 x = 1$$

The corresponding relationships to keep in mind here are

$$\frac{d}{dx}(\tan x) = \sec^2 x \qquad \frac{d}{dx}(\sec x) = \sec x \tan x$$

$$\sec^2 x = 1 + \tan^2 x$$

EXAMPLE 21 Evaluate $\int \tan^3 x \sec^2 x \, dx$.

SOLUTION The presence of $\sec^2 x$ (which is the derivative of tan x) in our integrand is our clue that we should make the substitution

$$u = \tan x$$

Thus

$$du = \sec^2 x \, dx$$

and so

$$\int \underbrace{\tan^3 x}_{u^3} \underbrace{\sec^2 x \, dx}_{du} \xrightarrow{\text{becomes}} \int u^3 \, du = \frac{u^4}{4} + c$$

which, in terms of x, says that

$$\int \tan^3 x \sec^2 x \, dx = \frac{\tan^4 x}{4} + c$$

EXAMPLE 22 Evaluate $\int (2 + \sec x)^{1/3} \sec x \tan x \, dx$.

SOLUTION This time we spot the appearance of sec x tan x, which is the derivative of sec x. Thus, if we let

$$u = 2 + \sec x$$

then

$$du = \sec x \tan x \, dx$$

and

$$\int (2 + \sec x)^{1/3} \sec x \tan x \, dx \xrightarrow{\text{becomes}} \int u^{1/3} \, du = \frac{3}{4} u^{4/3} + c$$

which, in terms of x, gives

$$\int (2 + \sec x)^{1/3} \sec x \tan x \, dx = \frac{3}{4}(2 + \sec x)^{4/3} + c$$

EXAMPLE 23 Compute $\int \sec^3 x \tan x \, dx$.

SOLUTION This time neither $\sec^2 x$ nor $\sec x \tan x$ stands out in the integrand. We can, however, regroup the terms in our integral so that it appears more manageable:

$$\int \sec^3 x \tan x \, dx = \int (\sec^2 x) \sec x \tan x \, dx$$

In fact, both $\sec^2 x$ and $\sec x \tan x$ are present. What substitution should we use? If we try

$$u = \tan x \qquad du = \sec^2 x \, dx$$

we find

$$\int \overbrace{(\sec^2 x)} \sec x \overbrace{\tan x} \, \overbrace{dx} \xrightarrow{\text{becomes}} \int u \sec x \, du$$

We have not eliminated all the x's from our integral. Thus we try instead to set

$$u = \sec x$$

$$du = \sec x \tan x \, dx$$

Now

$$\int (\sec^2 x) \sec x \tan x \, dx \xrightarrow{\text{becomes}} \int u^2 \, du$$

$$= \frac{u^3}{3} + c \xrightarrow{\text{becomes}} \frac{\sec^3 x}{3} + c$$

PRACTICE EXERCISE (a) Use the substitution $u = \tan x$ to show that

$$\int \tan x \sec^2 x \, dx = \frac{\tan^2 x}{2} + c$$

(b) Use the substitution $u = \sec x$ to show that

$$\int \tan x \sec^2 x \, dx = \frac{\sec^2 x}{2} + c$$

(c) How can you reconcile the different answers you obtained in parts (a) and (b) for the same integral?

EXAMPLE 24 Evaluate:

(a) $\int \tan x \, dx$ (b) $\int \tan^2 x \, dx$ (c) $\int \tan^3 x \, dx$ (d) $\int \tan^4 x \, dx$

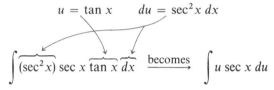

SOLUTION Parts (a) and (b) have already been done. In Example 13 we found that

$$\int \tan x\, dx = \ln|\sec x| + c$$

and in Example 15 we showed that

$$\int \tan^2 x\, dx = \tan x - x + c$$

(c) Now we turn our attention to $\int \tan^3 x\, dx$. Our integrand does not contain either $\sec^2 x$ or $\sec x \tan x$ to suggest an appropriate substitution to us. When this happens, it is often a good idea to try to rewrite it somehow by making use of the relationship

$$1 + \tan^2 x = \sec^2 x$$

This can also be expressed as

$$\tan^2 x = \sec^2 x - 1$$

which suggests that we might try

$$\int \tan^3 x\, dx = \int \tan x \tan^2 x\, dx$$

$$= \int \tan x(\sec^2 x - 1)\, dx = \int \tan x \sec^2 x\, dx - \int \tan x\, dx$$

We have reduced our integral to two other integrals, both of which we have done previously. In the practice exercise on page 426, we found that

$$\int \tan x \sec^2 x\, dx = \frac{\tan^2 x}{2} + c_1$$

whereas, from part (a) in Example 24,

$$\int \tan x\, dx = \ln|\sec x| + c_2$$

Combining these results, we get

$$\tan^3 x\, dx = \frac{\tan^2 x}{2} - \ln|\sec x| + c$$

where we have lumped the constants c_1 and c_2 into the single constant c.

(d) The key feature of part (c) was the way in which we were able to use the identity $\tan^2 x = \sec^2 x - 1$ to reduce the integral $\int \tan^3 x\, dx$ to integrals we had considered before. Because that idea worked so well, let us try it again.

$$\int \tan^4 x\, dx = \int \tan^2 x \tan^2 x\, dx = \int \tan^2 x(\sec^2 x - 1)\, dx$$

$$= \int \tan^2 x \sec^2 x\, dx - \int \tan^2 x\, dx$$

In order to compute the first of these integrals, we set

$$u = \tan x \qquad du = \sec^2 x\, dx$$

Then

$$\int \tan^2 x \sec^2 x\, dx \xrightarrow{\text{becomes}} \int u^2\, du$$

$$= \frac{u^3}{3} + c_1 \xrightarrow{\text{becomes}} \frac{\tan^3 x}{3} + c_1$$

The integral $\int \tan^2 x\, dx$ is given in part (b) as $\tan x - x + c_2$, so

$$\int \tan^4 x\, dx = \frac{\tan^3 x}{3} - \tan x + x + c$$

PRACTICE EXERCISE (a) Use the technique of the previous problem to evaluate $\int \tan^5 x\, dx$.

$$\left[\text{Your answer should be } \frac{\tan^4 x}{4} - \frac{\tan^2 x}{2} + \ln|\sec x| + c.\right]$$

(b) More generally, convince yourself that

$$\int \tan^n x\, dx = \frac{\tan^{n-1} x}{n-1} - \int \tan^{n-2} x\, dx$$

This formula reduces the problem of integrating $\tan^n x$ for any positive integer n to the problem of integrating $\tan^{n-2} x$. Formulas of this type are known as **reduction formulas**, and they are found in integration tables (see Appendix B, Formulas 61–66).

We remark that the methods we have just developed can also be used to integrate any expression of the form $\tan^p x \sec^q x$, provided that q is a nonnegative even integer. An illustration of this is given in Example 25.

EXAMPLE 25 Find $\int \tan^3 x \sec^4 x\, dx$.

SOLUTION

$$\int \tan^3 x \sec^4 x\, dx = \int \tan^3 x(\sec^2 x)(\sec^2 x)\, dx$$

$$= \int \tan^3 x(1 + \tan^2 x)\sec^2 x\, dx$$

$$= \int (\tan^3 x + \tan^5 x)\sec^2 x\, dx$$

Letting $u = \tan x$, so $du = \sec^2 x\, dx$ we find that

$$\int \tan^3 x \sec^4 x\, dx \xrightarrow{\text{becomes}} \int (u^3 + u^5)\, du = \frac{u^4}{4} + \frac{u^6}{6} + c$$

In terms of x,

$$\int \tan^3 x \sec^4 x\, dx = \frac{\tan^4 x}{4} + \frac{\tan^6 x}{6} + c$$

Having discussed in detail how we might integrate functions containing powers of $\tan x$, let us turn our attention to integrals involving the secant.

EXAMPLE 26 Evaluate:

(a) $\int \sec x\, dx$ (b) $\int \sec^2 x\, dx$ (c) $\int \sec^4 x\, dx$ (d) $\int \sec^3 x\, dx$

SOLUTION

(a) This first integral is a very hard one—no obvious substitution will work. We challenge you to try it.

In order to solve this problem, we employ a mathematical trick. We write

$$\sec x = \sec x \left(\frac{\sec x + \tan x}{\sec x + \tan x} \right)$$

and consider the modified integral

$$\int \sec x \left(\frac{\sec x + \tan x}{\sec x + \tan x} \right) dx = \int \frac{\sec^2 x + \sec x \tan x}{\sec x + \tan x}\, dx$$

Now we let

$$u = \sec x + \tan x$$

Thus $$du = (\sec x \tan x + \sec^2 x)\, dx$$

so

$$\int \sec x\, dx = \int \frac{\sec^2 x + \sec x \tan x}{\sec x + \tan x}\, dx \xrightarrow{\text{becomes}} \int \frac{du}{u} = \ln|u| + c$$

which, in terms of x, yields

(9.34) $$\int \sec x\, dx = \ln|\sec x + \tan x| + c$$

This answer is worth remembering, because we will have occasion to integrate $\sec x$ again. The trick we used to get our answer, however, is certainly *not* worth remembering. It is a very artificial and specialized mathematical trick that you would be very unlikely to think of on your own. Fortunately, you will not have to use it again.

(b) This problem is very easy.

$$\int \sec^2 x\, dx = \tan x + c$$

(c) The evaluation of $\int \sec^4 x\, dx$ is very similar to our solution of Example 25. We simply break up $\sec^4 x$ as follows:

$$\int \sec^4 x\, dx = \int (\sec^2 x)(\sec^2 x)\, dx = \int (1 + \tan^2 x)\sec^2 x\, dx$$

Now we make the substitution $u = \tan x$, and the rest is easy.

(d) We wish to compute $\int \sec^3 x\, dx$. Try it! Odd powers of the secant are very troublesome indeed. This problem does not yield to the techniques developed in this section. Instead, we will solve it in the next section by means of the method of integration by parts.

Appendix B contains a list of integration formulas involving trigonometric functions.

Section 9.3 EXERCISES

Use the technique of substitution developed in Section 5.4 and reviewed here to evaluate the integrals in Exercises 1–20.

1. $\int e^{\sin x}\cos x\, dx$
2. $\int \sin^5 x \cos x\, dx$
3. $\int \sin x \cos^4 x\, dx$
4. $\int \sin x\, e^{\cos x}\, dx$
5. $\int e^x \sin(e^x)\, dx$
6. $\int \sec^2 x \tan^3 x\, dx$
7. $\int \sec^4 x \tan x\, dx$
8. $\int \frac{\sin x}{1 + \cos x}\, dx$
9. $\int \frac{\cos x}{1 - \sin x}\, dx$
10. $\int \tan x \sec^6 x\, dx$
11. $\int \tan^4 x \sec^2 x\, dx$
12. $\int \frac{\cos \sqrt{x}}{\sqrt{x}}\, dx$
13. $\int \frac{\sec^2 x}{\tan x}\, dx$
14. $\int \frac{\sin x}{\cos^4 x}\, dx$
15. $\int \cos x \sqrt{\sin x}\, dx$
16. $\int \frac{\sin^3 x}{\cos^3 x}\, dx$
17. $\int \frac{\cos x}{\sin^3 x}\, dx$
18. $\int \frac{\sec^2 x}{\tan^3 x}\, dx$
19. $\int \frac{\sin \sqrt{x} \cos \sqrt{x}}{\sqrt{x}}\, dx$
20. $\int \frac{\cos(\ln x)}{x}\, dx$

Use the trigonometric identities $\sin^2 x + \cos^2 x = 1$, $\tan^2 x + 1 = \sec^2 x$, and $\cot^2 x + 1 = \csc^2 x$ and the methods of this section to evaluate the integrals in Exercises 21–40.

21. $\int \sin x \cos^2 x\, dx$

$\int \sin^4 x\, dx$

23. $\int \cos^2 x\, dx$
24. $\int \sin^3 x \cos^2 x\, dx$
25. $\int \tan^3 x \sec^2 x\, dx$
26. $\int \tan^3 x \sec^4 x\, dx$
27. $\int \cos^3 x \sin^2 x\, dx$
28. $\int \sin^2 x\, dx$
29. $\int \tan^2 x\, dx$
30. $\int \frac{\tan^2 x}{\sec x}\, dx$
31. $\int \cos^4 x\, dx$
32. $\int \sin^2 x \cos^2 x\, dx$
33. $\int \tan^4 x\, dx$
34. $\int \sec^4 x\, dx$
35. $\int \frac{\sin^3 x}{\cos^2 x}\, dx$
36. $\int \frac{\sin^3 x}{\sqrt{\cos x}}\, dx$
37. $\int \frac{\sec^4 x}{\tan^2 x}\, dx$
38. $\int \cot^3 x \csc^4 x\, dx$
39. $\int \csc^4 x\, dx$
40. $\int \cot x \csc^4 x\, dx$

Evaluate the definite integrals in Exercises 41–50.

41. $\int_0^{\pi/4} \sin x \cos^3 x \, dx$

42. $\int_0^{\sqrt{\pi}} x \cos \frac{x^2}{2} \, dx$

43. $\int_0^{\pi/4} \frac{\sin x}{\sqrt{\cos x}} \, dx$

44. $\int_0^{\pi/6} \sin^2 x \cos x \, dx$

45. $\int_0^{\pi} \sin x \cos x \, dx$

46. $\int_0^{\pi/3} \tan x \sec^2 x \, dx$

47. $\int_0^{\pi/3} \cos^3 x \sin x \, dx$

48. $\int_0^{\pi/6} \sin 2x \cos^3 2x \, dx$

49. $\int_0^{\pi/6} \tan x \sec^3 x \, dx$

50. $\int_{\pi/6}^{\pi/4} \tan^3 x \sec^2 x \, dx$

51. Can you find the correct mathematical trick to evaluate $\int \csc x \, dx$? (*Hint*: Look back at how we treated $\int \sec x \, dx$.)

52. Find the volume obtained by rotating $y = \sin x$, $0 \le x \le \pi$, about the x-axis.

53. Find the volume obtained by rotating $y = \cos x$, $0 \le x \le \pi/2$, about the x-axis.

54. Find the area between the curves $y = \sin x$, $y = \cos x$, $x = 0$, and $x = \pi/4$.

SECTION 9.4

Integration by Parts Revisited

The integrals in Section 9.3 all involved some type of substitution. Here we turn our attention to some trigonometric integrals that require the integration-by-parts formula (Section 7.1). We recall, then, that

(9.35) $$\int u \, dv = uv - \int v \, du$$

Let us begin with the last problem of the previous section.

EXAMPLE 27 Compute $\int \sec^3 x \, dx$.

SOLUTION The trick here is to rewrite the integrand as the product $\sec x \sec^2 x$. The functions $\sec x$ and $\sec^2 x$ are of comparable "sophistication," but it is certainly easier to find the antiderivative of $\sec^2 x$, because $(\tan x)' = \sec^2 x$. Thus we choose $dv = \sec^2 x \, dx$, $u = \sec x$, and apply the integration-by-parts formula.

$u = \sec x$	$v = \tan x$
$du = \sec x \tan x \, dx$	$dv = \sec^2 x \, dx$

$$\int \underset{\underset{u}{\uparrow}}{\sec x} \; \underset{\underset{dv}{\uparrow}}{\sec^2 x \, dx} = \underset{\underset{u}{\uparrow}}{\sec x} \; \underset{\underset{v}{\uparrow}}{\tan x} - \int \underset{\underset{v}{\uparrow}}{\tan x} (\underset{\underset{du}{\uparrow}}{\sec x \tan x \, dx})$$

$$= \sec x \tan x - \int \sec x \tan^2 x \, dx$$

Now $\tan^2 x = \sec^2 x - 1$, so we have

$$\int \sec^3 x \, dx = \sec x \tan x - \int \sec x [\sec^2 x - 1] \, dx$$

$$= \sec x \tan x - \int \sec^3 x \, dx + \int \sec x \, dx$$

Here is a new wrinkle! The integral we are trying to compute, $\int \sec^3 x\, dx$, appears on both sides of the equation. If we set

$$A = \int \sec^3 x\, dx$$

then
$$A = \sec x \tan x - A + \int \sec x\, dx$$

or
$$2A = \sec x \tan x + \int \sec x\, dx$$

Dividing both sides by 2, we find that

$$\int \sec^3 x\, dx = A = \tfrac{1}{2}\sec x \tan x + \tfrac{1}{2}\int \sec x\, dx$$
$$= \tfrac{1}{2}\sec x \tan x + \tfrac{1}{2}\ln|\sec x + \tan x| + c$$

EXAMPLE 28 Evaluate $\int x \cos x\, dx$.

SOLUTION We need to decide which term of $\int x \cos x\, dx$ should be denoted by $f(x)$ and which by $g'(x)$. Let us try

$$u = x, \qquad dv = \cos x\, dx$$

The information to be used in Equation (9.35) is summarized by the following scheme:

$u = x$	$v = \sin x$
$du = dx$	$dv = \cos x\, dx$

Thus
$$\int x \cos x\, dx = x \sin x - \int \sin x\, dx$$

and we conclude that

$$\int x \cos x = x \sin x + \cos x + c$$

We leave it for you to verify that trying

$$u = \cos x \qquad dv = x\, dx$$

is not satisfactory; it replaces our integral by one that looks even more difficult. ⊟

EXAMPLE 29 Evaluate $\int x^2 \sin x\, dx$.

SOLUTION The integrand is the product of x^2 and $\sin x$. The latter is "more sophisticated," and we do know its antiderivative, namely $-\cos x$.

So we set $dv = \sin x\,dx$, $u = x^2$.

$u = x^2$	$v = -\cos x$
$du = 2x\,dx$	$dv = \sin x\,dx$

Thus

$$\int \underset{u}{\underset{\uparrow}{x^2}} \underset{dv}{\underset{\uparrow}{\sin x\,dx}} = \underset{u}{\underset{\uparrow}{x^2}}\underset{v}{\underset{\uparrow}{(-\cos x)}} - \int \underset{v}{\underset{\uparrow}{(-\cos x)}}\underset{du}{\underset{\uparrow}{(2x\,dx)}}$$

$$= -x^2\cos x + 2\int x\cos x\,dx$$

We have succeeded in reducing the problem of evaluating $\int x^2 \sin x\,dx$ to that of evaluating $\int x\cos x\,dx$. In order to evaluate this latter integral, *we must use integration by parts again.* Actually, we did this problem earlier in Example 28, wherein we discovered that

$$\int x\cos x\,dx = x\sin x + \cos x + c$$

When we use this relationship in the expression we obtained above, our final answer is

$$\int x^2 \sin x\,dx = -x^2\cos x + 2[x\sin x + \cos x] + c$$

PRACTICE EXERCISE How many times do you think it will be necessary to use the integration-by-parts formula to evaluate each of the following?

(a) $\int x^3\cos x\,dx$ (b) $\int x^5 e^x\,dx$

[Your answers should be (a) three times and (b) five times.]

EXAMPLE 30 Evaluate $\int e^x\cos x\,dx$.

SOLUTION Here it is not at all clear whether e^x or $\cos x$ is "more sophisticated." Further, we know the antiderivative of both of them. Let us arbitrarily decide to select $dv = \cos x\,dx$, $u = e^x$. Thus

$u = e^x$	$v = \sin x$
$du = e^x dx$	$dv = \cos x\,dx$

and

(9.36) $$\int \underset{u}{\underset{\uparrow}{e^x}}\underset{dv}{\underset{\uparrow}{\cos x\,dx}} = \underset{u}{\underset{\uparrow}{e^x}}\underset{v}{\underset{\uparrow}{\sin x}} - \int \underset{v}{\underset{\uparrow}{(\sin x)}}\underset{du}{\underset{\uparrow}{e^x\,dx}}$$

We have converted the problem of computing $\int e^x \cos x\, dx$ to one of evaluating $\int e^x \sin x\, dx$, which certainly does not look any simpler. What shall we do? Well, let us try to use integration by parts again—this time to evaluate $\int e^x \sin x\, dx$. Let

$$dv = \sin x\, dx \quad \text{and} \quad u = e^x$$

to obtain

$u = e^x$	$v = -\cos x$
$du = e^x\, dx$	$dv = \sin x\, dx$

Therefore

$$\int \underset{u}{\underset{\uparrow}{e^x}} \underset{dv}{\underset{\uparrow}{\sin x\, dx}} = \underset{u}{\underset{\uparrow}{e^x}}\underset{v}{\underset{\uparrow}{(-\cos x)}} - \int \underset{v}{\underset{\uparrow}{(-\cos x)}}\underset{du}{\underset{\uparrow}{e^x\, dx}}$$

$$= -e^x \cos x + \int e^x \cos x\, dx$$

If we substitute this into Equation (9.36), we get

$$\int e^x \cos x\, dx = e^x \sin x - \left[-e^x \cos x + \int e^x \cos x\right]$$

which is the same as

$$\int e^x \cos x\, dx = e^x \sin x + e^x \cos x - \int e^x \cos x$$

We seem to be going around in circles! We started with the problem $\int e^x \cos x\, dx$ and, after using integration by parts twice, we have an expression that still involves $\int e^x \cos x\, dx$. Yet despite this, we have actually made substantial progress toward our solution. To see how, let us denote the value of the antiderivative we seek by A. Then

$$A = \int e^x \cos x\, dx$$

and the foregoing equation may be written as

$$A = e^x \sin x + e^x \cos x - A$$

If we add A to both sides, we have

$$2A = e^x \sin x + e^x \cos x$$

or

$$A = \frac{e^x \sin x + e^x \cos x}{2}$$

Recalling that $A = \int e^x \cos x\, dx$, we write our final answer as

$$\int e^x \cos x\, dx = \frac{e^x \sin x + e^x \cos x}{2} + c$$

Example 30 has several interesting aspects that are worthy of further discussion. At the beginning of our solution, we chose

$$u = e^x \quad \text{and} \quad dv = \cos x\, dx$$

Later on, when we applied the method of integration by parts the second time, we chose

$$u = e^x \quad \text{and} \quad dv = \sin x\, dx$$

What do you think would have happened if we had chosen

$$u = \sin x \quad \text{and} \quad dv = e^x\, dx$$

the second time? Let's see

$u = \sin x$	$v = e^x$
$du = \cos x$	$dv = e^x\, dx$

Now

$$\int \underset{u}{\underset{\uparrow}{\sin x}}\, \underset{dv}{\underset{\uparrow}{e^x\, dx}} = \underset{u}{\underset{\uparrow}{(\sin x)}} \underset{v}{\underset{\uparrow}{e^x}} - \int \underset{v}{\underset{\uparrow}{e^x}} \underset{du}{\underset{\uparrow}{\cos x\, dx}}$$

and, if we substitute this into Equation (9.36), we find that

$$\int e^x \cos x\, dx = e^x \sin x - \left[e^x \sin x - \int e^x \cos x\, dx\right]$$

or

$$\int e^x \cos x\, dx = \int e^x \cos x\, dx$$

This is certainly true, but it is not very helpful. The point is that, once we have made our original choice of

$$u = e^x \quad \text{and} \quad dv = \cos x\, dx$$

for purposes of using the integration-by-parts formula, our subsequent choices must be consistent with it. To use integration by parts again, we must again let $u = e^x$ and let $dv = \sin x\, dx$.

Section 9.4 EXERCISES

Use integration by parts to evaluate the integrals in Exercises 1–24. In some cases you may need to integrate by parts more than once to solve as you did in Examples 29 and 30.

1. $\int x \cos x\, dx$

2. $\int x \sin x\, dx$

3. $\int x^2 \sin 3x\, dx$

4. $\int x^2 \cos x\, dx$

5. $\int x^3 \cos x\, dx$

6. $\int x^3 \sin x\, dx$

7. $\int e^x \sin x\, dx$

8. $\int e^x \cos 4x\, dx$

9. $\int x \cos^4 x\, dx$

10. $\int x \sin 3x\, dx$

11. $\int x^5 \cos x^3\, dx$

12. $\int x^3 \cos x^2\, dx$

13. $\int x^3 \cos x^2\, dx$

14. $\int x^5 \sin x^3\, dx$

15. $\int x \sec^2 x\, dx$

16. $\int x \sec x \tan x\, dx$

17. $\int x \csc x \cot x\, dx$

18. $\int \cos(\ln x)\, dx$

19. $\int \sin(\ln x)\, dx$

20. $\int x \csc^2 x\, dx$

21. $\int e^{4x} \cos 3x\, dx$

22. $\int e^{5x} \sin x\, dx$

23. $\int x \sin(3x + 4)\, dx$

24. $\int x \cos(2x - 3)\, dx$

25. Refer to Example 27 and your answer to Exercise 51 in Section 9.3 to evaluate $\int \csc^3 x\, dx$.

Key Mathematical Concepts and Tools

Periodic functions
Cosine function
Sine function
Tangent function
Secant function
Cosecant function
Cotangent function
Degrees
Radian measure
Trigonometric identities

Applications Covered in This Chapter

Velocity of a beam of light cast by a rotating beacon

NINE Warm-Up Test

1. Find $\frac{dy}{dx}$ for each of the following.
   (a) $y = (\sin 2x)^{1/4}$
   (b) $y = \frac{\sin x}{x^3}$
   (c) $y = \tan^2 3x$

2. Without using a calculator, compute:
   (a) $\cos 225°$
   (b) $\sin^2 \pi/3$
   (c) $\sec\left(-\frac{5\pi}{6}\right)$
   (d) $\ln\left(\sin \frac{5\pi}{2}\right)$

3. Evaluate:
   (a) $\int (x + 2) \sin 2x\, dx$
   (b) $\int \cos^2 \frac{x}{3}\, dx$

4. Find the following limits if they exist.
   (a) $\lim_{x \to 0} \frac{\sin 3x}{4x}$
   (b) $\lim_{z \to \pi/2} \frac{\cos z}{z - \pi/2}$
   (c) $\lim_{x \to \pi/4} \tan 2x$
   (d) $\lim_{x \to 0} \frac{\csc 2x}{\cot x}$

5. Sketch the graph of $y = \sin 2x$ for $0 \le x \le \pi$. Use curve-sketching techniques to locate all maxima, minima, and inflection points, if any. Also discuss concavity.

6. The temperature in Boston over a certain 12-hr period has been found to vary according to the formula

$$T(t) = 65 + 10 \cos \frac{\pi}{24} t, \qquad 0 \le t \le 12$$

Here t is measured in hours, $t = 0$ corresponding to 12 A.M. and $t = 12$ to 12 P.M.

(a) Find the average temperature over this 12-hr period.

(b) Find the maximal and the minimal temperatures over this time interval.

7. Differentiate $f(t)$ with respect to t:

(a) $f(t) = \tan(t^3 + t)$

(b) $f(t) = \cos^2(t^2)$

(c) $f(t) = \ln(\sec t + \tan t)$

8. Evaluate:

(a) $f'\left(\frac{\pi^2}{4}\right)$ for $f(t) = \sin\sqrt{t}$

(b) $f'(0)$ for $f(t) = e^t \tan t$

(c) $\int_{\pi/4}^{\pi/2} \sin^2 x\, dx$

9. Find the area of the region between the curves $y = \cos 2x$ and $y = \sin 2x$, where x is restricted to the interval $-\pi/8 \le x \le \pi/8$.

10. Find the equation of the tangent line to $y = x + \sin x$ at the point where $x = \pi/6$.

NINE Final Exam

1. Evaluate each of the following indefinite integrals.

(a) $\int \sin^3 x \cos^2 x\, dx$

(b) $\int \sec^2 x \tan^2 x\, dx$

(c) $\int (t + 3)\sin(t - 1)\, dt$

2. Compute the derivative of each of the following:

(a) $\tan^5 2x$ (b) $\tan x^5$

(c) $\sin x \cos^2 x$

3. Find each of the following limits.

(a) $\lim_{x \to \pi} \frac{\cos x}{1 + \sin 2x}$

(b) $\lim_{h \to 0} \frac{\sin h}{\ln(e^{2h})}$

(c) $\lim_{z \to \pi/4} \frac{\sin z - \cos z}{z - \pi/4}$

(*Hint:* Substitute $x = z - \pi/4$, simplify, and let $x \to 0$.)

4. Evaluate:

(a) $f''(\pi/3)$ for $f(x) = \sin 2x$

(b) $f'(0)$ for $f(x) = \tan(e^x \pi/4)$

(c) $\int_{-\pi/3}^{+\pi/2} \cos^3 t\, dt$

5. Without using a calculator, compute:

(a) $\cos^2 390°$ (b) $\cos\left(-\frac{7\pi}{6}\right)$

(c) $e^{\sin 5\pi/4}$ (d) $\sin \pi/6 \cos \pi/6$

6. Sketch the function $f(x) = x + \sin x$ for $0 \le x \le 4\pi$. Use curve-sketching techniques to locate all maxima, minima and inflection points, if any. Also discuss concavity.

7. Find the area of the region under the curve $y = \sin x + \cos x$ for $|x| \le 3\pi/4$.

8. Ship A leaves the harbor at 12 P.M., sailing at 10 km/hr due east. Ship B leaves one hour later sailing northwesterly at 20 km/hr. How fast are the two ships separating at 3 P.M.? (*Hint:* Use the law of cosines.)

9. Find $\frac{dy}{dx}$ in each of the following.

(a) $y = (x^2 + \sin x)^3$

(b) $\sin(xy) + y^2 = 2$

(c) $y = \tan(e^{\sin x})$

10. Suppose the formula

$$P(t) = 115 - 10 \sin \frac{\pi t}{8}, \qquad 0 \le t \le 8$$

is a crude model of a factory worker's blood pressure over an 8-hr work day.

(a) At what time of the work day is his blood pressure lowest?

(b) What is his average blood pressure over the 8-hr work day?

TEN

Probability

Outline

In Context

Specialists in the branch of mathematics called *probability* study the likelihood of occurrence of different events. Probability theory was originally developed as a tool for the analysis of games of chance, such as cards, dice, and roulette, but it is now applied to many diverse fields. The well-known laws of genetics, first proposed by Gregor Mendel, were the first major application of probability to biology. A physicist might use this theory to estimate the life span of an atom. And an engineer employs probability to determine the likelihood that a particular telephone network will be adequate to handle the anticipated volume of telephone calls. There is a stimulating interplay between the theory of probability and its applications.

In the first two sections of this chapter, we introduce and motivate the basic concepts of probability by investigating some simple experiments in which there are a finite number of possible outcomes. In the later sections we examine situations in which there is a continuum of outcomes, necessitating the intertwining of probability with calculus.

SECTION 10.1

Discrete Random Variables

Consider the following simple probabilistic experiment.

EXPERIMENT 1 A single coin is tossed once. In this experiment the two possible **outcomes** are (i) a head, denoted by H, and (ii) a tail, denoted by T. The set S of possible outcomes of an experiment is called the **sample space**. Thus we write

$$S = \{H, T\}$$

where H and T are referred to as **simple events**.

If we assume that we have been dealing with a "fair" coin, the likelihood of flipping a head and that of flipping a tail should be the same. In this case we say that the probability of obtaining a head is 1/2 and that the probability of obtaining a tail is 1/2. Symbolically, we write

$$\Pr(H) = 1/2$$
$$\Pr(T) = 1/2$$

□

Now let us move on to a more complicated example.

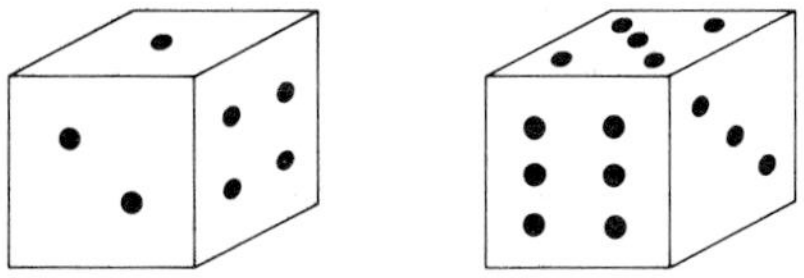

FIGURE 10.1

EXPERIMENT 2 A pair of dice is rolled. For example, Figure 10.1 shows a pair of dice displaying a 1 and a 5. It is easy to list all of the 36 possible outcomes for this experiment (see Table 10.1).

TABLE 10.1

Sample space when two dice are rolled

(1, 1)	(1, 2)	(1, 3)	(1, 4)	(1, 5)	(1, 6)
(2, 1)	(2, 2)	(2, 3)	(2, 4)	(2, 5)	(2, 6)
(3, 1)	(3, 2)	(3, 3)	(3, 4)	(3, 5)	(3, 6)
(4, 1)	(4, 2)	(4, 3)	(4, 4)	(4, 5)	(4, 6)
(5, 1)	(5, 2)	(5, 3)	(5, 4)	(5, 5)	(5, 6)
(6, 1)	(6, 2)	(6, 3)	(6, 4)	(6, 5)	(6, 6)

Because each outcome is equally likely, each has probability 1/36. Now in many games that use dice, the player's actions are determined by the sum of the numbers shown on the two faces. Thus we assign a single number to each simple event. For instance, the total of the dice shown in Figure 10.1 is 6. If we let X represent a variable whose value is the sum of the dice, then X can take on any of the following values:

$$2, 3, 4, 5, 6, 7, 8, 9, 10, 11, 12$$

X is an example of a **random variable**. □

DEFINITION 10.1 A **random variable** X is a function that assigns a numerical value to each simple event in our sample space S. For sample spaces that have a finite number of outcomes, the random variable is called a **discrete random variable**. ⊟

If there are n possible values that X can assume, we denote them by x_1, $x_2, \ldots, x_n$. The associated probabilities $\Pr(x_i)$ are denoted by p_i.

To illustrate these ideas, return to the dice-rolling experiment and consider Table 10.1, which shows the number of ways in which it is possible to roll a sum of 6: (1, 5), (2, 4), (3, 3), (4, 2), (5, 1). There are 5 pairs with probability 1/36, so $\Pr(6) = 5/36$. We display all of the possibilities in Table 10.2. Rolling a 6 is a **compound event**; there are several different simple events that can make up this compound event.

TABLE 10.2

Sum	2	3	4	5	6	7	8	9	10	11	12
Probability	1/36	2/36	3/36	4/36	5/36	6/36	5/36	4/36	3/36	2/36	1/36

It is worth noting that the random variable defined as the sum of the faces shown on the dice is not the only one possible, although it is the one that is most familiar. Alternatively, we might define a random variable Y to be the product of the numbers shown or another random variable Z to be the larger of the two values. The number of random variables that can be defined for this experiment is literally endless.

EXPERIMENT 3 Two coins are tossed simultaneously. Here the sample space S consists of four simple events.

$$S = \{\text{HH, HT, TH, TT}\}$$

where, for example, HT corresponds to obtaining an H on coin 1 and a T on coin 2. Note that HT and TH are both (different) simple events.

Now all of the four simple events in S are equally likely, so that

$$\Pr(\text{HH}) = 1/4 \qquad \Pr(\text{HT}) = 1/4 \qquad \Pr(\text{TH}) = 1/4 \qquad \Pr(\text{TT}) = 1/4$$

(a) Define a random variable X to be the number of heads displayed on the two coins. Then X can take on the values 0, 1, 2, and we have the probability values shown in Table 10.3.

TABLE 10.3

X	0	1	2
$\Pr(X)$	1/4	1/2	1/4

observe that $X = 1$ can arise in two ways: HT and TH. This is why $\Pr(X = 1) = \Pr(\text{HT}) + \Pr(\text{TH}) = 1/4 + 1/4 = 1/2$.

(b) Define another random variable Y for the same experiment to have the value 5 if the two coins agree and to have the value -2 if they are different. Then

$$Y(\text{HH}) = 5 \qquad Y(\text{HT}) = -2 \qquad Y(\text{TH}) = -2 \qquad Y(\text{TT}) = 5$$

and the probabilities are as shown in Table 10.4.

TABLE 10.4

Y	5	-2
$\Pr(Y)$	1/2	1/2

Intuitively, in any experiment, the probability of an event represents the relative frequency of its occurrence if the experiment is performed a great many times. It follows that the probability of any event must be less than or equal to 1, whereas the sum of all the probabilities must be 1. If you go back and examine all three experiments we have considered, you will see that this was the case. We state these observations as a formal definition.

DEFINITION 10.2 Let X be a discrete random variable for the sample space S that assumes values $x_1, x_2, \ldots, x_n$. A **probability distribution** for X is a function p for which

$$p(x_1) = p_1, p(x_2) = p_2, \ldots, p_n(x_n) = p_n$$

where the probabilities $p_1, p_2, \ldots, p_n$ satisfy

(10.1) $$0 \le p_i \le 1 \qquad i = 1, 2, \ldots, n$$

and

(10.2) $$p_1 + p_2 + \cdots + p_n = 1$$

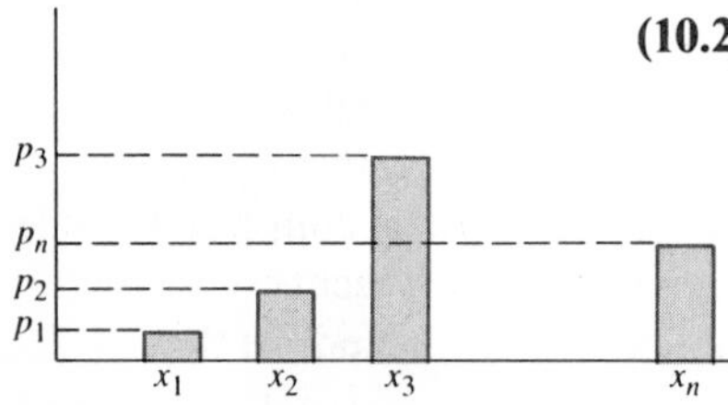

FIGURE 10.2

We have been displaying probability distributions in tabular form (see Table 10.5). It is frequently useful to display this information by means of a bar graph, or histogram, where a rectangle of height p_i is drawn over the value x_i. This is called a **probability histogram** (see Figure 10.2).

TABLE 10.5

X	x_1	x_2	$\cdots$	x_n
$\Pr(X)$	p_1	p_2	$\cdots$	p_n

We illustrate these ideas by returning to two experiments introduced earlier.

EXAMPLE 1 Construct the probability histogram for Experiment 2 for the random variable X whose value is the sum of the two dice.

SOLUTION The values of the random variable X, together with the associated probabilities, are shown in Table 10.2. We can easily transform the data into graphical form, as shown in Figure 10.3.

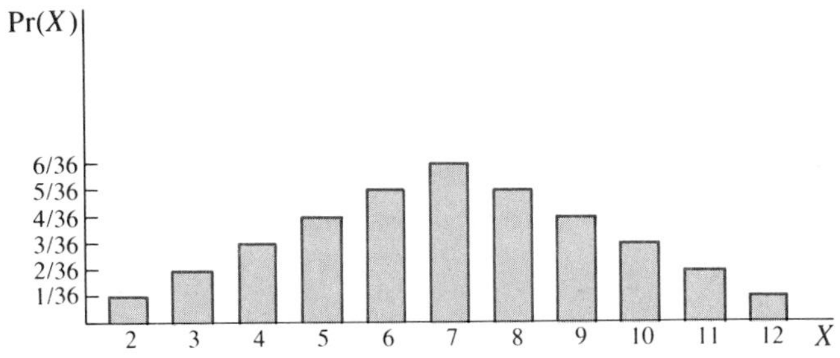

FIGURE 10.3

EXAMPLE 2 Construct the probability histogram for the random variable X representing the number of heads displayed when two coins are tossed simultaneously (see Experiment 3).

SOLUTION We easily construct the appropriate graph (Figure 10.4) from Table 10.3.

It is interesting to compare the theoretical probabilities displayed in Table 10.3 and Figure 10.4 with the actual results obtained by flipping two coins. One of the authors flipped two coins and obtained HH (two heads). For this single repetition of the experiment, the random variable X has value 2. Thus the empirically obtained relative frequency after one repetition is given by Figure 10.5. A second repetition yielded two tails (TT). Now our relative frequency histogram was as shown in Figure 10.6. After 100 trials, HH appeared 26 times; HT or TH appeared 45 times and TT appeared 29 times. Our relative frequency histogram at this point was as shown in Figure 10.7.

Note the similarity between Figure 10.7 and the theoretically obtained Figure 10.4. The greater the number of repetitions, the greater the correspondence we expect between the empirically obtained relative frequency histogram and the probability histogram. This provides confirmation that we have selected our probabilities correctly. For instance, suppose we had *incorrectly* reasoned that the flip of two coins must yield either two heads, one head, or no heads, and that each of these three events should therefore have probability 1/3—that is, $\Pr(X = 0) = 1/3$, $\Pr(X = 1) = 1/3$, $\Pr(X = 2) =$

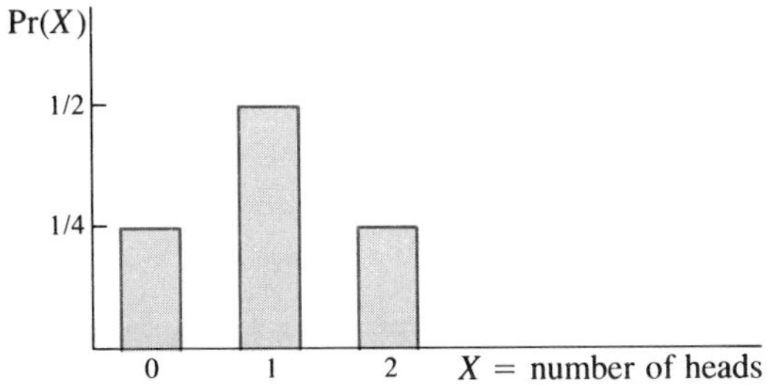

FIGURE 10.4

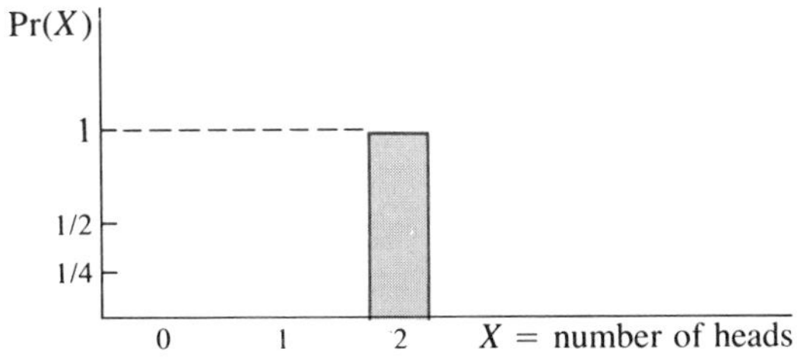

FIGURE 10.5
Relative frequencies after one repetition

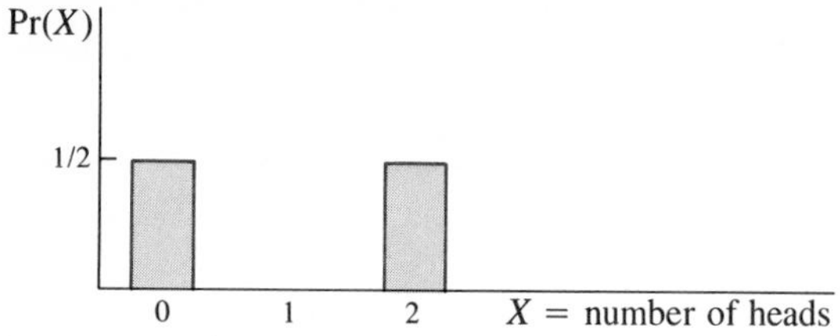

FIGURE 10.6
Relative frequency histogram after two trials

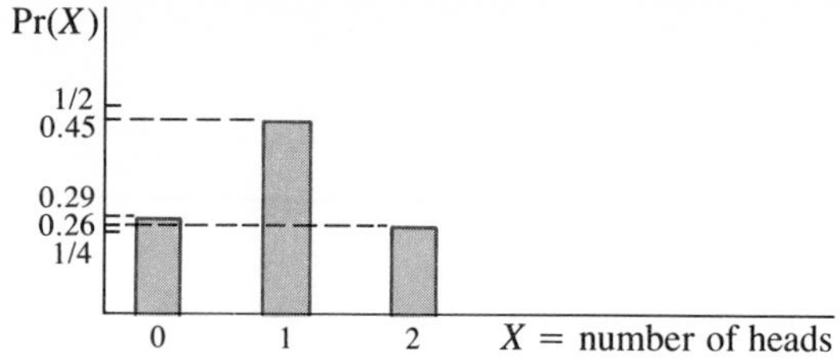

FIGURE 10.7
Relative frequency histogram after 100 trials

1/3. This is theoretically plausible, because each probability is less than 1 and the sum of the probabilities is 1. Performing the experiment a great many times reveals that the relative frequency histogram does *not* resemble the probability histogram proposed, so we would reject this proposed probability model for the tossing of two coins as not being an accurate reflection of reality. (For some complicated experiments, the choice of probabilities is not at all obvious, and there are sophisticated tests to decide which probability model most accurately represents a sample set of data.)

EXAMPLE 3 The price-to-earnings ratio for a stock, or P/E, is computed by dividing the current stock price by the preceding year's earnings. For example, if a company earns \$12 per share and each share sells for \$72, then its P/E is $72/12 = 6$. It is a widely used measure of value. The lower the P/E, the greater the amount of earnings we receive for each dollar invested in that company.

Table 10.6 shows the price-to-earnings ratio for Chase Manhattan Corp. for a fifteen-year period. In Table 10.7, we have grouped these data into a frequency table. It is now a simple matter to compute the relative frequencies and draw the relative frequency histogram (Figure 10.8).

TABLE 10.6
P/E for Chase Manhattan Corp., 1966–1980

Year	1966	1967	1968	1969	1970	1971	1972	1973	1974	1975	1976	1977	1978	1979	1980
P/E	12	13	14	14	11	11	13	10	4	6	9	8	6	4	4

TABLE 10.7

P/E	4	6	8	9	10	11	12	13	14
Frequency	3	2	1	1	1	2	1	2	2

Previously, we saw that, if we repeated our experiment a great many times, the relative frequency histogram closely resembled the probability histogram. In this example, we do not have probabilities available. Instead we utilize relative frequencies, provided that our sample is sufficiently large that the sample data are representative. (Actually, the data given in Table 10.6 seem to indicate a downward trend over time, which brings into question the fitness of a relative frequency model for this example. Looking at the table, we would probably not consider the likelihood that P/E $\geq$ 12 in 1981 to be 1/3, as can be seen from Figure 10.8.)

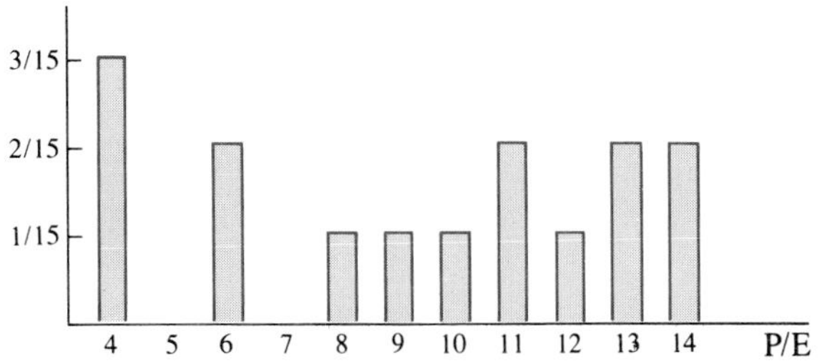

FIGURE 10.8

Section 10.1 EXERCISES

Three coins are tossed simultaneously. Develop a probability histogram to answer the questions posed in Exercises 1–3.

1. What is the probability that all three coins will turn up the same?
2. What is the probability of getting one head?
3. What is the probability of getting two tails?
4. Try tossing three coins 10 times, 20 times, and 30 times. Draw the relative frequency histograms. Do your results more closely approximate your probability model as you increase the number of throws?

In Exercises 5–10 assume that two dice are tossed simultaneously. Let the random variable X represent the larger of the two numbers. Let the random variable Y represent the product of the two dice.

5. List the possible outcomes for X.
6. List the possible outcomes for Y.
7. Draw a probability histogram for X.
8. Construct a table showing the probability distribution of y.
9. What is the probability that Y is 12?
10. What is the probability that X is 4?
11. Toss two dice simultaneously (or draw slips of paper bearing the numbers 1 through 6 out of two hats simultaneously) 20 times. Draw the relative frequency histogram for the sum. How well do your experimental results agree with the model?

In Exercises 12 and 13 assume that one green sock and two identical blue socks are placed in each of two dresser drawers.

12. If two socks are chosen, one from each drawer, what is the probability that they will match?
13. What is the probability of getting a blue pair?

In Exercises 14–17 assume that three identical red socks are added to each drawer that you worked with in Exercises 12 and 13.

14. List all the possible combinations of two socks drawn one from each drawer.
15. What are the chances of drawing a red pair?
16. What are the chances of drawing a green pair?
17. What is the probability that the socks will not match?

Draw a probability or relative frequency histogram from the probability or data tables given in Exercises 18–21.

18.

Outcome	2	4	6	8
Probability	0.2	0.3	0.3	0.2

19. *Attendance Record*

Days absent	0	1	2	3
Number weeks	20	15	9	6

20. *Orchard Production*

Bushels apples	3	4	5
Probability	1/3	1/2	1/6

21.

Outcome	up	down	right	left
Probability	1/5	3/10	1/10	2/5

In Exercises 22–25 construct a table corresponding to each histogram given.

22.

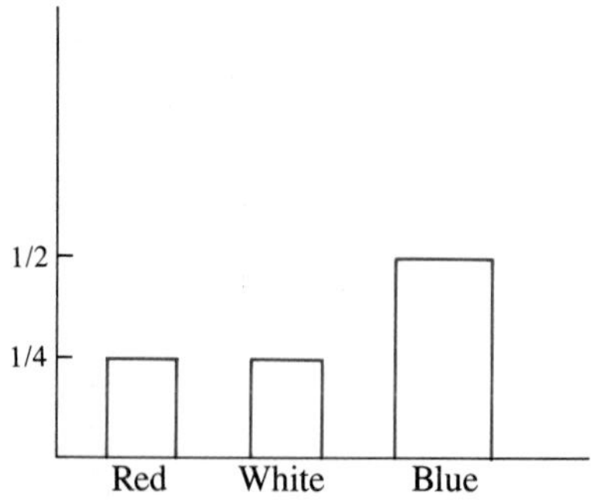

23.

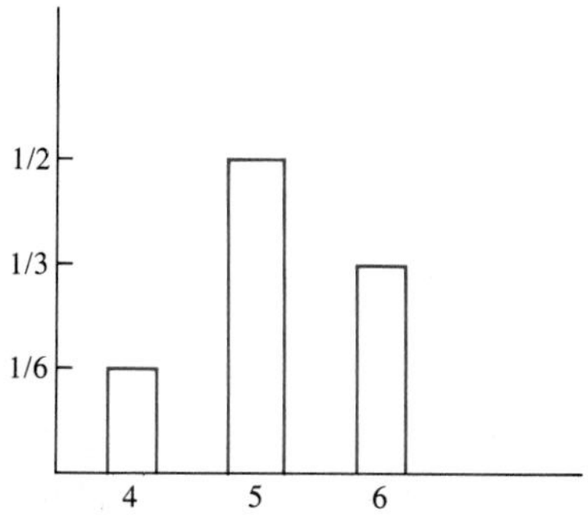

24.

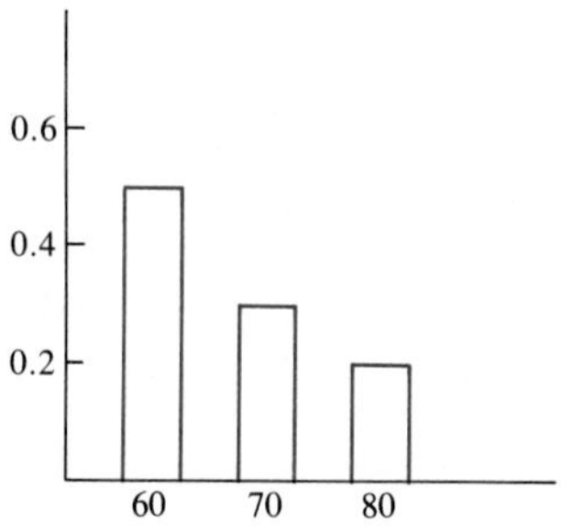

25.

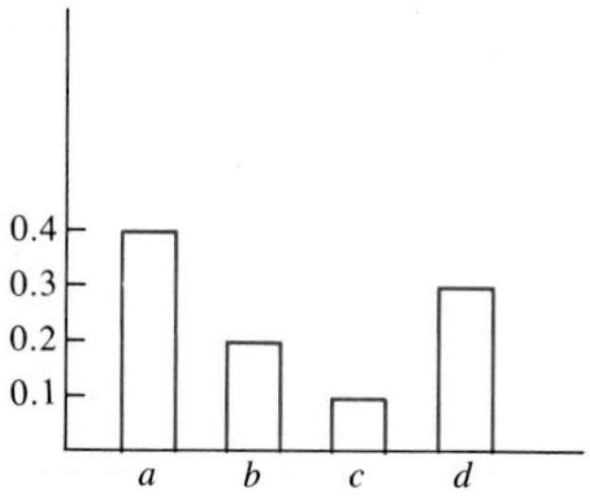

In Exercises 26–33 decide whether each of the tables or histograms could represent a probability distribution. If not, why not?

26.

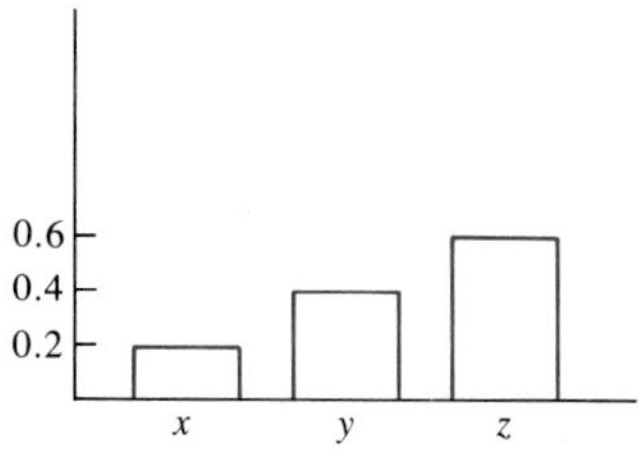

27.

Outcome	TIC	TAC	TOE
Probability	1/3	1/3	1/3

28.

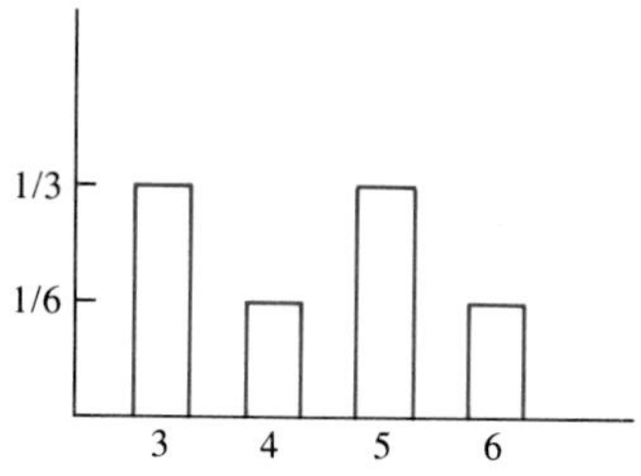

29.

Outcome	1	2	3	4
Probability	0.3	0.2	0.4	0.3

30.

Outcome	Red	Bl	Gr	Wh
Probability	0.4	0.2	0.1	0.4

31.

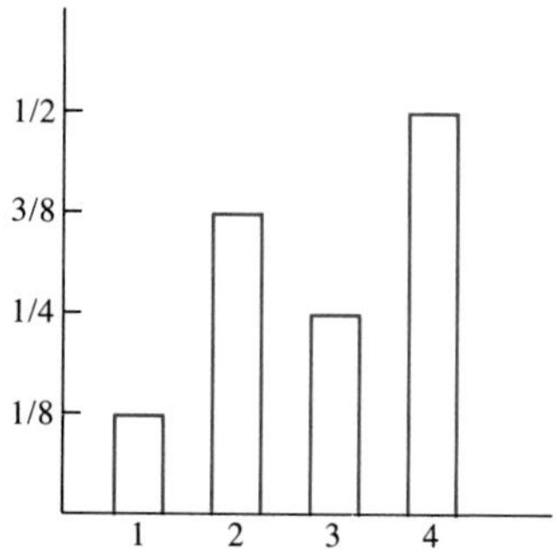

32.

Outcome	60	65	70	75
Probability	0.2	0.3	0.2	0.3

33.

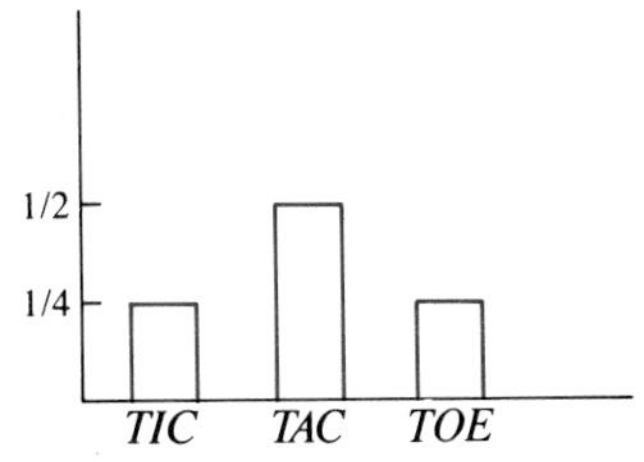

SECTION 10.2

Expected Value, Variance, and Standard Deviation

In Section 10.1 we introduced the concept of a probability distribution for a random variable and used the probability histogram as a means of graphically representing the distribution. In many situations it is useful to have analytical tools (rather than graphical tools) to help us summarize and describe probability distributions. By far the most important and well-known summarizing statistic is the **expected value**, or **mean value**, of the random variable.

Consider the data presented in Table 10.8, which shows the "savings rate" for the decade 1973–1982. This is the percentage of household income that goes into savings each year. It is a simple matter to compute the average (or **sample mean**) savings rate for the decade:

$$\bar{x} = \frac{9 + 9 + 9 + 7 + 6 + 6 + 6 + 6 + 6 + 7}{10} = \frac{71}{10} = 7.1$$

A bar above a variable is frequently used to denote its average value.

Alternatively, suppose we had chosen to group the data into a relative frequency table, such as that shown in Table 10.9. Then we could have obtained the same mean value by computing the weighted sum of the savings rates:

$$\bar{x} = 6 \cdot 5/10 + 7 \cdot 2/10 + 9 \cdot 3/10 = \frac{30}{10} + \frac{14}{10} + \frac{27}{10} = \frac{71}{10} = 7.1$$

By analogy, we can now define the **mean value** or **expected value** of a random variable X to be the weighted sum of the values of X. The expected value is

TABLE 10.8

U.S. savings rate as a percentage of household income

Year	1973	1974	1975	1976	1977	1978	1979	1980	1981	1982
Savings rate	9%	9%	9%	7%	6%	6%	6%	6%	6%	7%

TABLE 10.9

Savings rate	6	7	9
Relative frequency	5/10	2/10	3/10

frequently denoted by $E(X)$, or by the Greek letter μ (mu). Thus, given the probability distribution shown in Table 10.10, we define μ as follows:

$$\mu = E(X) = x_1p_1 + x_2p_2 + \cdots + x_np_n = \sum_{i=1}^{n} x_ip_i$$

TABLE 10.10

X	x_1	x_2	$\cdots$	x_n
$\Pr(X)$	p_1	p_2	$\cdots$	p_n

EXAMPLE 4 Compute the expected value for the random variable representing the number of heads that appear when two coins are tossed simultaneously. Refer to part (a) of Experiment 3.

SOLUTION We know that the probability distribution for this random variable is as given in Table 10.11. Therefore

$$\mu = E(X) = 0 \cdot 1/4 + 1 \cdot 1/2 + 2 \cdot 1/4 = 1$$

Not surprisingly, when two coins are tossed simultaneously, an average of one head will appear.

TABLE 10.11

X	0	1	2
$\Pr(X)$	1/4	1/2	1/4

EXAMPLE 5 Find the expected value for the random variable X that is the sum of the faces of a pair of dice.

SOLUTION From Table 10.2 we see that the probability distribution for X is as shown in Table 10.12. It follows that

$$\begin{aligned}\mu = E(X) &= 2(1/36) + 3(2/36) + 4(3/36) + 5(4/36) + 6(5/36) + 7(6/36) \\ &\quad + 8(5/36) + 9(4/36) + 10(3/36) + 11(2/36) + 12(1/36) \\ &= 252/36 = 7\end{aligned}$$

TABLE 10.12

Sum	2	3	4	5	6	7	8	9	10	11	12
Probability	1/36	2/36	3/36	4/36	5/36	6/36	5/36	4/36	3/36	2/36	1/36

EXAMPLE 6 A roulette wheel in Las Vegas has 38 numbers around its circumference. There are 2 green numbers, 18 red numbers, and 18 black numbers. A player can bet $1 on any of the numbers, and the casino will pay $35 if that number comes up; if it doesn't, the casino takes the player's dollar. Find the expected value of the amount won (or lost).

SOLUTION Suppose that each number on the roulette wheel is equally likely to come up, so that the probability of any particular number is 1/38. Let X be the random variable that represents the amount won. The outcomes for X, together with the associated probabilities, are shown in Table 10.13. Thus,

$$E(X) = (-1)(37/38) + (35)(1/38) = -2/38 \approx -0.0526$$

On average, a player will lose slightly over 5¢ out of every dollar he or she wagers.

TABLE 10.13

X	-1	$+35$
$\Pr(X)$	$37/38$	$1/38$

PRACTICE EXERCISE Consider again the roulette wheel in Example 6. A player can also place a \$1 bet that the number that comes up will be a "red" number. If a red number does come up, the player wins \$1. In the event that the number is either black or green, the wager is lost. If Y is the random variable that represents the amount won, find $E(Y)$.

[Your answer should be $-\frac{2}{38} \approx -0.0526$ again. Are you surprised?]

EXAMPLE 7 There are two basic types of oil wells that can be drilled. A *developmental well* is one drilled in an area that has already been proved to have oil and gas reserves. Typically, the likelihood of finding oil in a proven field is quite high: 65% of developmental well strike oil, and the remaining 35% are dry holes. Because of the fact that the oil and gas potential of the area is known in advance, however, most developmental wells do not have the potential for a dramatically large find. The average developmental well produces revenue equal to 4 times the cost of drilling the well.

A second type of oil well is an *exploratory well*. This is a well drilled in an area that lies above a geological formation that suggests the presence of hydrocarbons. Exploratory wells strike oil only 15% of the time, but the payoff is much larger, averaging a \$25 return for each \$1 invested in drilling the well.

Should an investor expect to make a better return for each dollar invested in a developmental well or an exploratory well?

SOLUTION Let X be the random variable that represents the investor's payoff for each \$1 invested in a developmental well, and let Y denote the corresponding random variable for an exploratory well. The probability distributions for X and Y are shown in Table 10.14.

TABLE 10.14

(a) Development well

X	-1	$+4$
$\Pr(X)$	0.35	0.65

(b) Exploratory well

Y	-1	$+25$
$\Pr(Y)$	0.85	0.15

$$E(X) = (-1)(0.35) + (4)(0.65) = 2.25$$

$$E(Y) = (-1)(0.85) + (25)(0.15) = 2.90$$

Thus, on average, an exploratory well has a higher payoff than a developmental well, even though the likelihood of drilling a dry hole is considerably greater.

The expected value is an important quantity for summarizing the nature of a probability distribution. It is a measure of **central tendency**; it tells us the "average" value of the distribution. We are also interested in studying the **variance** (and the **standard deviation**) of a distribution. These are measures of **dispersion**. They tell us whether the values of the random variable are generally clustered near the expected value or tend to be located far from the mean.

The significance of dispersion emerges when we compare two simple experiments. In the first experiment a coin is tossed 100 times, and the random variable X represents the total number of heads that appear. It is intuitively clear (and can easily be proved) that $E(X) = 50$. That is, we will average 50 heads in 100 tosses of a fair coin.

In the second experiment, the numbers 0 through 100 are written on slips of paper and placed in a hat. A single slip is drawn from the hat, and Y represents the value drawn. There are a total of 101 slips, so each slip has probability 1/101, and the expected value of Y is

$$E(Y) = 0(1/101) + 1(1/101) + 2(1/101) + \cdots + 100(1/101) = \frac{5050}{101} = 50$$

This is what we would expect, too.

Despite the fact that both random variables assume values from 0 through 100 and both have expected value 50, they are quite different. If we tossed a coin 100 times, we might expect to get 43 heads, or 48, or 55. We would be quite startled to obtain only 1 or 2 heads and would doubt the honesty of the coin.

On the other hand, in pulling a slip of paper from a hat, the number 1 or 2 is just as likely to be selected as 43, 48, or 55. The values in this experiment tend to be scattered further from the mean.

Now we are ready to make the appropriate definitions.

DEFINITION 10.3 The **variance** of a random variable X, denoted by $\mathrm{Var}(X)$, is the weighted sum of the squares of the differences between each value of X and the expected value.

$$\mathrm{Var}(X) = (x_1 - \mu)^2 p_1 + (x_2 - \mu)^2 p_2 + \cdots + (x_n - \mu)^2 p_n.$$

The **standard deviation**, σ, is the square root of the variance.

$$\sigma = \sqrt{\mathrm{Var}(X)}$$ ⊟

EXAMPLE 8 Compute the variance and standard deviation for the random variable X representing the number of heads that appear when two coins are tossed simultaneously (from Example 4).

SOLUTION We found that the expected value of X was 1. We now modify Table 10.11 by adding the values of $(X - \mu)^2$; see Table 10.15. Now,

$$\text{Var}(X) = (-1)^2(1/4) + (0)^2(1/2) + (1)^2(1/4) = 1/2$$

$$\sigma = \sqrt{\text{Var}(X)}$$
$$= \sqrt{1/2} = \frac{\sqrt{2}}{2}$$

TABLE 10.15

X	0	1	2
$(X - \mu)^2$	$(0 - 1)^2$	$(1 - 1)^2$	$(2 - 1)^2$
$\Pr(X)$	1/4	1/2	1/4

EXAMPLE 9 Compute the variance and standard deviation for the random variable X that represents the sum of the faces of a pair of dice.

SOLUTION In Example 5 we found that $E(X) = 7$. We add the values of $(X - \mu)^2$ to Table 10.12 to produce Table 10.16.

$$\begin{aligned}\text{Var}(X) &= (-5)^2(1/36) + (-4)^2(2/36) + (-3)^2(3/36) + (-2)^2(4/36) \\ &\quad + (-1)^2(5/36) + (0)^2(6/36) + (1)^2(5/36) + (2)^2(4/36) \\ &\quad + (3)^2(3/36) + (4)^2(2/36) + (5)^2(1/36) \\ &= 210/36 = 35/6 \approx 5.83\end{aligned}$$

$$\sigma = \sqrt{\text{Var}(X)} = \sqrt{35/6} \approx 2.42$$

TABLE 10.16

X	2	3	4	5	6	7	8	9	10	11	12
$(X - \mu)^2$	$(2 - 7)^2$	$(3 - 7)^2$	$(4 - 7)^2$	$(5 - 7)^2$	$(6 - 7)^2$	$(7 - 7)^2$	$(8 - 7)^2$	$(9 - 7)^2$	$(10 - 7)^2$	$(11 - 7)^2$	$(12 - 7)^2$
$\Pr(X)$	1/36	2/36	3/36	4/36	5/36	6/36	5/36	4/36	3/36	2/36	1/36

We can also compute the variance and standard deviation for data that emerge when we sample a population. These are called the **sample variance**, s^2, and the **sample standard deviation**, s. If we are given a collection of data $x_1, x_2, \ldots, x_n$, we first compute the sample mean, $\bar{x}$, and then define

$$s^2 = \frac{(x_1 - \bar{x})^2 + (x_2 - \bar{x})^2 + \cdots + (x_n - \bar{x})^2}{n - 1}$$
$$= \frac{\sum_{i=1}^{n} (x_i - \bar{x})^2}{n - 1}$$

Thus

$$s = \sqrt{\frac{\sum_{i=1}^{n} (x_i - \bar{x})^2}{n - 1}}$$

EXAMPLE 10 At one time, the sales tax rates for the six New England states were as follows:

Connecticut	7%
Maine	5%
Massachusetts	5%
New Hampshire	0%
Rhode Island	6%
Vermont	3%

Find the sample mean, variance, and standard deviation.

SOLUTION We begin by computing the sample mean.

$$\bar{x} = \frac{7 + 5 + 5 + 0 + 6 + 3}{6} = \frac{26}{6} = \frac{13}{3}$$

Next we find the sample variance.

$$\begin{aligned} s^2 &= \frac{(7 - 13/3)^2 + (5 - 13/3)^2 + (5 - 13/3)^2 + (0 - 13/3)^2}{5} \\ &\quad + \frac{(6 - 13/3)^2 + (3 - 13/3)^2}{5} \\ &= \frac{(8/3)^2 + (2/3)^2 + (2/3)^2 + (-13/3)^2 + (5/3)^2 + (-4/3)^2}{5} \\ &= \frac{282/9}{5} = \frac{94}{15} \approx 6.27 \end{aligned}$$

Finally,

$$s = \sqrt{94/15} \approx 2.50$$

We close with an example that illustrates the use of these concepts in decision making.

EXAMPLE 11 An East Coast student graduates from college and decides to accept a job offer in California. She must choose a moving company to transport her belongings across the country. A sampling of eight moves by each of the two leading movers revealed the information shown in Table 10.17 about the number of days from the time the furniture is picked up to the time it is delivered. Which mover should our graduate choose?

TABLE 10.17

	Number of days							
Rapid Movers	7	8	6	5	9	10	5	6
Speedy Van Lines	5	6	10	8	5	10	7	5

SOLUTION The sample mean delivery time for Rapid Movers is

$$\bar{x} = \frac{7 + 8 + 6 + 5 + 9 + 10 + 5 + 6}{8} = \frac{56}{8} = 7 \text{ days}$$

The corresponding computation for Speedy Van Lines is

$$\bar{y} = \frac{5 + 6 + 10 + 8 + 5 + 10 + 7 + 5}{8} = \frac{56}{8} = 7 \text{ days}$$

Our preliminary attempt to resolve the problem is a failure; the average delivery time for the two companies is the same. Next we compute the sample variance for each set of data.

$$s_x^2 = \frac{(7-7)^2+(8-7)^2+(6-7)^2+(5-7)^2+(9-7)^2+(10-7)^2+(5-7)^2+(6-7)^2}{7}$$

$$= \frac{0^2+1^2+(-1)^2+(-2)^2+2^2+3^2+(-2)^2+(-1)^2}{7}$$

$$= \frac{24}{7} \approx 3.43$$

$$s_y^2 = \frac{(5-7)^2+(6-7)^2+(10-7)^2+(8-7)^2+(5-7)^2+(10-7)^2+(7-7)^2+(5-7)^2}{7}$$

$$= \frac{(-2)^2+(-1)^2+3^2+1^2+(-2)^2+3^2+0^2+(-2)^2}{7}$$

$$= \frac{32}{7} \approx 4.57$$

Our graduate should hire Rapid Movers. Their average time is the same as that of Speedy Van Lines, but the lower variance indicates greater reliability.

Section 10.2 EXERCISES

The accompanying table gives samples of the number of French fries in a large order at the several fast-food places. Compute the sample mean, sample variance, and sample standard deviation for the restaurant given in each of Exercises 1–4.

1. Burger Doodle
2. MacBurger
3. Big Doodle
4. Doodle Chef

Burger Doodle	37	41	29	45	31	45	47
MacBurger	40	33	50	35	52	32	33
Big Doodle	22	43	41	28	45	34	37
Doodle Chef	35	26	37	43	39	48	34

Compute the expected value, variance, and standard deviation for each of the probability tables or histograms given in Exercises 5–13.

5.

Outcome	1	2	3	4	5
Probability	0.1	0.1	0.4	0.2	0.2

6.

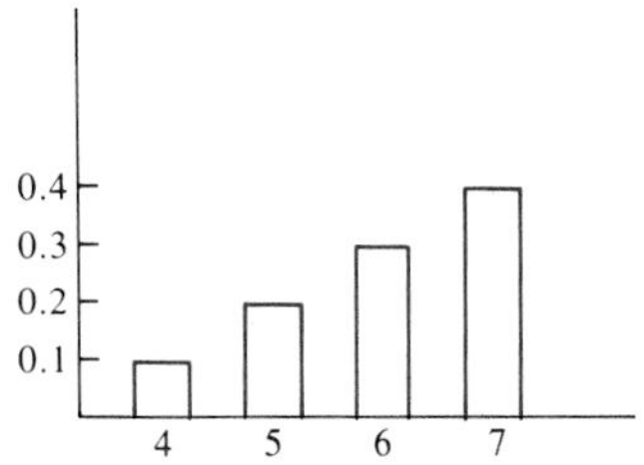

7.

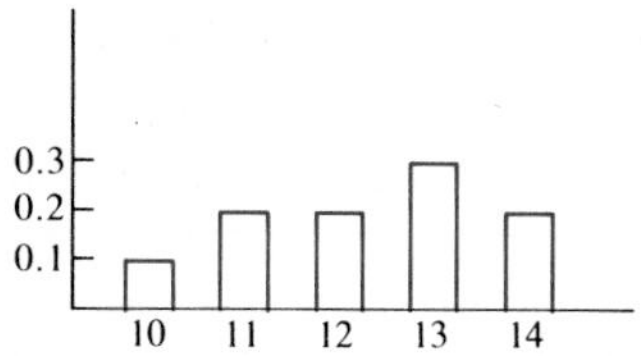

8.

High Temp.	60s	70s	80s
Probability	0.2	0.5	0.3

9.

Age	18	19	20	21
Probability	0.3	0.4	0.2	0.1

10.

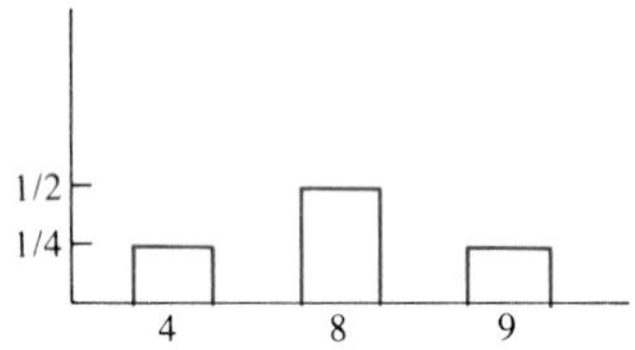

11.

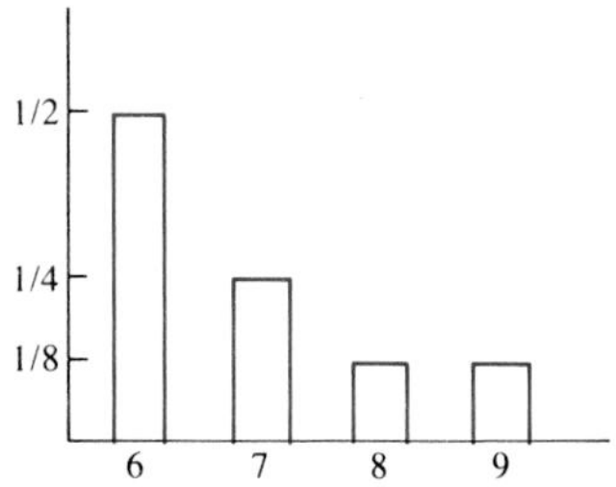

12.

Weight	120	130	140	150
Probability	0.2	0.3	0.4	0.1

13.

Outcome	37	47	41	31
Probability	0.3	0.4	0.2	0.1

In Exercises 13 and 14 assume that two dice are tossed simultaneously. Let *X* represent the smaller of the two numbers. Let *Y* represent the product of the two numbers.

14. Compute the expected value, variance, and standard deviation for X.

15. Compute the expected value, variance, and standard deviation for Y.

16. In a class of 40 students the number of students n_i of each age is shown in the accompanying table. If a student is to be selected at random from the class, what is the expected value of his or her age?

Age, i	n_i
18	10
19	17
20	8
21	3
23	2

17. Suppose that a word is selected at random from the clause "WHEN, IN THE COURSE OF HUMAN EVENTS, IT BECOMES NECESSARY FOR ONE PEOPLE TO DISSOLVE THE POLITICAL BONDS WHICH HAVE CONNECTED THEM WITH ANOTHER. . . ." If X denotes the number of letters in the word that is selected, what is the value of $E(X)$?

18. Suppose that one word is to be selected at random from the letters in the clause given in Exercise 17. If Y denotes the number of vowels in the word selected, what is the value of $E(Y)$?

19. In a town of 123 families, the number of families that have k children ($k = 0, 1, 2, \ldots$) is given in the table to the right. Determine the mean number of children per family.

Number of children	*Number of families*
0	19
1	32
2	35
3	23
4	14

20. Suppose that, in a gambling game, a person is paid \$15 if she draws a 2 or an ace and \$5 if she draws a queen or a king from an ordinary deck of 52 playing cards. If she draws any other card, she pays \$4. What is the expected gain when the game is played?

SECTION 10.3

Continuous Random Variables

In Section 10.1 we considered experiments in which the number of outcomes was finite. Many experiments, however, have an infinite number of possible outcomes. Typically the values assumed by a random variable for such an experiment lie in an interval along the real line. Such random variables are called **continuous random variables**.

As an illustration of this concept, consider an experiment in which we set up a radar device at a given location on a major highway, and define the random variable X to be the speed of each car as it passes. Thus X can assume any value on the real line between, say, 0 and 100 miles per hour.

We can now generalize the notion of a probability distribution (Definition 10.2) to the case of a continuous random variable X.

DEFINITION 10.4 Let X be a continuous random variable assuming any value x for which $A \leq x \leq B$. A **probability distribution** (or **density function**) for X on the interval $[A, B]$ is a function $p(x)$ for which

$$0 \leq p(x) \qquad A \leq x \leq B \tag{10.3}$$

and

$$\int_A^B p(x)\,dx = 1 \tag{10.4}$$

□

Property (10.3) states that the graph of a probability density function must lie above the x-axis. Property (10.4) says that there is probability 1 that the random variable X will assume a value between A and B. It is analogous to Property (10.2) of Definition 10.2, which states that the sum of the probabilities of a discrete random variable must be 1. We permit the possibility that either $A = -\infty$ or $B = \infty$. In this case, we interpret Property (10.4) as an improper integral (see Section 7.7).

There is a subtle difference between the use of the probability distribution for discrete random variables and for continuous random variables. In order to explain this difference, consider the experiment of rolling a single die and letting X represent the (discrete) random variable of the number that comes up. Then X can assume any of six different values and, because each is equally likely, each has probability $1/6$.

Similarly, if we spin a roulette wheel, there are 38 possible outcomes (0, 00, or the numbers 1 through 36), and each outcome has probability $1/38$.

A continuous random variable, by contrast, can assume an infinite number of values. It does not make sense to say that each outcome has probability $1/\infty = 0$, though. Instead, for continuous random variables we

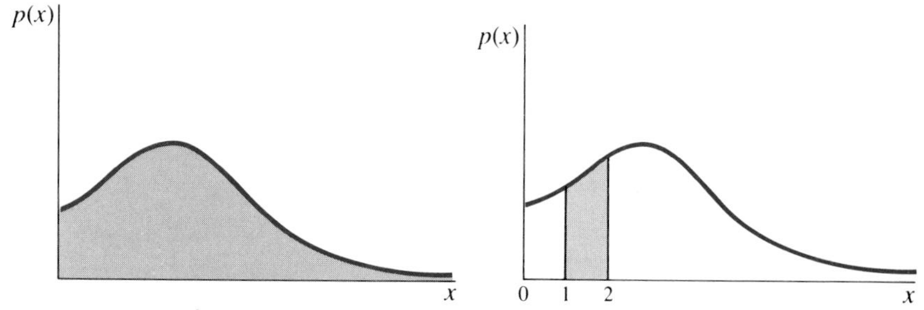

(a) $\int_0^\infty p(x) = 1$ (b) $\Pr(1 \leqslant x \leqslant 2) = \int_1^2 p(x)\,dx$

FIGURE 10.9

consider the probability that the variable assumes any value within an interval, rather than some specific value. The probability that X lies in the interval $[a, b]$, which we denote by $\Pr(a \le X \le b)$, is given by the area under the probability distribution function:

$$\Pr(a \le X \le b) = \int_a^b p(x)\,dx$$

Figure 10.9(a) shows a hypothetical probability density on $[0, \infty)$; in Figure 10.9(b) we have shaded in an area that represents the probability that $1 \le X \le 2$.

EXAMPLE 12 Let $p(x) = 3(1 - x^2)/4$, where $-1 \le x \le 1$.

(a) Show that $p(x)$ is a probability density function.

(b) Compute $\Pr(\frac{1}{2} \le X \le \frac{3}{4})$.

(c) Compute $\Pr(-\frac{1}{4} \le X \le 0)$.

SOLUTION

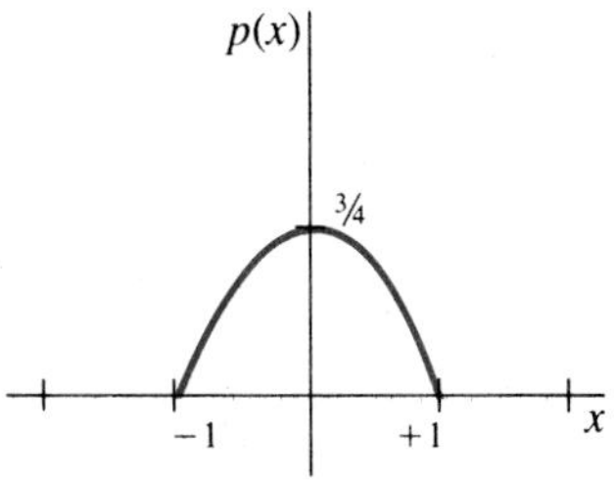

FIGURE 10.10

$$p(x) = \frac{3(1 - x^2)}{4}$$

(a) We begin by drawing a sketch of the graph of p (see Figure 10.10). Strictly speaking, this is unnecessary, but it helps us visualize the problem. It is clear from the graph that $p(x) \ge 0$ for $-1 \le x \le 1$. Next we evaluate

$$\int_{-1}^{1} \frac{3(1 - x^2)}{4}\,dx = \frac{3}{4}\int_{-1}^{1} (1 - x^2)\,dx = \frac{3}{4}\left(x - \frac{x^3}{3}\right)\Bigg|_{-1}^{1}$$
$$= \frac{3}{4}\left(\frac{2}{3}\right) - \frac{3}{4}\left(-\frac{2}{3}\right) = \frac{1}{2} - \left(-\frac{1}{2}\right) = 1$$

Thus p is a probability density function.

(b) We know that

$$\Pr\left(\frac{1}{2} \le X \le \frac{3}{4}\right) = \int_{1/2}^{3/4} \frac{3(1 - x^2)}{4}\,dx = \frac{3}{4}\left(x - \frac{x^3}{3}\right)\Bigg|_{1/2}^{3/4}$$
$$= \frac{3}{4}\left[\left(\frac{3}{4} - \frac{27}{192}\right) - \left(\frac{1}{2} - \frac{1}{24}\right)\right] = \frac{29}{256} \approx 0.113$$

(c) By analogy with part (b), we compute

$$\Pr\left(-\frac{1}{4} \le X \le 0\right) = \int_{-1/4}^{0} \frac{3(1 - x^2)}{4}\,dx = \frac{3}{4}\left(x - \frac{x^3}{3}\right)\Bigg|_{-1/4}^{0}$$
$$= \frac{3}{4}\left[0 - \left(-\frac{1}{4} + \frac{1}{192}\right)\right] = \frac{47}{256} \approx 0.184$$

EXAMPLE 13 Hodgkin's disease is a form of lymphoma—that is, cancer of the lymphatic system. Let X be the number of years that a person lives after receiving treatment, which is called the survival time. It has been determined that the density function for X is given by

$$p(x) = 0.12e^{-0.12x}$$

Find the probability that a patient survives at least 5 years (the 5-year survival rate), which is the commonly used criterion for cure.

SOLUTION The continuous random variable X can assume any value $0 \le x < \infty$. The probability that a patient survives at least 5 years is equivalent to the mathematical expression

$$\Pr(5 \le X < \infty)$$

Therefore

$$\begin{aligned}\Pr(5 \le X \le \infty) &= \int_5^\infty 0.12e^{-0.12x}\,dx = \lim_{b\to\infty} \int_5^b 0.12e^{-0.12x}\,dx \\ &= \lim_{b\to\infty}\left(-e^{-0.12x}\Big|_5^b\right) = \lim_{b\to\infty}(-e^{-0.12b} + e^{-0.12(5)}) \\ &= 0 + e^{-0.6} \approx 0.549\end{aligned}$$

This means that nearly 55% of all patients treated for Hodgkin's disease survive at least 5 years. (It is notable that 20 years ago, the 5-year survival rate for this disease was only 25%).

Let us now extend the notions of expected value, variance, and standard deviations to the case of continuous random variables.

DEFINITION 10.5 Let X be a continuous random variable on the interval $[A, B]$ with probability density function $p(x)$. We define the **expected value**, or **mean**, of X to be

(10.5) $$\mu = E(X) = \int_A^B xp(x)\,dx$$

The **variance** is given by

(10.6) $$\operatorname{Var}(X) = \int_A^B (x - \mu)^2 p(x)\,dx$$

and the **standard deviation**, σ, is

(10.7) $$\sigma = \sqrt{\operatorname{Var}(X)}$$

As in the discrete case, the expected value is the average outcome of the random variable X if the experiment is repeated a great many times, whereas both the variance and the standard deviation are measures of dispersion around the mean.

EXAMPLE 14 Consider again the random variable X of Example 12, for which

$$p(x) = \frac{3(1 - x^2)}{4} \qquad -1 \le x \le 1$$

Compute μ, $\operatorname{Var}(X)$, and σ.

SOLUTION By straightforward computation,

$$\mu = E(X) = \int_A^B xp(x)\,dx = \int_{-1}^{1} x\left[\frac{3(1-x^2)}{4}\right]dx$$
$$= \frac{3}{4}\int_{-1}^{1}(x - x^3)\,dx = \frac{3}{4}\left(\frac{x^2}{2} - \frac{x^4}{4}\right)\Bigg|_{-1}^{1}$$
$$= \frac{3}{4}\left[\left(\frac{1}{2} - \frac{1}{4}\right) - \left(\frac{1}{2} - \frac{1}{4}\right)\right] = 0$$

Similarly,

$$\text{Var}(X) = \int_A^B (x-\mu)^2 p(x)\,dx = \int_{-1}^{1}(x-0)^2\frac{3(1-x^2)}{4}\,dx$$
$$= \frac{3}{4}\int_{-1}^{1}(x^2 - x^4)\,dx = \frac{3}{4}\left(\frac{x^3}{3} - \frac{x^5}{5}\right)\Bigg|_{-1}^{1}$$
$$= \frac{3}{4}\left[\left(\frac{1}{3} - \frac{1}{5}\right) - \left(-\frac{1}{3} + \frac{1}{5}\right)\right] = \frac{3}{4}\left[\frac{2}{15} + \frac{2}{15}\right]$$
$$= \frac{1}{5}$$

Finally,

$$\sigma = \sqrt{\text{Var}(X)} = \sqrt{\frac{1}{5}} = \frac{\sqrt{5}}{5}$$

Before proceeding with more examples, it may be useful to explain why the definition of expected value in Definition 10.5 is analogous to that given in Section 10.2.

Consider a random variable X on $[A, B]$ whose probability distribution is shown in Figure 10.11. If we had to compute the expected value, or average value, of X, in the absence of Definition 10.5 we might try to *discretize* the function $p(x)$. That is, we would approximate the random variable X by a discrete random variable. To do this, we begin by partitioning the interval $[A, B]$ into subintervals of equal width Δx and by inserting subdivision points $x_0, x_1, \ldots, x_n$, as shown in Figure 10.12.

Imagine that we perform our experiment and that we agree to round the value of the random variable X up to the next higher of the discrete values

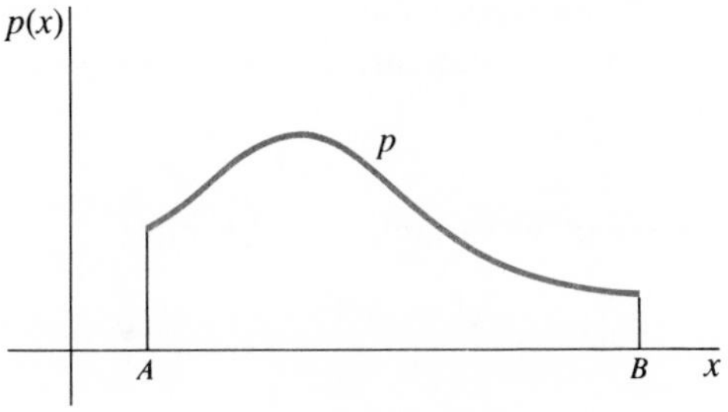

FIGURE 10.11

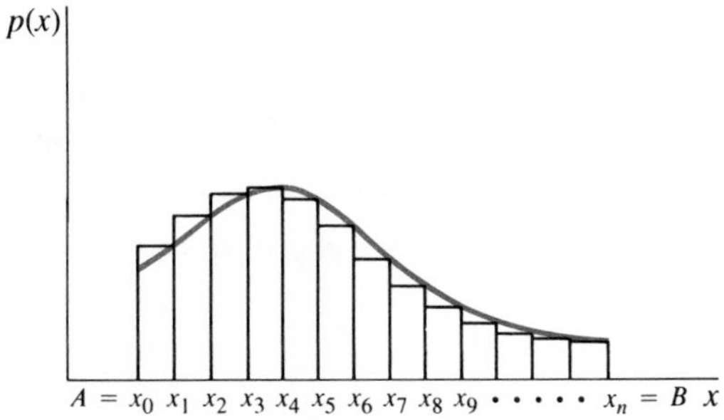

FIGURE 10.12

$x_1, x_2, \ldots, x_n$. Thus, for instance, if X assumes any value between x_{i-1} and x_i, we agree to consider this an occurrence of x_i. If we perform our experiment a great many times, we can construct a relative frequency table of the form shown in Table 10.18. The expected value of this discrete random variable is the weighted sum of the x_i terms.

TABLE 10.18

X	x_1	x_2	$\cdots$	x_n
$\Pr(X)$	p_1	p_2	$\cdots$	p_n

$$E(X) = x_1p_1 + x_2p_2 + \cdots + x_np_n = \sum_{i=1}^{n} x_ip_i$$

But what will be the values of the relative frequencies p_i? The number p_i measures the relative frequency of x_i. Figure 10.12 shows that the frequency of occurrence of x_i is given by the area of the rectangle from x_{i-1} to x_i with height $p(x_i)$. That is,

$$p_i = p(x_i)\,\Delta x$$

Substituting this expression into the sum for $E(X)$ yields

$$E(X) \approx \sum_{i=1}^{n} x_ip(x_i)\,\Delta x$$

We recognize this as a Riemann sum. If we partition the interval $[A, B]$ into finer and finer subintervals, then in the limit this Riemann sum approaches

$$E(X) = \int_A^B xp(x)\,dx$$

Similar reasoning can be used to obtain the formula for the variance.

There is an alternative formula for calculating variance that is often easier to apply than Equation (10.6). Let X be a continuous random variable on $[A, B]$ with probability density $p(x)$ and mean μ. Then

(10.8) $$\operatorname{Var}(X) = \int_A^B x^2p(x)\,dx - \mu^2$$

To see that this formula is equivalent to our previous definition, simply expand the original formula as follows:

$$\begin{aligned}\operatorname{Var}(X) &= \int_A^B (x-\mu)^2p(x)\,dx = \int_A^B (x^2 - 2x\mu + \mu^2)p(x)\,dx \\ &= \int_A^B x^2p(x)\,dx - 2\mu\int_A^B xp(x)\,dx + \mu^2\int_A^B p(x)\,dx\end{aligned}$$

Now, because $p(x)$ is a probability density function, $\int_A^B p(x)\,dx = 1$. Also,

$$\mu = \int_A^B xp(x)\,dx$$

from the definition of expected value, so

$$\mathrm{Var}(X) = \int_A^B x^2 p(x)\,dx - 2\mu^2 + \mu^2$$
$$= \int_A^B x^2 p(x)\,dx - \mu^2$$

as claimed.

EXAMPLE 15 Suppose the amount of time (in hours) that it takes a person to memorize a particular poem is a random variable with probability density function

$$p(x) = \frac{1}{50}x \qquad 0 \le x \le 10$$

Find the average time that it takes to memorize the poem, and compute the variance.

SOLUTION To compute $\mu = E(X)$ we simply evaluate

$$\mu = E(X) = \int_0^{10} x\left(\frac{1}{50}x\right)dx = \frac{1}{50}\int_0^{10} x^2\,dx$$
$$= \frac{1}{50}\frac{x^3}{3}\bigg|_0^{10} = \frac{1}{50}\cdot\frac{1000}{3} = \frac{20}{3}$$

To compute $\mathrm{Var}(X)$ we make use of Equation (10.8). This requires us to first compute

$$\int_0^{10} x^2 p(x)\,dx = \int_0^{10} x^2\left(\frac{1}{50}x\right)dx = \frac{1}{50}\int_0^{10} x^3\,dx$$
$$= \frac{1}{200}x^4\bigg|_0^{10} = 50$$

Then

$$\mathrm{Var}(X) = \int_0^{10} x^2 p(x)\,dx - \mu^2 = 50 - \left(\frac{20}{3}\right)^2 = \frac{50}{9}$$

Section 10.3 EXERCISES

Check whether $p(x)$ in each of Exercises 1–8 is a probability density function, or find the value of c (if any) that makes it so. If it is a probability density function, determine the indicated probability.

1. $p(x) = \begin{cases} \dfrac{c}{\sqrt{x}} & \text{for } 1 \le x \le 4 \\ 0 & \text{otherwise} \end{cases}$

$\Pr(X \le 3/2)$

2. $p(x) = \begin{cases} cx^2 & \text{for } 0 \le x \le 1 \\ 0 & \text{otherwise} \end{cases}$

$\Pr(X > 0.5)$

3. $p(x) = \begin{cases} 3(x-1) & \text{for } 1 \le x \le 2 \\ 0 & \text{otherwise} \end{cases}$

$\Pr(X \le 0.5)$

4. $p(x) = \begin{cases} e^{x-2} & \text{for } x \le 2 \\ 1 & \text{for } x > 2 \end{cases}$

$\Pr(X \ge 3)$

5. $p(x) = \begin{cases} \dfrac{c}{1+x} & \text{for } x \ge 0 \\ 0 & \text{otherwise} \end{cases}$

$\Pr(X \ge 0)$

6. $p(x) = \begin{cases} \dfrac{c}{x} & \text{for } 1 \le x < \infty \\ 0 & \text{otherwise} \end{cases}$

$\Pr(X \le 1)$

7. $p(x) = \begin{cases} 4(1-x^3)/3 & \text{for } 0 \le x \le 1 \\ 0 & \text{otherwise} \end{cases}$

$\Pr(1/4 < X < 1/2)$

8. $p(x) = \begin{cases} ce^{-3x} & \text{for } x \ge 0 \\ 0 & \text{otherwise} \end{cases}$

$\Pr(1 < X < 3)$

Suppose that the density function p of a random variable X is specified as follows:

$$p(x) = \begin{cases} 0 & \text{for } x < 0 \\ e^{-x} & \text{for } x \ge 0 \end{cases}$$

Find each of the probabilities in Exercises 9–17.

9. $\Pr(X \le -3)$
10. $\Pr(X > -2)$
11. $\Pr(X = 2)$
12. $\Pr(X \ge 0)$
13. $\Pr(X < 3)$
14. $\Pr(-1 \le X \le 1)$
15. $\Pr(1 \le X < 2)$
16. $\Pr(1 \le X \le 2)$
17. $\Pr(1 < X \le 2)$

18. A gas station has two pumps, each of which can pump up to 2000 gallons of gas in a week. The total amount of gas pumped at the station in a week is a random variable X (measured in thousands of gallons), whose probability density function is given by

$$p(x) = \begin{cases} x/4 & 0 \le x \le 2 \\ 1 - x/4 & 2 < x \le 4 \\ 0 & \text{elsewhere} \end{cases}$$

(a) Graph $p(x)$

(b) Find the probability that the station pumps between 1600 and 2500 gallons in a given week.

19. The proportion of time per day that all checkout counters at the local supermarket are busy is a random variable X with density function

$$p(x) = \begin{cases} cx(1-x)^2 & 0 \le x \le 1 \\ 0 & \text{elsewhere} \end{cases}$$

(a) Find the value of c that makes $p(x)$ a probability density function.

(b) Find $E(X)$.

Compute the mean, variance, and standard deviation for the probability distributions in Exercises 20–23.

20. $p(x) = 2(x-1), \quad 1 \le x \le 2$
21. $p(x) = 1/4, \quad 0 \le x \le 4$
22. $p(x) = 20x^3(1-x), \quad 0 \le x \le 1$
23. $p(x) = x^4/625, \quad 0 \le x \le 5$

SECTION 10.4

Uniform and Exponential Distributions

In this section we consider two important probability density functions that arise in statistical applications. We begin with the uniform distribution.

DEFINITION 10.6 Suppose that a number is selected "at random" from the interval $A \le X \le B$. The phrase *at random* means here that any value between A and B is equally likely to be chosen. Then we say that the continuous random variable X is **uniformly distributed** over the interval $A \le X \le B$.

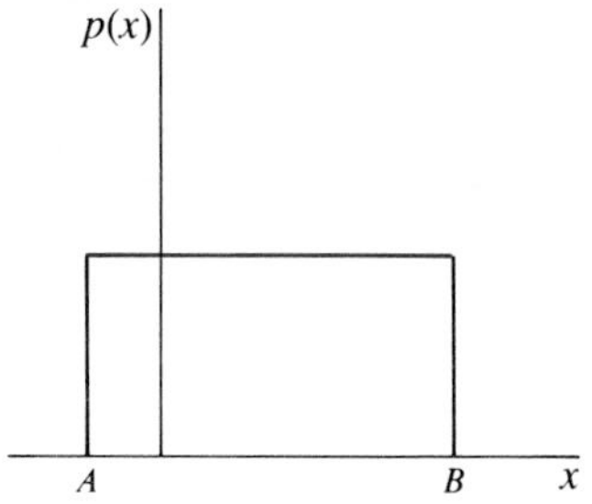

FIGURE 10.13
Uniform distribution on $A \leq x \leq B$

The corresponding probability density function is given by

$$p(x) = \frac{1}{B - A} \qquad A \leq x \leq B$$

and is shown in Figure 10.13. This density function has the property that subintervals of $[A, B]$ that have equal lengths have equal probabilities.

It is a simple matter to verify that p is, in fact, a probability density function. First we observe that

$$p(x) \geq 0 \qquad A \leq x \leq B$$

Second,

$$\begin{aligned}\int_A^B p(x)\,dx &= \int_A^B \frac{1}{B - A}\,dx = \frac{x}{B - A}\bigg|_A^B \\ &= \frac{B}{B - A} - \frac{A}{B - A} \\ &= \frac{B - A}{B - A} = 1\end{aligned}$$

We can now determine the mean, variance, and standard deviation of the uniform distribution. From Definition 10.5,

$$\begin{aligned}\mu = E(X) &= \int_A^B xp(x)\,dx = \int_A^B \frac{x}{B - A}\,dx \\ &= \frac{x^2}{2(B - A)}\bigg|_A^B = \frac{B^2 - A^2}{2(B - A)} = \frac{B + A}{2}\end{aligned}$$

This is what intuition would suggest. If we choose a number at random between A and B then, on average, we will select the midpoint of the interval. Next, from Equation (10.8),

$$\begin{aligned}\operatorname{Var}(X) &= \int_A^B x^2p(x)\,dx - \mu^2 = \int_A^B \frac{x^2}{B - A}\,dx - \left(\frac{A + B}{2}\right)^2 \\ &= \frac{x^3}{3(B - A)}\bigg|_A^B - \left(\frac{A + B}{2}\right)^2 = \frac{B^3 - A^3}{3(B - A)} - \left(\frac{A + B}{2}\right)^2 \\ &= \frac{B^2 + AB + A^2}{3} - \frac{(B^2 + 2AB + A^2)}{4} \\ &= \frac{B^2 - 2AB + A^2}{12} = \frac{(B - A)^2}{12}\end{aligned}$$

It now follows that

$$\begin{aligned}\sigma &= \sqrt{\operatorname{Var}(X)} = \sqrt{\frac{(B - A)^2}{12}} \\ &= \frac{B - A}{\sqrt{12}}\end{aligned}$$

In summary:

> For the uniform distribution $p(x) = \dfrac{1}{B - A}$, where $A \leq X \leq B$,
>
> $$\mu = \frac{A + B}{2}$$
>
> $$\sigma = \frac{B - A}{\sqrt{12}}$$

EXAMPLE 16 A number between 5 and 11 is picked at random so that any value in the interval $5 \leq X \leq 11$ is equally likely to be chosen.

(a) What is the probability that this number will lie between 8 and 10?

(b) What is the mean, μ?

(c) What is the standard deviation?

SOLUTION

(a) We know that

$$p(x) = \frac{1}{11 - 5} = \frac{1}{6} \qquad 5 \leq X \leq 11$$

Then

$$\Pr(8 \leq x \leq 10) = \int_8^{10} p(x)\, dx = \int_8^{10} \frac{1}{6}\, dx = \frac{x}{6}\bigg|_8^{10}$$

$$= \frac{10}{6} - \frac{8}{6} = \frac{1}{3}$$

(b)

$$\mu = \frac{A + B}{2} = \frac{5 + 11}{2} = 8$$

(c)

$$\sigma = \frac{B - A}{\sqrt{12}} = \frac{11 - 5}{\sqrt{12}} = \frac{6}{\sqrt{12}} = \frac{\sqrt{6}}{\sqrt{2}} = \sqrt{3}$$

SPECIAL PROBLEM FOR STUDENTS WHO KNOW PROGRAMMING These days, a good many students have access to either a personal or a university computer on which some form of the BASIC computer language is available.

The BASIC language has a built-in function "RND" that produces on call a so-called pseudorandom number x, in decimal form, with $0 \leq x < 1$. It is difficult to define *pseudorandom* precisely, but for our purposes we can say that, subject to certain statistical tests, the numbers produced appear to be uniformly distributed between 0 and 1. In particular, if we use the "RND" function successively n times and then compute the sample mean $\bar{x}$ and the sample standard deviation s, their values will approach 0.5 and $1/\sqrt{12} \approx 0.29$, respectively, as n grows large. (See μ and σ in the summary that precedes Example 16.)

A slight modification can pseudorandomly produce an *integer* uniformly distributed between 1 and K. For example, let $K = 6$. We wish to randomly select an integer between 1 and 6 inclusive. We proceed as follows:

Step 1 Use the "RND" function to select x, for $0 \le x < 1$.

Step 2 Set $Z = 6x + 1$.

Step 3 Let $W = [Z]$, where $[Z]$ denotes the largest integer that is $\le Z$

Then W is the desired random integer between 1 and 6.

For example, if "RND" produces $x = 0.38147$, then $Z = 6(0.38147) + 1 = 3.2888$, so the greatest integer of Z, $W = [Z]$ is 3. Similarly, if "RND" produces $x = 0.56243$, then $Z = 6(0.56243) + 1 = 4.3746$, so $W = [Z] = 4$.

If we now wish to simulate a simple coin-tossing experiment with the result either a head or a tail, we can simply modify the foregoing example to randomly pick an integer from the possibilities: 1, standing for a head H; or 2, standing for a tail T. To do this, all we need do is modify step 2 above by changing $Z = 6x + 1$ to $Z = 2x + 1$. We can now carry out some experiments.

PRACTICE EXERCISE

(a) Simulate the tossing of 5 coins simultaneously and record the number of heads obtained. Your answer must be 0H, 1H, 2H, 3H, 4H, or 5H. (You can easily carry out this simulation by calling the random-number generator 5 times. *Caution*: be sure that you read the instructions for using "RND" carefully so that you obtain a new random number each time you call for one.)

(b) Thinking of part (a) as the basic experiment, now repeat part (a) in its entirety 1000 times, keeping a running tally of the number of times the result was 0H, 1H, 2H, 3H, 4H, and 5H. (Of course, this part involves a looping procedure; the BASIC language FOR-NEXT construct could be useful.) Please present your results in a table.

We tried this experiment and obtained the following results:

0H	1H	2H	3H	4H	5H
35	150	304	311	171	29

Note that the sum crosswise is 1000, because we performed the experiment 1000 times. If we now divide the tabular values by 1000, statisticians know that we should obtain values close to the theoretical probabilities of the so-called **binomial distribution**, which is described in almost every statistics book.

The theoretical values are

0H	1H	2H
$\frac{1}{32} \approx 0.031$	$\frac{5}{32} \approx 0.156$	$\frac{10}{32} \approx 0.313$
3H	**4H**	**5H**
$\frac{10}{32} \approx 0.313$	$\frac{5}{32} \approx 0.156$	$\frac{1}{32} \approx 0.031$

and you can see that our experimental results are in fairly good agreement.

How well did the experiment you carried out agree?

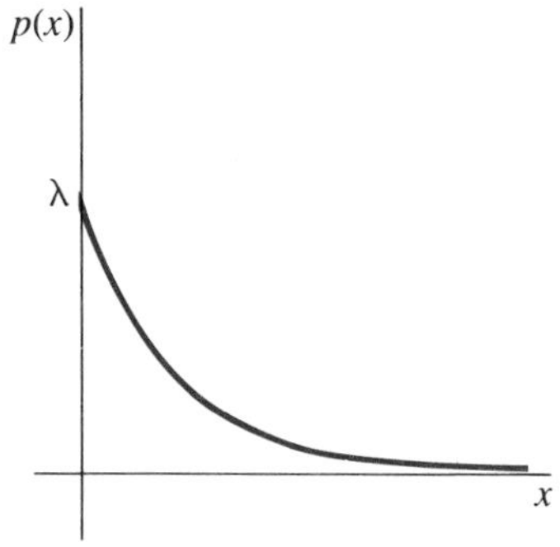

FIGURE 10.14
Exponential distribution function, $p(x) = \lambda e^{-\lambda x}$

Next we turn our attention to another important distribution function, the exponential density.

DEFINITION 10.7 Let the random variable X assume values on $0 \le x < \infty$. The density,

$$p(x) = \lambda e^{-\lambda x} \qquad x \ge 0$$

where λ is a positive constant, is called an **exponential distribution function.** Its graph is shown in Figure 10.14. ⊟

As before, we can check to see that p satisfies the requirements of a density.

(10.9)
$$p(x) \ge 0 \qquad 0 \le x < \infty$$

and

(10.10)
$$\int_0^\infty p(x)\,dx = \int_0^\infty \lambda e^{-\lambda x}\,dx = \lim_{b\to\infty} \int_0^b \lambda e^{-\lambda x}\,dx$$
$$= \lim_{b\to\infty} \left(-e^{-\lambda x}\Big|_0^b \right) = \lim_{b\to\infty} (-e^{-\lambda b} + 1) = 0 - (-1) = 1$$

Next let us compute the mean and standard deviation of X.

$$\mu = \int_0^\infty xp(x)\,dx = \int_0^\infty x(\lambda e^{-\lambda x})\,dx = \lim_{b\to\infty} \lambda \int_0^b xe^{-\lambda x}\,dx$$
$$= \lim_{b\to\infty} \left(-xe^{-\lambda x} - \frac{e^{-\lambda x}}{\lambda} \right)\Bigg|_0^b$$
$$= \lim_{b\to\infty} \left[\left(-be^{-\lambda b} - \frac{e^{-\lambda b}}{\lambda} \right) - \left(0 - \frac{1}{\lambda} \right) \right] = 0 + \frac{1}{\lambda} = \frac{1}{\lambda}$$

We have made use of the integration-by-parts formula to evaluate the integral.

In order to compute the variance, we first integrate by parts, letting $u = x^2$ and $dv = \lambda e^{-\lambda x}\,dx$.

$$\int_0^\infty x^2 p(x)\,dx = \lim_{b\to\infty} \int_0^b x^2 \lambda e^{-\lambda x}\,dx$$
$$= \lim_{b\to\infty} \left[x^2(-e^{-\lambda x})\Big|_0^b - \int_0^b 2x(-e^{-\lambda x})\,dx \right]$$
$$= \lim_{b\to\infty} \left[-b^2 e^{-\lambda b} + 0 + 2\int_0^b xe^{-\lambda x}\,dx \right]$$
$$= \lim_{b\to\infty} \left[-b^2 e^{-\lambda b} + \frac{2}{\lambda} \int_0^b x\lambda e^{-\lambda x}\,dx \right]$$

Now, as $b \to \infty$, $b^2 e^{-\lambda b} \to 0$. Moreover, from our computation of the expected value μ, we know that

$$\lim_{b \to \infty} \int_0^b x(\lambda e^{-\lambda x})\, dx = \frac{1}{\lambda}$$

Thus

$$\int_0^\infty x^2 p(x)\, dx = \frac{2}{\lambda} \cdot \frac{1}{\lambda} = \frac{2}{\lambda^2}$$

and, from Equation (10.8),

$$\text{Var}(X) = \int_0^\infty x^2 p(x)\, dx - \mu^2 = \frac{2}{\lambda^2} - \left(\frac{1}{\lambda}\right)^2 = \frac{2}{\lambda^2} - \frac{1}{\lambda^2} = \frac{1}{\lambda^2}$$

Finally,

$$\sigma = \sqrt{\text{Var}(X)} = \sqrt{\frac{1}{\lambda^2}} = \frac{1}{\lambda}$$

Thus we have

> For the exponential distribution $p(x) = \lambda e^{-\lambda x}$, where $0 \le x < \infty$,
>
> $$\mu = \frac{1}{\lambda}$$
>
> $$\sigma = \frac{1}{\lambda}$$

The exponential distribution has numerous applications. It is associated with problems involving waiting times for telephone calls and lines at the supermarket checkout counter. It also arises in the study of the time-to-failure of electrical components and in the study of the time required to service a machine. Actually, we encountered the exponential distribution in Example 13, where we had $\lambda = 0.12$. You may want to go back and review that example.

EXAMPLE 17 Suppose X is a random variable that represents the duration of a local telephone call. Measurements have shown that X is an exponential random variable. The average telephone conversation lasts 3 minutes.

(a) Find the probability density function for X.

(b) Find the probability that a telephone conversation will last between 6 and 8 minutes.

SOLUTION

(a) The exponential density has the form

$$p(x) = \lambda e^{-\lambda x} \qquad x \ge 0$$

It is determined completely by the parameter λ. We know that the mean value μ for the exponential density is

$$\mu = \frac{1}{\lambda}$$

Because we are told that the average telephone call lasts 3 minutes, we must have

$$\frac{1}{\lambda} = 3$$

or

$$\lambda = \frac{1}{3}$$

$$p(x) = \frac{1}{3} e^{-x/3} \qquad x \geqslant 0$$

(b) $$\Pr(6 \leq X \leq 8) = \int_6^8 p(x)\,dx = \int_6^8 \frac{1}{3} e^{-x/3}\,dx$$

$$= -e^{-x/3}\Big|_6^8 = -e^{-8/3} + e^{-2} \approx -0.0695 + 0.1353$$

$$\approx 0.066$$

Thus roughly 6.6% of the telephone calls last between 6 and 8 minutes.

EXAMPLE 18 In a certain bank there are several tellers to process transactions. Customers wait in a line until it is their turn, at which time they proceed to the next available window. The length of time that a customer waits is an exponentially distributed random variable. The average wait is 5 minutes.

(a) Find the probability density function.

(b) What is the probability that an individual will be served within 2 minutes?

(c) What is the probability that a customer will have to wait more than 10 minutes?

SOLUTION

(a) The average wait is 5 minutes, so

$$\frac{1}{\lambda} = 5 \qquad \lambda = 0.2$$

Thus

$$p(x) = 0.2e^{-0.2x} \qquad x \geq 0$$

(b) The probability that an individual will be served within 2 minutes is given by the value of $\Pr(0 \leq X \leq 2)$.

$$\Pr(0 \leq X \leq 2) = \int_0^2 0.2e^{-0.2x}\,dx = -e^{-0.2x}\Big|_0^2 = -e^{-0.4} + 1$$

$$\approx 0.33$$

(c) $$\Pr(10 \leq X < \infty) = \int_{10}^{\infty} 0.2e^{-0.2x}\,dx = \lim_{b\to\infty} \int_{10}^{b} 0.2e^{-0.2x}\,dx$$

$$= \lim_{b\to\infty}\left(-e^{-0.2x}\Big|_{10}^{b}\right) = \lim_{b\to\infty}\left(-e^{-0.2b} + e^{-2}\right)$$

$$\approx 0 + 0.135 = 0.135$$

Section 10.4 EXERCISES

Find the mean, variance, and standard deviation for the following density functions.

1. $p(x) = 4e^{-4x}, \quad x \geq 0$
2. $p(x) = 1/3, \quad 1 \leq x \leq 4$
3. $p(x) = 1/6, \quad 0 \leq x \leq 6$
4. $p(x) = 1/4, \quad 2 \leq x \leq 6$
5. $p(x) = 5e^{-5x}, \quad x \geq 0$
6. $p(x) = \frac{1}{3}e^{-x/3}, \quad x \geq 0$
7. $p(x) = \frac{1}{2}e^{-x/2}, \quad x \geq 0$
8. $p(x) = 7e^{-7x}, \quad x \geq 0$
9. $p(x) = 1/8, \quad 1 \leq x \leq 9$
10. $p(x) = 1/5, \quad 0 \leq x \leq 5$
11. Suppose that a calculus test is given and that the number of minutes required by any particular student to complete the test has an exponential distribution for which the mean is 70. Suppose the test begins at 9:00 A.M. Determine the probability that a student will finish the examination before 9:30 A.M.
12. A carpenter has found through experience that the low bid for a job (excluding his own bid) is a random variable that is uniformly distributed over the interval $(3C/4, 2C)$, where C is the carpenter's cost estimate (no profit or loss) of the job. What should the carpenter bid to maximize the expected profit?
13. If a parachutist lands at a random point on a line between markers A and B, find the probability that he is closer to B than to A. Find the probability that his distance to A is more than four times his distance to B.
14. The number of reams of paper used by a downtown business in one day is modeled by an exponential distribution with $\lambda = 4$. Find the probability that the business will use more than 4 reams in a given day.
15. The lifetime (in hours) of a cartridge for a turntable is an exponential random variable with $\lambda = 0.005$. Find the probability that the cartridge will last for at least 250 hr.
16. The time (in hours) it takes to apply for a driver's license has an exponential distribution with $\lambda = \frac{1}{2}$. Appointments are scheduled at $\frac{1}{4}$-hour intervals, beginning at 8:00 A.M., and applicants arrive on time. If you have an 8:15 A.M. appointment, what is the probability that you will have to wait?
17. The lifetime of a TV picture tube is found to be an exponential random variable with an expected value of 5 years. The manufacturer sells the tube for \$100 but will give a complete refund if the tube fails within 2 years. The revenue the manufacturer receives on each tube is thus a discrete random variable Y with values 100 and 0. Determine the expected revenue per tube.
18. At rush hour the time between the arrivals of successive buses is an exponential random variable with expected value 10 min. What is the probability that a bus will arrive within 8 min?
19. What is the probability that the bus in Exercise 18 will not arrive for 14 min?

In Exercises 20–23 assume that the time required to serve a customer at a supermarket is an exponential random variable with mean 4 min.

20. Find the probability that a customer will be served in less than 3 min.
21. Find the probability that a customer will be served within 1 min.
22. Find the probability that a customer will have to wait at least 8 min.
23. Find the probability that a customer will have to wait between 5 and 10 min.

SECTION 10.5

The Normal Distribution

In this section we study what is probably the most widely used distribution in statistics: the normal distribution. This is the familiar bell-shaped curve, used to analyze course grades, SAT scores, and so on.

DEFINITION 10.8 A continuous random variable X is said to be **normally distributed** with mean μ and standard deviation σ if its probability density function is given by

$$P_{\mu,\sigma}(x) = \frac{1}{\sigma\sqrt{2\pi}} e^{-\frac{1}{2}\left(\frac{x-\mu}{\sigma}\right)^2} \qquad -\infty < x < \infty$$

Figure 10.15 is a graph of the function $P_{\mu,\sigma}(x)$ with $\mu = 1$ and various values of σ. Note that each density function is centered symmetrically about the straight line $x = \mu$ and that a larger value of σ corresponds to a wider "spread" whereas a smaller value of σ corresponds to a narrower "spread" about the mean.

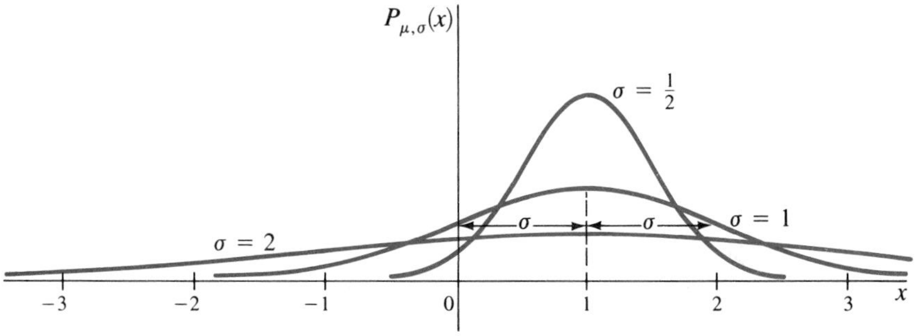

FIGURE 10.15
Probability density function of the normal distribution

In our definition, we have implied three important facts about the normal distribution. They are as follows:

1 $P_{\mu,\sigma}(x)$ is a probability density function for each value of μ, and $\sigma > 0$. Specifically,

$$P_{\mu,\sigma}(x) \geq 0 \qquad -\infty < x < \infty$$

which is easy to see, and

$$\int_{-\infty}^{\infty} P_{\mu,\sigma}(x)\,dx = \frac{1}{\sigma\sqrt{2\pi}} \int_{-\infty}^{\infty} e^{-\frac{1}{2}\left(\frac{x-\mu}{\sigma}\right)^2} dx = 1$$

2 The random variable X with probability density $P_{\mu,\sigma}(x)$ actually has mean μ:

$$\mu = E(x) = \int_{-\infty}^{\infty} xP_{\mu,\sigma}(x)\,dx = \frac{1}{\sigma\sqrt{2\pi}} \int_{-\infty}^{\infty} xe^{-\frac{1}{2}\left(\frac{x-\mu}{\sigma}\right)^2} dx$$

3 The variance of X is σ^2, so the standard deviation is σ:

$$\sigma^2 = \text{Var}(X) = \int_{-\infty}^{\infty} (x-\mu)^2 P_{\mu,\sigma}(x)\,dx$$

$$= \frac{1}{\sigma\sqrt{2\pi}} \int_{-\infty}^{\infty} (x-\mu)^2 e^{-\frac{1}{2}\left(\frac{x-\mu}{\sigma}\right)^2} dx$$

A careful verification of these three facts is beyond the scope of this book. We will simply accept them.

In order to motivate the discussion to follow, let us consider an example.

EXAMPLE 19 The heights of adult American males are normally distributed with mean 70 inches and standard deviation 4.3 inches. Formulate an integral whose value represents the probability that a male is over 72 inches (6 feet) in height.

SOLUTION We are given $\mu = 70$ and $\sigma = 4.3$. The corresponding normal distribution is

$$P_{\mu,\sigma}(x) = P_{70,4.3}(x) = \frac{1}{4.3\sqrt{2\pi}} e^{-\frac{1}{2}\left(\frac{x-70}{4.3}\right)^2}$$

The probability that a male is over 72 inches is represented symbolically by

$$\Pr(72 \le X < \infty)$$

Thus

(10.11) $$\Pr(72 \le X < \infty) = \int_{72}^{\infty} \frac{1}{4.3\sqrt{2\pi}} e^{-\frac{1}{2}\left(\frac{x-70}{4.3}\right)^2} dx.$$

The previous example raises two important questions. First, how were the numbers $\mu = 70$ and $\sigma = 4.3$ arrived at? Obviously it is impractical to record and tabulate the height of all adult males in the United States; there are too many of them, and it would take too long. Instead we record, at random, the heights of just a few cases, say n of them, where n might be 50, or 100, or any number that seems appropriate. This is called taking a sample, and n is referred to as the **sample size**. Needless to say, a certain amount of care must be exercised in selecting a sample; it would be bad practice only to measure the heights of males living on the West Coast, for example. We want to use our relatively small sample to deduce information about *all* males.

Suppose that our sample consists of the n numbers

$$x_1, x_2, \ldots, x_n.$$

We can then compute the sample mean $\bar{x}$ and the sample standard **deviation** introduced in Section 10.2.

(10.12) $$\bar{x} = \frac{x_1 + x_2 + \cdots + x_n}{n}$$

(10.13) $$s = \sqrt{\frac{(x_1 - x)^2 + (x_2 - x)^2 + \cdots + (x_n - x)^2}{n - 1}}$$

Statistical theory tells us that, if we have a normal distribution with mean μ and standard deviation σ, the sample mean $\bar{x}$ is an approximation to μ and the sample standard deviation s is an approximation to σ. The larger the sample size n, the better the approximations $\bar{x}$ and s.

Even more, suppose we take several samples, each of size n, and for each one compute the sample mean $\bar{x}$. Then taking the mean of these $\bar{x}$'s gives a much better approximation to μ than a single $\bar{x}$. Similarly, taking the mean of the sample standard deviations gives a better approximation to σ than a single

s. In any event, we can find μ and σ reasonably accurately without too much difficulty.

The second question that arises in the solution of Example 19 is how to evaluate the definite integral, Equation (10.11):

$$\Pr(72 \leq X < \infty) = \int_{72}^{\infty} \frac{1}{4.3\sqrt{2\pi}} e^{-\frac{1}{2}\left(\frac{x-70}{4.3}\right)^2} dx$$

As with the integral of $e^{-x^2/2}$, which was discussed in Section 7.6, the antiderivative

$$\int \frac{1}{4.3\sqrt{2\pi}} e^{-\frac{1}{2}\left(\frac{x-70}{4.3}\right)^2} dx$$

does not exist, in the sense that there is no combination of elementary functions that can be written down whose derivative is the integrand.

Despite this, the definite integral does correspond to the shaded area under the curve $P_{70,4.3}(x)$ beyond $x = 72$, as shown in Figure 10.16. How can we estimate its value? One approach is to use one of the numerical integration schemes we developed—the trapezoidal rule, or Simpson's rule, for instance. This will work, though it is a bit tedious to carry out. Actually, because of the importance of integrals of the general form

$$\int_a^b \frac{1}{\sigma\sqrt{2\pi}} e^{-\frac{1}{2}\left(\frac{x-\mu}{\sigma}\right)^2} dx$$

statisticians have prepared tables containing all the information necessary to evaluate such integrals. In order to explain how to use such a table and to explain why it works, we must digress and discuss the standard normal distribution.

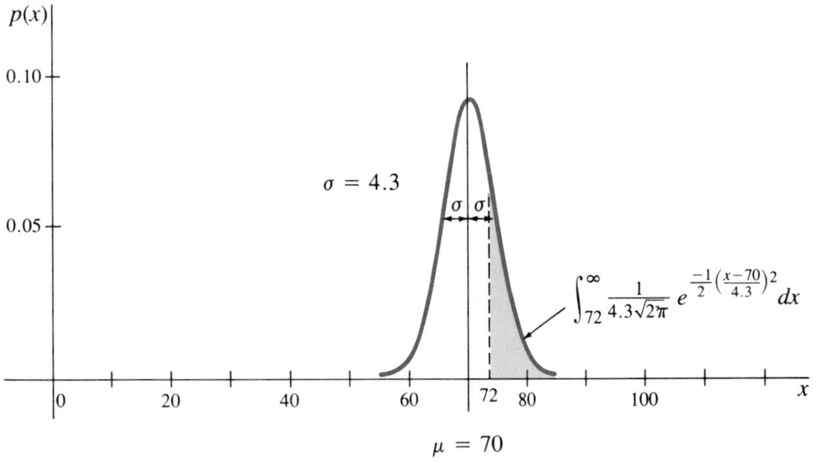

FIGURE 10.16

STANDARD NORMAL DISTRIBUTION The standard normal distribution is defined to be the special case of the normal distribution for which $\mu = 0$ and $\sigma = 1$. Specifically, it is the normal distribution whose probability

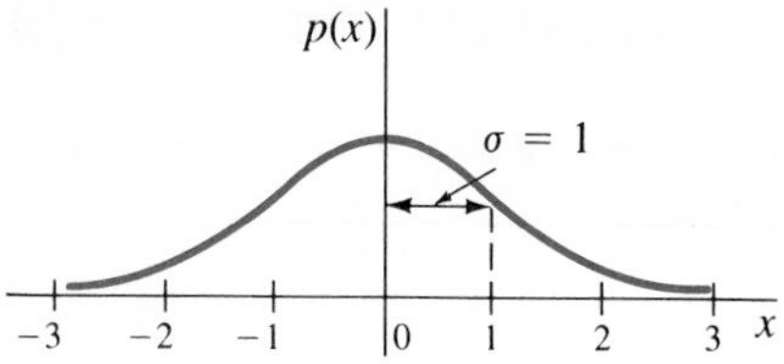

$\mu = 0$

FIGURE 10.17
Standard normal distribution

density function is given by

$$p(x) = \frac{1}{\sqrt{2\pi}} e^{-x^2/2} \qquad -\infty < x < \infty$$

The graph of the standard normal distribution is shown in Figure 10.17. Using the curve-sketching techniques of Chapter 3, it can be shown that

1. $p(x) > 0$ for $-\infty < x < \infty$.
2. The graph is symmetric with respect to the y-axis.
3. There is an absolute maximum at $x = 0$.
4. The curve is increasing on $(-\infty, 0)$.
5. The curve is decreasing on $(0, \infty)$.
6. There are inflection points at $x = 1$ and $x = -1$.
7. The curve is concave down on $(-1, 1)$.
8. The curve is concave up on $(-\infty, -1)$ and $(1, \infty)$.
9. The area under the curve is 1.
10. The x-axis is a horizontal asymptote.

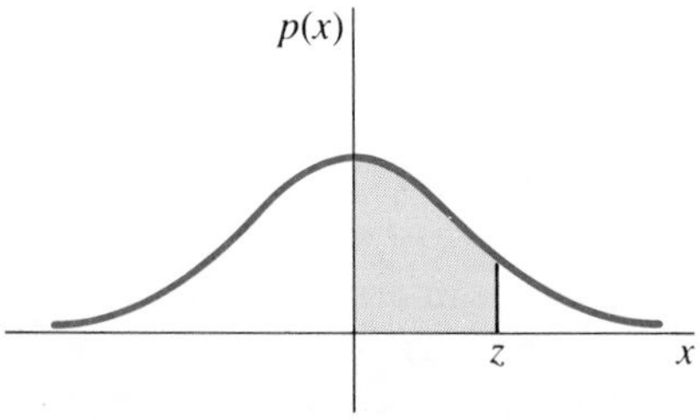

FIGURE 10.18

$$p(x) = \frac{1}{\sqrt{2\pi}} e^{-\frac{1}{2}x^2}$$

We have noted that the total area under the curve

$$p(x) = \frac{1}{\sqrt{2\pi}} e^{-x^2/2}$$

is 1. We want to be able to find the area under portions of this curve. The key to doing this is the table reproduced in Table 10.19. How does the table work? For $z \geq 0$ it gives the area under the curve $p(x)$ from 0 to z; that is,

$$\int_0^z p(x)\, dx$$

which is the shaded area in Figure 10.18. For example, to find $\int_0^{1.18} p(x)\, dx$ we must find the entry in the table corresponding to $z = 1.18$. This is easy. We look in the column on the far left until we come to the z-value 1.1. The entry where the row determined by 1.1 crosses the column headed by 0.08 is 0.3810, so

$$\int_0^{1.18} p(x)\, dx = 0.3810$$

To check your understanding of how to read the table, verify that

$$\int_0^{2.33} p(x)\, dx = 0.4901$$

Once we know how to use the table, it is easy to find the area under any piece of the curve—that is, to compute

$$\int_a^b p(x)\, dx$$

for all choices of real numbers $a < b$, even when $a = -\infty$ or $b = +\infty$. Let us illustrate this assertion with a few specific examples.

TABLE 10.19
Standard normal distribution

z	0.00	0.01	0.02	0.03	0.04	0.05	0.06	0.07	0.08	0.09
0.0	0.0000	0.0040	0.0080	0.0120	0.0160	0.0199	0.0239	0.0279	0.0319	0.0359
0.1	0.0398	0.0438	0.0478	0.0517	0.0557	0.0596	0.0636	0.0675	0.0714	0.0753
0.2	0.0793	0.0832	0.0871	0.0910	0.0948	0.0987	0.1026	0.1064	0.1103	0.1141
0.3	0.1179	0.1217	0.1255	0.1293	0.1331	0.1368	0.1406	0.1443	0.1480	0.1517
0.4	0.1554	0.1591	0.1628	0.1664	0.1700	0.1736	0.1772	0.1808	0.1844	0.1879
0.5	0.1915	0.1950	0.1985	0.2019	0.2054	0.2088	0.2123	0.2157	0.2190	0.2224
0.6	0.2257	0.2291	0.2324	0.2357	0.2389	0.2422	0.2454	0.2486	0.2517	0.2549
0.7	0.2580	0.2611	0.2642	0.2673	0.2704	0.2734	0.2764	0.2794	0.2823	0.2852
0.8	0.2881	0.2910	0.2939	0.2967	0.2995	0.3023	0.3051	0.3078	0.3106	0.3133
0.9	0.3159	0.3186	0.3212	0.3238	0.3264	0.3289	0.3315	0.3340	0.3365	0.3389
1.0	0.3413	0.3438	0.3461	0.3485	0.3508	0.3531	0.3554	0.3577	0.3599	0.3621
1.1	0.3643	0.3665	0.3686	0.3708	0.3729	0.3749	0.3770	0.3790	0.3810	0.3830
1.2	0.3849	0.3869	0.3888	0.3907	0.3925	0.3944	0.3962	0.3980	0.3997	0.4015
1.3	0.4032	0.4049	0.4066	0.4082	0.4099	0.4115	0.4131	0.4147	0.4162	0.4177
1.4	0.4192	0.4207	0.4222	0.4236	0.4251	0.4265	0.4279	0.4292	0.4306	0.4319
1.5	0.4332	0.4345	0.4357	0.4370	0.4382	0.4394	0.4406	0.4418	0.4429	0.4441
1.6	0.4452	0.4463	0.4474	0.4484	0.4495	0.4505	0.4515	0.4525	0.4535	0.4545
1.7	0.4554	0.4564	0.4573	0.4582	0.4591	0.4599	0.4608	0.4616	0.4625	0.4633
1.8	0.4641	0.4649	0.4656	0.4664	0.4671	0.4678	0.4686	0.4693	0.4699	0.4706
1.9	0.4713	0.4719	0.4726	0.4732	0.4738	0.4744	0.4750	0.4756	0.4761	0.4767
2.0	0.4772	0.4778	0.4783	0.4788	0.4793	0.4798	0.4803	0.4808	0.4812	0.4817
2.1	0.4821	0.4826	0.4830	0.4834	0.4838	0.4842	0.4846	0.4850	0.4854	0.4857
2.2	0.4861	0.4864	0.4868	0.4871	0.4875	0.4878	0.4881	0.4884	0.4887	0.4890
2.3	0.4893	0.4896	0.4898	0.4901	0.4904	0.4906	0.4909	0.4911	0.4913	0.4916
2.4	0.4918	0.4920	0.4922	0.4925	0.4927	0.4929	0.4931	0.4932	0.4934	0.4936
2.5	0.4938	0.4940	0.4941	0.4943	0.4945	0.4946	0.4948	0.4949	0.4951	0.4952
2.6	0.4953	0.4955	0.4956	0.4957	0.4959	0.4960	0.4961	0.4962	0.4963	0.4946
2.7	0.4965	0.4966	0.4967	0.4968	0.4969	0.4970	0.4971	0.4972	0.4973	0.4974
2.8	0.4974	0.4975	0.4976	0.4977	0.4977	0.4978	0.4979	0.4979	0.4980	0.4981
2.9	0.4981	0.4982	0.4982	0.4983	0.4984	0.4984	0.4985	0.4985	0.4986	0.4986
3.0	0.4987	0.4987	0.4987	0.4988	0.4988	0.4989	0.4989	0.4989	0.4990	0.4990

EXAMPLE 20 Compute the following areas:

(a) $\int_{-1.96}^{0} p(x)\,dx$ (b) $\int_{0.83}^{2.10} p(x)\,dx$

(c) $\int_{-1.23}^{1.57} p(x)\,dx$ (d) $\int_{-2.07}^{-1.74} p(x)\,dx$

(e) $\int_{2.33}^{\infty} p(x)\,dx$ (f) $\int_{-1.64}^{\infty} p(x)\,dx$

(g) $\displaystyle\int_{-\infty}^{2.58} p(x)\,dx$

SOLUTION In trying to find each integral, it is helpful to make a rough sketch of the area involved. This area will be shaded in each of our sketches.

(a) See Figure 10.19. Because the curve is symmetric with respect to the y-axis, it is clear that the area from -1.96 to 0 is the same as the area from 0 to 1.96. That is,

$$\int_{-1.96}^{0} p(x)\,dx = \int_{0}^{1.96} p(x)\,dx$$

The table tells us that the integral on the right equals 0.4750.

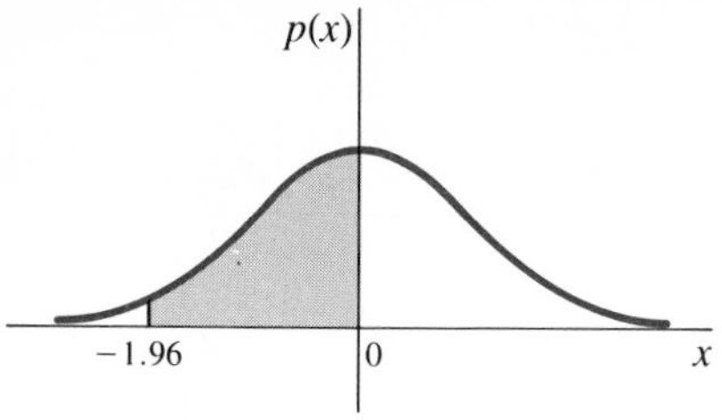

FIGURE 10.19

(b) See Figure 10.20. The area under the curve between 0.83 and 2.10 is

$$\begin{aligned}\int_{0.83}^{2.10} p(x)\,dx &= \int_{0}^{2.10} p(x)\,dx - \int_{0}^{0.83} p(x)\,dx\\ &= 0.4821 - 0.2967 \qquad \text{(from the table)}\\ &= 0.1854\end{aligned}$$

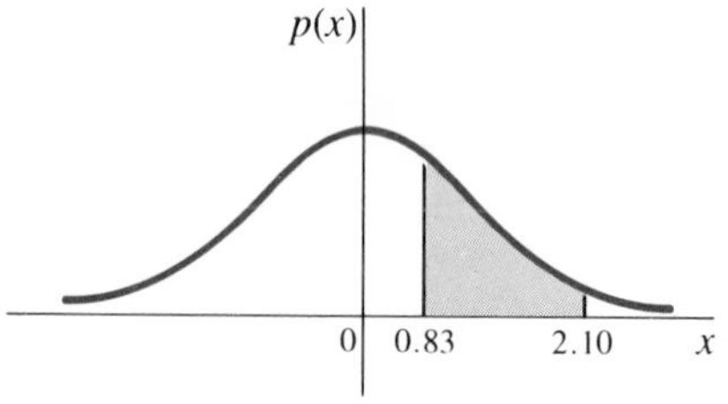

FIGURE 10.20

(c) See Figure 10.21. The area under the curve between -1.23 and 1.57 is

$$\begin{aligned}\int_{-1.23}^{1.57} p(x)\,dx &= \int_{-1.23}^{0} p(x)\,dx + \int_{0}^{1.57} p(x)\,dx\\ &= \int_{0}^{1.23} p(x)\,dx + \int_{0}^{1.57} p(x)\,dx\\ &= 0.3907 + 0.4418\\ &= 0.8325\end{aligned}$$

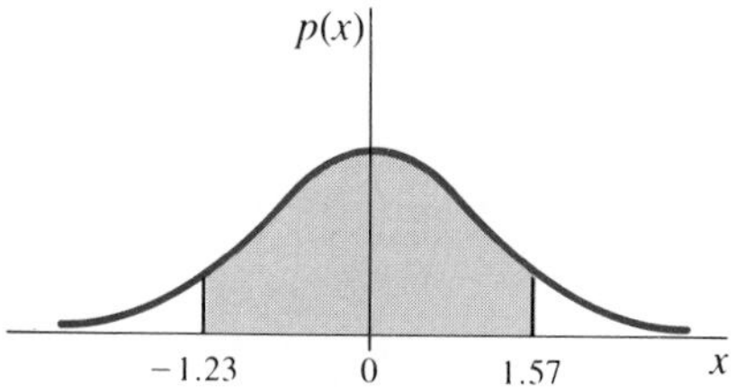

FIGURE 10.21

(d) See Figure 10.22. By symmetry, the area between -2.07 and -1.74 is

$$\begin{aligned}\int_{-2.07}^{-1.74} p(x)\,dx &= \int_{1.74}^{2.07} p(x)\,dx\\ &= \int_{0}^{2.07} p(x)\,dx - \int_{0}^{1.74} p(x)\,dx\\ &= 0.4808 - 0.4591\\ &= 0.0217\end{aligned}$$

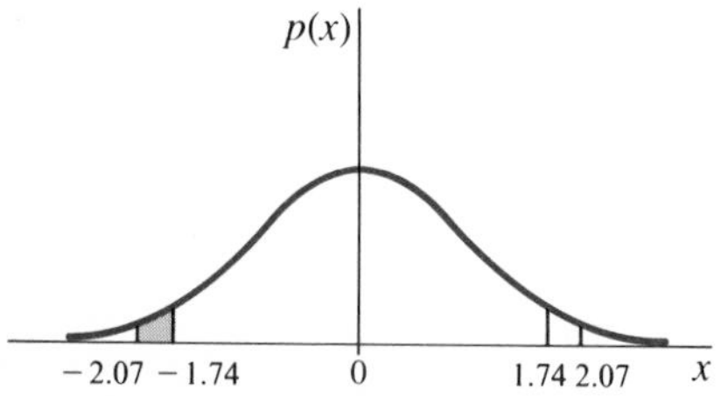

FIGURE 10.22

(e) See Figure 10.23. Because the total area under the curve is 1 and the curve is symmetric with respect to the vertical line $x = 0$, it is clear that

$$\int_{0}^{\infty} p(x)\,dx = \int_{-\infty}^{0} p(x)\,dx = \tfrac{1}{2} = 0.5000$$

Therefore the area under the curve to the right of $x = 2.33$ is

$$\begin{aligned}\int_{2.33}^{\infty} p(x)\,dx &= \int_{0}^{\infty} p(x)\,dx - \int_{0}^{2.33} p(x)\,dx\\ &= 0.5000 - 0.4901\\ &= 0.0099\end{aligned}$$

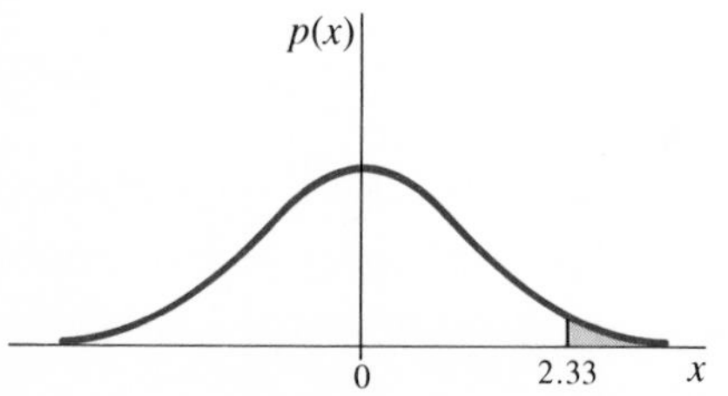

FIGURE 10.23

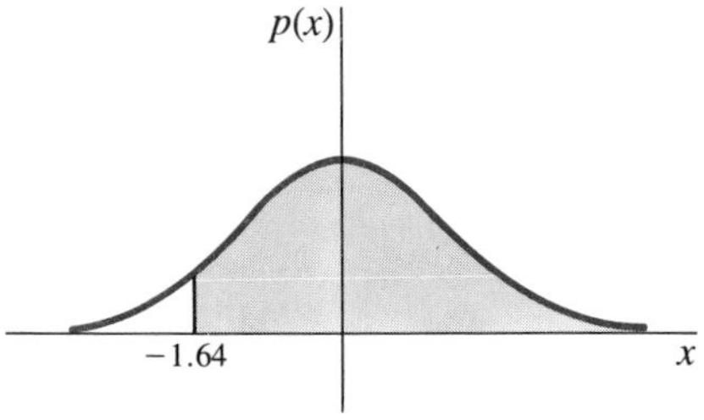

FIGURE 10.24

(f) See Figure 10.24. The area under the curve to the right of -1.64 is

$$\begin{aligned}\int_{-1.64}^{\infty} p(x)\,dx &= \int_{-1.64}^{0} p(x)\,dx + \int_{0}^{\infty} p(x)\,dx \\ &= \int_{0}^{1.64} p(x)\,dx + \int_{0}^{\infty} p(x)\,dx \\ &= 0.4495 + 0.5000 \\ &= 0.9495\end{aligned}$$

(g) This final situation is shown in Figure 10.25 and the area in question is

$$\begin{aligned}\int_{-\infty}^{2.58} p(x)\,dx &= \int_{-\infty}^{0} p(x)\,dx + \int_{0}^{2.58} p(x)\,dx \\ &= 0.5000 + 0.4951 \\ &= 0.9951\end{aligned}$$

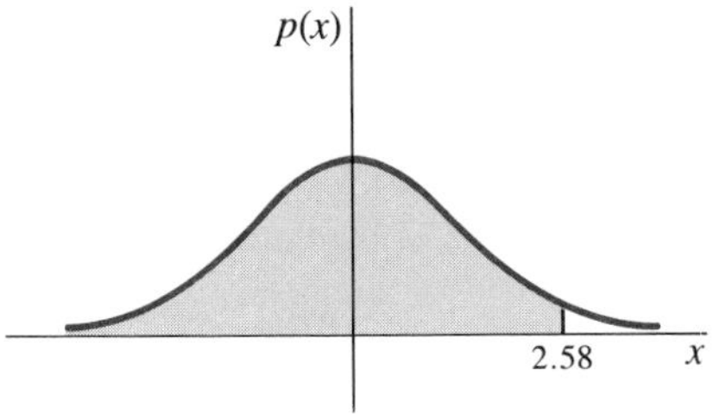

FIGURE 10.25

By using the techniques illustrated in Example 20, we can compute all areas $\int_a^b p(x)\,dx$, where $a < b$ (even when $a = -\infty$ or $b = +\infty$).

It should be noted that this statement is not entirely accurate. Because Table 10.19 goes only from $z = 0$ to $z = 3.09$, we will not be able to evaluate $\int_a^b p(x)\,dx$ when a or b or both are in $(-\infty, -3.09)$ or $(3.09, \infty)$. However, in practical terms this does not matter; the area under the curve between -3.09 and 3.09 (see Figure 10.26) is

$$\begin{aligned}\int_{-3.09}^{3.09} p(x)\,dx &= \int_{-3.09}^{0} p(x)\,dx + \int_{0}^{3.09} p(x)\,dx \\ &= 0.4990 + 0.4990 = 0.9980\end{aligned}$$

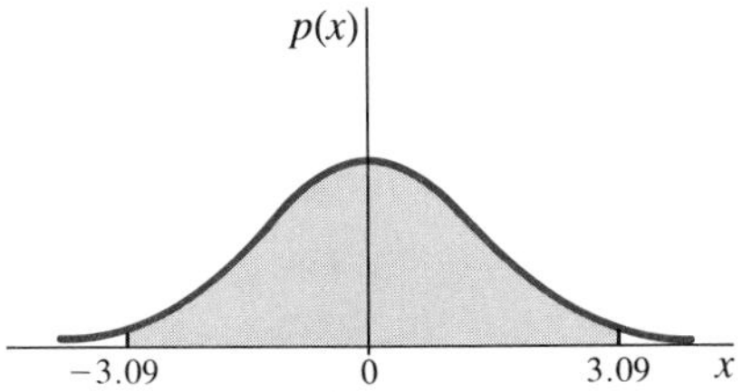

FIGURE 10.26

This is so close to 1, the total area $\int_{-\infty}^{\infty} p(x)\,dx$ under the curve, that we can pretty safely ignore any contributions to our area from the right of 3.09 or the left of -3.09. This is why the table goes only as far as $z = 3.09$—there isn't much contribution to the area for $z > 3.09$.

Finally, we are ready to return to the general situation of a normal distribution with mean μ and standard deviation σ:

$$P_{\mu,\sigma}(x) = \frac{1}{\sigma\sqrt{2\pi}}\,{}^2e^{-\frac{1}{2}\left(\frac{x-\mu}{\sigma}\right)^2} \qquad -\infty < x < \infty$$

Suppose we know that a certain random variable is normally distributed with mean μ and standard deviation σ. How do we find the probability that X will lie between two values α and β? We know that

$$\begin{aligned}\Pr(\alpha \le X \le \beta) &= \int_{\alpha}^{\beta} P_{\mu,\sigma}(x)\,dx \\ &= \frac{1}{\sigma\sqrt{2\pi}} \int_{\alpha}^{\beta} {}^2e^{-\frac{1}{2}\left(\frac{x-\mu}{\sigma}\right)^2}\,dx\end{aligned}$$

In this last definite integral, let us make the change of variables.

(10.14) $$z = \frac{x - \mu}{\sigma}$$

Then

$$dz = \frac{1}{\sigma}\, dx$$

or, equivalently,

$$\sigma\, dz = dx$$

After this transformation, our integral becomes

$$\Pr(\alpha \le X \le \beta) = \frac{1}{\sqrt{2\pi}} \int_a^b e^{-\frac{1}{2}z^2}\, dz$$

where

(10.15) $$a = \frac{\alpha - \mu}{\sigma} \quad \text{and} \quad b = \frac{\beta - \mu}{\sigma}$$

Remarkably, our change of variables has resulted in the problem of evaluating the area under a portion of the standard normal distribution between a and b, which is precisely the kind of problem we have just developed the techniques to solve. Thus the area under a portion of *any* normal curve may be found by computing the area under a portion of the standard normal curve with $\mu = 0$ and $\sigma = 1$.

EXAMPLE 21 Evaluate

$$\int_2^{14.2} P_{10,5}(x)\, dx = \int_2^{14.2} \frac{1}{5\sqrt{2\pi}}\, e^{-\frac{1}{2}\left(\frac{x-10}{5}\right)^2}\, dx$$

which represents the area under the bell-shaped curve with $\mu = 10$ and $\sigma = 5$ from $x = 2$ to $x = 14.2$.

SOLUTION We make the change of variables shown in Equations (10.14).

$$z = \frac{x - \mu}{\sigma} = \frac{x - 10}{5}$$

Then, from Equation (10.15),

$$a = \frac{2 - 10}{5} = -1.6 \quad \text{and} \quad b = \frac{14.2 - 10}{5} = 0.84$$

so the transformed picture is

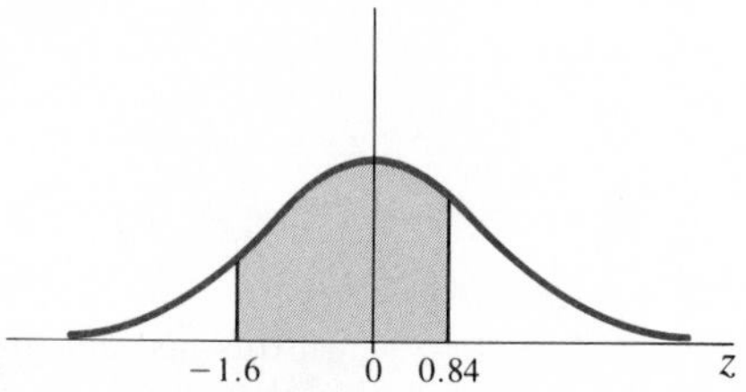

FIGURE 10.27

The area is given by

$$\begin{aligned}\int_{2}^{14.2} P_{10,5}(x)\,dx &= \int_{-1.6}^{0.84} \frac{1}{\sqrt{2\pi}} e^{-z^2/2}\,dz \\ &= \int_{-1.6}^{0} \frac{1}{\sqrt{2\pi}} e^{-z^2/2}\,dz + \int_{0}^{0.84} \frac{1}{\sqrt{2\pi}} e^{-z^2/2}\,dz \\ &= 0.4452 + 0.2995 \\ &= 0.7447\end{aligned}$$

We are finally able to go back and complete the solution to the problem posed in Example 19. There we were trying to determine the probability that an adult male will be over 6 feet in height. We were stopped by our inability to evaluate

(10.16) $$\int_{72}^{\infty} \frac{1}{4.3\sqrt{2\pi}} e^{-\frac{1}{2}\left(\frac{x-70}{4.3}\right)^2}\,dx$$

We do this by making the change of variables.

$$z = \frac{x-\mu}{\sigma} = \frac{x-70}{4.3}$$

Then

$$a = \frac{72-70}{4.3} = 0.47 \quad \text{and} \quad b = \infty$$

The transformed integral is

$$\int_{0.47}^{\infty} \frac{1}{\sqrt{2\pi}} e^{-\frac{1}{2}z^2}\,dz$$

From Table 10.19 we see that the value associated with 0.47 is 0.1808. Thus

$$\begin{aligned}\int_{0.47}^{\infty} \frac{1}{\sqrt{2\pi}} e^{-\frac{1}{2}z^2}\,dz &= 0.5000 - 0.1808 \\ &= 0.3192\end{aligned}$$

Almost 32% of all adult males are over 6 feet tall.

Let us consider one final example.

EXAMPLE 22 A college decides to administer a single final examination to all calculus classes. It is found that the test scores are normally distributed with mean $\mu = 72$ and standard deviation $\sigma = 9.2$.

(a) What percentage of the students receive a score between 80 and 89?

(b) It is decided that the top 10% of the scores will receive a grade of A. What is the minimal score necessary to receive an A?

(c) What is the probability that a student's score will lie within 2 standard deviations of the mean?

SOLUTION

(a) We are considering questions concerning the probability density function

$$P_{72,9.2}(x) = \frac{1}{9.2\sqrt{2\pi}} e^{-\frac{1}{2}\left(\frac{x-72}{9.2}\right)^2}$$

The probability that a score will lie between 80 and 89 is given by

$$\Pr(80 \le X \le 89) = \int_{80}^{89} \frac{1}{9.2\sqrt{2\pi}} e^{-\frac{1}{2}\left(\frac{x-72}{9.2}\right)^2} dx$$

We make the change of variables

$$z = \frac{x - 72}{9.2}$$

$$a = \frac{89 - 72}{9.2} \approx 1.85 \quad \text{and} \quad b = \frac{80 - 72}{9.2} \approx 0.87$$

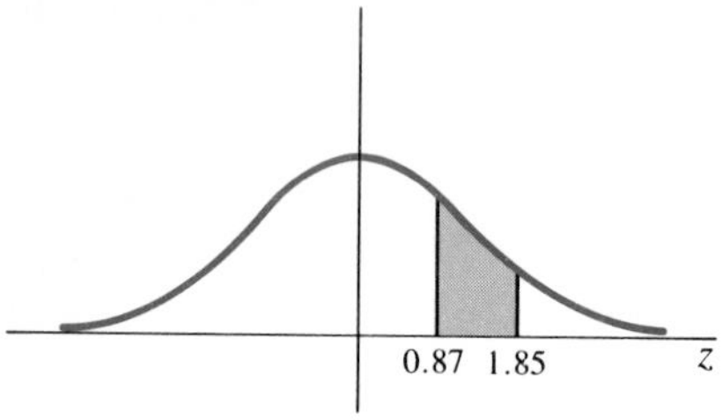

FIGURE 10.28

so that we must now evaluate

$$\int_{0.87}^{1.85} \frac{1}{\sqrt{2\pi}} e^{-\frac{1}{2}z^2} dz$$

(See Figure 10.28). This is given by

$$0.4678 - 0.3078 = 0.1600$$

Thus 16% of the scores are in the eighties.

(b) This question is somewhat different. Here we are asked to find a value α such that

$$\int_{\alpha}^{\infty} \frac{1}{9.2\sqrt{2\pi}} e^{-\frac{1}{2}\left(\frac{x-72}{9.2}\right)^2} dx = 0.10$$

The change of variables

$$z = \frac{x - 72}{9.2}$$

$$a = \frac{\alpha - 72}{9.2} \quad \text{and} \quad b = \infty$$

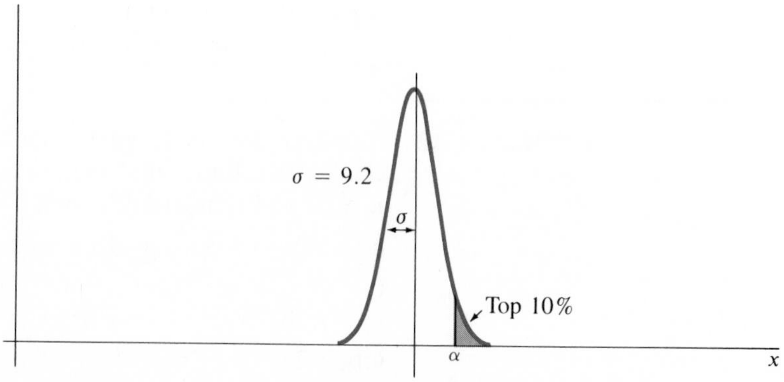

FIGURE 10.29

results in the problem of choosing a such that (Figure 10.30)

$$\int_a^{\infty} \frac{1}{\sqrt{2\pi}} e^{-\frac{1}{2}z^2}\, dz = 0.10$$

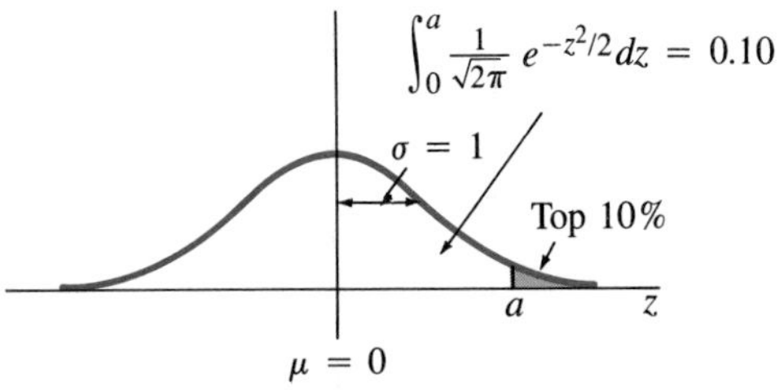

FIGURE 10.30

Since the area under the standard normal curve to the right of $\mu = 0$ is 0.5000, we see that the desired value of a must be such that, when we look it up in Table 10.19 we find the value 0.4000. This yields $a \approx 1.28$. In other words, a student whose score is roughly 1.28 standard deviations higher than the mean will be in the top 10% of all students. This allows us to find α:

$$1.28 = \frac{\alpha - 72}{9.2}$$

or

$$\alpha \approx 83.8$$

This is the minimal score a student must make to receive an A.

(c) Two standard deviations from the mean gives a range of

$$\mu - 2\sigma = 72 - 2(9.2) = 53.6$$

to

$$\mu + 2\sigma = 72 + 2(9.2) = 90.4$$

Then we seek

$$\int_{53.6}^{90.4} \frac{1}{9.2\sqrt{2\pi}} e^{-\frac{1}{2}\left(\frac{x-72}{9.2}\right)^2} dx$$

We let

$$z = \frac{x - 72}{9.2}$$

$$a = \frac{90.4 - 72}{9.2} = 2.0 \quad \text{and} \quad b = \frac{53.6 - 72}{9.2} = -2.0$$

and compute

$$\int_{-2}^{2} \frac{1}{\sqrt{2\pi}} e^{-\frac{1}{2}z^2} dz = \int_{-2}^{0} \frac{1}{\sqrt{2\pi}} e^{-\frac{1}{2}z^2} dz + \int_{0}^{2} \frac{1}{\sqrt{2\pi}} e^{-\frac{1}{2}z^2} dz$$
$$= 0.4772 + 0.4772 = 0.9544$$

The scores of approximately 95% of all the students lie within 2 standard deviations of the mean. This statement is true for any normal distribution.

For convenient reference, we conclude this chapter with a list of important formulas for both discrete and continuous probability distributions. For an arbitrary density function, we have the formulas shown in Table 10.20.

TABLE 10.20

	Discrete random variable	*Continuous random variable*
Probability density function, $p(x)$	$p_i \geq 0$, $\sum_{i=1}^{n} p_i = 1$	$p(x) \geq 0$ for $a \leq x \leq b$ and $\int_a^b p(x)\,dx = 1$
Expected value, mean, $\bar{x}$, $E(X)$, μ	$\sum_{i=1}^{n} x_i p_i$	$\int_a^b x p(x)\,dx$
$\text{Var}(X)$ *variance*, σ^2	$\sum_{i=1}^{n} (x_i - \mu)^2 p_i$	$\int_a^b (x - \mu)^2 p(x)\,dx$ or $\int_a^b x^2 p(x)\,dx - \mu^2$
Standard deviation, $\sigma = \sqrt{\text{Var}(X)}$	$\sqrt{\sum_{i=1}^{n} (x_i - \mu)^2 p_i}$	$\sqrt{\text{Var}(X)}$

We introduced three probability distributions which arise frequently in applications. For these specific distributions, Table 10.20 takes the form shown in Table 10.21.

TABLE 10.21

	Exponential distribution	*Standard normal distribution*	*Normal distribution*	*Uniform distribution* $A \leq x \leq B$
Probability density function, $p(x)$	$\lambda e^{-\lambda x}$	$\frac{1}{\sqrt{2\pi}} e^{-\frac{1}{2}x^2}$	$\frac{1}{\sigma\sqrt{2\pi}} e^{-\frac{1}{2}\left(\frac{x-\mu}{\sigma}\right)^2}$	$\frac{1}{B - A}$
Expected value, mean, $\bar{x}$, $E(X)$, μ	$\frac{1}{\lambda}$	$\mu = 0$	μ	$\frac{B + A}{2}$
$\text{Var}(X)$ *variance*, σ^2	$\frac{1}{\lambda^2}$	$\sigma^2 = 1$	σ^2	$\frac{(B - A)^2}{12}$
Standard deviation, $\sigma = \sqrt{\text{Var}(X)}$	$\frac{1}{\lambda}$	$\sigma = 1$	σ	$\frac{B - A}{\sqrt{12}}$

Section 10.5 EXERCISES

In Exercises 1–8 we assume that, on a given midsummer day, a few of the suitcases passing through Kennedy Airport are weighed, and it is found that the sample mean $\bar{x}$ is 38.2 lb and the sample standard deviation s is 5.7 lb. Assume the weights of all suitcases to be normally distributed.

1. Estimate the fraction of the total number of suitcases weighing between 36 and 42 lb.
2. Estimate the percentage of suitcases weighing over 45 lb.
3. Find that number of pounds below which 95% of all suitcases will weigh.
4. Find that number of pounds above which 65% of all suitcases will weigh.
5. Out of 1200 suitcases, how many would you expect to weigh between 32 and 35 lb?
6. Out of 1150 suitcases, how many would you expect to weigh between 1 and 2 standard deviations above the mean?
7. A new scale is brought in for testing on this day (the old one is known to be accurate) and it is found that, out of 840 suitcases, 55 weigh over 50 lb. Would you or would you not question the accuracy of the scale?
8. What percentage of the suitcases would you expect to weigh between 38 and 39 lb?

In Exercises 9–12 use Table 10.19 to evaluate the indicated integrals.

9. $\int_{-\infty}^{\infty} e^{\frac{-(x-2)^2}{8}}\,dx$

10. $\int_{0}^{\infty} e^{-\frac{(x-9)^2}{25}}\,dx$

11. $\int_{1}^{2} e^{-\frac{x^2}{18}}\,dx$

12. $\int_{-\infty}^{\infty} xe^{-x^2}\,dx$

13. Compute $\frac{1}{\sqrt{2\pi}\,\sigma}\int_{-\infty}^{\infty} xe^{-\frac{(x-\mu)^2}{2\sigma^2}}\,dx$

14. Compute $\frac{1}{\sqrt{2\pi}\,\sigma}\int_{-\infty}^{\infty} x^2e^{-\frac{(x-\mu)^2}{2\sigma^2}}\,dx$

15. A manufacturer is considering purchasing a machine that can produce 1-g tablets of a certain drug. Federal law permits no more than a total of 1% of the tablets produced to weigh less than 0.9 g or more than 1.1 g (in other words, at least 99% must weigh in the range 1 ± 0.1 g). Assuming that the machine produces a tablet of mean size 1 g, what must be the largest standard deviation in order to meet the required tolerance limits?
16. Do Exercise 15, assuming that the machine produces a tablet of mean size 1.05 g rather than 1 g.
17. If the machine in Exercise 15 is tested on a large sample and its sample standard deviation is found to be 0.4 g, what would be the weight range in which 95% of tablets would lie? In which 99% of the tablets would lie?

Key Mathematical Concepts and Tools

Outcomes
Simple event
Compound event
Random variable
Discrete random variable
Probability distribution (density function)
Probability histogram
Expected (mean) value
Variance
Standard deviation
Continuous random variable
Uniform distribution
Exponential distribution
Normal distribution

Applications Covered in This Chapter

Price-to-earnings ratio

Average U.S. savings rate as a percentage of household income

Roulette

Oil drilling—developmental versus exploratory wells

Selection of a reliable moving company

Survival time after cancer treatment

Time to memorize information

Duration of a telephone call

Waiting time on a line

Probability of growing to a given height

Distribution of test scores on an exam

TEN

Warm-Up Test

1. Compute the expected value, variance, and standard deviation for the following probability table.

Barometric pressure	23.9	29.1	29.3	29.5	29.7
Probability	0.08	0.26	0.32	0.27	0.07

2. A phone-in mail order company finds that the time between successive incoming switchboard calls is an exponential random variable with expected value 50 sec.

(a) What is the probability that a call will arrive within 35 sec?

(b) What is the probability that a call will arrive later than 60 sec after the last call?

(c) What is the probability that a call will arrive between 45 and 55 sec of the last call?

3. For the random variable X as given below with $f(x)$ the probability density function, compute the mean and the variance.

$$f(x) = \begin{cases} 0 & x < -1 \\ \frac{3}{10}(2 - x^2) & -1 \le x \le 1 \\ 0 & x > 1 \end{cases}$$

4. For the random variable of Exercise 3, find each of the following probabilities.

(a) $\Pr(|X| < 0.5)$ (b) $\Pr(X > -4)$

(c) $\Pr(-3 \le X \le 0.2)$

5. A ball bearing buyer who produces bicycles cannot use a ball bearing if its diameter is less than 3.8 mm or greater than 6.0 mm. How sure can he be of obtaining usable ball bearings from a factory that produces ball bearings with a mean diameter of 5.2 mm and with standard deviation of 0.6 mm?

6. Compute the expected value, variance, and standard deviation for the following probability table.

Outcome	22	31	19	8	40
Probability	0.2	0.2	0.2	0.3	0.1

7. Consider the following random variable X.

$$f(x) = \begin{cases} \dfrac{c}{5 + x} & 1 \le x \le 2 \\ 0 & \text{otherwise} \end{cases}$$

(a) Find the value of c that makes $f(x)$ into a probability density function.

(b) Given this value of c, find $\Pr(X < 1)$.

8. The lifetime of a new lightbulb is reputed to be 1000 hours on average (that is, this is the mean life). The lifetime is known to be an exponential random variable.

(a) Write down the probability density function for the lifetime of this new bulb.

(b) Find the probability that a new bulb will last between 1200 and 1500 h.

9. In a gambling game a person is paid \$10 if he draws a picture card and \$3 if he draws an ace from an ordinary deck of 52 playing cards. If he draws any other card he must pay \$4. What is the expected gain or loss when the game is played? Would you wish to play?

10. In a recent statistics final exam, the mean score was found to be $\mu = 68$ and the standard deviation $\sigma = 7.3$. Assume a normal distribution.

(a) What percentage of the students would you expect to receive a score between 65 and 82?

(b) If it is decided that 5% of the class is to fail, what should be the lowest passing grade?

TEN Final Exam

1. Compute $\int_{-2}^{5} e^{-\frac{(x-6)^2}{98}}\,dx$.

2. Three dice are tossed simultaneously. Let the random variable X represent the sum of the three numbers obtained.

(a) List the possible outcomes for X.

(b) What is the probability that $X \leq 6$?

(c) What is the probability that $X > 15$?

3. A study shows that the weights of adult females in the United States are normally distributed with mean 120 lb and standard deviation 6 lb.

(a) Write down the integral whose value represents the probability that a given adult female weighs between 118 and 123 lb.

(b) Evaluate this integral with the aid of the standard normal table.

4. In a class of 24 students, the numbers of students of each indicated weight are shown in the accompanying table. If a student is selected at random, what is the expected value of his weight?

Weight	120	130	140	150	160
Number	2	3	10	5	4

5. Assume that the number of defects per new car produced at a given factory is normally distributed with a mean of 28 and a standard deviation of 6.

(a) What percentage of the cars will have fewer than 20 defects?

(b) What percentage will have between 25 and 32 defects?

6. For the random variable X given below, with $f(x)$ the probability density function, compute the mean and variance if

$$f(x) = \begin{cases} 0 & x < 0 \\ 2(e^{-x} - e^{-2x}) & x \geq 0 \end{cases}$$

7. A number between 100 and 250 is picked at random in such a manner that any value in the interval $100 \leq x \leq 250$ is equally likely.

(a) Find the probability that this number will be between 108 and 173.

(b) Find the mean μ and the standard deviation σ.

8. In a popular ice cream store, a line is formed by patrons waiting for service. It is found that the average waiting time is an exponentially distributed random variable and that the average waiting time is 85 sec.

(a) Find the probability density function.

(b) What is the probability that a patron will have to wait longer than 3 min?

9. The following chart represents the weights of suitcases checked in at Boston's Logan Airport on a recent winter day.

Weight in pounds	10	15	20	25	30	35	40	45	50
Probability	0.01	0.05	0.06	0.11	0.25	0.20	0.18	0.13	0.01

(a) Find the mean and the standard deviation for the suitcase weights.

(b) What would you estimate as the probability that a given suitcase will weigh less than 28 lb?

10. For the random variable of Exercise 6, find each of the following probabilities:

(a) $\Pr(X > 4)$

(b) $\Pr(2 < X < 3)$

(c) $\Pr(X < 2.7)$

ELEVEN

Taylor Polynomials and Numerical Approximation Methods

Outline

In Context

This chapter begins with a discussion of the use of polynomial approximations for evaluating functions such as the exponential, logarithmic, and trigonometric functions. This leads to the representation of a function by means of a power series—its *Taylor series*—and then to a discussion of the properties possessed by such series in general. This may be viewed as a continuation of our brief introduction to series in Section 4.1. We also extend our study of differential equations by considering solution by *power series*.

In addition, we consider an efficient technique for solving equations that depends on the close approximation of a graph by its tangent line—*Newton's method*, also known as the Newton–Raphson method (after Joseph Raphson, who improved on Sir Isaac Newton's original formulation of the technique). Our final topic, *L'Hôpital's rule*, is an aid to evaluating limits of rational functions when the numerator and the denominator either both approach 0 or both approach infinity.

SECTION 11.1

Taylor Polynomials

Throughout the text we have introduced an array of different techniques used in calculus to solve various problems. Despite this variety of methods, however, nearly every technique we have seen is based on a common approach to problem solving.

1 *Approximate* the unknown quantity by something more easily computed.
2 Then use successively better approximations to obtain the unknown quantity exactly as a *limit*.

The question we wish to consider in this section is one of the most fundamental and simple to state: Given a function f, how can we compute $f(x)$ for various values of x?

Obviously if $f(x) = x^3 + 4$, we can easily compute $f(4)$; or if $f(x) = 2x^2 - x$, then $f(5)$ is readily found. After all, computation with polynomials is easy!

But what if $f(x) = \sin x$? What is $f(4) = \sin 4$? Or if $f(x) = \ln x$, what is $f(5) = \ln 5$? It is a pretty safe bet that, if you do not have either a calculator or a table of values for the trigonometric or natural logarithm function, you cannot determine these function values. (Come to think of it, where do the numbers in the tables come from in the first place, and how does the calculator evaluate these functions?)

In this section we will try to answer such questions. Our idea is to try to find a polynomial of degree n, call it p_n, that "closely resembles" the function f under consideration. Then, given any number $x = x_0$, we will use the value of $p_n(x_0)$ to approximate the value of $f(x_0)$. Polynomials are quickly evaluated.

For "nice" functions we will obtain better and better approximations by using polynomials of higher and higher degree. In the next section we will show that, as we allow the degree n of the polynomial p_n to increase, we can obtain the desired function exactly, in the limit, in the form of an "infinite-degree polynomial," called a **Taylor series**. Actually, if you reread Section 4.1 you will discover that we made use of these ideas when we developed the exponential function e^x by means of the infinite-degree polynomial

$$e^x = 1 + x + \frac{x^2}{2!} + \frac{x^3}{3!} + \cdots$$

Let us now move from generalities to specifics.

We begin by considering functions f that are defined in an interval (a, b) containing the origin and such that f possesses derivatives of all orders on that interval. Thus

$$f'(x), f''(x), f'''(x), \ldots$$

all exist for $x \in (a, b)$. Some examples of such "nice" functions are $\sin x$, $\cos x$, e^x, and any polynomial. We now ask the following question. Given f, how can we find the polynomial of degree n that does the "best job" of approximating f near 0? With the goal of answering this question, we first introduce the notion of ***n*th-order contact** for two functions f and g.

DEFINITION 11.1 Two "nice" functions f and g have **0th-order contact** at $x = 0$ if $f(0) = g(0)$. That is, the two functions have the same value at $x = 0$; see Figure 11.1(a).

We say that f and g have **1st-order contact** at $x = 0$ if both $f(0) = g(0)$ and $f'(0) = g'(0)$. Thus, not only do the graphs of $y = f(x)$ and $y = g(x)$ both go through the same point at $x = 0$, but also the slopes of these graphs coincide at $x = 0$. See Figure 11.1(b). ⊟

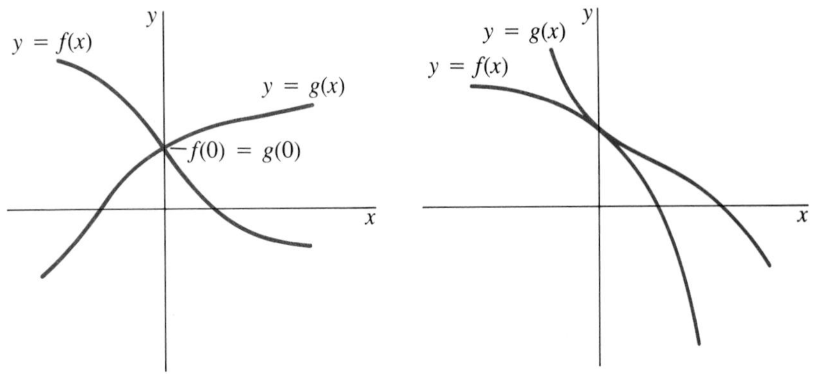

(a) 0th-order contact at $x = 0$; curves pass through the same point

(b) 1st-order contact at $x = 0$; curves pass through the same point and their slopes are equal when $x = 0$

FIGURE 11.1

Similarly, we shall say that f and g have 2nd-order contact at $x = 0$ if $f(0) = g(0)$, $f'(0) = g'(0)$, and $f''(0) = g''(0)$. In general, for any nonnegative integer n, we say that f and g have ***n*th order contact** at $x = 0$ if

$$f(0) = g(0),\ f'(0) = g'(0),\ f''(0) = g''(0), \ldots,\ f^{(n)}(0) = g^{(n)}(0)$$

In what way is the idea of nth-order contact a step toward answering our question?

We are trying to find a polynomial p_n, of degree n, that is very much like the function f near $x = 0$. Definition 11.1 is intended to provide a yardstick by which we can measure how nearly alike any two functions f and g are near $x = 0$. Certainly it is a minimal requirement to ask that f and g have 0th-order contact $[f(0) = g(0)]$. Then the curves $y = f(x)$ and $y = g(x)$ will pass through the same point, as shown in Figure 11.1(a).

EXAMPLE 1 The functions $f(x) = x + 2$ and $g(x) = 2\cos x$, as shown in Figure 11.2, have 0th-order contact at $x = 0$, because

$$f(0) = g(0) = 2$$

Nevertheless, these functions are not very much alike at all.

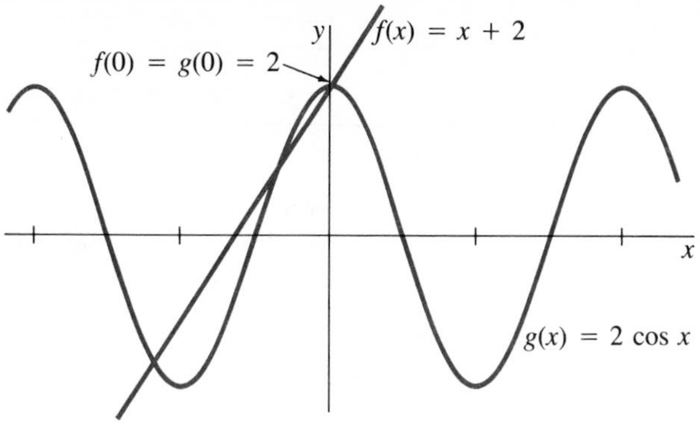

FIGURE 11.2

The requirement of 1st-order contact at $x = 0$ asks that $f(0) = g(0)$ and $f'(0) = g'(0)$. This assures us not only that the graphs of $y = f(x)$ and $y = g(x)$ pass through the same point, but also that their slopes coincide at $x = 0$.

EXAMPLE 2 The functions $f(x) = 2\sin x$ and $g(x) = 2x$, as shown in Figure 11.3, have 1st-order contact at $x = 0$, because

$$f(0) = g(0) = 0$$

and
$$f'(0) = g'(0) = 2$$

As you can see, these graphs look very similar right near $x = 0$. (Look at those segments of the graph contained in the box.) It is also worth noting that, the further x is from 0, the more the two graphs diverge and go their separate ways.

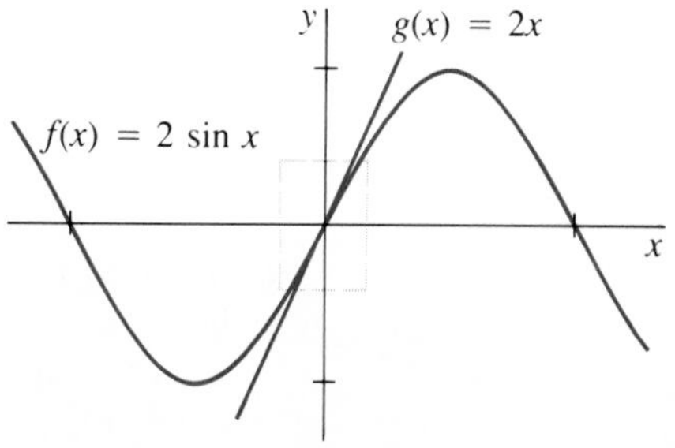

FIGURE 11.3

We know from Chapter 3 on curve sketching that the second derivative describes the concavity of a function. Thus we can say that, if two functions have 2nd-order contact at $x = 0$ [$f(0) = g(0)$, $f'(0) = g'(0)$, and $f''(0) = g''(0)$], not only do their graphs pass through the same point with the same slope at $x = 0$, but also they have the same concavity at $x = 0$. To make this clearer, we look at two examples.

EXAMPLE 3 The functions $f(x) = x^2 + 1$ and $g(x) = x^3 + 1$ satisfy

$$f(0) = g(0) = 1$$

and

$$f'(0) = g'(0) = 0$$

But $f''(0) = 2$, whereas $g''(0) = 0$. Thus these functions have 1st-order contact, but they fail to have 2nd-order contact at $x = 0$. Figure 11.4 shows how the concavity of the two curves differs at $x = 0$. Contrast this situation with the one in Example 4.

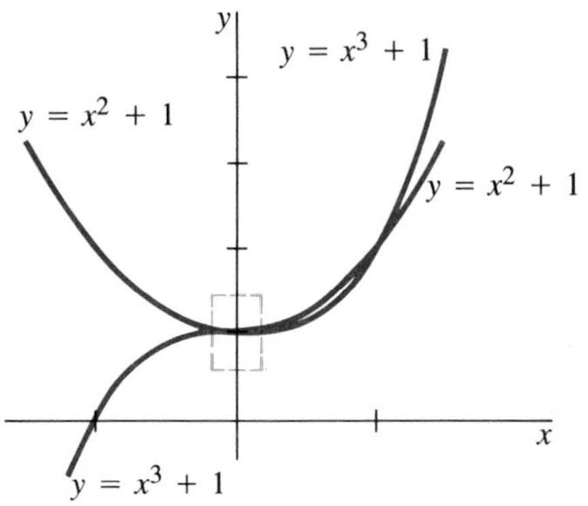

FIGURE 11.4

EXAMPLE 4 Let $f(x) = x^2 + 1$ and $g(x) = \sin^2 x + 1$. Then $f(0) = g(0) = 1$, $f'(0) = g'(0) = 0$, and $f''(0) = g''(0) = 2$. This is shown in Figure 11.5. In this case, we have 2nd-order contact at $x = 0$, and the curves agree more closely than in the last example.

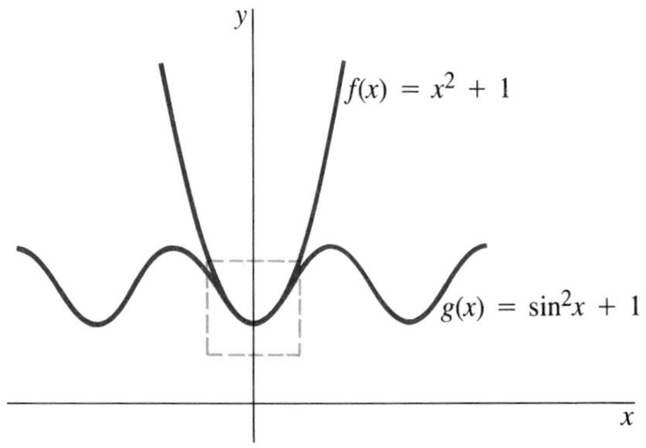

FIGURE 11.5

Unfortunately our eyes cannot make the fine distinctions necessary to see the difference in orders of contact much beyond the second. But if you imagine that you have in your possession a powerful microscope that you can use to blow up a small region of the graphs of $y = f(x)$ and $y = g(x)$ near $x = 0$, then with the aid of this instrument you would presumably be able to tell that f and g are a little more alike if they have, say, 10th-order contact than if they have only 9th. One other remark worth making before we move on is that it seems reasonable to expect that, for two functions f and g, the higher the order of contact they have at $x = 0$, the further away from 0 we can go before the graphs start to look very different.

With these rather lengthy remarks in mind, let's try to answer the question we posed on page 488.

First, let us try to find the polynomial of degree 1 that makes 1st-order contact with f at 0. If we denote the polynomial we seek by p_1, then p_1 has the form

$$p_1(x) = a_0 + a_1 x$$

where a_0, a_1 are constants to be determined. Now if we want f and p_1 to have 1st-order contact at 0, we must require that

$$p_1(0) = a_0 + a_1 \cdot 0 = a_0 = f(0)$$

and $$p_1'(0) = a_1 = f'(0)$$

Thus both a_0 and a_1 are determined: $a_0 = f(0)$ and $a_1 = f'(0)$. This in turn yields

(11.1) $$p_1(x) = f(0) + f'(0)x$$

Note that we were looking for the polynomial of *1st* degree that makes 1st-order contact with f at 0, and we had just enough undetermined constants (a_0 and a_1) to be able to do this. The polynomial p_1 best approximates f near 0, and there is only one such polynomial.

Actually we have encountered the polynomial $p_1(x)$ before—way back in Chapter 2. In that chapter we learned how to compute the tangent line to a curve $y = f(x)$ at a point $(a, f(a))$. The tangent line at the point $(0, f(0))$ must have slope $f'(0)$. From the point-slope form of the equation of a line,

$$y - y_1 = m(x - x_1)$$

we see that the tangent line is given by

$$y - f(0) = f'(0)(x - 0)$$

or $$y = f(0) + f'(0)x$$

which agrees with $p_1(x)$.

EXAMPLE 5 (a) Find the polynomial of degree 1 having 1st-order contact at 0 with

$$f(x) = 2 \sin 2x + 3$$

(b) Similarly, find the polynomial of degree 2 having 2nd-order contact at 0 with

$$f(x) = 2 \sin 2x + x^2 + 3$$

SOLUTION

(a) We easily find $f(0) = 2 \sin 0 + 3 = 3$, and

$$f'(0) = 4 \cos 2x|_{x=0} = 4 \cos 0 = 4$$

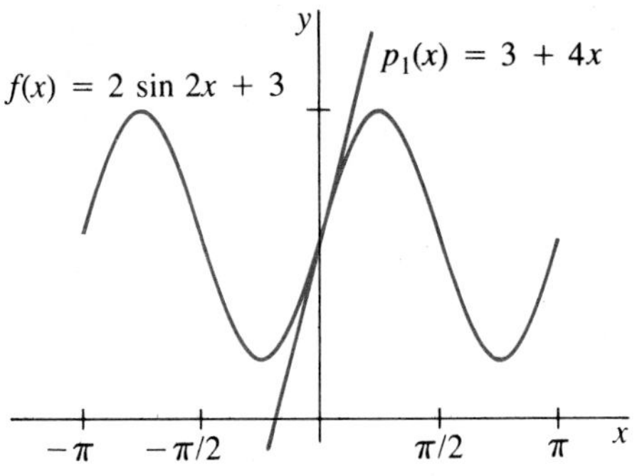

FIGURE 11.6

From Equation (11.1) we now have (see Figure 11.6)

$$p_1(x) = 3 + 4x$$

(b) Next let us construct a polynomial of degree 2, $p_2(x) = a_0 + a_1x + a_2x^2$, which makes *2nd*-order contact with f at 0. Such a polynomial must satisfy

$$p_2(0) = a_0 + a_1(0) + a_2(0)^2 = a_0 = f(0)$$
$$p_2'(0) = a_1 + 2a_2x|_{x=0} = a_1 + 2a_2(0) = a_1 = f'(0)$$
$$p_2''(0) = 2a_2|_{x=0} = 2a_2 = f''(0)$$

Thus we must have

$$a_0 = f(0) \qquad a_1 = f'(0) \qquad a_2 = \frac{f''(0)}{2}$$

and our polynomial is

(11.2) $$p_2(x) = f(0) + f'(0)x + \frac{f''(0)}{2}x^2 = 3 + 4x + x^2$$

because $f(0) = 3$, $f'(0) = 4$, $f''(0) = 2$, as you can check.

In general, given any positive integer n, there is a uniquely determined polynomial of degree n that has nth-order contact with f at 0. Let us try to find it. If we write

$$p_n(x) = a_0 + a_1x + a_2x^2 + a_3x^3 + \cdots + a_nx^n$$

then

$$\begin{aligned} p_n'(x) &= a_1 + 2a_2x + 3a_3x^2 + \cdots + na_nx^{n-1} \\ p_n''(x) &= 2a_2 + (3)(2)a_3x + \cdots + (n)(n-1)a_nx^{n-1} \\ p_n'''(x) &= (3)(2)(1)a_3 + \cdots + (n)(n-1)(n-2)a_nx^{n-3} \\ &\vdots \\ p_n^{(n)}(x) &= (n)(n-1)(n-2)\cdots(2)(1)a_n \end{aligned}$$

We must now require that

$$\begin{aligned} p_n(0) &= a_0 = f(0) \\ p_n'(0) &= a_1 = f'(0) \\ p_n''(0) &= 2a_2 = f''(0) \\ p_n'''(0) &= (3)(2)(1)a_3 = f'''(0) \\ &\vdots \\ p_n^{(n)}(0) &= (n)(n-1)(n-2)\cdots(1)a_n = f^{(n)}(0) \end{aligned}$$

Thus the coefficients are given by

(11.3) $$a_0 = f(0),\ a_1 = f'(0),\ a_2 = \frac{f''(0)}{2!},\ a_3 = \frac{f'''(0)}{3!},\ \ldots,\ a_n = \frac{f^{(n)}(0)}{n!}$$

and the polynomial p_n must be

(11.4)
$$p_n(x) = f(0) + f'(0)x + \frac{f''(0)}{2!}x^2 + \frac{f'''(0)}{3!}x^3 \cdots + \frac{f^{(n)}(0)}{n!}x^n$$

It is called the **Taylor polynomial of degree *n*** for f near 0.

EXAMPLE 6 (a) Compute the Taylor polynomial of degree 2 for $f(x) = \cos x$ near 0.
(b) Use this polynomial to approximate $\cos(\pi/12)$

SOLUTION

(a)
$$f(x) = \cos x$$
$$f'(x) = -\sin x$$
and
$$f''(x) = -\cos x$$
Thus
$$f(0) = 1$$
$$f'(0) = 0$$
$$f''(0) = -1$$

Utilizing Equation (11.4) with $n = 2$ gives (see Figure 11.7)

$$p_2(x) = 1 + 0x - \frac{1}{2!}x^2 = 1 - \frac{x^2}{2}$$

(b) If we now reason that $p_2(x)$ is a good approximation to $f(x) = \cos x$ near 0, we can try to use $p_2(\pi/12)$ to estimate $f(\pi/12) = \cos(\pi/12)$.

$$\cos(\pi/12) \approx p_2(\pi/12)$$
$$= 1 - \frac{(\pi/12)^2}{2}$$
$$= 0.9657$$

(We have retained only 4 decimal places in our computation.) The true value, correct to 4 decimal places, is 0.9659. Our answer is in very good agreement with this value, but it did not require us to look up anything in a table of trigonometric functions.

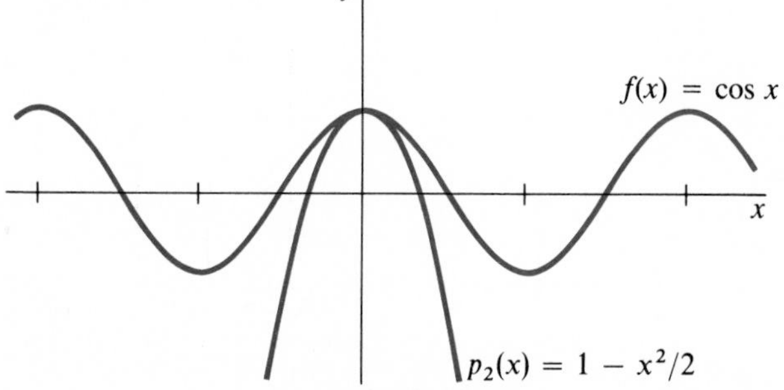

FIGURE 11.7

EXAMPLE 7 (a) Compute the Taylor polynomial of degree 3 for $f(x) = \sqrt{x + 1}$ near 0.
(b) Use this polynomial to approximate $\sqrt{2}$.

SOLUTION

(a)
$$f(x) = (x + 1)^{1/2}$$
$$f'(x) = \frac{1}{2}(x + 1)^{-1/2}$$
$$f''(x) = -\frac{1}{4}(x + 1)^{-3/2}$$
$$f'''(x) = \frac{3}{8}(x + 1)^{-5/2}$$

Therefore
$$f(0) = 1$$
$$f'(0) = \frac{1}{2}$$
$$f''(0) = -\frac{1}{4}$$
$$f'''(0) = \frac{3}{8}$$

From Equation (11.4) with $n = 3$, we find

$$p_3(x) = 1 + \frac{1}{2}x - \frac{\frac{1}{4}}{2!}x^2 + \frac{\frac{3}{8}}{3!}x^3$$
$$= 1 + \frac{x}{2} - \frac{x^2}{8} + \frac{x^3}{16}$$

(b) Because $f(1) = \sqrt{1 + 1} = \sqrt{2}$, we will use $p_3(1)$ to estimate its value. Thus

$$\sqrt{2} \approx p_3(1) = 1 + \frac{1}{2} - \frac{1}{8} + \frac{1}{16} = \frac{23}{16} = 1.438$$

The true value, correct to 3 decimal places, is 1.414. Thus our answer is a reasonable (though not an excellent) approximation to $\sqrt{2}$.

PRACTICE EXERCISE Show that the Taylor polynomial of degree 5 for $f(x) = \sqrt{x + 1}$ near 0 is

$$p_5(x) = 1 + \frac{x}{2} - \frac{x^2}{8} + \frac{x^3}{16} - \frac{5}{128}x^4 + \frac{7}{256}x^5$$

Then use this polynomial to estimate $\sqrt{2}$.

[Your answer should be $\sqrt{2} \approx \frac{365}{256} = 1.426$. This answer is in somewhat better agreement with the correct value than is the answer in Example 7. In general, the higher the degree of the Taylor polynomial that we use to approximate f, the more accuracy we expect.]

EXAMPLE 8 Compute the Taylor polynomial of degree 3 for $f(x) = e^x$ near $x = 0$.

SOLUTION As in our previous examples, we compute

$$f(x) = e^x$$
$$f'(x) = e^x$$
$$f''(x) = e^x$$
$$f'''(x) = e^x$$

Then

$$f(0) = e^0 = 1$$
$$f'(0) = 1$$
$$f''(0) = 1$$
$$f'''(0) = 1$$

and from Equation (11.4) with $n = 3$,

$$p_3(x) = 1 + 1x + \frac{1}{2!}x^2 + \frac{1}{3!}x^3$$

REMARK We have considered Taylor polynomials only for a function f near $x = 0$. That is due to the fact that we were trying to answer a question that asked how to approximate f near 0. We could just as easily have considered the problem of how to approximate f near $x = a$, for any particular number a. In that case the Taylor polynomial of degree n for f near $x = a$ takes the form

(11.5)

$$p_n(x) = f(a) + f'(a)(x - a) + \frac{f''(a)}{2!}(x - a)^2 + \frac{f'''(a)}{3!}(x - a)^3 + \cdots + \frac{f^{(n)}(a)}{n!}(x - a)^n$$

EXAMPLE 9 (a) Find the Taylor polynomial of degree 2 for $f(x) = \ln x$ near $x = 1$.
(b) Use this to estimate ln 1.5.

SOLUTION

(a)

$$f(x) = \ln x$$
$$f'(x) = \frac{1}{x}$$
$$f''(x) = -\frac{1}{x^2}$$

and

$$f(1) = \ln 1 = 0$$

$$f'(1) = \frac{1}{1} = 1$$

$$f''(1) = \frac{1}{1^2} = -1$$

If we now use these values in Equation (11.5) with $n = 2$ and $a = 1$, then

$$p_2(x) = 0 + 1(x-1) - \frac{1}{2!}(x-1)^2 = (x-1) - \frac{1}{2}(x-1)^2$$

(b) We wish to estimate $f(1.5) = \ln 1.5$ by computing $p_2(1.5)$. Thus

$$\ln 1.5 \approx p_2(1.5) = (1.5-1) - \frac{1}{2}(1.5-1)^2 = 0.5 - \frac{1}{2}(0.5)^2$$

$$= 0.3750$$

The correct value is $\ln 1.5 = 0.4055$. We would expect that the higher the degree of the Taylor polynomial we construct, the more closely our answer will approximate the true value.

Section 11.1 EXERCISES

Find the Taylor polynomial of degree 6 near $x = 0$ for each of the functions in Exercises 1–17.

1. $\cos x$
2. $\sin x$
3. $\cos 2x$
4. $\sin 2x$
5. $\cos x^2$
6. $\sin x^2$
7. e^x
8. e^{-x}
9. e^{3x}
10. $\dfrac{e^x + e^{-x}}{2}$
11. $\ln(1+x)$
12. $\sqrt{1+x}$
13. $\tan x$
14. x^4
15. e^{-x^2}
16. $x^7 + x^6 + x + 1$
17. $x^6 + x^5 + x + 1$
18. $x^5 + x^4 + x + 1$

Find the Taylor polynomial of degree 3 near $x = 1$ for each of the functions in Exercises 19–24.

19. $\cos x$
20. $\sin x$
21. e^x
22. $\tan x$
23. $\sqrt{1+x}$
24. $\ln(1+x)$
25. $\ln x$
26. $1/x$

27. Find a polynomial that makes 2nd-order contact with the function $f(x) = \sin x + \cos x$ at $x = 0$.
28. Find a polynomial that makes 3rd-order contact with the function $f(x) = x + \sin x$ at $x = 0$.
29. Find a polynomial that makes 3rd-order contact with the function $f(x) = (x-1)^3 + 2(x-1)^2 - 3(x-1) + 4$ at $x = 0$. Do the same at $x = 1$.
30. Use the result of Exercise 1 to approximate $\cos 3°$. (Keep in mind that angles need to be in radian measure.)
31. Use the result of Exercise 2 to approximate $\sin 3°$.
32. Use the result of Exercise 7 to approximate e.
33. Use the result of Exercise 7 to approximate $\sqrt{e}$.
34. Use the result of Exercise 13 to approximate $\tan 5°$.
35. Find the Taylor polynomial of degree 3, near $x = \pi/6$, for the function $f(x) = \sin x$. Use this to approximate $\sin 31°$.
36. Find the Taylor polynomial of degree 3, near $x = \pi/6$, for the function $f(x) = \cos x$. Use this to approximate $\cos 31°$.
37. Find the Taylor polynomial of degree 3, near $x = \pi/3$, for the function $f(x) = \sin x$. Use this to approximate $\sin 59°$.

SECTION 11.2

Taylor Series

In the previous section we saw how, given a "nice" function f, we could construct Taylor polynomials of any degree for f near 0. We then used these polynomials to approximate $f(x)$ for various values of x. Sometimes, as in Example 6 wherein we approximated $\cos(\pi/12)$, our estimate was a very good one; at other times, as in Example 7 wherein we estimated $\sqrt{2}$, it was not so good. Is there any way to tell, when we make an estimate, what the magnitude of our error might be? Fortunately, the answer is yes.

One measure of how close $p_n(x)$ is to $f(x)$ is their difference, which we denote by

(11.6) $$R_n(x) = f(x) - p_n(x)$$

If $R_n(x)$ is small, $f(x)$ and $p_n(x)$ are close and our approximation is a good one. If $R_n(x)$ is large, $p_n(x)$ is not a very good approximation. The function $R_n(x)$ is called the **remainder term** in the Taylor expansion of f about $x = 0$ after $(n + 1)$ terms. The size of the remainder term is described in the following theorem.

THEOREM 11.2 Let f be a function that has $(n + 1)$ continuous derivatives on an open interval I containing $x = 0$. Then for any $x \in I$, the remainder term $R_n(x) = f(x) - p_n(x)$ satisfies the inequality

(11.7) $$|R_n(x)| \le \frac{|x|^{n+1}}{(n+1)!}\left(\max_z |f^{(n+1)}(z)|\right)$$

Our notation $\max_z |f^{(n+1)}(z)|$ denotes the maximum value of the $(n + 1)$st derivative of f for z between 0 and x inclusive. ⊟

The proof of this theorem is somewhat involved, so we omit it here. But let us see how this theorem can be used.

EXAMPLE 10

(a) Use the Taylor polynomial of degree 3 for $f(x) = e^x$ near 0 to approximate $e^{1/2} = \sqrt{e}$.

(b) Estimate the error in performing this approximation.

SOLUTION

(a) In Example 8 we found that

$$p_3(x) = 1 + x + \frac{x^2}{2} + \frac{x^3}{3!}$$

Because we would like to estimate $f(1/2) = e^{1/2}$, we use

$$e^{1/2} \approx p_3\left(\frac{1}{2}\right) = 1 + \frac{1}{2} + \frac{\left(\frac{1}{2}\right)^2}{2} + \frac{\left(\frac{1}{2}\right)^3}{3!} = 1 + \frac{1}{2} + \frac{1}{8} + \frac{1}{48} = 1.6458$$

(b) Now we want to estimate the error inherent in using the value $e^{1/2} \approx 1.6458$. This is equivalent to determining the size of the remainder term R_3. According to Equation 11.7 with $n = 3$ and $x = \frac{1}{2}$, we have

$$\left|R_3\left(\frac{1}{2}\right)\right| \leq \frac{\left|\left(\frac{1}{2}\right)\right|^4}{4!}\left(\max_z |f^{(4)}(z)|\right)$$

Now $f(x) = e^x$, so $f^{(4)}(x) = e^x$. If z is allowed to vary between 0 and $\frac{1}{2}$, the largest value of e^z will occur when $z = \frac{1}{2}$ (because e^z is an increasing function). For the sake of simplicity we will use the fact that, because $e \leq 3$, we have

$$e^{1/2} < 2$$

Therefore $$\left|R_3\left(\frac{1}{2}\right)\right| \leq \frac{\left(\frac{1}{2}\right)^4}{4!}(e^{1/2}) < \frac{\frac{1}{16}}{24}(2) = \frac{1}{192} \approx 0.0052$$

Combined with part (a), this means that the true value of $\sqrt{e}$ lies between

$$1.6458 - 0.0052 = 1.6406$$

and $$1.6458 + 0.0052 = 1.6510$$

We sometimes write this as

$$\sqrt{e} = 1.6458 \pm 0.0052$$

The actual value to 4 decimal places happens to be $\sqrt{e} = 1.6487$.

EXAMPLE 11 What degree Taylor polynomial is required in order to estimate $\sin(0.1)$ within ± 0.0001?

SOLUTION This problem is asking us how large n must be in order to have

$$|R_n(0.1)| \leq 0.0001$$

If we set $f(x) = \sin x$, then

$$f'(x) = \cos x,\ f''(x) = -\sin x,\ f'''(x) = -\cos x,\ f^{(4)}(x) = \sin x,$$

$$f^{(5)}(x) = \cos x,\ f^{(6)}(x) = -\sin x,\ f^{(7)}(x) = -\cos x,\ f^{(8)}(x) = \sin x,$$

$$f^{(9)}(x) = \cos x, \ldots$$

Because $|\sin x| \leq 1$ and $|\cos x| \leq 1$ no matter what value z has, we must have

$$|f^{(n)}(z)| \leq 1$$

Thus, setting $x = 0.1$ in Equation (11.7), we desire

$$|R_n(0.1)| \leq \frac{(0.1)^{n+1}}{(n+1)!}\left(\max_z |f^{(n+1)}(z)|\right) \leq \frac{(0.1)^{n+1}}{(n+1)!}(1) \leq 0.0001$$

This will be true if $n = 3$, because

$$\frac{(0.1)^{3+1}}{(3+1)!} = \frac{0.0001}{24} < 0.0001$$

From the fact that $f(0) = 0$, $f'(0) = 1$, $f''(0) = 0$, and $f'''(0) = -1$, we easily (using Equation 11.4 with $n = 3$) obtain

$$p_3(x) = 0 + 1x + \frac{0}{2!}x^2 - \frac{1}{3!}x^3 = x - \frac{x^3}{6}$$

Setting $x = 0.1$ gives

$$p_3(0.1) = (0.1) - \frac{(0.1)^3}{6} = 0.0998$$

which we are confident is accurate to within ± 0.0001.

Let us return now to our general development. According to Equation (11.6), for any "nice" function f we can write

(11.8) $$f(x) = p_n(x) + R_n(x)$$

where $p_n(x)$ is the Taylor polynomial of degree n about $x = 0$, and $R_n(x)$ is the remainder term described in Theorem 11.2. To motivate what is to follow, let us focus our attention on the particular function $f(x) = e^x$. In this case

$$f(x) = f'(x) = f''(x) = \cdots = f^{(n)}(x) = e^x$$

and $$f(0) = f'(0) = f''(0) = \cdots = f^{(n)}(0) = 1$$

Thus the Taylor polynomial of degree n for e^x near $x = 0$ takes the form

$$p_n(x) = 1 + 1x + \frac{1}{2!}x^2 + \frac{1}{3!}x^3 + \cdots + \frac{1}{n!}x^n$$

How big is the remainder term $R_n(x)$? Well, according to Equation (11.7),

$$|R_n(x)| \leq \frac{|x|^{n+1}}{(n+1)!}\left(\max_z |f^{(n+1)}(z)|\right)$$

But $f^{(n+1)}(z) = e^z$, which assumes its maximum on the interval $[0, x]$ at the right hand end point $z = x$ (because e^x is an increasing function). See Figure 11.8.

Thus $$|R_n(x)| \leq \frac{|x|^{n+1}}{(n+1)!}e^x$$

If we can show that as $n \to \infty$ we have $R_n(x) \to 0$, it will follow from Equation (11.8) that, for large values of n, $p_n(x)$ will be a very good approximation to $f(x)$. In fact, we will have

(11.9) $$\lim_{n \to \infty} p_n(x) = f(x)$$

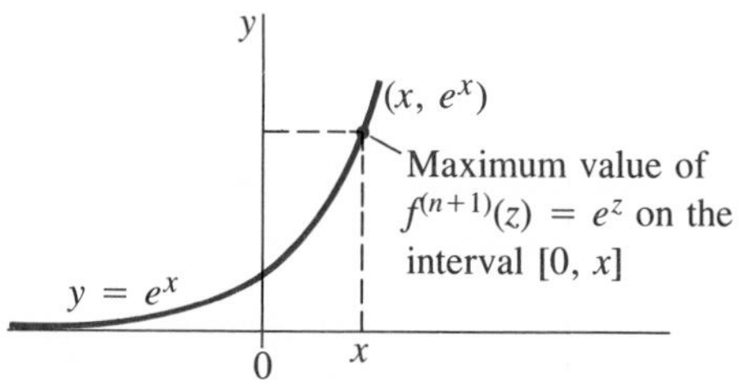

FIGURE 11.8

For definiteness, suppose we are interested in the value of

$$f(82.5) = e^{82.5}$$

Then, when we set $x = 82.5$,

$$|R_n(82.5)| \leq \frac{|82.5|^{n+1}}{(n+1)!} e^{82.5}$$

Because $e^{82.5}$ is some fixed (though large) number, the remainder term will go to 0 as $n \to \infty$, provided we can show that

$$\frac{(82.5)^{n+1}}{(n+1)!} \to 0 \quad \text{as} \quad n \to \infty$$

To see this, we observe that, as n grows larger, it will eventually exceed 82.5. Thus, when n is bigger than 83, we have

$$\frac{(82.5)^{n+1}}{(n+1)!} = \underbrace{\frac{\overbrace{(82.5)(82.5)\cdots(82.5)}^{83 \text{ times}}}{(1)(2)\cdots(83)}}_{A} \cdot \underbrace{\frac{(82.5)}{84} \frac{(82.5)}{85} \cdots \frac{(82.5)}{n+1}}_{B_n}$$

Now as n increases, the first term, which we have denoted by A, stays fixed. The second term, B_n, consists of an increasing number of fractions multiplied together, each of which is less than 1. The larger n grows, the smaller is the value of the fractions that are multiplied. When numbers that are less than 1 are multiplied together, their product is even smaller. So as n increases, the smaller B_n becomes, and hence the smaller the product AB_n. Therefore

$$\lim_{n \to \infty} \frac{(82.5)^{n+1}}{(n+1)!} = 0$$

and

$$R_n(82.5) \to 0 \quad \text{as} \quad n \to \infty$$

Actually there is nothing special about the number $x = 82.5$. The same reasoning could have been applied to *any* fixed value of x to demonstrate the following proposition.

PROPOSITION For any fixed x,

$$\lim_{n \to \infty} \frac{|x|^{n+1}}{(n+1)!} = 0 \qquad \qquad \square$$

Thus $R_n(x) \to 0$ as $n \to \infty$, and Equation (11.9) is valid. In words, $f(x)$ is the limit of $p_n(x)$ as n increases. But as n increases, the degree of our Taylor polynomial increases, and we obtain

$$e^x = \lim_{n \to \infty} \left[1 + x + \frac{x^2}{2!} + \cdots + \frac{x^n}{n!} \right]$$

$$= 1 + x + \frac{x^2}{2!} + \frac{x^3}{3!} + \cdots + \frac{x^n}{n!} + \cdots = \sum_{n=0}^{\infty} \frac{x^n}{n!}$$

This "infinite-degree polynomial" will have contact of every order with e^x. This expression is the same one we obtained intuitively in Section 4.1, and our discussion has established its validity.

More generally, if f is a "nice" function (so that its derivatives $f^{(n+1)}(z)$ don't grow too large), then, according to the foregoing proposition, we will have $R_n(x) \to 0$ as $n \to \infty$, so that allowing $n \to \infty$ in Equation (11.4) gives

(11.10)
$$f(x) = f(0) + f'(0)x + \frac{f''(0)}{2!}x^2 + \frac{f'''(0)}{3!}x^3 + \cdots + \frac{f^{(n)}(0)}{n!}x^n + \cdots$$
$$= \sum_{n=0}^{\infty} \frac{f^{(n)}(0)}{n!}x^n$$

This is called the **Taylor series** for f around $x = 0$. Observe that for each choice of x the right-hand side of Equation (11.10) is an infinite-degree polynomial, more generally referred to as an **infinite series**. Equation (11.10) can be interpreted as saying that the infinite series

$$\sum_{n=0}^{\infty} \frac{f^{(n)}(0)}{n!}x^n$$

converges to $f(x)$.

EXAMPLE 12 Compute the Taylor-series expansions for each of the following functions.

(a) $\sin x$ (b) $\ln(x + 1)$ (c) $\dfrac{1}{1 - x}$

SOLUTION

(a) If $f(x) = \sin x$ we have

f	f'	f''	f'''	$f^{(4)}$	$f^{(5)}$	$f^{(6)}$	$f^{(7)}$	$f^{(8)}$	···
$\sin x$	$\cos x$	$-\sin x$	$-\cos x$	$\sin x$	$\cos x$	$-\sin x$	$-\cos x$	$\sin x$	···

and the pattern repeats itself. Setting $x = 0$ gives

$f(0)$	$f'(0)$	$f''(0)$	$f'''(0)$	$f^{(4)}(0)$	$f^{(5)}(0)$	$f^{(6)}(0)$	$f^{(7)}(0)$	$f^{(8)}(0)$	···
0	1	0	-1	0	1	0	-1	0	···

From Equation (11.10) the Taylor series for $\sin x$ is now

$$\sin x = 0 + 1x + \frac{0}{2!}x^2 - \frac{1}{3!}x^3 + \frac{0}{4!}x^4 + \frac{1}{5!}x^5 + \frac{0}{6!}x^6 - \frac{1}{7!}x^7 + \cdots$$

or

(11.11)
$$\sin x = x - \frac{x^3}{3!} + \frac{x^5}{5!} + \frac{x^7}{7!} + \cdots = \sum_{n=0}^{\infty} \frac{(-1)^n x^{2n+1}}{(2n+1)!}$$

We may note that, for $f(x) = \sin x$, all derivatives of f are $\sin x$, $-\sin x$, $\cos x$, or $-\cos x$. Thus for any value of z, $|f^{n+1}(z)| \le 1$. If we combine this with the proposition given on page 501, we can see that, for any x, $R_n(x) \to 0$ as $n \to \infty$. This means that the Taylor series given in Equation (11.11) does, in fact, converge to $\sin x$ for each x, so we can approximate $\sin x$ to any desired accuracy simply by using enough of the terms in the Taylor series. For instance, in Example 11 we approximated $\sin(0.1)$ to within 0.0001 by using $p_3(x) = x - x^3/3!$, the first two nonzero terms in the Taylor series expansion.) By retaining as many terms as necessary to achieve our desired accuracy we could construct a table of values of the sine function. (In fact, this is precisely how such tables are computed!)

(b) This time $f(x) = \ln(x + 1)$, so we find

f	f'	f''	f'''	$f^{(4)}$	$f^{(5)}$	$f^{(6)}$
$\ln(x+1)$	$\dfrac{1}{x+1}$	$\dfrac{-1}{(x+1)^2}$	$\dfrac{2!}{(x+1)^3}$	$\dfrac{-3!}{(x+1)^4}$	$\dfrac{4!}{(x+1)^5}$	$\dfrac{-5!}{(x+1)^6}$

It is a good idea to compute enough derivatives to be able to discover what pattern, if any, is emerging. Here we can see that

$$f^{(n)}(x) = (-1)^{n-1}\frac{(n-1)!}{(x+1)^n}$$

Setting $x = 0$ gives

$f(0)$	$f'(0)$	$f''(0)$	$f'''(0)$	$f^{(4)}(0)$	$f^{(5)}(0)$	$f^{(6)}(0)\cdots$
0	1	-1	$2!$	$-3!$	$4!$	$-5!$

so that

$$f^{(n)}(0) = (-1)^{n-1}(n-1)!$$

The Taylor series is then

(11.12)
$$\begin{aligned}\ln(x+1) &= 0 + 1x - \frac{1}{2!}x^2 + \frac{2!}{3!}x^3 - \frac{3!}{4!}x^4 \\ &\quad + \frac{4!}{5!}x^5 - \frac{5!}{6!}x^6 + \cdots + (-1)^{n-1}\frac{(n-1)!}{n!}x^n + \cdots \\ &= x - \frac{x^2}{2} + \frac{x^3}{3} - \frac{x^4}{4} + \frac{x^5}{5} - \frac{x^6}{6} + \cdots \\ &= \sum_{n=1}^{\infty}\frac{(-1)^{n-1}x^n}{n}\end{aligned}$$

(c) Here $f(x) = \dfrac{1}{1-x}$ and

f	f'	f''	f'''	$f^{(4)}$	$f^{(5)}$	$\cdots$	$f^{(n)}$
$\dfrac{1}{1-x}$	$\dfrac{1}{(1-x)^2}$	$\dfrac{2!}{(1-x)^3}$	$\dfrac{3!}{(1-x)^4}$	$\dfrac{4!}{(1-x)^5}$	$\dfrac{5!}{(1-x)^6}$	$\cdots$	$\dfrac{n!}{(1-x)^n}$

$f(0)$	$f'(0)$	$f''(0)$	$f'''(0)$	$f^{(4)}(0)$	$f^{(5)}(0)$	$\cdots$	$f^{(n)}(0)$
1	1	$2!$	$3!$	$4!$	$5!$	$\cdots$	$(n)!$

The Taylor series is

$$\frac{1}{1-x} = 1 + 1x + \frac{2!}{2!}x^2 + \frac{3!}{3!}x^3 + \frac{4!}{4!}x^4 + \cdots \tag{11.13}$$

$$= 1 + x + x^2 + x^3 + x^4 + \cdots$$

If we set $x = r$ in (11.13) it becomes

$$\frac{1}{1-r} = 1 + r + r^2 + r^3 + r^4 + \cdots$$

The right-hand side of this is just a geometric series with first term $a = 1$ and common ratio r. We know that when $|r| < 1$ this series converges to

$$\frac{a}{1-r} = \frac{1}{1-r}$$

which is exactly what Equation (11.13) shows.

REMARK Just as in the case of Taylor polynomials, we have so far considered only Taylor series for a function f around $x = 0$. Such series are often called **Maclaurin series**. We could just as easily have extended our discussion to consider Taylor series expansions around $x = a$, for arbitrary a. In this more general case the series is obtained from Equation (11.5) by letting $n \to \infty$:

$$f(x) = f(a) + \frac{f'(a)}{1!}(x-a) + \frac{f''(a)}{2!}(x-a)^2 \tag{11.14}$$

$$+ \frac{f'''(a)}{3!}(x-a)^3 + \cdots + \frac{f^{(n)}(a)}{n!}(x-a)^n + \cdots$$

$$= \sum_{n=0}^{\infty} \frac{f^{(n)}(a)}{n!}(x-a)^n$$

EXAMPLE 13 Construct the Taylor-series expansion for $f(x) = \sqrt{x}$ about $x = 1$.

SOLUTION Setting $f(x) = \sqrt{x} = x^{1/2}$, we compute

f	f'	f''	f'''	$f^{(4)}$	$f^{(5)}$
$x^{1/2}$	$\frac{1}{2}x^{-1/2}$	$-\frac{1}{2}\left(\frac{1}{2}\right)x^{-3/2}$	$-\frac{3}{2}\left(-\frac{1}{2}\right)\left(\frac{1}{2}\right)x^{-5/2}$	$-\frac{5}{2}\left(-\frac{3}{2}\right)\left(-\frac{1}{2}\right)\left(\frac{1}{2}\right)x^{-7/2}$	$-\frac{7}{2}\left(-\frac{5}{2}\right)\left(-\frac{3}{2}\right)\left(-\frac{1}{2}\right)\left(\frac{1}{2}\right)x^{-9/2}$

$f(1)$	$f'(1)$	$f''(1)$	$f'''(1)$	$f^{(4)}(1)$	$f^{(5)}(1)$
1	$\frac{1}{2}$	$-\frac{1}{2}\left(\frac{1}{2}\right)$	$-\frac{3}{2}\left(-\frac{1}{2}\right)\left(\frac{1}{2}\right)$	$-\frac{5}{2}\left(-\frac{3}{2}\right)\left(-\frac{1}{2}\right)\left(\frac{1}{2}\right)$	$-\frac{7}{2}\left(-\frac{5}{2}\right)\left(-\frac{3}{2}\right)\left(-\frac{1}{2}\right)\left(\frac{1}{2}\right)$

From Equation (11.14) we obtain

$$\sqrt{x} = 1 + \frac{1}{2}(x-1) - \frac{1/4}{2!}(x-1)^2 + \frac{3/8}{3!}(x-1)^3 - \frac{15/16}{4!}(x-1)^4 + \frac{105/32}{5!}(x-1)^5 - \cdots$$

$$= 1 + \frac{1}{2}(x-1) - \frac{1}{8}(x-1)^2 + \frac{1}{16}(x-1)^3 - \frac{5}{128}(x-1)^4 + \frac{7}{256}(x-1)^5 - \cdots$$

Section 11.2 EXERCISES

1. Use the Taylor polynomial of degree 6 for $f(x) = e^x$ near 0 to approximate e. Estimate the error.
2. Use the Taylor polynomial of degree 4 for $f(x) = e^x$ near 0 to approximate $e^{1/3} = \sqrt[3]{e}$. Estimate the error.
3. Use the Taylor polynomial of degree 5 for $f(x) = e^{-x}$ near 0 to approximate $e^{-1} = 1/e$. Estimate the error.
4. Approximate cos 0.1 by using the Taylor polynomial of degree 6 for $f(x) = \cos x$ near 0. Estimate the error.
5. Approximate sin 0.02 by using the Taylor polynomial of degree 6 for $f(x) = \sin x$ near 0. Estimate the error.

In Exercises 6–26, write as many terms of the Maclaurin series (that is, the Taylor series near 0) as you can. If possible, find an expression for the general term (that is, the term of degree *n*).

6. $\sin 2x$
7. $\cos 3x$
8. $\sin x + \cos x$
9. $\sin x + \sin 2x$
10. $\cos^2 x$
11. $\tan x$
12. $\sin^3 2x$
13. $\sec x$
14. e^{-x}
15. e^{3x+2}
16. $e^{-x^2/2}$
17. $\sqrt{1+x}$
18. $(1+x)^{\sqrt{2}}$
19. $(1+\alpha)^{\infty}$, α real
20. $\tan^2 x$
21. $\dfrac{1}{1+\sin x}$
22. $\dfrac{e^x - e^{-x}}{2}$
23. $\dfrac{e^x + e^{-x}}{2}$
24. $(\sin x)e^x$
25. $\ln\left(\dfrac{1+x}{1-x}\right)$
26. $\dfrac{1}{1+x^2}$
27. What degree Taylor polynomial near 0 is needed to estimate cos 0.1 within ± 0.0001? Make the computation.
28. Compute sin 0.2 to 3 decimal places.
29. Compute tan 0.1 to 3 decimal places.
30. Compute tan 0.15 to an accuracy of $5(10)^{-4}$. What degree Taylor polynomial is needed?
31. What degree Taylor polynomial for $\ln(1+x)$ near 0 is needed to compute ln 1.2 to within 5×10^{-4}?
32. Compute e to an accuracy of 10^{-5}. What degree Taylor polynomial is needed?
33. Compute $\sqrt[3]{e}$ to 4 decimal places.
34. Compute $1/e$ to an accuracy of 10^{-4}.
35. Make use of the Taylor expansion of $\cos x$ near $x = \pi/3$ to find $\cos(\pi/3 + 0.1)$ with an accuracy of 10^{-4}.
36. Compute cos 29° to 3 places.
37. Compute $\sqrt{65}$ to 4 decimals.
38. Compute $\sqrt[3]{63}$ to 4 decimals.

SECTION 11.3

Manipulation of Taylor Series

In the last section we learned how to compute Taylor series for such familiar functions as e^x, $\sin x$, and $1/(1 - x)$. To do so we employed the series expansion

$$f(0) + f'(0)x + \frac{f''(0)x^2}{2!} + \frac{f'''(0)x^3}{3!} + \cdots$$

as given in Equation (11.10) for the Taylor series of a function $f(x)$ about the point $x = 0$.

This direct method for computing the Taylor series of a function is not always the only way the Taylor series for a function may be found. We will now describe an alternative approach based on the idea of differentiation and integration of Taylor series. The series manipulations we will do can be rigorously justified in a more advanced calculus course.

We view, as we have done before, a (Taylor) series as a sort of limiting case of a polynomial of increasingly larger degree. Then, because the derivative of a polynomial is easily obtained by differentiating it term by term, we can at least conjecture that, if a function $f(x)$ has a certain Taylor series

$$a_0 + a_1x + a_2x^2 + \cdots$$

then the derivative $f'(x)$ should have as its Taylor series the expression obtained by differentiating $a_0 + a_1x + a_2x^2 + \cdots$ term by term to obtain

$$a_1 + 2a_2x + 3a_3x^2 + \cdots$$

An analogous statement concerning the integral of $f(x)$ can also be made.

Let us try these ideas out on some simple examples.

EXAMPLE 14 Suppose we start with the geometric series

$$1 + x + x^2 + x^3 + \cdots \tag{11.15}$$

which we know from part (c) of Example 12 to be the Taylor series for

$$f(x) = \frac{1}{1 - x}$$

When we differentiate this $f(x)$ with respect to x, we obtain

$$f'(x) = \frac{1}{(1 - x)^2}$$

Thus, if $x = \frac{1}{2}$, then

$$f'\left(\frac{1}{2}\right) = \frac{1}{\left(1 - \frac{1}{2}\right)^2} = 4$$

Term-by-term differentiation of the geometric series

$$1 + x + x^2 + x^3 + \cdots$$

results in

$$1 + 2x + 3x^2 + \cdots$$

(As we have said, the idea of differentiating a series in this manner suggests itself because this is the way polynomials are differentiated.) If we now insert $x = \frac{1}{2}$ into the differentiated series, we get

$$1 + 2\left(\frac{1}{2}\right) + 3\left(\frac{1}{2}\right)^2 + 4\left(\frac{1}{2}\right)^3 + \cdots = 1 + \frac{2}{2} + \frac{3}{2^2} + \frac{4}{2^3} + \cdots$$

If our procedure of term-by-term differentiation is valid, this series should tend toward 4. In fact, this series *does* tend toward 4, as you may check with your calculator, and in general it can be shown that

$$f'(x) = \frac{1}{(1-x)^2} = 1 + 2x + 3x^2 + \cdots$$

for $-1 < x < 1$. The justification of term-by-term differentiation of power series, and also term-by-term integration, is given in more advanced courses.

EXAMPLE 15 We have already encountered the Taylor series for $\sin x$, which is

$$x - \frac{x^3}{3!} + \frac{x^5}{5!} - \frac{x^7}{7!} + \cdots$$

Differentiating term by term, we should have

$$f'(x) = \cos x = 1 - \frac{3x^2}{3!} + \frac{5x^4}{5!} - \frac{7x^6}{7!} + \cdots$$

$$= 1 - \frac{x^2}{2!} + \frac{x^4}{4!} - \frac{x^6}{6!} + \cdots$$

Here we have used the fact that

$$\frac{n}{n!} = \frac{n}{(n)(n-1)(n-2)\cdots(2)(1)} = \frac{1}{(n-1)!}$$

In fact, this is indeed the Taylor series expansion of $\cos x$ that we found in solving Exercise 4 in Section 11.2.

EXAMPLE 16 (Integration of a Series)

We start now with $f(x) = 1/(1 + x)$ and compute its Taylor series. We obtain

(11.16)
$$\frac{1}{1+x} = 1 - x + x^2 - x^3 + x^4 - \cdots$$

as you may verify. [Please also note that Equation (11.16) can be obtained by replacing x by $-x$ in Equation (11.15).] We may integrate both sides of this relation term by term to obtain

$$\int \frac{1}{1+x}\,dx = \ln(1 + x) = \int (1 - x + x^2 - x^3 + \cdots)\,dx$$

$$= x - \frac{x^2}{2} + \frac{x^3}{3} - \frac{x^4}{4} + \cdots + c$$

Because $\ln(1+0) = \ln 1 = 0$, we know that c must be 0. Therefore

$$\ln(1+x) = x - \frac{x^2}{2} + \frac{x^3}{3} - \frac{x^4}{4} + \cdots \tag{11.17}$$

This is the same series we obtained directly in part (b) of Example 12.

Using Equation (11.17) we may, for instance, compute ln 1.5. To do so, set $x = 0.5$ and use the relation

$$\begin{aligned}\ln(1.5) &= (0.5) - \left(\frac{0.5}{2}\right)^2 + \left(\frac{0.5}{3}\right)^3 - \left(\frac{0.5}{4}\right)^4 + \cdots \\ &= 0.4055\end{aligned}$$

Let us attempt one final example.

EXAMPLE 17 (a) Find the Taylor-series expansion of $e^{-x^2/2}$.

(b) Use the first two nonzero terms you obtained in (a) to approximate $\int_{-1}^{1} e^{-x^2/2}\,dx$.

SOLUTION

(a) Rather than computing the Taylor-series expansion of $e^{-x^2/2}$ directly, we will make use of the power series

$$e^u = 1 + u + \frac{u^2}{2!} + \frac{u^3}{3!} + \cdots$$

Setting $u = -x^2/2$ gives

$$\begin{aligned}e^{-x^2/2} &= 1 + \left(\frac{-x^2}{2}\right) + \frac{\left(\frac{-x^2}{2}\right)^2}{2!} + \frac{\left(\frac{-x^2}{2}\right)^3}{3!} + \cdots \\ &= 1 - \frac{x^2}{2} + \frac{x^4}{8} - \frac{x^6}{48} + \cdots\end{aligned}$$

(b) If we throw away all but the first two nonzero terms in the power series of part (a), we have

$$e^{-x^2/2} \approx 1 - \frac{x^2}{2}$$

We reason that

$$\begin{aligned}\int_{-1}^{1} e^{-x^2/2}\,dx &\approx \int_{-1}^{1}\left(1 - \frac{x^2}{2}\right)dx = x - \frac{x^3}{6}\Big|_{-1}^{1} \\ &= \left(1 - \frac{1}{6}\right) - \left((-1) - \frac{(-1)}{6}\right) = \frac{5}{6} - \left(-\frac{5}{6}\right) \\ &= \frac{10}{6} = \frac{5}{3} = 1.667\end{aligned}$$

We have already commented, in our discussion of the normal distribution, on the importance of being able to estimate integrals of the form $\int_a^b e^{-x^2/2}\,dx$. Here we see one way to approximate such integrals.

Section 11.3 EXERCISES

Manipulate the geometric series $\frac{1}{1-x} = 1 + x + x^2 + x^3 + \cdots$ by using term-by-term substitution, integration, differentiation, multiplication, or division to generate the given series in Exercises 1–5.

1. $\dfrac{1}{1+x}$

2. $\dfrac{1}{(1+x)^2}$

3. $\ln(1+x)$

4. $\dfrac{x}{1+x^2}$

5. $\ln(1+x^2)$

Approximate $\int_0^z e^{-x^2}\,dx$ to 3 decimal places for each of the values of z given in Exercises 6–11. How does your result compare with the entry corresponding to the given z in Table 10.19 for the standard normal distribution?

6. $z = 1$

7. $z = 0.1$

8. $z = 0.5$

9. $z = 1.96$

10. $z = 1.64$

11. $z = 2.33$

12. $z = 2.58$

Series techniques can be used to approximate definite integrals. For example,

$$\begin{aligned}\int_0^1 xe^x\,dx &= \int_0^1 x\left(1 + x + \frac{x^2}{2!} + \frac{x^3}{3!} + \cdots\right)dx\\ &= \int_0^1 \left(x + x^2 + \frac{x^3}{2!} + \frac{x^4}{3!} + \cdots\right)dx\\ &= \left(\frac{x^2}{2} + \frac{x^3}{3} + \frac{x^4}{2!4} + \frac{x^5}{3!5} + \cdots\right)\Bigg|_0^1\\ &= \left(\frac{1}{2} + \frac{1}{3} + \frac{1}{8} + \frac{1}{30} + \cdots\right) - 0\\ &= 1\end{aligned}$$

Estimate, to 3 decimal places, the area given by each of the integrals in Exercises 13–21. (Use series techniques.)

13. $\displaystyle\int_0^1 \sin x\,dx$

14. $\displaystyle\int_0^1 \cos x\,dx$

15. $\displaystyle\int_0^1 \frac{\sin x}{x}\,dx$

16. $\displaystyle\int_0^1 \frac{x - \sin x}{x}\,dx$

17. $\displaystyle\int_0^{0.1} \frac{x - \sin x}{x^2}\,dx$

18. $\displaystyle\int_0^1 \frac{1 - \cos x}{x}\,dx$

19. $\displaystyle\int_0^{0.1} \frac{1 - \cos x}{x^2}\,dx$

20. $\displaystyle\int_0^1 \frac{e^x - 1}{x}\,dx$

21. $\displaystyle\int_0^{0.2} \frac{e^x - e^{-x}}{x}\,dx$

22. Verify that $\dfrac{1}{1+t^2} = 1 - t^2 + t^4 + \cdots (-1)^{n-1}t^{2(n-1)} + (-1)^n \dfrac{t^{2n}}{1+t^2} + \cdots$

SECTION 11.4

Formal Power Series: A Last Look at Differential Equations

In the first two sections of this chapter we learned how, starting with a function f that has derivatives of all orders, we can write the Taylor-series expansion for f around $x = 0$.

$$f(0) + f'(0)x + \frac{f''(0)}{2!}x^2 + \frac{f'''(0)}{3!}x^3 + \cdots \tag{11.18}$$

As indicated in our examples, if the remainder term $R_n(x)$ tends to 0 as $n \to \infty$, the series given in Equation (11.18) represents $f(x)$, and it can be used to approximate $f(x)$ to any desired degree of accuracy. Indeed, for many basic functions such as $\sin x$, e^x, and $\sqrt{x}$, the only efficient way to compute their values numerically is to make use of their series expansions. The hand calculators that many students now own compute the values of these functions by using Taylor polynomials.

In this section we turn our point of view around. Rather than starting with a specific function f and using it to determine the series in Equation (11.18), we start with a series and use it to determine a function wherever the series converges. For definiteness we start with a series of the form

(11.19) $$a_0 + a_1x + a_2x^2 + a_3x^3 + \cdots$$

also written as $\sum_{n=0}^{\infty} a_nx^n$, where the a_i are constants. Such an infinite series, containing powers of the variable x, is called a **power series**.

How do power series arise? Well, we have already seen that for "nice" functions f we can write down a Taylor series, and this is one example of a power series. In particular, it is a power series in which the coefficients are of the form

$$a_n = \frac{f^n(0)}{n!}$$

There is another way, however, in which power series arise. We touched on it briefly in Section 4.1 when we discussed the exponential function. We asked whether there is a function that is equal to its own derivative. We wanted to know whether there is a function $y = y(x)$ for which

$$y'(x) = y(x)$$

We now recognize this as a differential equation. The solution that we obtained was

$$y(x) = 1 + x + \frac{x^2}{2!} + \frac{x^3}{3!} + \cdots$$

Basically, we constructed a power series that satisfied the differential equation. Let us illustrate this idea for some other differential equations.

EXAMPLE 18 Find a power series of the form

$$y(x) = a_0 + a_1x + a_2x^2 + a_3x^3 + \cdots$$

that satisfies the differential equation

$$y' = y + x^2$$

together with the initial condition

$$y(0) = 2$$

SOLUTION We reason in nonrigorous fashion as follows: Because

$$y(x) = a_0 + a_1x + a_2x^2 + a_3x^3 + a_4x^4 + a_5x^5 + \cdots$$

we would expect $y'(x)$ to be given by

$$y'(x) = a_1 + 2a_2x + 3a_3x^2 + 4a_4x^3 + 5a_5x^4 + \cdots$$

If we now substitute these expressions into the differential equation

$$y' = y + x^2$$

we obtain

$$a_1 + 2a_2x + 3a_3x^2 + 4a_4x^3 + 5a_5x^4 + \cdots$$
$$= (a_0 + a_1x + a_2x^2 + a_3x^3 + a_4x^4 + a_5x^5 + \cdots) + x^2$$

Collecting terms on the right-hand side of this equation yields

$$a_1 + 2a_2x + 3a_3x^2 + 4a_4x^3 + 5a_5x^4 + \cdots$$
$$= a_0 + a_1x + (a_2 + 1)x^2 + a_3x^3 + a_4x^4 + a_5x^5 + \cdots$$

Now it seems plausible that, if the power series on the left is to be equal to the power series on the right, the coefficients of like powers of x should coincide. Thus we have

(11.20)
$$\begin{aligned} a_1 &= a_0 && \text{(constant term on left and right)} \\ 2a_2 &= a_1 && \text{(coefficient of } x \text{ on left and right)} \\ 3a_3 &= a_2 + 1 && \text{(coefficient of } x^2 \text{ on left and right)} \\ 4a_4 &= a_3 && \text{(coefficient of } x^3 \text{ on left and right)} \\ 5a_5 &= a_4 && \text{(coefficient of } x^4 \text{ on left and right)} \\ &\vdots && \vdots \end{aligned}$$

From the initial condition $y(0) = 2$, we determine

$$y(0) = a_0 + a_10 + a_20^2 + a_30^3 + a_40^4 + \cdots = 2$$

so that $a_0 = 2$. It follows from Equation (11.20) that

$$\begin{aligned} a_1 &= a_0 &&= 2 && \text{so} \quad a_1 = 2 \\ 2a_2 &= a_1 &&= 2 && \text{so} \quad a_2 = 1 \\ 3a_3 &= a_2 + 1 &&= 2 && \text{so} \quad a_3 = \frac{2}{3} \\ 4a_4 &= a_3 &&= \frac{2}{3} && \text{so} \quad a_4 = \frac{1}{6} \\ 5a_5 &= a_4 &&= \frac{1}{6} && \text{so} \quad a_5 = \frac{1}{30} \end{aligned}$$

and so on. Thus our power series takes the form

(11.21)
$$y(x) = a_0 + a_1x + a_2x^2 + a_3x^3 + a_4x^4 + a_5x^5 + \cdots$$
$$= 2 + 2x + 1x^2 + \frac{2}{3}x^3 + \frac{1}{6}x^4 + \frac{1}{30}x^5 + \cdots$$

We could obtain as many terms as we want in the series expansion for y by continuing this process.

Is the series in Equation (11.21) really a solution of the differential equation $y' = y + x^2$? Well, it does have the property that, when we substitute it into the differential equation, we get an equality. Such an expression is called a **formal power series solution** to the differential equation

$y' = y + x^2$. The word *formal* refers to the fact that we have not concerned ourselves with the question of whether our series makes sense (converges) for any values of x but have simply followed steps that enabled us to determine the coefficients in the series expansion.

Note that we never really obtain any nice, simple expression for $y(x)$. Rather, the function y is determined only by its series. This is exactly the reverse of considerations in Section 11.2, in which the given function f determined a power series (namely, its Taylor series).

Let's work some more examples in which we find a formal series solution to a differential equation

EXAMPLE 19 Find the formal power series solution of the differential equation

$$y'' = xy$$

satisfying the initial conditions

$$y(0) = 1$$
$$y'(0) = 3$$

We try to express our solution in the form

$$y(x) = a_0 + a_1x + a_2x^2 + a_3x^3 + a_4x^4 + a_5x^5 + \cdots$$

Differentiation yields

$$y'(x) = a_1 + 2a_2x + 3a_3x^2 + 4a_4x^3 + 5a_5x^4 + \cdots$$

and $$y''(x) = 2a_2 + (3)(2)a_3x + (4)(3)a_4x^2 + (5)(4)a_5x^3 + \cdots$$

Substituting these series into the differential equation

$$y'' = xy$$

gives

$$\begin{aligned} 2a_2 + 6a_3x + 12a_4x^2 + 20a_5x^3 + \cdots &= x[a_0 + a_1x + a_2x^2 + a_3x^3 \\ &\quad + a_4x^4 + a_5x^5 + \cdots] \\ &= a_0x + a_1x^2 + a_2x^3 + a_3x^4 \\ &\quad + a_4x^5 + a_5x^6 + \cdots \end{aligned}$$

We reason, as in Example 18, that if the power series on the left is to equal the power series on the right, the coefficient of like powers of x should be equal. Thus

(11.22)
$$\begin{aligned} 2a_2 &= 0 && \text{(constant term on left and right)} \\ 6a_3 &= a_0 && \text{(coefficient of } x \text{ on left and right)} \\ 12a_4 &= a_1 && \text{(coefficient of } x^2 \text{ on left and right)} \\ 20a_5 &= a_2 && \text{(coefficient of } x^3 \text{ on left and right)} \\ &\vdots \quad \vdots \end{aligned}$$

Now, from the initial condition $y(0) = 1$, we find

$$y(0) = a_0 + a_1 \cdot 0 + a_2 \cdot 0^2 + a_3 \cdot 0^3 + a_4 \cdot 0^4 + \cdots = 1$$

so that $a_0 = 1$. Similarly, $y'(0) = 3$ implies that

$$y'(0) = a_1 + 2a_2 \cdot 0 + 3a_3 \cdot 0^2 + 4a_4 \cdot 0^3 + 5a_5 \cdot 0^4 + \cdots = 3$$

and $a_1 = 3$. Using these values together with Equation (11.22) gives

$$\begin{aligned}
& & a_0 &= 1 \\
& & a_1 &= 3 \\
2a_2 &= 0, & \text{so} \quad a_2 &= 0 \\
6a_3 &= a_0 = 1 & \text{so} \quad a_3 &= \frac{1}{6} \\
12a_4 &= a_1 = 3 & \text{so} \quad a_4 &= \frac{1}{4} \\
20a_5 &= a_2 = 0 & \text{so} \quad a_5 &= 0 \\
&\vdots & &\vdots
\end{aligned}$$

Our formal power series solution in this case is

$$\begin{aligned}
y(x) &= a_0 + a_1 x + a_2 x^2 + a_3 x^3 + a_4 x^4 + a_5 x^5 + \cdots \\
&= 1 + 3x + 0x^2 + \frac{1}{6}x^3 + \frac{1}{4}x^4 + 0x^5 + \cdots \\
&= 1 + 3x + \frac{1}{6}x^3 + \frac{1}{4}x^4 + \cdots
\end{aligned}$$

EXAMPLE 20 Find the formal power series solution of

$$y'' = -y$$

satisfying the initial conditions

$$y(0) = 2$$
$$y'(0) = 1$$

SOLUTION This problem looks very much like the previous one. We set

$$\begin{aligned}
y(x) &= a_0 + a_1 x + a_2 x^2 + a_3 x^3 + a_4 x^4 + \cdots \\
y'(x) &= \quad a_1 + 2a_2 x + 3a_3 x^2 + 4a_4 x^3 + \cdots \\
y'' &= \quad 2a_2 + (3)(2)a_3 x + (4)(3)a_4 x^2 + \cdots
\end{aligned}$$

Substituting into the differential equation

$$y'' = -y$$

gives

$$\begin{aligned}
&2a_2 + (3)(2)a_3 x + (4)(3)a_4 x^2 + (5)(4)a_5 x^3 + (6)(5)a_6 x^4 + \cdots \\
&\quad = -[a_0 + a_1 x + a_2 x^2 + a_3 x^3 + a_4 x^4 + \cdots] \\
&\quad = -a_0 - a_1 x - a_2 x^2 - a_3 x^3 - a_4 x^4 \cdots
\end{aligned}$$

Again we equate the coefficients of like powers of x.

$$
\begin{aligned}
2a_2 &= -a_0 \\
(3)(2)a_3 &= -a_1 \\
(4)(3)a_4 &= -a_2 \qquad (11.23) \\
(5)(4)a_5 &= -a_3 \\
(6)(5)a_6 &= -a_4
\end{aligned}
$$

From our initial condition $y(0) = 2$, we discover that

$$y(0) = a_0 + a_1 \cdot 0 + a_2 \cdot 0^2 + a_3 \cdot 0^3 + a_4 \cdot 0^4 + \cdots = 2$$

and $a_0 = 2$. Also, $y'(0) = 1$, which tells us that

$$y'(0) = a_1 + 2a_2 \cdot 0 + 3a_3 \cdot 0^2 + 4a_4 \cdot 0^3 ++ \cdots = 1$$

so $a_1 = 1$. Combining this with Equation (11.23) yields

$$
\begin{aligned}
a_0 &= 2 \\
a_1 &= 1
\end{aligned}
$$

$$
\begin{aligned}
2a_2 &= -a_0 = -2 & &\text{so} \quad a_2 = -\frac{2}{2} \\
(3)(2)a_3 &= a_1 = -1 & &\text{so} \quad a_3 = -\frac{1}{(3)(2)} \\
(4)(3)a_4 &= -a_2 = -(-1) & &\text{so} \quad a_4 = \frac{2}{(4)(3)(2)} \\
(5)(4)a_5 &= -a_3 = -\left[-\frac{1}{(3)(2)}\right] & &\text{so} \quad a_5 = \frac{1}{(5)(4)(3)(2)} \\
(6)(5)a_6 &= -a_4 = -\frac{2}{(4)(3)(2)} & &\text{so} \quad a_6 = -\frac{2}{(6)(5)(4)(3)(2)}
\end{aligned}
$$

and so on. Our formal power series solution is

$$
\begin{aligned}
y(x) &= a_0 + a_1 x + a_2 x^2 + a_3 x^3 + a_4 x^4 + \cdots \\
&= 2 + 1x - \frac{2}{2!}x^2 - \frac{1}{3!}x^3 + \frac{2}{4!}x^4 + \frac{1}{5!}x^5 - \frac{2}{6!}x^6 - \cdots
\end{aligned}
$$

Do not recognize this series? Do not be upset if you don't; it is not easy to see. Let us regroup even and odd terms separately, to obtain

$$
\begin{aligned}
y(x) &= (2 - \frac{2x^2}{2!} + \frac{2x^4}{4!} - \frac{2x^6}{6!} + \cdots) + (x - \frac{x^3}{3!} + \frac{x^5}{5!} - \frac{x^7}{7!} + \cdots) \\
&= 2(1 - \frac{x^2}{2!} + \frac{x^4}{4!} - \frac{x^6}{6!} + \cdots) + (x - \frac{x^3}{3!} + \frac{x^5}{5!} - \frac{x^7}{7!} + \cdots)
\end{aligned}
$$

Now you may recognize it. The first set of parentheses contains the Taylor expansion for $\cos x$, whereas the second contains the Taylor series for $\sin x$.

Thus

$$y(x) = 2 \cos x + \sin x$$

We leave it for you to check that this function does indeed satisfy $y'' = -y$ with the initial conditions $y(0) = 2$ and $y'(0) = 1$.

In most situations wherein we solve differential equations by series, we are *not* lucky enough to recognize the answer as the Taylor series of some function we already know. In those cases we must use the power series itself as our solution. In fact, some power series that first arose as the solution of certain differential equations have turned out to be so important in applications to physical problems that names have been given to them. The so-called Bessel functions are well-known examples of this. We shall not discuss such matters any further. If you want to know more about such series solutions of differential equations, see Boyce and DiPrima, *Elementary Differential Equations*, (New York: John Wiley, 1977).

Section 11.4 EXERCISES

Use the technique of formal power series to find one or more solutions of the following differential equations (some of which include initial conditions).

1. $y' = y$; $y(0) = 3$

2. $y' = 2y$; $y(0) = 2$

3. $y' = -y$; $y(0) = -1$

4. $y' = y - 2$

5. $y' = 2y + 3$

6. $y' = x + y$; $y(0) = -1$

7. $y' = xy$; $y(0) = 1$

8. $y' = y/x$

9. $y' = x/y$, $y(0) = 1$

10. $y' = x + x^2$; $y(0) = 8$

11. $y'' = -y$; $y(0) = 3$, $y'(0) = 2$

12. $y'' = -y$; $y(0) = 1$, $y'(0) = -1$

13. $y'' = -4y$; $y(0) = a$, $y'(0) = b$

14. $y' = y + \cos x$

15. $y'' + 2y' = -y$

16. $y'' = xy$; $y(0) = 1$, $y'(0) = -1$

SECTION 11.5

Numerical Approximation and Newton's Method

We have seen how to expand a function f in a Taylor series about $x = a$:

$$f(a) + f'(a)(x - a) + \frac{f''(a)}{2!}(x - a)^2 + \cdots$$

and also how such a series can often be used to approximate the value of $f(x)$. In general, the more terms of the Taylor series expansion of f that we use, the better our approximation will be.

It may surprise you, but quite often in practice a very good approximation can be obtained by using only the first two terms of the Taylor series. That is, we use the approximation

$$f(x) \approx f(a) + f'(a)(x - a) \qquad \textbf{(11.24)}$$

and this works well provided that x is sufficiently close to a.

Let us rewrite Equation (11.24) by letting

$$h = x - a$$

Then $x = a + h$, and we have

(11.25) $$f(a + h) \approx f(a) + f'(a)h$$

Let us see how to make use of this approximation.

EXAMPLE 21 Use Equation (11.25) to estimate $\sqrt[3]{8.01}$.

SOLUTION In order to use Equation (11.25) we must have a function f. It seems reasonable to set

$$f(x) = \sqrt[3]{x} = x^{1/3}$$

Our problem is now to estimate $f(x)$ when $x = 8.01$. What values should we choose for a and h? Well, because we will have to calculate $f(a) = \sqrt[3]{a}$, the number a should be chosen so that $f(a)$ is easy to compute. It should also be fairly close to the number of interest, $x = 8.01$. A moment's thought tells us that $\sqrt[3]{8} = 2$ exactly, so we let $a = 8$. Then $h = x - a = 8.01 - 8 = 0.01$. Now

$$f'(x) = \frac{1}{3}x^{-2/3}$$

so $$f'(a) = f'(8) = \frac{1}{3}(8)^{-2/3} = \frac{1}{12} = 0.083$$

Substituting these values into (11.25) yields

$$\underset{\substack{\uparrow \\ f(a+h)}}{\sqrt[3]{8.01}} \approx \underset{\substack{\uparrow \\ f(a)}}{2} + \underset{\substack{\uparrow \\ f'(a)}}{(0.083)}\underset{\substack{\uparrow \\ h}}{(0.01)} = 2.00083$$

EXAMPLE 22 A biologist wishes to measure the volume of a red blood cell, which is assumed to be spherical in shape. He measures the diameter of the cell as 7.50 microns, with a possible error of ± 0.02 microns. Estimate the maximum possible error in computing the volume of the cell.

SOLUTION If x is the diameter of a solid sphere, its radius is $x/2$. The volume is given by $V = (4/3)\,\pi r^3$, which in terms of x is

$$V = \frac{4}{3}\pi\left(\frac{x}{2}\right)^3 = \frac{\pi x^3}{6}$$

Let us put

$$f(x) = \frac{\pi x^3}{6}$$

so that

$$f'(x) = \frac{3\pi}{6}x^2 = \frac{\pi}{2}x^2$$

If we now set $a = 7.50$ and $h = \pm 0.02$, the approximation

$$f(a + h) \approx f(a) + f'(a)h$$

yields

$$\begin{aligned} f(7.50 \pm 0.02) &\approx f(7.50) + f'(7.50)(\pm 0.02) \\ &= \frac{\pi}{6}(7.50)^3 + \frac{\pi}{2}(7.50)^2(\pm 0.02) \\ &= 220.89 \pm 1.77 \end{aligned}$$

The maximum possible error is 1.77 out of 220.89, or roughly 0.8%.

The approximation formula (11.25) forms the basis of a very useful iterative technique for finding roots of a function. Before going into detail, let us familiarize ourselves with the idea of an iterative technique by discussing a somewhat surprising trick for approximating square roots.

Given a positive number a, proceed to carry out the following steps, the results of which we hope will approach $\sqrt{a}$.

1 Make a "rough guess" at an answer. Call your guess x_0. (Don't take $x_0 = 0$, though.)

2 Compute $\dfrac{x_0 + \dfrac{a}{x_0}}{2}$ and call it x_1.

3 Compute $\dfrac{x_1 + \dfrac{a}{x_1}}{2}$ and call it x_2.

4 Continue, successively, to compute

(11.26)
$$x_n = \frac{x_{n-1} + \dfrac{a}{x_{n-1}}}{2}$$

5 Stop whenever you feel that x_n^2 is sufficiently close to a. Let us see how this procedure works for $a = 2$. That is, we will approximate $\sqrt{2}$. For our initial guess we will try $x_0 = 1$. Then

$$x_1 = \frac{x_0 + \dfrac{a}{x_0}}{2} = \frac{1 + \dfrac{2}{1}}{2} = \frac{3}{2} = 1.5$$

$$x_2 = \frac{x_1 + \dfrac{a}{x_1}}{2} = \frac{1.5 + \dfrac{2}{1.5}}{2} = \frac{1.500 + 1.333}{2} = 1.417$$

$$x_3 = \frac{x_2 + \dfrac{a}{x_2}}{2} = \frac{1.417 + \dfrac{2}{1.417}}{2} = 1.414\ldots$$

Because $(1.414)^2 = 1.999\ldots$, our answer is sufficiently good and we stop. After only three repetitions of our technique, we have obtained a very fine approximation to $\sqrt{2}$.

PRACTICE EXERCISE Use the same procedure to approximate $\sqrt{2}$ by choosing other values for the "rough guess" x_0.

The type of procedure we have just illustrated is an **iterative** technique. Methods such as this one form the backbone of an important mathematical subject called **numerical analysis**. As you can see, our iterative technique utilizes the simple idea that at each stage we repeat the same type of calculation, using the value computed at the previous stage. As each step is performed, our results draw successively closer to the desired answer. The modern computer (and even some hand calculators) are ideal instruments for carrying out such iterative techniques.

We would like to explain how the rule, or algorithm, for finding $\sqrt{a}$:

$$\text{Let } x_n = \frac{x^{n-1} + \dfrac{a}{x_{n-1}}}{2}$$

comes about, and how you can make up similar rules yourself.

The basic for this square root algorithm is a technique known as **Newton's method**, which is used to locate the roots of a function f. Newton's method, in turn, is based on the simple approximation formula

$$f(x) \approx f(a) + f'(a)(x - a)$$

which we have already examined.

We begin with a function f and consider the problem of finding a root x_* of f—that is, a real number x_* for which

$$f(x_*) = 0$$

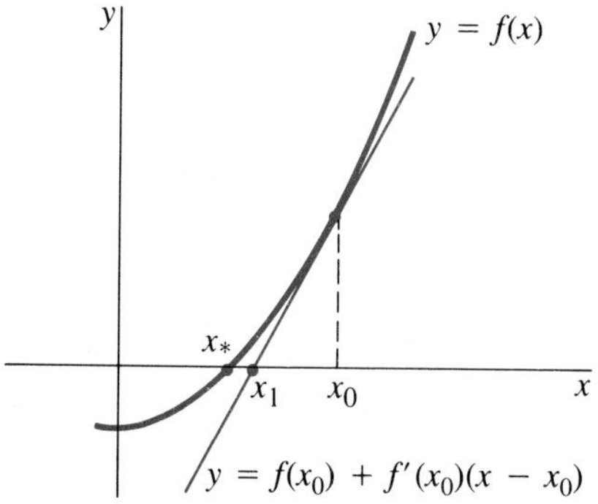

FIGURE 11.9

Suppose we make an initial guess of x_0 for the root. Then the point $(x_0, f(x_0))$ will lie on the graph of $y = f(x)$, not too far, we hope, from where the graph crosses the x-axis: $y = 0$. (The point where it crosses is the number x_* that we seek.)

At the point $(x_0, f(x_0))$ on the graph of $y = f(x)$, the tangent line will have slope $f'(x_0)$, and the equation of this line must be

$$y - f(x_0) = f'(x_0)(x - x_0)$$

or

$$y = f(x_0) + f'(x_0)(x - x_0)$$

See Figure 11.9. Note that, if we replace the number x_0 by a, the expression $f(x_0) + f'(x_0)(x - x_0)$ is just

$$f(a) + f'(a)(x - a)$$

which is the approximation formula for $f(x)$ derived in Equation (11.24), valid for values of x near a.

To put it another way, in the vicinity of $(x_0, f(x_0))$, the straight line $y = f(x_0) + f'(x_0)(x - x_0)$ looks a lot like the curve $y = f(x)$. In the language of Section 11.1, $y = f(x_0) + f'(x_0)(x - x_0)$ and $y = f(x)$ have 1st-order contact at $x = x_0$.

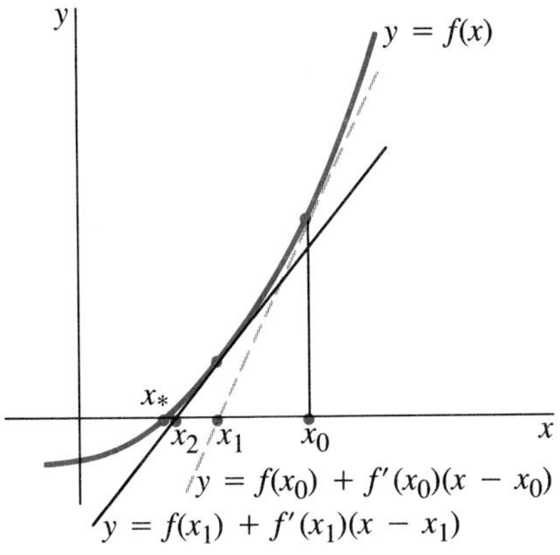

FIGURE 11.10

We might expect that the root of the equation

$$f(x_0) + f'(x_0)(x - x_0) = 0$$

(which is just the place where the tangent line crosses the x-axis) will be close to the root x_* of $f(x) = 0$.

Explicitly, we wish to solve

$$f(x_0) + f'(x_0)(x - x_0) = 0$$

This will be satisfied whenever

$$x - x_0 = \frac{f(x_0)}{f'(x_0)} \quad \text{or} \quad x = x_0 - \frac{f(x_0)}{f'(x_0)}$$

[Let us assume here that $f'(x_0) \neq 0$.] If we call this value of x, x_1, so that

$$x_1 = -x_0 - \frac{f(x_0)}{f'(x_0)}$$

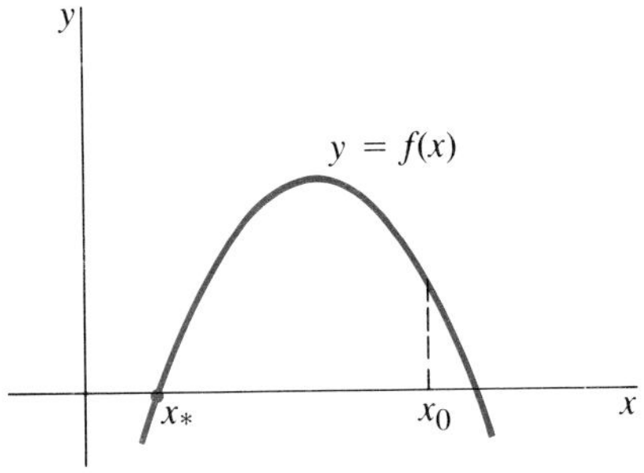

FIGURE 11.11
A bad initial guess x_0

we expect that x_1 will be a closer approximation to x_* than our original rough guess x_0 (see Figure 11.9). To improve on x_1, we can repeat our entire procedure. First we find the point $(x_1, f(x_1))$ on the graph of $y = f(x)$. Next we draw the tangent line to the curve at $(x_1, f(x_1))$ and see where it hits the x-axis. If we denote this point by x_2, then

$$x_2 = x_1 - \frac{f(x_1)}{f'(x_1)} \qquad \text{provided } f'(x_1) \neq 0$$

and we take x_2 as our next approximation to x_* (see Figure 11.10).

In general, given any approximation x_{n-1} we compute the next approximation x_n by

(11.27)
$$x_n = x_{n-1} - \frac{f(x_{n-1})}{f'(x_{n-1})}$$

provided $f'(x_{n-1}) \neq 0$. This iterative technique is known as **Newton's method**. Under most circumstances the sequence of approximate roots $x_0, x_1, x_2, \ldots$ will approach a real root x_* of $f(x) = 0$.

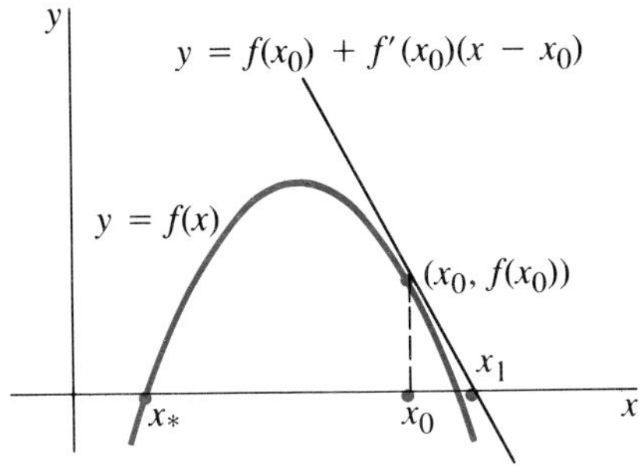

FIGURE 11.12

We say that Newton's method will work "under most circumstances," but there are several things that can go wrong. For example, suppose we wish to estimate the root x_* of the function f whose graph is shown in Figure 11.11. If the initial guess is too far to the right of x_*, our iteration procedure will lead us to the wrong root (see Figure 11.12). There are other undesirable things that can happen, too. Occasionally the sequence $x_0, x_1, x_2, \ldots$ generated by Newton's method will not only not converge to x_*, but it will not converge to anything. We will ignore such difficulties, however, and assume throughout that our choice of x_0, somehow or other, will be a good initial guess.

EXAMPLE 23 Derive the algorithm discussed previously for estimating $\sqrt{2}$ by means of Newton's method.

SOLUTION Let

$$f(x) = x^2 - 2$$

Then we see that a root of the equation $0 = f(x) = x^2 - 2$ is $x_* = \sqrt{2}$. If we estimate this root by means of Newton's method, we will simultaneously be approximating $\sqrt{2}$.

Assuming that we make an initial guess x_0 that is reasonably close to x_* and proceed to compute $x_1, x_2, x_3, \ldots$ by means of Equation (11.27):

$$x_n = x_{n-1} - \frac{f(x_{n-1})}{f'(x_{n-1})}$$

In our case $f(x) = x^2 - 2$ and $f'(x) = 2x$, so

$$\frac{f(x)}{f'(x)} = \frac{x^2 - 2}{2x}$$

Thus

$$x_n = x_{n-1} - \frac{x_{n-1}^2 - 2}{2x_{n-1}} = \frac{2x_{n-1}^2 - x_{n-1}^2 + 2}{2x_{n-1}}$$

$$= \frac{x_{n-1}^2 + 2}{2x_{n-1}}$$

which can be rewritten as

$$x_n = \frac{x_{n-1} + \dfrac{2}{x_{n-1}}}{2}$$

This is precisely Equation (11.26) with $a = 2$. Thus we iteratively should define

$$x_1 = \frac{x_0 + \dfrac{2}{x_0}}{2}$$

$$x_2 = \frac{x_1 + \dfrac{2}{x_1}}{2}$$

and so on.

PRACTICE EXERCISE Following the method of Example 23, derive an algorithm for estimating $\sqrt{13}$. (*Hint*: Set $f(x) = x^2 - 13$.)

[Your answer should be

$$x_n = \frac{x_{n-1} + \dfrac{13}{x_{n-1}}}{2}$$

Try $x_0 = 3$ as an initial guess and use the formula you've derived to approximate $\sqrt{13}$.]

EXAMPLE 24 Derive a method for estimating $\sqrt[3]{a}$ for any number a.

SOLUTION This time we consider the function

$$f(x) = x^3 - a$$

Then $x_* = \sqrt[3]{a}$ is a root of the equation $f(x) = x^3 - a = 0$. Now,

$$f(x) = x^3 - a \quad \text{and} \quad f'(x) = 3x^2$$

Thus, from Newton's method [Equation (11.27)],

(11.28)

$$x_n = x_{n-1} - \frac{f(x_{n-1})}{f'(x_{n-1})} = x_{n-1} - \frac{x_{n-1}^3 - a}{3x_{n-1}^2}$$

$$= \frac{3x_{n-1}^3 - x_{n-1}^3 + a}{3x_{n-1}^2} = \frac{2x_{n-1}^3 + a}{3x_{n-1}^2}$$

$$= \frac{2x_{n-1} + \dfrac{a}{x_{n-1}^2}}{3}$$

To see how this formula might work, let's try to approximate $\sqrt[3]{6}$, taking $x_0 = 2$ as our initial guess. Then, according to Equation (11.28),

$$x_1 = \frac{2x_0 + \dfrac{6}{x_0^2}}{3} = \frac{(2)(2) + \dfrac{6}{(2)^2}}{3}$$

$$= \frac{\dfrac{11}{2}}{3} = \frac{11}{6} \approx 1.8333$$

$$x_2 = \frac{2x_1 + \dfrac{6}{x_1^2}}{3} = \frac{2(1.8333) + \dfrac{6}{(1.8333)^2}}{3}$$

$$= \frac{3.6666 + 1.7852}{3} \approx 1.8173$$

$$x_3 = \frac{2x_2 + \dfrac{6}{x_2^2}}{3} = \frac{2(1.8173) + \dfrac{6}{(1.8173)^2}}{3}$$

$$= \frac{3.6346 + 1.8168}{3} \approx 1.8171$$

If you consult a table of cube roots or use your calculator, you will find that our approximation of $x_3 = 1.8171$, obtained after only three iterations, is accurate to four decimal places. Quite remarkable, isn't it? In the problems at the end of this section, you will be asked to develop algorithms for estimating quantities other than square roots and cube roots.

EXAMPLE 25 Use Newton's method to estimate the largest root of the equation $x^3 - 5x + 3 = 0$.

SOLUTION The function $f(x) = x^3 - 5x + 3$, shown in Figure 11.13, has a maximum at $x = -\sqrt{5/3}$ and a minimum at $x = \sqrt{5/3}$. The maximum is positive and the minimum is negative. Thus the function has three real roots.

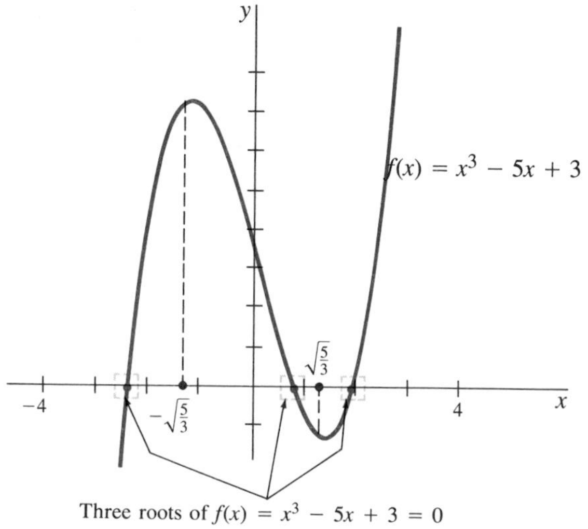

FIGURE 11.13

(This information is obtained by utilizing the curve-sketching ideas developed in Chapter 3.)

We are seeking the largest root, which is greater than $\sqrt{5/3} = \sqrt{1.66}$. Let us try an initial guess of $x_0 = 2$ and see what happens. Then

$$x_n = x_{n-1} - \frac{f(x_{n-1})}{f'(x_{n-1})} = x_{n-1} - \frac{x_{n-1}^3 - 5x_{n-1} + 3}{3x_{n-1}^2 - 5}$$

or

$$x_n = \frac{2x_{n-1}^3 - 3}{3x_{n-1}^2 - 5}$$

Thus, with $x_0 = 2$,

$$x_1 = \frac{2x_0^3 - 3}{3x_0^2 - 5} = \frac{2(2)^3 - 3}{3(2)^2 - 5} \approx 1.857$$

$$x_2 = \frac{2x_1^3 - 3}{3x_1^2 - 5} = \frac{2(1.857)^3 - 3}{3(1.857)^2 - 5} \approx 1.835$$

$$x_3 = \frac{2x_2^3 - 3}{3x_2^2 - 5} = \frac{2(1.835)^3 - 3}{3(1.835)^2 - 5} \approx 1.834$$

$$x_4 = \frac{2x_3^3 - 3}{3x_3^2 - 5} = \frac{2(1.834)^3 - 3}{3(1.834)^2 - 5} \approx 1.834$$

There seems to be no point in continuing. After four steps our approximation to the largest root of $x^3 - 5x + 3 = 0$ is $x \approx 1.834$.

EXAMPLE 26 Use Newton's method to find an approximation to the solution of the equation

$$e^x = \sin x + 2$$

SOLUTION Graphs of the functions $y = e^x$ and $y = \sin x + 2$ are both shown in Figure 11.14. They cross at only one point, and this is the point we are looking for. Set

$$f(x) = e^x - \sin x - 2$$

By Newtons' method,

$$x_n = x_{n-1} - \frac{f(x_{n-1})}{f'(x_{n-1})} = x_{n-1} - \frac{e^{x_{n-1}} - \sin x_{n-1} - 2}{e^{x_{n-1}} - \cos x_{n-1}}$$

If we make an initial guess of $x_0 = 1$, then

$$x_1 = 1 - \frac{e^1 - \sin 1 - 2}{e^1 - \cos 1} \approx 1.057$$

$$x_2 = 1.057 - \frac{e^{1.057} - \sin(1.057) - 2}{e^{1.057} - \cos(1.057)} \approx 1.054$$

$$x_3 = 1.054 - \frac{e^{1.054} - \sin(1.054) - 2}{e^{1.054} - \cos(1.054)} \approx 1.054$$

No improvement in our answer is taking place. After only three iterations, our approximation to the solution of $e^x = \sin x + 2$ is $x = 1.054$. (If our initial guess had not been such a good one, it would have taken us a little longer.)

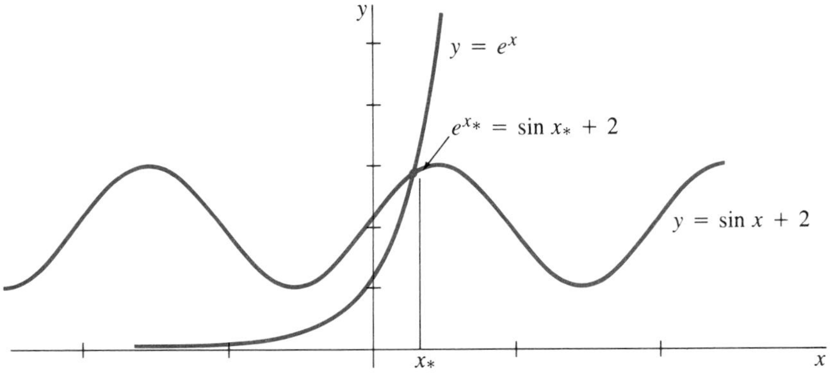

FIGURE 11.14

Section 11.5 EXERCISES

1. Use Equation (11.25) to approximate $\sqrt{5}$. Observe that all we are doing is using the Taylor polynomial of degree 1 for $f(x) = \sqrt{x}$ near $a = 4$ to approximate $f(5)$—this is called taking "the linear approximation."

2. Approximate $\sqrt{5}$ by using the Taylor polynomial of degree 2 for $f(x) = \sqrt{x}$ near $a = 4$ to approximate $f(5)$—this is called taking "the quadratic approximation."

3. Apply Newton's method to approximate the positive root of $f(x) = x^2 - 5$ and thus to get an approximation to $\sqrt{5}$.

4. Approximate $\sqrt{63}$ by using the linear approximation (as in Exercise 1).

5. Approximate $\sqrt{63}$ by using the quadratic approximation (as in Exercise 2).

6. Use Newton's method to approximate $\sqrt{63}$ (as in Exercise 3).

Use the linear approximation (as in Exercise 1) to approximate the radicals in Exercises 7–9.

7. $\sqrt[3]{2}$ 8. $\sqrt[4]{2}$ 9. $\sqrt[4]{17}$

Use the quadratic approximation (as in Exercise 2) to approximate the radicals in Exercises 10–12.

10. $\sqrt[3]{2}$ 11. $\sqrt[4]{2}$ 12. $\sqrt[4]{17}$

Use Newton's method (as in Exercise 3) to approximate the radicals in Exercises 13–15.

13. $\sqrt[3]{2}$ 14. $\sqrt[4]{2}$ 15. $\sqrt[3]{17}$

16. Sketch the graph of $f(x) = x^2 + 3x - 2$ and then use Newton's method to approximate the positive root.

17. Find an approximation to the negative root in Exercise 16.

18. Sketch the graph of $f(x) = x^3 + x - 3$ and show that this polynomial has just one real root. (That is, show that the graph crosses the x-axis once). Use Newton's method to approximate this root.

19. The graph of $f(x) = x^3 - 3x + 1$ has three real roots. Approximate each of these roots by using Newton's method.

20. Sketch the graph of $f(x) = x^4 - 4x^3 - 2x^2 + 12x + 1$ and observe that this polynomial has four real roots. Approximate these roots (as many of them as you wish) by using Newton's method.

21. Approximate all real roots of $f(x) = x^5 + x + 1$.

22. Approximate all real roots of $f(x) = x^5 - 5x - 1$.

Find, as best you can, a positive number x for which the relationships in Exercises 23–25 hold.

23. $x = 2 \sin x$ 24. $e^x = x^4$

25. $e^x = \cos x + 2$

26. Consider the function $f(x) = \sqrt[3]{x} = x^{1/3}$. Suppose we have failed to notice that the graph crosses the x-axis at the origin and want to apply Newton's method to approximate this root $x = 0$. If we start with $x_0 = 1$, what happens? What happens if we start with $x_0 = 1/8$? What happens if we start with an arbitrary $x_0 \neq 0$?

IF YOU WORK WITH A COMPUTER (OPTIONAL)

Use either one of the computer programs that follow to work Exercises 27–32.

27. Find the smallest positive number x satisfying $\tan x = 2x$.

28. Let $f(x) = x^3 - 7x^2 + 5x + 19$.
    (a) Apply Newton's method with the initial guess $x = -1$.
    (b) Apply Newton's method with the initial guess $x = 1$. Sketch the graph of f and explain why Newton's method failed.

29. A certain product has the following demand and supply equations:

$$\text{Demand:} \qquad p = 20 - \frac{x^2}{100}$$

$$\text{Supply:} \qquad p = \left(85 + \frac{x^2}{100}\right)^{1/2}$$

Find the demand and price at market equilibrium.

30. Let $f(x) = xe^{-x}$.
    (a) Apply Newton's method with the initial guess $x = \frac{1}{2}$.
    (b) Apply Newton's method with the initial guess $x = 2$. Sketch the graph of f and explain why Newton's method failed.
    (c) Apply Newton's method with the initial guess $x = 1$. Explain why Newton's method failed.

31. Redo Exercise 19.

32. Redo Exercise 20.

```
10   REM - Newton's Method - BASIC version

20   REM - define function and its derivative
30   DEF FNf(x) = EXP(x)-SIN(x)-2
40   DEF FNd(x) = EXP(x)-COS(x)
50   epsilon=.00001
```

```
(* Newton's Method - Pascal Version *)
program Newton;

 const
  epsilon = 0.00001;
```

(*Programs continue at top of next page*)

```
60    INPUT "Initial guess"; oldx

70    REM - Heading (Optional)
80    PRINT " x", " f(x)"

90    REM - compute next approximation
100   newx = oldx - FNf(oldx)/FNd(oldx)
110   REM - print successive approximations (optional)
120   PRINT newx, FNf(newx)
130   IF ABS(newx-oldx) < epsilon THEN 160
140   oldx=newx
150   GOTO 100

160   PRINT newx;"is an approximate root"
170   PRINT "f(";newx;") = ";FNf(newx)
180   END
```

```
var
  oldx, newx : real;

(* define function and its derivative *)
 function f (x : real) : real;
 begin
   f := x * x - 13
 end;

 function deriv (x : real) : real;
 begin
   deriv := 2 * x
 end;

begin
 write('Initial guess? ');
 readln(newx);
 writeln('  x', 'f(x)' : 16);   (*heading*)

 repeat
   oldx := newx;
   newx := oldx - f(oldx) / deriv(oldx);
   writeln(newx : 1 : 5, '      ', f(newx) : 1 : 5);
 until abs(newx - oldx) < epsilon;

 writeln;
 writeln(newx : 1 : 5, 'is an approximate root.');
 writeln('f(', newx : 1 : 5, ') = ', f(newx) : 1 : 5)
end.
```

SECTION 11.6

L'Hôpital's Rule

Suppose we are given two functions, f and g, and we wish to compute

$$\lim_{x \to a} \frac{f(x)}{g(x)}$$

As we have seen, in many cases this limit is evaluated without difficulty simply by substituting $x = a$:

$$\lim_{x \to 2} \frac{x^2 + 2}{x^3 + 1} = \frac{(2)^2 + 2}{(2)^3 + 1} = \frac{6}{9} = \frac{2}{3}$$

or

$$\lim_{x \to 2} \frac{x - 2}{x + 8} = \frac{2 - 2}{2 + 8} = \frac{0}{10} = 0$$

We can sometimes even assign a "value" if the denominator vanishes at $x = a$ but the numerator does not, as in

$$\lim_{x \to 2} \frac{x}{(x-2)^2} = \infty$$

If, however, at $x = a$ both $f(x)$ and $g(x)$ are 0, as in

$$\lim_{x \to 3} \frac{x^2 - 9}{x - 3} \quad \text{or} \quad \lim_{x \to 0} \frac{\sin x}{x}$$

we cannot find the limit by simply evaluating both numerator and denominator at $x = a$. In each of the last two examples we would obtain an ambiguous 0/0 by doing so.

We do know from a geometric argument in Chapter 9 that

$$\lim_{x \to 0} \frac{\sin x}{x} = 1$$

so in this case the 0/0 form actually stands for 1. Also $x^2 - 9 = (x - 3) \cdot (x + 3)$, so

$$\frac{x^2 - 9}{x - 3} = \frac{(x-3)(x+3)}{x - 3} = x + 3$$

for $x \neq 3$, which implies that

$$\lim_{x \to 3} \left(\frac{x^2 - 9}{x - 3} \right) = \lim_{x \to 3} (x + 3) = 6$$

This time 0/0 stands for 6.

In fact it is easy to cook up examples in which the form 0/0 will arise but where the true limiting value is anything at all. Because of this we refer to an expression $\lim_{x \to a} f(x)/g(x)$ with $f(a) = 0$ and $g(a) = 0$ as an **indeterminate form.** Our task in this section is to see how to evaluate indeterminate forms. The principal result that enables us to compute limits of this type is **L'Hôpital's rule**.

THEOREM 11.3 (L'Hôpital's Rule for 0/0) Let f and g be two functions that have Taylor series expansions on some interval about the point $x = a$. Suppose further that $f(a) = g(a) = 0$ [so that $\lim_{x \to a} f(x)/g(x)$ is an indeterminate form, 0/0] and that $g'(a) \neq 0$. Then

$$\lim_{x \to a} \frac{f(x)}{g(x)} = \lim_{x \to a} \frac{f'(x)}{g'(x)} \qquad \boxminus$$

Before proving this theorem, let us work a few examples to make sure we understand how to use it.

EXAMPLE 27 Find the indicated limits, using L'Hôpital's rule wherever it applies.

(a) $\lim_{x \to 3} \frac{x^2 - 9}{x - 3}$ (b) $\lim_{x \to 0} \frac{\sin 2x}{x}$ (c) $\lim_{x \to 0} \frac{\sin x}{e^x - e^{-x}}$

(d) $\lim_{x \to 0} \frac{1 - \cos x}{1 + \cos x}$ (e) $\lim_{x \to 0} \frac{1 - \cos 3x}{x^2}$

SOLUTION

(a) Let $f(x) = x^2 - 9$ and $g(x) = x - 3$. Then

$$f(3) = g(3) = 0$$

so $\lim_{x \to 3} f(x)/g(x)$ is the indeterminate form 0/0. We can easily compute $f'(x) = 2x$ and $g'(x) = 1$. Now, by L'Hôpital's rule,

$$\lim_{x \to 3} \frac{x^2 - 9}{x - 3} = \lim_{x \to 3} \frac{2x}{1} = \frac{6}{1} = 6$$

Note that we did *not* differentiate $\frac{x^2 - 9}{x - 3}$ by the quotient rule but rather took the *derivative* of the *numerator* and the *denominator separately.*

(b) This time $f(x) = \sin 2x$ and $g(x) = x$. Again

$$f(0) = g(0) = 0$$

so $\lim_{x \to 0} \frac{\sin 2x}{x}$ is of the form 0/0. By L'Hôpital's rule,

$$\lim_{x \to 0} \frac{\sin 2x}{x} = \lim_{x \to 0} \frac{2 \cos 2x}{1} = \frac{2 \cos 0}{1} = 2$$

(c) Here, $f(x) = \sin x$ and $g(x) = e^x - e^{-x}$. As in the previous examples,

$$f(0) = \sin 0 = 0 = e^0 - e^{-0} = g(0)$$

and $\lim_{x \to 0} f(x)/g(x)$ is of the form 0/0. From Theorem 11.3 we know that

$$\lim_{x \to 0} \frac{\sin x}{e^x - e^{-x}} = \lim_{x \to 0} \frac{\cos x}{e^x + e^{-x}} = \frac{\cos 0}{e^0 + e^{-0}} = \frac{1}{2}$$

(d) Let $f(x) = 1 - \cos x$ and $g(x) = 1 + \cos x$. Then

$$f(0) = 1 - \cos 0 = 0$$

but

$$g(0) = 1 + \cos 0 = 2$$

Thus $\lim_{x \to 0} f(x)/g(x)$ is not a 0/0 form—L'Hôpital's rule does not apply. This problem is too simple, really.

$$\lim_{x \to 0} \frac{1 - \cos x}{1 + \cos x} = \frac{1 - \cos 0}{1 + \cos 0} = \frac{0}{2} = 0$$

It is worth noting that, if we tried to apply L'Hôpital's rule in this problem where it really doesn't apply, we would obtain an erroneous answer:

$$f'(x) = \sin x, \qquad g'(x) = -\sin x$$

$$\lim_{x \to 0} \frac{f'(x)}{g'(x)} = \frac{\sin x}{-\sin x} = -1$$

which does not agree with the correct answer of 0.

(e) Set $f(x) = 1 - \cos 3x$ and $g(x) = x^2$. Then

$$f(0) = g(0) = 0$$

and $\lim_{x \to 0} f(x)/g(x)$ is the indeterminate form 0/0. Because

$$f'(x) = 3 \sin 3x \quad \text{and} \quad g'(x) = 2x$$

by L'Hôpital's rule we have

$$\lim_{x \to 0} \frac{1 - \cos 3x}{x^2} = \lim_{x \to 0} \frac{3 \sin 3x}{2x}$$

But when $x = 0$, $\dfrac{3 \sin 3x}{2x}$ is also of the form 0/0. Has L'Hôpital's rule failed us? Not at all. We simply have to apply it again to the new 0/0 form in order to resolve our difficulties.

$$\lim_{x \to 0} \frac{1 - \cos 3x}{x^2} = \lim_{x \to 0} \frac{3 \sin 3x}{2x} = \lim_{x \to 0} \frac{9 \cos 3x}{2} = \frac{9}{2}$$

We have seen how L'Hôpital's rule works. Now let's see *why* it works.

PROOF OF THEOREM 11.3 We are assuming that f and g have Taylor-series expansions on some interval containing a. Thus, according to Equation (11.14),

$$f(x) = f(a) + f'(a)(x - a) + \frac{f''(a)}{2!}(x - a)^2 + \frac{f'''(a)}{3!}(x - a)^3 + \cdots$$

and

$$g(x) = g(a) + g'(a)(x - a) + \frac{g''(a)}{2!}(x - a)^2 + \frac{g'''(a)}{3!}(x - a)^3 + \cdots$$

We have also assumed that $f(a) = g(a) = 0$, which tells us that the constant term in each of these series expansions is 0. Therefore

$$\begin{aligned} f(x) &= f'(a)(x - a) + \frac{f''(a)}{2!}(x - a)^2 + \frac{f'''(a)}{3!}(x - a)^3 + \cdots \\ &= (x - a)\left[f'(a) + \frac{f''(a)}{2!}(x - a) + \frac{f'''(a)}{3!}(x - a)^2 + \cdots\right] \end{aligned}$$

and similarly,

$$g(x) = g'(a)(x - a) + \frac{g''(a)}{2!}(x - a)^2 + \frac{g'''(a)}{3!}(x - a)^3 + \cdots$$

$$= (x - a)\left[g'(a) + \frac{g''(a)}{2!}(x - a) + \frac{g'''(a)}{3!}(x - a)^2 + \cdots\right]$$

Now, if $x \neq a$, then

$$\frac{f(x)}{g(x)} = \frac{(x - a)\left[f'(a) + \frac{f''(a)}{2!}(x - a) + \frac{f'''(a)}{3!}(x - a)^2 + \cdots\right]}{(x - a)\left[g'(a) + \frac{g''(a)}{2!}(x - a) + \frac{f'''(a)}{3!}(x - a)^2 + \cdots\right]}$$

$$= \frac{f'(a) + \frac{f''(a)}{2!}(x - a) + \frac{f'''(a)}{3!}(x - a)^2 + \cdots}{g'(a) + \frac{g''(a)}{2!}(x - a) + \frac{g'''(a)}{3!}(x - a)^2 + \cdots}$$

As $x \to a$, each of the terms containing a power of $(x - a)$ will become smaller and smaller, leaving only the terms $f'(a)$ in the numerator and $g'(a)$ in the denominator:

$$\lim_{x \to a} \frac{f(x)}{g(x)} = \frac{f'(a)}{g'(a)} = \lim_{x \to a} \frac{f'(x)}{g'(x)}$$ ⊟

L'Hôpital's rule actually works in more situations than we have indicated. It is actually not necessary that f and g be sufficiently nice that they have Taylor-series expansions about $x = a$. It would suffice that they be differentiable (with $g'(x) \neq 0$) for x near a.

Another indeterminate form that frequently arises is ∞/∞. L'Hôpital's rule can be used to evaluate limits of this form, too, though we shall not prove this. We will content ourselves with looking at a few examples.

EXAMPLE 28 Evaluate:

(a) $\displaystyle\lim_{x \to \infty} \frac{\ln x}{x}$ (b) $\displaystyle\lim_{x \to \infty} \frac{5x^2 + 2x}{3x^2 - 8}$

SOLUTION

(a) Because $\lim_{x \to \infty} \ln x = \lim_{x \to \infty} x = \infty$, we have the indeterminate form ∞/∞. Differentiating numerator and denominator separately yields

$$\lim_{x \to \infty} \frac{\ln x}{x} = \lim_{x \to \infty} \frac{\frac{1}{x}}{1} = \frac{0}{1} = 0$$

(b) If we let $f(x) = 5x^2 + 2x$ and $g(x) = 3x^2 - 8$, then

$$\lim_{x \to \infty} f(x) = \lim_{x \to \infty} g(x) = \infty$$

Because $f'(x) = 10x + 2$ and $g'(x) = 6x$, by L'Hôpital's rule we have

$$\lim_{x\to\infty} \frac{5x^2 + 2x}{3x^2 - 8} = \lim_{x\to\infty} \frac{10x + 2}{6x}$$

which is still the indeterminate form ∞/∞. We use L'Hôpital's rule a second time.

$$\lim_{x\to\infty} \frac{5x^2 + 2x}{3x^2 - 8} = \lim_{x\to\infty} \frac{10x + 2}{6x} = \lim_{x\to\infty} \frac{10}{6} = \frac{5}{3}$$

EXAMPLE 29 Evaluate

$$\lim_{x\to 0+} \left(\frac{1}{x} - \frac{1}{\sin x}\right)$$

Note: The symbol "$\lim\limits_{x\to 0+}$" means that x is tending towards 0 through positive values.

SOLUTION Because $\lim\limits_{x\to 0+} \left(\frac{1}{x}\right) = \infty$ and $\lim\limits_{x\to 0+} \left(\frac{1}{\sin x}\right) = \infty$, our limit takes the form $\infty - \infty$, which is another indeterminate form (though somewhat different from the previous ones). If we combine fractions:

$$\frac{1}{x} - \frac{1}{\sin x} = \frac{\sin x - x}{x \sin x}$$

we find that $\lim\limits_{x\to 0+} \left(\frac{\sin x - x}{x \sin x}\right)$ is an indeterminate form 0/0. Thus, by L' Hôpital's rule,

$$\lim_{x\to 0+} \left(\frac{1}{x} - \frac{1}{\sin x}\right) = \lim_{x\to 0+} \left(\frac{\sin x - x}{x \sin x}\right) = \lim_{x\to 0+} \left(\frac{\cos x - 1}{x \cos x + \sin x}\right)$$

But $\lim\limits_{x\to 0+} (\cos x - 1) = 0$ and $\lim\limits_{x\to 0+} (x \cos x + \sin x) = 0$, so this is again the indeterminate form 0/0. Applying L' Hôpital's rule a second time yields

$$\lim_{x\to 0+} \left(\frac{\cos x - 1}{x \cos x + \sin x}\right) = \lim_{x\to 0+} \left(\frac{-\sin x}{(-x \sin x + \cos x) + \cos x}\right) = \frac{0}{2} = 0$$

Sometimes we are faced with the problem of evaluating an expression of the form

$$\lim_{x\to a} [f(x)]^{g(x)}$$

This can often be done by the following trick.

1 Set $y = [f(x)]^{g(x)}$ and take the natural logarithm of both sides of the relation to obtain

$$\ln y = \ln\{[f(x)]^{g(x)}\} = g(x)(\ln f(x))$$

2 Find $\lim_{x\to a} (\ln y) = \lim_{x\to a} \{g(x) \ln f(x)\}$. Denote this limit by K. It may be necessary to use L'Hôpital's rule to do this.

3 Then exponentiate on both sides.

$$\lim_{x\to a} e^{\ln y} = \lim_{x\to a} y = \lim_{x\to a} [f(x)]^{g(x)} = e^K$$

EXAMPLE 30 Evaluate:

(a) $\lim_{x\to 0+} (1 + x)^{1/x}$ (b) $\lim_{x\to 0+} x^{2x}$ (c) $\lim_{x\to\infty} x^{1/x^2}$

SOLUTION

(a) Let

$$y = (1 + x)^{1/x}$$

Then $$\ln y = \ln(1 + x)^{1/x} = \frac{1}{x}\ln(1 + x) = \frac{\ln(1 + x)}{x}$$

Now, $\lim_{x\to 0+} \frac{\ln(1 + x)}{x}$ is of the form 0/0. By L'Hôpital's rule,

$$\lim_{x\to 0+} \ln y = \lim_{x\to 0+} \frac{\ln(1 + x)}{x} = \lim_{x\to 0+} \frac{\frac{1}{1 + x}}{1} = 1$$

Thus $$\lim_{x\to 0+} y = \lim_{x\to 0+} (1 + x)^{1/x} = e^1 = e$$

Some books actually use the formula $\lim_{x\to 0+} (1 + x)^{1/x} = e$ to define the number e, rather than defining it as we did to be the base of the exponential function.

(b) Setting $y = x^{2x}$ and taking logarithms gives

$$\ln y = 2x(\ln x) = \frac{2(\ln x)}{\frac{1}{x}}$$

As $x \to 0+$, this takes the form $\frac{-\infty}{\infty}$, which is an indeterminate form. By L'Hôpital's rule,

$$\lim_{x\to 0+} \frac{2(\ln x)}{\frac{1}{x}} = \lim_{x\to 0+} \frac{\frac{2}{x}}{-\frac{1}{x^2}} = \lim_{x\to 0+} (-2x) = 0$$

Thus $\lim_{x\to 0+} \ln y = 0$, and

$$\lim_{x\to 0+} y = \lim_{x\to 0+} x^{2x} = e^0 = 1$$

(c) If $y = x^{1/x^2}$, then $\ln y = \dfrac{1}{x^2}(\ln x) = \dfrac{\ln x}{x^2}$

Now $$\lim_{x\to\infty} \frac{\ln x}{x^2} = \lim_{x\to\infty} \frac{\frac{1}{x}}{2x} = \lim_{x\to\infty} \frac{1}{2x^2} = 0$$

Thus $$\lim_{x\to\infty} y = \lim_{x\to\infty} x^{1/x^2} = e^0 = 1$$

Section 11.6 EXERCISES

Use L'Hôpital's rule or any other method to evaluate the limits.

1. $\lim\limits_{x\to 0} \dfrac{\sin 3x}{2x}$
2. $\lim\limits_{x\to 0} \dfrac{\cos x - 1}{x^2}$
3. $\lim\limits_{x\to 0} \dfrac{\sin x}{x^2}$
4. $\lim\limits_{x\to 0} \dfrac{\sin x^2}{x}$
5. $\lim\limits_{x\to 0} \dfrac{\sin x^2}{x^2}$
6. $\lim\limits_{x\to 0} \dfrac{\sin x^2}{\sin^2 x}$
7. $\lim\limits_{x\to 0} \dfrac{e^x - e^{-x}}{x}$
8. $\lim\limits_{x\to 0} \dfrac{e^x - 1}{x}$
9. $\lim\limits_{x\to 0} \dfrac{e^x - (1 + x)}{x^2}$
10. $\lim\limits_{x\to 0} \dfrac{x - \sin x}{x}$
11. $\lim\limits_{x\to 0} \dfrac{x - \sin x}{x^2}$
12. $\lim\limits_{x\to 0} \dfrac{x - \sin x}{x^3}$
13. $\lim\limits_{x\to 0} \dfrac{\tan x - \sin x}{x^3}$
14. $\lim\limits_{x\to 0} \dfrac{\sqrt{1 + x} - \left(1 + \dfrac{x}{2}\right)}{x^2}$
15. $\lim\limits_{x\to 0} \dfrac{(1 + x)^{1/3} - (1 + \frac{x}{3})}{x^2}$
16. $\lim\limits_{x\to 0} \dfrac{\ln(1 + x)}{x}$
17. $\lim\limits_{x\to 0} \dfrac{\sqrt{9 + x} - 3}{x}$
18. $\lim\limits_{x\to 0} \dfrac{e^x - \cos x}{x}$
19. $\lim\limits_{x\to 0} \dfrac{e^x - 1 - \sin x}{x}$
20. $\lim\limits_{x\to 0} \dfrac{\cos x - \left(1 + \dfrac{x^2}{2}\right)}{x^3}$
21. $\lim\limits_{x\to 0} \sin \dfrac{1}{x}$
22. $\lim\limits_{x\to 0} x \sin \dfrac{1}{x}$
23. $\lim\limits_{x\to \infty} \dfrac{3x^2 - 5x + 1}{7x^2 - x - 1}$
24. $\lim\limits_{x\to \infty} \dfrac{3x^2 - 5x + 1}{7x^3 - x - 1}$
25. $\lim\limits_{x\to \infty} \dfrac{3x^3 - 5x + 1}{7x^3 - x - 1}$
26. $\lim\limits_{x\to \infty} \dfrac{3x^3 - 5x + 1}{7x^2 - x - 1}$
27. $\lim\limits_{x\to 0} \dfrac{a_0 + a_1 x + a_2 x^2 + \cdots + a_n x^n}{b_0 + b_1 x + b_2 x^2 + \cdots + b_m x^m}$, $b_0 \neq 0$
28. $\lim\limits_{x\to \infty} \dfrac{a_0 + a_1 x + a_2 x^2 + \cdots + a_n x^n}{b_0 + b_1 x + b_2 x^2 + \cdots + b_m x^m}$, $a_n \neq 0$, $b_m \neq 0$ (There are several cases to consider.)
29. $\lim\limits_{x\to \infty} (\sqrt{x^2 + 3} - x)$
30. $\lim\limits_{x\to \infty} (\sqrt{x^2 + x} - x)$
31. $\lim\limits_{x\to 0+} x \log x$
32. $\lim\limits_{x\to \infty} \dfrac{(\ln x)^2}{x}$
33. $\lim\limits_{x\to \infty} \dfrac{(\ln x)^m}{x}$, $m > 0$
34. $\lim\limits_{x\to \infty} \dfrac{x^2}{e^x}$
35. $\lim\limits_{x\to \infty} \dfrac{x^{75}}{e^x}$
36. $\lim\limits_{x\to \infty} \dfrac{x^m}{e^x}$, $m > 0$
37. $\lim\limits_{x\to \infty} \left(1 + \dfrac{1}{x}\right)^x$
38. $\lim\limits_{x\to 0+} (1 + 2x)^{1/x}$
39. $\lim\limits_{x\to \infty} \left(1 + \dfrac{3}{x}\right)^x$
40. $\lim\limits_{x\to 0+} (1 + nx)^{1/x}$
41. $\lim\limits_{x\to \infty} \left(1 + \dfrac{n}{x}\right)^x$
42. $\lim\limits_{x\to 0+} x^x$
43. $\lim\limits_{x\to \infty} x^{1/x}$
44. $\lim\limits_{x\to 0} (1 - 2x)^{1/x}$
45. $\lim\limits_{x\to 0} (1 - 3x)^{2/x}$
46. $\lim\limits_{x\to 0} (\cos x)^{1/x}$
47. $\lim\limits_{x\to 0} \dfrac{\ln(1 + x)^2}{x}$
48. $\lim\limits_{x\to 0} [\ln\{(1 + x)^{1/x}\}]$
49. $\lim\limits_{x\to 0} x \ln|\sin x|$

Key Mathematical Concepts and Tools

Taylor polynomials
*n*th-order contact
Taylor series
Estimating errors by using the remainder term
Maclaurin series
Formal power series
Numerical approximation—Newton's method
Iterative methods
L'Hôpital's rule

Applications Covered in This Chapter

Approximation of trigonometric functions
Approximation of square roots
Approximation of the exponential and natural logarithm functions
Power-series solution of differential equations
Estimating the error in approximating the volume of a red blood cell
Approximation of cube root
Approximating roots of equations
Approximating the intersection of a pair of curves

ELEVEN

Warm-Up Test

1. (a) Find the Taylor polynomial of second degree, p_2, for $f(x) = e^{-2x}$ expanded about $x = 0$.
   (b) Use your answer to part (a) to approximate $e^{-0.6}$.

2. Evaluate the following limits. Use L'Hôpital's rule where needed.
   (a) $\lim_{x\to 0} \dfrac{5\sin x - x}{x^2}$ (b) $\lim_{x\to\infty} \dfrac{x+3}{x^2-4}$
   (c) $\lim_{x\to\pi/2} \dfrac{\cos x}{\pi/2 - x}$ (d) $\lim_{x\to+\infty} \dfrac{3x^2-4}{5-8x^2}$

3. Calculate the Taylor-series expansions of the following functions about $x = 0$.
   (a) $\cos 2x$ (b) $\left(\dfrac{1}{1-x}\right)^2$

4. Let $f(x) = x^3 - 7$. Estimate $\sqrt[3]{7}$ by applying Newton's method to find a root of the equation $f(x) = 0$. Explicitly write down the iterative scheme you are using and, starting with $x_0 = 2$, compute two successive approximations, x_1 and x_2.

5. Let $g(x) = \sqrt[3]{x} = x^{1/3}$. Estimate $\sqrt[3]{8.1}$ by approximating $g(x)$ about $a = 8$ with its Taylor polynomial of degree 3.

6. (a) Estimate e^{-x^2} by the Taylor polynomial about $x = 0$ of minimal degree from which you can be assured that your estimate is within ± 0.001 for $0 \leq x \leq 0.02$.
   (b) Estimate $e^{-(0.016)^2}$ from your approximation in part (a) above.

7. Use the technique of formal power series to find the first three terms of the series solution to the differential equation
$$y' = -y + x - x^2 \qquad y(0) = 4$$

8. Use Newton's method to approximate the unique real root of the polynomial $f(x) = x^3 + x - 3$ by computing three successive approximations. Start with $x_0 = 1$.

9. At the given value of x, find the highest-order contact of the functions $f(x)$ and $g(x)$.

(a) $x = 0$; $f(x) = \sin x^2$, $g(x) = (\sin x)^2$

(b) $x = 1$; $f(x) = \ln x$, $g(x) = \dfrac{x^3}{3} - \dfrac{3}{2}x^2 + 3x - \frac{11}{6}$

10. Use the formal power series method to solve the differential equation

$$y'' = -y \qquad y(0) = 0,\ y'(0) = 1$$

(a) Carry out your expansion until you obtain three nonzero terms.

(b) Use the series you found in part (a) to approximate $y(0.06)$.

(c) What is the "name" of the function that you approximated in part (a)?

ELEVEN Final Exam

1. Calculate the Taylor-series expansions of the following functions about $x = 0$.

(a) $f(x) = \dfrac{x}{(1 + x^2)^2}$ (b) $f(x) = x \ln(1 + x)$

2. Let $g(x) = x^{2/5}$. Estimate $(-31)^{2/5}$ by approximating $g(x)$ by its Taylor polynomial of degree 2 about $x = -32$.

3. (a) Find the Taylor polynomial of third degree, p_3, for

$$f(x) = \sin[\ln(1 + x)]$$

expanded about $x = 0$.

(b) Use your answer to part (a) to approximate $\sin[\ln(1.1)]$.

4. Evaluate the following limits. Use L'Hôpital's rule where needed.

(a) $\displaystyle\lim_{h\to 0} \frac{-\cos h}{1 + \sin h^2}$ (b) $\displaystyle\lim_{x\to 0} \frac{x \sin 2x}{x^2}$

(c) $\displaystyle\lim_{x\to\infty} \frac{3x^2 - 6}{5x^2 + 7}$ (d) $\displaystyle\lim_{z\to\pi/2} \frac{\sin^2 z - 1}{\cos 2z}$

5. By applying Newton's method, find the unique positive value of x that satisfies $e^{-x} = x$ to within the tolerance 0.001.

(a) Explicitly write down the iterative scheme you will be using.

(b) Start with $x_0 = 1$ and carry your procedure out until your answer is within the given tolerance. You will need your calculator for this.

6. Use the formal power series technique to find a power-series solution to the differential equation

$$y'' = (1 + x)y \qquad y(0) = 1,\ y'(0) = 1$$

Explicitly compute the first four terms of this series solution.

7. Find the highest-order contact at $x = 0$ of $f(x)$ and $g(x)$.

(a) $f(x) = e^{2x} - e^{-x}$, $g(x) = 3x + \frac{3}{2}x^2$

(b) $f(x) = 2\cos x^2$, $g(x) = 2 - 3x^2$

8. (a) What minimal-degree Taylor polynomial about $x = 0$ is needed in order to estimate $(1 + x)^{1/2}$ to within 0.01 for $-0.1 \le x \le 0.1$?

(b) Use the Taylor polynomial you found in part (a) to estimate $\sqrt{1.05}$ within a tolerance of 0.01.

9. Use L'Hôpital's rule to compute

(a) $\displaystyle\lim_{x\to 0} x^{x^2}$ (b) $\displaystyle\lim_{x\to\infty} \left(x - \frac{1}{e^{1/x} - 1}\right)$

10. (a) Compute $\sin^2 31^\circ$ to within 1% of its exact value.

(b) Compute $\cos 62^\circ$ to within 1% of its exact value.

Appendixes

A Review of Algebra

SECTION A.1

Elementary Algebraic Expressions

In Appendix A we provide a summary of several topics from high school algebra. In particular, we focus on rules for combining algebraic expressions. These are important prerequisites for studying calculus.

We begin by recalling a few familiar rules about the properties of negative numbers and multiplication. Our list is not intended to be complete but simply to recall the most frequently used facts.

For any numbers a, b, and c,

(A.1)

1 $(-a)b = a(-b) = -(ab)$

2 $-(-a) = a$

3 $(-a)(-b) = ab$

4 $(-1)a = -a$

5 $a(b \pm c) = ab \pm ac$

6 $-(a + b) = -a - b$

EXAMPLE 1 As numerical illustrations of these rules, we offer the following:

1 $(-2)(3) = (2)(-3) = -(2 \cdot 3) = -6$

2 $-(-5) = 5$

3 $(-5)(-7) = (5)(7) = 35$

4 $(-1)(\sqrt{2}) = -\sqrt{2}$

5 This rule is known as the **distributive law**. We have used the notation $\pm$ to denote the two cases

$$a(b + c) = ab + ac$$

and

$$a(b - c) = ab - ac$$

Thus for instance,

$$2(3 + 5) = 2 \cdot 3 + 2 \cdot 5 = 16$$

and $$2(3 - 5) = 2 \cdot 3 - 2 \cdot 5 = -4$$

6 $-(-1 + 4) = -(-1) - (4) = 1 - 4 = -3$

Next we turn our attention to the standard rules for arithmetic operations with fractions. For any numbers a, b, c, d, and k,

(A.2)

1 $\dfrac{a}{b} = \dfrac{ka}{kb}$

2 $\dfrac{a}{b} \cdot \dfrac{c}{d} = \dfrac{ac}{bd}$

3 $\begin{cases} \dfrac{a}{b} \pm \dfrac{c}{d} = \dfrac{ad \pm bc}{bd} \\[2ex] \dfrac{a}{c} \pm \dfrac{b}{c} = \dfrac{a \pm b}{c} \end{cases}$

4 $\dfrac{\dfrac{a}{b}}{\dfrac{c}{d}} = \dfrac{a}{b} \cdot \dfrac{d}{c} = \dfrac{ad}{bc}$

5 $\dfrac{a}{\dfrac{b}{c}} = \dfrac{ac}{b}$

6 $\dfrac{\dfrac{a}{b}}{c} = \dfrac{a}{bc}$

It is understood that these are valid only when the denominators are not equal to 0; for example, it is understood that, in Equations (A.2), part 1, we must have $k \neq 0$ and $b \neq 0$.

EXAMPLE 2 Let us illustrate the foregoing rules with some specific numbers.

1 $\dfrac{5}{6} = \dfrac{5\sqrt{2}}{6\sqrt{2}}$ and $\dfrac{4\sqrt{2}}{4\sqrt{3}} = \dfrac{\sqrt{2}}{\sqrt{3}}$, or $\dfrac{\sqrt{6}}{3}$

2 $\left(\dfrac{-2}{3}\right)\left(\dfrac{\sqrt{5}}{7}\right) = \dfrac{-2\sqrt{5}}{21}$

3 $\dfrac{2}{3} + \dfrac{4}{5} = \dfrac{(2)(5) + (3)(4)}{3 \cdot 5} = \dfrac{22}{15}$

$\dfrac{2}{3} - \dfrac{4}{5} = \dfrac{(2)(5) - (3)(4)}{3 \cdot 5} = \dfrac{-2}{15}$

Observe that this rule involved the use of a **common denominator** for the fractions a/b and c/d. In the event that the denominators are already equal, we have $\frac{2}{5} + \frac{4}{5} = \frac{6}{5}$.

4 $\dfrac{\frac{2}{3}}{\frac{5}{9}} = \dfrac{2}{3} \cdot \dfrac{9}{5} = \dfrac{18}{15} = \dfrac{6}{5}$

5 $\dfrac{\sqrt{6}}{\left(\frac{-2}{5}\right)} = \dfrac{(\sqrt{6})(5)}{-2} = -\dfrac{5\sqrt{6}}{2}$

6 $\dfrac{\frac{4}{5}}{3} = \dfrac{4}{5 \cdot 3} = \dfrac{4}{15}$

REMARK According to the rules for adding fractions,

$$\frac{a+b}{c} = \frac{a}{c} + \frac{b}{c}$$

A common error that many students make is to assume that $\dfrac{a}{b+c}$ is the same as $\dfrac{a}{b} + \dfrac{a}{c}$. This is simply not the case. In general,

$$\frac{a}{b+c} \neq \frac{a}{b} + \frac{a}{c}$$

For example, if $a = 2$, $b = 3$, and $c = 5$, we have $\dfrac{a}{b+c} = \dfrac{2}{3+5} = \dfrac{1}{4}$, whereas $\dfrac{a}{b} = \dfrac{2}{3}$ and $\dfrac{a}{c} = \dfrac{2}{5}$, so

$$\frac{a}{b} + \frac{a}{c} = \frac{2}{3} + \frac{2}{5} = \frac{2 \cdot 5 + 2 \cdot 3}{15} = \frac{16}{15}$$

which is not the same as $\frac{1}{4}$.

Another variation of this error occurs when $b = c$. It is, of course, true that $\dfrac{a+b}{b} = \dfrac{a}{b} + 1$ $\left(\text{because } \dfrac{b}{b} = 1\right)$. But

$$\frac{b}{a+b} \neq \frac{b}{a} + 1$$

To see this, simply set $a = 1$ and $b = 2$. ⊟

Thus far we have been performing computations using real numbers. Now let us practice manipulating algebraic expressions in one or more variables. The rules expressed in Equations (A.1) and (A.2) are valid for such expressions.

EXAMPLE 3 Simplify

$$\frac{x + \frac{1}{y}}{y + \frac{1}{x}}$$

SOLUTION We begin by combining the terms in the numerator. The common denominator is y. Thus we have

$$x + \frac{1}{y} = \frac{xy}{y} + \frac{1}{y} = \frac{xy + 1}{y}$$

Similarly, the denominator can be simplified:

$$y + \frac{1}{x} = \frac{xy}{x} + \frac{1}{x} = \frac{xy + 1}{x}$$

Therefore

$$\frac{x + \frac{1}{y}}{y + \frac{1}{x}} = \frac{\frac{xy + 1}{y}}{\frac{xy + 1}{x}} = \left(\frac{xy + 1}{y}\right)\left(\frac{x}{xy + 1}\right) = \frac{x}{y}$$

EXAMPLE 4 Simplify $\dfrac{3}{x - 2} - \dfrac{4}{x + 1}$.

SOLUTION The common denominator is $(x - 2)(x + 1) = x^2 - x - 2$ Thus

$$\frac{3}{x - 2} - \frac{4}{x + 1} = \frac{3(x + 1) - 4(x - 2)}{(x - 2)(x + 1)} = \frac{3x + 3 - 4x + 8}{x^2 - x - 2} = \frac{-x + 11}{x^2 - x - 2}$$

We will have occasion in the text to make use of the notion of the absolute value of a number. The **absolute value** of a real number, a, denoted by $|a|$, is defined to be

$$|a| = \begin{cases} a & \text{if } a \geq 0 \\ -a & \text{if } a < 0 \end{cases}$$

In words, the absolute value of a positive number is simply the number itself, while for a negative number we take its negative (which is a positive number). Thus, for instance, $|2| = 2$, $|\pi| = \pi$, and $|-4| = 4$.

EXAMPLE 5 Evaluate

$$\left|\frac{\frac{1}{2} + \frac{1}{3}}{\frac{2}{3} - \frac{4}{5}}\right|$$

SOLUTION

$$\frac{1}{2}+\frac{1}{3}=\frac{(1)(3)+(1)(2)}{(2)(3)}=\frac{5}{6}$$

while

$$\frac{2}{3}-\frac{4}{5}=\frac{(2)(5)-(4)(3)}{(3)(5)}=\frac{-2}{15}$$

Therefore,

$$\left|\frac{\frac{1}{2}+\frac{1}{3}}{\frac{2}{3}-\frac{4}{5}}\right|=\left|\frac{\frac{5}{6}}{\frac{-2}{15}}\right|=\left|\left(\frac{5}{6}\right)\left(-\frac{15}{2}\right)\right|$$

$$=\left|-\frac{75}{12}\right|=\left|-\frac{25}{4}\right|=\frac{25}{4}$$

Section A.1 EXERCISES

In Exercises 1–16 simplify.

1. $\frac{2}{3}+\frac{5}{6}$ **2.** $\frac{5}{8}-\frac{4}{5}$ **3.** $\frac{3}{4}\div\frac{7}{8}$

4. $\frac{3}{5}\div\frac{6}{7}$ **5.** $\frac{2}{3}\div 4$ **6.** $\frac{3}{4}\div 6$

7. $5\div\frac{3}{4}$ **8.** $7\div\frac{2}{3}$ **9.** $\frac{1}{4}+\frac{2}{3}-\frac{5}{6}$

10. $3+4\sqrt{2}-8+6\sqrt{2}$ **11.** $5+3\sqrt{5}+6-2\sqrt{5}$

12. $\sqrt{3}(5+2\sqrt{3})$

13. Does $a+(b\cdot c)=(a+b)(a+c)$?

14. Use the distributive law to show

$$(a+b)(c+d)=ac+ad+bc+bd$$

15. $(1-\sqrt{3})(5+2\sqrt{3})$ **16.** $(6-\sqrt{5})(2+3\sqrt{5})$

Write Exercises 17–28 without absolute value bars.

17. $|\pi-3|$ **18.** $|-2-\pi|$

19. $|2-\pi|$ **20.** $|\sqrt{7}-3|$

21. $-|-8|+|-3|$ **22.** $|-5|-|4|$

23. $|36-12m|$ if $m>3$ **24.** $|-x-y|$

25. $|x-y|$ if $x<y$ **26.** $|-4-3m|$ if $m>-\frac{4}{3}$

27. $\frac{|x|}{x}$ **28.** $\frac{|2-x|}{x-2}$

The rules governing the order of operations indicate that multiplication and division should be performed *before* addition and subtraction. With this in mind, simplify Exercises 29–46.

29. $3+7\cdot 2-8$ **30.** $(6+4)\div 5+3\cdot 2$

31. $8+14\div 2-3$ **32.** $6-12\div 6+4\cdot 2$

33. $\frac{1}{x}+\frac{x}{4}$ **34.** $\frac{x}{8}-\frac{3}{x}$

35. $\frac{1}{x-2}+\frac{3}{4}$ **36.** $\frac{3}{x-2}+\frac{5}{x+2}$

37. $\frac{x}{x-1}+\frac{x+1}{x}$ **38.** $\frac{x+4}{x}-\frac{3x}{x+2}$

39. $\frac{1}{x-1}+\frac{1}{x-1}$ **40.** $\frac{\frac{x}{y}-1}{\frac{y}{x}-1}$

41. $\frac{\frac{1}{x}-\frac{1}{y}}{\frac{1}{x}+\frac{1}{y}}$ **42.** $\frac{3-\frac{1}{x+2}}{\frac{1}{x}-2x}$

43. $\left|\dfrac{\frac{3}{5}-\frac{5}{6}}{\frac{1}{2}+\frac{1}{3}}\right|$

44. $\left|\dfrac{\frac{1}{4}-\frac{5}{8}}{\frac{3}{5}+\frac{2}{3}}\right|$

45. $\dfrac{1}{1+\dfrac{1}{x+\dfrac{1}{x}}}$

46. $\dfrac{\dfrac{1}{x^2}+\dfrac{1}{y^2}}{\dfrac{1}{xy}}$

SECTION A.2

Exponents

In this section we turn our attention to the rules for manipulating exponents. Let a, b, m, n be real numbers, $a \neq 0$, $b \neq 0$. Then

(A.3)

1 $a^0 = 1$

2 $a^{-n} = \dfrac{1}{a^n}$

3 $(ab)^n = a^n b^n$, $\left(\dfrac{a}{b}\right)^n = \dfrac{a^n}{b^n}$

4 $a^m a^n = a^{m+n}$

5 $\dfrac{a^m}{a^n} = a^{m-n}$

6 $(a^n)^m = a^{nm}$

Before we illustrate the use of these rules, let us remark that there are cases in which some of the rules are meaningless. For example, you know that for any positive number a, $\sqrt{a} = a^{1/2}$. On the other hand, if a is negative (such as $a = -2$) then $\sqrt{a} = \sqrt{-2}$ has no meaning. That is, there does not exist any real number whose square is -2. For simplicity, however, we have chosen to focus on the rules as they generally apply, not on the special cases that create a problem.

EXAMPLE 6 Evaluate $3^2 \cdot 3^4 + (-2)^2 \cdot (-2)^3$

SOLUTION According to Equations (A.3), part 4,

$$3^2 \cdot 3^4 = 3^6 = 729$$

and

$$(-2)^2 \cdot (-2)^3 = (-2)^5 = -32$$

Thus

$$3^2 \cdot 3^4 + (-2)^2 \cdot (-2)^3 = 729 - 32 = 697$$

EXAMPLE 7 Evaluate $(\frac{3}{4})^2(\frac{1}{6})^3(\frac{-5}{2})^{-1}(\frac{1}{3})^{-2}$

SOLUTION We have

$$\left(\frac{-5}{2}\right)^{-1} = \frac{1}{\frac{-5}{2}} = -\frac{2}{5}$$

and

$$\left(\frac{1}{3}\right)^{-2} = \frac{1}{\left(\frac{1}{3}\right)^2} = \frac{1}{\frac{1}{9}} = 9$$

Thus

$$\left(\frac{3}{4}\right)^2\left(\frac{1}{6}\right)^3\left(\frac{-5}{2}\right)^{-1}\left(\frac{1}{3}\right)^{-2} = \left(\frac{3^2}{4^2}\right)\left(\frac{1}{6^3}\right)\left(-\frac{2}{5}\right)(9) = \left(\frac{9}{16}\right)\left(\frac{1}{216}\right)\left(-\frac{2}{5}\right)(9)$$
$$= -\frac{162}{17{,}280} = -\frac{3}{320}$$

EXAMPLE 8 Simplify $\dfrac{3^{7/5} \cdot 3^{5/7}}{3^{9/7}}$.

SOLUTION

$$\frac{3^{7/5} \cdot 3^{5/7}}{3^{9/7}} = \frac{3^{7/5+5/7}}{3^{9/7}} = \frac{3^{49/35} \cdot 3^{25/35}}{3^{45/35}}$$
$$= \frac{3^{49/35+25/35}}{3^{45/35}} = \frac{3^{74/35}}{3^{45/35}} = 3^{74/35-45/35} = 3^{29/35}$$

EXAMPLE 9 Evaluate $(\sqrt{2})^{5/3}(2)^{5/4}(\sqrt{2})^{-3/4}$.

SOLUTION

$$(\sqrt{2})^{5/3}(2)^{5/4}(\sqrt{2})^{-3/4} = (2^{1/2})^{5/3}(2)^{5/4}(2^{1/2})^{-3/4} = (2^{1/2})^{5/3}(2)^{5/4}\frac{1}{(2^{1/2})^{3/4}}$$
$$= (2^{5/6})(2^{5/4})\frac{1}{2^{3/8}} = \frac{(2^{20/24})(2^{30/24})}{(2^{9/24})} = \frac{2^{50/24}}{2^{9/24}}$$
$$= 2^{41/24}$$

As we did in the previous section, let us apply our ideas to expressions involving variables.

EXAMPLE 10 Simplify $(3x^3y^4z)(4x^5y^8z^3) + (2x^2y^3z)^4$.

SOLUTION Using our exponent rules (A.3) we obtain

$$(3x^3y^4z)(4x^5y^8z^3) + (2x^2y^3z)^4 = 12x^8y^{12}z^4 + 2^4x^8y^{12}z^4$$
$$= 28x^8y^{12}z^4$$

EXAMPLE 11 Simplify $(x^{-2}y^3)^{-3}$

SOLUTION In straightforward fashion we have

$$(x^{-2}y^3)^{-3} = \left(\frac{y^3}{x^2}\right)^{-3} = \frac{1}{\left(\frac{y^3}{x^2}\right)^3} = \frac{1}{\left(\frac{y^9}{x^6}\right)} = \frac{x^6}{y^9}$$

EXAMPLE 12 Simplify $\frac{2+\sqrt{3}}{2-\sqrt{3}}$.

SOLUTION The procedure to use is to **rationalize the denominator**. This is accomplished by multiplying both numerator and denominator by the **conjugate** of the denominator—namely, by $2 + \sqrt{3}$. This serves to eliminate the square root from the denominator as follows:

$$\frac{2+\sqrt{3}}{2-\sqrt{3}} = \frac{2+\sqrt{3}}{2-\sqrt{3}} \cdot \frac{2+\sqrt{3}}{2+\sqrt{3}} = \frac{4+4\sqrt{3}+3}{4-3} = 7 + 4\sqrt{3}$$

Notice that we made use of the fact that for any real numbers a and b, $(a+b)(a-b) = a^2 - b^2$.

PRACTICE EXERCISE Convince yourself that the following hold:

1 $(x+y)^2 = x^2 + 2xy + y^2$

2 $(x+y)^3 = x^3 + 3x^2y + 3xy^2 + y^3$

3 $(x+y)^4 = x^4 + 4x^3y + 6x^2y^2 + 4xy^3 + y^4$

4 $(x+y)^5 = x^5 + 5x^4y + 10x^3y^2 + 10x^2y^3 + 5xy^4 + y^5$

All of these are special cases of the expansion of $(x+y)^n$, where n is any positive integer. The general expression is known as the **binomial theorem**.

Let us now restrict ourselves to expressions involving only one variable. These are the kind we will encounter most often in our study of calculus. Expressions such as $3x^2 - 5x + 7$ or $2 + 8x^2 - 7x^5 + 9x^7$ are examples of the general expression

$$a_0 + a_1x + a_2x^2 + \cdots + a_nx^n$$

where $a_0, a_1, \ldots, a_n$ are numbers and x is a variable. This is called a **polynomial** in x. An expression that is the ratio of two polynomials—that is, an expression of the form

$$\frac{a_0 + a_1x + a_2x^2 + \cdots + a_nx^n}{b_0 + b_1x + b_2x^2 + \cdots + b_nx^n}$$

is called a **rational function** of x. All of the rules in (A.1), (A.2), and (A.3) hold for such expressions. Let's practice a few computations.

EXAMPLE 13 Multiply $(x^2 + 2x + 3)$ and $(x^3 - x^2 + 4x + 7)$.

SOLUTION We obtain

$$\begin{aligned}(x^2 + 2x + 3)(x^3 - x^2 + 4x + 7) &= x^2(x^3 - x^2 + 4x + 7)\\ &\quad + 2x(x^3 - x^2 + 4x + 7)\\ &\quad + 3(x^3 - x^2 + 4x + 7)\\ &= (x^5 - x^4 + 4x^3 + 7x^2)\\ &\quad + (2x^4 - 2x^3 + 8x^2 + 14x)\\ &\quad + (3x^3 - 3x^2 + 12x + 21)\\ &= x^5 + x^4 + 5x^3 + 12x^2 + 26x + 21\end{aligned}$$

PRACTICE EXERCISE Verify the following:

1 $(x - 1)(x^6 + x^5 + x^4 + x^3 + x^2 + x + 1) = x^7 - 1$

2 $(x + 1)(x^6 - x^5 + x^4 - x^3 + x^2 - x + 1) = x^7 + 1$

EXAMPLE 14 Simplify (that is, express as a rational function):

$$\frac{2}{x - 3} + \frac{x}{x^2 - 9} + \frac{3x - 1}{x^2 + 6x + 9}$$

SOLUTION Because $x^2 - 9 = (x - 3)(x + 3)$ and $x^2 + 6x + 9 = (x + 3)^2$, we obtain

$$\frac{2}{x - 3} + \frac{x}{x^2 - 9} + \frac{3x - 1}{x^2 + 6x + 9} = \frac{2}{x - 3} + \frac{x}{(x - 3)(x + 3)} + \frac{3x - 1}{(x + 3)^2}$$

Then we take the common denominator $(x - 3)(x + 3)^2$ and obtain

$$\frac{2(x + 3)^2 + x(x + 3) + (3x - 1)(x - 3)}{(x - 3)(x + 3)^2}$$

which becomes

$$\frac{2(x^2 + 6x + 9) + (x^2 + 3x) + (3x^2 - 10x + 3)}{(x - 3)(x + 3)^2}$$

So the final result is

$$\frac{6x^2 + 5x + 21}{(x - 3)(x + 3)^2}$$

EXAMPLE 15 Divide $(x - 1)$ into $x^4 + 8x^3 - 9$.

SOLUTION The procedure for doing this is called **long division** and is exactly parallel to the familiar long division procedure for numbers. Here is the technique:

$$
\begin{array}{r}
x^3 + 9x^2 + 9x \ \ + 9 \\
(x - 1)\overline{\big)\, x^4 + 8x^3 + 0x^2 + 0x - 9} \\
\underline{x^4 - x^3} \qquad\qquad\qquad\quad \\
9x^3 + 0x^2 \qquad\quad \\
\underline{9x^3 - 9x^2} \qquad\quad \\
9x^2 + 0x \quad \\
\underline{9x^2 - 9x} \quad \\
9x - 9 \\
\underline{9x - 9} \\
0 \quad
\end{array}
$$

We see that the **quotient** is $x^3 + 9x^2 + 9x + 9$ and the **remainder** is 0, so

$$\frac{x^4 + 8x^3 - 9}{x - 1} = x^3 + 9x^2 + 9x + 9$$

or, equivalently,

$$x^4 + 8x^3 - 9 = (x - 1)(x^4 + 8x^3 - 9)$$

EXAMPLE 16 Divide $x^2 + 3x - 2$ into $x^5 + 4x^4 + 3x^3 + 2x^2 + x + 1$; in other words, express

$$\frac{x^5 + 4x^4 + 3x^3 + 2x^2 + x + 1}{x^2 + 3x - 2}$$

as a quotient plus a remainder.

SOLUTION We carry out the long division explicitly:

$$
\begin{array}{r}
x^3 + \ \ x^2 + 2x \ \ - 2 \qquad\quad \\
x^2 + 3x - 2\overline{\big)\, x^5 + 4x^4 + 3x^3 + 2x^2 + \ \ x + 1} \\
\underline{x^5 + 3x^4 - 2x^3} \qquad\qquad\qquad\quad \\
x^4 + 5x^3 + 2x^2 \qquad\qquad \\
\underline{x^4 + 3x^3 - 2x^2} \qquad\qquad \\
2x^3 + 4x^2 + \ \ x \qquad \\
\underline{2x^3 + 6x^2 - 4x} \qquad \\
-2x^2 + 5x + 1 \\
\underline{-2x^2 - 6x + 4} \\
11x - 3
\end{array}
$$

Thus $x^3 + x^2 + 2x - 2$ is the quotient and $11x - 3$ is the remainder. We have (as you may check by multiplying out)

$$
\begin{aligned}
&x^5 + 4x^4 + 3x^3 + 2x^2 + x + 1 \\
&\qquad = (x^3 + x^2 + 2x - 2)(x^2 + 3x - 2) + (11x - 3)
\end{aligned}
$$

or equivalently:

$$\frac{x^5 + 4x^4 + 3x^3 + 2x^2 + x + 1}{x^2 + 3x - 2} = x^3 + x^2 + 2x - 2 + \frac{11x - 3}{x^2 + 3x - 2}$$

Section A.2 EXERCISES

In Exercises 1–10 evaluate.

1. $\dfrac{3^2 \cdot 5^3 \cdot 7 \cdot 11^0}{3 \cdot 5^2 \cdot 7^3}$

2. $(2^{-1} + 3^{-1})^2$

3. $(2^{-1} - 9^{-1/2})^{-1}$

4. $2^{-1} + 3^{-1} + 2^{-1} \cdot 3^{-1}$

5. $(3 \cdot 4)^2 - (3 \cdot 4)^{-1} + (3 \cdot 4)^{-2}$

6. $\left|\dfrac{\left(\frac{1}{2}\right)\left(-\frac{1}{3}\right)^{-1}}{\left(\frac{2}{3}\right)\left(\frac{4}{5}\right)^{-1}}\right|$

7. $(\sqrt{2})^3(2^{1/3})$

8. $(-3)^4(3)^{-2} - (-8)^{2/3}(-2)^3(2)^{1/3}$

9. $\dfrac{(-7)^0(-7)^2(-7)^3}{(25)^{1/2} - (25)^{-1/2}}$

10. $(\frac{1}{3})^{-1}(9)^{1/2} + (\frac{1}{9})^{-1/2}(-3)^{-1}$

11. Verify that $(2^3)^{-1} \neq 2^{(3^{-1})}$

12. Verify that $(3^3)^3 \neq 3^{(3^3)}$

13. Verify that $(2^2)^2 = 2^{(2^2)}$ but $(-2^{-2})^{-2} \neq (-2)^{(-2^{-2})}$

14. Are there any other numbers $a \neq 0$ for which $(a^a)^a = a^{(a^a)}$?

Simplify Exercises 15–29.

15. $\dfrac{8x^3y^{-5}}{4x^{-1}y^2}$

16. $\left(\dfrac{2x^3}{y}\right)^2\left(\dfrac{y}{x^3}\right)^3$

17. $(x^{1/3} - x^{-5/3})x^{2/3}$

18. $\left(\dfrac{2x^{2/3}}{y^{1/2}}\right)^2\left(\dfrac{y^{1/3}}{3x^{-5/6}}\right)^{-1}$

19. $\dfrac{x^{-1}}{y^{-1}} + \dfrac{x}{y}$

20. $\dfrac{x^{-1}}{y^{-1}} - \left(\dfrac{x}{y}\right)^{-1}$

21. $\dfrac{x^{-2} + y^{-2}}{(xy)^{-1}}$

22. $\dfrac{(x + y)^{-1}}{x^{-1} + y^{-1}}$

23. $\dfrac{\sqrt{a} - \sqrt{b}}{\sqrt{a} + \sqrt{b}}$

24. $\dfrac{px^{p-1}}{q(x^{p/q})^{q-1}}$

25. $(x - y)(x^2 + xy + y)^2$

26. $\left(x^2 - \dfrac{3}{x}\right)^3$

27. $(x - y - z)(x - y + z)$

28. $(x^{2/3} + y^{2/3})^3$

29. $\dfrac{x^3 - y^3}{x^2 - y^2}$

30. Show that

$$(x + y)^3 = x^3 + 3x^2y + 3xy^2 + y^3$$

and that

$$(x + y)^4 = x^4 + 4x^3y + (\text{assorted stuff})y^2$$

What would you expect $(x + y)^n$ to be?

31. Multiply out $(2x^3 + 4x - 1)(3x^2 - x + 5)$.

32. Multiply out $(3x^4 + x^3 - x + 2)(x^3 - 4x^2 + 2x - 3)$.

33. Multiply out $(x - 1)(x + 1)(x^4 + x^2 + 1)$.

34. (a) Show $x^2 - 1 = (x - 1)(x + 1)$,
$x^4 - 1 = (x - 1)(x^3 + x^2 + x + 1)$
How can you write $x^n - 1$ if n is even?
(b) Show $x^3 + 1 = (x + 1)(x^2 - x + 1)$,
$x^5 + 1 = (x + 1)(x^4 - x^3 + x^2 - x + 1)$
How can you write $x^n + 1$ if n is odd?

35. Simplify $\dfrac{3x^2 - 7x - 6}{x^2 - 9}$.

36. Multiply out $(x^{-2} + x^2)^3$.

37. Multiply out $\left(\dfrac{1}{x} - x\right)^4$.

Perform long division in Exercises 38–45.

38. $\dfrac{x^2 - 5x + 6}{x - 2}$

39. $\dfrac{x^2 - 5x + 6}{x - 1}$

40. $\dfrac{x^3 + x - 10}{x + 3}$

41. $\dfrac{x^3 + x - 10}{x - 2}$

42. $\dfrac{x^4 + x^3 - 3x^2 + 2x - 5}{x^2 - x - 1}$

43. $\dfrac{x^4 + x^3 - 3x^2 + 2x - 5}{x + 1}$

44. $\dfrac{x^7 + x + 1}{x^2 + x + 1}$

45. $\dfrac{x^7 + 1}{x^2 + 1}$

SECTION A.3

Solving Simple Equations

We now turn our attention to the problem of solving equations. As before, our approach will be to review the principal concepts by working a large number of examples.

If p and q are two algebraic expressions involving a variable, x, then the statement $p = q$ is called an **algebraic equation** in x. A **root**, or a **solution** of the equation is a number c that makes the statement true when it is substituted for the variable. For example, if we consider the polynomial $p = x^3 - x^2 + x + 14$, then $x = -2$ is a root of the equation

$$x^3 - x^2 + x + 14 = 0$$

because

$$(-2)^3 - (-2)^2 + (-2) + 14 = (-8) - (4) + (-2) + 14 = 0$$

For obvious reasons, $x = -2$ is also called a **zero** of the polynomial p.

To **solve an equation** means to find all the roots of the equation; that is, to find all values of the variable for which the equation is satisfied.

We are interested primarily in equations that involve polynomials. For such equations, the location of roots is closely associated with the **factorization** of the polynomial. A polynomial is said to be **factored** if it is written as a product of polynomials other than itself; the polynomials appearing in this product are called **factors**. Factorization is important because of the following. Suppose we are given a polynomial p. Then the real number c will be a root of the given polynomial if and only if $(x - c)$ is a factor of the polynomial. This means that when we divide $(x - c)$ into the polynomial—which we can do via long division—the remainder is zero.

EXAMPLE 17 Solve the equation $\dfrac{2x + 3}{x + 4} = 7$.

SOLUTION Multiply both sides of the equation by $x + 4$. This gives $2x + 3 = 7(x + 4)$, or $2x + 3 = 7x + 28$. Collecting terms gives $-5x = 25$, so $x = -5$. We can check any solution by substituting it back into the original equation in place of the variable. In this example we would have

$$\frac{2(-5) + 3}{(-5) + 4} = \frac{-7}{-1} = 7$$

so $x = -5$ is correct.

EXAMPLE 18 Solve the equation $x^2 - 5x + 6 = 0$.

SOLUTION By inspection we see that we have the factorization $x^2 - 5x + 6 = (x - 2)(x - 3)$. Note that the product of -2 and -3 is $+6$, the constant term, and the sum of -2 and -3 is -5, the coefficient of x. Settling $x - 2 = 0$, and $x - 3 = 0$ we see that $x = 2$ and $x = 3$ are the solutions of the given equation. (Another way to express this is to say that $x = 2$ and $x = 3$ are the *roots* of the polynomial $(x^2 - 5x + 6)$.

EXAMPLE 19 Solve $(x - 1)(x + 4) + (2x - 5)(x - 1) = 0$.

SOLUTION We may factor $(x - 1)$ out of the left side. This gives $(x - 1)[(x + 4) + (2x - 5)] = 0$, which is $(x - 1)(3x - 1) = 0$. Setting each factor equal to 0, we have $x = 1, x = \frac{1}{3}$.

EXAMPLE 20 Solve $(x + 1)(x + 4) + (2x - 5)(x - 2) = 0$.

SOLUTION Because $(x + 1)(x + 4) = x^2 + 5x + 4$ and $(2x - 5)(x - 2) = 2x^2 - 9x + 10$, our given equation becomes

$$3x^2 - 4x + 14 = 0$$

If you try to factor the left side, you will not succeed. Instead, it is necessary to turn to the **quadratic formula**. This says that the two roots of the general quadratic polynomial $ax^2 + bx + c$ are given by

$$\frac{-b \pm \sqrt{b^2 - 4ac}}{2a}$$

The expression $b^2 - 4ac$ is known as the **discriminant**. If $b^2 - 4ac = 0$, then the two roots are equal (and both equal $-b/2a$). If $b^2 - 4ac > 0$, then there are two real and unequal roots. If, however, $b^2 - 4ac < 0$, then there are no real roots, because there is no real square root of a negative number. In our case, $a = 3$, $b = -4$ and $c = 14$. Because

$$b^2 - 4ac = (-4)^2 - 4(3)(14) < 0$$

we see that our equation has no real solutions.

EXAMPLE 21 Solve the equation $\dfrac{2x - 3}{x - 4} = \dfrac{7x - 1}{x + 2}$.

SOLUTION By clearing of fractions (which here means multiplying both sides by $(x - 4)(x + 2)$ or by cross-multiplication (which says that if we have $a/b = c/d$ then $ad = bc$), we obtain

$$(2x - 3)(x + 2) = (x - 4)(7x - 1)$$

Multiplication gives $2x^2 + x - 6 = 7x^2 - 29x + 4$, and therefore,

$$5x^2 - 30x + 10 = 0$$

Division by 5 yields

$$x^2 - 6x + 2 = 0$$

According to the quadratic formula, the two roots of this equation are

$$\frac{6 \pm \sqrt{36 - 8}}{2} = \frac{6 \pm \sqrt{28}}{2} = \frac{6 \pm 2\sqrt{7}}{2} = 3 \pm \sqrt{7}$$

EXAMPLE 22 Solve $\dfrac{3}{x + 1} - \dfrac{2}{x} = \dfrac{1}{10}$.

SOLUTION We multiply through by $10x(x + 1)$, and we get $3(10x) - 2[10(x + 1)] = x(x + 1)$ or $30x - 20x - 20 = x^2 + x$. Thus $x^2 - 9x + 20 = 0$. The solutions of this equation are easily seen to be $x = 4$ and $x = 5$.

EXAMPLE 23 Solve $\sqrt{x^2 - 8} = 4 - x$.

SOLUTION By squaring both sides, we get $x^2 - 8 = 16 - 8x + x^2$. So $8x = 24$ and $x = 3$. It is particularly important to check the answer in the original equation because squaring may introduce extraneous solutions.

EXAMPLE 24 Solve

$$\frac{\dfrac{3}{x + 1} - \dfrac{2}{x - 3}}{\dfrac{1}{x - 1} + \dfrac{2}{x + 2}} = \frac{2}{5}$$

SOLUTION We have in the numerator

$$\frac{3}{x + 1} - \frac{2}{x + 3} = \frac{x + 7}{(x - 1)(x + 3)}$$

and in the denominator

$$\frac{1}{x - 1} + \frac{2}{x + 2} = \frac{3x}{(x - 1)(x + 2)}$$

Therefore, by (A.2), part 4,

$$\frac{(x + 7)(x - 1)(x + 2)}{(x + 1)(x + 3)(3x)} = \frac{2}{5}$$

Multiplication in both the numerator and denominator gives

$$\frac{x^3 + 8x^2 + 5x - 14}{3x^3 + 12x^2 + 9x} = \frac{2}{5}$$

This reduces (upon cross-multiplying and combining terms) to

$$x^3 - 16x^2 - 7x + 70 = 0$$

By trial and error, you might discover that 2 is a root. (A useful hint, applicable when the leading coefficient is one, is to test the factors of the constant term 70 as possible roots — ± 2, ± 5, ± 7, ± 14, ± 35.) Thus $(x - 2)$ is a factor, and long division produces the factorization

$$(x - 2)(x^2 - 14x - 35) = 0$$

Thus $x - 2 = 0$ and $x^2 - 14x - 35 = 0$. So the solutions of the original equation are $x = 2$ and also

$$x = \frac{14 \pm \sqrt{196 + 140}}{2}$$

$$= \frac{14 \pm \sqrt{336}}{2} = 7 \pm 2\sqrt{21}.$$

Sometimes we will want to solve more than one equation simultaneously. In this case we seek values of the variables that satisfy all of the given equations.

EXAMPLE 25 Solve the simultaneous equations

$$x - y = 4$$

$$3x + 2y = 15$$

SOLUTION There are two principal approaches to solving this problem.

(a) We multiply the first equation by 2 and then add the resulting equation to the second given equation

$$\begin{aligned} 2x - 2y &= 8 \\ 3x + 2y &= 15 \\ \hline 5x &= 23 \\ x &= \frac{23}{5} \end{aligned}$$

Now, substituting this value into the first equation we have

$$\frac{23}{5} - y = 4$$

so that

$$y = \frac{23}{5} - 4 = \frac{3}{5}$$

Of course, we could just as easily have substituted the value of x into the second equation.

(b) Let us solve the first equation for y:

$$y = x - 4$$

Next, we substitute this expression into the second equation

$$3x + 2(x - 4) = 15$$

or

$$5x - 8 = 15$$

$$5x = 23$$

$$x = \frac{23}{5}$$

As previously, $y = \frac{3}{5}$. In many problems, especially if the equations involve expressions of higher degree than the first degree, this second approach is the only one available.

EXAMPLE 26 Solve the simultaneous equations

$$3x + 2y = 15$$

$$2x - 3y = -3$$

SOLUTION To eliminate y, multiply the first equation by 3 and the second equation by 2. This yields

$$9x + 6y = 45$$

$$4x - 6y = -6$$

Adding these two equations results in

$$13x = 39 \quad \text{or} \quad x = 3$$

Substituting back (for x) in either of the original equations gives $y = 3$. Thus the only solution is $x = 3$, $y = 3$.

EXAMPLE 27 Solve the simultaneous equations

$$y = x^2 - 4$$

$$x + y = 2$$

SOLUTION The first equation is already solved for y in terms of x. Thus we substitute this expression into the second equation to obtain

$$x + (x^2 - 4) = 2$$

or

$$x^2 + x - 6 = 0$$

This equation can be factored to yield

$$(x + 3)(x - 2) = 0$$

so that the roots are $x = -3$ and $x = 2$ We substitute each of the values, in turn, into either equation. If $x = -3$, then $y = (-3)^2 - 4 = 5$, while if $x = 2$, then $y = 2^2 - 4 = 0$. Our solutions are $x = -3$, $y = 5$ and $x = 2$, $y = 0$.

EXAMPLE 28 Solve the simultaneous equations

$$x^2 + y^2 = 25$$
$$-2x + y = 10$$

SOLUTION It is relatively simple to solve the second equation for y:

$$y = 2x + 10$$

We substitute this into the first equation. This produces

$$x^2 + (2x + 10)^2 = 25$$
$$x^2 + 4x^2 + 40x + 100 = 25$$
$$5x^2 + 40x + 75 = 0$$
$$x^2 + 8x + 15 = 0$$

Again, factoring produces

$$(x + 3)(x + 5) = 0$$

so that our roots are $x = -3$ and $x = -5$. Working with the second equation (because it is simpler) produces the complete solution. If $x = -3$, then $y = 2(-3) + 10 = 4$, while if $x = -5$, then $y = 2(-5) + 10 = 0$. The solutions of our system of equations are $x = -3$, $y = 4$, and $x = -5$, $y = 0$.

Section A.3 EXERCISES

Factor Exercises 1–16.

1. $x^2 - x - 12$

2. $3x^2 + 7x - 6$

3. $4x^2 - 12x + 9$

4. $3x^2 - 14x + 8$

5. $6x^2 - 16x - 32$

6. $6x^2 + 7x - 3$

7. $x^2 - 25$

8. $x^2 - 36$

9. $x^4 - 13x^2 + 36$

10. $x^4 - 10x^2 + 9$

11. $x^3 - 1$

12. $x^3 + 1$

13. $x^3 - 8$

14. $x^3 + 27$

15. $x^3 + x^2 - 3x - 6$

16. $3x^3 + x^2 + 3x + 1$

Solve Exercises 17–40.

17. $3x - 8 = 5x + 6$

18. $-6 + 5x = 3x^2$

19. $20 - x - x^2 = 0$

20. $\dfrac{2x^2 + 3x - 4}{x - x^2} = -1$

21. $\dfrac{(x - 1)(2x + 3)}{x - 5} = x + 1$

22. $\dfrac{3}{2x - 5} - \dfrac{8}{6x - 15} = 0$

23. $\dfrac{3}{2x - 5} + \dfrac{2}{3x + 5} = 1$

24. $\sqrt{x-5} + 1 = \sqrt{x}$

25. $x + 3y = 7,\ 3x - 2y = 1$

26. $5x - 2y = 3,\ 15x - 6y = 7$

27. $y = x^2 - x - 6,\ y = -x - 5$

28. $x^2 + y^2 = 25, x + y = 7$

29. $y = x^2 + 2x - 3,\ y = x - 1$

30. $y = x^3 + 2x^2 - 4, y = x^2 + 3x - 1$

31. $y = x^2 - x - 6,\ y = x^2 + x - 2$

32. $y = 8 - x^2,\ y = x^2$

33. $y = \dfrac{x}{x-2} - 2,\ y = \dfrac{2}{x-2}$

34. $y = 8x^2 + 14x + 99,\ y = 7x^2 - 12x - 70$

35. $y = 3x + 4,\ y = 6(7x + 2) - 9$

36. $y = x^4 - 11x^2,\ y = 11x^2 + 75$

37. $y = x^2 + x,\ y = 3 - x$

38. $y = 2x^2,\ y = x^3 - 3x$

39. $y = x^2,\ y = 1/x^2$

40. $y = x^3 - 5x^2,\ y = 2x - 24$

B Table of Integrals

INTEGRALS INVOLVING $a + bu$

1. $$\int (a + bu)^n \, du = \frac{(a + bu)^{n+1}}{b(n + 1)} + c, n \neq -1$$

2. $$\int \frac{du}{a + bu} = \frac{1}{b} \ln|a + bu| + c$$

3. $$\int \frac{u \, du}{a + bu} = \frac{1}{b^2} [bu - a \ln|a + bu|] + c$$

4. $$\int \frac{u \, du}{(a + bu)^2} = \frac{1}{b^2} \left[\frac{a}{a + bu} + \ln|a + bu| \right] + c$$

5. $$\int \frac{u \, du}{(a + bu)^n} = \frac{1}{b^2} \left[\frac{a}{(n - 1)(a + bu)^{n-1}} - \frac{1}{(n - 2)(a + bu)^{n-2}} \right] + c,$$ $n \neq 1, 2$

6. $$\int \frac{u^2 \, du}{a + bu} = \frac{1}{b^3} \left[a^2 \ln|a + bu| + \frac{bu}{2}(bu - 2a) \right] + c$$

7. $$\int \frac{u^2 \, du}{(a + bu)^2} = \frac{1}{b^3} \left[bu - \frac{a^2}{a + bu} - 2a \ln|a + bu| \right] + c$$

8. $$\int \frac{u^2 \, du}{(a + bu)^3} = \frac{1}{b^3} \left[\ln|a + bu| + \frac{2a}{a + bu} - \frac{a^2}{2(a + bu)^2} \right] + c$$

9. $$\int \frac{du}{u(a + bu)} = \frac{1}{a} \ln\left| \frac{u}{a + bu} \right| + c$$

10. $$\int \frac{du}{u^2(a + bu)} = -\frac{1}{au} + \frac{b}{a^2} \ln\left| \frac{a + bu}{u} \right| + c$$

11. $$\int \frac{du}{u^2(a + bu)^2} = -\frac{1}{a^2} \left[\frac{a + 2bu}{u(a + bu)} + \frac{2b}{a} \ln\left| \frac{u}{a + bu} \right| \right] + c$$

12. $$\int \frac{u \, du}{(a + bu)^2} = \frac{1}{b^2} \left(\ln|a + bu| + \frac{a}{a + bu} \right) + c$$

13. $\displaystyle\int \frac{du}{(a + bu)(a' + b'u)} = \frac{1}{ab' - a'b} \ln\left|\frac{a' + b'u}{a + bu}\right| + c$

14. $\displaystyle\int \frac{u\,du}{(a + bu)(a' + b'u)} = \frac{1}{ab' - a'b}\left[\frac{a}{b}\ln|a + bu| - \frac{a'}{b'}\ln|a' + b'u|\right] + c$

INTEGRALS INVOLVING $\sqrt{a + bu}$

15. $\displaystyle\int u\sqrt{a + bu}\,du = -\frac{2(2a - 3bu)(a + bu)^{3/2}}{15b^2} + c$

16. $\displaystyle\int u^n\sqrt{a + bu}\,du = \frac{2}{b(2n + 3)}\left[u^n(a + bu)^{3/2} - na\int u^{n-1}\sqrt{a + bu}\,du\right]$

17. $\displaystyle\int \frac{\sqrt{a + bu}}{u}\,du = 2\sqrt{a + bu} + a\int \frac{du}{u\sqrt{a + bu}}$

18. $\displaystyle\int \frac{\sqrt{a + bu}}{u^n}\,du = -\frac{(a + bu)^{3/2}}{a(n - 1)u^{n-1}} - \frac{b(2n - 5)}{2a(n - 1)}\int \frac{\sqrt{a + bu}\,du}{u^{n-1}}$

19. $\displaystyle\int \frac{du}{u\sqrt{a + bu}} = \frac{1}{\sqrt{a}}\ln\left|\frac{\sqrt{a + bu} - \sqrt{a}}{\sqrt{a + bu} + \sqrt{a}}\right| + c \qquad (a > 0)$

20. $\displaystyle\int \frac{du}{u^n\sqrt{a + bu}} = -\frac{\sqrt{a + bu}}{a(n - 1)u^{n-1}} - \frac{b(2n - 3)}{2a(n - 1)}\int \frac{du}{u^{n-1}\sqrt{a + bu}}$

21. $\displaystyle\int \frac{u\,du}{\sqrt{a + bu}} = \frac{2(bu - 2a)\sqrt{a + bu}}{3b^2} + c$

22. $\displaystyle\int \frac{u^2\,du}{\sqrt{a + bu}} = \frac{2(3b^2u^2 - 4abu + 8a^2)\sqrt{a + bu}}{15b^3} + c$

23. $\displaystyle\int \frac{u^n\,du}{\sqrt{a + bu}} = \frac{2u^n\sqrt{a + bu}}{b(2n + 1)} - \frac{2an}{b(2n + 1)}\int \frac{u^{n-1}\,du}{\sqrt{a + bu}}$

INTEGRALS INVOLVING $u^2 - a^2$, $a > 0$

24. $\displaystyle\int \frac{du}{u^2 - a^2} = \frac{1}{2a}\ln\left|\frac{u - a}{u + a}\right| + c$

25. $\displaystyle\int \frac{du}{a^2 - u^2} = \frac{1}{2a}\ln\left|\frac{u + a}{u - a}\right| + c$

INTEGRALS INVOLVING $\sqrt{u^2 \pm a^2}$, $a > 0$

26. $\displaystyle\int \sqrt{u^2 \pm a^2}\,du = \tfrac{1}{2}[u\sqrt{u^2 \pm a^2} \pm a^2\ln|u + \sqrt{u^2 \pm a^2}|] + c$

27. $\int u^2\sqrt{u^2 \pm a^2}\,du = \frac{1}{8}[u(2u^2 \pm a^2)\sqrt{u^2 \pm a^2} - a^4\ln|u + \sqrt{u^2 \pm a^2}|] + c$

28. $\int \frac{\sqrt{u^2 + a^2}}{u}\,du = \sqrt{u^2 + a^2} - a\ln\left|\frac{a + \sqrt{u^2 \pm a^2}}{u}\right| + c$

29. $\int \frac{\sqrt{u^2 \pm a^2}}{u^2}\,du = -\frac{\sqrt{u^2 \pm a^2}}{u} + \ln|u + \sqrt{u^2 \pm a^2}| + c$

30. $\int \frac{du}{\sqrt{u^2 \pm a^2}} = \ln|u + \sqrt{u^2 \pm a^2}| + c$

31. $\int \frac{du}{u\sqrt{u^2 + a^2}} = \frac{1}{a}\ln\left|\frac{u}{a + \sqrt{u^2 + a^2}}\right| + c$

32. $\int \frac{u^2\,du}{\sqrt{u^2 \pm a^2}} = \frac{1}{2}[u\sqrt{u^2 \pm a^2} \mp a^2\ln|u + \sqrt{u^2 \pm a^2}|] + c$

33. $\int \frac{du}{u^2\sqrt{u^2 \pm a^2}} = \mp\frac{\sqrt{u^2 \pm a^2}}{a^2u} + c$

34. $\int \frac{du}{(u^2 \pm a^2)^{3/2}} = \frac{\pm u}{a^2\sqrt{u^2 \pm a^2}} + c$

35. $\int \frac{u^2\,du}{(u^2 \pm a^2)^{3/2}} = \frac{-u}{\sqrt{u^2 \pm a^2}} + \ln|u + \sqrt{u^2 \pm a^2}| + c$

INTEGRALS INVOLVING $\sqrt{a^2 - u^2}$

36. $\int \frac{du}{u\sqrt{a^2 - u^2}} = -\frac{1}{a}\ln\left|\frac{a + \sqrt{a^2 - u^2}}{u}\right| + c$

37. $\int \frac{du}{u^2\sqrt{a^2 - u^2}} = -\frac{\sqrt{a^2 - u^2}}{a^2u} + c$

38. $\int \frac{\sqrt{a^2 - u^2}\,du}{u} = \sqrt{a^2 - u^2} - a\ln\left|\frac{a + \sqrt{a^2 - u^2}}{u}\right| + c$

39. $\int \frac{du}{(a^2 - u^2)^{3/2}} = \frac{u}{a^2\sqrt{a^2 - u^2}} + c$

INTEGRALS INVOLVING e^u

40. $\int e^u\,du = e^u + c$

41. $\int ue^u\,du = (u - 1)e^u + c$

42. $\int u^n e^u \, du = u^n e^u - n \int u^{n-1} e^u \, du$

43. $\int \frac{du}{1 + e^u} = u - \ln(1 + e^u) + c$

44. $\int \frac{du}{1 + e^{nu}} = u - \frac{1}{n} \ln(1 + e^{nu}) + c$

INTEGRALS INVOLVING ln u

45. $\int \ln u \, du = u \ln u - u + c$

46. $\int u \ln u \, du = \frac{u^2}{2} \ln u - \frac{u^2}{4} + c$

47. $\int u^n \ln u \, du = \frac{u^{n+1}}{n+1} \ln u - \frac{u^{n+1}}{(n+1)^2} + c, \; n \neq -1$

48. $\int (\ln u)^n \, du = u(\ln u)^n - n \int (\ln u)^{n-1} \, du, \; n \neq -1$

INTEGRALS INVOLVING TRIGONOMETRIC FUNCTIONS

49. $\int \sin u \, du = -\cos u + c$

50. $\int \cos u \, du = \sin u + c$

51. $\int \tan u \, du = -\ln|\cos u| + c = \ln|\sec u| + c$

52. $\int \cot u \, du = \ln|\sin u| + c$

53. $\int \sec u \, du = \ln|\sec u + \tan u| + c$

54. $\int \csc u \, du = \ln|\csc u - \cot u| + c = \ln\left|\tan \frac{u}{2}\right| + c$

55. $\int \sec^2 u \, du = \tan u + c$

56. $\int \csc^2 u \, du = -\cot u + c$

57. $\int \sec u \tan u \, du = \sec u + c$

58. $\int \csc u \cot u \, du = -\csc u + c$

59. $\int \sin^2 u \, du = \frac{1}{2} u - \frac{1}{4} \sin 2u + c$

60. $\int \cos^2 u \, du = \frac{1}{2} u + \frac{1}{4} \sin 2u + c$

61. $\int \sin^n u \, du = -\frac{\sin^{n-1} u \cos u}{n} + \frac{n-1}{n} \int \sin^{n-2} u \, du$

62. $\int \cos^n u \, du = \frac{\cos^{n-1} u \sin u}{n} + \frac{n-1}{n} \int \cos^{n-2} u \, du$

63. $\int \tan^n u \, du = \frac{\tan^{n-1} u}{n-1} - \int \tan^{n-2} u \, du$

64. $\int \cot^n u \, du = -\frac{\cot^{n-1} u}{n-1} - \int \cot^{n-2} u \, du$

65. $\int \sec^n u \, du = \frac{\tan u \sec^{n-2} u}{n-1} + \frac{n-2}{n-1} \int \sec^{n-2} u \, du$

66. $\int \csc^n u \, du = -\frac{\cot u \csc^{n-2} u}{n-1} + \frac{n-2}{n-1} \int \csc^{n-2} u \, du$

C Answers to Odd-Numbered Exercises

CHAPTER 1

Section 1.1 (page 14)

1.

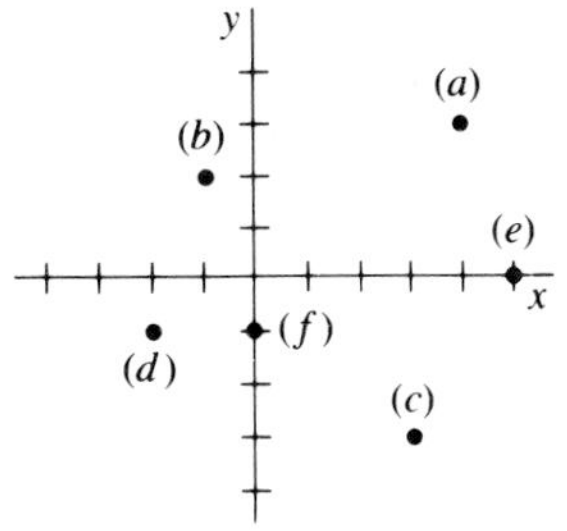

3. (a) $\frac{5}{2}$; (b) -2; (c) 0; (d) Undefined
5. $y = 3x + 7$ **7.** $y = -x + 5$ **9.** $y = 4x - 1$
11. $x = 3$ **13.** $y = 2x$ **15.** $5y - 2x = 23$
17. $(\frac{2}{5}, \frac{29}{5})$ **19.** $(4, -1)$
21. Infinite solution set—same line
23. $2y - 3x = 17$ **25.** $3y - x = 2$ **27.** $x = 3$
29. (a) 4; (b) 7; (c) $\sqrt{2}$; (d) 25; (e) 10; (f) $\sqrt{37}$; (g) $\sqrt{29 + 2\sqrt{2}}$; (h) $\sqrt{(x - a)^2 + (y - b)^2}$
31. $(x + 2)^2 + (y + 5)^2 = 2$
33. $(x + 5)^2 + (y - 2)^2 = 25$
35. Center (2, 3); radius 4 **37.** Center (2, 3); radius 4
39. Center $(-2, -1)$; radius $\sqrt{10}$ **41.** Yes; yes
43. Yes **45.** No **47.** Yes **49.** Yes **51.** Yes
53. *Hint*: Check that $(a, 0)$ and $(0, b)$ satisfy this equation.
55. (a) $m = \frac{4}{3}$; (b) $m = -\frac{3}{4}$ **57.** $3y + 4x = 5$
59. $q_d = -2000p + 49{,}000$ **61.** $q_s = 100p - 4600$
63. (a) It rises by 0.71 cents; (b) It rises by 1.6 cents.
65. $C(x) = 240 + 5x$; $R(x) = 20x$; 16 hr

Section 1.2 (page 21)

1. (a) 7; (b) -21; (c) 0; (d) 17.5; (e) 7π; (f) $-7\sqrt{2}$; (g) $7t$
3. (a) 2; (b) 2; (c) 1; (d) 3.5; (e) $\pi + 1$; (f) $\sqrt{2} - 1$; (g) $t + 1, t \geq -1$ or $-t + 1, t < -1$
5. (a) 1; (b) $\sqrt{5}$; (c) $\sqrt{2}$; (d), (e) Not permissible; (f) $\sqrt{2 + \sqrt{2}}$; (g) $\sqrt{2 - t}, t \leq 2$
7. (a) $4 + 3x$; (b) $4 - 3(-2 + h) = 10 - 3h$; (c) $4 - 3(h^2 - 1) = 7 - 3h^2$; (d) $4 - 3(x + h)$; (e) $4 - \dfrac{3}{x - 1}$
9. (a) $x^2 - x + 1$; (b) $(-2 + h)^2 + (-2 + h) + 1$; (c) $(h^2 - 1)^2 + (h^2 - 1) + 1$; (d) $(x + h)^2 + (x + h) + 1$; (e) $\left(\dfrac{1}{x - 1}\right)^2 + \dfrac{1}{x - 1} + 1$
11. (a) $-x^3 + x + 1$; (b) $(-2 + h)^3 - (-2 + h) + 1$; (c) $(h^2 - 1)^3 - (h^2 - 1) + 1$; (d) $(x + h)^2 - (x + h) + 1$; (e) $\left(\dfrac{1}{x - 1}\right)^3 - \dfrac{1}{(x - 1)} + 1$
13. D: all reals **15.** D: $\{x | x \neq 3\}$
17. D: $\{x | x \leq 3\}$ **19.** R: all reals
21. R: $\{y | y \geq 0\}$ **23.** R: $\{y | y \neq 0\}$
25. D, R: all reals **27.** D, R: all reals **29.** D, R: all reals
31. D: $\{x | x \geq 2$ or $x \leq -2\}$; R: $\{y | y \geq 0\}$
33. D, R: all reals
35. $g(r) = 2\pi r$, D: $\{r | r > 0\}$; R: $\{y | y > 0\}$
37. (a) 0.22; 0.56; 0.73; (b) D: $\{x | x > 0\}$; R: $\{y | y = 0.22 + 0.17a, a = 0, 1, 2, \ldots\}$
39. $C(x) = 45 + 0.1x$
41. $V = x(12 - 2x)^2$; $D = \{x | 0 < x < 6\}$
43. Yes; $a = 0$ and $a = -2$ **45.** (a) $2 + h$; (b) $-6 + h$
47. (a) $\dfrac{1}{2h + 4}$ (b) $\dfrac{-1/2}{-2 + h}$
49. (a) $8h^2 + 36h + 54$; (b) $8h^2 - 60h + 150$
51. (a) $\dfrac{m(x + h) + b - mx - b}{h}$; (b) m
53. (a) $\dfrac{3(x + h)^2 - (x + h) + 1 - 3x^2 + x - 1}{h}$; (b) $6x + 3h - 1$
55. (a) $\dfrac{(x + h)^3 - x^3}{h}$ (b) $3x^2 + 3xh + h^2$

57. (a) $\dfrac{\sqrt{x+h}-\sqrt{x}}{h}$ (b) $\dfrac{1}{\sqrt{x+h}+\sqrt{x}}$

Section 1.3 (page 29)

1.
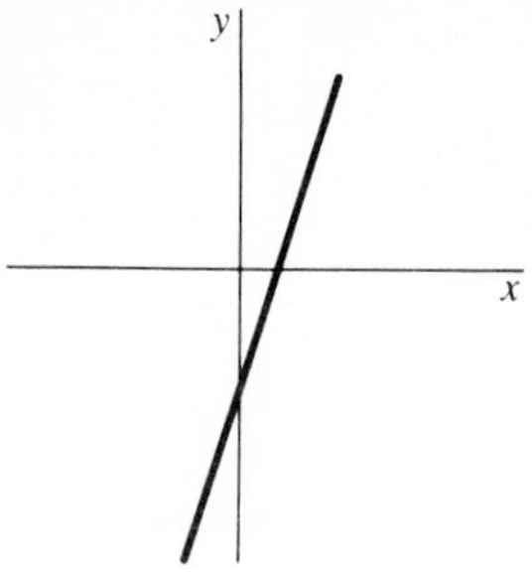

3.
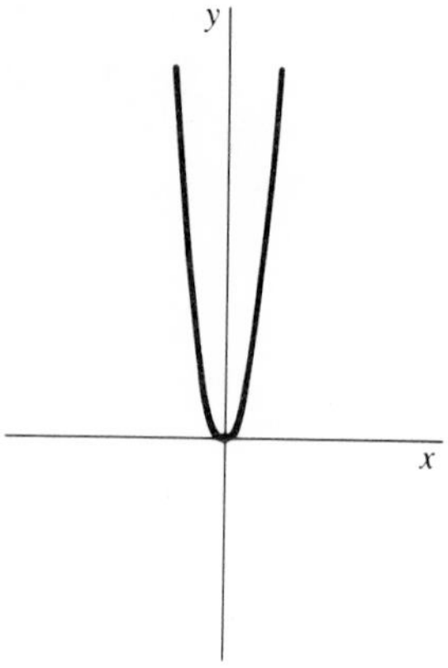

5.
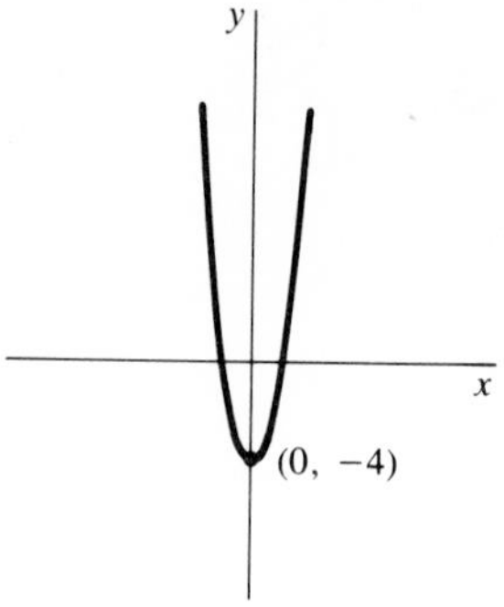

7.
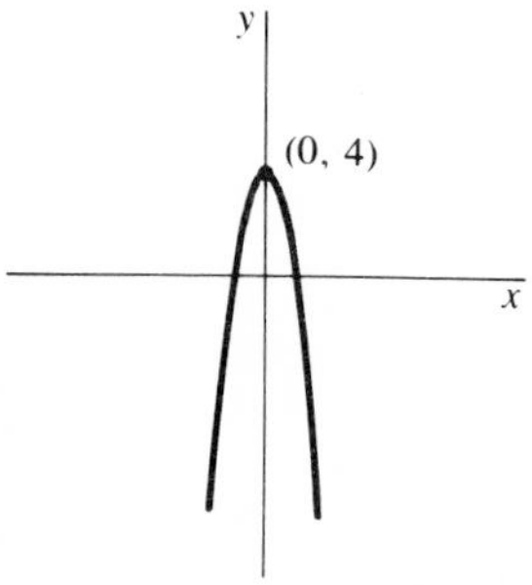

9.
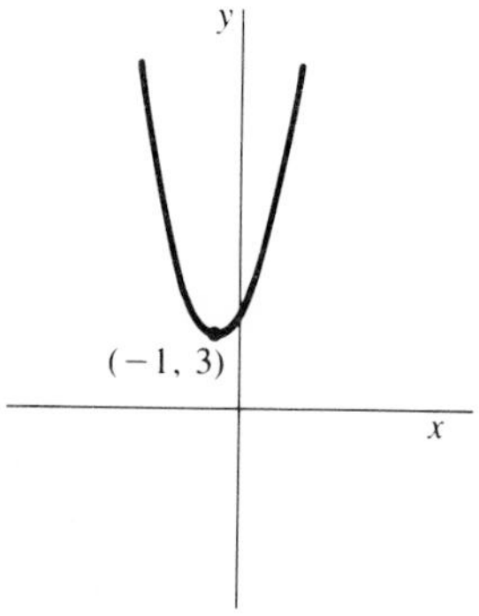

11.
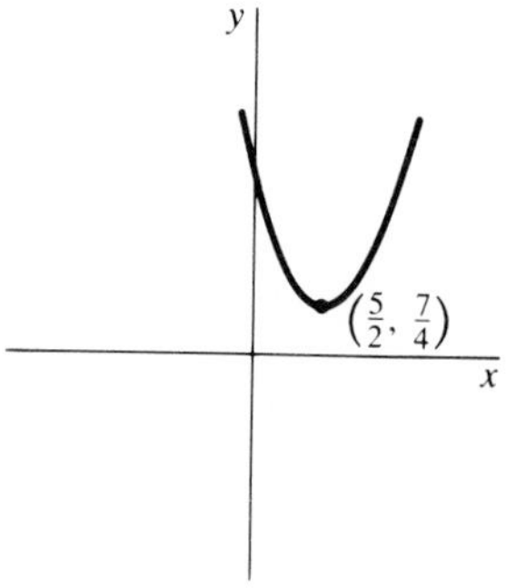

13.
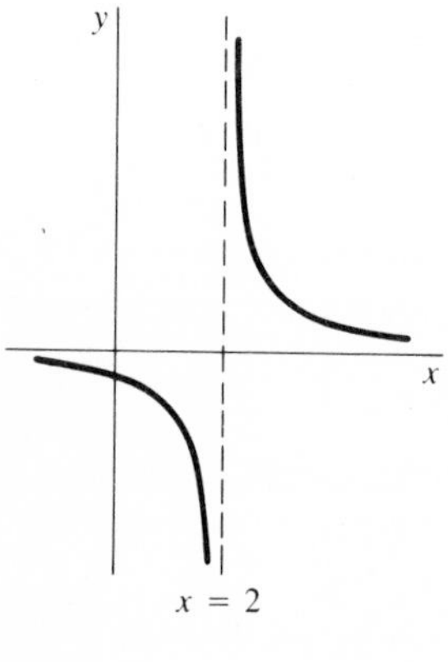

15.
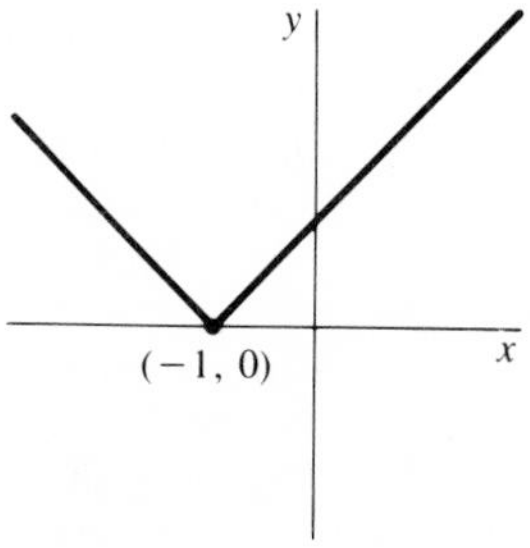

17.
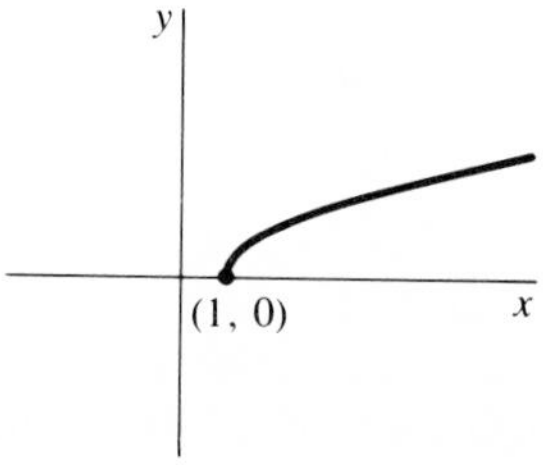

19. $f(x) = 11 - 8x$ 21. $f(x) = 1 + 4x - x^2$
23. $f(x) = 1$

25.

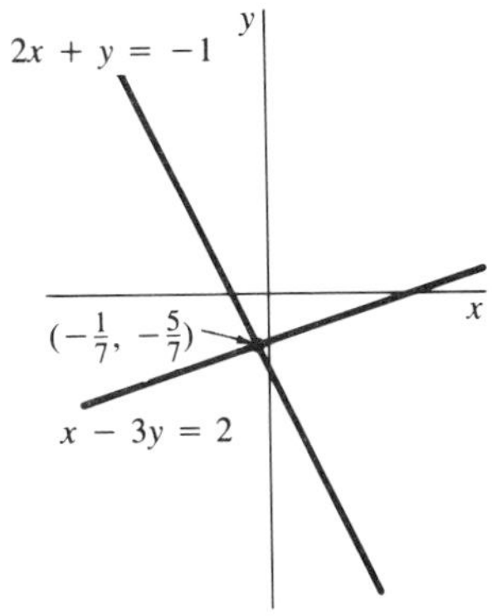

27.

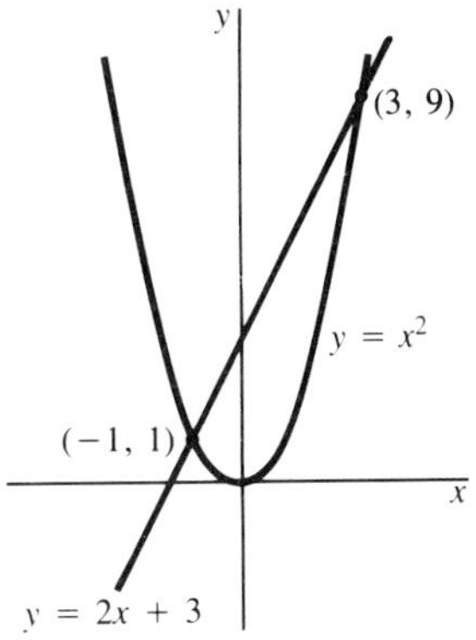

29.

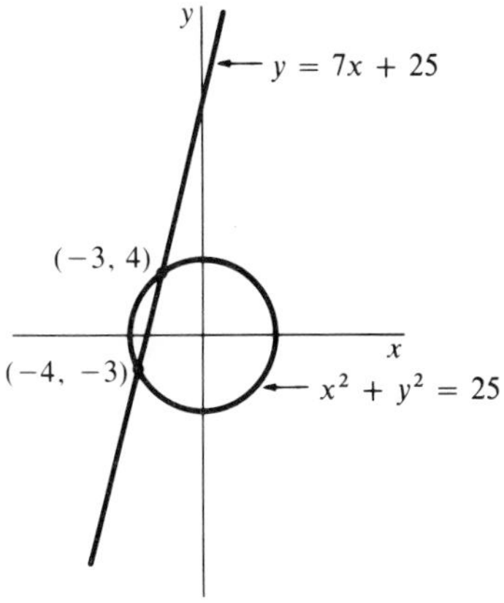

31. Not a graph of a function.
Parabola

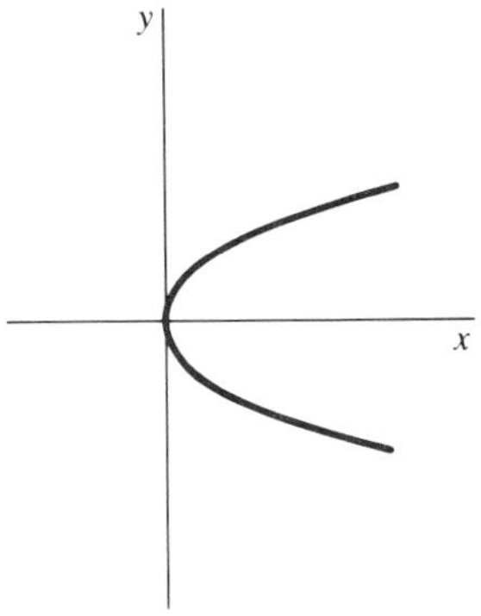

32. Not a graph of a function.
Hyperbola

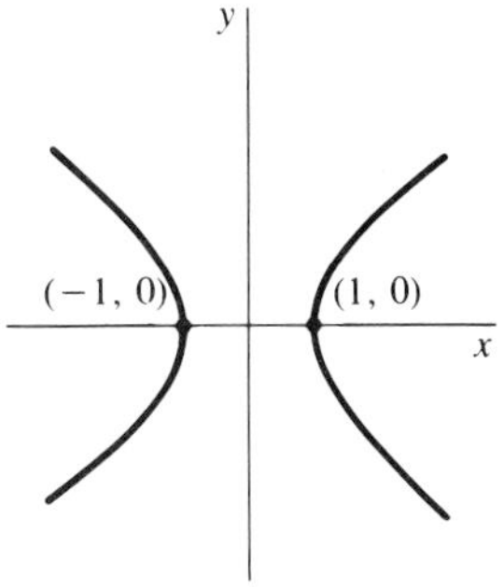

33. Not a graph of a function.
Circle

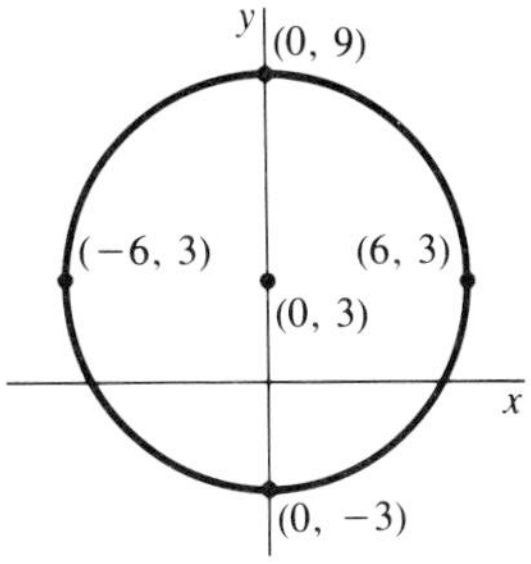

37.

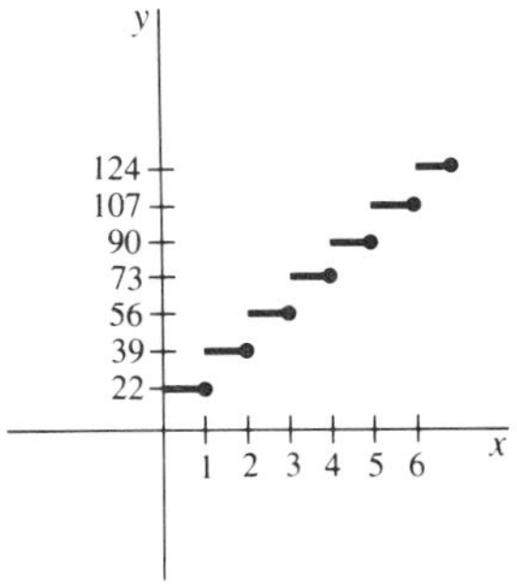

39. (a) No; (b) Yes. D: all reals; R: $\{y|y \leq 3\}$;
(c) No; (d) Yes. D: $\{x|x \neq \pm 1\}$; R: all reals;
(e) Yes. D: all reals; R: $\{y|y \geq 1\}$ (f) No; (g) No;
(h) Yes. D: $\{x|x \neq 0\}$; R: $\{y|y \neq 0\}$;
(i) Yes. D, R: all reals; (j) No; (k) Yes. D, R: all reals;
(l) No

41. (a) Yes; (b) No; (c) No; (d) Yes; (e) No

43. (a) No; (b) No; (c) No; (d) No; (e) No

45.

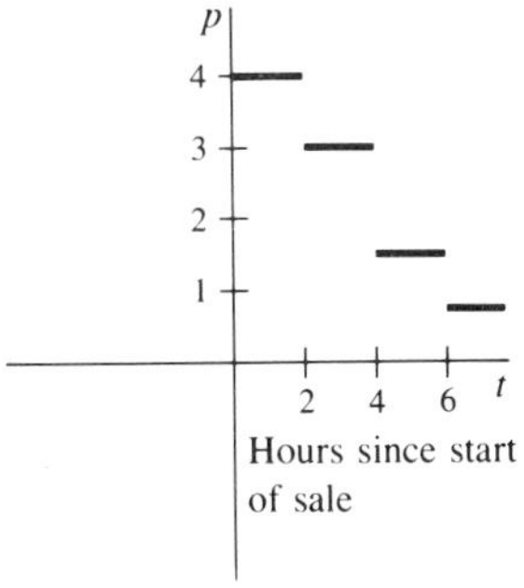

47. A square with sides of length 60 ft

49. (a) $R(x) = x(40 - x)$; (b)
(c) $R = 400$; (d) \$20

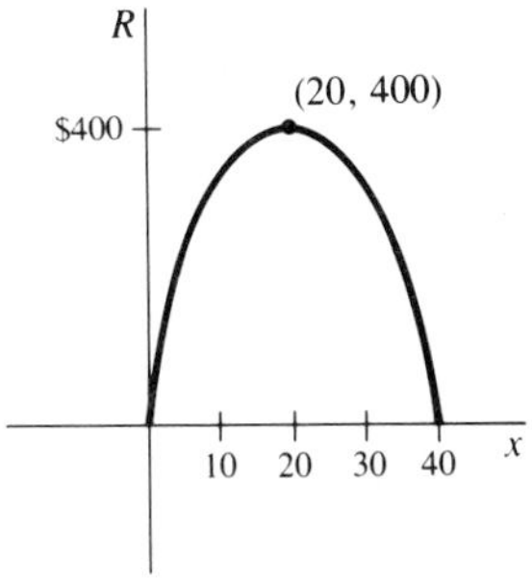

51.

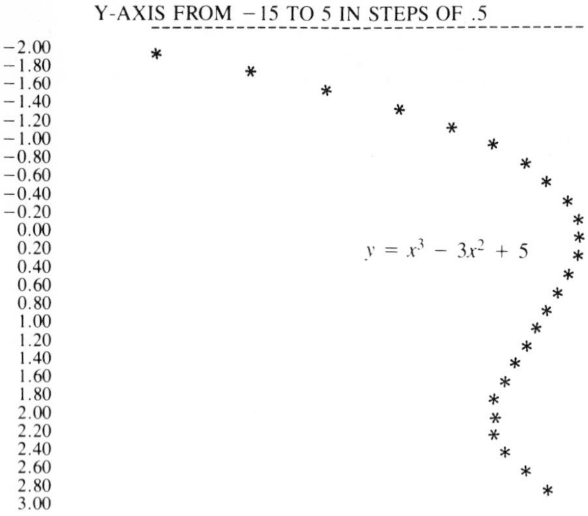

53.

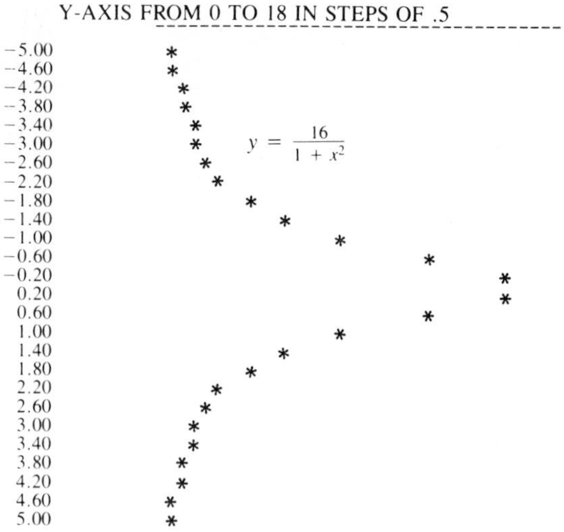

55.

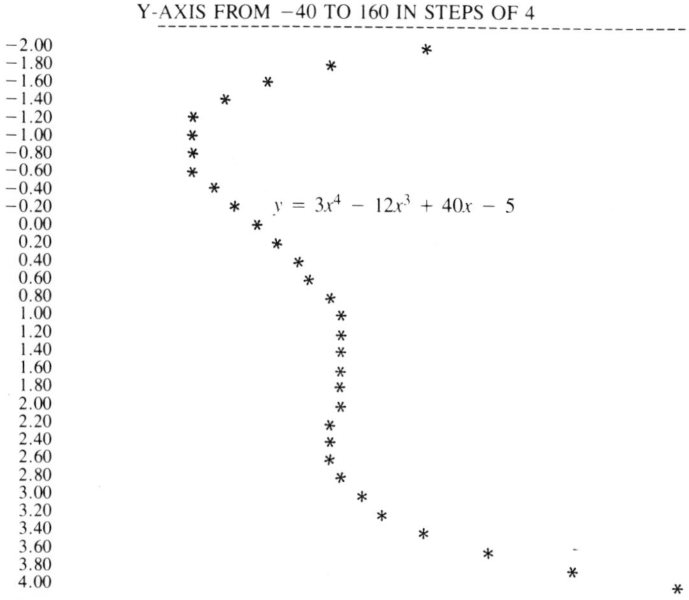

57.

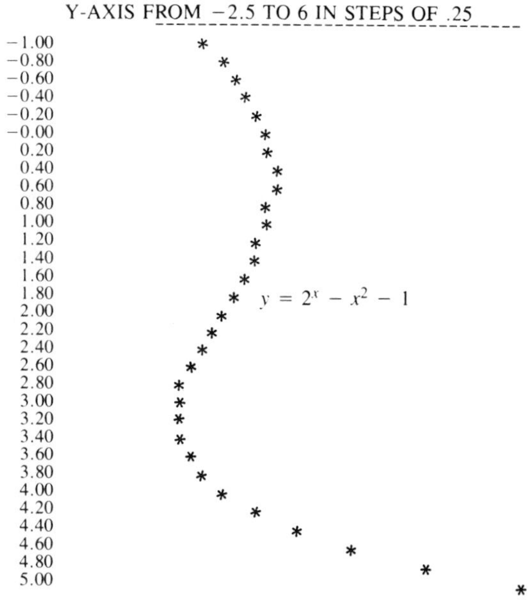

Section 1.4 (page 35)

1. $(f \circ g)(x) = \dfrac{4}{x + 2} - 2 = \dfrac{-2x}{x + 2}$, D: $\{x | x \neq -2\}$;

$(g \circ f)(x) = \dfrac{1}{4x}$, D: $\{x | x \neq 0\}$

3. $(f \circ g)(x) = x + 2 - \sqrt{x + 2}$, D: $\{x \geq -2\}$;
$(g \circ f)(x) = \sqrt{x^2 - x + 2}$, D: all reals

5. $f(x) = (g \circ h)(x)$; $g(x) = x^2$, $h(x) = x^2 - x + 3$

7. $f(x) = (g \circ h)(x)$; $g(x) = \dfrac{1}{x}$, $h(x) = \sqrt{3x^2 + 4}$

9. (a) $C(t) = 200 + 5(60t - \frac{3}{2}t^2)$; (b) $2120

CHAPTER 2

Section 2.1 (page 44)

1. No **3.** 3 **5.** 18 **7.** $v(t) = 3$
9. $v(t) = 6t + 1$ **11.** $v(t) = -1/t^2$ **13.** $v(t) = -2/t^3$
15. (a) $v(t) = 3t^2, a(t) = 6t$; (b) $v(t) = 6t, a(t) = 6$
17. $v(t) = 15t^2, a(t) = 30t$ **19.** $v(t) = 5, a(t) = 0$
21. \$1.24 per year, \$1.24 per year
23. (a) $-3.7 - 0.3h$ gal/min; (b) -3.7 gal/min

Section 2.2 (page 50)

1. $m = 3$ **3.** $m = 27$ **5.** $m = -5$ **7.** -1
9. $m = a$ **11.** $m = 2ax + b$ **13.** $m = 2x + 2$
15. $m = -2/x^2$ **17.** $m = 3x^2$ **19.** $m = \dfrac{-1}{(x-3)^2}$
21. (a) \$1550; (b) \$70 per additional unit;
(c) \$80 per additional unit

Section 2.3 (page 57)

1. $v(t) = 3t^2, y'(x) = 3x^2$
3. $v(t) = 6t - 1, y'(x) = 6x - 1$
5. $v(t) = 2at + b, y'(x) = 2ax + b$
7. $f'(7) = 14$ **9.** $f'(-2) = 12$ **11.** $f'(x) = 6x - 5$
13. $4, f(x) = x^2, f'(2)$ **15.** $3, f(x) = 3x, f'(2)$
17. $-\frac{1}{4}, f(x) = 1/x^2, f'(2)$ **19.** $3x^2, f(x) = x^3, f'(x)$
21. (a) $16, 12 + 3h + h^2$; (b) $4, 3h + h^2$;
(c) $4, 3 + h$; (d) 3
23. (a) $\frac{119}{4}, 29, 3h^2 - 3h + \frac{119}{4}$; (b) $-\frac{3}{4}, 3h^2 - 3h$;
(c) $-\frac{3}{2}, 3h - 3$; (d) -3; (e) $-\frac{9}{2}, 0, 3$; (f) $6t_0 - 6$

(g)

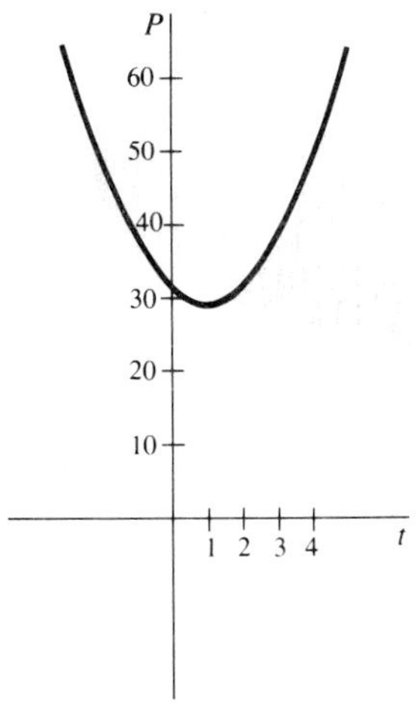

25. (a) 21,000, 18,750, 2250; (b) 45; (c) $40 + 0.1h$; (d) 40
27. $-1, -1.298, -0.740, 0.740, 1.298, 1$
29. $-16, -2, -0.593, -0.25, -0.128, -0.074$

Section 2.4 (page 64)

1. $12x^3$ **3.** $-18\frac{1}{9}$ **5.** 28 **7.** 0
9. $6x^2 - 2x + 5$ **11.** $-1/x^2 + 2/x^3$
13. $y = 13x + 20$ **15.** $y = -10x - 9$
17. $n, f(x) = x^n, f'(1)$
19. $f(x) = x^2 + 5x, f'(x) = 2x + 5$ **21.** Yes
23. $f'(x) = 256x^{255}$ **25.** $f'(x) = -10/x^3$
27. $f'(x) = 169x^{12}$ **29.** $f'(x) = -2/x^2$
31. (a) 200; (b) -64 **33.** 240

Section 2.5 (page 69)

1. Not differentiable at $x = 0$; $f'(x) = 2$ if $x > 0$; $f'(x) = 0$ if $x < 0$
3. $f'(x) = 2x$; $f'(0) = 0$
5. Not differentiable at $x = 0$; $f'(x) = 2$ if $x > 0$; $f'(x) = -2$ if $x < 0$
7. Differentiable at $x = 0$; $f'(x) = 3x^2$ if $x \geq 0$; $f'(x) = -3x^2$ if $x < 0$
9. Not differentiable at $x = 4$

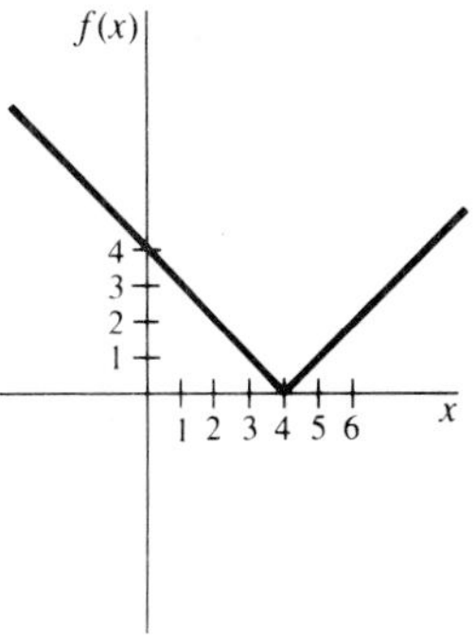

11. Not differentiable at $x = \frac{1}{2}$

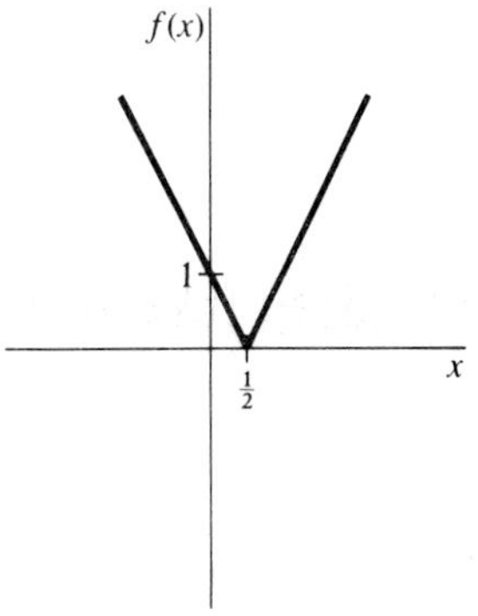

13. Not differentiable at $x = 3$

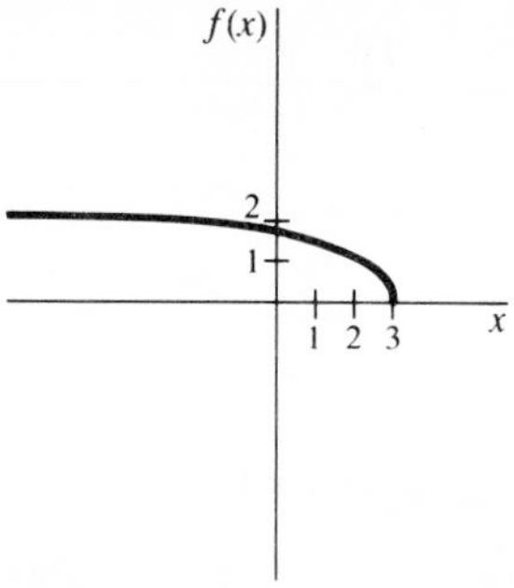

15. Not differentiable at $x = \frac{1}{2}$

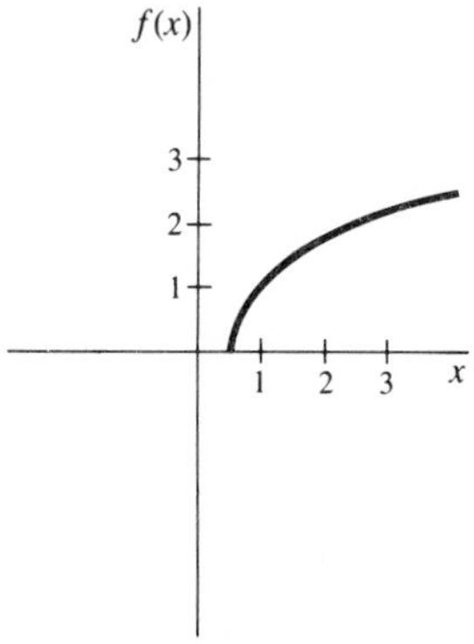

17. Not differentiable at $x = 0$ and $x = 2$

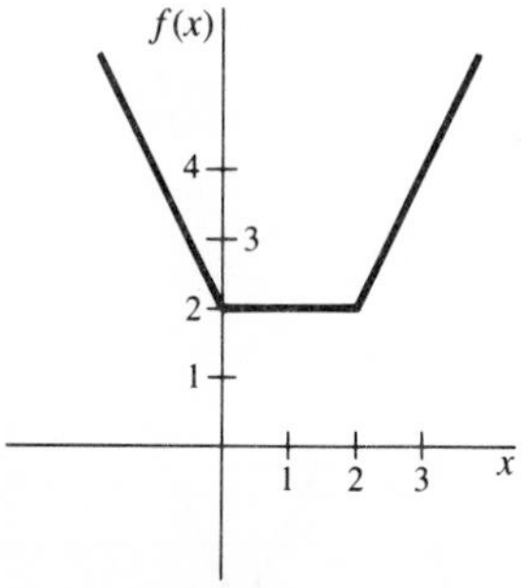

19. The slope of the tangent line is infinite at $x = 0$. The function is not differentiable at $x_0 = 0$.

Section 2.6 (page 72)

1. 5 **3.** 0 **5.** -2 **7.** 11 **9.** 2 **11.** -9
13. $3m + b$ **15.** -6 **17.** bd **19.** 2 **21.** $\sqrt{3}$
23. 1 **25.** 1 **27.** $-\frac{1}{5}$ **29.** $\sqrt{\frac{3}{7}}$ **31.** 2
33. -1 if n odd, 1 if n even **35.** -3 **37.** -3
39. -1 **41.** $\frac{2}{25}$ **43.** 5^{13} **45.** x^2 **47.** x^3
49. $x^3 - 5$ **51.** $\frac{1}{2}$

Section 2.7 (page 84)

1. $15x^2 - 3$ **3.** $f'(x) = 5x^4 + 4x^3$ **5.** $-2/x^3$
7. $3/(x + 3)^2$
9. $(-x^4 - 2x^3 - 4x^2 + 2x + 2)/(x^3 - x + 1)^2$
11. $82x^{81} - 81x^{80} + 80x^{79}$ **13.** $4x$ **15.** $\dfrac{4(x-1)}{(x+1)^3}$
17. $(-10x - 5)/(x^2 + x + 1)^2$ **19.** $30x + 46$ **21.** 0
23. $(-2x^4 - 4x^3 - 6x^2 - 4x - 2)/(x^3 - 1)^2$
25. $[19(25x^2 - 10x + 1) - 36(9x^2 + 30x + 25)]/$ $(3x + 5)^2(5x - 1)^2$
27. $(-4x + 6)/x^4$ **29.** $6x(x^2 + 1)^2$ **31.** $\frac{4}{25}$
33. $(-11\sqrt{2})/100$ **35.** $-\frac{8}{3}, \frac{10}{3}$ **37.** $\frac{1}{2}, -\frac{16}{25}$
39. -13 **41.** 48 **43.** $9y + x = 6$
45. $4y - x = 4$ **47.** $y = 2x_0 x - x_0^2$
49. $(-2, \frac{13}{3}), (1 - \frac{1}{6})$ **51.** $(\frac{3}{2}, \frac{13}{4})$ **53.** $(\frac{5}{2}, -\frac{19}{4})$
55. (a) $s(0) = 160$; (b) 192;
(c) No physical meaning, not in domain;
(d) $t = -2, t = 5$; (e) $t = 5$; (f) $0 \le t \le 5$;
(g) $v(t) = 48 - 32t$; (h) $t = \frac{3}{2}$
57. 5 **59.** $2x + 15x^2$ **61.** $1 - 6/x^3$ **63.** $12x^{1/2}$
65. (a) $R = 50x(30 - \sqrt{x})$; (b) (i) 1000, 750 (ii) 500, 0
67. (a) 62.5; (b) 5.625; (c) 77.17, 1.361
69. (a) 8.9%; (b) 9%

Section 2.8 (page 90)

1. $f'(x) = 8x^7 - 18x^5 + 2x$; $f''(x) = 56x^6 - 90x^4 + 2$; $f^{(3)}(x) = 336x^5 - 360x^3$; $f^{(4)}(x) = 1680x^4 - 1080x^2$
3. $f'(x) = 6x - 4$; $f''(x) = 6$; $f^{(3)}(x) = 0$; $f^{(4)}(x) = 0$
5. $f'(x) = 25x^{24}$; $f''(x) = 600x^{23}$; $f^{(3)}(x) = 13800x^{22}$; $f^{(4)}(x) = 303600x^{21}$
7. $3024x^5$ **9.** $7920x^7$ **11.** $3024x^5 + 840x^3$
13. $-7920x^7 + 1680x^4 + 24$ **15.** 0 **17.** $100!x$
19. 0 **21.** 0 **23.** 84
25. $-1/x^2$; $2/x^3$; $-6/x^4$; $24/x^5$; $-120/x^6$; $\dfrac{(-1)^n n!}{x^{n+1}}$
27. $1, \frac{4}{9}, -\frac{7}{64}$ **29.** $-\frac{5}{36}, \frac{26}{27}$
31. (a) 64 ft/sec; (b) 256 ft; (c) 128 ft/sec

CHAPTER 3

Section 3.1 (page 101)

1.

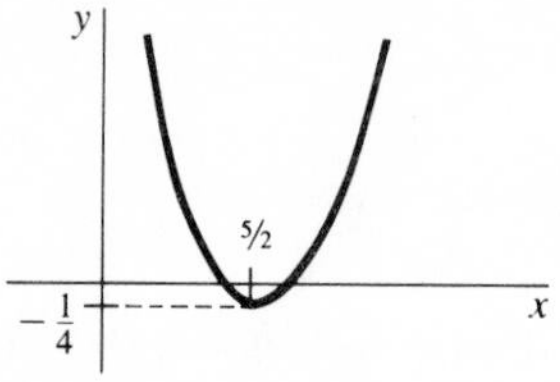

3.

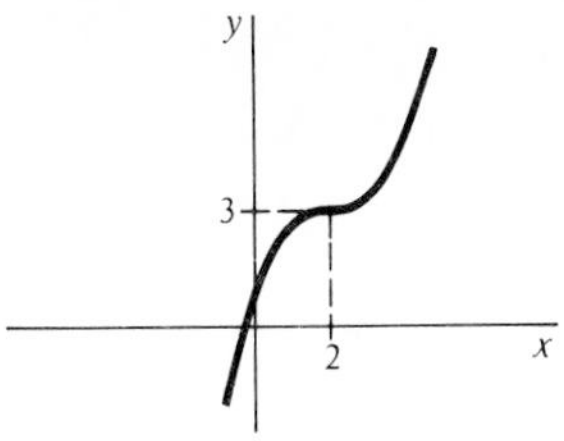

5.

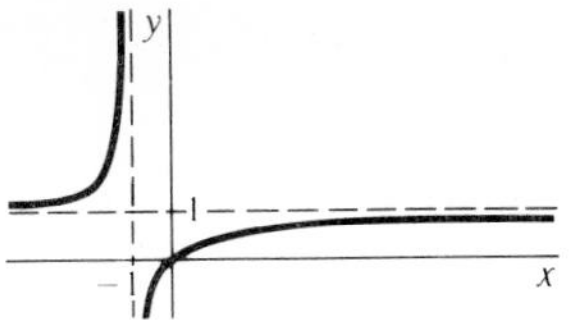

7.

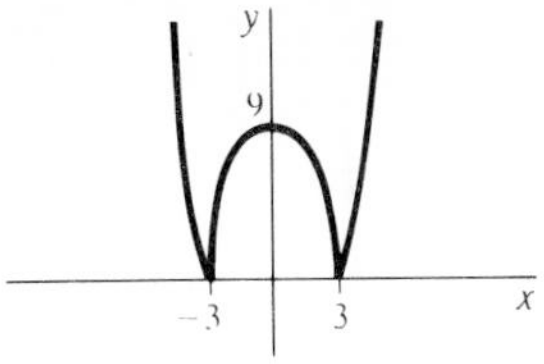

9.

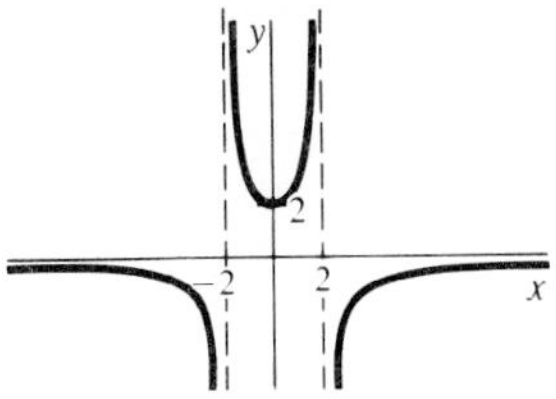

11.

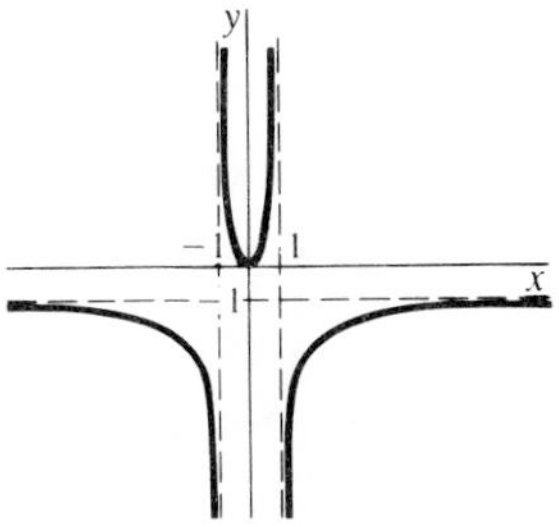

13.

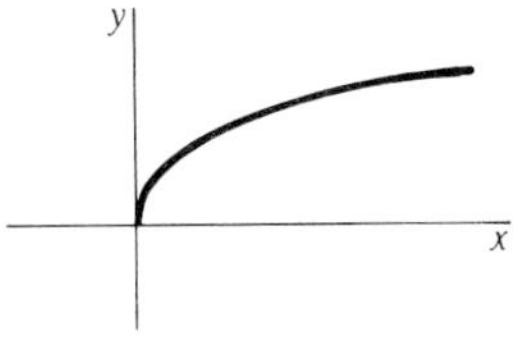

15.

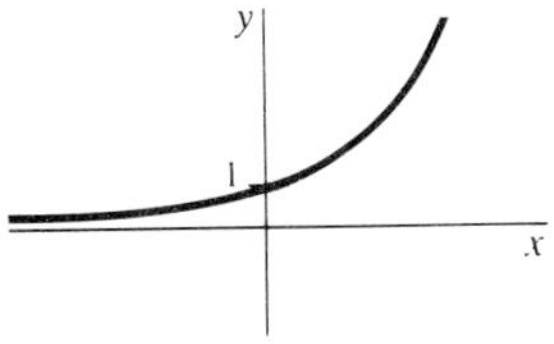

17.

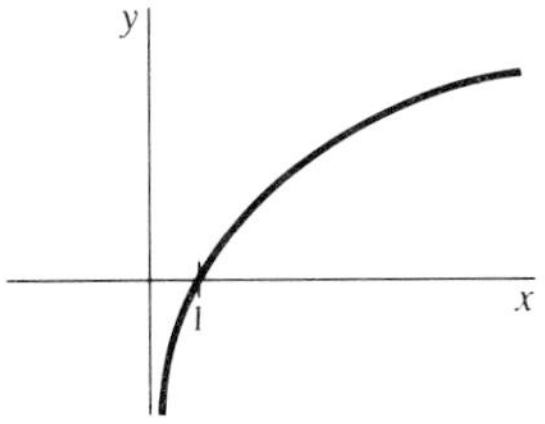

19.

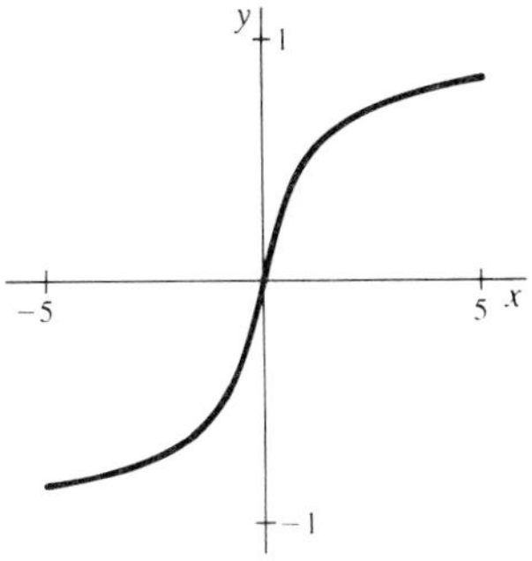

Section 3.2 (page 108)

1. $x = 1$ **3.** $x = 0, x = \frac{1}{\sqrt{2}}, x = -\frac{1}{\sqrt{2}}$

5. No critical points

7. Decreasing for $0 < x < \frac{2}{3}$, increasing for $x < 0$ and $x > \frac{2}{3}$

9. Increasing for $x < 0$, decreasing for $x > 0$.

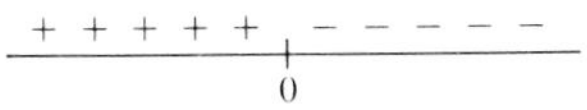

11. Decreasing for $x < 0$, increasing for $x > 0$

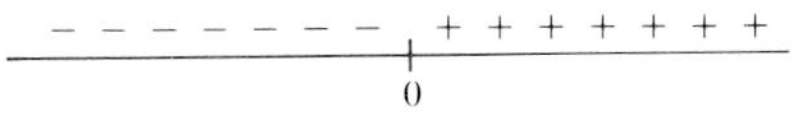

13. $-\infty, -\infty$ **15.** $+\infty, -\infty$ **17.** $\frac{3}{5}$

19.

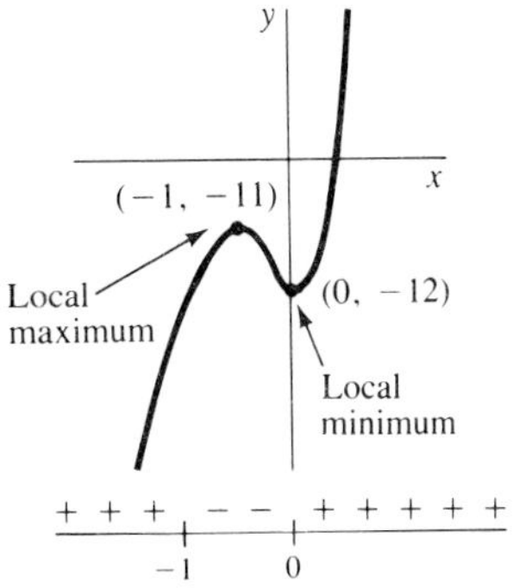

21.

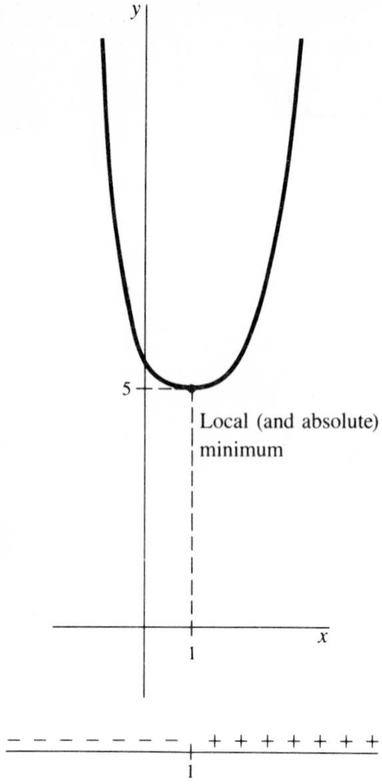

23.

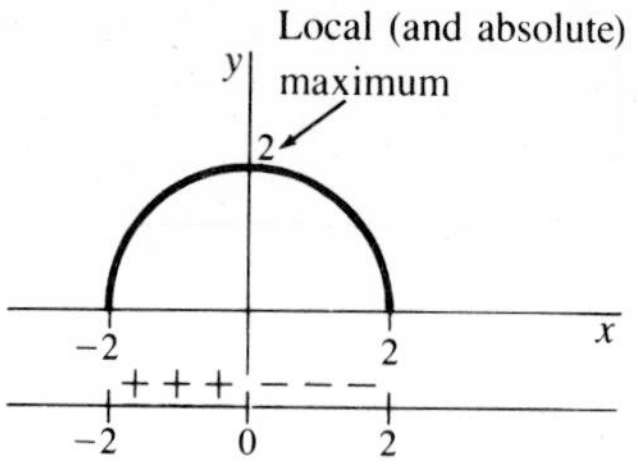

25.

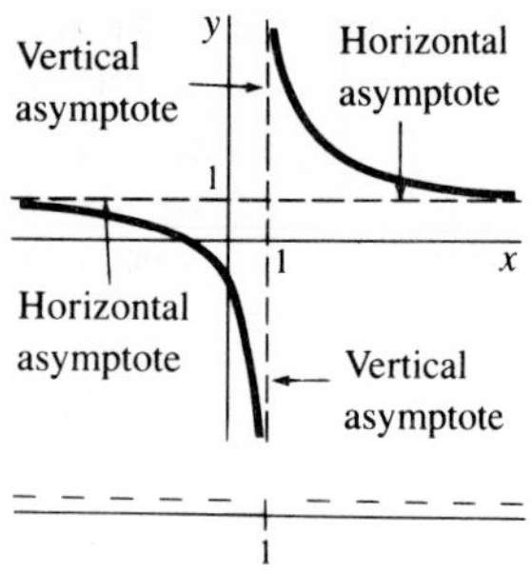

Undefined at $x = 1$

27.

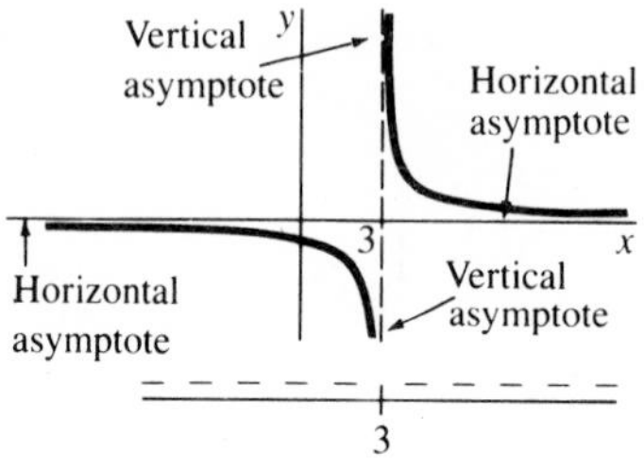

Undefined at $x = 3$

29.

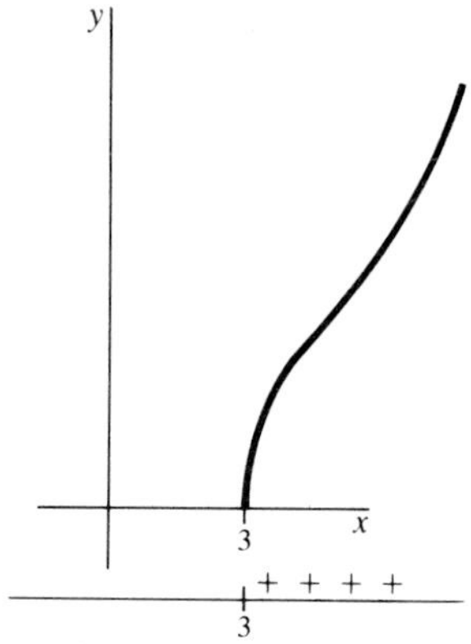

Undefined for $x \leq 3$

Section 3.3 (page 119)

1. $(-2, 10)$ local max., $(1, -17)$ local min., $(-\frac{1}{2}, -\frac{7}{2})$ pt. of inflection. Rises on $-\infty < x < -2$, falls on $-2 < x < +1$, rises on $+1 < x < \infty$. Concave down on $-\infty < x < -\frac{1}{2}$, concave up on $-\frac{1}{2} < x < +\infty$.

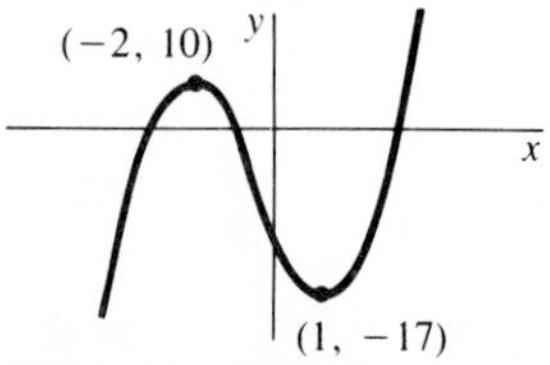

3. $(-\frac{1}{3}, \frac{88}{9})$ local min., $(\frac{1}{3}, \frac{92}{9})$ local max., $(0, 10)$ pt. of inflection. Falls on $-\infty < x < -\frac{1}{3}$, rises on $-\frac{1}{3} < x + \frac{1}{3}$, falls on $+\frac{1}{3} < x < \infty$. Concave up on $-\infty < x < 0$, concave down on $0 < x < \infty$.

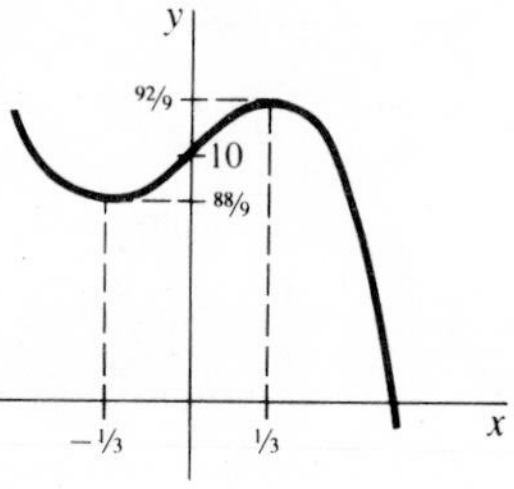

5. $(-1.82, 5.09)$ local max., $(-0.18, 2.91)$ local min., $(-1, 4)$ pt. of inflection. Rises on $-\infty < x < -1.82$, falls on $-1.82 < x < -0.18$, rises on $-0.18 < x < \infty$. Concave down on $-\infty < x < -1$, concave up on $-1 < x < \infty$.

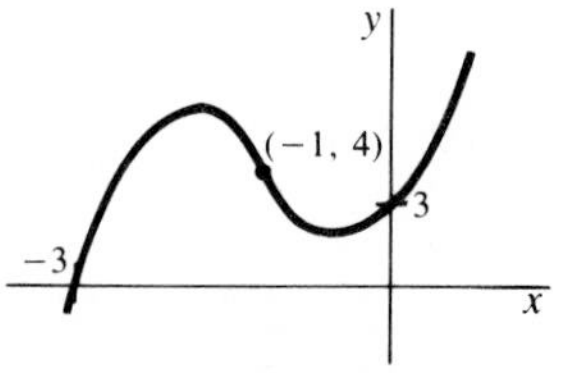

7. Rises for all x, pt. of inflection at $(-2, -1)$, concave down on $-\infty < x < -2$, concave up on $-2 < x < \infty$.

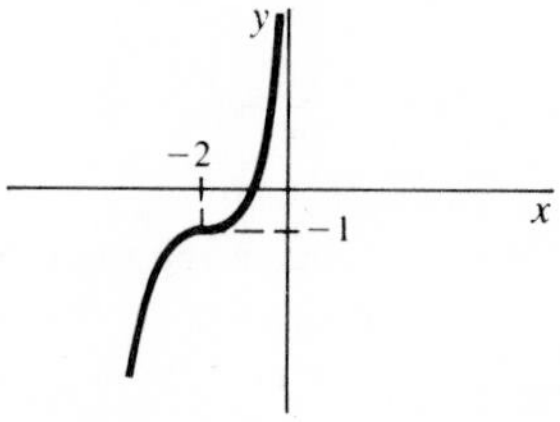

9. $(-1, 13)$ and $(+1, 13)$ maxima, $(0, 12)$ local min. $(-\frac{1}{\sqrt{3}}, 12\frac{5}{9})$ and $(+\frac{1}{\sqrt{3}}, 12\frac{5}{9})$ pts. of inflection. Rises on $-\infty < x < -1$, falls on $-1 < x < 0$, rises on $0 < x < 1$ and falls on $1 < x < \infty$. Concave down on $-\infty < x < -\frac{1}{\sqrt{3}}$ and $\frac{1}{\sqrt{3}} < x < \infty$, and concave up on $-\frac{1}{\sqrt{3}} < x < \frac{1}{\sqrt{3}}$.

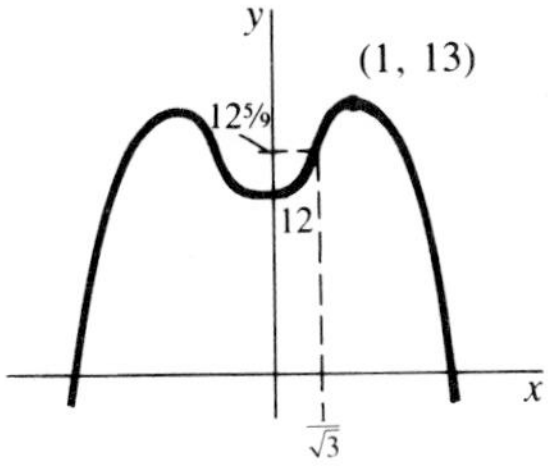

11. $(-1, -2)$ local min. $(+1, 2)$ local max. $(0, 0)$, $(-\frac{1}{\sqrt{2}}, -\frac{7}{4\sqrt{2}})$ and $(\frac{1}{\sqrt{2}}, \frac{7}{4\sqrt{2}})$ pts. of inflection. Falls on $-\infty < x < -1$, rises on $-1 < x < +1$ and falls on $1 < x < \infty$. Concave up on $-\infty < x < -\frac{1}{\sqrt{2}}$ and $0 < x < \frac{1}{\sqrt{2}}$, concave down on $-\frac{1}{\sqrt{2}} < x < 0$ and $\frac{1}{\sqrt{2}} < x < \infty$.

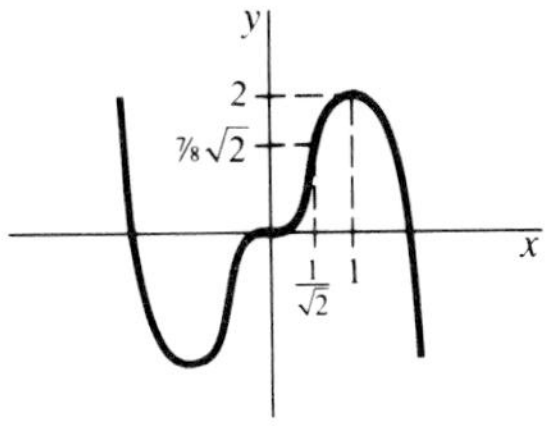

13. Rises for all x, horizontal asymptote at $y = 1$, vertical asymptote at $x = -1$.

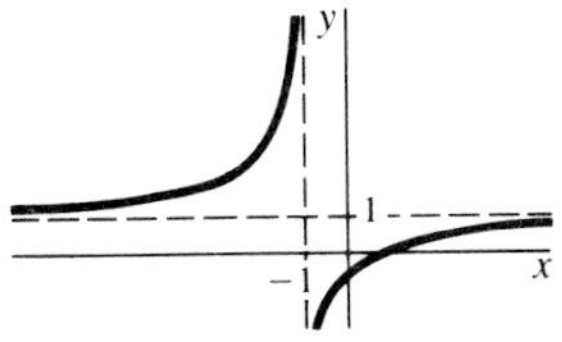

15. $(1, 0)$ local max. $(\frac{5}{3}, \frac{-4}{27})$ local min. $(\frac{4}{3}, -\frac{2}{27})$ pt. of inflection. Rises on $-\infty < x < 1$, falls on $1 < x < \frac{5}{3}$, and rises on $\frac{5}{3} < x < \infty$. Concave down on $-\infty < x < \frac{4}{3}$, concave up on $\frac{4}{3} < x < \infty$.

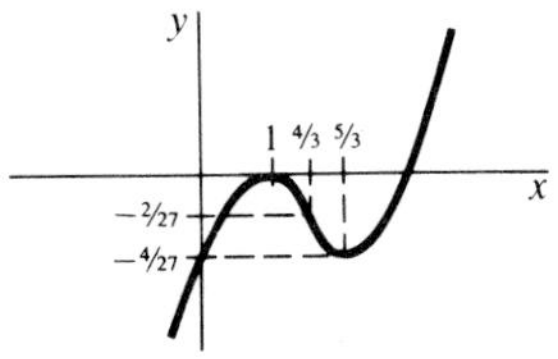

17. $(\frac{14}{3}, -\frac{25}{16})$ minimum, $(6, -\frac{3}{2})$ pt. of inflection. Rises on $-\infty < x < 2$, falls on $2 < x < \frac{14}{3}$, and rises on $\frac{14}{3} < x < \infty$. Concave up on $-\infty < x < 2$ and $2 < x < 6$, concave down on $6 < x < \infty$. Vertical asymptote at $x = 2$, horizontal asymptote at $y = -1$.

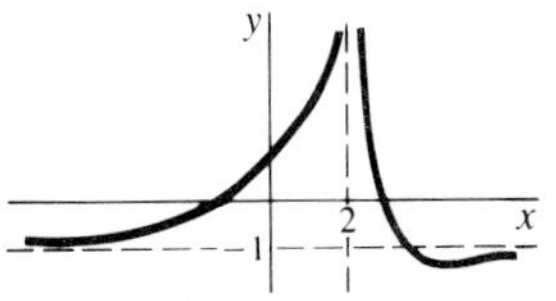

19. $(0, -1)$ local max. Rises on $-\infty < x < -1$ and on $-1 < x < 0$. Falls on $0 < x < 1$ and on $1 < x < \infty$. Concave up on $-\infty < x < -1$ and $1 < x < \infty$, concave down on $-1 < x < 1$. Vertical asymptotes at $x = -1$ and $x = 1$. Horizontal asymptote at $y = 1$.

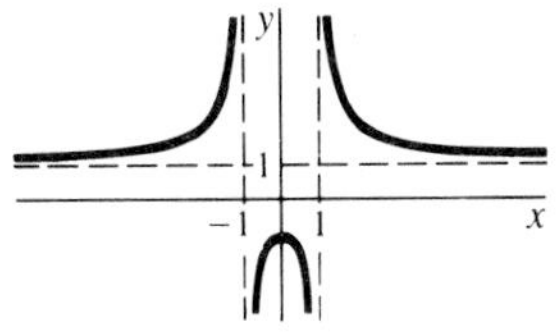

21. $(2, 0)$ minimum, $(\frac{3}{2}, \frac{1}{9})$ pt. of inflection. Falls on $-\infty < x < 2$, rises on $2 < x < 3$ and falls on $3 < x < \infty$. Concave down on $-\infty < x < \frac{3}{2}$, concave up on $\frac{3}{2} < x < 3$ and $3 < x < \infty$. Vertical asymptote at $x = 3$. Horizontal asymptote at $y = 1$.

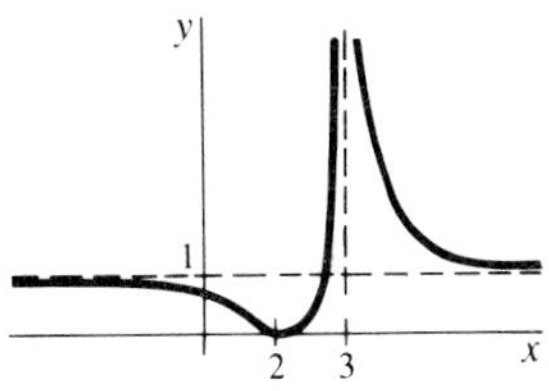

23. $(0, 8)$ local min. Falls on $-\infty < x < -1$ and $-1 < x < 0$. Rises on $0 < x < 1$ and $1 < x < \infty$. Concave down on $-\infty < x < -1$ and $1 < x < \infty$, concave up on $-1 < x < 1$. Vertical asymptotes at $x = \pm 1$, horizontal asymptote at $y = 0$.

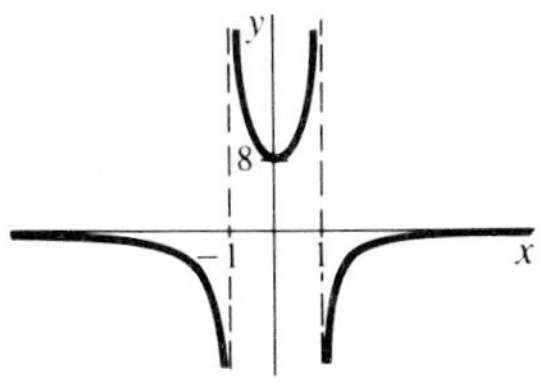

25. $(2^{-1/3}, 2^{1/3} + 2^{-2/3}) \approx (0.79, 1.89)$ local min. $(-1, 0)$ pt. of inflection. Falls on $-\infty < x < 0$ and on $0 < x < 2^{-1/3}$ and rises on $2^{-1/3} < x < \infty$. Concave up on $-\infty < x < -1$

and $0 < x < \infty$, concave down on $-1 < x < 0$. Vertical asymptote at $x = 0$.

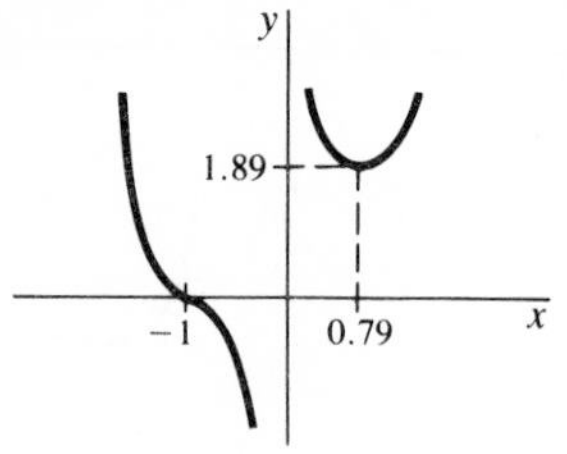

27. Undefined for $-1 < x < 1$. Falls on $-\infty < x < -1$ and rises on $1 < x < \infty$. Concave down on $-\infty < x < -1$ and on $1 < x < \infty$. As $x \to -\infty$ there is a horizontal asymptote at $y = 0$.

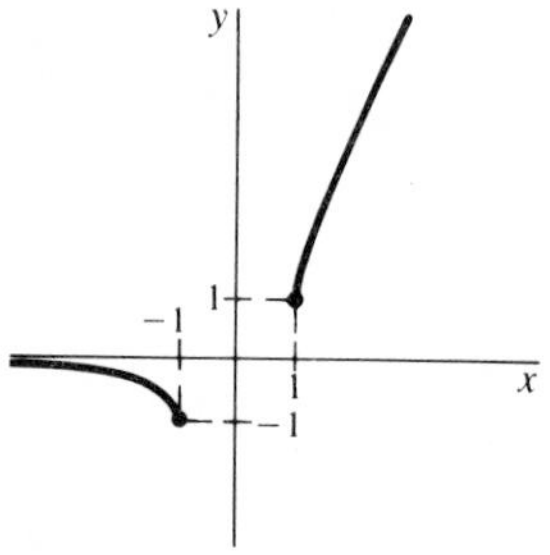

29. $(-2^{1/3}, 3 \cdot 2^{2/3}) \approx (-1.26, 4.76)$ local min., $(4^{1/3}, 0) \approx (1.59, 0)$ pt. of inflection. Falls on $-\infty < x < -2^{1/3}$ and rises on $-2^{1/3} < x < 0$ and $0 < x < \infty$. Concave up on $-\infty < x < 0$ and $4^{1/3} < x < \infty$, concave down on $0 < x < 4^{1/3}$. Vertical asymptote at $x = 0$.

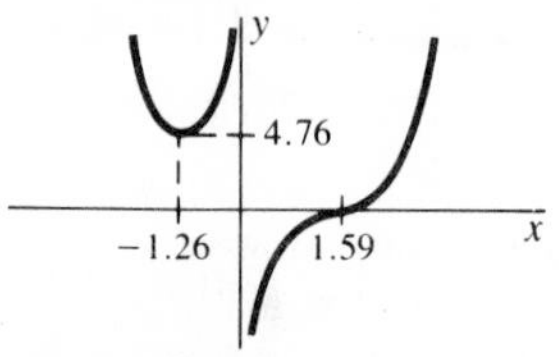

31. $y = \dfrac{1}{x-2} + x$; Asymptote $y = x$

33. $y = \dfrac{1}{x} + x$; Asymptote $y = x$

35.*

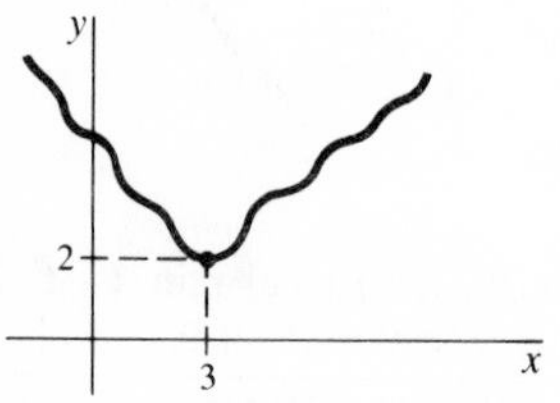

37.*

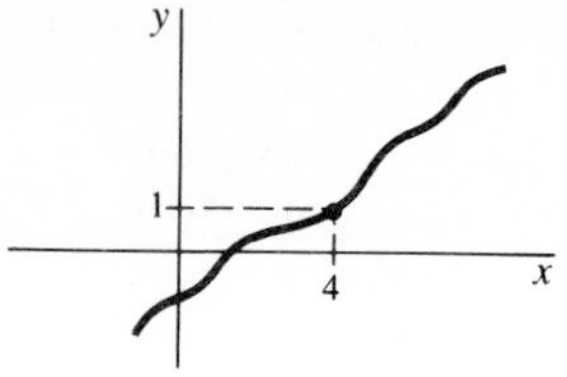

39.

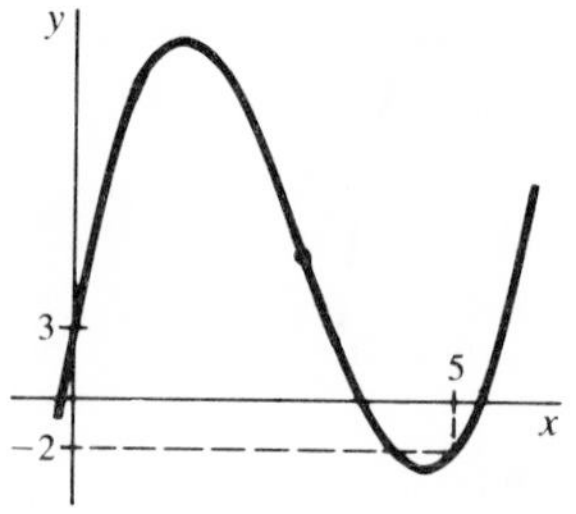

41.

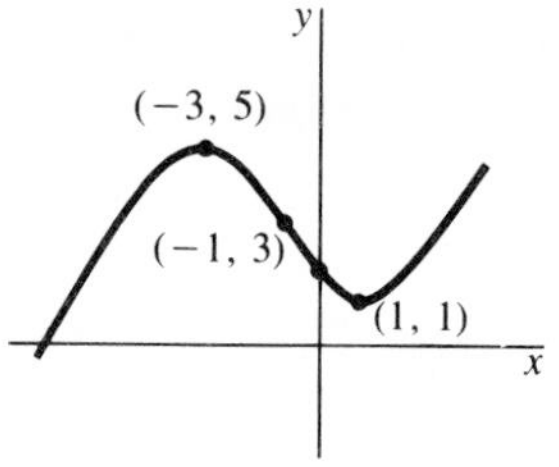

Section 3.4 (page 127)

1. 144; 12 and 12 **3.** $\frac{1}{2}$

5. No maximum; minimum $= 2\sqrt{a}$ **7.** 31,250 yd^2

9. Show that length of a side is $\frac{1}{4} \cdot$ (perimeter).

11. $\dfrac{150}{4 + \pi}$ inches

13. The minimal area is 0, corresponding to $r = 0$.

15. $a\sqrt{2}$ by $a\sqrt{2}$, $2a^2$ **17.** $r = \sqrt[3]{\frac{3}{2}} \approx 1.14$, $h \approx 2.92$

19. For a maximum the entire string should be used for circle. For a minimum the length of circumference of circle should be $\dfrac{\pi l}{4 + \pi} \approx 0.44l$.

21. For a maximum use entire string on circle. For a minimum the length of circumference of circle should be $\dfrac{\pi l\sqrt{3}}{9 + \pi\sqrt{3}} \approx 0.38l$.

23. $\dfrac{3\sqrt{10}}{10}$

25. (a) $(0, 0)$, $d = 1$; (b) $\left(\pm\dfrac{1}{\sqrt{2}}, \dfrac{1}{2}\right)$, $d = \dfrac{\sqrt{3}}{2}$;

* In Answers 35 and 37 waviness is due to lack of information on second derivative.

(c) (0, 0), $d = \frac{1}{2}$; (d) (0, 0), $d = \frac{1}{4}$; (e) For $b \le \frac{1}{2}$, (0, 0), with $d = |b|$. For $b > \frac{1}{2}$, $(\pm\sqrt{b - \frac{1}{2}},\ b - \frac{1}{2})$ with $d = \sqrt{b - \frac{1}{4}}$

27. 18.22 ft. **29.** A cube of sides $\sqrt{a/6}$

31. (a) $\frac{\sqrt{2a}}{2} \times \frac{\sqrt{2a}}{3} \times \frac{\sqrt{2a}}{5}$ (b) $\frac{\sqrt{a}}{2} \times \frac{\sqrt{a}}{3} \times \frac{2\sqrt{a}}{5}$

Section 3.5 (page 134)

1. $4(2)^{1/3} \times 4(2)^{1/3} \times 2(2)^{1/3}$ **3.** \$350
5. Use all 10 presses. **7.** 25 ft **9.** $p = 46¢$
11. 10% **13.** 2.59
15. Production level $93\frac{1}{3}$; unit price \$$103\frac{1}{3}$
17. 6350 items
19. $s = \sqrt{a/c}$ mph
21. The height should be 5 times the radius.
23. (a) -2 (b) $1 - \frac{3000}{q}$ (c) $2 - \frac{100}{\sqrt{q}}$

Section 3.6 (page 141)

1. (a) 4.5 (b) 5.5 **3.** $y = 2.65x$ **5.** $y = 0.508x$
7. $y = 3x - 4$
13. If it starts crawling on the $l \times h$ wall it will pass to the $w \times h$ wall at a corner point $\frac{lh}{l + w}$ units above the floor.
15. (a) Point A; (b) $\frac{7\sqrt{15}}{15} \approx 1.81$ miles from point P; (c) $\frac{7\sqrt{15}}{15} \approx 1.81$ miles from point P; (d) Point A; (e) $\frac{7\sqrt{15}}{15} \approx 1.81$ miles from point P

CHAPTER 4

Section 4.1 (page 157)

1. 0.3681 **3.** 19.4125 **5.** 1.3956
7. $1/2.718 = 0.3679$ **11.** $e^x - e^{-x}$ **13.** $xe^{-x}(2 - x)$
15. $e^{x+2}(x - 2)/x^3$ **17.** $4/(e^x + e^{-x})^2$ **19.** $-2xe^{-x^2}$
21. $(1 + 2x)e^{x+x^2}$ **23.** $\frac{1}{2\sqrt{x}}e^{\sqrt{x}}$ **25.** $4e^{4x}$
27. $4x^3e^{x^4}$ **29.** *Hint*: Reflect graph of $y = e^x$ about y-axis.
31. $\lim_{h\to 0} \frac{e^{x+h} - e^x}{h} = e^x$; set $x = 0$
33. 11.63% **35.** \$13,534
37. (a) 2.7183; (b) 7.3991; (c) 0.3679; (d) 1.3956

39. A 30-year mortgage at 14% has a larger monthly payment (0.0118 times face amount) than 25-year mortgage at 12% (0.0105 times face amount).

Section 4.2 (page 161)

1. $(2x - 3)^5$; $10(2x - 3)^4$ **3.** $(2x - 3)^{-5}$; $-10(2x - 3)^{-6}$
5. $\sqrt{x^2 - 2x + 3}$; $(x - 1)/\sqrt{x^2 - 2x + 3}$
7. $u = 3x - 2$; $f = u^{17}$; $g' = 51(3x - 2)^{16}$
9. $u = 3x^2 - x + 1$; $f = u^{1/2}$;
$g' = \frac{6x - 1}{2}(3x^2 - x + 1)^{-1/2}$
11. $440(5x + 7)^{87}$ **13.** $49(2x + 1)(x^2 + x - 3)^{48}$
15. $-51(3x - 2)^{-18}$ **17.** $(5x + 2)(5x^2 + 4x - 1)^{-1/2}$
19. $(1 - 6x)(3x^2 - x + 1)^{-2}$
21. $[(e^x + e^{2x})^2(6x + 3)(x^2 + x + 1)^2 - (x^2 + x + 1)^3 2(e^x + e^{2x})(e^x + 2e^{2x})]/(e^x + e^{2x})^2$
23. $e^{\sqrt{x^5+x+1}}[\frac{1}{2}(x^5 + x + 1)^{-1/2}(5x^4 + 1)]$
25. $-2xe^{-x^2}(x^2 + 2)/(x^2 + 1)^2$
27. $8(3x^5 + x^4 - 2)^7(15x^4 + 4x^3) + 18(2x - 7)^8$; Sum Rule, Generalized Power Rule or Chain Rule, Sum Rule, Power Rule
29. $[(2x - 7)^9 \frac{1}{2}(3x^5 + x^4 - 2)^{-1/2}(15x^4 + 4x^3) - \sqrt{3x^5 + x^4 - 2}(18)(2x - 7)^8]/(2x - 7)^{18}$; Quotient Rule, Chain Rule, Generalized Power Rule, Sum Rule, Power Rule
31. $\frac{1}{2}\left(\frac{(x^2 + x + 1)^5 - 2}{(3x - 2)^7 + 1}\right)^{-1/2} \cdot \frac{((3x - 2)^7 + 1)5(x^2 + x + 1)^4 \cdot (2x + 1) - ((x^2 + x + 1)^5 - 2)(21)(3x - 2)^6}{((3x - 2)^7 + 1)^2}$;
Chain Rule, Quotient Rule, Generalized Power Rule or Chain Rule, Sum Rule, Power Rule
33. $(6x - 5)e^{3x^2 - 5x + 4}$; Chain Rule, Sum Rule, Power Rule
35. $[(e^x - 3e^x)^2(2xe^{-x^3} - 3x^4e^{-x^3}) - x^2e^{-x^3}(2)(e^x - e^{3x})(e^x - 3e^{3x})]/(e^x - e^{3x})^4$; Quotient Rule, Product Rule, Chain Rule, Power Rule, Generalized Power Rule, Sum Rule.
37. $n(x^2 + x + 1)^{n-1}(2x + 1)$ **39.** $-nx^{n-1}e^{-x^n}$
41. $\frac{(e^{mx}m - e^{nx}n)(e^{mx} + e^{nx}) - (e^{mx}m + e^{nx}n)(e^{mx} - e^{nx})}{(e^{mx} + e^{nx})^2}$
43. Rising \$400 per month.

Section 4.3 (page 169)

1. (5, 2) **3.** $(-3, -2)$ **5.** (6, 0) **7.** (i) $y = \frac{x + 2}{3}$

(iii)

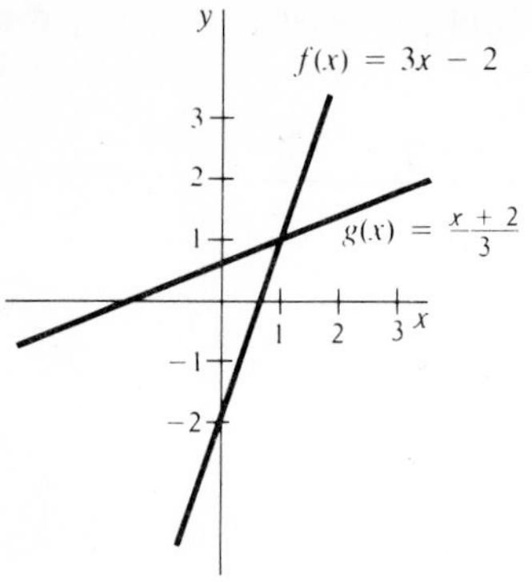

9. (i) $y = x - 1$ (iii)

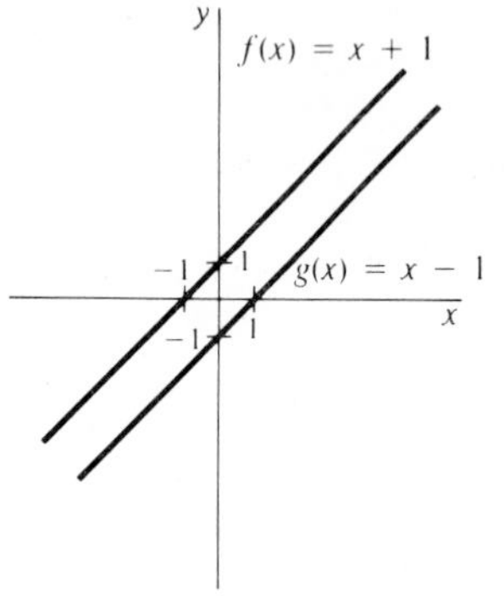

11. (i) $y = -x$
(iii) One graph; the function is its own inverse.

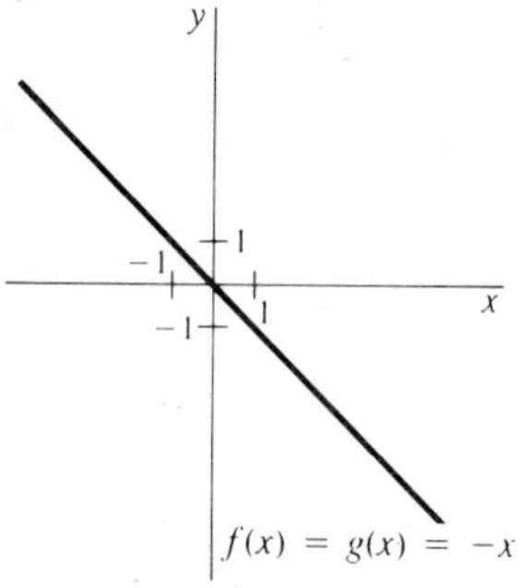

13. (i) $y = (x - 2)^{1/3}$ (iii)

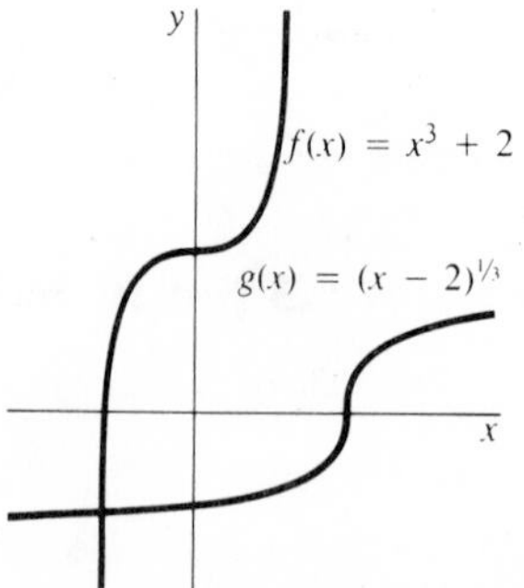

15. No inverse exists. **17.** $y = (x + 1)^5, -\infty < x < \infty$

19. $y = x^{1/4}, x \geq 0$ **21.** $y = \dfrac{1}{x}, x \neq 0$

23. No inverse exists. **25.** $y = \dfrac{x}{1 - x}, x \neq 1$

27. $y = x^2 + 2, x \geq 0$ (The domain of the inverse must equal the range of the original function.)

29. $y = x^{1/4} + 1, x \geq 0$ **31.** $y = \dfrac{1}{x} - 3, x \neq 0$

33. $y = x^3 - 4, -\infty < x < \infty$

35. $x = 2500\left(1 - \dfrac{p}{24}\right)^2, 0 \leq p \leq 24$

Section 4.4 (page 174)

1. $(\frac{2}{3})x^{-1/3}$ **3.** $(\frac{30}{11})x^{-6/11} - (\frac{1}{9})x^{-7/9}$
5. $(\frac{3}{5})(3x - 2)^{-4/5}$
7. $(-8x^4 - 22x^2 + 10x + 6)/[15(x^3 - x + 1)^{7/5}(x^2 + 1)^{2/3}]$.
9. $2(3e^{3x} - 2xe^{x^2})/[3(e^{3x} - e^{x^2})^{1/3}]$ **11.** $(\frac{3}{4})5^{-1/4}$
13. $\sqrt{3}(8)^{\sqrt{3}-1}$ **15.** $y = \frac{1}{2}(9x - 27)$
17. $y = (\frac{20}{3})x + \frac{64}{3}$ **19.** $4x$ **21.** $1/y$
23. $-(2 + 5x)/y$ **25.** $(4x - 4)/(3y^2)$ **27.** $-(y/x)^{2/3}$
29. $3x^2/(1 - 4y)$
31. $(6xy^2 - 6x^2y - 1)/(2x^3 - 6x^2y - 1)$
33. $-(y + 2x)/(2y + x)$ **35.** $(2x - y)/(x - 2y)$
37. $y = \frac{1}{4}(-3x + 25)$ **39.** $y = -\frac{3}{2}x - 6$
41. $y = \frac{3}{2}x + \frac{7}{2}$
47. *Hint*: $y = x^{p/q}$ is equivalent to $y^q = x^p$. Differentiate the latter implicitly.

Section 4.5 (page 182)

1. $3 \ln 2 = 2.0793$ **3.** $2 \ln 5 = 3.2189$
5. $\ln 5 - \ln 2 = 0.9163$ **7.** $2 \ln 3 - 2 \ln 5 = -1.022$
9. $\frac{2}{3} \ln 2 = 0.4621$ **11.** $-\ln 2 - \ln 5 = -2.303$
13. $2x$ **15.** $1/x$ **17.** x^{-3} **19.** x^2 **21.** $3x^2$

23. $-2 \ln x$ **25.** $\dfrac{1}{x}$ **27.** $1 + \ln x$

29. $2x\sqrt{\ln 3x} + \dfrac{x}{2\sqrt{\ln 3x}}$ **31.** $\dfrac{2}{x} + \dfrac{1}{2x - 2}$

33. $e^{bx}\left(\dfrac{1}{x} + b \ln ax\right)$ **35.** $e^{1/x}(1 - 1/x)$

37. $2e^{-x}/(e^x + e^{-x})$ **39.** $\dfrac{1}{x \ln x}$ **41.** $e^x e^{e^x}$

43. $y = (3 - x)e^{-2}$ **45.** $y = x$ **47.** $y = 3x - e^2$
49. $y = x - 1$ **51.** $10^x \ln 10$ **53.** $\sqrt{2}x^{\sqrt{2}-1}$
55. $2(\ln x)x^{-1+\ln x}$ **57.** $(2 \ln x + 1)x^{x^2+1}$
59. $y = -3(x + 1) \ln 3 + 3$ **61.** $y = \pi(x - 1) \ln \pi + \pi$
75. $(6x^2 - 12x + 4)x(x - 1)(x - 2)$

77. $2\left[\sum_{i=1}^{n} \frac{1}{x - a_i}\right] \prod_{i=1}^{n} (x - a_i)^2$

Section 4.6 (page 196)

1. (a) $y = x^2 - 3x + c$; (b) $y = x^2 - 3x + 5$
3. (a) $y = -1/x - x^2/2 + x + c$;
(b) $y = -1/x - x^2/2 + x + 11/2$
5. (a) $y = -3x^2/2 + c$; (b) $y = -3x^2/2 + 8$
19. $f(x) = x^3 + x^2 - 4x + 5$ 21. $f(x) = \ln x + 2$
23. $s(t) = \frac{2}{3}t^3 + 16t + 4$ 25. $s(t) = 2t^3 - \frac{1}{2}t^2 - 10$
27. $s(t) = -16t^2 + 60$; 1.94 sec; 61.97 ft/sec
29. 33.75 ft; 18.98 ft 31. 4.4 ft/sec
33. 352 ft 35. (a) 0.905 g (b) 0.368 g
37. (a) \$1126.49; (b) \$1127.16; (c) \$1127.49;
(d) \$1127.50
39. \$54.23
41. (a) 2560; (b) In the 8th week without advertising
43. 19.93 hr 45. 12.44 hr; 16.25 hr
47. 1034.55 years; 6873.39 years 49. 69.7 years
51. (a) 66.44 hr; (b) 23.77 hr 53. $k = 0.077$
55. (a) 132°; (b) 76.92°; (c) 70.02°; (d) 70°
57. $T - A = 130e^{-0.1466t}$; $T - A = ce^{-kt}$
59. (a) 321.4 mm; (b) 16.5 km 61. \$2904.58

Section 4.7 (page 204)

1. 700 in.^{2}/sec 3. 150π in.^{2}/min 5. $7/40\pi$ in/hr
7. 1.25 in./sec 9. 45.89 mph 11. $\frac{1}{18\pi}$ ft/sec
13. When radius $= \frac{5}{3\pi}$ in. 15. $\frac{3}{10{,}000\pi}$ ft/sec
19. 3.636 ft/sec
21. (a) $\left(\frac{5\sqrt{2}}{2}, -\frac{5\sqrt{2}}{2}\right)$ and $\left(-\frac{5\sqrt{2}}{2}, \frac{5\sqrt{2}}{2}\right)$;
(b) $\left(\frac{5\sqrt{2}}{2}, \frac{5\sqrt{2}}{2}\right)$ and $\left(-\frac{5\sqrt{2}}{2}, -\frac{5\sqrt{2}}{2}\right)$;
(c) (0, 5) and (0, −5); (d) $2\sqrt{5}, -\sqrt{5})$ and $(-2\sqrt{5}, \sqrt{5})$
23. (a) 8/3 ft/sec; (b) $\frac{20}{3}$ ft/sec. These results do not depend on distance to lamp-post.
25. −\$30/week

CHAPTER 5

Section 5.2 (page 224)

1. 20 3. 16 5. Cannot evaluate exactly yet
7. Cannot evaluate exactly yet 9. 10 11. 14, 13
13. $\frac{43}{4}, \frac{155}{16}$ 15. 100, 81 17. 1.39, 1.80
19. 84.79, 68.11 21. $\int_3^5 x^4\,dx$ 23. $\int_{-2}^{2} (4 - x^2)\,dx$
25. $1 + 2 + 3 + 4 + \cdots + 18 + 19 + 20$
27. $1^2 + 2^2 + \cdots + 14^2 + 15^2$
29. $1 + \frac{1}{2} + \frac{1}{3} + \cdots + \frac{1}{(n-1)} + \frac{1}{n}$
31. $\ln 1 + 2\ln 2 + \cdots + 7\ln 7 + 8\ln 8$
33. $f(x_1) + f(x_2) + \cdots + f(x_{n-1}) + f(x_n)$
35. $\sum_{i=1}^{17} \frac{1}{i}$ 37. $\sum_{i=1}^{25} \frac{1}{i(i+1)}$
39. (i) 20; (ii) 20; (iii) 20; (iv) 20
41. (i) 40; (ii) 32; (iii) 38; (iv) 34
43. (i) 11; (ii) 9; (iii) $\frac{21}{2}$; (iv) $\frac{19}{2}$
45. (i) $\frac{23}{4}$; (ii) $\frac{15}{4}$; (iii) $\frac{83}{16}$; (iv) $\frac{67}{16}$
47. (i) $\frac{99}{8}$; (ii) $\frac{117}{8}$; (iii) $\frac{207}{16}$; (iv) $\frac{225}{16}$
49. $\frac{85}{2}, \frac{75}{2}$ 51. 0.669, 0.719 53. 24.503, 16.402
55. 1.806, 1.634 57. 1.896, 1.896 59. 84.7910, 31.1929
61. 64.963 63. 1.494 65. 63.043, 0.793, 1.494, 19.5625

Section 5.3 (page 223)

1. $x^6/6 + c$ 3. $x^4/4 + x^2/4 + x + c$
5. $2x^{3/2}/3 + c$ 7. $2\sqrt{x} + c$ 9. $\frac{1}{3}\ln x + c$
11. $\frac{1}{2}e^{x^2} + c$ 13. $ax^3/3 + bx^2/2 + cx + c$
15. Area under the curve $y = 2x + 3$ from 1 to 5
17. Area under the curve $y = x^3 + 1$ from 0 to 3
19. Area under the curve $y = 3x^2 + 8x + 1$ from 0 to 5
21. $\frac{125}{6}$ 23. $\frac{64}{5}$ 25. $1 - e^{-3}$

Section 5.4 (page 239)

1. $2x^2 - 7x + c$ 3. $\frac{5}{3}x^3 + 4x^2 + c$
5. $\frac{5}{6}x^2 + \frac{8}{3}x + c$ 7. $\frac{x^{18}}{18} - \frac{1}{16x^{16}}$
9. $\frac{2}{5}(1 + x)^{5/2} + c$ 11. $-\frac{1}{48}(2 - 3x)^{16} + c$
13. $\frac{1}{12}(2 - 3x)^{-4} + c$ 15. $-e^{-x} + \frac{x^3}{3} + c$
17. $\frac{1}{4}(1 + x^2)^2 + c$ 19. $\frac{1}{66}(4 + 3x^2)^{11} + c$
21. $\frac{1}{3}\ln|3x + 2| + c$ 23. $\frac{1}{2}\ln(1 + x^2) + c$
25. $\frac{4x^2 - 3x}{4} + c$ 27. $2x^2 - 5x + 3\ln|x| + c$
29. $\frac{15}{7}x^{7/5} - 2x^{5/2} + c$ 31. $\frac{1}{5}e^{5x} + c$
33. $5\ln|x - 2| + c$ 35. $-\frac{1}{2}e^{-x^2} + c$
37. $\frac{2}{3}(x^2 + x + 1)^{3/2} + c$ 43. $x^4 - \frac{x^3}{3} + \frac{5}{2}x^2 + 2x + \frac{7}{6}$
45. $\frac{2}{3}(x + 1)^{3/2} - \frac{28}{3}$ 47. $x + c$ 49. $\frac{1}{2}e^{x^2} + c$
51. $\frac{2}{15}\sqrt{4 + 5x^3} + c$ 53. $-\frac{2}{\sqrt{x}} + c$
55. $\frac{1}{3}\ln|3x - 5| + c$ 57. $-\frac{1}{5}e^{-5x} + c$
59. $\frac{1}{21}(1 + x)^{21} + c$ 61. $3(x^3 - x + 7)^{1/3} + c$
63. $3(e^2 - 1)$ 65. $\frac{1}{2}\left(1 - \frac{1}{e}\right)$
67. $\frac{1}{2}e^{2x} + 2x - \frac{1}{2}e^{-2x} + c$ 69. $\frac{1}{8}\ln 2$

71. $\frac{5}{2}(\ln x)^2 + c$ **73.** $3 \ln|x^2 - x^5| + c$

75. $\frac{3}{4}(x^2 + 2x + 2)^{2/3} + c$ **77.** $-\dfrac{1}{x} - \dfrac{1}{6x^6} + c$

79. $\dfrac{-1}{\ln|x|} + c$ **81.** $\frac{2}{3}e^{(3/2)x} + c$

83. $R(x) = 1000x - 0.01x^2 - x^3$

85. $W(t) = 50t + t^2 - \dfrac{t^3}{30}$

CHAPTER 6

Section 6.1 (page 249)

1. 7/3 **3.** $\frac{1}{2} \ln \frac{7}{5}$ **5.** $(e^3 - 1)/3$ **7.** $(e^4 - 1)/4$
9. 1.28 grains **11.** 265 **13.** -0.002 **15.** 28.2

Section 6.2 (page 259)

1. $\frac{9}{2}$

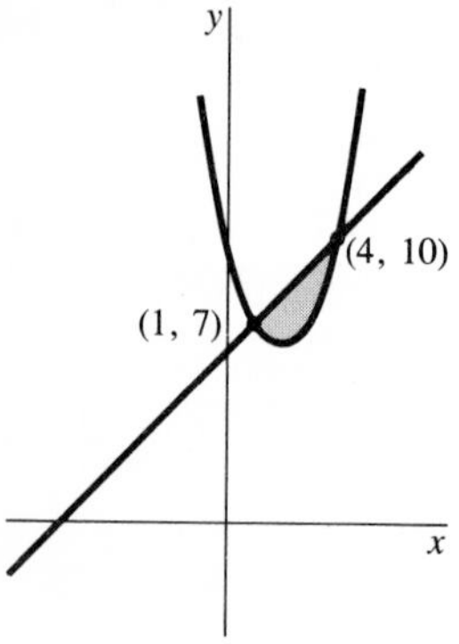

3. $\frac{9}{2}$

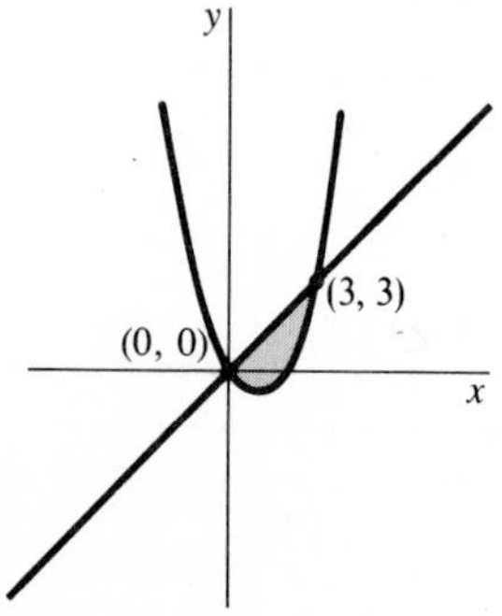

5. 2.7

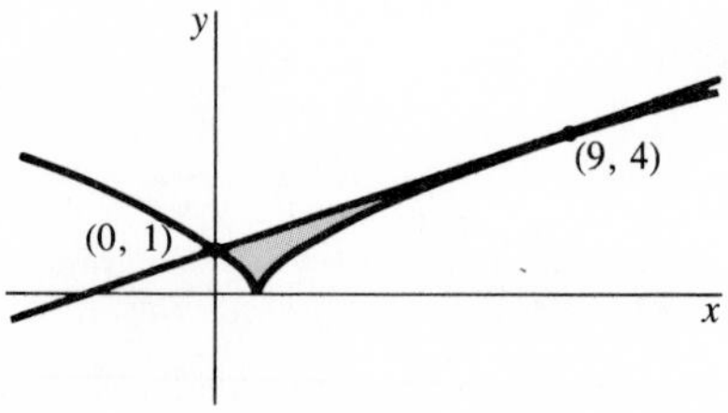

7. $\frac{1}{12}$

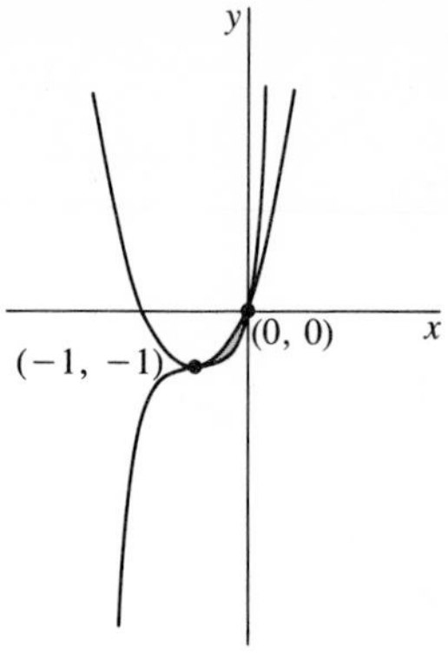

9. 32

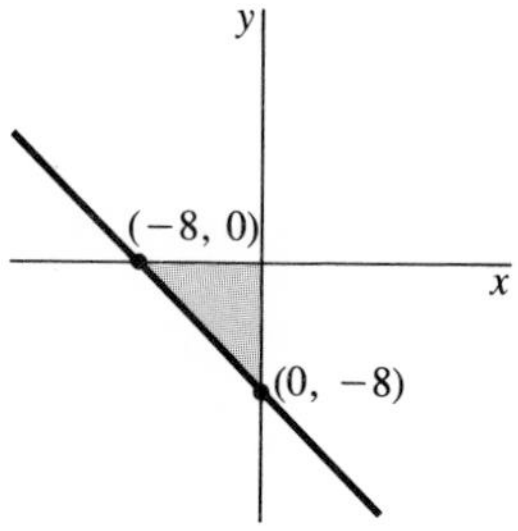

11. $\frac{1}{12}$

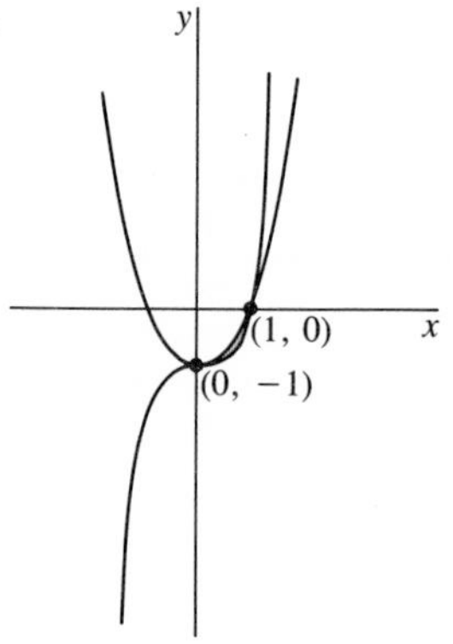

13. $\frac{512}{5}$

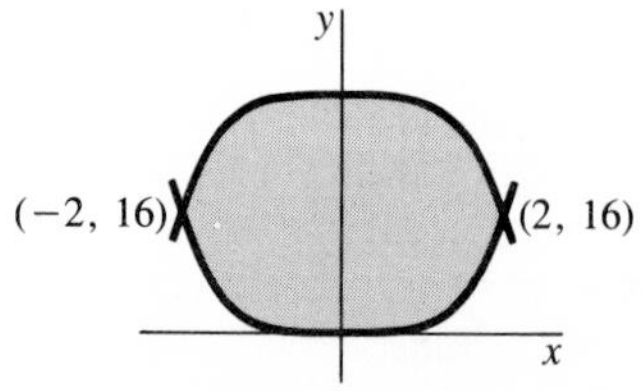

15. $2(1 - e^{-1})$

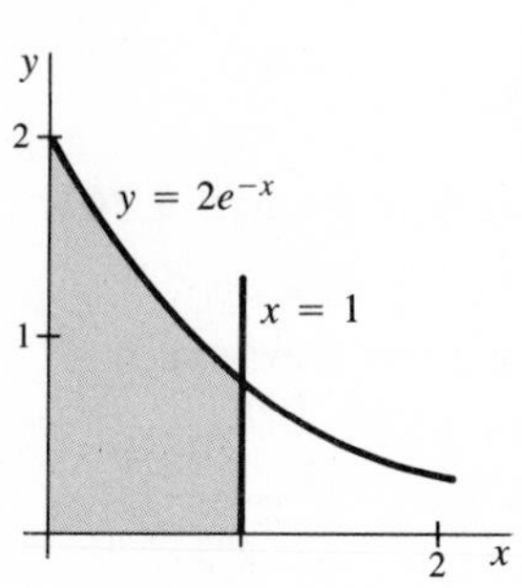

17. $y = (-b/a^2)x^2 + b$ **19.** 13 **21.** Loss of \$670,000
23. (15,60), 225, 450 **25.** (1,4), 3.09, 0.5 **27.** 1466
29. 16 years, \$64,000 **31.** (a) 20%; (b) 0.40

Section 6.3 (page 267)

1. $(496\pi)/15$ **3.** $\pi(77/4 - 7\ln 2)$ **5.** $\pi \ln 2$
7. $\pi/5$ **9.** π **11.** $(3\pi)/10$ **13.** $32\pi/3$
15. $4\pi ab^2/3$

Section 6.4 (page 273)

1. (a) Revenue increases \$600, cost increases \$400;
(b) profit increases \$200
3. (a) −27.53 (b) −12.37

5. (a)

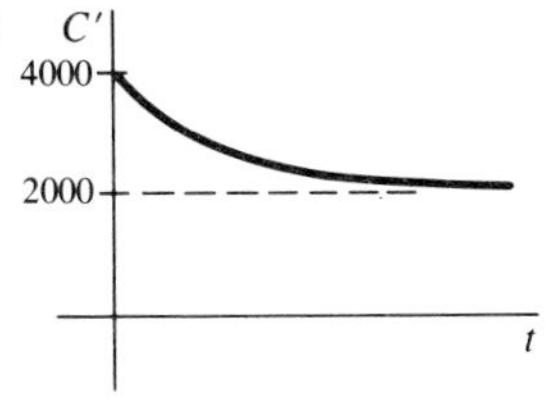

(b) \$105,118

7. \$18,706 **9.** 3.72 sec **11.** 144 ft **13.** 24 ft

CHAPTER 7

Section 7.1 (page 283)

1. $xe^x - e^x + c$ **3.** $(x^3 - 3x^2 + 6x - 6)e^x + c$

5. $x(\ln x)^2 - 2x\ln x + 2x + c$ **7.** $\dfrac{x^3}{3}\ln x - \dfrac{x^3}{9} + c$

9. $-\frac{1}{2}e^{-x^2} + c$ **11.** $\frac{2}{3}(1 + e^x)^{3/2} + c$

13. $2\sqrt{1 + e^x} + c$ **15.** $-\frac{1}{2}e^{-2x} - e^{-x} + c$

17. $(x^2 - 2x + 2)e^x - \dfrac{x^3}{3} + c$ **19.** $-\dfrac{1 + 2\ln x}{4x^2} + c$

21. $-\frac{1}{15}(3x^2 + 8)(4 - x^2)^{3/2} + c$ **23.** $\frac{1}{2}(x\ln x - x) + c$

25. $5\ln 5 - 4$ **27.** (a) Let $u = x^5, v' = e^x$;
(b) $(x^5 - 5x^4 + 20x^3 - 60x^2 + 120x - 120)e^x + c$,
$(x^6 - 6x^5 + 30x^4 - 120x^3 + 360x^2 - 720x + 720)e^x + c$

Section 7.2 (page 293)

1. $3\ln|x - 7| + c$ **3.** $x - 3\ln|x + 3| + c$

5. $\dfrac{x^2}{2} + 2x + c$ **7.** $x + \ln|2x - 1| + c$

9. $\dfrac{x^2}{2} - x + 2\ln|x + 1| + c$ **11.** $\dfrac{x^3}{3} - \dfrac{x^2}{2} + x + c$

13. $\dfrac{x^3}{3} - \dfrac{x^2}{2} + 3x - 5\ln|x + 2| + c$

15. $-x^3 + x^2 - x + c$ **17.** $-x - 9\ln|3 - x| + c$

19. $\dfrac{x^3}{3} + \dfrac{x^2}{2} + x + c$ **21.** $\frac{5}{7}(\ln|x - 4| - \ln|x + 3|) + c$

23. $x + \frac{1}{7}(17\ln|x - 4| - 10\ln|x + 3|) + c$
25. $\frac{1}{3}(4\ln|x - 3| - \ln|x|) + c$

27. $\dfrac{1}{9}\ln|x| - \dfrac{1}{9}\ln|x - 3| - \left(\dfrac{1}{3}\right)\dfrac{1}{x - 3} + c$

29. $-\dfrac{4}{9}\ln|x| + \dfrac{1}{3x} + \dfrac{13}{9}\ln|x - 3| + c$

31. $\frac{1}{4}(\ln|x - 2| - \ln|x + 2|) + c$
33. $-\frac{1}{2}\ln|x| + \frac{1}{3}\ln|x + 1| + \frac{7}{6}\ln|x - 2| + c$

35. $\ln|x - 2| - \dfrac{4}{x - 2} - \left(\dfrac{5}{2}\right)\dfrac{1}{(x - 2)^2} + c$

37. $\frac{6}{5}(\ln|x - 4| - \frac{1}{5}\ln|x + 1|) + c$
39. $-\frac{1}{2}\ln|x| + \frac{2}{3}\ln|x - 1| + \frac{5}{6}\ln|x + 2| + c$

41. $\dfrac{x^2}{2} + x + \frac{1}{3}(11\ln|x - 2| - 2\ln|x + 1|) + c$

43. $\frac{1}{3}\ln|x^3 - 1| + c$
45. $x + \ln|x + 1| - 4\ln|x + 2| + c$

47. $\dfrac{x^2}{2} - 3x - \ln|x + 1| + 8\ln|x + 2| + c$

49. $\dfrac{x^2}{2} + x + 2\ln|x + 1| - 4\ln|x + 2| + c$

51. $\frac{11}{15}\ln|x + 2| - \frac{1}{3}\ln|x - 1| + \frac{17}{20}\ln|x - 3|$
$- \frac{1}{4}\ln|x + 1| + c$

53. $\dfrac{7}{9}\ln|x + 2| + \dfrac{2}{9}\ln|x - 1| - \left(\dfrac{4}{3}\right)\dfrac{1}{x - 1} + c$

55. $\dfrac{x^2}{2} + 2x + 3\ln|x - 1| - \dfrac{1}{x - 1} + c$

57. $\frac{1}{9}\ln|x + 1| - \frac{5}{9}\ln|x + 4| + \frac{4}{9}\ln|x - 2| + c$
59. $\frac{9}{10}\ln|x - 3| - \frac{1}{2}\ln|x + 1| + \frac{3}{5}\ln|x + 2| + c$

Section 7.3 (page 296)

1. $\dfrac{1}{4}\ln\left|\dfrac{x}{x + 2}\right| + \left(\dfrac{1}{2}\right)\dfrac{1}{x + 2} + c$

3. $\sqrt{4t^2 + 1} - \ln\left|\dfrac{1 + \sqrt{4t^2 + 1}}{2t}\right| + c$

5. $\sqrt{9 - x^2} - 3\ln\left|\dfrac{3 + \sqrt{9 - x^2}}{x}\right| + c$

7. $\dfrac{1}{4}\ln\left|\dfrac{z}{z + 4}\right| + c$

9. $\frac{2}{5}(x + 7)^{5/2} - \frac{28}{3}(x + 7)^{3/2} + 98(x + 7)^{1/2} + c$

13. $e - 2$ **15.** $\sqrt{3}/2$ **17.** $= 300\ln 5 - 400 \approx 82.83$

19. $\dfrac{1 - 3e^{-2}}{2}$

Section 7.4 (page 306)

1. $y = \dfrac{3}{6 - 3t - t^3}$ **3.** $y = \sqrt{1 - (1/4)e^{t^2}}$

5. $y = \dfrac{12x + 1}{3x + 1}$ **7.** $(-1/2)e^{-y^2} = 1 - (1/2)e^{-1} - e^{-x}$

9. $y = 7e^{[(5t^2 + 2t)/6]}$ **11.** $y \ln y - y = \frac{3}{2}t^2 - 4t - 1$

13. $y \ln y - y = e^t + 2t - 1$ **15.** $y = \dfrac{4}{1 + 2t^2 - t^4} - 1$

17. (a) 96.7° (b) 27.1 min

19. (a) 28,302 birds; (b) 6.85 years

21. (a) $M = \dfrac{c}{\sqrt{c^2 - v^2}} M_0$; (b) $(\sqrt{3}/2)$ (speed of light)

23. $VP = c$ **25.** $r = -1 - \dfrac{4}{(2\theta - 1)e^{2\theta} + c}$

27. $4x^2 + 9y^2 + c$

Section 7.5 (page 315)

1. (a) $40e^{-(3t/200)}$; (b) 25.5 lb; (c) 65.4 min

3. (a) $(2.5 \times 10^{-8})(t - 200)^4$; (b) 20.9 lb; (c) 43.5 min

7. 12.5 g, 1.97 g, 6.5 years

9. (a) $100 - 80e^{-t/50}$; (b) 21.6 lb; (c) 23.5 min

11. 11,449; 8192; the population will fall below 1 in 1938.

13. \$4.48

Section 7.6 (page 323)

1. (a) 12; (b) 12 **3.** (a) 2.75; (b) 2.688

5. (a) 0.6970; (b) 0.6941 **7.** (a) 0.6970; (b) 0.6941

9. (a) 226; (b) 210.125 **11.** (a) 1.248; (b) 1.2495

13. (a) 1.727; (b) 1.721 **15.** (a) 4.288; (b) 4.276

17. (a) 29155.15; (b) 23059.05 **19.** (a) 8; (b) 8

21. (a) 9.333; (b) 9.333 **23.** (a) 0.8374; (b) 0.8048

25. (a) 81.248; (b) 81.050 **27.** $n = 13$

29. (a) Because $f^{(4)}(x)| = \left|\dfrac{48}{x^5}\right|$; (b) $n = 18$

31. 1.2499 **33.** 6.1975 **35.** 1.5681 **37.** 2.393

39. 1.473 **41.** 3.14159

Section 7.7 (page 331)

1. $\frac{1}{5}$ **3.** ∞ **5.** $\frac{1}{7}\ln 8$ **7.** ∞ **9.** e^{-2} **11.** $2/e$

13. $\frac{1}{4}$ **15.** ∞ **17.** ∞ **19.** -120

CHAPTER 8

Section 8.1 (page 339)

1. 11, 9, 0 **3.** −3630.86, 0, 29.56

5. −3.246, 1, 54.598 **7.** Not defined, not defined, 5.386

9. Not defined, 1, −0.6 **11.** −367.2

13. (a) $6, 6 + 2e^3, 6 + e^6$;
(b) $7 + e^9, 10 + 2e^9, 13 + 3e^9, 16 + 4e^9$;
(c) line with slope $(e^9 + 3)$, z intercept $= 4$

15. (a) 136, 364; (b) 81, 818; (c) 187,500, 112,500;
(d) 107,143, 64,286

17. (a) $Y(V, r) = Vr$; (b) a yield greater than \$2400/year

Section 8.2 (page 345)

1. $f_x = \ln y + y, f_y = x/y + x$

3. $f_x = 1/y^2, f_y = -(2x + y)/y^3$

5. $f_x = x/\sqrt{x^2 + 10y}, f_y = 5/\sqrt{x^2 + 10y}$

7. $f_x = xye^{x+y}(x + 2), f_y = x^2e^{x+y}(y + 1)$

9. $f_x = y/x + \ln y, f_y = \ln x + x/y$ **11.** 1, 4.5, 1.5

13. −1, 0 **15.** $-\frac{25}{216}$ **17.** 1 **19.** 1.646

21. (a) $2x + 1/y$; (b) $-(x + 1)/y^2$; (c) 2;
(d) $2(x + 1)/y^3$

23. (a) $1.6L^{-0.2}K^{0.2}$ (b) $0.4L^{0.8}K^{-0.8}$

25. (a) $0.45L^{-0.55}K^{0.55}$ (b) $0.55L^{0.45}K^{-0.45}$

27. (a) $z = 100L^{0.6}K^{0.4}$ (b) 56.91 (c) 36.03

29. (a) $0.42L^{-0.58}K^{0.58}$ (b) $0.58L^{0.42}K^{-0.42}$

33. Neg., pos., pos., neg.

35. $R(x, y) = 150x - 5x^2 + 3xy + 240y - 6y^2$,
$C(x, y) = 400 + 40x + 20y$,
$P(x, y) = 110x - 5x^2 + 3xy + 220y - 6y^2 - 400$

Section 8.3 (page 356)

1. (2, −3) **3.** $(\frac{10}{13}, \frac{9}{13})$ **5.** (1, 2), (−1, 2)

7. (−1, 0) **9.** (1, −2) **11.** Minimum

13. Saddle point.

15. (1, 2) saddle point; (−1, 2) maximum

17. Saddle point **19.** Maximum

21. Box with maximum volume: dimensions 1.826 in., 1.826 in., 0.913 in.

23. $x = \frac{49}{13}, y = \frac{24}{13}$ **27.** There is no maximal profit.

29. $x = 1, y = 1$, profit = \$9 **31.** $x, y, z = 10$

33. 5.10 in. by 2.55 in. by 7.65 in.

Section 8.4 (page 363)

1. (a) $y = 2.9x + 4.1$; (b) 0.03

3. (a) $y = -0.35x + 4.31$; (b) 47.98

5. (a) $y = 2x + 1$; (b) 0 **7.** (a) $y = 0.5x + 1$; (b) 0

9. (a) $3y = 12x - 10$; (b) 0.667 **11.** $y = 15.7$

13. (a) $F = 0.5I + 48.8$;. (b) 136.3

15. (a) $E = 0.031Y + 2.34$; (b) \$2.59

19. (a) x = number of years since 1980, y = price,
$y = 3.804x + 19.512$; (b) \$49.94

21. \$46.26

Section 8.5 (page 373)

1. 10 at (1, 2) **3.** $(-\frac{8}{13}, \frac{38}{13})$ **5.** 27 at (−6, 9)

7. $20\pi/(4 + \pi), 80/(4 + \pi)$ **9.** $K/L = \frac{60}{11}$

11. $K/L = \frac{75}{22}$ 13. (0.53, 0.85)
15. $x = 1.31, y = 3.18, z = 5.55$
17. $x = 1, y = 2.048, z = 4.635$
19. $x = \frac{76}{21}, y = \frac{152}{21}, z = -\frac{38}{21}$ 21. 3456 in.^{3}

Section 8.6 (page 387)

1. 2   3. $\frac{16}{3}$ 5. $\frac{\sqrt{5}-2}{2}$ 7. $2\sqrt{2}-2$
9. $1 - e^{-1} - e^{-3} + e^{-4}$ 11. $1 - \frac{1}{2}\ln 3$
13. $\int_0^2\int_0^{2-y} 2xy\,dx\,dy, \frac{4}{3}$
15. $\int_1^e\int_0^{\ln y} y\,dx\,dy, \frac{e^2+1}{4}$
17. $\frac{4}{3}$ 19. $\frac{1}{3}$ 21. 18   23. $\frac{3(e-1)}{2}$
25. $\bar{f} = 0$ 27. $\bar{f} = \frac{\ln 82}{72}$

CHAPTER 9

Section 9.1 (page 409)

1. $\sqrt{3}$ 3. $-1/2$ 5. Undefined   7. $-2/\sqrt{3}$
9. $-1/\sqrt{3}$ 11. 1   13. $\sqrt{3}/2$ 15. Undefined
17. $\sqrt{7}/4$ 19. 12/5   21. 17/15   23. 8/15
25. 13/5   27. $2\pi/3$ 29. 210°   31. $3\pi/2$
33. $7\pi/4$ 35. 0.259   37. 0.966

39.
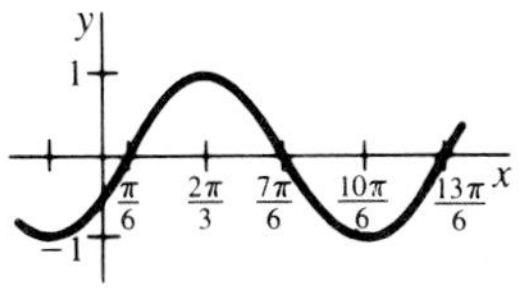

41.
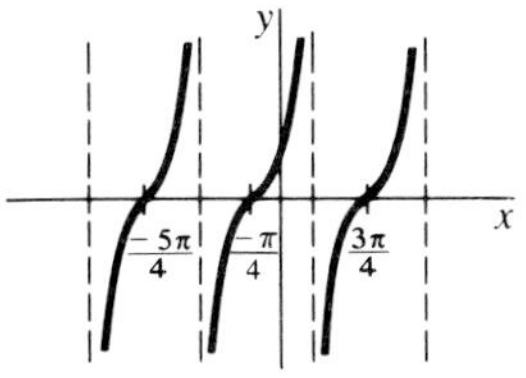

43.
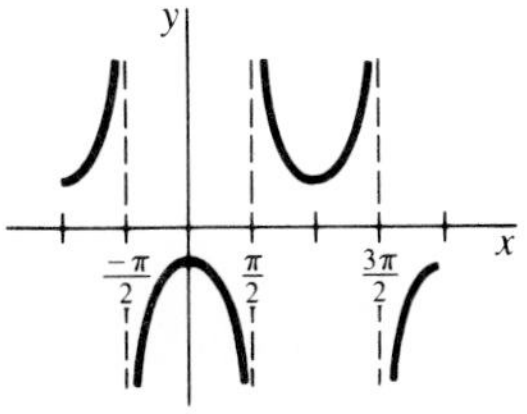

45.
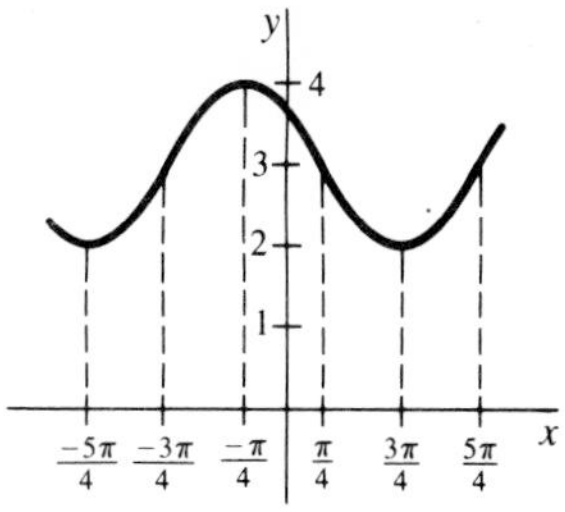

Section 9.2 (page 417)

1. $x^2(3\cos x - x\sin x)$ 3. $3\sin^2 x\cos x$
5. $(-2\cos x - x\sin x)/x^3$ 7. $e^x(\sec^2 x + \tan x)$
9. $-2\cos x\sin x$ 11. $(\cos x)e^{\sin x}$
13. $-12\csc^4(3x)\cot(3x)$ 15. $\sec^2 x/(2\sqrt{\tan x})$
17. $20\tan^4(4x)\sec^2(4x)$ 19. $-\csc^2(\sqrt{x})/(2\sqrt{x})$
21. $\cot x$ 23. $-\sec^2(1/x)/x^2$
25. $e^x\cos(e^x)$ 27. $\csc x$
29. $(4x^3 + \frac{1}{2}x^{-1/2})\sec(x^4 + x^{1/2})\tan(x^4 + x^{1/2})$
31. 1   33. 2   35. $\frac{1}{4}$ 37. 551 mph

Section 9.3 (page 430)

1. $e^{\sin x} + c$ 3. $-\frac{1}{5}\cos^5 x + c$ 5. $-\cos(e^x) + c$
7. $\frac{1}{4}\sec^4 x + c$ 9. $-\ln|1 - \sin x| + c$
11. $\frac{1}{5}\tan^5 x + c$ 13. $\ln|\tan x| + c$
15. $\frac{2}{3}(\sin x)^{3/2} + c$ 17. $-\frac{1}{2}\csc^2 x + c$
19. $\sin^2\sqrt{x} + c$ 21. $-\frac{1}{3}\cos^3 x + c$
23. $\frac{1}{2}x + \frac{1}{4}\sin(2x) + c$ 25. $\frac{1}{4}\tan^4 x + c$
27. $\frac{1}{3}\sin^3 x - \frac{1}{5}\sin^5 x + c$ 29. $\tan x - x + c$
31. $\frac{3}{8}x + \frac{1}{4}\sin(2x) + \frac{1}{32}\sin(4x) + c$
33. $\frac{1}{3}\tan^3 x - \tan x + x + c$ 35. $\sec x + \cos x + c$
37. $\tan x - \cot x + c$ 39. $-\cot x[1 + (\cot^2 x)/3] + c$
41. $\frac{3}{16}$ 43. $2 - 8^{1/4}$ 45. 0   47. $\frac{15}{64}$
49. $\frac{8\sqrt{3}}{27} - \frac{1}{3}$ 51. $\ln|\csc x - \cot x| + c$ 53. $\frac{1}{4}\pi^2$

Section 9.4 (page 435)

1. $x\sin x + \cos x + c$
3. $\frac{2}{27}\cos 3x - \frac{1}{3}x^2\cos 3x + \frac{2x}{9}\sin 3x + c$
5. $(x^3 - 6x)\sin x + (3x^2 - 6)\cos x + c$
7. $\frac{1}{2}e^x(\sin x - \cos x) + c$
9. $\frac{3}{16}x^2 + \frac{x\sin 4x}{32} + \frac{\cos 4x}{128} + \frac{x\sin 2x}{4} + \frac{\cos 2x}{8} + c$
11. $\frac{1}{3}[x^3\sin(x^3) + \cos(x^3)] + c$
13. $\frac{1}{2}[x^2\sin(x^2) + \cos(x^2)] + c$
15. $x\tan x - \ln|\sec x| + c$
17. $\ln|\csc x - \cot x| - x\csc x + c$
19. $\frac{1}{2}x[\sin(\ln|x|) - \cos(\ln|x|)] + c$
21. $\frac{1}{25}e^{4x}[3\sin(3x) + 4\cos(3x)] + c$
23. $\frac{1}{9}[\sin(3x + 4) - 3x\cos(3x + 4)] + c$
25. $-\frac{\csc x\cot x}{2} - \frac{1}{2}\ln|\csc x + \cot x| + c$

CHAPTER 10

Section 10.1 (page 445)

1. 1/4 **3.** 3/8 **5.** 1, 2, 3, 4, 5, 6

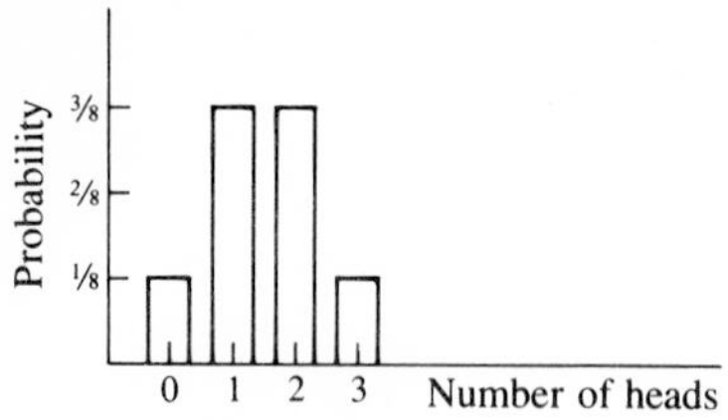

7.

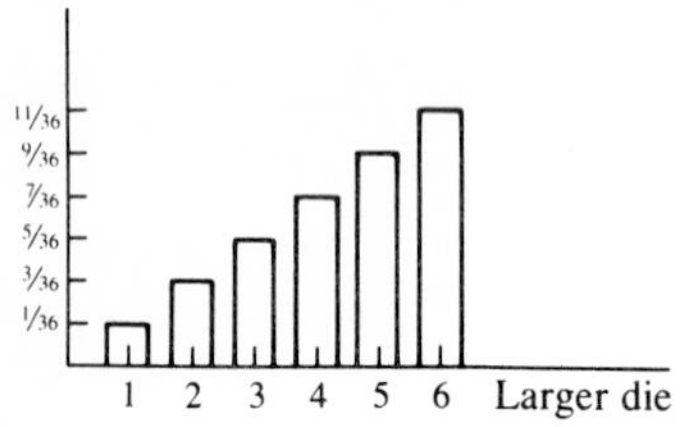

9. 1/9 **13.** 4/9 **15.** 1/4 **17.** 11/18

19.

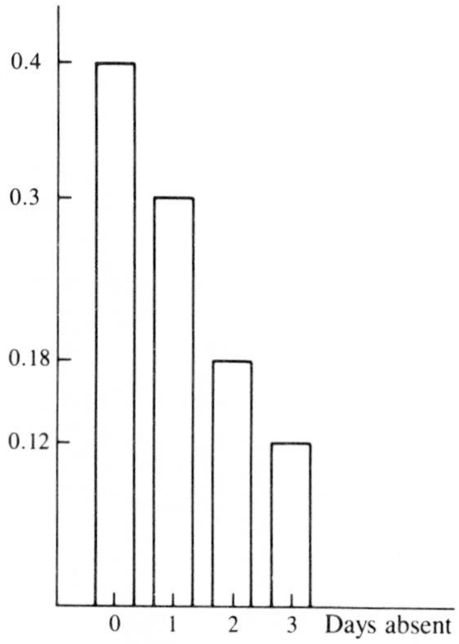

21.

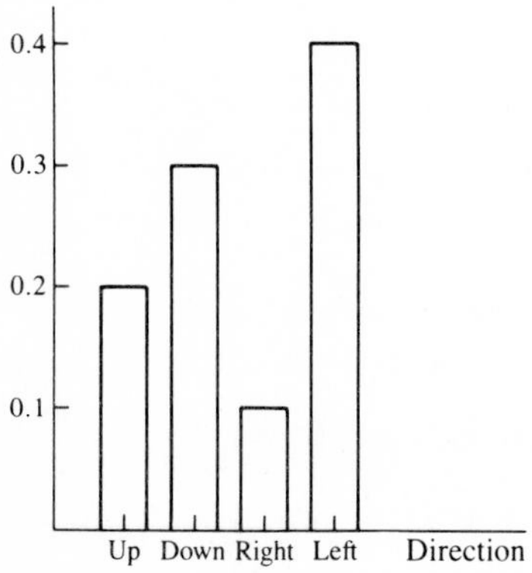

23.

Number	4	5	6
Prob.	1/6	1/2	1/3

25.

Letter	*a*	*b*	*c*	*d*
Prob.	0.4	0.2	0.1	0.3

27. Yes **29.** No **31.** No **33.** Yes

Section 10.2 (page 453)

1. 39.3, 51.2, 7.2 **3.** 35.7, 69.9, 8.4 **5.** 3.3, 1.41, 1.19
7. 12.3, 1.61, 1.27 **9.** 19.1, 0.89, 0.94
11. 6.875, 1.11, 1.05 **13.** 41.2, 29.16, 5.4
15. 12.25, 79.97, 8.94 **17.** 4.92 **19.** 1.85

Section 10.3 (page 460)

1. $c = 1/2$; $\Pr = \frac{1}{2}\sqrt{6} - 1$ **3.** Not a pdf
5. Not a pdf **7.** 241/768 **9.** 0 **11.** 0
13. $1 - e^{-3}$ **15.** $e^{-1} - e^{-2}$ **17.** $e^{-1} - e^{-2}$
19. (a) 12; (b) 2/5 **21.** 2, 4/3, $2\sqrt{3}/3$
23. 25/6, 125/252, 0.704

Section 10.4 (page 468)

1. 1/4, 1/16, 1/4 **3.** 3, 3, $\sqrt{3}$ **5.** 1/5, 1/25, 1/5
7. 2, 4, 2 **9.** 5, 16/3, $4\sqrt{3}/3$ **11.** 0.349
13. 1/2, 1/5 **15.** 0.287 **17.** $67.03 **19.** 0.247
21. 0.221 **23.** 0.204

Section 10.5 (page 481)

1. 0.400 **3.** 47.6 **5.** 180
7. Yes, question the accuracy. **9.** $2\sqrt{2\pi}$ **11.** 0.8971
13. μ **15.** 0.0388 g **17.** 1 g ± 0.784 g; 1 g ± 1.03 g

CHAPTER 11

Section 11.1 (page 497)

1. $1 - \frac{x^2}{2} + \frac{x^4}{24} - \frac{x^6}{720}$

3. $1 - 2x^2 + \frac{2}{3}x^4 - \frac{4}{45}x^6$ **5.** $1 - \frac{x^4}{2}$

7. $1 + x + \frac{x^2}{2!} + \frac{x^3}{3!} + \frac{x^4}{4!} + \frac{x^5}{5!} + \frac{x^6}{6!}$

9. $1 + 3x + \frac{9}{2}x^2 + \frac{9}{2}x^3 + \frac{27}{8}x^4 + \frac{81}{40}x^5 + \frac{81}{80}x^6$

11. $x - \frac{x^2}{2} + \frac{x^3}{3} - \frac{x^4}{4} + \frac{x^5}{5} - \frac{x^6}{6}$

13. $x + \frac{x^3}{3} + \frac{2}{15}x^5$ **15.** $1 - x^2 + \frac{x^4}{2} - \frac{x^6}{6}$

17. $1 + x + x^5 + x^6$

19. $\cos 1 - (\sin 1)(x-1) - (\cos 1)\dfrac{(x-1)^2}{2} + (\sin 1)\dfrac{(x-1)^3}{6}$

21. $e\left(1 + (x-1) + \dfrac{(x-1)^2}{2} + \dfrac{(x-1)^3}{6}\right)$

23. $\sqrt{2}\left(1 + \dfrac{(x-1)}{4} - \dfrac{(x-1)^2}{32} + \dfrac{(x-1)^3}{128}\right)$

25. $(x-1) - \dfrac{(x-1)^2}{2} + \dfrac{(x-1)^3}{3}$ 27. $1 + x - \dfrac{x^2}{2}$

29. Because $f(x)$ itself is of degree 3, *it* is the desired polynomial at $x = 0$ and at $x = 1$.

31. 0.05234 33. 1.6487

35. $\frac{1}{2} + \dfrac{\sqrt{3}}{2}\left(x - \dfrac{\pi}{6}\right) - \dfrac{1}{4}\left(x - \dfrac{\pi}{6}\right)^2 - \dfrac{\sqrt{3}}{12}\left(x - \dfrac{\pi}{6}\right)^3$, 0.51504

37. $\dfrac{\sqrt{3}}{2} + \dfrac{1}{2}\left(x - \dfrac{\pi}{3}\right) - \dfrac{\sqrt{3}}{4}\left(x - \dfrac{\pi}{3}\right)^2 - \dfrac{1}{12}\left(x - \dfrac{\pi}{3}\right)^3$, 0.8572

Section 11.2 (page 505)

1. 2.71805, $|\text{error}| \le 0.0006$
3. 0.366667, $|\text{error}| \le 0.00139$
5. 0.0199987, $|\text{error}| \le 2.54 \times 10^{-16}$

7. $\displaystyle\sum_{n=0}^{\infty} (-1)^n \frac{(3x)^{2n}}{(2n)!}$

9. $\displaystyle\sum_{n=0}^{\infty} (-1)^n \frac{(2^{2n+1}+1)x^{2n+1}}{(2n+1)!}$

11. $x + \dfrac{x^3}{3} + \dfrac{2x^5}{15} + \dfrac{17x^7}{315} + \dfrac{62x^9}{2835} + \cdots$

13. $1 + \dfrac{x^2}{2} + \dfrac{5}{24}x^4 + \dfrac{61}{720}x^6$ 15. $\displaystyle\sum_{n=0}^{\infty} \frac{e^2(3x)^n}{n!}$

15. Does not exist; $e^{1/x}$ is not continuous at $x = 0$.

17. $1 + \dfrac{x}{2} - \dfrac{x^2}{8} + \cdots + (-1)^{n+1}\dfrac{(1/2)[(1/2)-1]\cdots[(1/2)-n+1]}{n!} + \cdots$

19. $1 + \alpha x + \dfrac{\alpha(\alpha-1)x^2}{2!} + \cdots \dfrac{\alpha(\alpha-1)\cdots(\alpha-n+1)}{n!}x^n + \cdots$

21. $1 - x + x^2 - \dfrac{5}{6}x^3 + \cdots$

23. $\displaystyle\sum_{n=0}^{\infty} \frac{x^{2n}}{(2n)!}$ 25. $\displaystyle 2\sum_{n=0}^{\infty} \frac{x^{2n+1}}{2n+1}$

27. Degree 2, 0.995; Note that $p_2(x) = p_3(x)$ for $\cos x$ expanded about $x = 0$.
29. 0.100 31. Degree 3 33. 1.396 35. 0.4110
37. 8.0623

Section 11.3 (page 509)

1. $1 - x + x^2 - x^3 + x^4 \cdots$

3. $x - \dfrac{x^2}{2} + \dfrac{x^3}{3} - \dfrac{x^4}{4} + \cdots$

5. $x^2 - \dfrac{x^4}{2} + \dfrac{x^6}{3} - \dfrac{x^8}{4} + \cdots$

7. 0.100. It is $\approx \sqrt{\pi}$ times table value.
9. 1.191. It is $\approx \sqrt{2\pi}$ times table value.
11. 1.229. It is $\approx \sqrt{2\pi}$ times table value.
13. 0.460 15. 0.946 17. 0.001 19. 0.050
21. 0.401

Section 11.4 (page 515)

1. $y = 3\left(1 + x + \dfrac{x^2}{2!} + \dfrac{x^3}{3!} + \cdots\right) = 3e^x$

3. $y = -1\left(1 - x + \dfrac{x^2}{2!} - \dfrac{x^3}{3!} + \cdots\right) = -e^{-x}$

5. $y = (-\frac{3}{2} + a) + a\left(1 + 2x + 2x^2 + \dfrac{4x^3}{3} + \cdots\right) = -\frac{3}{2} + ae^{2x}$

7. $y = 1 + \dfrac{x^2}{2} + \dfrac{x^4}{8} + \dfrac{x^6}{48} + \cdots = e^{x^2/2}$

9. $y = 1 + \dfrac{x^2}{2} - \dfrac{x^4}{8} + \cdots = \sqrt{1 + x^2}$

11. $y = 3 + 2x - \dfrac{3x^2}{2} - \dfrac{x^3}{3} + \cdots = 3\cos x + 2\sin x$

13. $y = a\left(1 - x^2 + \dfrac{2x^4}{3}\cdots\right) + b\left(x - \dfrac{2}{3}x^3 + \cdots\right) = a\cos 2x + \dfrac{b}{2}\sin 2x$

15. $y = a\left(1 - x + \dfrac{x^2}{2!}\cdots\right) + b\left(x - x^2 + \dfrac{x^3}{2!}\cdots\right) = ae^{-x} + bxe^{-x}$

Section 11.5 (page 523)

1. 2.25 3. 2.236 5. 7.937 7. 1.333 9. 2.031
11. 1.156 13. 1.26 15. 2.031 17. −3.562
19. 0.347, 1.53, −1.88 21. Only one real root, −0.755
23. 1.895 25. 0.949 27. 1.1656
29. $x = 32$, $p = 9.76$ 31. 0.34730, 1.532, −1.879

Section 11.6 (page 532)

1. $\frac{3}{2}$ **3.** Infinite **5.** 1 **7.** 2 **9.** $\frac{1}{2}$ **11.** 0
13. $\frac{1}{2}$ **15.** $-\frac{1}{9}$ **17.** $\frac{1}{6}$ **19.** 0
21. Undefined **23.** $\frac{3}{7}$ **25.** $\frac{3}{7}$
27. a_0/b_0
29. 0 **31.** 0 **33.** 0 **35.** 0 **37.** e **39.** e^3
41. e^n **43.** 1 **45.** e^{-6} **47.** 1 **49.** 0

APPENDIX A

Section A.1 (page 540)

1. $\frac{3}{2}$ **3.** $\frac{6}{7}$ **5.** $\frac{1}{6}$ **7.** $\frac{20}{3}$ **9.** $\frac{1}{12}$
11. $11 + \sqrt{5}$ **13.** no **15.** $-1 - 3\sqrt{3}$ **17.** $\pi - 3$
19. $\pi - 2$ **21.** -5 **23.** $12m - 36$ **25.** $y - x$
27. $+1$ if $x > 0$, -1 if $x < 0$, undefined if $x = 0$
29. 9 **31.** 12 **33.** $\dfrac{x^2 + 4}{4x}$
35. $\dfrac{3x - 2}{4x - 8}$ **37.** $\dfrac{2x^2 - 1}{x^2 - x}$ **39.** $\dfrac{2}{x - 1}$
41. $\dfrac{y - x}{y + x}$ **43.** $\frac{7}{25}$ **45.** $\dfrac{x^2 + 1}{x^2 + x + 1}$

Section A.2 (page 546)

1. $\frac{15}{49}$ **3.** 6 **5.** $143\frac{133}{144}$ **7.** $2^{11/6}$ **9.** $\dfrac{-5 \cdot 7^5}{24}$
11. $(2^3)^{-1} = \frac{1}{8}$ but $2^{(3^{-1})} = 2^{1/3} \approx 1.26$
13. $(2^2)^2 = 4^2 = 16$ and $2^{(2^2)} = 2^4 = 16$ but $(-2^{-2})^{-2} = (-\frac{1}{4})^{-2} = 16$ while $(-2)^{(-2^{-2})} = (-2)^{-1/4} = \dfrac{1}{(-2)^{1/4}}$, which is undefined
15. $\dfrac{2x^4}{y^7}$ **17.** $x - \dfrac{1}{x}$ **19.** $\dfrac{y^2 + x^2}{xy}$
21. $\dfrac{y^2 + x^2}{xy}$ **23.** $\dfrac{a - 2\sqrt{ab} + b}{a - b}$ **25.** $x^3 - y^3$
27. $(x - y)^2 - z^2$ **29.** $\dfrac{x^2 + xy + y^2}{x + y}$
31. $6x^5 - 2x^4 + 22x^3 - 7x^2 + 21x - 5$ **33.** $x^6 - 1$
35. $\dfrac{3x + 2}{x + 3}$ **37.** $x^4 - 4x^2 + 6 - \dfrac{4}{x^2} + \dfrac{1}{x^4}$
39. Quotient $x - 4$, remainder 2
41. Quotient $x^2 + 2x + 5$, remainder 0
43. Quotient $x^3 - 3x + 5$, remainder -10
45. Quotient $x^5 - x^3 + x$, remainder $-x + 1$

Section A.3 (page 552)

1. $(x - 4)(x + 3)$ **3.** $(2x - 3)^2$ **5.** $2(x - 4)(3x + 4)$
7. $(x + 5)(x - 5)$ **9.** $(x + 3)(x - 3)(x + 2)(x - 2)$
11. $(x - 1)(x^2 + x + 1)$ **13.** $(x - 2)(x^2 + 2x + 4)$
15. $(x - 2)(x^2 + 3x + 3)$ **17.** $x = -7$
19. $x = 4, -5$ **21.** $x = \dfrac{-5 \pm \sqrt{17}}{2}$
23. $x = \dfrac{3 \pm \sqrt{29}}{2}$ **25.** $x = \frac{17}{11}$, $y = \frac{20}{11}$
27. $x = 1, y = -6$ or $x = -1, y = -4$
29. $x = 1, y = 0$ or $x = -2, y = -3$
31. $x = -2, y = 0$ **33.** no solution
35. $x = \frac{1}{39}$, $y = \frac{53}{13}$
37. $x = -3, y = 6$ or $x = 1, y = 2$
39. $x = 1, y = 1$ or $x = -1, y = 1$

D Complete Solutions to Warm-Up Tests

CHAPTER 1 (page 36)

1. (a) $\boxed{m = \dfrac{5-2}{3-(-1)} = \dfrac{3}{4}}$ (b) $\boxed{m = \dfrac{0-5}{2-1} = -5}$

   (c) $\boxed{m = \dfrac{0-0}{5-(-3)} = 0}$ (d) $\boxed{m = \dfrac{8-8}{4-1} = 0}$

2. (a) $y = 4x + 5$ so $m = 4$

   $4 = \dfrac{5-y}{2-x}$

   $8 - 4x = 5 - y$

   $\boxed{y = 4x - 3}$

   (b) $m = \dfrac{-22-3}{-4-1} = 5$

   $5 = \dfrac{3-y}{1-x}$

   $\boxed{y = 5x - 2}$

   (c) $m = -\dfrac{1}{2}, b = 6$

   $y = -\dfrac{1}{2}, b = 6$

   $\boxed{y = -\dfrac{1}{2}x + 6}$

   (d) $m_1 = -2$ so $m_2 = \dfrac{1}{2}$

   $\dfrac{1}{2} = \dfrac{3-y}{2-x}$

   $\boxed{y = \dfrac{1}{2}x + 2}$

3. $p_1 = 86, q_{s_1} = 1500$
   $p_2 = 70, q_{s_2} = 1200$

   $m = \dfrac{1500 - 1200}{86 - 70} = \dfrac{300}{16} = 18.75$

   $1200 = 18.75(70) + b$
   $b = -112.5$

   $\boxed{q_s = 18.75p - 112.5}$

4. (a) $x^2 - 9 \geq 0, x^2 \geq 9$

   $\boxed{\{x | x \leq -3 \quad \text{or} \quad x \geq 3\}}$

   (b) $x - 4 \neq 0$

   $\boxed{\{x | x \neq 4\}}$

   (c) $\boxed{\text{All real } x}$

   (d) $5 - x^2 \neq 0$

   $\boxed{\{x | x \neq \pm\sqrt{5}\}}$

5. (a) $x^2 - 8x + y^2 + 10y = 128$
   $x^2 - 8x + 16 + y^2 + 10y + 25 = 128 + 16 + 25$
   $= 169$
   $(x-4)^2 + (y+5)^2 = 169$

   $\boxed{\text{Circle with center at } (4, -5) \text{ and radius } 13}$

   (b) $x^2 - 8x + y^2 + 10y = -41$
   $x^2 - 8x + 16 + y^2 + 10y + 25 = -41 + 16 + 25$
   $(x-4)^2 + (y+5)^2 = 0$

   $\boxed{\text{The point } (4, -5)}$

6. (a) $\boxed{\begin{array}{l} g(x) = x^2 - 4x + 5 \quad \text{and} \quad h(x) = \sqrt{x} \\ h(g(x)) = \sqrt{x^2 - 4x + 5} \end{array}}$

(b) $g(x) = x^2 + 1$ and $h(x) = \frac{1}{x^3}$

$h(g(x)) = \frac{1}{(x^2 + 1)^3}$

or $g(x) = (x^2 + 1)^3$ and $h(x) = \frac{1}{x}$

$h(g(x)) = \frac{1}{(x^2 + 1)^3}$

(c) $g(x) = 8x^2 + 5$ and $h(x) = x^{1/4}$

$h(g(x)) = (8x^2 + 5)^{1/4}$

7. (a) $32°F \to 0°C$

$212°F \to 100°C$

$$m = \frac{\Delta C}{\Delta F} = \frac{100 - 0}{212 - 32} = \frac{100}{180} = \frac{5}{9}$$

$$C - 0 = \frac{5}{9}(F - 32)$$

$$\boxed{C = \frac{5}{9}(F - 32)}$$

(b) $C = \frac{5}{9}(F - 32) = \frac{5}{9}\left[\left(70 + \frac{(t - 8)^2}{30}\right) - 32\right]$

$$\boxed{C = \frac{5}{9}\left[38 + \frac{(t - 8)^2}{30}\right] \quad 0 \le t \le 24}$$

8. $f(g(x)) = 2(3x + B) - 3, \quad g(f(x)) = 3(2x - 3) + B$

$2(3x + B) - 3 = 3(2x - 3) + B$

$6x + 2B - 3 = 6x - 9 + B$

$$\boxed{B = -6}$$

9. The original line is $3y - 2x = 6$ or $y = \frac{2}{3}x + 2$. Its slope is $\frac{2}{3}$, so a line perpendicular to it will have slope $-\frac{3}{2}$. Such a perpendicular line will have an equation

$$y = -\frac{3}{2}x + b$$

and, if it is to pass through $(-1,4)$, we must have $b = \frac{5}{2}$, so

$$y = -\frac{3}{2}x + \frac{5}{2}$$

Next we seek the intersection of the two lines by equating

$$-\frac{3}{2}x + \frac{5}{2} = \frac{2}{3}x + 2$$

So $x = \frac{3}{13}$, $y = \frac{28}{13}$, and

$$d = \sqrt{(x_2 - x_1)^2 + (y_2 - y_1)^2}$$
$$= \sqrt{(\tfrac{28}{13} - 4)^2 + (\tfrac{3}{13} - (-1))^2}$$

$$\boxed{\text{Distance} = 8\sqrt{13}/13 = 2.22}$$

10. $C = (22)(5) + (0.14)x + 18$

$$\boxed{C = 0.14x + 128}$$

CHAPTER 2 (page 91)

1. $v(t_0) = \lim_{h \to 0} \frac{s(t_0 + h) - s(t_0)}{h}$

(a) $v(3) = \lim_{h \to 0} \frac{4(3 + h) + 2 - (4 \cdot 3 + 2)}{h}$

$= \lim_{h \to 0} \frac{4h}{h} = \boxed{4}$

(b) $v(2) = \lim_{h \to 0} \frac{(2 + h)^2 + (2 + h) - (2^2 + 2)}{h}$

$= \lim_{h \to 0} \frac{5h + h^2}{h} = \lim_{h \to 0} (5 + h) = \boxed{5}$

(c) $v(1) = \lim_{h \to 0} \frac{\frac{3}{2 + h} - \frac{3}{2}}{h} = \lim_{h \to 0} \left(-\frac{3h}{2(2 + h)h}\right)$

$= \lim_{h \to 0} \left(-\frac{3}{2(2 + h)}\right) = \boxed{-\frac{3}{4}}$

2. Average rate of change (5 years) $= \frac{155.17 - 108.21}{1985 - 1980}$

$= \boxed{\$9.39 \text{ per year}}$

Average rate of change (3 years) $= \frac{123.81 - 108.21}{1983 - 1980}$

$= \boxed{\$5.20 \text{ per year}}$

3. (a) $f'(x) = 8, f'(2) = 8$

$\boxed{\text{The slope is 8.}}$

(b) $f'(x) = 2x - 4$

$f'(-1) = -2 - 4$

$f'(-1) = -6$

$\boxed{\text{The slope is } -6.}$

(c) $f'(x) = 2x - 2$

$f'(0) = -2$

$\boxed{\text{The slope is } -2.}$

4. (a) $f'(x) = \boxed{6x - 4}$

(b) $f'(3) = 6 \cdot 3 - 4 = \boxed{14}$

(c) $f'(2 + z) = 6(2 + z) - 4 = 12 + 6z - 4$
$= \boxed{8 + 6z}$

(d) $\left.\dfrac{dy}{dx}\right|_{x=-3} = f'(-3) = 6(-3) - 4$
$= -18 - 4$
$= \boxed{-22}$

5. $P(x) = 75x^3 - 84$ so $P'(x) = 225x^2$

(a) $P'(7) = 225 \cdot 49$
$= \boxed{\$11{,}025 \text{ per unit}}$

(b) $P'(11) = 225 \cdot 121$
$= \boxed{\$27{,}225 \text{ per unit}}$

6. (a) $$f'(3) = \lim_{h\to 0} \frac{\sqrt{4+h} - 2}{2}$$
$$= \lim_{h\to 0} \frac{\sqrt{4+h} - 2}{h} \cdot \frac{\sqrt{4+h} + 2}{\sqrt{4+h} + 2}$$
$$= \lim_{h\to 0} \frac{h}{h(\sqrt{4+h} + 2)}$$
$$= \lim_{h\to 0} \frac{1}{\sqrt{4+h} + 2} = \boxed{\frac{1}{4}}$$

(b) $$f'(3) = \lim_{h\to 0} \frac{\dfrac{1}{4+h} - \dfrac{1}{4}}{h}$$
$$= \lim_{h\to 0} \frac{-h}{h(4)(4+k)}$$
$$= \lim_{h\to 0} \frac{-1}{4(4+h)} = \boxed{\frac{-1}{16}}$$

7. (a) $\lim_{h\to 0} \sqrt{h^2 + 5} = \boxed{\sqrt{5}}$

(b) $\displaystyle\lim_{x\to 4} \frac{x^2 - 16}{x - 4} = \lim_{x\to 4} (x + 4) = \boxed{8}$

(c) $\displaystyle\lim_{x\to 4} \frac{x + 7}{(2x - 1)^2} = \frac{4 + 7}{(2 \cdot 4 - 1)^2} = \boxed{\frac{11}{49}}$

8. (a) Newton: $(f(x)g(x))' = f(x)g'(x) + f'(x)g(x)$
Leibniz: $\dfrac{d}{dx}(uv) = u\dfrac{dv}{dx} + \dfrac{du}{dx}v$

(b) $\dfrac{dy}{dx} = (x^2 + x - 1)(3x^2) + (2x + 1)(x^3 + 4)$
$= \boxed{5x^4 + 4x^3 - 3x^2 + 8x + 4}$

9. (a) $\dfrac{dy}{dx} = 74x^{73} - 32x^{15}$ Power rule

(b) $\dfrac{dy}{dx} = 2(3x^4 - 2x^2 + 1)(12x^3 - 4x)$
Generalized power rule

(c) $$\frac{dy}{dx} = \frac{(2x - 1)(3x^2 - 1) - (x^3 - x)(2)}{(2x - 1)^2}$$
Quotient rule
$$= \frac{6x^3 - 3x^2 - 2x + 1 - 2x^3 + 2x}{(2x - 1)^2}$$
$$= \boxed{\frac{4x^3 - 3x^2 + 1}{(2x - 1)^2}}$$

10. (a) $y' = \dfrac{-1}{(x - 3)^2}$.

$y'(4) = \dfrac{-1}{(4 - 3)^2} = -1$, so slope $= -1$
$y - 1 = -1(x - 4)$
$y - 1 = -x + 4$

$\boxed{y = -x + 5}$

(b) $y'(5) = -\dfrac{1}{(5 - 3)^2} = -\dfrac{1}{4}$, so slope $= -\dfrac{1}{4}$

$y - \dfrac{1}{2} = -\dfrac{1}{4}(x - 5)$

$y - \dfrac{2}{4} = -\dfrac{1}{4}x + \dfrac{5}{4}$

$\boxed{y = -\dfrac{1}{4}x + \dfrac{7}{4}}$

CHAPTER 3 (page 143)

1. $f(x) = \dfrac{x - 1}{(x + 1)^2}$, $f'(x) = \dfrac{3 - x}{(x + 1)^3}$, $x \neq -1$

There is a critical point at $x = 3$ because $f'(3) = 0$. The function and its derivative are undefined at $x = -1$.

sign chart $f'(x)$

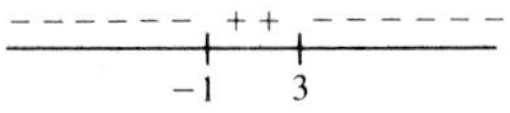

Maximum at (3, 1/8). The curve falls on $-\infty < x < -1$, rises on $-1 < x < 3$, and falls on $3 < x < \infty$. Vertical asymptote at $x = -1$; horizontal asymptote at $y = 0$.

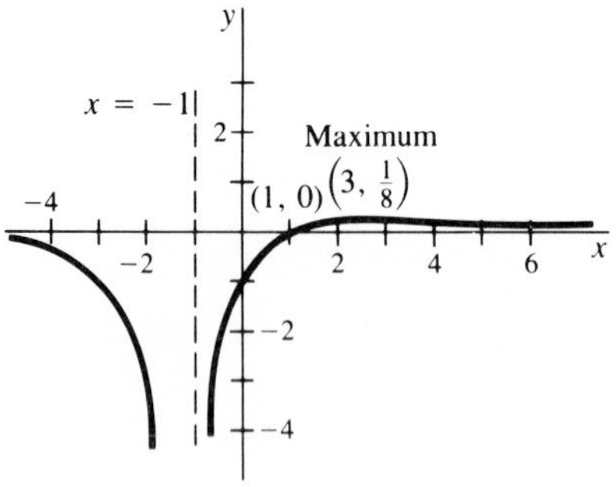

2. $y = -\dfrac{x^3}{8} + \dfrac{3}{2}x^2 + 6$

$$y' = -\frac{3}{8}x^2 + 3x = -\frac{3}{8}x(x - 8)$$

Critical points at $x = 0$ and $x = 8$.

sign y'

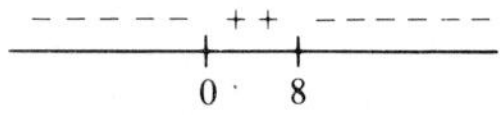

Minimum at (0, 6), maximum at (8, 38). The curve falls on $-\infty < x < 0$, rises on $0 < x < 8$, and falls on $8 < x < \infty$.

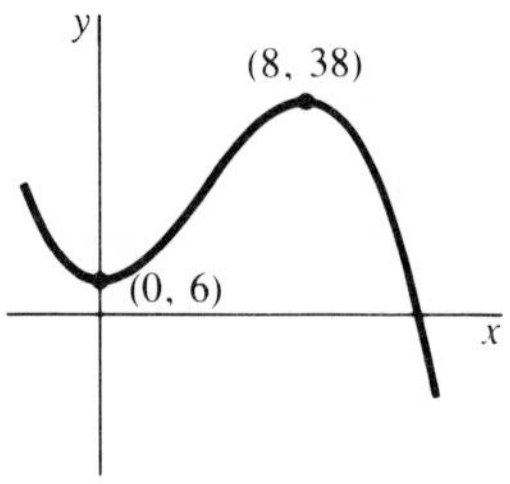

3. (a) approaches $= \infty$

(b) $$\lim_{x\to-\infty} \frac{2x^3 + 4x + 5}{x^3 - 8x} = \lim_{x\to-\infty} \frac{2 + \dfrac{4}{x^2} + \dfrac{5}{x^3}}{1 - \dfrac{8}{x^2}} = \boxed{2}$$

(c) $$\lim_{x\to-\infty} \frac{3 + 2x}{x + 7} = \lim_{x\to-\infty} \frac{2 + \dfrac{3}{x}}{1 + \dfrac{7}{x}} = \boxed{2}$$

(d) $$\lim_{x\to-\infty} \frac{1 + x^3}{x^4 + x^2} = \lim_{x\to-\infty} \frac{\dfrac{1}{x^4} + \dfrac{1}{x}}{1 + \dfrac{1}{x^2}} = \boxed{0}$$

4. (a) $$m = \frac{6 - 7}{20{,}000 - 10{,}000} = -0.0001$$

p	x
6	20,000
7	10,000

The equation has the form $p = -0.0001x + b$. Because $x = 20{,}000 \to p = 6$ we find that $b = 6 + (0.0001)(20{,}000) = 6 + 2 = 8$

$$\boxed{p = (-0.0001)x + 8}$$

(b) We are given $R(x) = px + 1x = (p + 1)x$. Thus

$$R(x) = ((-0.0001)x + 9)x = -0.0001x^2 + 9x$$

Next, differentiating gives

$$R'(x) = -0.0002x + 9$$

and setting equal to 0 yields the only critical point

$$x = \frac{9}{0.0002} = \boxed{45{,}000}$$

Because $R''(x) = -0.0002 < 0$, this value of x must be a maximum.

Finally, corresponding to 45,000 fans, the price $p = (-0.0001)(45{,}000) + 8 = \$3.50$.

A price of \$3.50 will maximize revenue and at this price 45,000 fans will attend.

5. Let x = length of side parallel to the school and let y equal lengths of two ends. We are given that $5x + 2(2y) = 1000$ so that $y = \dfrac{1000 - 5x}{4}$. We are asked to maximize area $= xy$. Thus

$$A(x) = x\left(\frac{1000 - 5x}{4}\right) = 250x - \frac{5x^2}{4}$$

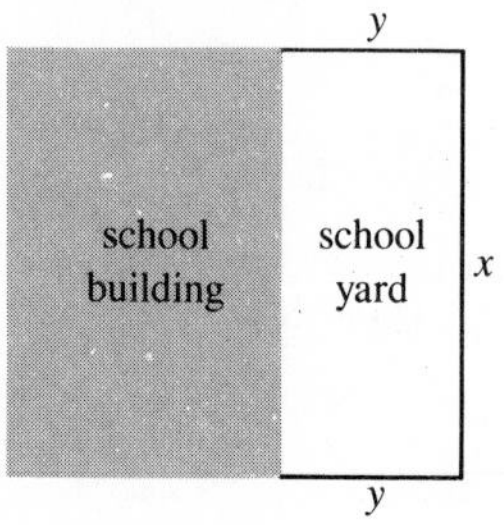

Differentiating, we obtain

$$A'(x) = 250 - \frac{5}{2}x$$

and $$A''(x) = -\frac{5}{2} < 0$$

The only critical point is the solution of $A'(x) = 0$,

$$250 - \frac{5}{2}x = 0 \quad \text{or} \quad x = 100$$

and this must be a maximum because $A''(100) < 0$.

$$y = \frac{1000 - 5x}{4} = \frac{1000 - 500}{4} = 125$$

We can state:

For largest area, take parallel side 100 feet and two ends 125 feet each.

6. $y = \dfrac{x}{x + 2} \qquad x \neq -2$

$$y' = \frac{2}{(x + 2)^2}$$

$$y'' = -\frac{4}{(x + 2)^3}$$

$y' > 0$ for all x; $y'' < 0$ for all $x > -2$ and $y'' > 0$ for all $x < -2$

The curve has no maxima, minima, or inflection points. It has a vertical asymptote at $x = -2$ and a horizontal asymptote at $y = 1$. It rises for $-\infty < x < -2$ and $-2 < x < \infty$. It is concave up for $-\infty < x < -2$ and concave down for $-2 < x < \infty$.

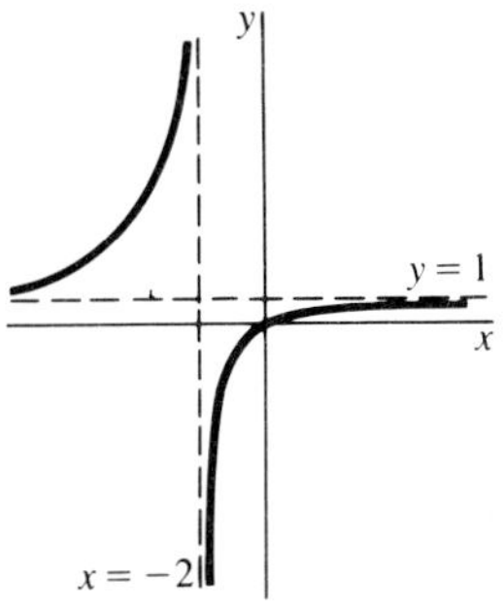

7. Here the demand function is $p = f(p) = 450 - 0.2q$, $0 < q < 2250$. By definition

$$E = \frac{p/q}{dp/dq} = \frac{450 - 0.2q}{-0.2q} = 1 - \frac{2250}{q}$$

(a) $E(180) = 1 - \dfrac{2250}{180} = \boxed{-11.5}$

(b) $E(2000) = 1 - \dfrac{2250}{2000} = \boxed{-0.125}$

(c) $\boxed{E(q) = 1 - \dfrac{2250}{q} \qquad 0 < q < 2250}$

8. We must minimize the distance squared.

$$E(m) = (m - 3)^2 + (3m - 11)^2 + (5m - 17)^2$$

Computing the derivative gives

$$\begin{aligned} E'(m) &= 2(m - 3) + 6(3m - 11) + 10(5m - 17) \\ &= 70m - 242 \end{aligned}$$

Setting $E'(m) = 0$, we find the only critical point, $m = \dfrac{242}{70} = 3.46$. This must correspond to a minimum of $E(m)$ as $E'' > 0$.

The best-fit line is $y = 3.46x$.

9. Set $f(x) = (x + 5)^2 + 2(x - 3)^2 + 8(x - 4)^2$. $f(x)$ will be minimized when $f'(x) = 0$. Differentiating gives

$$\begin{aligned} f'(x) &= 2(x + 5) + 4(x - 3) + 16(x - 4) \\ &= 22x - 66 \end{aligned}$$

Setting this equal to 0 and solving gives $x = 3$ as the only critical point. Since $f'' = 22 > 0$, we are assured that

$x = 3$ gives the minimum

10. $y = 2x^3 - 3x^2 - 12x + 12$

$y' = 6x^2 - 6x - 12 = 6(x - 2)(x + 1)$

so $\quad$ $x = 2$, $x = -1$ are critical points

sign y'

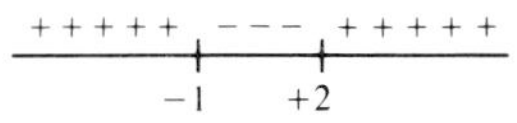

Here $(-1, 19)$ is a local maximum; $(2, -8)$ is a local minimum. The curve rises on $-\infty < x < -1$, falls on $-1 < x < 2$, and rises on $2 < x < \infty$.

We next compute the second derivative

$$y'' = 12x - 6$$

sign y''

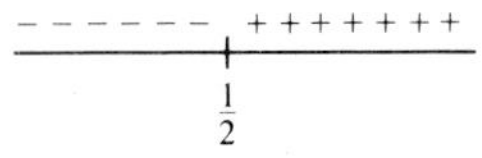

Thus $(\frac{1}{2}, \frac{11}{2})$ is a point of inflection. The curve is concave downward on $-\infty < x < 1/2$ and concave upward on $\frac{1}{2} < x < \infty$.

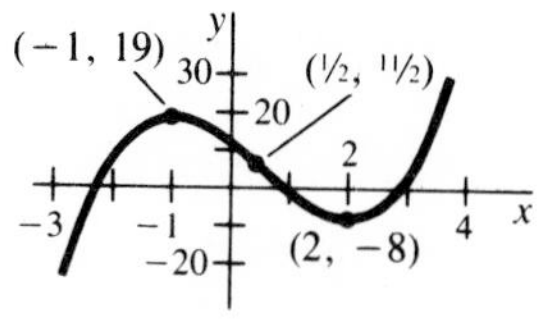

CHAPTER 4 (page 206)

1. (a) $\dfrac{dy}{dx} = \dfrac{(x^3+5)^4(2) - (2x)4(x^3+5)^3(3x^2)}{(x^3+5)^8} = \boxed{\dfrac{10-22x^3}{(x^3+5)^5}}$

(b) $\dfrac{dy}{dx} = x\,e^{3x^2}(6x) + e^{3x^2} = \boxed{(6x^2+1)e^{3x^2}}$

(c) $\boxed{\dfrac{dy}{dx} = x^2\left(\dfrac{1}{x+1}\right) + 2x\ln(x+1)}$

(d) $\dfrac{dy}{dx} = e^{3e^{2x+1}}(3e^{2x+1}(2)) = \boxed{6e^{2x+1}e^{3e^{2x+1}}}$

2. $p + xp = 3$

$$\frac{dp}{dt} + \frac{dx}{dt}(p) + x\frac{dp}{dt} = 0$$

$$(1+x)\frac{dp}{dt} = -p\frac{dx}{dt}$$

$$\frac{dp}{dt} = \frac{-p\dfrac{dx}{dt}}{1+x}$$

$$= \frac{-1(-0.1)}{1+2}$$

$$= 1/30 = \boxed{0.033 \text{ dollars/day}}$$

3. (a) $\ln y = \ln(x+1) + \dfrac{1}{3}\ln(x^2-6x) - \dfrac{1}{2}\ln(2x-3)$

$$\frac{1}{y}y' = \frac{1}{x+1} + \frac{1}{3}\frac{1}{x^2-6x}(2x-6) - \frac{1}{2}\frac{1}{2x-3}(2)$$

$$\boxed{y' = \left(\frac{1}{x+1} + \frac{2x-6}{3x^2-18x} - \frac{1}{2x-3}\right) \times \frac{(x+1)(x^2-6x)^{1/3}}{\sqrt{2x-3}}}$$

(b) $\ln y = \dfrac{1}{5}\ln(x^3-4) - \dfrac{3}{2}\ln(3x+7)$

$$\frac{1}{y}y' = \frac{1}{5}\frac{1}{x^3-4}(3x^2) - \frac{3}{2}\frac{1}{3x+7}(3)$$

$$\boxed{y' = \left(\frac{3x^2}{5x^3-20} - \frac{9}{6x+14}\right)\frac{(x^3-4)^{1/5}}{(3x+7)^{3/2}}}$$

4. $y' = 2xe^{x^2} \quad y'|_{x=0} = 0$

so $m = 0$

$y - 5 = 0(x - 0)$

$$\boxed{y = 5}$$

5. (a) Let P_0 = initial principal = \$35,000, r = rate = 0.13, t = time, $P(t)$ = accumulation after t years. Then

$$P(t) = P_0e^{rt}$$

$$P(10) = 35{,}000e^{(0.13)10}$$

$$= \boxed{\$128{,}425}$$

(b) In one year, \$1 accumulates to $e^{0.13}$ = \$1.139, so it is equivalent to

$$\boxed{13.9\% \text{ simple interest}}$$

6. (a) $\boxed{\dfrac{dy}{dx} = \dfrac{dy}{du}\cdot\dfrac{du}{dx}}$

(b) $\boxed{\dfrac{dy}{dx} = e^{(7-2x+x^3)}(-2+3x^2)}$

7. (a) Given $x^3 - 4xy + y^3 = 0$

$$3x^2 - 4\left(x\frac{dy}{dx} + y\right) + 3y^2\frac{dy}{dx} = 0$$

$$\frac{dy}{dx}(-4x+3y^2) + 3x^2 - 4y = 0$$

$$\boxed{\frac{dy}{dx} = \frac{-3x^2+4y}{3y^2-4x}}$$

(b) Given $x + ye^{xy} = 7$

$$1 + ye^{xy}\left(x\frac{dy}{dx} + y\right) + e^{xy}\frac{dy}{dx} = 0$$

$$\frac{dy}{dx}(xy\,e^{xy} + e^{xy}) = -y^2e^{xy} - 1$$

$$\boxed{\frac{dy}{dx} = -\frac{(y^2e^{xy}+1)}{e^{xy}(xy+1)}}$$

(c) Given $e^x \ln(xy+3) = 4$

$$e^x \frac{1}{xy+3}\left(x\frac{dy}{dx} + y\right) + \ln(xy+3)e^x = 0$$

$$\frac{dy}{dx}\left(\frac{xe^x}{xy+3}\right) = -\ln(xy+3)e^x - \frac{ye^x}{xy+3}$$

$$\frac{dy}{dx} = \frac{-(xy+3)\ln(xy+3)}{x} - \frac{y}{x}$$

$$\boxed{\frac{dy}{dx} = \frac{-(xy+3)\ln(xy+3) - y}{x}}$$

8. A_0 = initial amount of C^{11}, t = time.
$A(t) = A_0 e^{-kt}$
Given $A(0) = 200$, $A(20.5) = 100$
First solve for k as follows:

$$100 = 200e^{-20.5k}$$

$$0.5 = e^{-20.5k}$$

$$\ln 0.5 = -20.5k$$

$$k = -\frac{\ln 0.5}{20.5} = 0.034$$

so $A(t) = 200e^{-0.034t}$.
Next we solve for t_* such that

$$40 = 200e^{-0.034t_*}$$

Thus

$$0.2 = e^{-0.034t_*}$$

$$\boxed{t_* = -\frac{\ln 0.2}{0.034} = 47.3 \text{ min}}$$

9. $x^2 - 2xy - y = 3$. Differentiating with respect to t gives

$$2x\frac{dx}{dt} - 2y\frac{dx}{dt} - 2x\frac{dy}{dt} - \frac{dy}{dt} = 0$$

Thus

$$(2x+1)\frac{dy}{dt} = (2x - 2y)\frac{dx}{dt}$$

When $x = 5$, $y = 2$ and $\frac{dx}{dt} = 0.5$,

$$\left.\frac{dy}{dt}\right|_{\substack{x=5\\y=2}} = \frac{6\cdot(0.5)}{11} = \frac{3}{11} = 0.27 \text{ parts per million per year}$$

$$\boxed{0.027 \text{ parts per million per year}}$$

10. Let t be the number of years since 1962.

$$P(0) = 10{,}000,\ P(28) = ?$$

$$P(13) = 15{,}000$$

$$P(t) = P_0 e^{kt}$$

$$15{,}000 = 10{,}000e^{13k}$$

$$\frac{15{,}000}{10{,}000} = e^{13k}$$

$$1.5 = e^{13k}$$

$$\ln 1.5 = 13k$$

$$k = \frac{\ln 1.5}{13} = 0.0312$$

(a) $P(28) = 10{,}000e^{(0.0312)(28)} = \boxed{23{,}955}$

(b) $P(t) = 18{,}000$, $t_* = ?$
$18{,}000 = 10{,}000e^{0.0312t_*}$
$1.8 = e^{-0.0312t_*}$

$$t_* = -\frac{\ln 1.8}{0.0312}$$

$$t_* = 18.8 \approx \text{between 18th and 19th years after 1962,}$$

or

$$\boxed{\text{in 1980}}$$

CHAPTER 5 (page 241)

1. (a) $$\int_1^4 (x - x^2)\,dx \sim \frac{4-1}{3}[(2-2^2) + (3-3^2) + (4-4^2)]$$

$$\sim 1\cdot[-2 - 6 - 12]$$

$$\sim \boxed{-20}$$

(b) $$\int_0^1 \frac{1}{(x+1)^3}\,dx \sim \frac{1-0}{3}\left[\frac{1}{(1/3+1)^3} + \frac{1}{(2/3+1)^3} + \frac{1}{(1+1)^3}\right]$$

$$\sim \frac{1}{3}\left[\frac{27}{64} + \frac{27}{125} + \frac{1}{8}\right]$$

$$\sim (0.333)(0.763)$$

$$\sim \boxed{0.254}$$

2. (a) $$\sum_{i=3}^{i=7}(i + i^2) = (3+9) + (4+16) + (5+25) + (6+36) + (7+49)$$

$$= 12 + 20 + 30 + 42 + 56 = \boxed{160}$$

(b) $\sum_{i=0}^{i=4} (1 + (-1)^i) = (1 + 1) + (1 - 1) + (1 + 1) + (1 - 1) + (1 + 1)$

$= \boxed{6}$

(c) $\sum_{i=1}^{4} \frac{i-1}{i} = (0) + \left(\frac{1}{2}\right) + \left(\frac{2}{3}\right) + \left(\frac{3}{4}\right)$

$= \frac{6}{12} + \frac{8}{12} + \frac{9}{12} = \boxed{\frac{23}{12}}$

3. In each of parts (a)–(c), the area is the shaded region pictured.

(a)

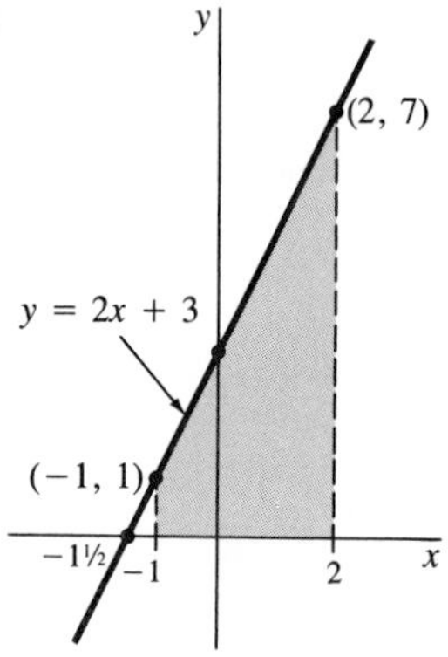

$$\int_{-1}^{2} (2x + 3)\, dx$$

$A = (\tfrac{1}{2}h_1b_1) - (\tfrac{1}{2}h_2b_2)$
$= \tfrac{1}{2}(7 \cdot \tfrac{7}{2}) - \tfrac{1}{2}(1 \cdot \tfrac{1}{2})$
$= \tfrac{49}{4} - \tfrac{1}{4}$
$= \boxed{12}$

(b)

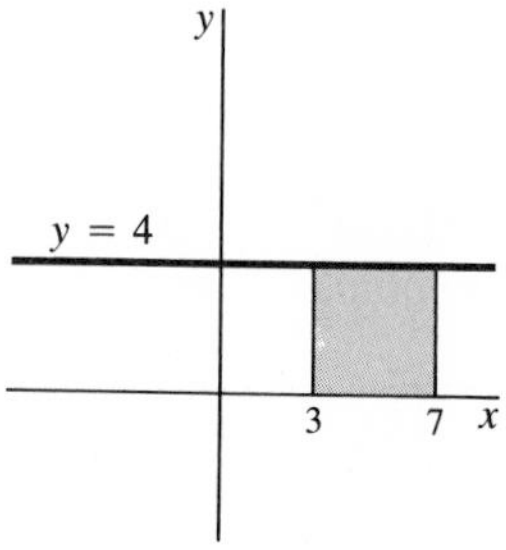

$$\int_{3}^{7} 4\, dx$$

$A = lw$
$= 4 \cdot 4$
$= \boxed{16}$

(c)

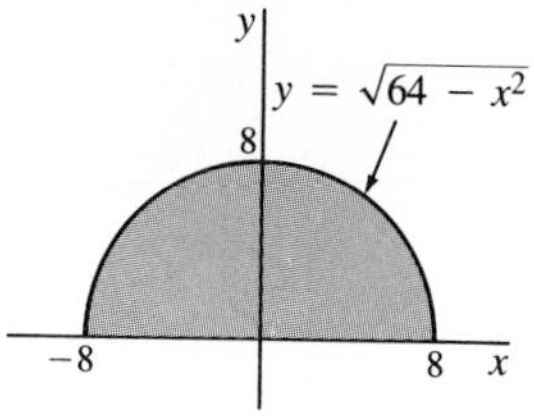

$$\int_{-8}^{8} \sqrt{64 - x^2}\, dx$$

$x^2 + y^2 = 64$

$A = \frac{1}{2}(\pi r^2)$
$= \frac{1}{2}(\pi 64)$
$= \boxed{32\pi}$

4. First, find intersections with x-axis.

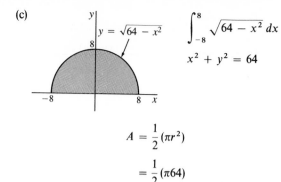

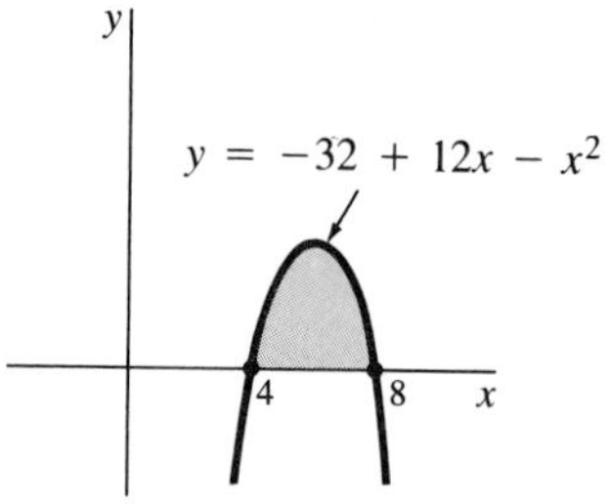

$0 = -32 + 12x - x^2$
$= x^2 - 12x + 32$
$= (x - 4)(x - 8)$

so $x = 4, 8$

Then

$$A = \int_{4}^{8} (-32 + 12x - x^2)\, dx$$

$= -32x + 6x^2 - \left.\frac{x^3}{3}\right|_4^8$

$= -32(8) + 6(8)^2 - (8)^3/3$
$-(-32(4) + 6(4)^2 - (4)^3/3)$

$= \boxed{10\tfrac{2}{3}}$

5. (a) $\int xe^{-x^2}\, dx$. Let $u = -x^2$. Then $du = -2x\, dx$, yielding

$$-\frac{1}{2}\int e^u\, du = -\frac{e^u}{2} + c$$

or, in terms of x,

$$\boxed{-\frac{1}{2}e^{-x^2} + c}$$

(b) $\int \frac{x^2}{4+x^3}\,dx$. Let $u = 4 + x^3$. Then $du = 3x^2\,dx$ giving

$$\left(\frac{1}{3}\right)\int \frac{du}{u} = \left(\frac{1}{3}\right)\ln|u| + c$$

or, in terms of x,

$$\boxed{\left(\tfrac{1}{3}\right)\ln|4 + x^3| + c}$$

(c) $$\int\left(\frac{5}{x^2} + 6 + 3x\right)dx = \boxed{-\frac{5}{x} + 6x + \frac{3}{2}x^2 + c}$$

6. (a) $$\int\left(3x^{-1/3} - x^{5/2} + \frac{2}{x}\right)dx$$

$$= \boxed{\frac{9}{2}x^{2/3} - \frac{2}{7}x^{7/2} + 2\ln|x| + c}$$

(b) $\int \frac{x}{(4-x^2)^{3/2}}\,dx$. Let $u = 4 - x^2$. Then $du = -2x\,dx$, yielding

$$-\tfrac{1}{2}\int \frac{du}{u^{3/2}} = u^{-1/2} + c$$

or, in terms of x,

$$\boxed{\frac{1}{\sqrt{4-x^2}} + c}$$

(c) $$\int 6dx = \boxed{6x + c}$$

7. *Method 1*: $f'''(x) = 0$ implies that $f''(x) = a$, constant, Therefore $f'(x) = ax + b$ for some constant b, and $f(x) = ax^2/2 + bx + c$ for some constant c. Substituting the three conditions

$$f'(2) = 8, \quad f(1) = 0, \quad f(-1) = 0$$

gives three simultaneous equations

$$2a + b = 8$$

$$a/2 + b + c = 0$$

$$a/2 - b + c = 0$$

which have the solution $a = 4, b = 0, c = -2$. Therefore $f(x) = 2x^2 - 2$.

Method 2: $f'''(x) = 0$ for all x tells us that $f(x)$ is a polynomial of degree 2. Because $f(1) = 0$ and $f(-1) = 0$, we must have

$$f(x) = c(x-1)(x+1) = cx^2 - c$$

for some constant c. But $f'(x) = 2cx$ and $f'(2) = 8$ gives $4c = 8$, so $c = 2$ and

$$\boxed{f(x) = 2x^2 - 2}$$

8. $C'(x) = 3x + 2\sqrt{x+1} = 3x + 2(x+1)^{\frac{1}{2}}$, $C(0) = 1500$. Thus

$$\boxed{C(x) = \tfrac{3}{2}x^2 + \tfrac{4}{3}(x+1)^{\frac{3}{2}} + c}$$

for some constant c. Substituting $x = 0$ and setting $C(0) = 1500$, we find $c = \frac{4496}{3}$. Thus

$$\boxed{C(x) = \tfrac{3}{2}x^2 + \tfrac{4}{3}(x+1)^{\frac{3}{2}} + \tfrac{4496}{3}}$$

9. (a) $y = \frac{1}{x}$, $1 \le x \le 3$

$$\boxed{\int_1^3 \frac{1}{x}\,dx} = \Big[\ln|x|\Big]_{x=1}^{x=3} = \ln|3| - \ln|1| = \ln 3$$

(b) $$\ln 3 = \int_1^3 \frac{1}{x}\,dx \sim \frac{3-1}{4}\left[\frac{1}{1} + \frac{1}{1.5} + \frac{1}{2} + \frac{1}{2.5}\right]$$

$$\sim 0.5[1 + 0.667 + 0.5 + 0.4]$$

$$\sim 0.5[2.567]$$

$$\sim \boxed{1.283}$$

10. We are given that $v(0) = 96$ ft/sec and that $s(0) = 7$ ft. From physics,

$$a = -32 \text{ ft/sec}^2$$

Integrating with respect to time yields

$$v(t) = \int a\,dt = \int -32\,dt$$

$$= -32t + C$$

Then $96 = v(0) = -32\cdot 0 + C$, so $C = 96$ and $v(t) = -32t + 96$. Integrating again, we get

$$s = \int v\,dt = \int(-32t + 96)\,dt$$

$$= -16t^2 + 96t + K$$

Now $7 = s(0) = -16(0)^2 + 96(0) + K$, so $K = 7$ and

$$s(t) = -16t^2 + 96t + 7$$

Finally, at time $t = 2$, the height of the ball above the ground will be

$$s(2) = -16(2)^2 + 96(2) + 7$$

$$= -64 + 192 + 7 = \boxed{135 \text{ ft}}$$

1. First find points of intersection of the two curves by equating

$$9 - x^2 = x^2 - 4x - 21$$

We obtain $2x^2 - 4x - 30 = 0$ or

$$x^2 - 2x - 15 = 0$$

$$(x - 5)(x + 3) = 0$$

which gives $x = 5$ and $x = -3$. Thus the desired area is

$$\int_{-3}^{5} (9 - x^2) - (x^2 - 4x - 21)\,dx$$

$$= \int_{-3}^{5} 30 + 4x - 2x^2\,dx$$

$$= \left(30x + 2x^2 - \left(\frac{2}{3}\right)x^3\right)\Bigg|_{-3}^{5} = \boxed{\frac{512}{3}}$$

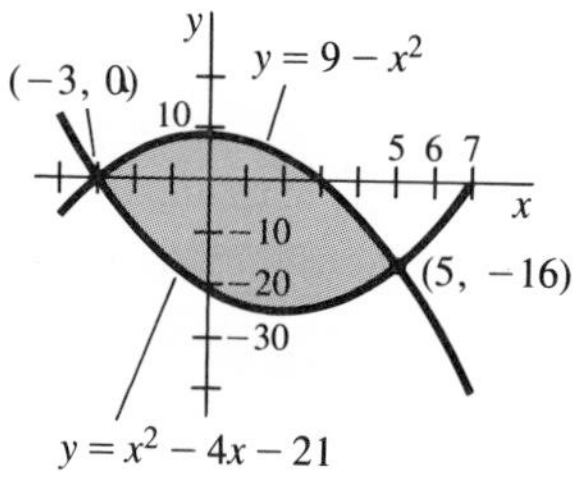

2. (a) Average $= \dfrac{1}{3 - 2}\displaystyle\int_2^3 e^{-0.41t}\,dt$

$$= -\frac{1}{0.41}(e^{-1.23} - e^{-0.82})$$

$$= -\frac{1}{0.41}(0.2923 - 0.4404) \approx \boxed{0.36}$$

(b) Let initial dose $= I_0$. Then we must solve

$$I_0\left(\frac{1}{8 - 5}\right)\int_5^8 e^{-0.41t}\,dt = 0.7$$

Now

$$\frac{1}{8-5}\int_5^8 e^{-0.41t}\,dt = \frac{1}{3(-0.41)}[e^{-0.41(8)} - e^{-0.41(5)}]$$

$$= \frac{-1}{1.23}[e^{-3.28} - e^{-2.05}]$$

$$= -\frac{1}{1.23}[0.0376 - 0.1287]$$

$$\approx 0.074$$

Thus $\quad \boxed{I_0 = \dfrac{0.7}{0.074} = 9.46 \text{ grams}}$

3. (a) $\bar{f} = \dfrac{1}{(4 - 1)}\displaystyle\int_1^4 \frac{x^2}{x^3 + 4}\,dx$

$$= \frac{1}{3}\left[\frac{1}{3}\ln(x^3 + 4)\right]\Bigg|_1^4$$

$$= \frac{1}{9}[\ln 68 - \ln 5]$$

$$= \frac{1}{9}[2.61]$$

$$= \boxed{0.290}$$

(b) $\bar{f} = \dfrac{1}{e - 2}\displaystyle\int_2^e \left(2x^2e^{-x^3} + \frac{4}{x}\right)dx$

$$= \frac{1}{e - 2}\left[\left(-\frac{2}{3}\right)e^{-x^3} + 4\ln x\right]\Bigg|_2^e$$

$$= \frac{1}{e - 2}\left[-\frac{2}{3}e^{-e^3} + 4\ln e + \frac{2}{3}e^{-8} - 4\ln 2\right]$$

$$\approx \boxed{1.7091}$$

4. (a) The market equilibrium (x_*, p_*) must satisfy simultaneously

$$p_* = -4x_* + 39$$

$$p_* = \frac{x_*^2}{2} + 15$$

Thus $\quad -4x_* + 39 = \dfrac{x_*^2}{2} + 15$

or $\quad x_*^2 + 8x_* - 48 = 0$

yielding

$$(x_* - 4)(x_* + 12) = 0$$

so $\quad x_* = 4 \text{ or } -12$

Discarding the negative root we obtain

$$\boxed{(x_*, p_*) = (4, 23)}$$

(b) The consumers' surplus is given by

$$\text{CS} = \int_0^4 [(-4x + 39) - 23]\,dx$$

$$= \int_0^4 (-4x + 16)\,dx$$

$$= (-2x^2 + 16x)\Bigg|_0^4 = \boxed{32}$$

The producers' surplus is given by

$$PS = \int_0^4 \left[23 - \left(\frac{x^2}{2} + 15\right)\right] dx$$

$$= \int_0^4 \left(8 - \frac{x^2}{2}\right) dx$$

$$= \left(8x - \frac{x^3}{6}\right)\Bigg|_0^4 = \boxed{21\tfrac{1}{3}}$$

5. (a) We must find t_* for which

$$R'(t_*) = C'(t_*)$$

That is,

$$10 + 0.3t_* = 7 + 0.4t_* \quad \text{or} \quad 0.1t_* = 3$$

Thus $\qquad t_* = 30$

The business should close down after 30 months.

(b) $\int_0^{30} (R'(t) - C'(t))\, dt$ = total profit, so

$$\text{Total profit} = \int_0^{30} [(10 + 0.3t) - (7 + 0.4t)]\, dt$$

$$= \int_0^{30} (3 - 0.1t)\, dt$$

$$= (3t - 0.05t^2)\Big|_0^{30} = 3(30) - 0.05(30)^2$$

$$= 45$$

The total profit is $45,000

6. A wafer-like cross section of thickness Δx has volume

$$\Delta V \approx \pi(2e^x + 3)^2\, \Delta x$$

so total volume

$$V = \Sigma \Delta V \approx \Sigma \pi (2e^x + 3)^2\, \Delta x$$

and in limit as $\Delta x \to 0$

$$V = \int_0^1 \pi(2e^x + 3)^2\, dx$$

$$= \pi \int_0^1 (4e^{2x} + 12e^x + 9)\, dx$$

$$= \pi(2e^{2x} + 12e^x + 9x)\Big|_0^1$$

$$= \pi[(2e^2 + 12e + 9) - (2 + 12)]$$

$$= \boxed{\pi(2e^2 + 12e - 5)}$$

7. In the first 3 years, the salvage value changes

$$\int_0^3 -3200(7 + 6t - t^2)\, dt \quad \text{dollars}$$

This is equal to

$$-3200\left(7t + 3t^2 - \frac{t^3}{3}\right)\Bigg|_0^3 = -3200(21 + 27 - 9)$$

$$= -124{,}800.$$

The minus sign indicates a decrease, so the salvage value drops

$124,800

The change in the second 3 years is

$$\int_3^6 -3200(7 + 6t - t^2)\, dt \quad \text{dollars}$$

This is equal to

$$-3200(7t + 3t^2 - t^3/3)\Big|_3^6 = -3200[(78) - (39)]$$

$$= -3200(39) = -124{,}800$$

Thus the salvage value drops, in the second 3 years,

$124,800

8. Acceleration due to gravity = -32 ft/sec, so

$$v'(t) = -32$$

$$s'(t) = v(t) = -32t + v_0$$

$$s(t) = -16t^2 + v_0 t + s_0$$

But $s_0 = 6$, which gives

$$s(t) = -16t^2 + v_0 t + 6$$

The ball reaches its highest point when

$$v(t_*) = 0 \text{ or } -32t_* + v_0 = 0$$

that is, at $t_* = \frac{v_0}{32}$. Thus we must have

$$s(t_*) = 40$$

or $$-16\left(\frac{v_0}{32}\right)^2 + v_0\left(\frac{v_0}{32}\right) + 6 = 40$$

We solve this quadratic for v_0.

$$\frac{v_0^2}{64} = 34$$

so $$v_0^2 = 34 \cdot 64$$

and finally we obtain

$$\boxed{v_0 = 46.65 \text{ ft/sec}}$$

9. $$\frac{1}{3-0}\int_0^3 (1 + 12x - x^2)\,dx = \frac{1}{3}\left(x + 6x^2 - \frac{x^3}{3}\right)\Bigg|_0^3$$

$$= \frac{1}{3}(3 + 54 - 9) = 16$$

The average cash reserve is \$16,000.

10. The two curves intersect at x-values for which

$$x^2 = 4x - x^2$$

or

$$2x(x - 2) = 0$$

That is, they intersect at $x = 0$ and $x = 2$. Also the curve $y = 4x - x^2$ is above $y = x^2$ for this range of x's. By Equation (6.10), the volume must be

$$\pi\int_0^2 (4x - x^2)^2 - (x^2)^2\,dx$$

$$= \pi\int_0^2 (16x^2 - 8x^3 + x^4 - x^4)\,dx$$

$$= \pi\int_0^2 (16x^2 - 8x^3)\,dx$$

$$= \pi\left(\frac{16x^3}{3} - 2x^4\right)\Bigg|_0^2$$

$$= \boxed{\frac{32}{3}\pi}$$

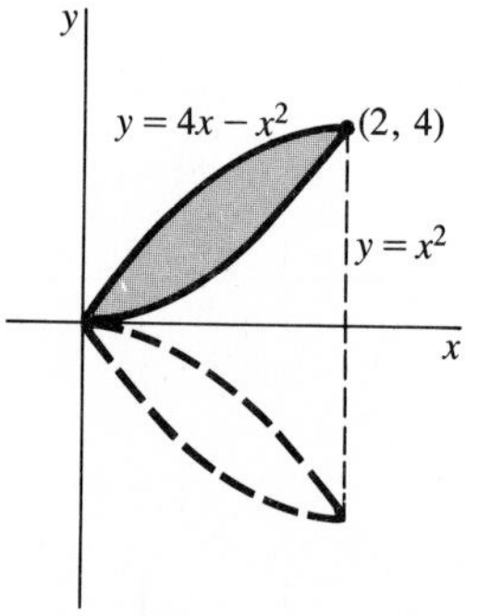

CHAPTER 7 (page 332)

1. (a) Let $u = x + 2$, $dv = e^{3x}\,dx$. Then $du = dx$, $v = \frac{1}{3}e^{3x}$ and

$$\int (x + 2)e^{3x}\,dx = \frac{x + 2}{3}e^{3x} - \frac{1}{3}\int e^{3x}\,dx$$

$$= \frac{x + 2}{3}e^{3x} - \frac{1}{9}e^{3x} + c$$

$$= \boxed{e^{3x}\left(\frac{3x + 5}{9}\right) + c}$$

(b) Making the change of variables $u = 4 + 3e^{-x}$ gives $du = -3e^{-x}\,dx$ and

$$\int \frac{e^{-x}}{\sqrt{4 + 3e^{-x}}}\,dx \to \left(-\frac{1}{3}\right)\int \frac{1}{\sqrt{u}}\,du = -\frac{2}{3}\sqrt{u} + c$$

or, in terms of x,

$$\boxed{-\frac{2}{3}\sqrt{4 + 3e^{-x}} + c}$$

(c) Let $u = \ln[(2x + 1)^2]$, $dv = x\,dx$. Then $du = \dfrac{4}{2x + 1}\,dx$, $v = \dfrac{x^2}{2}$, and

$$\int x\ln[(2x + 1)^2]\,dx$$

$$= x^2\ln(2x + 1) - \int \frac{2x^2}{2x + 1}\,dx$$

$$= x^2\ln(2x + 1) - \int\left((x - 1/2) + \frac{1/2}{2x + 1}\right)dx$$

$$= x^2\ln(2x + 1) - \frac{1}{4}\ln(2x + 1) - \frac{x^2}{2} + \frac{x}{2} + c$$

$$= \boxed{\left(x^2 - \frac{1}{4}\right)\ln(2x + 1) - \frac{x^2}{2} + \frac{x}{2} + c}$$

2. (a) Partition $[a, b]$ into n equal parts. Define $\Delta x = \dfrac{b - a}{n}$ and $x_i = a + i\Delta x$, $0 \le i \le n$. Let $y_i = f(x_i)$, $0 \le i \le n$. Then

$$\boxed{\int_a^b f(x)\,dx \approx \Delta x\left[\frac{y_0}{2} + y_1 + y_2 + \cdots + y_{n-1} + \frac{y_n}{2}\right]}$$

(b) We have

x_i	y_i
0	1.414
0.167	1.424
0.333	1.453
0.500	1.500
0.667	1.563
0.833	1.641
1	1.732

$$\Delta x = \frac{1 - 0}{6} = \frac{1}{6}$$

so

$$\int_0^1 \sqrt{2 + x^2}\,dx \sim \frac{1}{6}[0.707 + 1.424 + 1.453 + 1.500 + 1.563 + 1.641 + 0.866]$$

$$\sim \frac{1}{6}(9.154)$$

$$= \boxed{1.526}$$

3. (a) $\dfrac{x^2 + 2}{4 - x} = -x - 4 + \dfrac{18}{4 - x}$. Thus

$$\int \frac{x^2 + 2}{4 - x}\,dx = \boxed{-\frac{x^2}{2} - 4x - 18\ln|4 - x| + c}$$

(b) $\dfrac{1 - 2x}{(x - 2)(x - 3)} = \dfrac{3}{x - 2} - \dfrac{5}{x - 3}$. Thus

$$\int \frac{1 - 2x}{(x - 2)(x - 3)}\,dx = \boxed{3\ln|x - 2| - 5\ln|x - 3| + c}$$

(c) $\dfrac{x + 3}{x^2 + 6x} = \dfrac{1/2}{x} + \dfrac{1/2}{x + 6}$. So

$$\int \frac{x + 3}{x^2 + 6x}\,dx = \boxed{\frac{1}{2}\ln|x| + \frac{1}{2}\ln|x + 6| + c}$$

4. (a) Let $u = e^x$, $du = e^x\,dx$. Then

$$\int \frac{e^{3x} + 4e^x}{e^{2x} - 4}\,dx \to \int \frac{u^2 + 4}{u^2 - 4}\,du$$

Now $$\frac{u^2 + 4}{u^2 - 4} = 1 + \frac{-2}{u + 2} + \frac{2}{u - 2}$$

So

$$\int \frac{u^2 + 4}{u^2 - 4}\,du = u - 2\ln|u + 2| + 2\ln|u - 2| + c$$

Converting back to x gives

$$\boxed{e^x - 2\ln(e^x + 2) + 2\ln|e^x - 2| + c}$$

(b) Let $u = e^x + 2$, $du = e^x\,dx$. Then

$$\int e^x \ln(e^x + 2)\,dx \to \int \ln u\,du$$

which by integration by parts, as we saw in Example 3, equals $u\ln u - u + c$. Converting back to x now gives

$$\boxed{(e^x + 2)\ln(e^x + 2) - e^x + c}$$

5. (a) $\dfrac{dy}{dx} = (y + 3)e^{-x}$, so separating variables gives

$$\int \frac{dy}{y + 3} = \int e^{-x}\,dx$$

or $$\ln|y + 3| = -e^{-x} + c$$

Exponentiating both sides gives

$$|y + 3| = K\,e^{-e^{-x}}$$

Now when $x = 0$, $y(0) = -2$. Thus $1 = Ke^{-1}$, so $K = e$, and we obtain

$$|y + 3| = e^{1 - e^{-x}}$$

Assuming $y + 3 > 0$, we obtain $y + 3 = e^{1 - e^{-x}}$ and

$$\boxed{y = e^{1 - e^{-x}} - 3}$$

[The case $y + 3 \le 0$ is not allowable. It is inconsistent with the initial condition $y(0) = -2$.]

(b) Separating variables gives

$$\int \frac{y}{6 - y}\,dy = \int -\frac{2}{5}t\,dt$$

so that

$$-y - 6\ln|6 - y| = -\frac{t^2}{5} + c$$

Imposing the condition $y(5) = 5$ gives $-5 - 6\ln 1 = -5 + c$. Thus $c = 0$ and we obtain

$$\boxed{y + 6\ln|6 - y| = \frac{t^2}{5}}$$

6. (a) We make the change of variables $u = x + 2$, $du = dx$. Then

$$\int \frac{dx}{(x + 2)^2\sqrt{x + 3}} \to \int \frac{du}{u^2\sqrt{u + 1}}$$

This integral in u now appears as Equation 20 in Appendix B (with $a = b = 1$ and $n = 2$). Thus

$$\int \frac{du}{u^2\sqrt{u + 1}} = -\frac{\sqrt{u + 1}}{u} - \frac{1}{2}\int \frac{du}{u\sqrt{u + 1}}$$

Next we use Equation 19 of Appendix B to find that

$$\int \frac{du}{u\sqrt{u + 1}} = \ln\left|\frac{\sqrt{u + 1} - 1}{\sqrt{u + 1} + 1}\right| + c$$

Thus

$$\int \frac{du}{u^2\sqrt{u + 1}} = -\frac{\sqrt{u + 1}}{u} - \frac{1}{2}\ln\left|\frac{\sqrt{u + 1} - 1}{\sqrt{u + 1} + 1}\right| + c$$

Converting back to x we obtain

$$\boxed{-\frac{\sqrt{x+3}}{x+2} - \frac{1}{2}\ln\left|\frac{\sqrt{x+3}-1}{\sqrt{x+3}+1}\right| + c}$$

(b) Let $u = 2x$, $du = 2\,dx$. Then

$$\int \frac{dx}{(9+4x^2)^{3/2}} \rightarrow \frac{1}{2}\int \frac{du}{(9+u^2)^{3/2}}$$

This integral in u now is given in Equation 34 of the integral table (with $a = 3$). We have

$$\frac{1}{2}\int \frac{du}{(9+u^2)^{3/2}} = \frac{1}{2}\left(\frac{u}{9\sqrt{9+u^2}}\right) + c$$

Converting back to x gives

$$\boxed{\frac{x}{9\sqrt{9+4x^2}} + c}$$

7. The equation with $k = 2$, and $c = 5$ is $\frac{dx}{dt} = 2(5 - x)$.

Separating variables gives

$$\int \frac{dx}{5-x} = \int 2\,dt$$

so

$$-\ln|5 - x| = 2t + m$$

for some constant m. The condition $x(0) = 0$ yields

$$-\ln 5 = m$$

Thus

$$\ln|5 - x| = -2t + \ln 5$$

Exponentiating yields

$$|5 - x| = 5e^{-2t}$$

If we assume $5 - x \geq 0$ we obtain $5 - x = 5e^{-2t}$, or

$$\boxed{x(t) = 5(1 - e^{-2t})}$$

(The situation $5 - x < 0$ would lead to $x = 5 + 5e^{-2t}$, which is inconsistent with the initial condition $x(0) = 0$).

8. The area is given by

$$\int_0^\infty (x+3)e^{-x}\,dx = \lim_{b\to\infty}\left[-(x+4)e^{-x}\Big|_0^b\right]$$

$$= \lim_{b\to\infty}[-(b+4)e^{-b} + 4]$$

$$= \boxed{4}$$

9. (a) Let $u = x^2$, $dv = (3x^2 - 2)^{1/2}x\,dx$. Then

$$du = 2x\,dx \qquad v = \frac{1}{9}(3x^2 - 2)^{3/2}$$

Thus

$$\int x^3(3x^2 - 2)^{1/2}\,dx$$

$$= \frac{x^2}{9}(3x^2 - 2)^{3/2} - \frac{2}{9}\int x(3x^2 - 2)^{3/2}\,dx$$

$$= \boxed{\frac{x^2}{9}(3x^2 - 2)^{3/2} - \frac{2}{135}(3x^2 - 2)^{5/2} + c}$$

(b) Let $u = 3x^2 - 2$, $du = 6x\,dx$. Then

$$\int x^3(3x^2 - 2)^{1/2}\,dx \rightarrow \int\left(\frac{u+2}{18}\right)u^{1/2}\,du$$

$$= \int\left(\frac{u^{3/2}}{18} + \frac{1}{9}u^{1/2}\right)du$$

$$= \frac{1}{45}u^{5/2} + \frac{2}{27}u^{3/2} + c$$

Converting back to x gives

$$\boxed{\frac{1}{45}(3x^2 - 2)^{3/2} + \frac{2}{27}(3x^2 - 2)^{3/2} + c}$$

The answers to parts (a) and (b) are equivalent (as they should be), but it takes a bit of algebraic manipulation to show it.

10. The amount remaining at time t is given by $A(t) = A_0e^{-kt}$. We are given that the half-life is 11 days. This determines k via the formula

$$k = \frac{\ln 2}{11} = 0.063$$

Also we are given that $A(20) = 6$. That is, we know that

$$A_0e^{-(0.063)20} = 6$$

This equation determines $A_0 = 21.15$. Thus we have explicitly

$$A(t) = 21.15e^{-0.063t}$$

(a) At time $t = 40$

$$\boxed{A(40) = 1.70 \text{ g}}$$

(b) After 5 days a fraction $e^{-(0.063)5}$ remains, so the answer must be

$$1 - e^{-(0.063)5} = 0.270$$

$$\boxed{27\% \text{ disappears every 5 days}}$$

CHAPTER 8 (page 389)

1. $g(x, y) = \dfrac{x^2}{y} e^{-3x}$

$$\left.\frac{\partial g}{\partial x}\right|_{(-1,-2)} = \left.\left(\frac{2x}{y} e^{-3x} - \frac{3x^2}{y} e^{-3x}\right)\right|_{(-1,-2)}$$

$$= \frac{2(-1)}{-2} e^{-3(-1)} - \frac{3(-1)^2}{-2} e^{-3(-1)}$$

$$= e^3 + \frac{3}{2} e^3$$

$$= \boxed{\frac{5}{2} e^3}$$

$$\left.\frac{\partial g}{\partial y}\right|_{(-1,-2)} = \left.-\frac{x^2}{y^2} e^{-3x}\right|_{(-1,-2)}$$

$$= \frac{-(-1)^2}{(-2)^2} e^{-3(-1)}$$

$$= \boxed{-\frac{1}{4} e^3}$$

$$\frac{\partial^2 g}{\partial x \partial y} = \frac{\partial}{\partial x}\left(\frac{\partial g}{\partial y}\right) = \frac{\partial}{\partial x}\left[-\frac{x^2}{y^2} e^{-3x}\right]$$

$$= \frac{-2x}{y^2} e^{-3x} + \frac{3x^2}{y^2} e^{-3x}$$

$$= \frac{xe^{-3x}}{y^2}(-2 + 3x)$$

$$\left.\frac{\partial^2 g}{\partial x \partial y}\right|_{(-1,-2)} = \left.\frac{xe^{-3x}}{y^2}(-2 + 3x)\right|_{(-1,-2)}$$

$$= \frac{-1e^{-3(-1)}}{(-2)^2}(-2 + 3(-1))$$

$$= -\frac{e^3}{4}(-5)$$

$$= \boxed{\frac{5e^3}{4}}$$

2. $z = f(L, K) = 2L^{0.7} \cdot K^{0.3}$

(a) $\boxed{\dfrac{\partial z}{\partial L} = 1.4L^{-0.3}K^{0.3} = \text{marginal productivity of labor}}$

(b) $\boxed{\dfrac{\partial z}{\partial K} = 0.6L^{0.7}K^{-0.7} = \text{marginal productivity of capital}}$

3. $f(x, y) = x^3 + y^3 - 9xy + 27$. Compute first partials.

$$f_x = 3x^2 - 9y \qquad f_y = 3y^2 - 9x$$

Set equal to 0 to solve for critical points.

$$3x^2 - 9y = 0 \Rightarrow x^2 = 3y$$

$$3y^2 - 9x = 0 \Rightarrow y^2 = 3x$$

$$\text{so} \quad x^4 = 9y^2 = 9(3x) = 27x$$

$$\text{or} \quad x^4 - 27x = 0$$

Factoring yields $\quad x(x^3 - 27) = 0$

Thus $x = 0$, $x = 3$ are the only real solutions.

$\boxed{\text{The critical points are } (0, 0) \text{ and } (3, 3).}$

Now $f_{xx} = 6x$, $f_{yy} = 6y$ and $f_{xy} = -9$.

At (3, 3) we have

$$f_{xx}(3, 3) = 18 > 0$$

$$f_{xy}(3, 3) = 18 > 0$$

$$f_{xy}(3, 3) = -9$$

and $D = f_{xx}f_{yy} - f_{xy}^2 = 18 \cdot 18 - (-9)^2 = 243 > 0$. Thus

$\boxed{(3, 3) \text{ is a minimum.}}$

At (0, 0)

$$f_{xx}(0, 0) = 0$$

$$f_{xy}(0, 0) = 0$$

$$f_{xy}(0, 0) = -9$$

and $D = f_{xx}f_{yy} - f_{xy}^2 = -(-9)^2 = -81 < 0$. Thus

$\boxed{(0, 0) \text{ is a saddle point.}}$

4. We must maximize

$$P(x, y) = x(64 - 4x) + y(77 - 3y) - (3x^2 + 8y^2 + 11xy)$$

$$= 64x - 7x^2 + 77y - 11y^2 - 11xy$$

Thus

$$\frac{\partial P}{\partial x} = 64 - 14x - 11y$$

$$\frac{\partial P}{\partial y} = 77 - 11x - 22y$$

Setting these derivatives equal to 0, we compute the critical point as a solution of

$$14x + 11y = 64$$
$$11x + 22y = 77$$

The solution $x = 3$, $y = 2$ yields (3, 2) as the only critical point. Because $\dfrac{\partial^2 P}{\partial x^2} = -14 < 0$, $\dfrac{\partial^2 P}{\partial y^2} = -22 < 0$ and $\dfrac{\partial^2 P}{\partial x \partial y} = -11$.

$$D = \frac{\partial^2 P}{\partial x^2}\frac{\partial^2 P}{\partial y^2} - \left(\frac{\partial^2 P}{\partial x \partial y}\right)^2 = (-14)(-22) - (-11)^2$$
$$= 187 > 0$$

We are assured that

$x = 3$, $y = 2$ is the production level maximizing the profit.

5. The "best-fit" line $y = mx + b$ has m and b satisfying

$$m(\Sigma x_i^2) + b\Sigma x_i = \Sigma x_i y_i$$
$$m\Sigma x_i + bn = \Sigma y_i$$

For the current data, these equations yield

$$5.01m + 1.1b = 22.41$$
$$1.1m + 4b = -1.3$$

with solution $m = 4.84$, $b = -1.66$.

The "best-fit" line is $y = 4.84x - 1.66$.

6. (a) Using the same formula as in the previous solution, we obtain (with $x_1 = 0$, $x_2 = 1$, $x_3 = 2$)

$$5m + 3b = 7.17$$
$$3m + 3b = 6.85$$

with solution $m = 0.16$, $b = 2.12$, so

Regression line is $y = 0.16x + 2.12$.

(b) To predict the 1990 share earnings, use $y = 0.16x + 2.12$ with $x = 6$ to obtain

$0.16(6) + 2.12 =$ \$3.08 per share

7. We minimize $x^2 + y^2$ subject to the constraint $y - 7x + 1 = 0$. Set $F(x, y, \lambda) = x^2 + y^2 + \lambda(y - 7x + 1)$. Then solve

$$F_x = 2x - 7\lambda = 0$$
$$F_y = 2y + \lambda = 0$$
$$F_\lambda = y - 7x + 1 = 0$$

We obtain $x = \frac{7}{2}\lambda$, $y = -\lambda/2$.

$$y - 7x + 1 = 0 \Rightarrow +\lambda/2 + \frac{49\lambda}{2} = 1$$

or $25\lambda = 1$. Thus $\lambda = \frac{1}{25}$, $x = \frac{7}{50}$, $y = -\frac{1}{50}$.

The point closest to the origin on the line $y = 7x - 1$ is $(\frac{7}{50}, -\frac{1}{50})$.

8. (a) $$\int_0^2 \int_{x^3}^8 y^{-\frac{1}{3}}\, dy\, dx = \int_0^2 \left(\frac{3}{2} y^{2/3}\Big|_{x^3}^8\right) dx$$
$$= \frac{3}{2}\int_0^2 (4 - x^2)\, dx$$
$$= \frac{3}{2}\left(4x - \frac{x^3}{3}\right)\Big|_0^2 = \frac{3}{2}\left(8 - \frac{8}{3}\right) = 8$$

$$\int_0^8 \int_0^{y^{\frac{1}{3}}} y^{-\frac{1}{3}}\, dx\, dy = \int_0^8 \left(y^{-\frac{1}{3}} x\Big|_0^{y^{\frac{1}{3}}}\right) dy$$
$$= \int_0^8 dy = \boxed{8}$$

(b) $$\int_0^1 \int_x^{2-x} 3\, dy\, dx = \int_0^1 \left(3y\Big|_x^{2-x}\right) dx$$
$$= 3\int_0^1 (2 - 2x)\, dx$$
$$= 3(2x - x^2)\Big|_0^1 = 3$$

$$\int_0^1 \int_0^2 3\, dx\, dy + \int_1^2 \int_0^{2-y} 3\, dx\, dy$$
$$= \int_0^1 \left(3x\Big|_0^y\right) dy + \int_1^2 \left(3x\Big|_0^{2-y}\right) dy$$
$$= 3\int_0^1 y\, dy + 3\int_1^2 (2 - y)\, dy = \boxed{3}$$

(c) $$\int_0^1 \int_{e^y}^e \frac{1}{x}\, dx\, dy = \int_0^1 \left(\ln x\Big|_{e^y}^e\right) dy = \int_0^1 (1 - y)\, dy$$
$$= \left(y - \frac{y^2}{2}\right)\Big|_0^1 = \frac{1}{2}$$

$$\int_1^e \int_0^{\ln x} \frac{1}{x}\, dy\, dx = \int_1^e \left(\frac{y}{x}\Big|_0^{\ln x}\right) dx = \int_1^e \frac{\ln x}{x}\, dx$$
$$= \frac{(\ln x)^2}{2}\Big|_1^e = \boxed{\frac{1}{2}}$$

9. We must maximize $zp_z - xp_x - yp_y$ or $8z - 4x - 3y$, subject to the constraint $z - 8 + \dfrac{2}{x^{3/2}} + \dfrac{1}{y^{1/2}} = 0$. We introduce

$$F(x, y, z, \lambda) = 8z - 4x - 3y + \lambda\left(z - 8 + \frac{2}{x^{3/2}} + \frac{1}{y^{1/2}}\right)$$

and then solve

$$F_x = -4 - \frac{3\lambda}{x^{5/2}} = 0$$

$$F_y = -3 - \tfrac{1}{2}\frac{\lambda}{y^{3/2}} = 0$$

$$F_z = 8 + \lambda = 0$$

$$F_\lambda = z - 8 + \frac{2}{x^{3/2}} + \frac{1}{y^{1/2}} = 0$$

From the foregoing we obtain

$$\lambda = -8$$

$$-4 - \frac{3(-8)}{x^{5/2}} = 0$$

$$\text{so} \quad 4x^{5/2} = 24$$

$$x^{5/2} = 6$$

$$x = (36)^{1/5} \approx 2.048$$

$$-3 - \tfrac{1}{2}\left(\frac{-8}{y^{3/2}}\right) = 0$$

$$\text{so} \quad 3y^{3/2} = 4$$

$$y^{3/2} = \frac{4}{3}$$

$$y = (\tfrac{4}{3})^{2/3} \approx 1.211$$

and $$z = 8 - \frac{2}{(2.048)^{3/2}} - \frac{1}{(1.211)^{1/2}}$$

$$\approx 6.409$$

The optimal production schedule is $x = 2.048$, $y = 1.211$, $z = 6.409$.

10. Let the height of this rectangle be h, and let the length of the base be x. Then $\left(\frac{x}{2}\right)^2 + h^2 = 16$, and we must

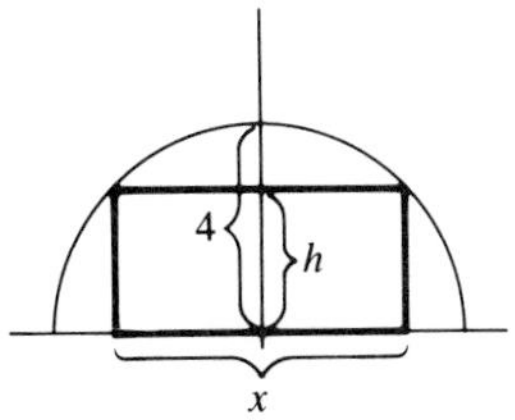

maximize xh subject to this constraint. Letting

$$F(x, h, \lambda) = xh + \lambda\left(\frac{x^2}{4} + h^2 - 16\right)$$

we solve

$$F_x = h + \frac{\lambda x}{2} = 0$$

$$F_h = x + 2\lambda h = 0$$

$$F_\lambda = \frac{x^2}{4} + h^2 - 16 = 0$$

which yields

$$h = -\frac{\lambda x}{2} \qquad x = -2\lambda h$$

so $$h = -\frac{\lambda}{2}(-2\lambda h) = \lambda^2 h \text{ or } h(1 - \lambda^2) = 0$$

Thus either $h = 0$ or $\lambda = \pm 1$. We exclude $h = 0$ as not corresponding to a rectangle. Also $\lambda = +1$ will give $h = -x/2$, which is impossible, because we must have $x > 0$ and $h > 0$ for a rectangle. Thus we must have $\lambda = -1$ and $h = \frac{x}{2}$. Then $\frac{x^2}{4} + h^2 = 16$ yields

$$2x^2 = 64 \Rightarrow x^2 = 32$$

so $x = 4\sqrt{2}$ and $h = 2\sqrt{2}$.

The base has length $4\sqrt{2}$ and the height length $2\sqrt{2}$.

CHAPTER 9 (page 436)

1. (a) $\dfrac{dy}{dx} = \frac{1}{4}(\sin 2x)^{-3/4}(\cos 2x)2 = \boxed{\frac{1}{2}(\sin 2x)^{-3/4}(\cos 2x)}$

(b) $\dfrac{dy}{dx} = \dfrac{x^3 \cos x - 3x^2 \sin x}{x^6} = \boxed{\dfrac{x \cos x - 3 \sin x}{x^4}}$

(c) $\dfrac{dy}{dx} = (2 \tan 3x)(\sec^2 3x)(3) = \boxed{6 \tan 3x \sec^2 3x}$

2. (a) $\cos 225^\circ = \cos \dfrac{5\pi}{4} = \boxed{-\dfrac{\sqrt{2}}{2}}$

(b) $\sin^2 \dfrac{\pi}{3} = \left(\dfrac{\sqrt{3}}{2}\right)^2 = \boxed{\dfrac{3}{4}}$

(c) $\sec\left(-\dfrac{5\pi}{6}\right) = \dfrac{1}{\cos\left(-\dfrac{5\pi}{6}\right)} = \boxed{-\dfrac{2\sqrt{3}}{3}}$

(d) $\ln\left(\sin \frac{5\pi}{2}\right) = \ln 1 = \boxed{0}$

3. (a) Use integration by parts. Let

$$u = x + 2 \qquad dv = \sin 2x\, dx$$
$$du = dx \qquad v = -\tfrac{1}{2}\cos 2x$$

Then

$$\int (x+2)\sin 2x\, dx = (x+2)(-\tfrac{1}{2}\cos 2x) + \tfrac{1}{2}\int \cos 2x\, dx$$
$$= \boxed{(-\tfrac{1}{2}\cos 2x)(x+2) + \tfrac{1}{4}\sin 2x + c}$$

(b) To evaluate $\int \cos^2 \frac{x}{3}\, dx$, use the fact that

$\cos^2 \frac{x}{3} = \dfrac{1 + \cos \frac{2x}{3}}{2}$. Thus

$$\int \cos^2 \frac{x}{3}\, dx = \int \frac{1 + \cos \frac{2x}{3}}{2}\, dx = \boxed{\frac{x}{2} + \frac{3}{4}\sin \frac{2x}{3} + c}$$

4. (a) $\lim\limits_{x\to 0} \dfrac{\sin 3x}{4x} = \lim\limits_{x\to 0} \dfrac{\sin 3x}{3x} \cdot \dfrac{3}{4} = \dfrac{3}{4} \lim\limits_{x\to 0} \dfrac{\sin 3x}{3x} = \boxed{\dfrac{3}{4}}$

(b) Let $x = z - \pi/2$. Then

$$\lim_{z\to\pi/2} \frac{\cos z}{z - \pi/2} = \lim_{x\to 0} \frac{\cos\left(x + \frac{\pi}{2}\right)}{x}$$
$$= \lim_{x\to 0} \frac{\cos x \cos \frac{\pi}{2} - \sin x \sin \frac{\pi}{2}}{x}$$
$$= -\lim_{x\to 0} \frac{\sin x}{x} = \boxed{-1}$$

(c) Does not exist

(d) $$\lim_{x\to 0} \frac{\cos 2x}{\cot x} = \lim_{x\to 0} \frac{\sin x}{\sin 2x \cos x}$$
$$= \lim_{x\to 0} \frac{\sin x}{x} \cdot \lim_{x\to 0} \frac{2x}{\sin 2x} \cdot \lim_{x\to 0} \frac{1}{2\cos x}$$
$$= \boxed{\frac{1}{2}}$$

5. $y = \sin 2x$
$y' = 2\cos 2x$
$y'' = -4\sin 2x$

Critical points are at $x = \dfrac{\pi}{4}$ and at $x = \dfrac{3\pi}{4}$

the solutions of $y' = 0$.

At $x = \frac{\pi}{4}$, $y''\left(\frac{\pi}{4}\right) = -4 < 0$, so

$\left(\frac{\pi}{4}, +1\right)$ is a maximum

At $x = \frac{3\pi}{4}$, $y''\left(\frac{3\pi}{4}\right) = +4 > 0$, so

$\left(\frac{3\pi}{4}, -1\right)$ is a minimum

In addition, by observing the second derivative, we see that

the curve is concave up on $\frac{\pi}{2} < x < \pi$ and concave down on $0 < x < \pi/2$, with $(\frac{\pi}{2}, 0)$ a point of inflection.

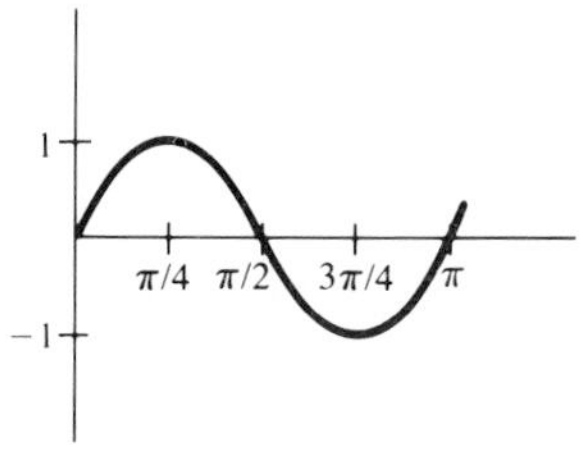

6. $T(t) = 65 + 10 \cos \frac{\pi}{24} t$, $0 \le t \le 12$ where $t\ 0 \leftrightarrow 12$ A.M. and $t = 12 \leftrightarrow 12$ P.M.

(a) $$\text{Average T} = \frac{1}{12}\int_0^{12} \left(65 + 10\cos \frac{\pi}{24} t\right) dt$$
$$= \frac{1}{12}\left[65t + 10 \cdot \frac{24}{\pi} \sin \frac{\pi t}{24}\right]_0^{12}$$
$$= \frac{1}{12}\left(65 \cdot 12 + \frac{240}{\pi} \sin \frac{\pi}{2}\right) - \frac{1}{12}\left(\frac{240}{\pi} \sin(0)\right)$$
$$= 65 + \frac{20}{\pi}$$

(b) At $t = 0$ the temperature is $65 + 10 = 75$, whereas at $t = 12$ the temperature is 65. These must represent the maximal and minimal temperatures, respectively, as $T'(t) \le 0$ over the time interval.

7. (a) $\boxed{f'(t) = (3t^2 + 1)\sec^2(t^3 + t)}$

(b) $f'(t) = 2\cos(t^2)(-\sin(t^2))2t$

$= \boxed{-4t\cos(t^2)\sin(t^2)}$

(c) $f'(t) = \dfrac{\sec t \tan t + \sec^2 t}{\sec t + \tan t} = \boxed{\sec t}$

8. (a) $f'(t) = \dfrac{1}{2\sqrt{t}}\cos\sqrt{t}$; thus $\boxed{f'\left(\dfrac{\pi^2}{4}\right) = 0}$

(b) $f'(t) = e^t\sec^2 t + e^t\tan t$; thus $\boxed{f'(0) = 1}$

(c) $$\int_{\pi/4}^{\pi/2}\sin^2 x\,dx = \int_{\pi/4}^{\pi/2}\frac{1-\cos 2x}{2}\,dx$$

$$= \left(\frac{x}{2} - \frac{\sin 2x}{4}\right)\Bigg|_{\pi/4}^{\pi/2}$$

$$= \left(\frac{\pi}{4} - \frac{\sin\pi}{4}\right) - \left(\pi/8 - \frac{\sin\pi/2}{4}\right)$$

$$= \boxed{\frac{\pi}{8} + \frac{1}{4}}$$

9. The two curves $y = \cos 2x$ and $y = \sin 2x$ intersect at $x = \frac{\pi}{8}$. In addition, for $-\frac{\pi}{8} < x < \frac{\pi}{8}$, the curve $y = \cos 2x$ lies above the curve $y = \sin 2x$. Thus the area enclosed is given by

$$\int_{-\pi/8}^{+\pi/8}(\cos 2x - \sin 2x)\,dx$$

$$= \frac{1}{2}\sin 2x\Bigg|_{-\pi/8}^{\pi/8} + \frac{1}{2}\cos 2x\Bigg|_{-\pi/8}^{\pi/8}$$

$$= \frac{1}{2}\left(\frac{\sqrt{2}}{2} + \frac{\sqrt{2}}{2}\right) + \frac{1}{2}\left(\frac{\sqrt{2}}{2} - \frac{\sqrt{2}}{2}\right)$$

$$= \boxed{\frac{\sqrt{2}}{2}}$$

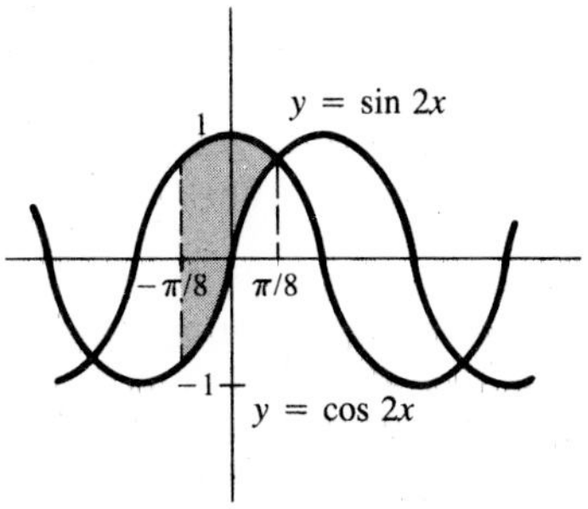

10. Because $y = x + \sin x$, $y' = 1 + \cos x$, so $y'(\pi/6) = 1 + \cos\frac{\pi}{6} = \frac{3}{2}$. The tangent line must therefore have slope $\frac{3}{2}$ and must pass through $(\frac{\pi}{6}, \frac{\pi}{6} + \frac{1}{2})$. An equation for the tangent line is obtained from

$$\frac{y - \left(\frac{\pi}{6} + \frac{1}{2}\right)}{x - \frac{\pi}{6}} = \frac{3}{2}$$

which yields

$$y = \frac{3}{2}\left(x - \frac{\pi}{6}\right) + \frac{\pi}{6} + \frac{1}{2}$$

or

$$\boxed{y = \frac{3}{2}x + \frac{6-\pi}{12}}$$

CHAPTER 10 (page 482)

1. $E(X) = (23.9)(0.08) + (29.1)(0.26) + (29.3)(0.32) + (29.5)(0.27) + (29.7)(0.07)$

$\boxed{E(X) = 28.90}$

$$\text{Var}(X) = (23.9 - 28.9)^2(0.08) + (29.1 - 28.9)^2(0.26) + (29.3 - 28.9)^2(0.32) + (29.5 - 28.9)^2(0.27) + (29.7 - 28.9)^2(0.07)$$

$= \boxed{2.20}$

$\sigma = \sqrt{\text{Var}(X)}$

$= \boxed{1.48}$

2. Because $p = 50 \Rightarrow \lambda = 1/50$, the probability density function must be $p(x) = (1/50)e^{-x/50}$, $x > 0$.

(a) $$\Pr(X \le 35) = \int_0^{35}\frac{1}{50}e^{-x/50}\,dx$$

$$= -e^{-x/50}\Big|_0^{35}$$

$$= 1 - e^{-0.7} \sim \boxed{0.50}$$

(b) $$\Pr(X \ge 60) = \int_{60}^{\infty}\frac{1}{50}e^{-x/50}\,dx$$

$$= -e^{-x/50}\Big|_{60}^{\infty}$$

$$= 0 + e^{-1.2} \sim \boxed{0.30}$$

(c) $$\Pr(45 \le X \le 55) = \int_{45}^{55} \frac{1}{50} e^{-x/50}\, dx$$

$$= -e^{-x/50}\Big|_{45}^{55}$$

$$= e^{-0.9} - e^{-1.1}$$

$$= 0.407 - 0.333$$

$$\sim \boxed{0.07}$$

3. $$\mu = \frac{3}{10}\int_{-1}^{+1} x(2 - x^2)\, dx$$

$$= -\frac{3}{40}(2 - x^2)^2\Big|_{-1}^{+1}$$

$$= -\frac{3}{40} - \left(-\frac{3}{40}\right)$$

$$= \boxed{0}$$

$$\mathrm{Var}(X) = \frac{3}{10}\int_{-1}^{+1} x^2(2 - x^2)\, dx$$

$$= \frac{3}{10}\int_{-1}^{+1} (2x^2 - x^4)\, dx$$

$$= \frac{3}{10}\left(\frac{2x^3}{3} - \frac{x^5}{5}\right)\Big|_{-1}^{+1}$$

$$= \frac{3}{10}\left[\left(\frac{2}{3} - \frac{1}{5}\right) - \left(-\frac{2}{3} + \frac{1}{5}\right)\right]$$

$$= \boxed{\frac{7}{25}}$$

4. (a) $$\Pr(|X| < 0.5) = \frac{3}{10}\int_{-0.5}^{+0.5} (2 - x^2)\, dx$$

$$= \frac{3}{10}\left(2x - \frac{x^3}{3}\right)\Big|_{-0.5}^{0.5}$$

$$= 0.3\left[\left(1 - \frac{0.125}{3}\right) - \left(-1 + \frac{0.125}{3}\right)\right]$$

$$= \boxed{0.575}$$

(b) $$\Pr(X > -4) = \frac{3}{10}\int_{-1}^{+1} (2 - x^2)\, dx$$

$$= \boxed{1}$$

(There is no need to carry out the computation because the integral is over the full range of values for which $f(x) \neq 0$.)

(c) $$\Pr(-3 \le X \le 0.2) = \frac{3}{10}\int_{-1}^{0.2} (2 - x^2)\, dx$$

$$= \frac{3}{10}\left(2x - \frac{x^3}{3}\right)\Big|_{-1}^{0.2}$$

$$= \frac{3}{10}\left[\left(0.4 - \frac{0.008}{3}\right) - \left(-2 + \frac{1}{3}\right)\right]$$

$$= \boxed{0.619}$$

5. Here we assume that the ball bearings have diameters distributed normally with $\mu = 5.2$ and $\sigma = 0.6$. Then

$$\Pr(3.8 \le X \le 6.0) = \frac{1}{\sqrt{2\pi}\sigma}\int_{3.8}^{6.0} e^{-\frac{1}{2}\left(\frac{x-\mu}{\sigma}\right)^2}\, dx$$

Making the change of variables $z = \dfrac{x - \mu}{\sigma} = \dfrac{x - 5.2}{0.6}$, we find that our problem is equivalent to

$\Pr\left(\dfrac{3.8 - 5.2}{0.6} \le Z \le \dfrac{6.0 - 5.2}{0.6}\right)$, where Z represents the standard normal distribution. Thus we must compute

$$\Pr(-2.33 \le Z \le 1.33) = 0.4901 + 0.4082$$

$$= 0.898$$

He is almost 90% sure.

6. $$E(X) = (22)(0.2) + (31)(0.2) + (19)(0.2) + (8)(0.3) + (40)(0.1)$$

$$= \boxed{20.8}$$

$$\mathrm{Var}(X) = (22 - 20.8)^2(0.2) + (31 - 20.8)^2(0.2) + (19 - 20.8)^2(0.2) + (8 - 20.8)^2(0.3) + (40 - 20.8)^2(0.1)$$

$$= \boxed{107.8}$$

$$\sigma = \boxed{10.4}$$

7. (a) We require

$$\int_1^2 \frac{c}{5+x}\,dx = 1$$

Now

$$\int_1^2 \frac{c}{5+x}\,dx = c\ln(5+x)\Big|_1^2 = c(\ln 7 - \ln 6)$$

$$= 0.154c \overset{must}{=} 1$$

Thus $\boxed{c = \frac{1}{0.154} = 6.49}$

(b) $$\Pr(|X| < 0.5) = \int_{-0.5}^{+0.5} f(x)\,dx$$

$$= \boxed{0}$$

because the density function is equal to 0 outside of $1 \le x \le 2$.

8. (a) Because $\mu = 1000$ and $\lambda = \frac{1}{1000}$, we have

$$\boxed{p(x) = \frac{1}{1000}e^{-\frac{x}{1000}} \qquad 0 \le x < \infty}$$

(b) $$\Pr(1200 \le X \le 1500) = \int_{1200}^{1500} \frac{1}{1000}e^{-\frac{x}{1000}}\,dx$$

$$= -e^{-\frac{x}{1000}}\Big|_{1200}^{1500}$$

$$= e^{-1.2} - e^{-1.5}$$

$$= \boxed{0.078}$$

9. $$\text{Value of game} = 10\left(\frac{12}{52}\right) + 3\left(\frac{4}{52}\right) - 4\left(\frac{36}{52}\right)$$

$$= -\frac{12}{52} = -\frac{3}{13}$$

$$\approx 0.231$$

$\boxed{\text{The expected loss is \$0.23 per game. I would not wish to play.}}$

10. (a) $\Pr(65 \le X \le 82)$ can be related to probabilities for standard normal distribution Z by setting

$$z = \frac{x-\mu}{\sigma} = \frac{x-68}{7.3}$$

(See Problem 5 of this test.)

Thus

$$\Pr(65 \le X \le 82) = \Pr\left(\frac{65-68}{7.3} \le Z \le \frac{82-68}{7.3}\right)$$

$$= \Pr(-0.41 \le Z \le 1.92)$$

$$= 0.1591 + 0.4726$$

$$= \boxed{0.6317}$$

(b) We must find c such that $\Pr(X \ge c) = 0.95$. Relating to the standard normal distribution this means that

$$\Pr\left(Z \ge \frac{c-68}{7.3}\right) = 0.95$$

Because $\Pr(Z \ge 0) = 0.50$, we must find c such that

$$\Pr\left(0 \ge Z \ge \frac{c-68}{7.3}\right) = 0.45$$

or, equivalently,

$$\Pr\left(0 \le Z \le \frac{68-c}{7.3}\right) = 0.45$$

From the standard normal table, we take $\frac{68-c}{7.3} = 1.645$ (interpolating the last digit). Thus $c = 68 - (7.3)(1.645) = 56$.

$\boxed{\text{The lowest passing grade should be 56.}}$

CHAPTER 11 (page 533)

1. (a) $p_2(x)$ must be of the form

$$p_2(x) = f(0) + f'(0)x + \frac{f''(0)}{2}x^2$$

Because $f(x) = e^{-2x}$, $f'(x) = -2e^{-2x}$, and $f''(x) = 4e^{-2x}$, we obtain

$$\boxed{p_2(x) = 1 - 2x + 2x^2}$$

(b) $e^{-0.6} \sim p_2(0.3) = 1 - 2(0.3) + 2(0.3)^2$

$$= \boxed{0.58}$$

2. (a) $\lim\limits_{x\to 0} \dfrac{5\sin x - x}{x^2} = \lim\limits_{x\to 0} \dfrac{5\cos x - 1}{2x} = \boxed{\text{undefined}}$

(b) $\lim\limits_{x\to\infty} \dfrac{x+3}{x^2-4} = \lim\limits_{x\to\infty} \dfrac{1+3/x}{x-4/x} = \boxed{0}$

(c) $\lim\limits_{x\to\pi/2} \dfrac{\cos x}{\pi/2 - x} = \lim\limits_{x\to\pi/2} \dfrac{-\sin x}{-1} = \boxed{1}$

(Compare this result using L'Hôpital's rule with your solution to the similar part (b) of Exercise 4 in the Chapter 9 Warm-Up Test.)

(d) $\lim_{x\to\infty} \dfrac{3x^2 - 4}{5 - 8x^2} = \lim_{x\to\infty} \dfrac{3 - 4/x^2}{5/x^2 - 8} = \boxed{-\dfrac{3}{8}}$

3. (a) Because

$$\cos x = 1 - \frac{x^2}{2!} + \frac{x^4}{4!} - \frac{x^6}{6!} \cdots$$

we replace x by $2x$ to obtain

$$\cos 2x = 1 - \frac{2^2x^2}{2!} + \frac{2^4x^4}{4!} - \frac{2^6x^6}{6!} + \cdots$$

$$= \boxed{\sum_{n=0}^{\infty} (-1)^n \frac{2^{2n}x^{2n}}{(2n)!}}$$

(b) Because $\left(\dfrac{1}{1-x}\right)^2$ is the derivative of $\dfrac{1}{1-x}$, we first write the Taylor series for $\dfrac{1}{1-x}$:

$$1 + x + x^2 + x^3 \cdots$$

and then differentiate term by term to obtain

$$\frac{1}{(1-x)^2} = 1 + 2x + 3x^2 - 4x^3 \cdots$$

$$= \boxed{\sum_{n=0}^{\infty} (n+1)x^n}$$

4. Newton's iteration is described by the recursion

$$\boxed{x_{n+1} = x_n - \frac{f(x_n)}{f'(x_n)} \qquad \mu = 0, 1, 2, \ldots}$$

For $f(x) = x^3 - 7$, $f'(x) = 3x^2$ and we obtain, on simplification,

$$\boxed{x_{n+1} = \frac{2x_n^3 + 7}{3x_n^2} \qquad n = 0, 1, 2}$$

Starting with $x_0 = 2$, we compute

$$x_1 = \frac{2(2)^3 + 7}{3(2)^2} = \frac{23}{12}$$

$$= \boxed{1.917}$$

$$x_2 = \frac{2(1.917)^3 + 7}{3(1.917)^2}$$

$$= \boxed{1.913}$$

(We note that $(1.913)^3 = 7.00075$).

5. We compute

$$g(x) = x^{1/3} \qquad g'(x) = 1/3x^{-2/3} \qquad g''(x) = -\frac{2}{9}x^{-5/3}$$

and $g'''(x) = \frac{10}{27}x^{-8/3}$
Evaluating at $x = 8$, we obtain

$$g(8) = 2 \qquad g'(8) = \frac{1}{12} \qquad g''(8) = -\frac{1}{144}$$

and $g'''(8) = \frac{5}{3456}$.
So the Taylor polynomial of degree 3 about $a = 8$ is

$$p_3(x) = 2 + \frac{1}{12}(x-8) - \frac{1}{144}\frac{(x-8)^2}{2!} + \frac{5}{3456}\cdot\frac{(x-8)^3}{3!}$$

Thus

$$p_3(8.1) = 2 + \frac{0.1}{12} - \frac{0.01}{288} + \frac{5(0.001)}{(6)(3456)} = 2.0083$$

Our approximation yields

$$\boxed{\sqrt[3]{8.1} \approx 2.0083}$$

6. (a) We compute

$$f'(x) = -2xe^{-x^2}$$

We note that, for $0 \le x \le 0.02$,

$$|f'(x)| \le 2(0.02) = 0.04$$

We differentiate again and note that $f''(x) = 4x^2e^{-x^2} - 2e^{-x^2}$ so that, for $0 \le x \le 0.02$,

$$|f''(x)| \le 4(0.02)^2 + 2 = 2.0016$$

Similarly we could find that, for $0 \le x \le 0.02$,

$$f'''(x) = -8x^3e^{-x^2} + 12xe^{-x^2}$$

so that, for $0 \le x \le 0.02$,

$$|f'''(x)| \le 8(0.02)^3 + 12(0.02) < 0.241$$

From Theorem (11.2) we seek the smallest n for which

$$\max_x |f^{(n+1)}(x)| \frac{(0.02)^{n+1}}{(n+1)!} < 0.001$$

Trying $n = 1$, we test whether

$$(2.0016)\frac{(0.02)^2}{2} \overset{?}{<} 0.001 \qquad \text{The answer is no!}$$

Trying $n = 2$, we test

$$(0.241)\frac{(0.02)^3}{6} \overset{?}{<} 0.001 \qquad \text{The answer is yes!}$$

We conclude that the second-order Taylor polynomial for e^{-x^2} can provide the desired accuracy.

(b) The second-order Taylor polynomial for e^{-x^2} is

$$p_2(x) = 1 - x^2$$

Thus

$$e^{-(0.016)^2} \sim p_2(0.016) = 1 - (0.016)^2 = \boxed{0.9997}$$

7. We write

$$y = a_0 + a_1x + a_2x^2 + a_3x^3 + \cdots$$

so $$y' = a_1 + 2a_2x + 3a_3x^2 + 4a_4x^3 + \cdots$$

Plugging into the equation

$$y' = -y + x - x^2$$

gives

$$a_1 + 2a_2x + 3a_3x^2 + \cdots = -a_0 - a_1x - a_2x^2 \cdots + x - x^2$$

Equating like powers of x gives the relations

$$a_1 = -a_0$$
$$2a_2 = -a_1 + 1$$
$$3a_3 = -a_2 - 1$$

and so on

The condition $y(0) = 4$ tells us that $a_0 = 4$. Then $a_1 = -a_0$ gives $a_1 = -4$, and $2a_2 = -a_1 + 1$ gives $2a_2 = -(-4) + 1$, so $a_2 = \frac{5}{2}$. Thus to three terms

$$\boxed{y = 4 - 4x + \frac{5}{2}x^2 + \cdots}$$

8. As in Exercise 4, the recursion is

$$x_{n+1} = x_n - \frac{f(x_n)}{f'(x_n)} \qquad n = 0, 1, 2 \cdots$$

For $f(x) = x^3 + x - 3$, $f'(x) = 3x^2 + 1$ and, on simplifying,

$$x_{n+1} = \frac{2x_n^3 + 3}{3x_n^2 + 1}$$

Starting with $x_0 = 1$, we compute

$$x_1 = \frac{2(1)^3 + 3}{3(1)^2 + 1} = \frac{5}{4} = \boxed{1.25}$$

$$x_2 = \frac{2(1.25)^3 + 3}{3(1.25)^2 + 1} = \boxed{1.214}$$

$$x_3 = \frac{2(1.214)^3 + 3}{3(1.214)^2 + 1} = \boxed{1.213}$$

9. (a) Because $\sin x = x - \frac{x^3}{6} + \cdots$, it follows that

$$\sin x^2 = x^2 - \frac{x^6}{6} + \cdots$$

whereas

$$(\sin x)^2 = \left(x - \frac{x^3}{6} + \cdots\right)\left(x - \frac{x^3}{6} + \cdots\right)$$
$$= x^2 - \frac{x^4}{3} + \cdots$$

Thus $\sin x^2$ and $(\sin x)^2$ have 3rd-order contact but not 4th-order contact at $x = 0$.

(b) $$f(x) = \ln x \quad g(x) = \frac{x^3}{3} - \frac{3}{2}x^2 + 3x - \frac{11}{6}$$

Now $f(1) = 0$ and $g(1) = 0$

Also $f'(x) = \frac{1}{x}$ and $g'(x) = x^2 - 3x + 3$

so $f'(1) = 1$ and $g'(1) = 1$

Differentiating again,

$$f''(x) = -\frac{1}{x^2} \quad \text{and} \quad g''(x) = 2x - 3$$

so $f''(1) = -1$ and $g''(1) = -1$

Once again,

$$f'''(x) = \frac{2}{x^3} \quad \text{and} \quad g'''(x) = 2$$

so $f'''(1) = 2$ and $g'''(1) = 2$

Finally, however,

$$f^{(iv)}(x) = -\frac{6}{x^4} \quad \text{with} \quad f^{(iv)}(1) = -6$$

But $g^{(iv)}(1) = 0$.

We conclude that $f(x)$ and $g(x)$ have 3rd-order but not 4th-order contact.

10. (a) We write

$$y = a_0 + a_1x + a_2x^2 + a_3x^3 + a_4x^4 + a_5x^5 + \cdots$$

and note that the conditions $y(0) = 0$ and $y'(0) = 1$ give $a_0 = 0$ and $a_1 = 1$. Differentiating twice gives

$$y'' = 2a_2 + 3 \cdot 2a_3x + 4 \cdot 3a_4x^2 + 5 \cdot 4a_5x^3 + \cdots$$

and setting $y'' = -y$ and equating like powers of x

give the recursive equations

$$2a_2 = -a_0 = 0 \Rightarrow a_2 = 0$$

$$(3 \cdot 2)a_3 = -a_1 = -1 \Rightarrow a_3 = \frac{-1}{3 \cdot 2} = -\frac{1}{3!}$$

$$4 \cdot 3a_4 = -a_2 \Rightarrow a_4 = 0$$

$$5 \cdot 4a_5 = -a_3 \Rightarrow a_5 = -\frac{a_3}{5 \cdot 4} = \frac{1}{5!}$$

Thus $\boxed{y(x) = x - \frac{x^3}{3!} + \frac{x^5}{5!} + \cdots}$

(b) $$y(0.06) \sim (0.06) - \frac{(0.06)^3}{6} + \frac{(0.06)^5}{120}$$

$$\sim \boxed{0.059964}$$

(c) $\boxed{\sin x}$

Index

Index of Applications

BUSINESS AND ECONOMICS